Biodiversity in India

Volume 4

T. Pullaiah

2024

Regency Publications

A Division of

Astral International Pvt. Ltd.

New Delhi – 110 002

First Published, 2006
Reprinted, 2024

ISBN: 9788189233204 (HB)

Publisher's Note:

Every possible effort has been made to ensure that the information contained in this book is accurate at the time of going to press, and the publisher and author cannot accept responsibility for any errors or omissions, however caused. No responsibility for loss or damage occasioned to any person acting, or refraining from action, as a result of the material in this publication can be accepted by the editor, the publisher or the author. The Publisher is not associated with any product or vendor mentioned in the book. The contents of this work are intended to further general scientific research, understanding and discussion only. Readers should consult with a specialist where appropriate.

Every effort has been made to trace the owners of copyright material used in this book, if any. The author and the publisher will be grateful for any omission brought to their notice for acknowledgement in the future editions of the book.

Published by : **Regency Publications**
A Division of
Astral International Pvt. Ltd.
– ISO 9001:2015 Certified Company
4736/23, Ansari Road, Darya Ganj
New Delhi-110 002
Ph. 011-43549197, 23278134
E-mail: info@astralint.com
Website: www.astralint.com

PREFACE

Since Rio Earth Summit Biodiversity has gained importance. Biodiversity bill has been approved by Parliament of India. Several countries are preparing national Biodiversity Strategy and Action Plans. Even after this not many journals and books are coming out on Biodiversity in India. We intend to bring out a series of Edited volumes on Biodiversity in India encompassing all aspects. This includes Species diversity, Genetic diversity and Ecosystem diversity. This volume is the fourth in this attempt. Authors who intend to publish their articles are requested to send them to the editor. We seek the opinion of the readers for the coming volumes.

CONTENTS

LIST OF CONTRIBUTORS

B. Viswanath
Department of Microbiology,
Sri Krishnadevaraya University,
Ananatapur-515 003
Andhra Pradesh

Dr. K.Venkateswara Rao
Department of Biosciences,
Sri Satya Sai Institute of Higher Learning,
Prasanthi Nilayam-515 134
Andhra Pradesh

S. Sai Kiran Kumar
Department of Biosciences,
Sri Satya Sai Institute of Higher Learning,
Prasanthi Nilayam-515 134
Andhra Pradesh

Dr. D. Someswari
Department of Botany,
T.J.P.S. College,
Guntur-522 006

Prof. T.N. Mary
Department of Botany,
Acharya Nagarjuna University,
Nagarjunanagar-522 510

Dr. G.T. Dabhade
Hon. Professor in Botany
Birla College,
Kalyan-421 304

Dr. V.K. Hile
Lecturer in Botany,
Bhavan's H. Somani College,
Mumbai-400 007
Maharashtra

Dr. M. Chandrasekhar
Central Sericultural Germplasm Resource Centre,
Central Silk Board, P.B. No. 44,
Thally Road,
Hosur-635 109
Tamil Nadu

Dr. R.P. Saraswat
Central Sericultural Germplasm Resource Centre,
Central Silk Board, P.B. No. 44,
Thally Road,
Hosur-635 109
Tamil Nadu

Dr. A. Tikader
Central Sericultural Germplasm Resource Centre,
Central Silk Board, P.B. No. 44,
Thally Road,
Hosur-635 109
Tamil Nadu

Dr. A. Anantha Rao
Central Sericultural Germplasm Resource Centre,
Central Silk Board, P.B. No. 44,
Thally Road,
Hosur-635 109
Tamil Nadu

Dr. B.P. Choudhury
P.G. Department of Botany,
Utkal University,
Bhubaneswar-751 002

Dr. Mihir Kumar Das
Department of Botany,
G.M. College,
Sambalpur

Dr. Priti D. Mahekar
Department of Botany,
Shivaji University,
Kolhapur-416 004
Maharashtra

Prof. S.R. Yadav
Department of Botany,
Shivaji University,
Kolhapur-416 004
Maharashtra

CHAPTER 1

DIFFERENCES IN THE INCIDENCE OF FUNGI ISOLATED FROM THE SOILS OF *PHYLLANTHUS EMBLICA* L. AND *OCIMUM SANCTUM* L. OF DHARWAD, KARNATAKA, INDIA

*Buddolla Viswanath**

ABSTRACT

The life in soil is highly diverse and consists of interacting population of microorganisms and soil fauna, and their activities affect physical, chemical and biological characteristics of soil. The present work deals with prevalence of soil fungi in both rhizosphere and non-rhizosphere soils of *Phyllanthus emblica* and *Ocimum sanctum* L. plants. A total of 90 species of fungi were isolated from rhizosphere and non-rhizosphere regions of both plants of Dharwad over a period of 12 months. Differences in the diversity pattern of soil fungi in both rhizosphere and non-rhizosphere regions of these plants occurred in different seasons. The most common genera viz., *Rhizopus, Aspergillus, Penicillium, Trichoderma, Rhizoctonia* etc., showed their adaptability for all the seasons and were isolated throughout the year from the soils of both plants. The experimental data will be discussed.

Introduction

The microbial ecology involves the study of the relationship of microorganisms to their surroundings and to each other. These surroundings may be major global ecosystems such as soils or aquatic bodies or other living organisms, such as plants or animals. The life in soil is highly diverse and consists of interacting population of microorganisms and soil fauna, and

* Department of Microbiology, Sri Krishnadevaraya University, Anantapur-515 003. A P. buddolla@yahoo.co.in

their activities affect physical, chemical and biological characteristics of soil (Dwivedi, 1997).

The present estimate of global fungal diversity is 1.5 million species while our routine efforts of exploring and reporting this diversity in more than a century and half has only yielded information on approximately 5% of this treasure (Jain, 2004). Fungi are dominant in soils in terms of their biomass. In some soils, fungal biomass may exceed that of all other soil microbes (Anderson and Domsch, 1978). Many factors influence the fungal distribution in soil, which include soil atmosphere, soil precipitation, soil water, soil pH that may vary according to seasons. By keeping all these in view the present paper endeavors to focus beam on diversity and seasonal variations in the soil fungi of *Phyllanthus emblica* and *Ocimum sanctum* plants from Dharwad, which is also known as Chotomahabaleshwar. The climate of this region is cool and the soil type is red laterite soil.

Materials and Methods

The soil samples were collected from several selected localities of both *Phyllanthus emblica* L., *Ocimum sanctum* L. plants. The localities were selected in such a way that where there is good vegetation. Soils were collected from both the rhizosphere and non-rhizosphere regions of these plants. The sampling was done during three seasons viz., winter, summer and rainy seasons. Soil samples from superficial layers were taken after removing the litter covering the soil and were picked up by sterile spatula. Samples at the depth of 10–15cm were also collected along with the superficial ones.

Soil dilution plate method (Warcup, 1950) and the soil plate method (Waksman, 1927) were used for primary screening of soil fungi. Fungal diversity study was carried out using different cultural media to get a maximum number of fungi, as suggested by Booth (1971) in his Methods of Microbiology. All the media were sterilized at 15 lbs pressure for 20 min. Antibiotic streptomycin was used in the media to avoid bacterial colonies. The inoculated plates were incubated for four days at 25–29°C temperature and daily observations were done. Isolated fungi were identified first on the basis of colony characters and morphological structures later with the help of direct mount from culture and their slide culture.

Different forms were identified up to species level with the help of recent literature (Watanabe 1994; Subramanian 1990; Ellis and Ellis 1985; Von Arx 1981; Bilgrami *et al.*, 1991; Domsch *et al.*, 1980, Sarabhoy *et al.*, 1975 Dennis 1960; Gilman 1957).

Results

The rhizosphere region of *Phyllanthus emblica* soil had a maximum flora with (Table 1) *Mucor hiemalis, Rhizopus arrhizus, Aspergillus niger, Curvularia pallescenes, Penicillium islandicum, Torula herbarum, Trichoderma koeningii* etc., as dominant forms; *Absidia butleri, Cunninghamella echinulata, Chaetomium cochliodes, Neurospora sitophila, Alternaria alternate, Curvularia maculans, Fusarium moniliforme, Trichoderma glacum* etc., are less dominant. *Absidia glauca, Chaetomium spirale, Alternaria tenuissima, Beltrania rhombica, Scolecobasidium verruculosum, Ulocladium atrum* were with less frequency. A total of 56 species were isolated from the rhizosphere of *Phyllanthus emblica* plant.

The non-rhizosphere region of *Phyllanthus emblica* soil had *Mucor hiemalis, Aspergillus sydowi, Aspergillus ustus, Curvularia pallescens* etc., as dominant forms (Table 1). Species like *Mucor racemosus, Aspergillus nidulans, Curvularia maculans, Penicillium brefeldianum, Penicillium mineo-luteum, Trichoderma viridae* etc., were less dominant. And the forms like *Absidia butleri, Acaetomium globosum, Alternaria tenuissima* etc., were with less frequency. A total of 48 species were recorded from the non-rhizosphere of *Phyllanthus emblica* plants.

The rhizosphere region of *Ocimum sanctum* soil had a maximum flora with (Table 2) *Mucor hiemalis, M. racemosus, Rhizopus arrhizus, Aspergillus fumigatus, A. niger* etc., as dominant forms; and the forms like *Syncephalastrum racemosum, Neurospora sitophila, Aspergillus flavus, A. restrictus, Curvularia penniseti, Drechslera australiensis, Torula herbarum* etc., were recorded as less dominant. *Absidia glauca, Chaetomium indicum, Thielavia terricola, Alternaria deniisi, Cordana pauciseptata, Veronia coprophila* etc., were recorded less frequently. A total of 53 species were isolated from the rhizosphere of *Ocimum sanctum* plant.

The non-rhizosphere region of *Ocimum sanctum* soil had *Mucor hiemalis, Rhizopus arrhizus, Aspergillus fumigatus, Aspergillus unguis, A.ustus, Curvularia penniseti, Torula herbarum* etc., as dominant forms (Table 2). Species like *Mucor racemosus, Neurospora sitophila, Aspergillus niveus, Curvularia maculans,C. lunata, Drechslera australiensis, Nigrospora oryzae, Penicillium corylophilum, P. islandicum, Trichoderma glaucum* etc., were less dominant. And the forms like *Absidia glauca, Chaetomium indicum, Alternaria tenuissima, Arthrobotrys oligospora, Corynespora cassicola* etc., were with less frequency. A total of 49 species were recorded from the non-rhizosphere of *Ocimum sanctum* plant.

In the course of this study a total of 90 species belonging to 47 genera were isolated from rhizosphere and non-rhizosphere regions of both plants. These can be divided as follows:

Zygomycotina – 10 species
Ascomycotina – 9 species
Deuteromycotina – 71 species

In this study, soil samples were collected at two different layers viz., superficial layer and about 15 cm depth layer. Higher fungal population was recorded at superficial layer than the 15 cm depth layer in summer and winter seasons. But in rainy season fungal population was predominantly found at 15cm depth, because of washing off of the fungal spores in superficial layers.

Different media like Czapek Sucrose Agar, Malt Extract Agar, Potato Dextrose Agar, Yeast Extract Agar and Dextrose Peptone Agar were used to get maximum number of fungi.

Table 1: Seasonal Occurrence and Percentage Occurrence of Fungi in the Rhizosphere and Non-rhizosphere of *Phyllanthus emblica* L.

Sr. No.	*Name of the Fungus*	*Rhizosphere region*			*% of Fungal Occurrence*	*Non-Rhizosphere region*			*% of Fungal Occurrence*
		W	S	R		W	S	R	
	ZYGOMYCOTINA								
1	*Absidia butleri* Lendner	112	—	—	1.24	61	—	—	0.80
2	*Absidia glauca* Hagem	42	—	—	0.46	—	—	—	—
3	*Cunninghamella echinulata* Thaxter	—	114	—	1.26	—	77	—	1.02
4	*Mucor hiemalis* Wehmeyer	241	159	40	4.89	122	215	128	6.16
5	*Mucor racemosus* Fres.	—	—	—	—	91	148	—	3.16
6	*Rhizopus arrhizus* Fischer	100	125	83	3.42	61	154	100	4.17
7	*Rhizopus* sp.	23	126	—	1.65	—	—	—	—
	ASCOMYCOTINA								
8	*Achaetomium globosum* Rai & Tewari	103	—	—	1.14	62	—	—	0.82
9	*Ascotricha chartarum* Berk	—	105	—	1.16	—	—	—	—
10	*Chaetomium cochliodes* Palliser	75	—	100	1.94	35	—	57	1.22
11	*Chaetomium globosum* Kunze & Schm.	71	—	73	1.60	84	—	62	1.93
12	*Chaetomium spirale* Zopt	—	35	—	0.38	—	—	—	—
13	*Emericella* sp.	—	50	—	0.55	—	53	—	0.70
14	*Neurospora sitophila* Shear & Dodge	60	—	75	1.50	—	—	—	—
	DEUTEROMYCOTINA								
15	*Alternaria alternata* Keissler	90	48	—	1.53	—	—	—	—
16	*Alternaria brassicola* Wiltshire	—	83	—	0.92	33	69	—	1.27
17	*Alternaria tenuis* Auct.	—	—	—	—	—	63	—	0.83
18	*Alternaria tenuissima* Wiltshire	26	48	—	0.71	38	17	—	0.73
19	*Aspergillus candidus* Link	183	96	—	3.10	164	136	—	3.68
20	*Aspergillus flavus* Link	78	87	56	2.45	86	119	100	4.04
21	*Aspergillus nidulans* Wingate	175	100	—	3.05	105	83	—	2.49
22	*Aspergillus niger* van Tieghem	120	177	172	5.21	124	65	111	3.68
23	*Aspergillus sydowi* Thom & Church	75	143	130	3.86	147	139	74	4.77
24	*Aspergillus unguis* Thom & Raper	100	190	90	4.22	—	—	—	—

Sr. No.	*Name of the Fungus*	*Rhizosphere region*			*% of Fungal Occurrence*	*Non-Rhizosphere region*			*% of Fungal Occurrence*
		W	S	R		W	S	R	
25	*Aspergillus ustus* Thom & Church	83	189	133	4.50	119	115	86	4.24
26	*Beltrania rhombica* Penzig	—	50	—	0.55	—	73	—	0.97
27	*Cladosporium cladosporioides* de Vries	75	18	—	1.03	—	—	—	—
28	*Cordana pauciseptata* Preuss	—	—	80	0.88	—	—	75	0.99
29	*Curvularia clavata* Jain	—	78	—	0.86	—	90	—	1.19
30	*Curvularia geniculata* Boedijin	96	—	—	1.06	—	—	—	—
31	*Curvularia maculans* Boedijn	106	—	112	2.42	86	—	106	2.54
32	*Curvularia pallescens* Boedijn	55	68	127	2.77	90	120	73	3.75
33	*Cylindrocladium parvum* Anderson	—	—	—	—	—	80	—	1.06
34	*Drechslera hawaiiensis* Subram. & Jain	55	—	85	1.55	54	—	58	1.48
35	*Fusarium moniliforme* Sheld.	120	—	—	1.33	91	—	—	1.20
36	*Fusarium nivale* (Fr.) Ces.	94	—	—	1.04	—	—	—	—
37	*Fusarium oxysporum* Schlecht	150	—	—	1.66	122	—	—	1.62
38	*Humicola fusco-atra* Traaen	—	95	—	1.05	—	72	—	0.95
39	*Humicola grisea* Traaen	—	80	—	0.88	—	79	—	1.04
40	*Monilia sitophila* Sacc	—	130	90	2.44	—	59	78	1.81
41	*Monodictys levis* Hughes	70	—	—	0.77	—	—	—	—
42	*Myrothecium roridum* Tode ex Fr.	—	—	—	—	—	62	36	1.29
43	*Nigrospora oryzae* Petch	—	—	—	—	49	—	70	1.57
44	*Paecilomyces marquandii* Hughes	—	65	—	0.72	—	—	—	—
45	*Paecilomyces varioti* Bainer	—	—	—	—	44	34	—	1.03
46	*Paecilomyces brefeldianum* Dodge	—	—	—	—	90	—	103	2.42
47	*Penicillium corylophilum* Dierckx	36	—	54	1.00	—	—	108	1.43
48	*Penicillium islandicum* Sopp.	85	—	110	2.16	—	—	—	—
49	*Penicillium mineo-luteum* Dierckx	35	95	50	2.00	101	54	95	3.31
50	*Penicillium nigricans* Thom	—	—	—	—	61	64	86	2.79
51	*Penicillium purpurogenum* Stoll	50	35	—	0.94	—	—	—	—
52	*Penicillium roseo-purpureum* Dierckx	—	—	—	—	51	35	80	2.20
53	*Penicillium rubrum* Stoll	30	100	25	1.72	30	78	63	2.27
54	*Peziotrichum corticola* Subram.	—	90	60	1.66	—	—	—	—
55	*Pithomyces atro-olivaceus* Ellis	—	90	35	1.38	—	—	—	—
56	*Rhizoctonia solani* Kuhn	—	95	41	1.51	—	101	37	1.83
57	*Scolecobasidium verruculosum* Roy	—	45	—	0.50	—	—	—	—
58	*Spegazzinia sundara* Subram.	50	—	62	1.24	81	—	53	1.77
59	*Stachybotrys cartarum* Hughes	—	—	—	—	—	58	—	0.77
60	*Stachybotrys cylindrospa* Jenson	—	73	—	0.81	—	—	—	—
61	*Tetracoccosporium paxianus* Szaba	—	—	—	—	70	—	55	1.65
62	*Thelviopsis paradoxa* Hoehn	—	75	—	0.83	—	—	—	—
63	*Torula graminis* Desm	90	—	56	1.62	—	—	75	0.99
64	*Torula herbarum* Link	127	83	160	4.11	60	67	68	2.58
65	*Trichoderma glaucum* Abbott	97	60	80	2.63	96	62	84	3.20
66	*Trichoderma koeningii* Oudem	—	133	120	2.81	—	—	—	—
67	*Trichoderma viride* Pres. ex Fr.	—	—	—	—	70	70	58	2.62
68	*Ulocladium atrum* Preuss	—	—	96	1.06	—	—	—	—
69	*Volutina concentrica* Penzig & Sacc	—	—	—	—	78	—	—	1.03

W = Winter; S = Summer; R = Rainy Season

Table 2: Seasonal Occurrence and Percentage Occurrence of Fungi in the Rhizosphere and Non-rhizosphere of *Ocimum sanctum* L.

Sr. No.	Name of the Fungus	Rhizosphere region			% of* Fungal Occurrence	Non-Rhizosphere region			% of* Fungal Occurrence
		W	S	R		W	S	R	
	ZYGOMYCOTINA								
1	*Absidia glauca* Hagem	63	—	—	0.79	55	—	—	0.77
2	*Cunninghamella echinulata* Thaxter	—	127	—	1.59	—	—	—	—
3	*Cunninghamella elegans* Lendner	—	132	—	1.65	—	97	—	1.35
4	*Mucor hiemalis* Wehmeyer	142	186	30	4.49	131	144	55	4.62
5	*Mucor racemosus* Fres.	137	203	35	4.70	85	90	—	2.45
6	*Phycomyces nitens* Agardh	81	—	59	1.75	—	—	—	—
7	*Rhizopus arrhizus* Fischer	75	239	186	6.27	50	150	105	4.27
8	*Syncephalastrum racemosum* Schr.	—	112	136	3.11	—	—	—	—
	ASCOMYCOTINA								
9	*Chaetomium cochlioides* Palliser	90	—	70	2.00	—	—	90	1.26
10	*Chaetomium indicum* Corda.	—	—	66	0.82	—	—	51	0.71
11	*Emericella* sp.	—	—	—	—	—	35	20	0.77
12	*Neurospora sitophila* Shea. & Dodg.	84	—	60	1.80	60	—	80	1.96
13	*Thielavia terricola* Emmons	—	70	—	0.87	—	87	—	1.21
	DEUTEROMYCOTINA								
14	*Alternaria alternata* Keissler	—	—	—	—	37	58	—	1.33
15	*Alternaria deniisi* Ellis	—	55	—	0.69	42	64	—	1.48
16	*Alternaria tenuis* Auct	52	38	—	1.13	—	—	—	—
17	*Alternaria tenuissima* Wiltshire	55	32	—	1.09	56	51	—	1.50
18	*Arthrobotrys oligospora* Fres.	—	—	65	0.81	—	—	70	0.98
19	*Aspergillus flavus* Link	77	68	85	2.88	94	75	61	3.22
20	*Aspergillus fumigatus* Fres.	—	167	197	4.57	—	130	175	4.27
21	*Aspergillus niger* van Tieghem	55	145	108	3.86	110	55	120	3.99
22	*Aspergillus niveus* Blochwitz	100	—	85	2.32	125	—	90	3.01
23	*Aspergillus restrictus* Smith	61	96	68	3.20	—	—	—	—
24	*Aspergillus unguis* Thom & Raper	—	—	—	—	125	150	105	5.32
25	*Aspergillus ustus* Thom & Church	—	—	—	—	99	141	95	4.69
26	*Cladosporium herbarum* Link	52	—	74	1.58	—	—	—	—
27	*Cordana pauciseptata* Preuss	—	—	80	1.00	—	—	70	0.98
28	*Corynespora cassiicola* Wei.	50	—	—	0.62	30	—	—	0.42
29	*Curvularia clavata* Jain	—	80	—	1.00	—	55	—	0.77
30	*Curvularia lunata* Boedijn	80	—	80	2.00	65	—	60	1.75
31	*Curvularia maculans* Boedijn	100	—	100	2.51	90	—	85	2.47
32	*Curvularia pallescens* Boedijn	—	—	—	—	20	40	95	2.17
33	*Curvularia penniseti* Boedijn	108	77	68	3.17	80	50	100	3.22
34	*Drechslera australiensis* Subr. & Jain	80	—	80	2.00	50	—	68	1.65
35	*Drechslera specifera* Nicot	88	—	80	2.10	50	—	35	1.19
36	*Fusarium nivale* (Fr.) Ces.	107	—	—	1.34	90	—	—	1.26
37	*Fusarium oxysporum* Schlecht	90	—	—	1.13	—	—	—	—
38	*Graphium putridinis* Hughes	—	95	—	1.19	—	70	—	0.98
39	*Humicola fusco-atra* Traaen	—	70	—	0.87	—	—	—	—
40	*Humicola grisea* Traaen	—	—	—	—	—	90	—	1.26

Sr. No	Name of the Fungus	Rhizosphere region			% of* Fungal Occurrence	Non-Rhizosphere region			% of* Fungal Occurrence
		W	S	R		W	S	R	
41	*Memnoniella echinata* Gall.	80	—	—	1.00	—	—	—	—
42	*Monilia sitophila* Sacc	—	—	—	—	—	—	95	1.33
43	*Monodictys levis* Hughes	67	—	—	0.84	—	—	—	—
44	*Myrothecium roridum* Tode ex Fr.	—	40	40	1.00	—	—	—	—
45	*Nigrospora oryzae* Petch.	—	—	—	—	43	—	84	1.78
46	*Paecilomyces marquandii* Hughes	—	68	—	0.85	—	—	—	—
47	*Paecilomyces varioti* Bainer	—	—	—	—	50	55	—	1.47
48	*Penicillium corylophilum* Dierckx	60	—	95	1.94	70	—	90	2.24
49	*Penicillium islandicum* Sopp.	—	—	—	—	90	—	100	2.66
50	*Penicillium nigricans* Thom	85	—	100	2.32	—	—	—	—
51	*Penicillium purpurogenum* Stoll	—	—	—	—	50	25	75	2.10
52	*Penicillium roseo-purpureum* Dierckx	75	48	70	2.42	—	—	—	—
53	*Penicillium rubrum* Stoll	—	—	—	—	55	95	60	2.94
54	*Peziotrichum corticola* Subram	85	—	—	1.06	—	—	—	—
55	*Pithomyces atro-olivaceus* Ellis	—	—	67	0.83	—	—	65	0.91
56	*Rhizoctonia solani* Kuhn	—	65	70	1.69	—	50	65	1.61
57	*Scolecobasidium verruculosum* Roy	—	80	—	1.00	—	—	—	—
58	*Scytalidium thermophilum* Austwick	—	55	110	2.07	—	—	65	0.91
59	*Stachybotrys cartarum* Hughes	—	—	—	—	—	64	—	0.89
60	*Stachybotrys cylindrospa* Jenson	—	70	—	0.87	—	—	—	—
61	*Tetracoccosporium paxianus* Szabo	—	—	—	—	75	—	—	1.05
62	*Thelviopsis paradoxa* Hoehn	—	80	—	1.00	—	77	—	1.07
63	*Torula herbarum* Link	80	91	90	3.27	130	100	140	5.18
64	*Trichoderma glaucum* Abbott	53	70	60	2.29	60	70	76	2.88
65	*Trichoderma koeningii* Oudem	—	130	115	3.07	—	—	—	—
66	*Trichoderma viride* Press. ex Fries	—	—	—	—	61	57	81	2.78
67	*Veronaea coprophila* Ellis	59	—	—	0.74	—	—	—	—
68	*Verticillium glaucum* Bonorden	—	75	—	0.94	—	54	—	0.75

W = Winter; S = Summer; R = Rainy Season

$$\text{* \% of Fungal Occurrence} = \frac{\text{Number of particular fungal colonies of the year}}{\text{Total number of fungal colonies occurred in the year}}$$

Discussion

Currently, there is a welcome awareness about biodiversity almost all around the world. Biodiversity is the total variability with in all the living organisms and the ecological complexes they inhibit, and forms very basis of human survival. Diversity can be recognized at ecosystem, species and genetic levels i.e. biodiversity is the totality of genes, species and ecosystems in a region and thus it represents the total variability within all the living organisms and the ecological complexes they inhabit. The soil is a natural body, differentiated into horizons of loose minerals and organic matter of variable depth on the earth crust. Along with minerals, organic matter, air, water the soil is also composed of innumerable organisms interrelated with each other and creates an atmosphere called soil biota.

The present investigation deals with the study of soil fungi of *Phyllanthus emblica* L. and *Ocimum sanctum* L. Of course it is difficult to estimate their number but attempts have been made to study as much as possible the soil mycoflora of both plants. The study was taken up mainly to understand soil mycoflora of the area and the variations occurring in these soils. The comparative account of the population of fungi has been worked out with a view of finding out population variations in different soils in relation to climatic conditions and soil types. In all, 90 species belonging to 47 genera have been isolated.

Maximum number of species were isolated from the rhizosphere of *Phyllanthus emblica* plants and least number of species were isolated from the non-rhizosphere of *Ocimum sanctum* plants. Thus, the fungal spora associated with different soils varies considerably. Most of the numbers of Zygomycotina were isolated from the rhizosphere soils of *Ocimum sanctum* plant. The most common members were the species of *Mucor* and *Rhizopus* and these were isolated from both the plants and both rhizosphere and non-rhizosphere regions.

The different genera of Ascomycotina occurred rather consistently in different localities. The species of *Chaetomium* were isolated in the soils of both plants. Species of *Achaetomium, Neurospora* were isolated from the soils of *Phyllanthus emblica* but absent in the soils of *Ocimum sanctum.* Maximum species of Deuteromycotina were recorded from the soils of both plants. During the course of study, the species of *Aspergillus* constituted a major group. Next to *Aspergillus* species of *Penicillium, Alternaria, Fusarium, Trichoderma* etc., were the dominant members.

Seasonal influence would appear to be reflected in isolation. Most of the members of Zygomycotina were isolated during all the seasons from different localities of the two plants. Among Ascomycotina, species of *Chaetomium* occurred rather consistently throughout the year. But the species of *Acaetomium* were isolated only during winter season and the species of *Thielavia* were isolated only in summer season. Maximum genera from *Deuteromycotina* were found to be occurring in winter season as against summer and rainy seasons. This data indicates that with the changes in soil moisture and fluctuations in temperature during different seasons, fungal population also varies.

REFERENCES

Anderson, J.P.E. and K.H. Domsch. 1978. A physiological method for the quantitative measurement of microbial biomass in soils. *Soil Biol. & Biochem.* **10**:215–221.

Bilgrami K.S., Jamaluddin and M.A. Rizwi. 1991. Fungi of India, Part I & II. Today and Tomorrows Printers and Publishers, New Delhi.

Booth, C. 1971. Methods in Microbiology. Academic press, London and Newyork, Vol. 4.

Dennis, R.W.G. 1960. British cup fungi and their allies. An introduction to the Ascomycetes. Adlord and son Ltd. Barthelomew press, Dorking, .

Domsch, K.H., and Gams, W. and Anderson T.H. 1980. Compendium of soil fungi, Vol. 1, Academic press, London,

Dwivedi R.S. 1997. Perspectives of Microbial interactions. *J. Indian bot. Soc.* **76**:145–156.

Ellis M.B. and J.P. Ellis. 1985. Microfungi and Land plants (An identification Handbook). Croom Helm Ltd., Sydney, Australia.

Gilman, J.C. 1951. A manual of soil fungi, 2nd Ed. Iowa state College Press, Ames, Iowa, p. 450.

Jain, P.C. 2004. Metabolic dynamism in fungal morphospecies: significance in determining biodiversity. National symposium on Recent trends in Biology, Biodiversity and Biotechnology of Fungi, CAS in Botany, University of Madras, Chennai–Feb. 5–6, 2004.

Sarabhoy, A.K., Girdharilal and J.L.Varshney. 1975. Fungi of India, Navayug traders publications, New Delhi.

Subramanian, C.V. 1971. Hyphomycetes, I.C.A.R. Publ. New Delhi.

Von Arx J.A. 1981. The Genera of Fungi sporulating in pure culture. III Edi. J. Cramer, Varduz, Germany.

Waksman, S.A. 1927. Principles of soil microbiology, Williams and Wilkins Co. Baltimore.

Warcup, J.H. 1951. The ecology of soil fungi. *Trans. Brit. Mycol. Soc.* **36**:376–400

Watanabe, T. 1994. Pictorial Atlas of soil and seed fungi. Lewis publishers, Boca Raton.

CHAPTER 2

STUDIES ON INTRASPECIFIC VARIATIONS OF *ANDROGRAPHIS PANICULATA* NEES IN PRASANTHINILAYAM

*K. Venkateswara Rao** and *S. Sai Kiran Kumar**

ABSTRACT

Andrographis paniculata Nees belongs to family Acanthaceae, the Kalmegh of Ayurveda and also known in northeastern India as 'Maha-tita', literally 'king of bitters' is used as a wonder drug in traditional Siddha and Ayurveda systems of medicine as well as in tribal medicine in India for treating a variety of ailments (Snakebite, stomach complaints, wounds, itches, malaria, fever, diarrhoea, dysentery etc.)

In the present study analysis of intraspecific variations in *Andrographis paniculata* Nees was carried out. Three local populations were sampled from different parts of Prasanthi Nilayam (AP 1: Institute campus; AP 2: Hilly ranges; AP 3: Local farm lands). They were raised under uniform growth conditions and analyzed at morphological, cytological, phytochemical and biochemical levels to reveal extent of intraspecific variability, with special reference to Andrographolide content.

Andrographolide ($C_{20}H_{30}O_5$; MW–350.44) the plant specific bicyclic diterpeniod lactone of the leaf samples were analyzed. Andrograpaholide content on the dry weight basis was highest in AP 2 (1.18%), with a mean value of 0.93% for all the populations. In general AP 1 recorded the maximum of plant height with a mean height of 81.36 cms and AP 2 and AP 3 showed a mean height of 55.6 cms and 72.5 cms respectively. Two enzyme systems (Esterase, Glutamate Oxaloacetate tranaminase) were used for the purpose of isozyme analysis. The isozyme analysis revealed that *A. paniculata* of the hilly regions was genetically variable when compared to other two populations.

In the light of the results discussed in the paper, it is a well considered suggestion, that a two way approach viz., selection of an improved natural variant for the development as cultivar and isolation of high-yielding cultivar from the

* Department of Biosciences, Sri Sathya Sai Institute of Higher Learning, Prasanthi Nilayam-515 134 (A.P).

selected genotype through mutagenesis would be most fruitful for achieving the twin objectives of conservation of existing genetic diversity and sustainable utilization of this medicinal resource, so unique with its wide range of variation, in India.

Introduction

India is a known mega-diversity center harboring a multitude of medicinal plant species each presumably studded with as yet unknown genetic and chemical variations of economic importance. Out of an estimated 17,000 higher plant species occurring in India, more than 1000 species are used over several centuries in the traditional systems of medicine viz. Ayurveda, Siddha, Unani and Amchi. The villagers and tribal folks spread across the length and breadth of the country make use of more than 7000 plant species through oral traditions.

Knowledge of the genetic diversity is also a prerequisite for any *in situ* and *ex situ* conservation schemes as it is not practical to conserve all genotypes of a given species against the mass extinction spasm projected for the 21st century (Raven, 1999).

Andrographis paniculata Nees (Acanthaceae), the Kalmegh of Ayurveda selected for the present investigation is an erect annual herb, extremely bitter in taste in each part of the plant body. It is also known as '*Bhui-neem*' since the plant, though much smaller in size, shows similar appearance and has bitter taste as that of Neem (*Azadirachta indica).* Incidentally, the genus *Andrographis* consist of 28 species of small annual shrubs essentially distributed in tropical Asia. Only a few species are medicinal, of which *A. paniculata* is the most popular. Since time immemorial, villagers and ethnic communities in India have been using this herb for treating a variety of ailments (Snakebite, stomach complaints, wounds, itches, malaria, fever, diarrhoea, dysentery etc.). It is one of the important ingredients in Ayurvedic preparations recommended for fever and liver diseases. These include 'Liv 52' for treating hepatoxicity; 'TBR-002' for fever, 'Tephroli' for treating viral hepatitis etc.

Andrographolide, the chief constituent extracted from the leaves of the plants, is a bitter water-soluble lactone exhibiting protective effects in carbon tetrachloride induced hepatopathy in rats. This bitter principle was isolated in pure form. Despite its enormous medicinal importance, attempts to cultivate *A. paniculata* have seldom been undertaken in any part of the country. Hence the local vaidyas as well as drug companies depend on wild source for supply for raw material.

The current work was undertaken with an objective to study intraspecific variations of *Andrographis paniculata.* Nees found in and around Prasanthi Nilayam. The intraspecific variations of *A. paniculata* was worked out well by Sabu (2001) in Kerala, but taking into consideration the extensively varying climatic and geographic conditions of Prasanthi Nilayam and its environs the authors felt necessary to conduct a study of the plant and prepare the Chemical and DNA (Genetic) profiles, after surveying the area of study and identifying the plants for study. In this process, variability existing in the wild populations of the species purported to have ecological and more importantly genetic basis, has been addressed at morphological, cytological, phytochemical and biochemical levels. It is needless to say that in the wilderness of nature, the populations of a sexually reproducing plant like *A. paniculata* are presumably so varied, that selection of high biomass and product yielding genotypes in distinctly possible and holds much promise for upgradation into cultivars.

Materials and Methods

Plant Materials

Samples of *Andrographis paniculata* Nees were collected from various sources, after a survey of areas in and around Prasanthi Nilayam. Three main areas were identified for final plant collection viz, the Institute Campus, the Hilly range and the local farmlands (Table 1). Plants from these areas were collected and raised in pots in an experimental area in the Institute. These plants were used for various experimental purposes during the course of the present study.

Table 1: List *Andrographis paniculata* population studies. Appropriate Ids were given with the prefix of AP derived from the botanical name, *Andrographis paniculata*

S. No.	*Locality*	*Status*	*ID*
1.	Institute Campus	Wild	AP 1
2.	Local Hills	Wild	AP 2
3.	Local Farm Lands	Wild	AP 3

Plants of *Andrographis paniculata* collected from these localities were raised in pots containing fertile soil and vermin compost in 3:1 ratio. Plants were raised under identical conditions and were irrigated regularly on alternate days except during the rainy days.

Methods

Ploidy Check and Cytological Studies

To check the ploidy level of the collected plants, excised young flower buds were fixed in Carnoy's fixative (Sharma and Sharma, 1980) as described by

Ziaciddin *et al.* (1997). Cytological observations were made under a microscope with photographic attachment.

Phenology

The flowering of plants was studied by observing the appearance of flowers in both the natural as well as the experimental plants, regularly.

Moisture Content of the Seeds

To study the effect of moisture content on seed germination, lots of 25 seeds were sown on filter paper discs moistened with 7.5 ml of deionized water in petri plates. The seeds were germinated in the dark at a temperature regime of 33°C/23°C (day/night) under laboratory conditions. The seed moisture content of the samples was determined by incubating seeds at 103$\pm$2°C for 17 hours in an electric oven.

Phytochemical Studies

Plants grown under uniform conditions were screened for the major active principle Andrographolide ($C_{20}H_{30}O_5$; MW 350.44), a bicylic diterpenoid lactone. The methanolic extract of the shade-dried leaves was analysed sprectrophotometrically at 223 nm to estimate its concentration.

Physiological Studies

Young (2 month-old) and mature (6 month-old) plants were selected for analyzing age-related changes in chlorophyll, carotenoid and anthocyanin content among the populations. Chlorophyll and carotenoids were extracted using 80% acetone (Sadasivam and Manikkam, 1996). For estimating variation in anthocyanin content of leaves, the absorbance of the supernatant was measured at 530 nm (Janda *et al.,* 1996).

Chemical Profiling

Fresh leaves (10gms.) cut in small pieces and were refluxed with ethanol (50 ml used, generally the amount of ethanol taken should be enough to submerge all the leaves) for 6 hours. The extract obtained was filtered and concentrated. The extract thus obtained was subjected to TLC using varying proportions of the solvent systems Ethyl acetate and Petroleum ether. The plates were exposed to iodine treatment for varying time periods depending on the bands/streaks on the TLC plates. The Rf values for all the plates were calculated.

Depending upon the separation obtained, PTLC was run with all the solvent systems that showed a very good separation. These plates were run with a larger volume of the solvent system (~ 90 ml.). The Rf values were calculated. The separate bands that were obtained were characterized using UV spectroscopy. The bands were carefully scraped out and were dissolved in the respective solvent. This separates the compound from the silica (as silica is insoluble). The compound thus obtained was taken the UV spectroscopy was done. The chemical profile of the sample was prepared.

Paper Chromatography and Shinoda test was done for confirming the presence of flavonoids.

Isozyme Analysis

Fully opened leaves of mature plants (3–4 months old) were used to prepare crude enzyme extracts for analyzing variation between the three populations using two enzyme systems Esterase (Est) and Glutamate oxaloacetate transaminse (Got). Vertical discontinuous Sodium Dodecyl Sulphate Polyacrylamide Gel Electrophoresis (SDS-PAGE) was carried out for separation of isozyme(s) using crude extracts, with appropriate modifications (Anonymous, 1997). Two enzyme systems were resolved in the three populations using Tris-Glycine electrophoresis buffer system. Staining of the gels was done by using standard methods.

Results and Discussion

Three populations of *A. paniculata* were sampled from regions in and around Prasanthi Nilayam. The plant was found distributed in wild lands, as a weed in cultivated farmlands and as a weed in the Institute campus.

Cytology

The microscopic examination of flower bud squashes from the three populations (AP 1, AP 2 and AP 3) revealed no change in ploidy level (Table 2). The haploid chromosome number (n) was uniform in all the three populations, i.e. n = 25. Reports in *A. paniculata,* confirm the ploidy (n) as 25 (Govindarajan and Subramanian, 1983; Roy and Dutta, 1988).

Table 2: Ploidy in *A. paniculata*

Population	*Ploidy*
AP 1	25
AP 2	25
AP 3	25

Phenology and Flowering

Initiation of flowering was noticed from December onwards. Plants growing under favorable conditions flowered throughout the year. In AP 1 (plants in drier hilly region) flowers started appearing from the last week of January onwards. Size of the flower was same in all the three populations.

Morphology

Considerable diversity in the morphology was found. AP 2 was shorter than the other two. In general AP 1 recorded the maximum for plant height with a mean height of 81.36 cms. AP 3 showed a mean height of 72.5 cms and AP 2 showed the least mean height of 55.6 cms. (Table 3).

Table 3: Mean heights of various populations of *A.paniculata*

Population	*Mean Height (cms)*
AP 1	81.36
AP 2	55.6
AP 3	72.5

Seed Moisture Content

The seeds showed average moisture content of 13.7%. There was no significant variation of the moisture content in the three populations (Table 4). Studies have shown that a content in the range of 17.4% to 13.5% is required for optimum germination (Singh and Ahmad, 1997).

Table 4: Moisture Content

Population ID	*Moisture content*
AP 1	15.34%
AP 2	11.24%
AP 3	14.52%

Variation in Pigment Content

The quantitative analysis of chlorophylls. Carotenoids and anthocyanins was carried out. Pigment content of leaves of 6 month old plants revealed considerable difference [Table 5(a) and 5(b)]. Total chlorophyll content varied from 0.042 (AP 2) to 0.057 (AP 1), with a mean value of 0.0476. Absorbance of carotenoids was lowest in AP 2 (3.711) and highest in AP 1

(5.393), with a mean value of 4.407. The values obtained for anthocyanins was maximum in AP 1 (1.631) and least in AP 2 (0.688), with a mean value of 1.285.

Table 5(a): Concentration of Carotenoids and Anthocyanins (Absorbance at 530)

Population	*Carotenoids (Abs)*	*Anthocyanins (Abs)*
AP 1	5.393	1.631
AP 2	3.711	0.688
AP 3	4.117	1.536

Table 5(b): Concentration of chlorophyll *A. paniculata*

Population	*Total Chlorophyll (mg/Plant tissue)*	*Chlorophyll a (mg/Plant tissue)*	*Chlorophyll b (mg/Plant tissue)*
AP 1	0.057	0.042	0.015
AP 2	0.042	0.031	0.011
AP 3	0.044	0.033	0.011

Plant pigments, particularly Anthocyanins are considered as genetic markets in many instances (Hahlbrock, 1981)

Andrographolide Content

The andrographolide content from the three populations was estimated using UV spectrophotometer. The andrographolide content in the leaves varied from 0.73 to 1.18%, with a mean value of 0.93%, in the three populations (Table 6).

Table 6: Andrographoplide content of individual populations

Populations ID	*Andrographolide Content (%)*
AP 1	0.73
AP 2	1.18
AP 3	0.90

The data presented in table 6, revealed that there is intraspecific diversity among populations of *A. paniculata* in the concentration of the active principle andrographolide. It is known that extreme climate conditions induce production of secondary metabolites in more quantity than the normal climatic conditions (Sabu, 2001). Based on the present investigation, the population AP 2 has the potential to be grown as a plant for economic benefits, in the draught prone areas like Anantapur.

Chemical Profiles

The ethanol extract was subjected to TLC with 11 solvent systems, starting from least polar to fully polar solvent systems. The solvent system (SS) 10 showed a complete streak when treated in iodine chamber. Plates with solvent systems 11, 9 and 8 showed discontinuous streaks. Solvent systems 1, 2, 3 did not show any streaks or spots. Solvent systems 4, 5, 6 and 7 showed single spots. The solvent systems that showed streaks were used again in the PTLC analysis for more separation of the compounds. SS-9 showed 5 distinct bands, (Table 7 for Rf values), whereas SS-8 showed two bands. The UV spectrophotometric data of these bands showed characteristic resemblance with already well established standard spectral data.

Table 7: Rf values of PTLC (Solvent Systems: 9)

Population	*Distance Moved by the compound (cms)*	*Rf Values*
AP 1	0.5	0.029
	6	0.352
	9.5	0.558
	13	0.764
AP 2	0.5	0.029
	6	0.352
	9.5	0.558
	13	0.764
AP 3	0.5	0.029
	6	0.352
	9.5	0.558
	13	0.764

The single spots obtained in SS-4 to 7, were also subjected to paper chromatography, confirming the nature of secondary metabolites (Table 8). The flavonoids were analysed by paper chromatography and later confirmed by Shinoda's test. There was no difference found in all the three populations regarding the nature of the secondary metabolites.

Table 8: Flavonoids from *A. paniculata*

UV	*UV/NH 3*	*Likely Flavonoid type*
Dark Purple	Yellowish Green	5, 4–Flavones/Flavonols with no free 3, 3–or 5-OH
Blue Florescence	Blue-green	Flavones lacking free 5-OH group
Dull Yellow	Dull Yellow	Flavonol Glycones

Andrographin, panicotin, aplegenin-4, 7-dimethyl ether, mono-o-methyl wightin, and hydroxy-7, 8, 2', 3'-trimethoxy flavone and β-sitosterol, were reported in *A. paniculata.* (Encyclopedia of Chemicals and Drugs, 1958).

Isozyme Analysis

Two enzyme systems were used for the purpose of isozyme analysis. In the isozyme-Esterase (Est), the bands on the PAGE gels were three in number, and these were present uniformly in all three populations. However, lane AP2 contained an extra band, which was unique for this population. Similar results were reported previously by Sabu (2001), in *A. paniculata* of Pazhakulam (a high altitude place), Kerala. In the Glutamte Oxaloacetate transaminase (Got) gels, two distint bands were observed uniformly in all the three populations. Both the enzymes systems used showed less amount of variation except for Est of AP 2.

REFERENCES

Anonymous, 1997. Laboratory workshop manual on "Purification and Characterization of plant proteins and basic recombinant DNA techniques" held at Department of Biochemistry and Department of Molecular and Cellular Biology, Indian Institute of Science, Bangalore, India, February 24–March 3, 1997.

Govindarajan, T. and D. Subramanian. 1983. Karyomorphological studies in South Indian Acanthaceae. *Cytologia,* **48:**491–504.

Hahlbrock, K. 1981. Flavonoids. In: E.E.Conn (ed.)The Biochemistry of Plants. Vol. 7. Secondary Plant Products. Academic Press. London.

Janda, T., G. Szalai and E. Paldi, 1996. Chlorophyll fluorescence and anthocyanin content in chilled maize plants after return to a non-chilling temperature under various irradiances. *Biologia Plantarum,* **38:**625–627.

Raven, P.H. 1999. World's biodiversity becoming extinct at levels rivaling Earth's past "Mass extinctions." *Botanical Garden conservation news,* **3:**31–32.

Roy, S.K. and P.C. Dutta, 1988. Production of a variant line by tissue culture of *Andrographis paniculata* with enhanced yield of andrographolide. *Bangladesh J. Bot.,* **17:**117–122.

Sabu, K.K. 2001. Intraspecific Variations in *Andrographis paniculata* Nees. Ph.D. Thesis, The Kerala University, Trivendrum, India.

Sadasivam, S. and A. Manikkam, 1996. Biochemical methods. 2nd ed. New Age International (P) Ltd., Publishers, New Delhi.

Sharma, A.K. and A. Sharma 1980 Chromosome techniques: Theory and praetice. 3rd ed.; (Butterworths, UK).

Singh, N.B. and Z. Ahmad, 1997. Genotypic differences in wheat under moisture stress conditions. *Indian J. Plant Physiol.,* **2:**163–165.

Ziaciddin A., M. Peng and D.J. Wolyn, 1997. Improved nuclear staining of *Asparagus* microspores for cytological analysis; *Hort. Science,* **32:**735–736.

CHAPTER 3

STUDIES ON MORPHOLOGY OF TEN GRASS SPECIES OF ANDROPOGONEAE

*D. Someswari** and *T.N. Mary***

ABSTRACT

Ten grass species belonging to four genera viz., *Eremopogon, Dichanthium, Cymbopogon* and *Iseilema* of Andropogoneae in Poaceae were studied from areas of Guntur district for various morphological traits. Morphology of the inflorescence was found to be interesting due to its complexity. Therefore, based on the details of spike variation existing in the ten grass species, the trends of elaboration and their adaptive significance in Adropogoneae were discussed.

Introduction

The value of grasses to the human race is incalculable as they affect and support virtually every facet of human existence. Grasses exhibit greater morphological, reproductive and adaptive variations due to hybridization, polyploidy and apomixis etc. The investigations on grasses of Poaceae can make a better understanding of plant evolution in general.

The grass tribe, Andropogoneae of the present study is considered to be the most highly specialized by many workers (Bews, 1929; Hartley, 1950). It is a well defined monophyletic tribe with spikelets and inflorescences of considerable complexity. The available morphological and geographical information suggests that the tribe is of comparatively recent origin.

Though extensive studies were conducted on grasses all over the world (Myers, 1947; Tateoka, 1960; Watson *et al.*, 1985; Soderstrom *et al.*, 1986), there is practically little work on the grasses of Andropogoneae from India.

* Department of Botany, T.J.P.S. College, Guntur-522 006.

** Department of Botany, Acharya Nagarjuna University, Nagarjunanagar-522 510.

Taxonomically, some of the genera in Andropogoneae are still in a confused state; revisionary work on these taxa is the most immediate requirement for a better understanding of the tribe (Clayton and Renovoize, 1986). Hence, ten species growing in nearby areas of Guntur district belonging to four genera viz., *Eremopogon* Stapf, *Dichanthium* Willemet, *Cymbopogon* Sprengel and *Iseilema* Anderson were chosen in this investigation.

Materials and Methods

Ten species belonging to four genera, i.e. *Eremopogon, Dichanthium Cymbopogon, Iseilema* chosen for study and are listed below:

1. *Eremopogon foveolatus* (Del.) Stapf Namalachettu
2. *Dichanthium annulatum* (Forsk.) Stapf Marvel grass
3. *Dichanthium pertusum* (L.) Clayton
4. *Cymbopogon coloratus* (Hook.f.) Stapf Doddanchi grass
5. *Cymbopogon caesius* (Hook. & Arn.) Stapf Kachi grass
6. *Cymbopogon jwarancusa* Nees. ex Steud.–Kovigrass
7. *Cymbopogon flexuosus* (Jones) Schult.– East Indian Lemon grass
8. *Iseilema laxum* Hack. Chengali gadde
9. *Iseilema anthephoroides* Hack.
10. *Iseilema venkateswarlui* Satyavathi

Among the ten species of the present study, species of *Iseilema, Dichanthium* and *Eremopogon* are fodder grasses while those of *Cymbopogon* are aromatic grasses. The morphological characters such as length of culm, root, number of tillers, leaves, spikes, spikelets, inflorescences, length and breadth of third leaf, flag leaf etc., were determined. For each parameter, the average value was calculated and the individual readings were analysed for mean and standard deviation. Statistical methods of Snedecor and Cochron (1967) were followed for F-test and CD analyses. DMR test was applied to the data when F-values were found significant.

Taxonomy

The tribe Andropogoneae belongs to the subfamily Panicoideae of Poaceae. The Panicoideae is characterised by 2-flowered spikelets which fall on maturity and usually with bisexual upper floret and lower barren or male floret, the latter often reduced to lemma. The three tribes of Panicoideae can be identified by the following characters.

1. Spikelets both male and female in separate inflorescence or in different parts of the same inflorescence and of different appearance..Maydeae
1. Spikelets bisexual or with male or barren and bisexual spikelets mixed in the same inflorescence:

2. Spikelets often paired, one sessile and the other pedicelled, rarely solitary and all alike; upper lemma usually awned...Andropogoneae
2. Spikelets solitary or paired, more or less similar, usually unawned ....Paniceae

The key characters of the genera and species selected for the present study are as follows:

Genus: Eremopogon

Inflorescence of solitary racemes. Spikelets awned. All spikelets heterogamous. Racemes supported by spathes, sessile spikelets pitted, perennial nodes densely bearded. Species–*Eremopogon foveolatus* (Del.) Stapf.

Genus: Dichanthium

Inflorescence not a panicle, digitate racemes. All spikelets both sessile and pedicelled in the racemes more or less alike, glumes herbaceous and greenish.

Key to Species

1. Lower glumes of the sessile spikelet pitted; nodes sparsely bearded .... *D. pertusum*
1. Lower glumes of the sessile spikelets not pitted, nodes with long hairs...................... ....*D. annulatum*

Genus: Cymbopogon

Involucral spikelets absent. Aromatic grasses, racemes in pairs, supported by spathes, often collected into huge panicles, one pair of spikelet in each raceme, homogamous male or neuter, all other pairs heterogamous.

1. Panicles narrow of pairs of racemes..*C. jwarancusa*
1. Panicles very large, compound:
   2. Lower glume of sessile spikelet with a narrow grove in the lower half......*C. caesius*
   2. Lower glume of sessile spikelet flat, wrinkled:
      3. Pedicel of lowest pedicelled spikelet in the sessile raceme not swollen................ ....*C. flexuosus*
      3. Pedicel of lowest pedicelled spikelet in the sessile raceme swollen...................... ....*C. coloratus*

Genus: Iseilema

Involucral spikelets present. Racemes solitary at the end of the branches at the base by an involucre of homogamous spikelets. Involucre of 4–1 flowered or sterile spikelets. Involucral spikelets pedicellate, deciduous, callus fertile, spikelets very short truncate or obtuse.

1. Pedicels of the involucral spikelets almost as broad as long................ *I. anthephoroides*
1. Pedicels of the involucral spikelets otherwise:
   2. Pedicelled spikelets shorter than and hidden by the involucral spikelets........................ .... *I. venkateswarlui*
   2. Pedicelled spikelets projecting beyond the tips of the involucral spikelets........*I. laxum*

Results and Discussion

Interspecific variations based on several morphological characters have been presented among ten species of four genera studied in the present investigation through bar graphs (Figs.1–6). 'F' values for the length of culm, root, leaf, number of tillers, leaves, inflorescences, spikelets etc., were found to be significant among them (Tables 1–6).

Table 1: Vegetative characters of species of *Dichathium* and *Eremopogon*

Character	*D. annulatum*	*D. pertusum*	*E. foveolatus*	*F-value*	*CD-value*
Culm length (cm)	76.27±2.31	14.19±0.41	33.03±1.41	169.37	3.43
Root length (cm)	4.10±0.15	3.02±0.07	2.83±0.17	12.12	0.09
S/R ratio	18.93±1.53	4.93±0.09	11.05±0.05	56.20	1.32
Internode length (cm)	6.62±0.13	5.06±1.80	2.85±0.28	28.03	0.27
Number of tillers	10.73±0.44	11.06±0.85	4.40±0.44	371.57	0.30
Number of branches	7.40±0.33	5.80±0.27	5.00±0.60	2.71	0.85
Number of leaves	68.26±2.38	76.73±2.69	31.86±2.71	39.87	5.33
Leaf length (cm)	5.98±0.12	3.12±0.05	6.43±0.16	411.79	0.12
Leaf breadth (cm)	0.10±0.00	0.10±0.00	0.10±3.00	12.30	0.02
Number of stomata	1.42±0.05	1.34±0.03	1.27±0.01	3.59	0.05

Table 2: Floral characters of species of *Dichanthium* and *Eremopogon*

Character	*D. annulatum*	*D. pertusum*	*E. foveolatus*	*F-value*	*CD-value*
Flag leaf length (cm)	3.34±0.09	3.66±0.09	3.70±0.08	4.81	0.13
Flag leaf breadth (cm)	0.01±0.00	0.09±0.10	0.07±0.60	2.90	0.19
Inflorescence length (cm)	9.74±0.16	9.06±0.81	3.86±0.23	237.05	0.35
Number of inflorescences	10.06±1.18	10.33±1.14	10.93±1.21	3.05	5.35
Number of spikes	7.31±0.81	3.80±0.34	8.20±0.21	8.54	1.81
Number of spikelets per spike	62.13±1.62	19.86±0.26	30.13±0.51	170.92	2.37
Seed L/B Ratio	1.80±0.27	2.09±0.40	2.86±0.46	2.33	0.66
Pollen fertility (%)	74.28	84.38	82.32		

All the species of four genera, *Eremopogon, Dichanthium, Cymbopogon* and *Iseilema* were noted as perennial grasses (Fig. 1), except *Iseilema anthephoroides* and *I. venkateswarlui* as annuals (Plate I). *Eremopogon foveolatus, Dicanthium annulatum* and *D. pertusum* are sympatric in growth on University campus grass level. In both the species of *Dichanthium* and *Eremopogon faveolatus*, the presence of hairs at the nodes extending to the base of the inflorescence was found as a common trait. The inflorescence is solitary raceme in *Eremopogon* (Plate I) digitate raceme in species of

Fig. 1. Vegetative characters of species of *Dichanthium* and *Eremopogon*

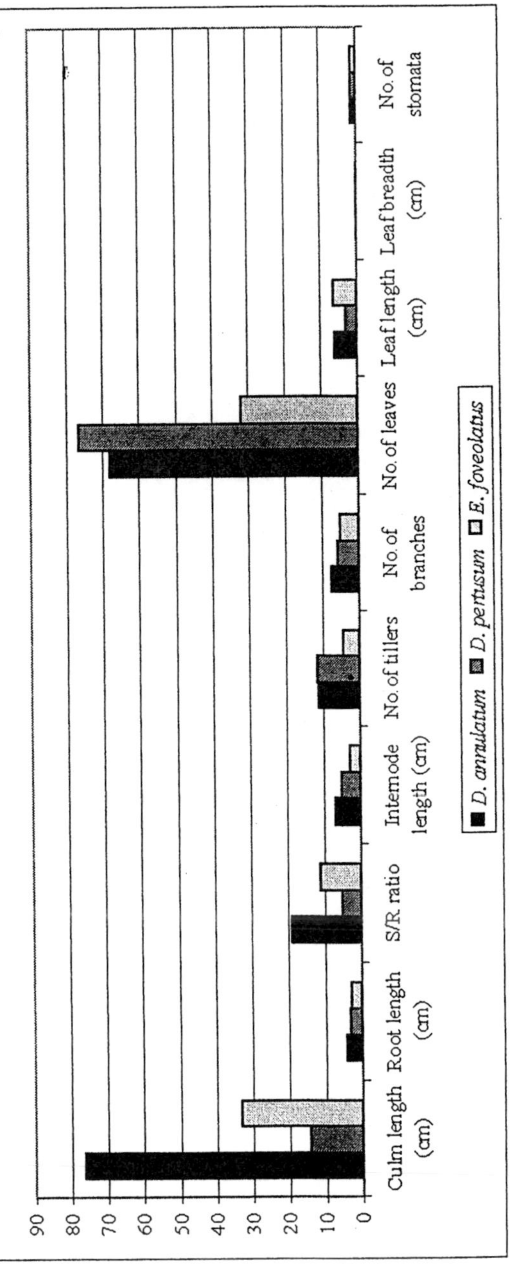

Fig. 2. Floral characters of species of *Dichanthium* and *Eremopogon*

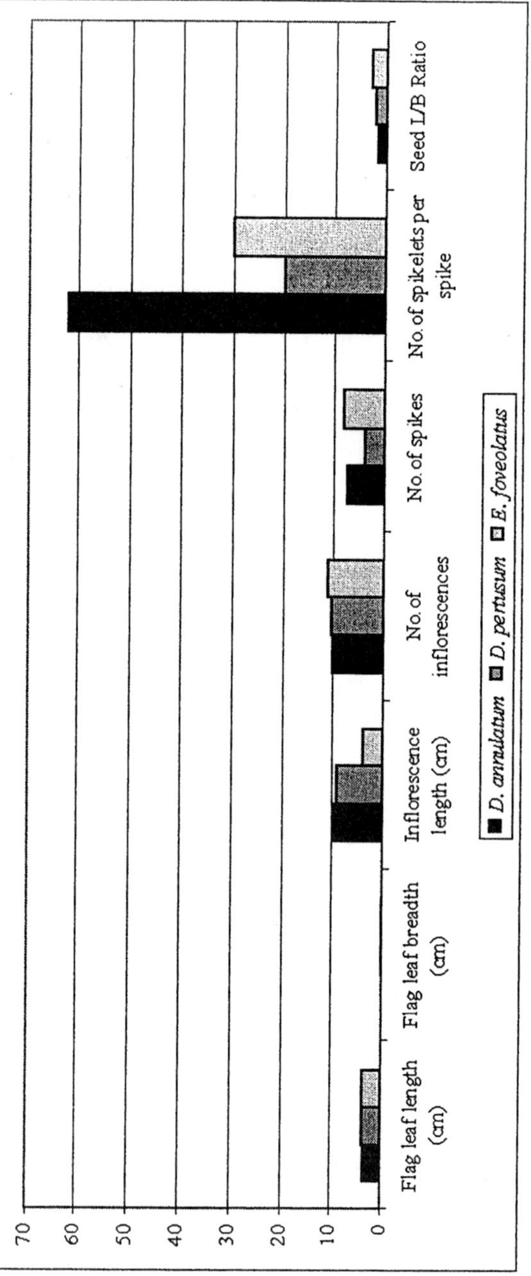

Species of Four Genera in Andropogoneae
(a) *Dichanthium pertusum*; (b) *Eremopogon foveolatus*; (c) *Dichanthium annulatum*; (d) *Iseilema laxum*; (e) *Cymbopogon caesius*

Plate I

Morphology of *Iseilema* Species (Variation in Spikes is shown)
(a) *I. anthephoroides*; (b) *I. venkateswarlui*; (c) *I. laxum*

Plate II

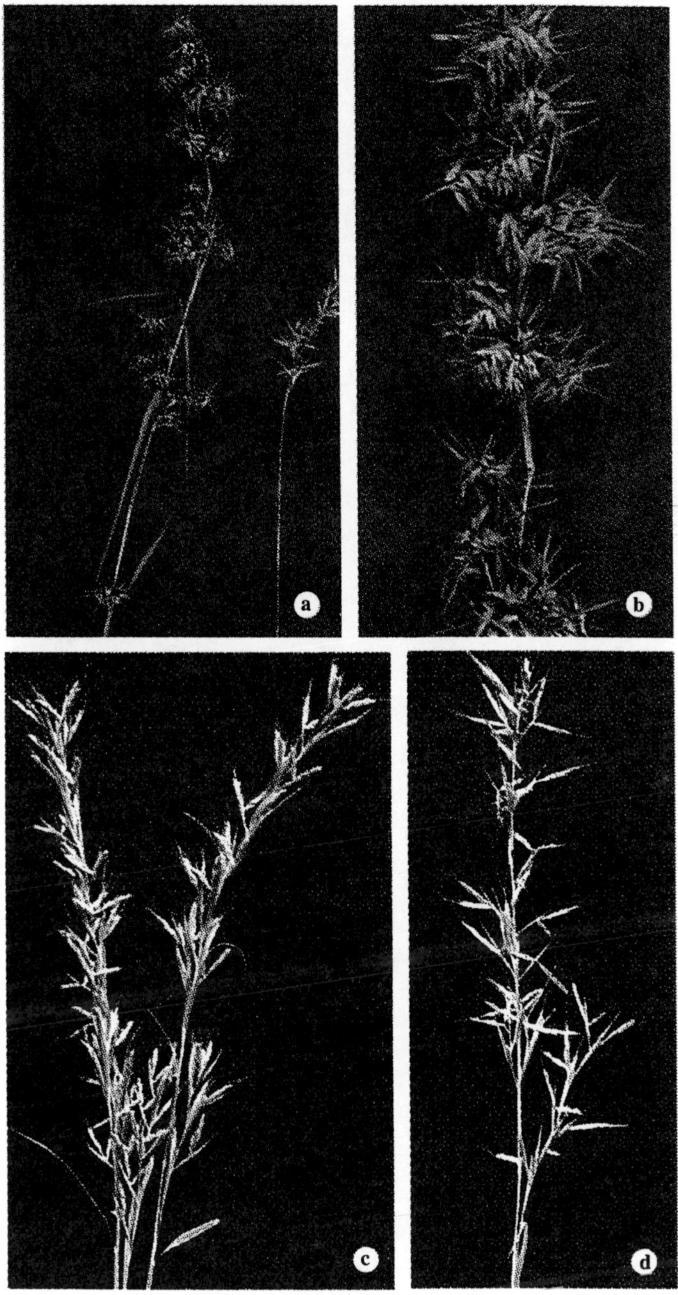

Inflorescence in Species of *Cymbopogon*
(a) *C. jwarancusa*; (b) *C. flexuosus*; (c) *C. coloratus*; (d) *C. caesius*

Plate III

Fig. 3. Vegetative characters of ***Cymbopogon*** species

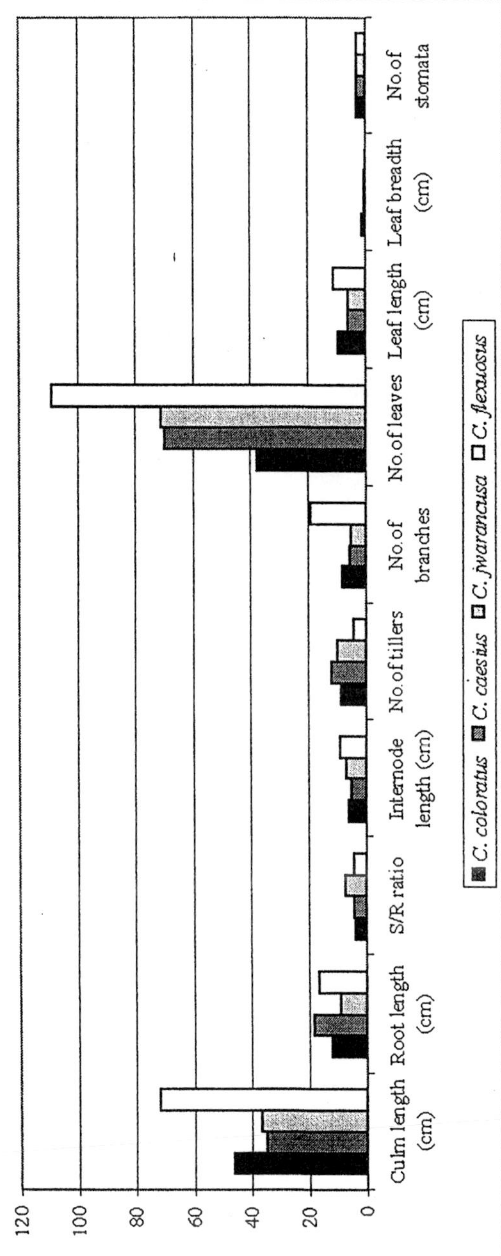

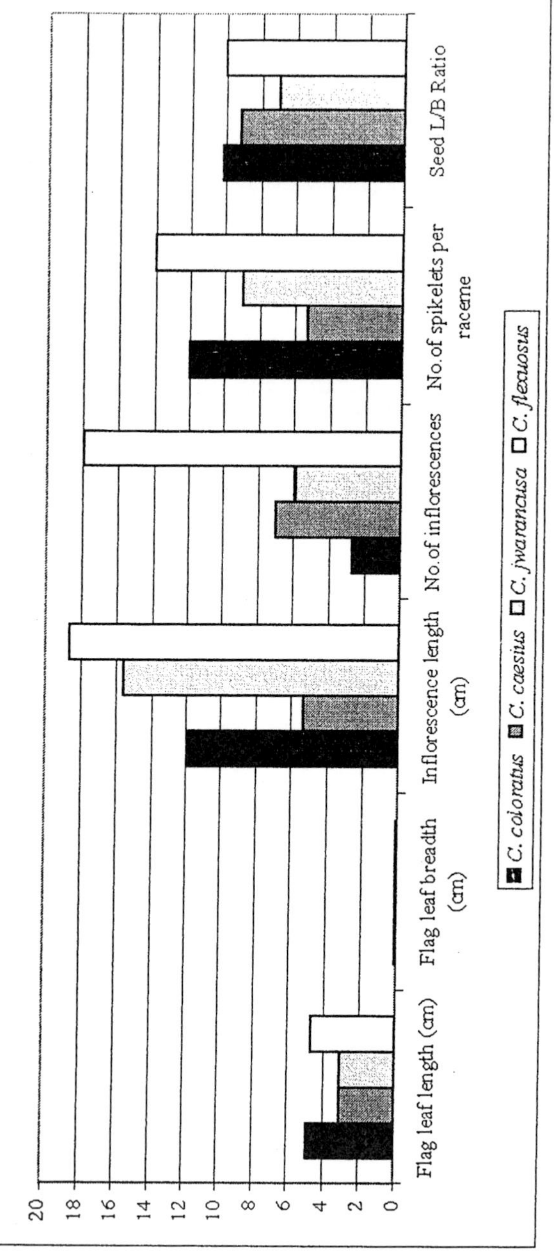

Fig. 4. Floral characters of *Cymbopogon* species

Fig. 5. Vegetative characters of species of *Iseilema*

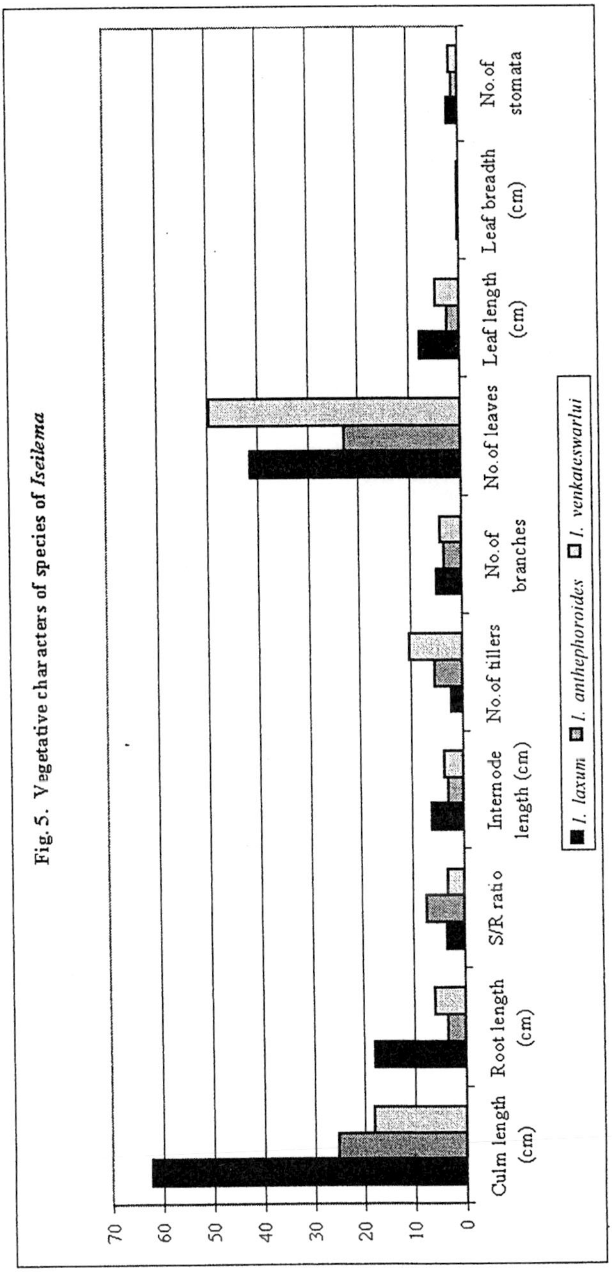

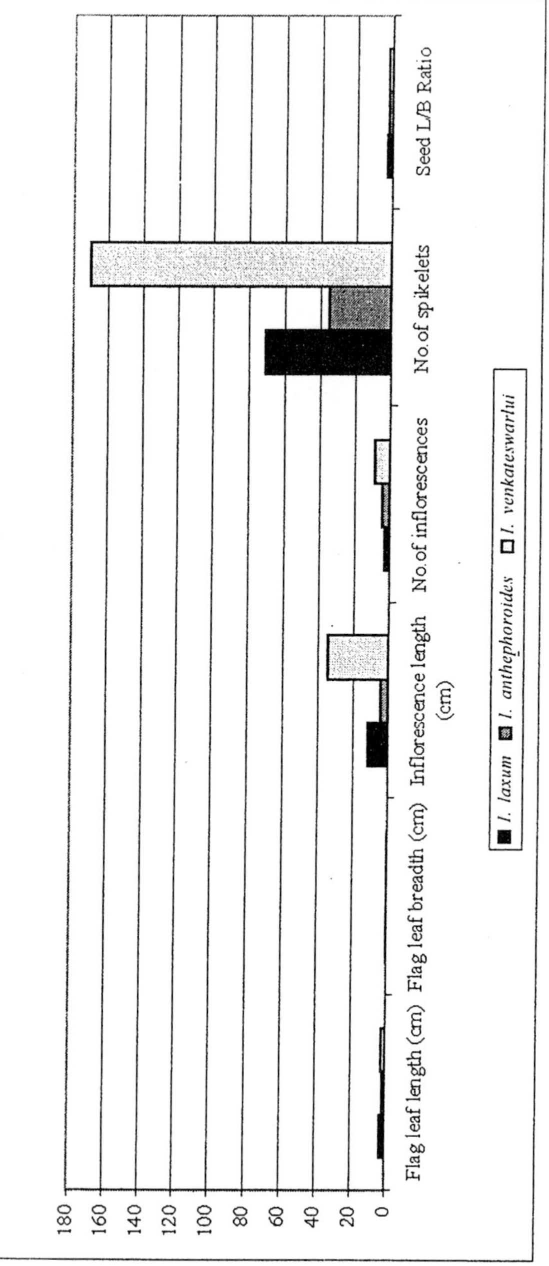

Fig. 6. Floral characters of three species of *Iseilema*

Dicanthium with sessile spikelets pitted in *D. pertusum* and absence of pits in *D. annulatum*.

Among the species of *Cymbopogon* from hills, *C. flexuosus* showed maximum values for length of culm, inflorescence, number of branches, leaves, inflorescence and racemes (Tables 3 and 4). In *C. jwarancusa*, rosette type of arrangement was found though leaves are arranged at nodes. The inflorescence in *C. jwarancusa* showed compact arrangement of racemes, while in *C. flexuosus* it was loosely arranged (Plate IIIa & b).

Table 3: Vegetative characters of *Cymbopogon* species

Character	*C. coloratus*	*C. caesius*	*C. jwarancusa*	*C. flexuosus*	*F-value*	*CD-value*
Culm length (cm)	45.86±1.48	34.75±1.41	36.33±4.09	71.92±0.31	109.08	2.23
Root length (cm)	11.95±0.86	18.54±0.46	9.04±0.49	16.68±0.74	35.71	1.12
S/R ratio	3.93±3.02	4.11±0.19	7.22±0.26	4.35±0.26	23.40	0.45
Internode length (cm)	5.85±0.50	4.98±0.20	6.91±0.20	9.03±0.10	1.52	0.50
Number of tillers	8.40±0.56	11.90±0.84	9.70±0.73	4.10±0.53	15.86	1.16
Number of branches	8.00±0.30	5.70±0.60	5.20±0.54	19.30±2.00	11.98	2.72
Number of leaves	37.60±1.69	70.00±4.26	71.30±4.49	108.90±5.29	33.53	7.23
Leaf length (cm)	9.39±0.28	5.86±0.21	5.99±0.23	11.33±0.26	120.66	0.34
Leaf breadth (cm)	1.18±0.04	0.20±0.01	0.18±0.01	0.10±0.01	0.63	0.10
Number of stomata	2.97±0.06	3.07±0.11	2.82±0.06	2.81±0.10	1.95	0.12

Table 4: Floral characters of *Cymbopogon* species

Character	*C. coloratus*	*C. caesius*	*C. jwarancusa*	*C. flexuosus*	*F-value*	*CD-value*
Flag leaf length (cm)	4.92±0.19	3.05±0.06	3.09±0.08	4.70±0.11	40.66	0.22
Flag leaf breadth (cm)	0.09±0.01	0.10±0.00	0.09±0.01	0.10±0.00	1.52	0.50
Inflorescence length (cm)	11.85±0.78	5.30±0.26	15.58±0.25	18.67±0.21	34.17	1.01
Number of inflorescences	2.70±0.89	7.00±3.16	5.87±0.78	17.90±1.98	20.85	1.96
Number of spikelets per raceme	12.00±1.45	5.30±0.26	9.00±0.61	14.00±0.79	2.69	1.03
Seed L/B Ratio	10.20±0.44	9.20±0.85	7.02±0.80	10.00±0.07	0.14	7.78
Pollen fertility (%)	83.89	84.28	79.39	80.75		

A comparison of the morphological characters of species of plains viz., *C. coloratus* and *C. caesius* revealed that quantitative morphological variability was not significant (Table 3 and 4) except for number of racemes per inflorescence. However, the species of *C. caesius* was distinguished in nature by light green colour of the leaf ending into curls, whereas *C. coloratus* is differentiated by thick green leaves, racemes are in pairs, appearing very much similar in both *C. coloratus* and *C. caesius* (Plate IIIc & d), but differ in the presence of narrow slit like groove in the lower glume of the sessile spikelet in *C. caesius* and its absence in *C. coloratus*.

Table 5: Vegetative characters of species of *Iseilema*

Character	*I. laxum*	*I. anthephoroides*	*I. venkateswarlui*	*F-value*	*CD-value*
Culm length (cm)	62.30±2.03	25.22±0.91	18.15±0.72	18.67	15.27
Root length (cm)	17.79±1.05	3.35±0.14	5.95±0.23	115.72	1.00
S/R ratio	3.50±0.24	7.34±0.54	3.08±0.26	43.06	0.52
Internode length (cm)	6.07±0.23	3.02±0.10	3.66±0.20	33.30	0.39
Number of tillers	2.23±0.16	5.50±0.38	10.40±1.42	9.27	2.13
Number of branches	4.83±0.25	3.35±0.23	4.10±0.53	2.16	0.93
Number of leaves	41.40±2.25	22.85±1.18	49.50±4.65	6.46	7.30
Leaf length (cm)	7.76±0.30	2.32±0.09	4.69±0.11	114.28	0.36
Leaf breadth (cm)	0.28±0.01	0.20±0.13	0.35±0.01	0.41	0.17
Number of stomata	2.22±0.17	1.29±0.07	1.74±0.06	1.15	0.61

Table 6: Floral characters of three species of *Iseilema*

Character	*I. laxum*	*I. anthephoroides*	*I. venkateswarlui*	*F-value*	*CD-value*
Flag leaf length (cm)	2.98±0.15	1.56±0.16	2.34±0.17	2.36	0.75
Flag leaf breadth (cm)	0.19±0.23	0.19±0.01	0.18±2.66	6.58	0.20
Inflorescence length (cm)	11.54±0.47	3.99±0.10	34.30±3.86	58.96	0.71
Number of inflorescences	2.66±0.16	4.26±0.30	8.60±0.75	13.61	1.20
Number of spikelets	70.50±2.56	35.00±4.57	170.50±18.80	15.49	2.68
Seed L/B Ratio	2.92±0.02	2.26±0.09	2.29±0.04	38.55	0.08
Pollen fertility (%)	93.46	82.34	95.35		

Among the species of *Iseilema, I. laxum* is a perennial grass showing wide range of habitat conditions. *I. anthephoroides* and *I. venkateswarlui* are annual grasses common in black clay field soils. *I. laxum* is well identified in nature due to greater length of culms (Plate IIc), whereas *I. venkateswarlui* and *I. anthephoroides* due to short culms (Plate IIa). In species of *Iseilema* though involucral spikelets are present in racemes they were found to be more in *I. venkateswarlui* (Plate IIb). In addition much variation was seen in the length of their pedicels (Plate II).

Thus, the morphology of the inflorescence among these species is found to be a special interesting character. The details of spike morphological variation existing among these species are described under individual taxonomic studies.

The species of *Dichanthium* and *Eremopogon* appear similar to each other by having spike like racemes. However, they differ by the presence (*Eremopogon foveolatus* and *Dicanthium pertusum*) or absence (*D. annulatum*) of pits in the sessile spikelets. Another important character is the pedicels of the pedicellate spikelets being slightly longer than the length of sessile spikelets in all the above three species. Since longer pedicels are considered as primitive than shorter pedicels, *Eremopogon faveolatus* and both the species of *Dichanthium* viz., *D. annulatum* and *D. pertusum* are found primitive for this trait. Perhaps, by possessing some such common morphological characters *Eremopogon* and *Dichanthium* were considered close allies of the tribe Andropogoneae that constitute the section Amphilophiastrae of Stapf (1917–34).

Among the wild aromatic grasses of *Cymbopogon*, the species from plains, i.e. *C. coloratus* and *C. caesius*, were found difficult to distinguish from each other, since both of them have false panicles. The only distinctive feature between them is the presence (*C. caesius*) or absence (*C. coloratus*) of a groove in the glume of sessile spikelet.

The hilly polyploid races of *Cymbopogon*, viz., *C. flexuosus* and *C. jwarancusa* were found robust in their growth. The arrangement of spikelets was crowded in *C. jwarancusa* while of lax-type in *C. flexuosus*. *C. jwarancusa* is clearly identified by its awns in sessile spikelet. Here, a trend towards elaboration of the inflorescence is followed by the condensation of the raceme segment in both the species.

The above-said traits, viz., basal reduction of spikelets in an inflorescence is considered an advanced feature (Clayton and Renvoize, 1986). From the present observations and from existing literature, it was understood that the genus *Cymbopogon* is a polymorphic group of Andropogoneae, distributed in regions with abundant rainfall and growing in a wide range of habitats from sea level to about 6,000 feet (Jagadishchandra and Narayan, 1971; Jagadishchandra, 1975). Natural hybridization seems to have been responsible for the formation of numerous transitional forms in this genus (Bor, 1960). Thus, during the evolutionary process of these grasses, the trend towards morphological advancement of the spike might have taken place to facilitate their dispersal under different climates in which they became dominant.

According to Sharma (1979), the morphologists take the simplicity in the method of seed dispersal and unreduced and unspecialized spikelet structure as primitive characters, while the converse represent advanced stages in the evolutionary path. Thus, he felt that grasses have been characteristically specialized by reduction which lead to complexity in structure rather than to simplicity. The present observation on spike variation among these grasses focus the pivotal theme that "Andropogoneae is the most advanced and specialized of all grasses".

In *Iseilema*, though the inflorescence is false panicles, it is distinct from the racemes of other genera owing to the presence of four involucral spikelets enclosing both sessile and pedicelled spikelets. The differentiation between the species of *Iseilema* was due to varying length of pedicelled spikelets. The pedicels of spikelets are shorter in *I. venkateswarlui* while longer in *I. laxum* and *I. anthephoroides*. From the earlier studies on *Iseilema* complex by Satyavathi and Murthy (1973), it was understood that *I. venkateswarlui* was found more advanced than *I. anthephoroides* based on the reduced pedicels of the spikelets. The present studies on spike morphology corroborate the earlier view that *I. venkateswarlui* is an advanced species in *Iseilema* complex.

The reduction in the length of the pedicel in the spike of *I. venkateswarlui* was followed by increased number of spikelets, thus enhancing the reproductive efficiency of the plant. The reproductive efficiency is an adaptive strategy for survival under selection pressure. Such r-type of strategic response is common among the natural populations (Abrahamson and Gadgil, 1973; Warwick, 1980; Hickman, 1975).

Cytological findings of Satyavathi and Murthy (1973) also supported the hybrid origin of *Iseilema venkateswarlui*. Thus, it can be understood how natural hybridization results in novel forms with a combination of morphological characters, having an adaptive significance. The perennial fodder grass, *I. laxum* survived wide ecological conditions unlike *I. venkateswarlui*, an annual species having restricted growth. *Dicanthium annulatum*, a grass with widespread growth also showed long pedicels. The increased intensity of purple color of the culm among all the three species of *Iseilema* was found very prominent during the summer months, an adaptation to xeric climate. Reduced pedicels seen in *I. venkateswarlui*, a field grass, can be judged by the fact that the reduced floral parts facilitate rapid flowering and seed development, a favourable feature for annual field grasses of *Iseilema*.

Referring in general to certain morphological features of the tribe Andropogoneae, as reviewed by Stebbins (1986), the presence of bicellular hairs is primitive, while basal reduction of spikelets; 2 lodicules; 3 stamens; large embryos; Kranz syndrome; dumbell-shaped siliceous cells are considered as advanced.

The trends of elaboration in Andropogoneae can be interpreted as adaptations to the very different climates in which the taxa are dominant, particularly with respect to reproductive characters, reduction aids in rapid development and maturation, a rhythm that is of particular value in highly seasonal climates. On the other hand, elaboration of complexity may promote greater phenotypic plasticity, which in warm climates, may help to adapt the plant to irregularities in the onset and end of moist seasons (Stebbins, 1986).

Thus among Poaceae, as in higher plants and animals generally, different characteristics have evolved at different rates within the same evolutionary line. Nearly every taxon is a mosaic of relatively primitive and more advanced character states (Stebbins, 1986). The degree of advancement of a particular taxon is expressed not in terms of the degree of advancement of every character, but in the proportion of primitive versus advanced states. With respect to advanced character states, the tribe Andropogoneae is distinctive in that advancement affects reproductive characters, much unlike vegetative characters as in Bambusoideae. The adaptive significance, if any, of a different combination of advanced characters found in this tribe is hard to interpret. Many of them might have become established by chance in small populations during the early evolution. Hence, the tribe Andropogoneae cries out for explanation of its function and presents an opportunity to relate the course of evolution to the selection pressures driving it.

The authors are thankful to the Head, Botany for necessary facilities and first author is grateful to U.G.C. for financial assistance under F.I.P.

REFERENCES

Abrahanson, W.G. and M.D. Gadgil. 1973. Growth form and reproductive efforts in golden rods (*Solidago* Compositae). *American naturalist,* **107**:651–661.

Bewis, J.W. 1929. The World's Congress, London.

Bor, N.L. 1960. Grasses of Burma, Ceylon, India and Pakistan. Pergamon Press, London.

Clayton, W.D. and S.A. Renovoize. 1986. Genera Graminium. *Kew Bulletin* Additional Series, **13**:389, 23–27.

Hartley. 1950. The global distribution of tribes of the Gramineae in relation to historical and environmental factors. *Aust. J. Agr. Res.,* **2**:355–373.

Hickman, J.C. 1975. *J. Ecol.,* **63**:689.

Jagadischandra, K.S. and K.N. Narayan. 1971. Studies on supernumerary chromosomes in the genus *Cymbopogon* (Poaceae). Proc. first All India Congr. of Cytol. and Genetics. *J. Cytol.* and *Genet. Congr. Suppl.,* 104–112.

Jagadischandra, S. 1975. Recent studies on *Cymbopogon* Spreng (Aromatic grasses) with special reference to Indian Taxa: Taxonomy Cytogenetics Chemistry and Scope. *J. Plantation Crops.,* **3(2)**:43–57.

Myers, W.N. 1947. Cytology and genetics of Forage grasses. *Bot. Rev.,* **13**:319–442.

Satyavathi, D. and G.R. Murthy. 1974. Aneuploid complex of *Iseilema* Anders. I Chromosome Pairing in F1 hybrids *Nucleus.,* **17(1)**:9–14.

Sharma, M.L. 1979. Some considerations on the phylogeny and chromosomal evolution in grasses. *Cytologia.,* **49**:679–685.

Snedecor, George W. and William G. Cochron, 1967. Statistical methods 5th ed. The Iowa State Univ. Press Amer., IOWA.

Soderstrom, T.R., W. Kiridir, Hilu, S. Christopher Campbell and E. Mary, Bark Worth, 1986. Grass Systematics and Evolution. An International Symposium held at the Smithsonian Institution. Washington DC, 27–31.

Stapf, O. 1917–34. Gramineae. In: D. Prain (ed.). Flora of tropical Africa, Vol. 9. Kent.

Stebbins, G.L. 1986. Grass Systematics and Evolution past, present and future. Grass syntematics and evolution, edited by L. Thomas Sodersfrom, K.W. Hilu, Christopher. S. Camphell and Mary E. Bark Worth.

Tateoka, 1960. Cytology in grass systematics. A critical review: *The Nucleus.,* **3(1)**:81–110.

Warwick, S.I. 1980. *New Phytol.,* **85**:461.

Watson, L., H.T. Clifford and M.J. Dallwitz. 1985. The classification of Poaceae. Subfamilies and supertribes. *Australian J. Botany,* **33**:433–484.

CHAPTER 4

PRELIMINARY STUDY ON THE "DIVERSITY OF MOSSESS OF KASARA GHATS IN SAHYADRI (WESTERN GHATS)"

*G.T. Dabhade** and *V.K. Hile***

Mosses which play a key role in the formation of natural biotic community and indicative of forest condition were abundant in the Western Ghats (India). Eversince the work of Kirtikar (1896 in Birdwood 1897), Sedgwick (1910, 1911, 1913), Blatter (1909), Dixon (1909), Bruhl (1931) and Dabhade (1988) etc. research on Mosses was neglected. In addition to this scenario, the area under study has undergone tremendous physiognomic changes. It had been observed that the forest have succumbed to heavy deforestration and urbanisation pressures and pollution, survival of many plant species including Mosses have been threatened and many mosses have become endangered species. Therefore an attempt has been made to undertake the task of the investigation of the Diversity of Mosses of Kasara Ghats of Western Ghats (Sahyadri) lying in Maharashtra State.

Kasara is a terminus of central railway between Mumbai (Bombay) and Nasik, situated on the Northern ranges of the Western Ghat 121 km by road from Bombay. It belongs to Thane dist. and extends between 18° 42′ and 20° 20′ N and 72° 45′ and 73° 45′ E. The highway from Bombay to Nasik passes through the Thalghat of Kasara which passes through Ghatandevi Temple. Elevation of Kasara Ghat is about 590 M above sea level. It stands on dark basalt rock. The area experiences maximum and minimum temperature of 34° C and 25° C respectively, and an average annual rainfall about 2850 mm mostly during the monsoon (June–Sept.). Winds are moderate during the July but their velocity is significantly reduced in Aug.–Sept. The relative

* Hon. Professor in Botany, Birla College, Kalyan-421 304.
** Lecturer in Botany, Bhavan's H. Somani College, Mumbai-400 007.

humidity is high (70–80%) during monsoon month but it seldom exceed 30% during summer month. Topography of the area is hilly.

The angiospermic vegetation in the area is dominated by *Tectona grandis* (Teak) and is ± of dry deciduous type. Variety of bryophytic vegetation appears during June–Sept. when the condition are highly conducive to their growth and development. Several liverworts such as species of *Riccia, Cyathodium, Targionia, Fimbraria, (Astrella), Plagiochasma* and leafy hepatics such as *Lejeunea* and hornworts, e.g. *Anthoceros erectus, A. laevis,* grow luxuriently in the area. Earlier, Sedgwick (1910, 1911, 1913) reported many mosses from the adjacent area (Trymbakeshwar of this region). The mosses of Kasara Ghat are not investigated so far. Therefore the present attempt is made to investigate them.

Material and Methods

Frequent surveys were made in and around Kasara Ghat during the monsoon period in 2000–2001 with a view to collecting mosses. The collections were examined and identified with the help of standard monographs. (Gangulee 1969, 1971, 1972 & Dabhade, 1998).

Observations

Several collection of mosses were made from the area. Seventeen taxa of Mosses belonging to thirteen genera under eleven Families were investigated and in what follows information is provided on the salient features, habitat, preference and site and date of collection of these species. (A species with an asterisk represents the first record from the area).

1. *Archidium birmannicum* Mitt. ex Dix. (Generic name is derived from the Greek *Archidium*–primitive) (Text Figs. 1–6)

Family I: **Archidiaceae**

This taxon collected in sterile conditions was characterised by having pyramidal (dome shaped) apical cell in ovate lanceolate leaves. Leaf cells were subrectangular to irregularly rhombiodal where as rectangular cells at the leaf base. Leaf base margin was bordered by two rowed rectangular cells. This terricolous moss was found to grow luxurienty on moist lateratic soil near the stream at Kasara (Sept. 2002, Leg. G.T. Dabhade and V.K. Hile). The species is endemic to Indo-burmese subcontinent and it has been reported from West Bengal, Upper Assam, Nilgiri, Palni Hill, Khandala and Burma.

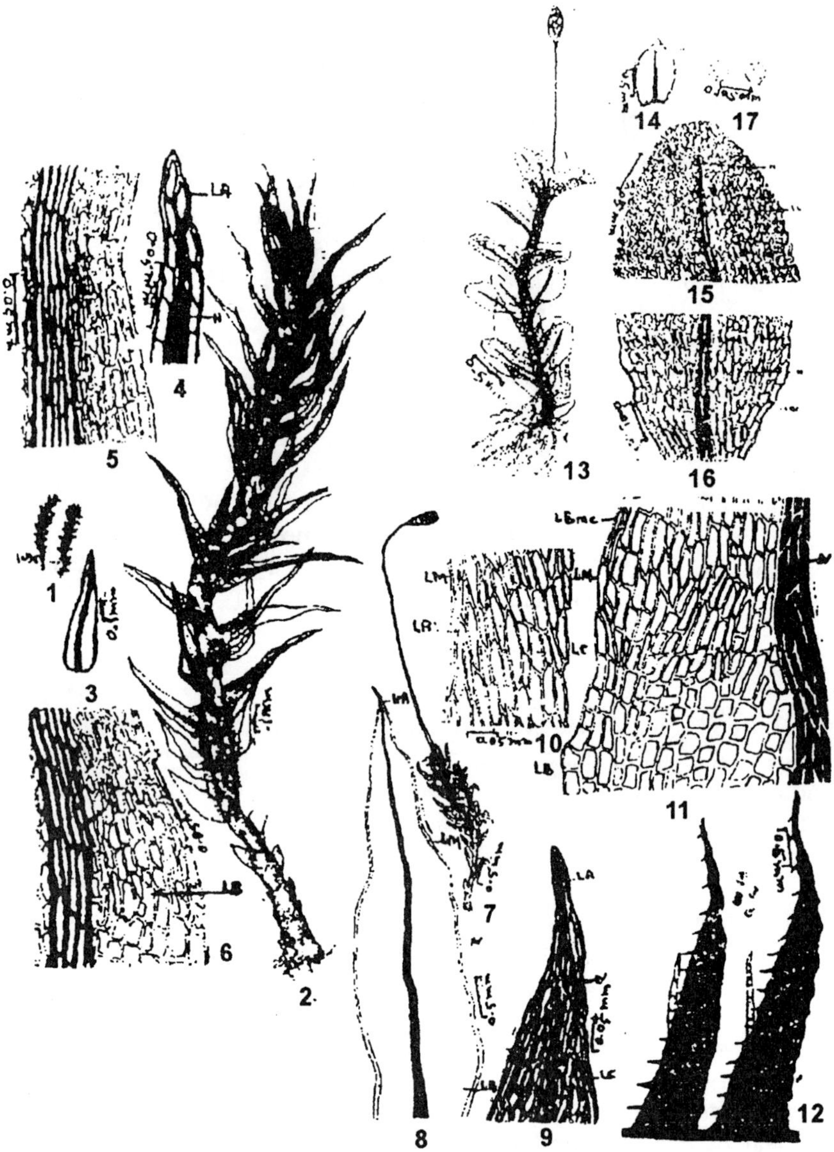

Archidium birmannicum Mitt. ex Dix.

1. Entire plant, 2. Enlarged view of entire plant, 3. Entire leaf, 4. Leaf apex & nerve, 5. Median Leaf cells, 6. Leaf base cells; *Funaria hygrometrica* Hedw., 7. Entire plant with sporophyte, 8. Entire leaf, 9. Leaf apex, 10. Median leaf cell, 11. Leaf base cells, 12. Outer & inner peristome teeth along with spores; *Gymnostomiella vernicosa* (Hook.) Fleisch., 13. Entire plant with capsule, 14. Entire leaf, 15. Leaf based cells, 17. Spores.

2. *Funaria hygrometrica* Hedw. (Text Figs. 7–12)

Family II: **Funariaceae**

This taxon was characterised by having obliquely placed capsule with reddish tinge at the tip of flexuose, twisted seta. It was also recognised by lanceolate, shortly pointed leaves with percurrent nerve. Leaf cells elongated, hexagonal, polygonal, but rectangular at the base. Leaf margin entire and bordered.

Plants of this terricolous taxon were found on semi exposed sandy brick walls and ground at road near Ghatandevi Temple and dome shaped stony well, Kasara Ghat (Sept. 2002, Leg G.T. Dabhade and V.K. Hile). This species was very common at different places in Western Ghat Mahabaleswar (Kirtikar in Sedgwick 1910) and Arthur seat point, Mahabaleshwar (Dabhade 1966 in Dabhade 1998) and Khandala (Dabhade 1967 in Dabhade 1988) and Mahur (Bhonsle 1996). This cosmopolitian species is very commonly seen at different places in Western Ghat, Kodaikanal, (South India). In North West and Eastern Himalayan regions in India. It was reported from Burma, Phillippines, Africa and other countries in the world. It was very easily recognised from other species by obliquely placed red brown asymetrical wide mouthed capsule.

3. *Gymnostomiella vernicosa* (Hook.) Fleisch. (Text Figs. 13–17)

Family III: **Splachnaceae**

Plants of this taxon were very small (5 to 7 mm) minute forming velvately coating on calcarious brick wall of houses and temples. They were characterised by their spathulate-obovate leaves with papillose, crenate margin. Leaf cells 4–6 sided hexagonal where as leaf base cells were elongated rectangular and hyaline. Nerve short faint and percurrent reaching up to midleaf. Plant terminates into red ovoid capsule, at the tip of erect, seta. Smaller male plant better represented than female ones. This terricolous and calciphilous plant was collected at Ghatandevi temple, Kasara (August 2001, leg. V.K. Hile). This taxon was reported from Thana by Sedgwick (1910 as a *Pottia vernicosa* Hampe.) and from Vihar (Bombay) by Blatter (1905 in Sedgwick 1911) as a *Gymnostomillum vernicosum* (Hook.) Fl.

This taxon was also reported from Khandala and Mahabaleshwar (Dabhade 1998) and Mahur (Bhosale 1996).

This cosmopolitan species was widely scattered in tropical to subtropical countries like Pakistan, Myanmar (Burma), Sri Lanka (Ceylon), Singapore, Malaya Archipelago, Java, Philipines, Amboina, Africa etc.

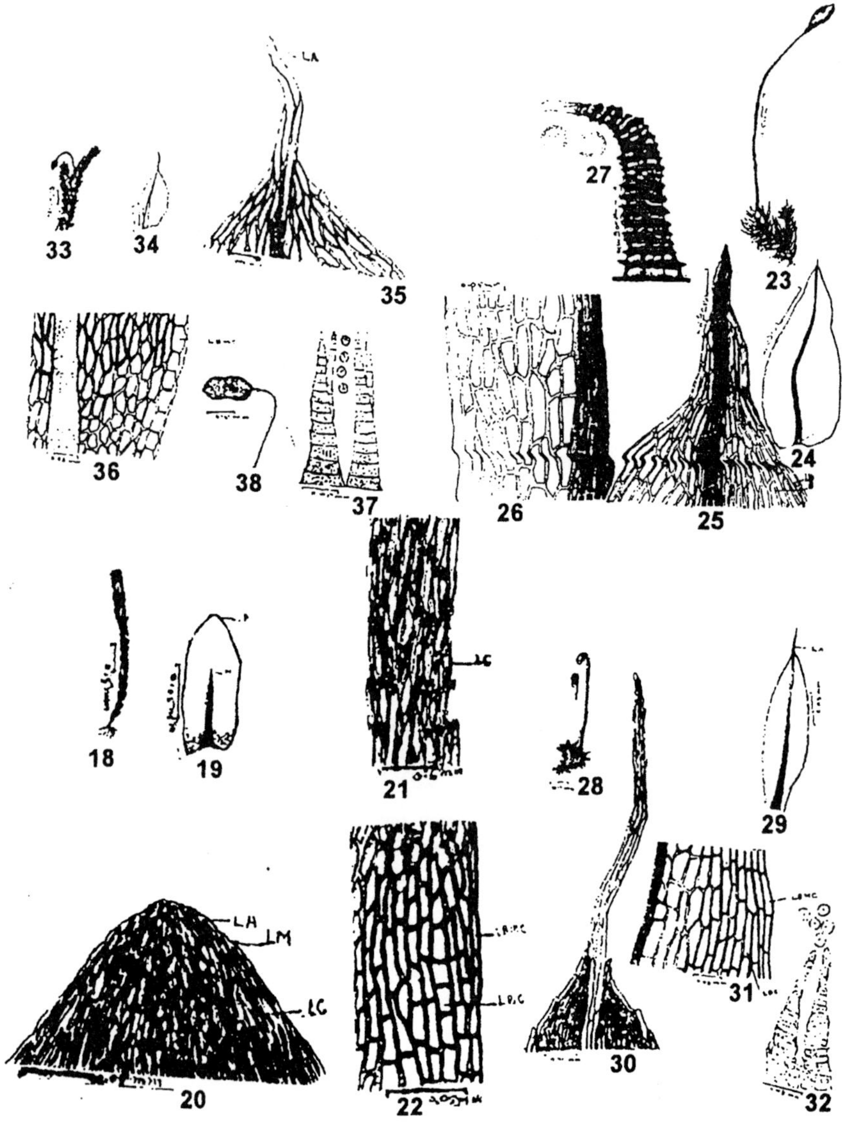

Anomobryum auratum (Mitt.) Jaeg.
18. Entire plant, 19. Entire Leaf, 20. Median leaf cells, 21. Leaf apex, 22. Leaf base cells; *Brachymenium nepalense* Hook., 23. Entire plant, 24. Entire Leaf, 25. Leaf apex, 26. Leaf base cells, 27. Peristome teeth with spores; *Bryum coronatum* Schwaegr., 28. Plants with pendulous capsule, 29. Entire leaf, 30. Leaf apex, 31. Leaf base cells, 32. Peristomal teeth with spores; *Bryum argenteum* Hedw., 33. Entire fertile plant, 34. Entire Leaf, 35. Enlarged view of leaf apex, 36. Enlarged view of leaf base, 37. Capsule, 38. Peristome teeth with spores.

4. *Anomobryum auratum* (Mltt.) Jaeg. Derived from Greek *a*-not, *nomous*-rule, means and nonconformist *Bryum*. (Text Figs. 18–22)

Family IV: **Bryaceae**

Plants of this taxon were very small with shining light green tuft, julaceous to highly form. Obtuse leaves with rounded apex and crenate margine; were closely appressed to the stem of the plant. Nerve reaching $^1/_3$ length of the leaf. Laminar cells were linear to rhomboidal where as basal cell rectangular. This tufted Julaceous moss was investigated on lateratic soil to the side of road, Jawahar Phata, Kasara Ghat (Sept. 2002 leg. V.K. Hile). It was also reported from various places in Khandala, Mahabaleshwar (Dabhade, 1998), Kodaikanal, Palani Hill, etc. in India.

This Julaceous moss was easily recognised by shining light green tufted plants with closely appressed obtuse leaves with rounded apex and crenate margin.

5. *Brachymenium nepalense* Hook. (Derived from Greek *brachus*-close celled, *Menium*-tissue, probable with reference to the anatomy of the plant.) (Text Figs. 23–27)

Plants of this species were robust, pale green in dense tuft slightly glossy in appearance. Leaves crowded in comal tuft, spirally twisted when dry, oblong, lingulate, acuminate and distinctly border. Leaf margin narrowly recurved toothed toward apex. Nerve brownish, excurrent in a long denticulate arista. Leaf cell rhomboidal, marginal cell thin walled linear forming a distinct border, leaf base cell lax, rectangular. Sub-errect red–brown pyriform capsule present at the tip of red seta. Seta is 2–3 cm long. Spores rounded, brownish, papilose. Pale green corticolous fruiting plant in dense tuft were found on the bark of *Thespesia populnea* Soland at road side on way to Ghatandevi temple, Kasara (Sept. 2002, Leg. V.K. Hile). This moss also reported by Sedgwick (1910) as a terricolous taxon (on wall, Lonawala). It was also reported from Khandala, Mahabaleshwar (Dabhade 1998) in Western ghat. It was also confined to various places in India, Pakistan, Srilanka (Ceylon), Myanmar (Burma), Siam, Indo-China, Malaya, Phillipine, Celebes in the world.

The striking feature of this species are as robust tufted plant with dentate apical margin, excurrent nerve and bordered margin of linear cells and pyriform red brown capsule on long seta.

6. *Bryum coronatum* Schwaegr. (Derived from Greek *bryon*-which means a cryptogamic plant). (Text Figs. 28–32)

Family V: **Bryaceae**

Dull yellowish green plants of this terricolous as well as saxicolous moss were present in the form of tuft. Leaves ovate lanceolate with denticulate

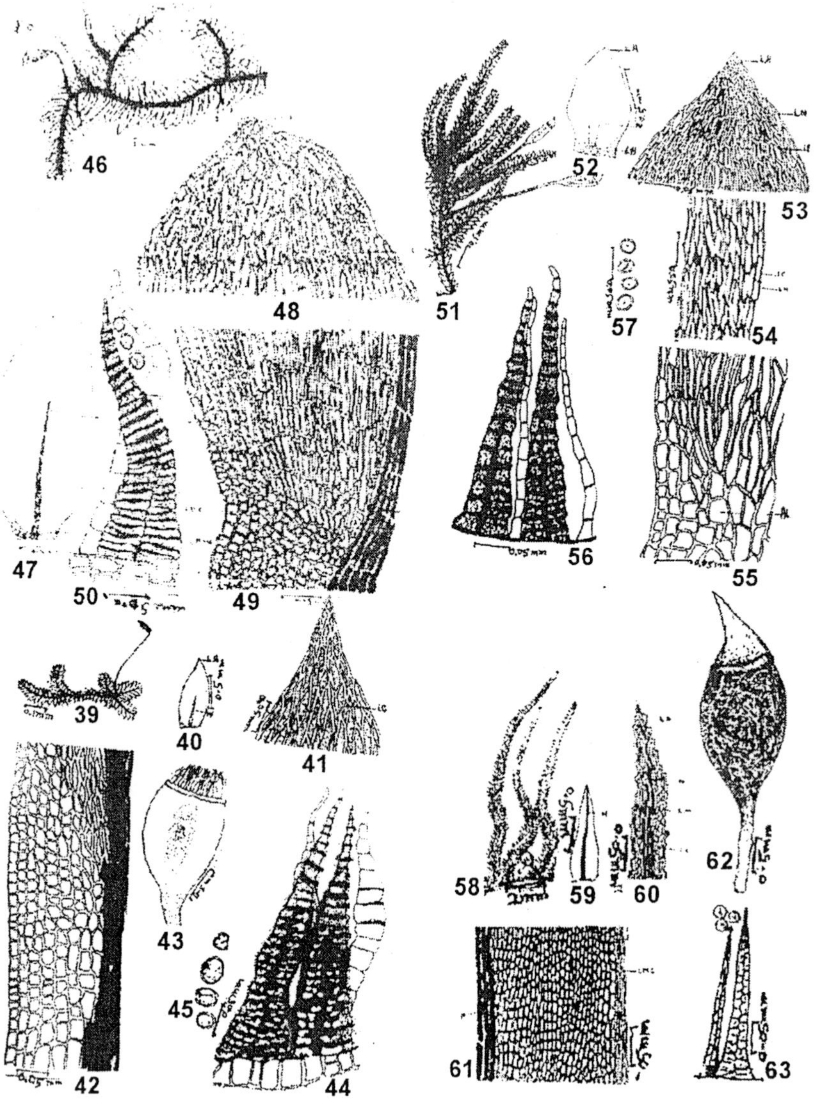

Stereophyllum tavoyense (Hook) Jaeg.

39. Entire plant with a sporophyte, 40. Entire leaf, 41. Leaf apex, 42. Leaf base cells, 43. Capsule, 44. Peristomal teeth, 45. Spores; *Stereophyllum ligulatum* Jaeg.: 46. Entire plant with sporophyte, 47. Entire leaf, 48. Leaf apex, 49. Leaf base cells, 50. Peristomal teeth & spores; *Entodon prorepens* (Mitt.) Jaeg., 51. Entire plant with sporophyte, 52. Entire leaf, 53. Leaf apex, 54. Median Leaf cells, 55. Leaf base cells, 56. Peristomal teeth, 57. Spores; *Diaphanodon procumbens* (C. Muell) Ren. et Card.: 58. Entire fertile plant, 59. Entire leaf, 60. Leaf apex, 61. Leaf base cells, 62. Capsule, 63. Peristome teeth & spores.

bordered margin at the apex. Nerve excurrent ending into long arista. Capsule of this was pear shaped, cylindrical or oblong, pendulous.

This taxon was very commonly growing on the calcarious walls of old houses at Kasara village as well as on the basaltic rock at Kasara Ghat (Aug. 2001, Leg. V.K. Hile). Said taxon was reported from Poona (Wood, 1895 in Sedgwick, 1910) and Khandala, Mahabaleshwar (Dabhade 1998).

7. *Bryum argenteum* Hedw. (Silver thread Moss) (Text Figs. 33–38)

The plants of this taxon were small silvery white, glossy with short reddish brown erect stem with julaceous branches. Leaves crowded acuminate with percurrent nerve ending below apex. Leaf cell rhomboidal colourless but leaf base cell subrectangular and chlorophyllose. Red pendulous capsule at the tip of red erect seta. Dense cushion of silvery white glossy plant were growing on the sandy bricks of compound wall (terricolous) and on large basaltic rock (saxicolous) in Kasara Ghat (Aug. 2002 Leg. V.K. Hile). Said taxon was reported from Castle rock (Gammie 1892 in Sedgwick 1913). Khandala and Mahabaleshwar (Dabhade 1998) and at Kodai Kanal, Eastern and Western Himalaya in India and tropical and subtropical and temparate countries of the world. Said terricolous as well as saxicolous taxon is cosmopolitan and popularly called "Silver thread moss" due to its silvery white glossy appearance.

8. *Sterophyllum tavoyense* (Hook.) Jaeg. (Derived from the Greek word steros-rigid, Phyllum-with reference to the leaf) (Text Figs. 39–45)

Family VI: **Plagiotheciateae**

Fruiting plants of this corticolous taxon with creeping primary stem, slender, tufted and light green to glossy coloured were growing on the bark of tree. Leaves ligulate to lanceolate, acuminate with incomplete nerve ending in midleaf. Leaf cell rhomboidal to linear with quadrate to rectangular, alar cell at the base. leaf margin entire ending into acute leaf apex. Cylindrical capsule at the tip of long brown seta possess double peristome along with colourless cilia. Spores brown, rounded to oval in shape.

This taxon was collected from the bark of several trees abodes near Ghatandevi Temple, (Sept. 2001 Leg. G.T. Dabhade and V.K. Hile). It was also reported from Vasind, Thane Dist. (Sedgwick 1910), Khandala and Mahabaleshwar (Dabhade 1998) and from various places such as Botampalli and Palni Hills, Deharadoon, near Tavoy, Bihar in India.

9. *Stereophyllum ligulatum* Jaeg. (Text Figs. 46–50)

Plants of said corticlous taxon were dull green, prostrate fruiting on the bark of the tree. Leave appressed, ligulate with rounded apex.

Nerve incomplete reaching up to ¾th region of the leaf cells rhombiodal linear. Alar cells quadrate to rectangular, capsule ovate at the tip of slender flexious seta.

Peristome teeth double with cilia and spores circular to oval brown colour. Said corticlours taxon was on the bark of tree at the side of Hiwala bridge of Kasara Ghat (Sept. 2001 Leg. V.K. Hile) and also reported from various places such as Khandala, Mahabaleshwar (Dabhade 1998) and Botampalli, Palni Hill, Kanan deva hill, etc. in India.

10. *Entodon Prorepens* (Mitt.) Jaeg. (Text Figs. 51–57)

Family VII: **Entodontaceae**

Said corticolous plants of this moss were growing in the form of glossy, greenish brown, cylindrical stem forming mats. Leaves were oblong acute with elongated linear laminar cell and subquadrate alar cells. Nerve incomplete. Capsule erect, cylindrical with double peristome along with hyline cilia. Spores smooth, brownish and circular. Said corticolous as well as saxicolous taxon reported from Kasara Ghat on the bark of tree and on basalt rock (Sept. 2001, Leg. Hile). It was also confined to Khandala and Mahabaleswar (Dabhade 1998), Western Himalaya (Kumaon, Simla), Eastern Himalaya (Sikkim, Darjeeling, Assam, Khasia Hills) and Bhutan, Burma, China (Yunnan).

11. *Diaphanodon procumbens* (C. Muell.) Ren. et Card. (Text Figs. 58–63)

Family VIII: **Trachypodaceae**

Plant of this taxon were slender greenish brown densly tufted with very small narrow thread like branches at the tip. Leaves ovate with acuminate apex and serrate border. Nerve percurrent ceasing below apex. Laminar cells elongated along with quadrate to rectangular, alar cells at the leaf base. Plants of this taxon were growing on the bark of tree at Hiwala bridge near tunnel at Kasara Ghat, (Sept. 2001, Leg. V.K. Hile) and also reported from Poona (Wood 1895 in Sedgwick 1970 and 1913) and Matheran, Mahabaleshwar (Kirtikar 1896 in Sedgwick 1910), Lonawala, Triambakeshwar, and also from Khandala, Mahabaleshwar (Dabhade 1998), Jog falls, Ooty, North–West Himalayas, Eastern Himalayas, Calcutta and also in South–East Asian Tropical Countries.

12. *Hyophila involuta* (Hook.) Jaeg. Generic name is derived from Greek *Hyo*-rain *philos*-loving with reference to the habitat of the plant. (Text Figs. 64–68)

Family IX: **Pottiaceae**

Dark green coloured plants of this terricolous as well as saxicolous moss were commonly growing luxuriently in the form of tuft. Leaves long

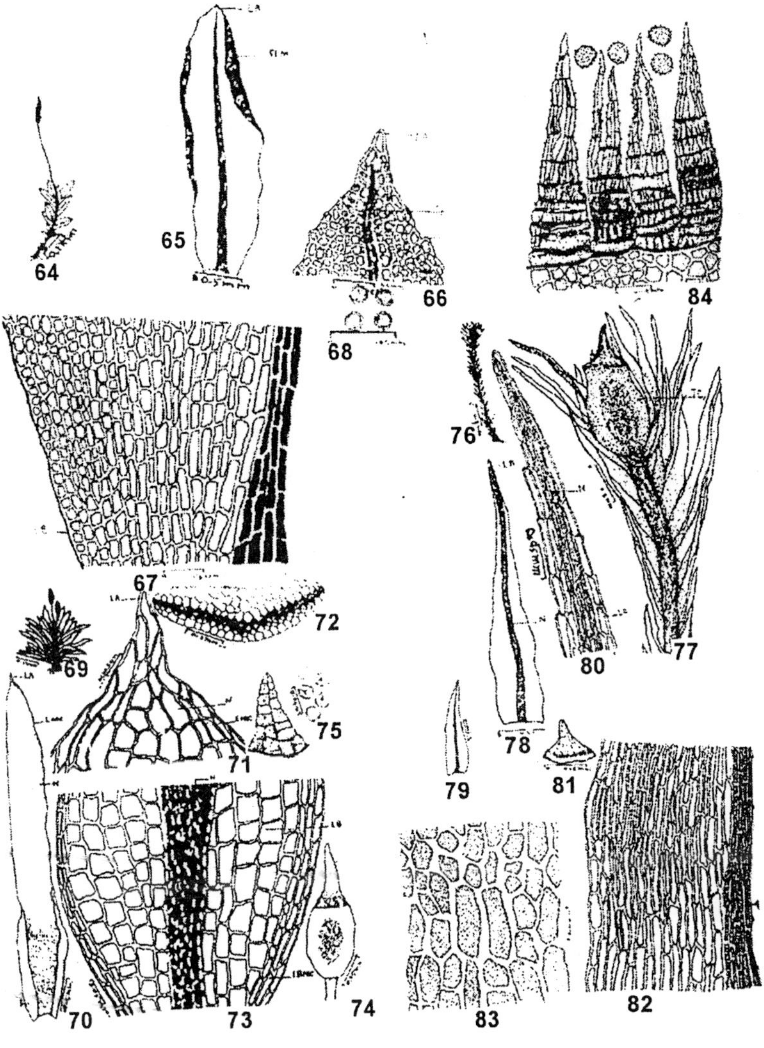

Hyophila involuta (Hook) Jaeg.
64. Entire plant with sporophyte, 63. Entire leaf, 66. Leaf apex, 67. Leaf base cells, 68. Spores; *Octoblepharum albidum* Hedw.: 69. Plant with sporophyte, 70. Entire leaf, 71. Leaf apex, 72. Leaf base cells, 73. T.S. of leaf showing median leucocyst & chlorocyst, 74. Sporophyte, 75. Dicranoid peristome tooth with spores; *Garckea phascoides* (Hook.) C. Muell., 76. Entire Plant, 77. Plant with immersed sporophyte, 78. Entire leaf, 79. Entire leaf at base, 80. Leaf apex, 81. Leaf base cells, 82. Calyptra, 83. Peristomal teeth with spores, 84. Exothecial cells.

spathulate and involute. Leaf margin mostly wavy, entire but serrate to the leaf apex. Nerve percurrent. Leaf cell quadrate to hexagonal, mammilose and leaf base cells large rectangular, cylindrical or needle like brownish capsule present at the tip of erect seta. Spores light brown smooth and circular. Earlier Surgeon-General Dr. Kirtikar reported this taxon as *Hyophilla subflaccida* Broth. & Dix. from Andheri, Bombay, where as Blatter (1905) reported it on a basalt rock of Khandala as a *Hyophilla stenocarpa* Ren. & Card. (Sedgwick 1911) and as a *H. cylindrica* Hook. by Sedgwick (1910) from Poona & Bandra, Lonavala, Mahabaleshwar (Dabhade 1968) and also from most of the part of India and from Burma, Vietnam, East China, Manchuria, Korea, Japan, Phillippines, Indonesia, North Central America, Europe, etc.

13. *Octoblepharum albidum* Hedw. (Text Figs. 69–75)

Family X: **Leucobryaceae**

Plants of this taxon were dense whitish green, corticolous as well as saxicolous forming tufts on tree trunks as well as on moist soil. Leaves crowded, erect ligulate spreading apex pointed with entire margin. Nerve broad, thick, multilayered with a median row of triangular chlorocyst and leucocyst on both side. Leaf base flanked by 5–6 row of hyaline rectangular cells. Leaf margin bordered by linear to rhomboidal cell. Greenish brown erect oblong capsule present at a tip of erect seta. Eight clefted peristome teeth present, spot light brown and circular. This terricolous taxon was collected on the moist ground near Jawahar phata of the Kasara Ghat (Sept. 2001 Leg. Hile). Earlier said taxon was reported as a corticolous from Khandala by Blatter (1905 in Sedgwick 1911). There after it was also reported from Khandala and Mahabaleshwar (Dabhade 1998). This cosmopolitan taxon is widely distributed in tropical and subtropical countries of world.

14. *Garckea phascoides* (Hook.) C. Muell. (Text Figs. 76–84)

Family XI: **Dictrichaceae**

The said interesting terricolous, slender, dull brown or yellowish green plants were growing in loose tuft. Leaves crowded at the apex in a distinct comal tuft, lanceolate, long slenderly, acuminate with entire margin. Leaf cell linear but rectangular to polygonal at the base. Nerve brownish and excurrent. Immersed, oblong, cylindrical capsule present at the tip of short seta. Peristome teeth red brown clefted about 2/3 down. Spore yellow brown and papillose. This taxon can be easily recognised by immersed capsule at the tip of plant.

As a terricolous taxon from the moist soil to the side of road of Kasara Ghat on way of Ghatandevi temple (Sept. 2001 Leg. V.K. Hile). It was also reported from Mahabaleshwar (Dabhade 1998) and West Bengal, Khasia hills and South India in India. It was also widely distributed in the countries of British India, China, Japan, Korea, Mangolia, S.E. Asian countries,

Madagascar, Mauritius, Africa, Panama, Australia and Pacific oceanic islands in the world.

15. *Fissidens splachnobryoides* Broth. (Text Figs. 85–89)

Family XII: **Fissidentaceue**

This terricolous taxon of the genus *Fissidens* Hedw. was present in the form of robust gregarious, full yellow green, with brownish red stem. Leaves distichous, 10–11 pairs, oblong lanceolate, shortly acuminate boarded all around by rows of narrow linear cell. Network like leaf cells–oval, rhomboidal to polygonal reticulate and prosenchymatous. Along with rectangular cells at the base. Nerve incomplete ending far below the apex. Earlier said taxon was reported on stone at Thane and Trimbakeshwar (Sedgwick, 1910). It was also reported from Khandala and Mahabaleshwar (Dabhade 1998) and from Western and Eastern Himalaya in India, Pakistan, Srilanka (Ceylon), Myanmar (Burma), Thailand, Farmosa, Indonesia, Malaya, Phillippine, New Guinea, Japan etc.

16. *Fissidens curvatoxiphoides* Dix. et Vard. (Text Figs. 90–96)

This minute terricolous, distichous green coloured plant possess 8–10 pairs of leaves. Leaves ovate, lanceolate, narrow, acute with bordered margin from base to apex, Nerve excurrent, along with quadrate to hexagonal, laminar cell and rectangular leaf base cells obliquely placed. Cylindrical capsule on erect brown seta. Spores circular to oval, light brown. This minute taxon of the said species was collected in the form of terricolous habitat on moist soil near the side of road at Latif wadi, Kasara Ghat (Sept. 2000, Leg. V.K. Hile). It was also reported as a saxicolous taxon from Khandala and Mahabaleshwar (Dabhade 1998). This South East Asiatic species was mainly confied to the Western Himalaya, S. India, (Palani hill in India), Myanmar (Burma), East Nepal, Yunnan, Japan etc. in the world.

17. *Fissidense crenulatus* Mitt. (Text Figs. 97–102)

Very small plant of red brown stem possesses oblong, lanceolate leaves with acute apex, vaginant lamina reaches to midleaf and bordered by row of elongated cells. Nerve light brown, excurrent, leaf cells mammiose, irregularly hexagonal, crenulate margin toward the leaf apex. Oval, erect brown capsule at the tip of erect seta. Spores smooth, circular, brown in colour.

This Indo-Burmese species was growing on the moist soil (Terricolous) to the side of the road of Kasara Ghat (Sept. 2002 Leg. Hile). Earlier it was reported at Mahabaleshwar (Sedgwick 1910; Dabhade, 1998). It was confined to S. India, Orissa, E. Nepal & Upper Burma.

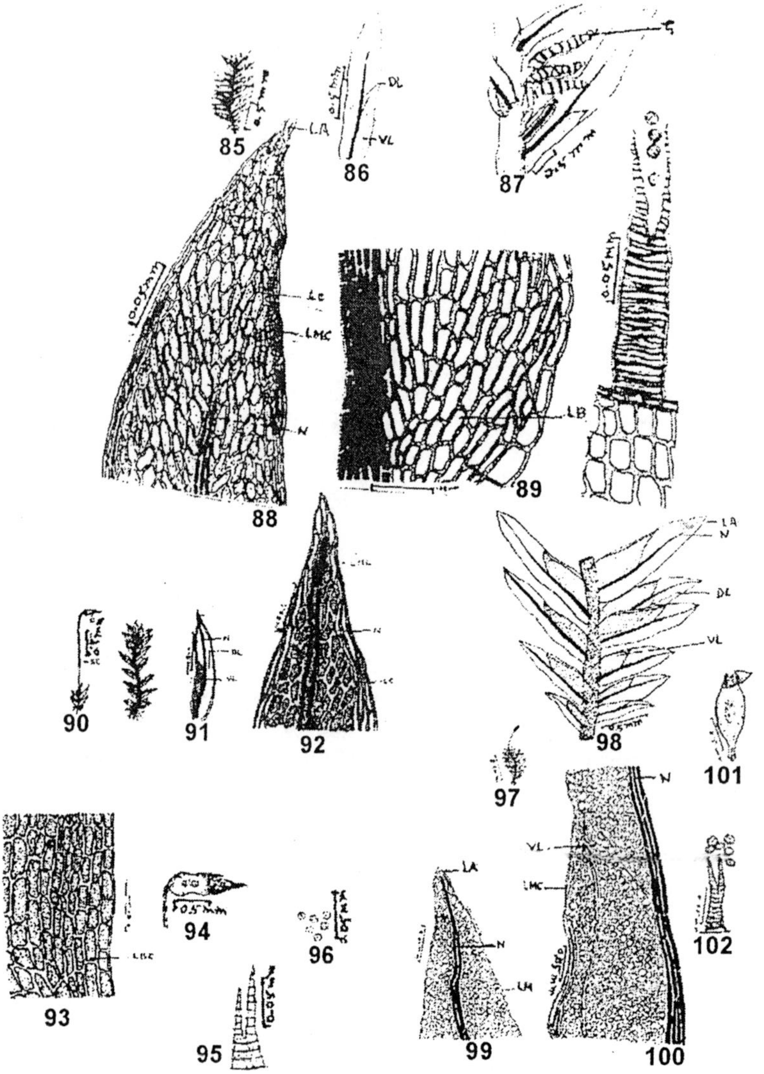

Fissidens splachnobryoides Broth.

85. Entire plant, 86. Entire leaf, 87. Clavate, septate gemmae, 87. Leaf apex, 88. Leaf base cells and spores with peristome teeth; *Fissidens curvatoxiphoides* Dix. & Vard.: 90. Entire fertile and sterile plant, 91. Entire leaf, 92. Leaf apex, 93. Leaf base cell, 94. Capsule with seta, 95. Peristome teeth, 96. Spores; *Fissidens crenulatus* Mitt.: 97. Entire fertile plant, 98. A branch showing arrangement and shape of leaves, 99. Enlarged view of leaf apex, 100. Leaf base cells showing bordered cells in leaf base margin and vaginant lamina margin, 101. Capsule, 102. Peristome teeth with spores.

Discussion

Presently seventeen species were collected and identified and they include *Archidium birmannicum* Mitt., *Funaria hygrometrica* Hedw., *Gymnostomiella vernicosa* (Hook.) Fleisch., *Anomobryum auratum* (Mitt.) Jaeg., *Brachymenium nepalense* Hook., *Bryum coronatum* Schwaegr., *Bryum argenteum* Hedw., *Stereophyllum tavoyense* (Hook) Jaeg., *Stereophyllum ligulatum* Jaeg., *Entodon prorepens* (Mitt.) Jaeg., *Diaphanodon procumbens* (C. Muell.) Ren. et Card., *Hyophila involuta* (Hook.) Jaeg., *Octoblepharum albidum* Hedw., *Garckea phascoides* (Hook.) C. Muell., *Fissidens splachnobryoides* Broth., *F. curvatoxiphoides* Dix. & Vard., *Fissidens crenulatus* Mitt. which has been recorded for first time from Kasara Ghat (Thal ghat).

Based on their habitat preference, seven species were strictly terricolous *(Archidium birmannicum, Funaria hygrometrica, Gymnostomiella vernicosa, Anomobryum auratum, Octoblepharum albidum, Garckea phascoides, Fissidens crenulatus* and three were purely corticolous *Brachymenium nepalense, Sterophyllum tavoyense, Stereophyllum liguatum,* where as five were both terricolous and saxicolous *(Bryum coronatum, Bryum argenteum, Hyophila involuta, Fissidens splachnobryoides, Fissidens curvatoxiphoides)* and two were both corticolous and saxicolous *(Entodon prorepens, Diaphanodan procumbens).*

REFERENCES

Birdwood, H.M. 1897. Catalogue of the Flora of Matheran and Mahabaleshwar; *J. Bom. Nat. Hist. Soc.*, **1**:203–214.

Blatter, E. 1909. The Flora of Panchgani, *J. Bom. Nat. Hist. Soc.*, **19**:314–332.

Bhonsle, U.S. 1996. Bryophytes of Mahur. Ph.D. Thesis Univ. of Bombay. (Unpublished).

Bruhl, P. 1931. A census of Indian Mosses. *Record Bot. Sur. India,* **13(10)**:50.

Chaudhari, K.K. 1982. Gazetter of Maharashtra (Thane District) Govt. of Maharashtra, Bombay.

Dabhade, G.T. 1988. Mosses of Khandala Mahabaleshwar in the Western Ghats (India). Publ. A.S. Dalvi, Kharigaon, Thane.

Dixon, H.N. 1908(-a, b). Mosses from Western India. *J. Bombay Nat. Hist. Soc.*, **19**:536–536.

Dixon, H.N. 1909 (b). Mosses from the Western Ghat. *J. Bot.*, **45**: 157–164.

Gangulee, H.C. 1969, 1971, 1972. Mosses of Eastern India Fascile 1–3 Calcutta.

Sedgwick, L.J. 1910. A First list of Mosses of Western India. *J. Bom. Nat. Hist. Soc.*, **19(4)**:938–942.

Sedgwick, L.J. 1911. A Second list of Mosses from Western India. *J. Bom. Nat. Hist. Soc.*, **20**:4, 1043–1045.

Sedgwick, J.J. 1913. A third list of Mosses from Western India. *J. Bom. Nat. Hist. Soc.*, **22**:370–371.

CHAPTER 5

DISTRIBUTION AND CONSERVATION OF MULBERRY (*MORUS* SPP.) GENETIC RESOURCES IN INDIA

*M. Chandrasekhar**, *R.P. Saraswat**, *A. Tikader** and *A. Ananda Rao**

ABSTRACT

Mulberry, the genus *Morus* (Moraceae), is an exclusive host plant of silkworm (*Bombyx mori* L.). Besides, it has multipurpose utility as fodder, fuel, timber, fruit and medicine. The distribution pattern and diversity of mulberry indicate that four species viz., *M. indica* L., *M. alba* L., *M. laevigata* Wall. and *M. serrata* Roxb., are found in India, especially in sub-Himalayan range under natural condition. *M. serrata* is endemic to North West Himalayas, where as *M. laevigata* is naturally distributed through out India including Andaman & Nicobar islands. *M. alba* and *M. indica* are mostly available under cultivated conditions. *Morus* diversity is also available in public domain for which awareness is to be created for its conservation.

Presently this centre conserve 1023 accessions (indigenous-761 and exotic-262) in its field gene bank (living stand) representing collections from geographical and genetic origin, representing 26 countries including India and 13 mulberry species. Different conservation strategies including collection, introduction, characterization, evaluation, documentation and supply of mulberry genetic resources are discussed. Certain safety back-up of mulberry genetic resources available other than CSGRC, Hosur are also enumerated.

Biodiversity and Conservation

Biodiversity, particularly of plant species, is the base for plant improvement. Genetic diversity of species is in need for protection or preservation. Endangered species in their habitats acquire great significance in conservation.

* Central Sericultural Germplasm Resources Centre, Central Silk Board, P.B. No. 44, Thally road, Hosur-635 109 (Tamil Nadu).

Genes contained in the natural populations is the focus of conservation interest and considered as a 'resource'. In this context wild genes have become prominent today. Plant breeders have to have greater range of genetic diversity for crop improvement. Usefulness of collections to plant breeders and other users, including evolutionists, plant pathologists, taxonomists and other experimental biologists depends on the extent to which the collections are geographically and ecologically representative and on the presence of genes of particular interest to plant breeders. Hence, conservation of biodiversity is fundamental to success of development process. Biodiversity conservation is an instrument that yields substantial local, national and global benefits. Primary objective of conservation is to preserve at least one copy of each of different alleles than the accurate representation of allelic frequencies in the sample (Bennett, 1970). This unique heritage serving human kind since ages has to be protected and conserved for the present and posterity.

Genetic stock conserved often harbour rare and unexploited genes for productivity as well as resistance to biotic and abiotic stresses. Conservation of any genetic resource ensures easy access to desired genes, a resource to revitalization, a "guard file" for disaster management, safe guard against gene erosion and comparative profiling of genes/traits etc. The choice of conservation strategy depends mainly on the nature of the material to be conserved i.e. life cycle, mode of reproduction, size and the ecological status (OCED, 1999). The major approaches for conservation are (i) *in situ* (ii) *ex situ.*

Mulberry

Mulberry, the genus *Morus,* belongs to the family Moraceae. Mulberry leaf is an indispensable food for silkworm, *Bombyx mori* L. Hence it is an important plant for sericulture industry. Mulberry comprises about 68 recognized species in different countries mainly from Japan, China, India, Korea and Taiwan. Today mulberry is grown in many countries, which practice sericulture, in a wide area from tropics to sub-arctic regions, from wetlands to semi-arid regions and from flat lands to upper lands up to an altitude of 11,000 feet MSL.

Mulberry fruits are eaten, refrigerant, used also for sore throat, dyspepsia and melancholia. Leaves eaten as vegetable, also as cattle fodder, rich in calcium and a good source of vitamin C. Fruits also used in the form of juices, jam, stews and tarts and fermented to yield spirituous liqueurs. Wood is used for hockey sticks, rackets and their presses, cricket bats, stumps and other sports goods. Wood is also suitable for house building, agricultural implements, and furniture, spokes and for shafts, bent parts of carriages and carts and for turnery. Bark is used for paper making, yields a textile fibre and also as a purgative and vermifuge. Roots are anthelmintic in nature (Anonymous, 1986).

Distribution of Mulberry Genetic Resources in India

Though mulberry is naturally distributed in Sub-Himalayan belt in India, it is also widely cultivated for sericulture activities and for other uses throughout India. Mulberry is known in different Indian languages as Sanskrit–*tula*; Hindi and Punjabi–*tut*, *tutri*; Bengali and Kashmiri–*tut*; Marathi–*tut, ambut*; Gujarati–*shetur*; Telugu–*rashme chettu, pippalipandu chettu*; Kannada–*hipnerle*; Tamil–*musukette, kambli chedi*; Oriya–*tuti, tuticoli.*

Genus *Morus* is a native of Indo-China and found widely distributed in the lower sub-Himalayan region up to an elevation of 2100 m covering both the temperate and sub-tropical regions of the northern hemisphere. Hooker (1885) and Brandis (1906) have reported occurrence of four species of genus *Morus* L. namely *Morus alba, Morus indica, Morus laevigata* and *Morus serrata.* Parkinson (1923) reported occurrence of *Morus laevigata* from Andaman and Nicobar islands. Osmoston (1927) reported distribution of *M. indica, M. serrata* and *M. laevigata* in the forest flora of Kumaon hill region of Uttaranchal state. Mention of *M. rubra* has been made by Hooker (1885) in Bhaderwah and Ramson areas of Jammu province. Presence of mulberry in Ladakh is recorded as early as 1853 by Cunningham. Wazir (1942) has reported mulberry wealth scattered both in private as well as State Government land in Ladakh province as a whole.

Kanjilal *et al.* (1940) enumerated three species from Assam and northeast viz. *Morus serrata*, *Morus laevigata* and *Morus acidosa*. Gamble and Fischer (1957) recorded two species viz. *Morus alba* and *Morus indica*. Nair (1977) documented occurrence of *Morus serrata* from Himalayas in the Sub Himalayan belt up to an altitude of 3300 m MSL. Botanical Survey of India recorded the distribution of *M. serrata* in altitudes between 1300 m to 1900 m in the locations of Sitapur, Joshimath, Chamoli Garhwal region and Tehri Garhwal region of Uttaranchal state. In Himachal Pradesh, *M. serrata* is found in Chamba and Sarahan at an altitude range between 1100 to 2400 m. *M. indica* is recorded in Chamoli Garhwal and Kumaon hills. *M. laevigata* has been reported from localities of Gopeshwar in Chamoli Garhwal region at an altitude of 1300 m. Foot hills of J&K to Assam up to the elevation of 7000' have a rich flora of Moraceae (Dandin, 1989).

M. serrata is native to India known as Himalayan mulberry and found only in Himalayan region up to an altitude of 2200 meters MSL particularly in North-West Himalayas. *M. laevigata* is found throughout the country; concentrated areas of natural mulberry trees of *Morus laevigata* are found in Island of Andaman & Nicobar, Arunachal Pradesh, Sikkim, West Bengal, Assam, Meghalaya, Manipur, Uttaranchal and Uttar Pradesh, the trees of which are mainly used for its fruits and timber. *M. alba* and *M. indica* are mostly found in cultivated form. Presence of *M. serrata* up to an altitude of 1800 m and mostly in association with oak was noticed in Uttar Pradesh and Uttaranchal states (Balakrishna. and Ramesh, 1987). Trees of *M. laevigata* were found around Dehradun. Distribution of *M. laevigata* in Manipur, Nagaland, Meghalaya and Darjeeling (West Bengal) was noticed (Jain and Kumar, 1988). *M. indica* and *M. alba* are common in N-E region and different varieties of these two species are found in cultivated form for sericulture purpose. *M. serrata* is sparsely distributed in the region. Mulberry plants in North-West Himalayan region are mostly found as huge and massive trees growing on hill sides, road sides and on river beds. *M. indica, M. alba* and *M. laevigata* are common in the region and *M. serrata* is not a common species. *M. nigra* found in the region was an introduced species from Iran (Dhar and Ahsan, 1989).

Ravindran (1995) reported occurrence of *M. laevigata* in Jalapaiguri and Kalimpong area (West Bengal), Sikkim, Assam and Manipur. Ananda Rao (1995) noticed huge trees of *M. laevigata* in Madhya Pradesh, Chathisgarh states. The age of these trees appear to be less than 100 years, indicating that the trees might have been introduced. Occurrence of natural population of *M. laevigata, M. indica* and *M. alba* was also noticed in Rajasthan (Ananda Rao, 2003), Gujarat and Uttar Pradesh (Chandrasekhar, 2001, 2002, 2003), Jharkhand (Ananda Rao, 1996) and Maharashtra (Saraswat, 2002, 2003). Thangavelu and Naik (1997) noticed occurrence of *M. laevigata* in Andaman & Nikobar islands and four collections were made among the population, which are conserved at CSGRC, Hosur. *M. serrata* is naturally growing in Dehradun, Pauri Garhwal and Haridwar areas of Uttaranchal (Tikader, 1997). Availability of *M. alba, M. indica, M. laevigata* and *M. serrata* under natural conditions in Punjab, Himachal Pradesh and Kashmir was confirmed (Tikader, 2001).

The higher altitudes of Western Ghats (Kerala) or even Eastern Ghats (Yercaud, Tamil Nadu) would be the primary home of introduction of the wild mulberry from its native habitat in the Himalayas (Nair, 1997). Naik (1995) confirmed that the Western Ghat region of Kerala and Tamil Nadu do not possess any wild mulberry. However, the Britishers during their tenure as Estate Managers of tea plantation have selectively introduced female genotypes of mulberry from Europe for their ornamental value and also for its fruits in wine making. The introduced mulberry has spread in different parts of Kerala and Tamil Nadu. Thangavelu and Muthu (2002) reported occurrence of naturally growing *M. indica* tress in Kerala. Herbarium collection at Botanical Survey of India indicate occurrence of *M. laevigata, M. alba* in Dharwad (Karnataka), Rajasthan and Maharashtra (Ananda Rao and Girish Naik, 1997).

Need for Mulberry Genetic Resources Conservation

Due to fast changing environment, fragmentation of forests, destruction of natural habitats and depletion of water resources, the genetic resources of mulberry are depleting very fast. Since breeders mostly depend on diverse germplasm resources, the conservation of mulberry becomes imperative for improvement of the crop.

Rich *Morus* biodiversity exists both under natural (Plate I) and managed habitats. In India, efforts have been made to use primary gene pools, constituted predominantly cultivars belong to *Morus indica* and *Morus alba,* to tailor the desired varieties. However, the promising genes exist in the secondary and tertiary gene pool in the wild species of *Morus laevigata* and *Morus serrata* and other wild species can be utilized for gene transfer by applying biotechnological and molecular techniques.

Conservation of Mulberry Genetic Resources

Realizing the importance of mulberry germplasm for sustainable development of sericulture in India, Central Sericultural Germplasm Resource Centre (CSGRC) was established in the year 1990 under the aegis of Central Silk Board (CSB), Ministry of Textiles, Government of India. It is the nodal agency for mulberry germplasm management in India and recognized as National Active Germplasm Site (NAGS) for mulberry by National Bureau of Plant Genetic Resources (NBPGR), New Delhi, under Indian National Plant Genetic Resources System (INPGRS).

Survey, Exploration and Collection of Mulberry Genetic Resources

Phyto-geographical information is a prerequisite for effective exploration and collection of natural genetic resources. Work on survey, exploration and collection of mulberry genetic resources was initiated from 1983. The main objective of exploration is to identify the areas where significant resources of *Morus* spp. occur as also most threatened with deforestation. Survey, exploration and collection of mulberry genetic resources was continued from 1993 onwards by CSGRC, Hosur. So far, 55 survey and exploration tours were conducted covering different agro-climatic and geographical regions of the country through which 666 mulberry accessions were collected from different eco-geographic zones. Besides, the introduced mulberry germplasm resources from different temperate and tropical countries, which are available in the Research institutes, Universities, and State departments, are also collected to enrich the mulberry field gene bank.

Strategies of Conservation: *Ex situ* Conservation Field Gene Bank

Mulberry being perennial out-breeding tree, exhibits high degree of heterozygosity. Hence, for conservation of mulberry outside its natural habitat, *ex situ* field gene bank (clonal repository) has been developed by planting vegetative clones of mulberry accessions for maintaining the genetic integrity of the conserved material. Rooted cuttings (saplings) after six-month establishment in the nursery were transplanted to the field gene bank. Exotic accessions and wild *Morus* species which do not respond for vegetative propagation through cuttings were raised through bud grafting to the local scions. The field lay-out of the mulberry field gene bank is indicated in the Table 1. *Morus* species represented in the field gene bank is presented in Table 2. National Accession Numbers (Indigenous collections: IC. No: 313662–314262 and Exotic collections: EC. No: 493758–493928) have been provided for majority of the mulberry accessions conserved in the field gene bank, by NBPGR, New Delhi for protection of mulberry genetic resources at global level under the present IPR regime.

Table 1: Layout of mulberry field gene bank at CSGRC, Hosur

Total accessions	*Indigenous–761; Exotic - 262; TOTAL –1023*
Area	5.0 ACRES
Plantation period	1993–2003
Plant type	Dwarf tree (1.35–1–5 m)
Spacing	• 2.44 × 2.44 m between plants. • 3.05 m between accessions.
Pruning	• Twice in a year (for data recordal) (Feb. and June) • Once in year (for conservation)
Cultural practices	• Neem oil cake: 500 g. per plant. • FYM: 10 kg/plant/year. • Green manure raising and mulching. • Chemical fertilizers: NPK–75 g/plant/year. • Urea–75 g/plant after leaf harvest. • Other cultural operations.
Irrigation	Once in 8–10 days during dry season.
Plant protection	• Bavistin (0.2%) for control of fungal diseases. • Plantomycin (0.2%) for control of bacterial diseases. • Insecticide for control of pests.

Table 2: *Morus* species in the field gene bank

Sl. No.	*Morus species*	*Origin/habitat*	*Economic importance*
1.	*M. alba* L.	China	Sericulture, fruits, fodder, fuel.
2.	*M. australis* Poirt. (*M. acidosa* Grifith.)	Japan	Sericulture, fruits.
3.	*M. bombycis* Koidz.	South Sakhaline, Japan, Korea and China	Sericulture.
4.	*M. cathayama* Hemsl.	Central China	Fruits.
5.	*M. indica* L.	India	Sericulture, fruits, fodder, fuel.
6.	*M. laevigata* Wall.	Himalayas, West China	Fruits, timber.
7.	*M. multicaulis* Poir.	Japan	Sericulture.
8.	*M. nigra* L.	West Asia, Caucasus, Persia	Fruits.
9.	*M. rotundiloba* Koidz.	Thailand	Sericulture, fruits.
10.	*M. rubra* L.	North America	Fruits.
11.	*M. serrata* Roxb.	Himalayas, Kashmir	Timber.
12.	*M. sinensis* Hort.	—	Sericulture.
13.	*M. tiliafolia* Makino.	South Japan, Korea	Sericulture.

Presently 1023 accessions of diverse mulberry germplasm of different origin and geographical region (indigenous-761 and exotic-262), representing 26 countries and 13 mulberry species are conserved in the ex *situ* field gene bank (Table 3) (Plate I). Number of germplasm collection in each category such as landraces, hybrids, ploidy level etc. is presented in Table 4. The germplasm information is documented in the Mulberry Germplasm Information System (MGIS) with query-based retrieval, which is available in the website *WWW.SILKGERMPLASM.COM* for easy access of the mulberry germplasm material for promoting utilization at global level.

Conservation Using *in vitro* and Cryopreservation Techniques

Cryopreservation technology forms an alternative to *ex situ* conservation strategy for crop like mulberry, which is vegetatively propagated. Cryopreservation method ensures the genetic stability of the germplasm and provides an alternate cost effective non-dependence on electricity, long-term strategy for conservation of mulberry germplasm. Application of *in vitro* methods of plant genetic conservation has several advantages such as disease free stocks can be maintained, which has implications for germplasm exchange and quarantine and large number of genotypes can be stored in a relatively small area at cost efficient management. Keeping this in view, *in vitro* and cryopreservation facilities were developed at CSGRC, Hosur Efforts are being made to standardize the techniques on cryopreservation of winter buds and pollen of mulberry accessions, comprising different species, landraces, local cultivars, wild and polyploid accessions. A total of 37 mulberry accessions representing different species are conserved in *in vitro*

Table 3: *Ex situ* conservation mulberry germplasm in the CSGRC, Hosur (Country and state wise representation)

Country	*No. of collections*	*States of India*	*No of collections*
Afghanistan	2	Andaman & Nicobar	15
Australia	2	Andhra Pradesh	3
Bangladesh	5	Arunachal Pradesh	9
Myanmar	7	Assam	10
China	53	Bihar	18
Cyprus	1	Gujarat	26
Egypt	3	Haryana	2
Spain	2	Himachal Pradesh	16
France	32	Jammu & Kashmir	32
Hungary	1	Karnataka	99
Indonesia	6	Kerala	68
Italy	7	Manipur	12
Japan	69	Meghalaya	21
Pakistan	8	Maharastra	16
Papua New Guinea	1	Madhya Pradesh	12
Paraguay	4	Nagaland	1
Philippines	1	New Delhi	2
Portugal	1	Punjab	11
Russia	1	Rajasthan	43
South Korea	6	Sikkim	3
Thailand	11	Tamil Nadu	53
Turkey	1	Uttar Pradesh	135
Venezuela	1	Uttaranchal	8
Vietnam	3	West Bengal	146
Zimbabwe	11		
Unidentified	23		
Total	262	Total	761

Table 4: Classification of mulberry germplasm in the *ex situ* field gene bank at CSGRC, Hosur

Sl. No.	*Genetic origin*	*No. of accessions*
1.	Selections	86
2.	Locals/Landraces	347
3.	Cross pollinated hybrids (CPH)	33
4.	Open pollinated hybrids (OPH)	43
5.	Triploids	23
6.	Tetraploids	30
7.	Hexaploid	1
8.	Decosoploid	1
9.	Mutants	7
10.	Backcross	1
11.	Wild	190
12.	Exotics	261
	Total	1023

culture using MS media supplemented with BAP 1mg/lit and maintained under 25 ± 1°C with daily photoperiod regime of 16 h light and 8 h dark. Thirty mulberry genetic resources representing different species, local landraces, wild and cultivated varieties are stored under –1.5°C to 15°C to test the cold hardiness and storage potentiality.

Characterization and Evaluation of Mulberry Genetic Resources

Objective of conservation of mulberry genetic resources is for preservation of gene pool, selection of parental stock for breeding, exploitation for multipurpose utility of mulberry, registration and supply of germplasm material for users for research. The conserved germplasm can be utilized only after documenting the qualitative and quantitative characters of the germplasm. Characterization of mulberry genetic resource determines the structural and functional attributes of the accessions and is highly essential for their genetic identity under IPR regime. The comprehensive information on variability recorded helps breeders and users for effective crop improvement. The detailed information is required to describe an accession. Molecular characterization of mulberry genetic resources provides information on genetic relationship among them.

Keeping the above in view, CSGRC, Hosur attempted to generate the information on mulberry genetic resources by characterization of morphological traits (22 characters), reproductive biological traits (26 characters), growth behaviour (16 parameters), leaf histology (14 characters), cytological status to assess the ploidy level of the germplasm (2n = 2x = 28), leaf quality assessment through biochemical content of leaf (total chlorophyll, carbohydrate and protein), evaluation of mulberry germplasm for propagation traits (vegetative propagation through cuttings–17 parameters), growth, leaf yield and quality of leaf through silkworm rearing (bio-assay) (16 parameters), screening of germplasm for major foliar fungal diseases (3 diseases) and pest infestation (mealy bug) under natural conditions, isozyme estimation (peroxidase) in leaf and molecular characterisastion. The data generated so far has been published in the form of three catalogues viz. Catalogue on Mulberry (*Morus* spp.) *Germplasm,* Vol. I. 1997, contain data on characterization of 328 mulberry germplasm accessions (indigenous-212 and exotic-116) and Catalogue on Evaluation of Mulberry (*Morus* spp.) *Germplasm,* Vol. I. 2003, contain data on evaluation of 316 mulberry germplasm accessions evaluated (indigenous-234 and exotic-82) for propagation traits, growth and yield parameters and leaf quality through silkworm rearing.

Upon evaluation for various parameters, selected (outstanding) germplasm accessions are further evaluated under varied agro-climatic conditions, abiotic

Mulberry field gene bank

Diversity in mulberry leaf

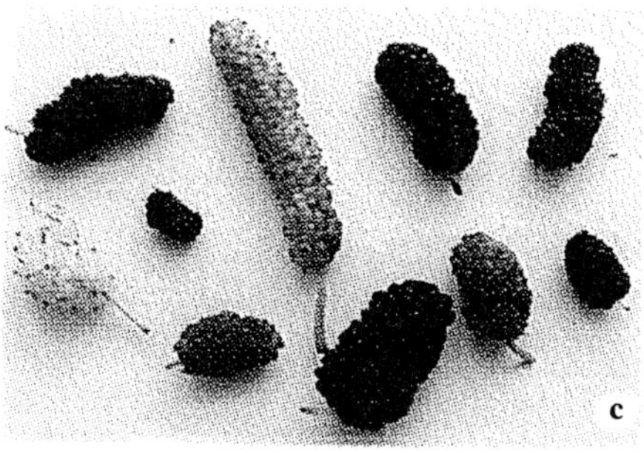

Variation in mulberry fruits

The oldest mulberry tree in Joshimutt

Wild mulberry in Rajasthan

Wild mulberry in UP

Plate I

and biotic stress conditions, so as to identify suitable mulberry accession for sustainable sericulture development in the country.

Supply of Mulberry Genetic Resources

The most effective way to conserve mulberry germplasm resources is to put them in continuous use, meeting competing demands of the society. Hence, multipurpose uses of mulberry germplasm have to be greatly exploited for their sustainable conservation. Realizing the increased use of mulberry germplasm resources for crop improvement programmes, CSGRC, Hosur being the nodal Institute for mulberry germplasm conservation and utilization in India. From 1997–2004, a total of 973 mulberry accessions (indidgenous-571 and exotic-402) were supplied to various research institutions/universities for research, conservation, evaluation, propagation and for raising mulberry plantation for fruit purpose.

Registration, Quarantine and Exchange of Mulberry Genetic Resources

Registration of germplasm is an important aspect for protection of breeders right under national and international agricultural/sericultural research systems. Outstanding varieties of mulberry are released for commercial exploitation after authorization. However, many promising germplasm do not get authorized due to one or other limitation in spite of their outstanding performance. Registration of such germplasm is required. Moreover, there is urgent need to document such materials in wake of GATT and IPR regime at national and international level. Keeping these points in view, CSGRC, Hosur has developed a set of guidelines for registration of sericultural germplasm under National Sericultural Research System of CSB. Phytosanitory certificate which is to be met by the supplier along with the sample ensuring that the material is not coated/treated with chemical pesticide is also issused at the time of supply of mulberry germplasm.

Mulberry Germplasm Database — Management and Utilization of Database

Docùmentation acts as sources of information to assist in the setting of the priorities, planning and operation of germplasm management. Realizing the importance of information database in the germplasm management, CSGRC, Hosur developed an information system named Mulberry Germplasm Information System (MGIS), incorporating various facilities (Sekar *et al.*, 1997). This system will work on MS-DOS platform and can be viewed on a VGA with resolution of 640 × 480 resolutions. While developing the system, the standard packages in the use at NBPGR, New Delhi and CYMMIT were taken into consideration so as to make the packages friendlier with scope for broad-spectrum of applications. In order to MGIS more meaningful and user friendly, the standard data format in terms of terminology and measurements with integrated database techniques were used. User-friendly menu driven facilities have been incorporated for easy data viewing, information retrieval, updating and report generation. Each file is linked to the National accessioning number which is a unique number assigned to each accession serving as primary key. The query module has the facility for identification and selection of accessions on the basis of certain desirable criteria up to 15 quantitative/qualitative characters has been developed as a separate module to facilitate breeders/users for utilization of information. The system is flexible enough to allow modification and insertion of new modules as per the changing needs of the scientists.

CSGRC, Hosur being the nodal agency for mulberry germplasm in India, maintain the entire mulberry germplasm available in the country. However, the Sericulture Research Institutes, like Central Sericultural Research and Training Institute (CSRTI) at Berhampore (West Bengal), Mysore (Karnataka) and Pampore (Jammu and Kashmir), Universities, State Sericulture Departments also maintain the mulberry germplasm for research purpose, which also serves as back-up conservation centers (Table 5).

Table 5: Mulberry germplasm conservation centers other than CSGRC

Sl. No.	*Centre*	*Exotic*	*Indigenous*	*Other*	*Total*
1.	CSR&TI, Mysore	138	188	101	427
2.	CSR&TI, Berhampore	91	210	–	301
3.	KSSR&DI, Bangalore	181	170	–	351
4.	SKUAST, J&K	26	31	18	75
5.	CSR&TI, Pampore	33	23	9	65
6.	ERRC, Kerala	–	71	–	71
7.	Bangalore University, Bangalore	21	21	10	52
8.	RSRS, Kalimpong	12	31	–	43
9.	RSRS, Sahaspur	6	33	–	39
10.	RSRS, Jorhat	1	10	13	24
11.	RSRS, Coonoor	3	18	–	21
12.	Padmavathi Mahila University, Thirupathi	2	26	–	28
13.	Mysore University, Mysore	5	5	–	10
14.	APSSRDI, Hindupur	–	5	–	5
15.	Other centers	31	45	1	77

REFERENCES

Ananda Rao, A. 1995. Technical Report on Survey, Exploration and Collection of Mulberry Genetic Resources in Central India, CSGRC, Hosur.

Ananda Rao, A. 2003. Technical Report on Survey, Exploration and Collection of Mulberry Genetic Resources in Western India, CSGRC, Hosur.

Ananda Rao, A. and V. Girish Naik. 1997. Technical Report on Survey and Exploration for Mulberry Genetic Resources in Western India, CSGRC, Hosur. pp. 4–10.

Annonymous, 1986. The useful plants of India, CSIR, 1986, p. 381.

Balakrishna, R. and S.R. Ramesh. 1989. Mulberry species and their distribution in Central Himalaya. In: K. Sengupta, and S.B. Dandin (eds.). Genetic Resources of Mulberry and Utilization. CSR&TI, Mysore. pp. 33–35.

Bennett, E. 1970. Tactics of plant exploration. In: Frankel, C.H. and E. Bennett (eds.). Genetic Resources in Plants: Their exploration and conservation. IBH Handbook No. 11, Blackwell Scientific Publications. Oxford. pp. 157–159.

Chandrasekhar, M. 2001. Technical Report on Survey, Exploration and Collection of Mulberry Genetic Resources in Gujarat, CSGRC, Hosur.

Chandrasekhar, M. 2002. Technical Report on Survey, Exploration and Collection of Mulberry Genetic Resources in Uttar Pradesh, CSGRC, Hosur.

Chandrasekhar, M. 2003. Technical Report on Survey, Exploration and Collection of Mulberry Genetic Resources in Uttar Pradesh, II tour, CSGRC, Hosur.

Cunningham, A. 1853. Ladakh: physical, statistical and historical, pp. 219.

Dandin, S.B. 1989. Indian collection of mulberry germplasm. In: K. Sengupta, and S.B. Dandin (eds.). Genetic Resources of Mulberry and Utilization. CSR&TI, Mysore. pp. 9–31.

Dhar, A. and M.M. Ahsan. 1989. Mulberry species and their distribution in North West Himalayas. In: K. Sengupta, and S.B. Dandin (eds.). Genetic Resources of Mulberry and Utilization. CSR&TI, Mysore. pp. 111–124.

Hooker, J.D. 1885. Flora of British India, Vol. V; P. 492. B.S.M.P. Singh Publications, Dehradoon.

Kanjilal and Das. 1940. Flora of Assam, pp. 233–275.

Naik, V.G. 1995. Technical report on survey, exploration and collection of mulberry genetic resources in Idukki district (Kerala state). CSGRC, Hosur, p. 6.

Nair, P.K.K. 1997. Exploration and characterization studies on wild mulberry genetic resources with reference to various specific ecosystems. Project Report of Environmental Resources Research Centre, Thiruvananthapuram submitted to the Central Silk Board. p. xvii.

Osmostan, A.E. 1927. A Forest Flora of Kumaon. pp. 502–503. Govt. Press.

Parkinson, C.E., 1923. A forest flora of Andaman Islands, p. 155, xvii.

Ravindran, S. 1995. Technical Report on Survey and Exploration of Mulberry Genetic Resources in North East India, CSGRC, Hosur. pp. 20–21.

Saraswat, R.P. 2002 Technical Report on Survey, Exploration and Collection of Mulberry Genetic Resources in Maharashtra, CSGRC, Hosur.

Saraswat, R.P. 2003 Technical Report on Survey, Exploration and Collection of Mulberry Genetic Resources in Maharashtra, CSGRC, Hosur.

Thangavelu, K. and V. Girish Naik, 1997. Technical report on survey, exploration and collection of mulberry germplasm in Andaman and Nicobar islands, CSGRC, Hosur, pp. 7–8.

Thangavelu, K. and E. Muthu, 2002. Technical Report on Survey, Exploration and Collection of Mulberry Genetic Resources in Kerala, CSGRC, Hosur.

Tikader, A. 1997. Technical Report on Survey and Exploration for Mulberry Genetic Resources in North-West India, CSGRC, Hosur.

Tikader, A. 2001. Technical Report on Survey, Exploration and Collection of Mulberry Genetic Resources in Jammu & Kashmir, Punjab and Himachal Pradesh, CSGRC, Hosur. p. 11.

Wazir, T.C. 1942. Report on conditions of the sericulture industry in Jammu & Kashmir.

CHAPTER 6

A GLIMPSE INTO THE BIODIVERSITY OF ORISSA MANGROVES AND STRATEGIES FOR THEIR GERMPLASM CONSERVATION

*B.P. Choudhury** and *Mihir Kumar Das***

ABSTRACT

The state of Orissa is rich in various natural resources of which the vegetable wealth deserves special mention. The forest cover of yore which was gregarious as well as luxuriant embracing broad spectrum of biodiversity has degraded at an alarming rate in recent time and is now restricted to only 20–30% against its total landmass. Consequently, the mangrove vegetation along the coastal belt has come down to an area of 195 sq. kms. as per satellite data (1995) and practically this plant community is restricted in the estuary of Dhamra river and Mahanadi delta while those in the Budhabalanga, Jambu, Devi, Rusikulya and along the fringes of Chilka lake are found in very denuded condition while at other places these have totally disappeared. Total/partial disappearance of this plant community has enhanced oceanic cyclones, unusual floods and other environmental hazards which have affected the sustenance of human life as well as other life forms resulting in the miserable financial backbone of coastal populace. Super Cyclone in the year 1999 is a bright example in this connection. In fact, as experienced now, the fury of environmental hazards in the costal belt of Orissa is in ascending order. Hence, there is an urgent need for improvement of coastal environment of this state for which shrinkage of mangrove vegetation should be immediately checked, while effective strategies should be implemented for proliferation of this plant group by adequate aforestation through protection, regeneration as well as propagation of this sensitive vegetation, development of artificial mangrove vegetation as well as promoting eco-friendly mangrove vegetation along the coastal zone of Orissa. Mangrove Genetic Resource Conservation Centre (MGRCC) should be established in Bhitarkanika National park as suggested by the scientists of M.S. Swaminathan Research Foundation (Chennai) and International Tropical Timber Organisation (Japan) after extensive survey works.

* P.G. Department of Botany, Utkal University, Bhubaneswar-751 002.

** Department of Botany, G.M. College Sambalpur.

Introduction

Of the 32 forest types, as described by Champion and Seth (1968) mangrove vegetation extant along 480 km sea shore deserves special mention. Mangroves of Orissa, in remote past, were very luxuriant as well as gregarious in the estuaries of the river systems as well as islets and fringes of Chilka lagoon (Map 1). However, due to constant operation of various anthropogenic activities the mangroves have disappeared leaving behind those present only in the estuaries of Dhamra (confluence of the rivers Brahmani and Baitarani)

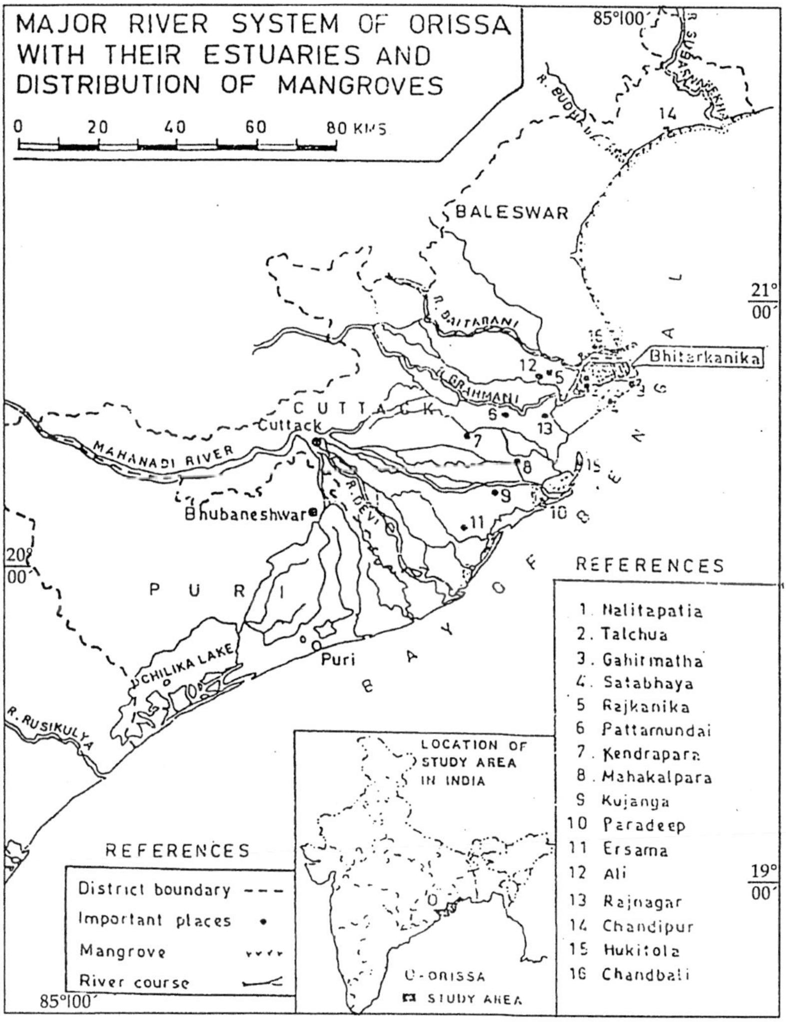

Map 1

and Mahanadi delta. These have disappeared partly/totally from the river mouths of Budhabalanga, Devi, Jatadhar, Jambu, Rusikulya (Choudhury 1990). Due to establishment of port (cf. Paradeep), factories, reclamation and conversion of forest lands for paddy cultivation, human habitation, prawn culture and felling of woody elements for fuel and other domestic usages this plant community is degrading at an alarming rate and is now restricted to 195 sq. kms. only according to satellite data 1995 (Misra, 1999).

Due to disappearance of mangroves from the coastal terrain the inhabitants are experiencing furious oceanic cyclones, unusual floods, encroachment of sea towards land mass, washing away of various life forms as well as widespread destruction of crops. The wide-scale devastation of lives, properties along with domestic and wild animals during the last week of 1999 in the coastal zone of Orissa by super cyclone is beyond one's imagination (Choudhury, 2002). According to a report by Committee for Legal Aid [Sunday Times of India, Bhubaneshwar, National July 11, 2004] for disappearance of mangroves from coastal belt of Orissa coupled with global warming, cyclone washed away the shelter at once submerged the village of Jambu, the Kendrapara canal to the North-West of False Point. The worst cyclone to hit the state, was the False Point cyclone of 1885. This cyclone burst upon the coast in the early morning of September, 1885. It was accompanied, by a storm-wave rising to height of about 6.5 m above sea level, which at once submerged the village of Jambu on the Kendrapara Canal (Map 3) to the north-west of False Point and rolled on it a north westerly direction till it lost itself in Brahmani river. Shortly after midnight of October 29, 1971 while the people were in their homes on the (coastal zone) north of Paradeep up to Basudevpur was lashed by a furious cyclone which lasted till the evening of next day. The tidal wave up to height of 4 m swept over 75 mile long stretch of coastal area which caused death of innumerable life forms. This was followed by super cyclone of October 29, 1999.

Despite, the drastric shrinkage of this plant community mangroves in the Bhitarkanika (Map 2) wildlife sanctuary (Chadha and Kar 2001) and those in the Mahanadi delta are rich in species composition and hence exhibit biodiversity in tangible degree. Mangroves of Bhitarkanika wildlife sanctuary was under the pestilent anthropogenic spell during the time of Kanika Zamindary (Choudhury, *et al.,* 1995). The mangrove forest was infested with many savage animals both in water bodies as well as in terrestrial regions. The humid atmosphere, mud flats studded with numerous pneumatophores and impenerable thorny scrub jungle was not suitable for human habitation. Haines (1921–1925) in the introductory part of his treatise has rightly mentioned the obstacles for exploration of such mangrove terrains.

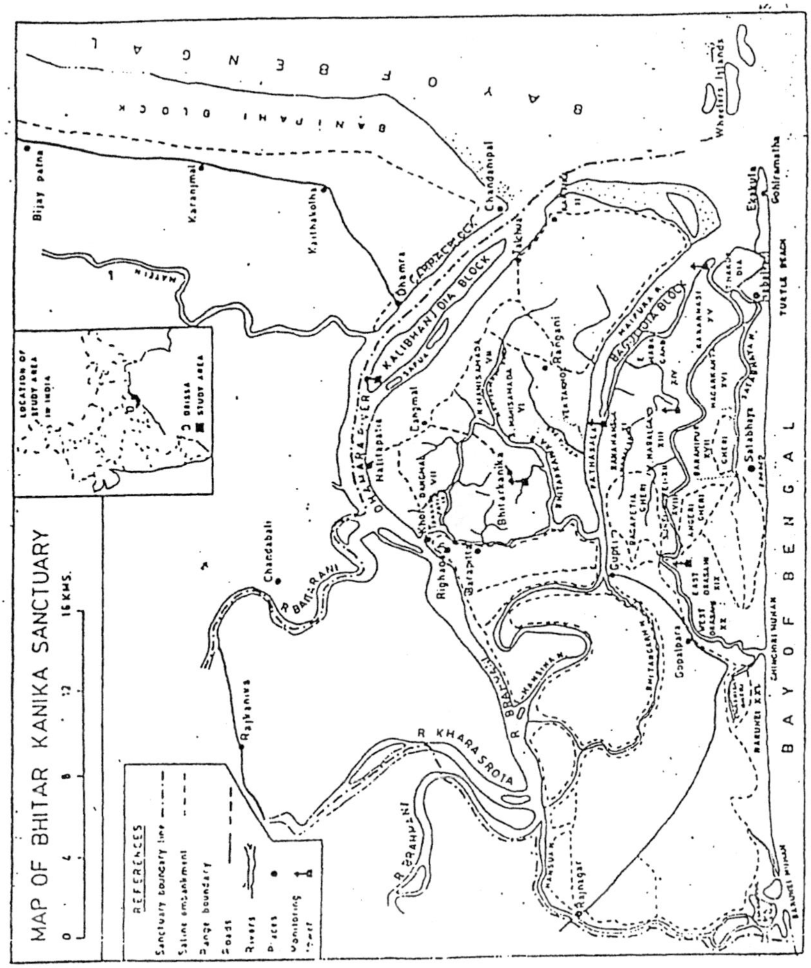

Map 2

For this reason, both Haines (l.c.) and Mooney (1950), the later plant explorer could not botanize Bhitarkanika and adjoining reagions. Of course, they have sporadically launched upon floristic studies in the mangroves extant in Mahanadi delta. Infact, some new reports have been established by Mooney (l.c) in his treatise. However, many aspects such as dominance, rarity, distributional pattern, socio-economic aspects of mangroves have not been highlighted by them. At present, the existing mangroves are also not free from danger although Bhitarkanika was declared as a wild life sanctuary in the year 1975. Later on (1998) a portion of it has been upgraded to the National park for effective management as well as better conservation of rare and vulnerable life forms (Choudhury, 1990). The mangroves of

Mahanadi delta (Nayak and Choudhury, 2001) though come under reserve forest category, mostly this plant community is under the detrimental effects of prawn culture as well as illegal felling of woody elements by the immigrants as well as local people (Choudhury, 1994). At present, there are *Ca* 67 species (both mangroves and their associates) belonging to 51 genera under 29 families of vascular plants in the state of Orissa. According to status report (Anonymous, 1994) 39 mangrove species are present in Bhitakanika which is one of the four sites studied by the experts of team constituted in collaboration with the scientists of M.S. Swaminathan Research Foundation (Madras) and International Tropical Timber Organisation (Japan). According to their opinion, of the sites visted in India, Bhitarkanika has the largest number of mangrove species. Genus *Xylocarpus* shows remarkable variation, representing three species viz. *Xylocarpus granatum, X. mekongensis* and *X.* sp. Prestine stands of *Heritiera fomes,* a threatened species in the mangrove forest of Sundarban, are also observed. Bhitarkanika is belived to be a primary centre of diversity in both the *Heritiera* species. Substantial variation is observed within a number of species. Nutritional status of the land is satisfactory and the general health of the forest seems to be very good. Diverse species and their genotypes are likely to thrive well at Bhitarkanika under conditions of minimum inter-specific and inter-genotype condition. The area under mangrove forest is wide, showing clear zonation. The association flora is rich. As a matter of fact, *Heritiera fomes,* in remote past i.e. 5000 years ago (Naskar and Guha Baxi, 1987) which was plentiful in and around present Calcutta including Sundarbana is now on the verge of disappearance due to obstacle of fresh water flow to the Bay of Bengal, created due to the settlement of the immigrants. On the contrary, in Bhitarkanika this taxon is plentiful and gregarious also (Choudhury, 1990). Beside this species, *Heritiera littoralis* and *H. kanikensis* are also present here. The later one has been identified as a new taxon (*sp. novum*) and endemic one (Majumdar and Banerjee, 1985). Species like *Cerbera manghas, Acanthus volubilis, Sonneratia caseolaris* etc. exhibit a trend towards vulnerability. Interestingly, *Nypa fruticans* is abundant in Sundarban. It is singularly absent from Orissa Mangrove forest. Haines (1921–1925) collected some fruits of *Nypa fruticans* from Orissa coast and remarked that this taxon may be present in Mahanadi delta. Choudhury (2001) has also collected the fruits of this species from Gahirmatha sea beach within the boundary of Bhitarkanika wildlife sanctuary. However, during the floristic studies in the mangrove forests of Orissa for the last twenty years this plant could not be traced. Quite possibly, these fruits might have been arrested in Orissa coast being drifted by the oceanic current from Sundarban. However, recently Gupta and Khandwala (1990) through palyno-stratigraphical study of profiles of islets of Chilka and its neighbourhood are of opinion that *Nypa fruticans* were plentiful in this region about six thousand years ago. Hence, possibility of its occurrence in the mangrove forests of Orissa can not be ruled out. It has been predicted that in remote past mangroves of Sunderbans and those of Orissa were in a continuous belt which is evidenced from the comparative floristic study of the above-said region. However, in course of time, these have been degraded largely due to operation of various factors and now they are occurring in discontinuous patches.

A Glimpse into Floristic Composition of Orissa Mangroves: Expressive of Biodiversity

As stated earlier, at present mangroves and their associated are largely extant in Bhitarkanika and Mahanadi delta (Choudhury, 1990, 1991, 1994, 1999, Patnaik and Choudhury, 1989). Members of *Sonneratia apetala* are the most dominant mangrove species found along the river banks and creeks of Bhitarkanika river. At places, pure, luxuriant and extensive distribution of this species are attractive. Differential water salinity mainly controls the distribution of *Sonneratia.* In less saline water *S. caseolaris* is restricted in Khola (artificially dug creek by the Kanika Raja) creek. Incidentally, this taxon is not found in any other mangrove forests of Orissa. Next to *S. apetala,* species of *Avicinnea* are aboundant. Other notable tree species constituting the top canopy are: *Heritiera fomes, H. littoralis, Excoecaria agallocha, Instia bijuga,* species of *Rhizophora* etc. The second canopy comprising of shrubs and under trees are: *Kandelia candel, Brownlowia tersa, Cynometra iripa, Tamarix troupii, Phoenix paludosa, Clerodendrum inerme, Acanthus ilicifolius, Acrostichum aureum, Ceriops tagal, Hibiscus tiliaceous, Aegiceras corniculatum* etc. *Derris trifoliata, Caesalpinia crista, Mucuna gigantea, Finlaysonia obovata* etc. are notable climbers. In the mudflats, *Myriostachya wightiana* are found in pure formation. Next to this species, *Cyperus malaccensis* deserves mention. In the same habitat patches of *Porteresia coarctata* are found which is comparatively less gregarious along the creeks and channels of the river Mahanadi, extending from Atharbanki to Hukitola (Map 3) while other two taxa are occasionally found in this terrain. Other mangrove associates like *Suaeda maritima, S. nudiflora, Sesuvium portulacastrum, Salicornia brachiata* are present in Bhitarkanika, Mahanadi delta and Jambu. *Drynaria quercifolia* and *Vanda tassellata* are the only epiphytes restricted in the mangrove forest of Bhitarkanika. Similarly, *Bruguiera gymnorrhiza,* species of *Xylocarpus, Cerbera manghas* are extant only in the Bhitarkanika (Choudhury *et al.,* 1991, 1995).

Majority of plants occurring between Devi to Dhamra estuaries as enumerated by Rao and Shastri (1974) have disappeared due to constant operation of various anthropogenic factors leaving behind some remnants

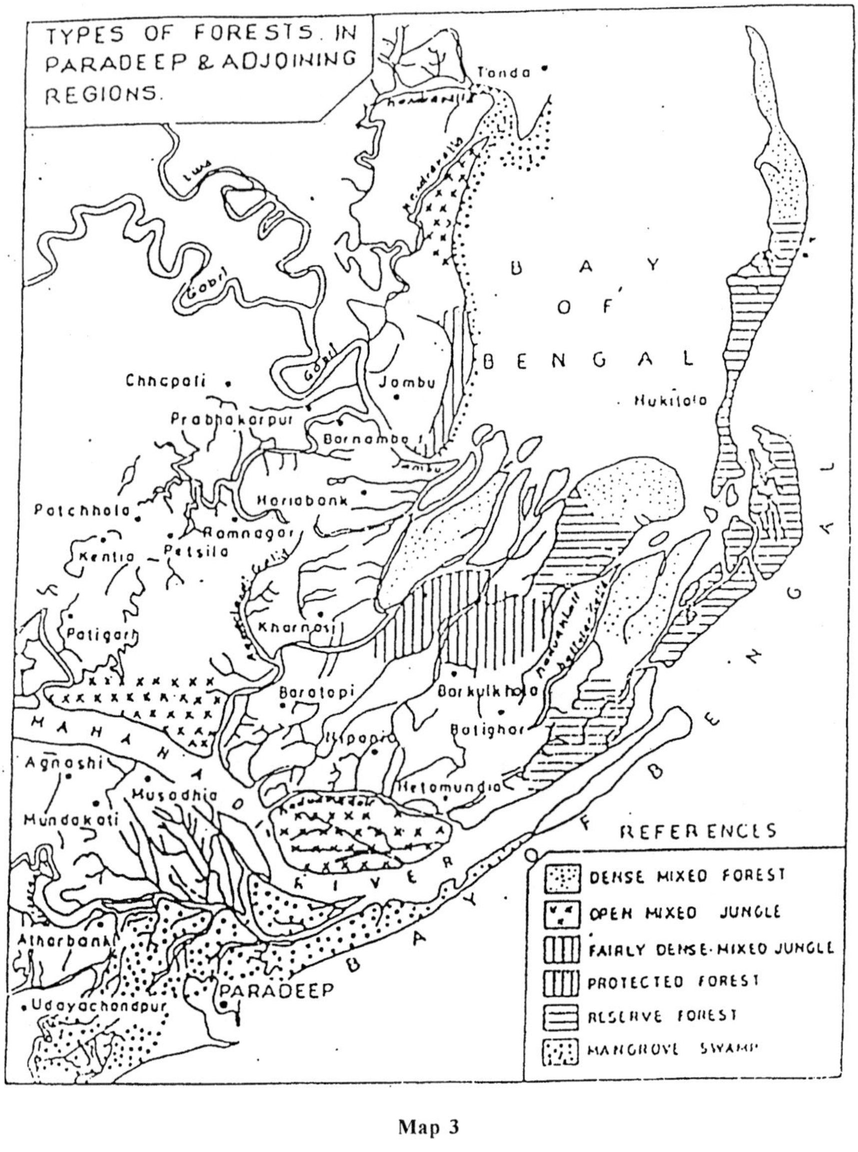

Map 3

such as *Tamarix troupii, Acrostichum aureum, Suaeda nudiflora, Acanthas ilicifolius* and a few tree species.

A newly formed island known as 'Udabali' near Chandinipal is crowded with species of *Avicennia* in denuded state since the branches are being regularly exploited by the local people for leaves which are good fodder for the buffalos.

In remote past, Chilka (Map 4) the largest brakish water lake harboured quite a good number of mangroves and their associates as evidenced from the palenostratigraphical studies leading to reconstruction of palaenoflora of this lagoon and its adjoining region (Choudhury *et al.*, 2004). Haines (1921–1925) and Narayan Swami and Carter (1922) had botanised this water body and its neighbourhood enumerating some mangrove species and their associates. However, in course of time probably due to change of ecological conditions coupled with constant operation of various biotic factors these have almost disappeared.

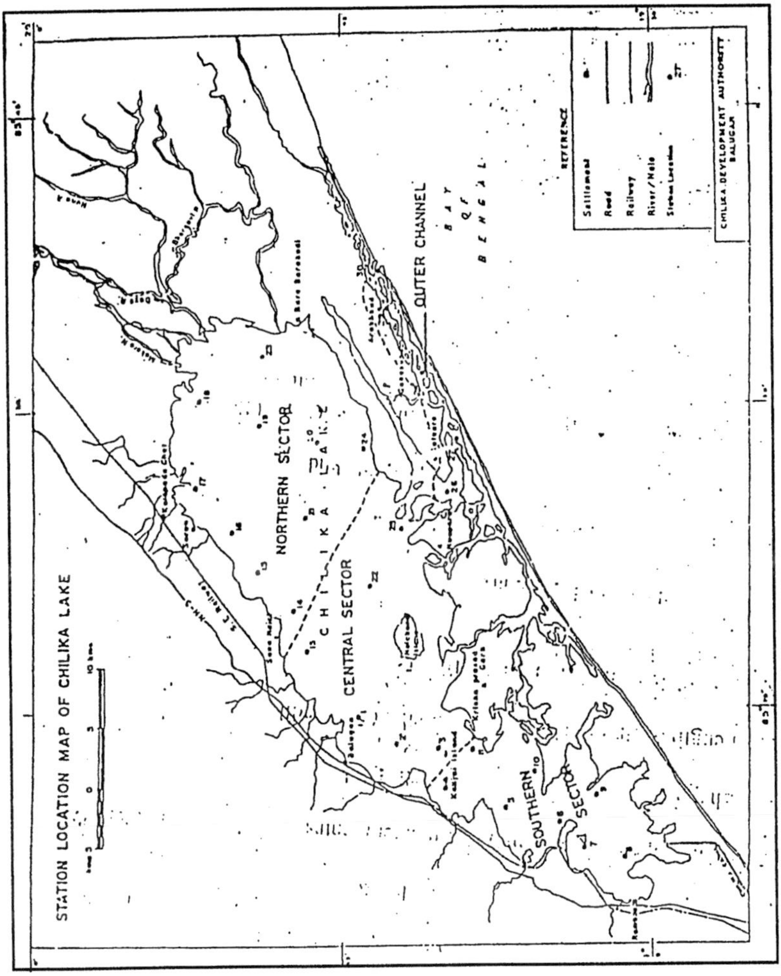

Map 4

Habitat improvement of Nalaban island, plantation of *Phragmites karka,* construction of earthen mounds, renovation of creeks have been undertaken to augment the habitat improvement for water foul. These activities are being implemented in collaboration with Wildlife wing of State government. Nalabana has been declared as "Nalabana Wildlife Sanctuary" which is an ideal abode of migratory birds (Panda and Patnaik, 1988). This wildlife sanctuary is managed by the wildlife wing of Forest Department, Government of Orissa, Chilka Development Authority (CDA) supports the 'Wildlife Division for restoration of habitat and water foul conservation.

Recently, Panda and Patnaik (1998), Saxena and Brahmam (1994–1996) have reported species like *Cressa cretica, Derris trifoliata, Salvadora persica, Azima tetracantha, Sesuvium portulacastrum, Acanthus ilicifolius, Luminitzera racemosa, Suaeda maritima, Excoecaria agallocha, Heliotropium curassavicum, Casipoura ceylanica* etc. from the islets and fringes of Chilka lake.

Due to unfavourable as well as disturbed ecological niche the number as well as variety of migratory birds coming from Siberia and Himalayan regions exhibit declination. On the other hand, due to Siltation of outer channel (Magarmukha) of Chilka (Map 4) there is decrease of water salinity which has adversely affected the various aquatic life forms of this water body. Realising this problem of Chilka lagoon, Government of Orissa constituted CDA in 1991 to formulate integrated resources management plan and conserve lagoon ecosystem with its genetic diversity and undertake measures for management of the lagoon in collaboration with various departments and institutions both within and outside the state. However, recently CDA has taken up a stable step to eradicate this problem. On the sandy soil, drazzed out from this water body extensive *Casuarina equisetefolia* plantation has been made which has grown up luxuriantly forming a thick green belt. Along with this, *Calophyllm inophyllum, Prosopis spicigera, Anacardium occidentale* can also be planted. This will enhance the biomass of this region and will improve the economic condition of the local people by proper exploitation of fruits of *Calophyllum inophyllum* which yields oil and cashew nut. This will also provide source of income by harvesting the nuts as well as fleshy thalamus of *Acacardium occidentale.* Almost regular blood-shed combat among the inhabitants who mainly depend upon pisciculture will be minimized. The polluted environment will be improved and harmonious social life may develop in course of time.

Spectrum of Biodiversity in the Fauna

The animals those are associated with mangroves cover a wide range of invertebrates and vertebrate fauna (Kar and Satapathy, 1996).

Bhitarkanika wildlife sanctuary, harbouring abundant mangrove elements and their associates, innumerable water bodies, along with sea beach provides an ideal niche for both aquatic as well as terrestrial organisms of diverse categories.

The Olive Ridley sea turtle (*Lepidochelys olivacea*) is an annual migrant reptile to Gahirmatha beach during their breeding season (December–March). Visit of about 100,000 to 200,000 sea turtle every year (Raju and Dev, 1982) for mass nesting provides a spectacular sight.

Salt water crocodile (*Crocodylus porosus*) is another important reptile found in the rivers and creeks of Bhitarkanika. They attain a size of 6–7 meters in length. It has been reported that the world's largest crocodile was killed in the year 1926 from the Bhitarkanika river system. Partial albino variety, locally known as 'Sankhua' is also found in the water bodies. Due to large-scale hunting *Crocodylus porosus* was fast disappearing and was on the threat of extinction. However, after 1975 this endangered species has been enjoying adequate protection and care due to declaration of wildlife sanctuary. At Dangmal, a Crocodile Research Centre was established for research and conservation of this salt water crocodile where active research works are being conducted and artificially hatched juveniles are being regularly released into the creeks and channels of Bhitarkanika. As per a report 550 crocodiles consisting of 27 males and 523 females have been released into the water bodies of this sanctuary. Their movement, behaviour as well as survival rate are being regularly monitored. It has been observed that they prefer hyposalinity condition of the water as well as areas which are thickly populated with mangrove elements and are less interfered by human activities. Further, from time to time juvenile salt water crocodiles are being sent to various research centres, Zoological/Biological parks of India for research and display. Captive breeding programme is now a fascinating facet of crocodile research which will provide some interesting data on breeding behaviour of this species. As a matter of fact, due to adequate care and protection its population has attained stability and therefore, this species is now out of danger.

However, rehabilitation of mangrove taxa in the partially degraded areas within this terrain will provide better environment for the growth and development of these crocodiles.

In addition to these, sea turtles and crocodiles, other aquatic animals are water monitor lizards, mud skippers, limulus, red crabs and varieties of fishes. Migratory birds like openbilled storks, white ibis, grey pelicans, barheaded goose, Bramhiny ducks etc. come in large number during winter season of everyyear and these are special attractions for the ornithologists as well as nature lovers. Besides these migratory birds, the resident birds also add to the diversity of wildlife. Night herons, cattle egrets, cormorants, grey herons, three varities of king fishers, water fowls, common sand piper, collard dove, Alexandrian parakeet etc. come under this category.

The terrestrial wild animals such as python, king cobra, spotted deer, wild boars, chital, jackel, hyena, porcupine, fishing cat, mangoose etc. are notable in this sanctuary.

Strategies for Conservation of Biodiversity

As indicated earlier the biodiversity of Orissa mangroves at present exhibits a tangible shrinkage and is still in declining state. This has posed many environmental hazards specially, in the coastal belt. Ultimately it has affected the socio-economic status of populace of Orissa in general and that of coastal terrain in particular. The fury of the environmental hazards has reached its climax in the end of Dec. 1999 causing widespread devastation of various life forms, loss of human habitations as well as properties. This super cyclone has also affected the neighbouring regions. Restoration of habitats, avenue plantation, and croplands has not yet been completed (Choudhury, 2002). But the imbalance of coastal environment is still continuing. As a matter of fact frequent oceanic cyclones are still occurring in the state of Orissa.

Hence, it is high time for conservation of coastal zone biodiversity of Orissa, atleast to improve the life style of coastal dwellers and to strengthen the socio-economic status of the state in general (Choudhury, 1987).

(a) In the first place, although Bhitarkanika mangroves are well protected due to its status of being a National Park, germplasm conservation of some endangered/vulnerable/endemic taxa should be attempted on priority basis. The population of mangrove species such as *Xylocarpus granatum, Cerbera manghas, Acanthus volubils, Bruguiera gymnorrhiza, Heritiera kanikensis* etc. which are restricted to Bhitarkanika only exhibiting shrinkage of population (Choudhury *et al.,* 1995, 1999) should be proliferated by modern biotechnological methods.

(b) Regeneration, rehabilitation as well as adequate protection of rare/ threatended mangrove taxa should be attempted. Prawn culture and conversion of Mangrove forests for paddy cultivation and human habitation as practiced by the immigrants and inhabitants specially in the Mahanadi delta, Paradeep, Astaranga, Ramnagar etc. should be prohibited.

(c) Simultaneously, felling of woody elements by the local people for fuel purpose and to meet the other commercial usages specially in the mangrove denuded area (cf. Mahanadi delta, Jambu, Hukitola, Astaranga etc.) should be checked strictly.

(d) Alternative fuel energy should be implemented by popularizing the biogas chula and plantation of fast-growing trees/shrubs in order to provide fuel wood to the general public inhabiting the mangrove forest regions.

(e) Mangrove nursery should also be developed in Hukitola island and Bhitarkanika where under adequate protection and care mangrove taxa and associates such as *Excoecaria agallocha, Instia bijuga, Cynometra inpa, Clerodendrum inerme* should be planted following recommendation given by Untawale (1986). After successful stabilization of population of these species other typical mangroves may be planted in the denuded areas so that genepool of mangrove species will be ideally maintained.

(f) The inhabitants, specially school/college students should be educated for conservation of mangrove species through organisation of eco-friendly camps/seminars/symposia/workshops using audio-visual aids.

(g) Government as well as Non Governmental Organizations (NGO's) should take active role in developing awareness towards judicial exploitation and proper conservation of mangrove species. Unless otherwise grass-root level attempts are made in this direction, conservation of mangroves will not be successful (Choudhury, 1996).

(h) Finally, development of artificial mangrove forests in the mangrove-denuded areas is need of the hour. In this connection, man-made mangrove forest in Hochiminh City is a bright example (Choudhury, 1999). By spray of poisonous chemicals a vast mangrove forest was destroyed. However, through successful plantation programme over an area of 20,000 ha. in south of Hochiminh City (Vietnam) a pure stand of *Rhizophora apiculata* has come into existence which provides characoal of high calorific value, honey, woody elements which are being exploited by the nearby inhabitants. This type of artificial mangrove forests are also made with in India (cf. Coringa). Development of such forests selecting some mangrove elements may be followed in Chilka area Mahanadi delta, Paradeep, Hukitola etc. This will promote eco-friendly tourism which will enhance the economic conditions of the inhabitants. Various aspects of biodiversity conservation have also been high lighted by Choudhury (1988) Patnaik *et al.* (2000).

Government of India have been implementing adequate strategies for the conservation of mangroves (vide notification No. J. 22014/22/83-CSC/ dt 14th December, 1988). The National mangrove committee has been reconstituted (vide letter ministry OM.No. 23/1/84/CSC/dt. 3rd Sept. 1986) which are adopting effective strategies for conservation of mangroves of 15 selected mangrove chunks (Anonymous 1989). Bhitarkanika and Mahanadi delta have been allotted to Utkal University and Central Soil Salinity Institute (Port Canning) and Utkal University in collaboration with Regional Plant

Resource Centre (RPRC), Bhubaneshwar, respectively. These organisations have been adopting various work plans and research activities for germplasm conservation (Anonymous, 1989). These are further augmented by various activities of the Department of Forests, Government of Orissa (Anonymous, 2001).

In recent time, M.S. Swaminathan Research Foundation Chennai in Collaboration with International Tropical Timber Organisation (Japan) had provided the final report on the "Travelling workshops to mangrove forests in South and Central Africa, Oceania, West and Central Africa". Out of 23 sites recommended from 9 countries India has been selected as one for the establishment of (MGRCC) Mangrove Genetic Resource Conservation Centre.

The summary of evaluation report for four sites in India has been highlighted. Four mangrove chunks such as (i) Choras, (ii) Pichavarum, (iii) Bhitarkanika and (iv) Coringa have been studied by the mangrove experts. Various parameters like (a) mean temperature (b) mean annual rainfall (c) mean relative humidity (d) tidal conditions (e) water pH (f) salinity and (g) total available area have been taken into account. Other salient features of Bhitarkanika have also been critically evaluated. From such study it has been concluded that, of the four sites in India, Bhitarkanika is the most suitable one for establishment of MGRCC.

Adequate steps should be taken up for implementation of this report given by the experts of the above-said organizations so that the germplasm as well as luxurience of Orissa mangroves will be restored. This will provide salubrious environment in the Orissa coast and the lost glory of mangroves will be revived.

REFERENCES

Anonymous. 1989. Conservation of Mangroves of India, Govt. of India, Ministry of Env. and Forest, New Delhi.

Anonymous. 1994. Establishment of International Network for the Conservation and sustainable utilization of mangrove forest genetic Resources (eds. M.S. Swaminathan, Sanjay, V. Deshmukh and V. Balaji), M.S. Swaminathan Research foundation, Madras and International Tropical Timber Organisation (Japan).

Anonymous. 2001. Bibliography of wildlife of Orissa. (eds. S.K. Kar, C.S. Kar and L.A.K. Singh). Wildlife Wing, Forest Department, Government of Orissa, Bhubaneshwar.

Chadha, S. and C.S. Kar. 2001. Bhitarkanika: Myth and Reality, Nataraj Publishers, Dehradun.

Champion, H.G. and S.K. Seth. 1968. A revised survey of the forest types of India. CSIR, New Delhi.

Choudhury, B.P. 1987. Socio-economic aspects of Bhitarkanika Wildlife Sanctuary in the state of Orissa. *Orissa Review*, **43(9)**:45–47.

Choudhury, B.P. 1990. Bhitarkanika: Mangroves Swamps. *J. Env. Sci.*, **3(1)**:1–16.

Choudhury, B.P. 1991. Mangroves of Orissa with speical references to those Bhitarkanika. In: Botanical Research in India. (ed. N.C. Aery et B.L. Chaudhury) 32–39. Himansu Publication, Udayapur.

Choudhury, B.P. 1994. A comparative study of mangrove vegetation in Bhitarkanika and Paradeep area in the state of Orissa in. Forest, Wildlife & Environment (ed. M.V. Subba Rao). 33–41. Andhra University, Visakhapatam.

Choudhury, B.P. 1996. Necessity of conservation of Mangroves in Bhitarkanika in the district of Cuttack. In: Taxonomy and Plant conservation (eds. K.S. Manilal and A.K. Pandey): 7–12. CBS Publisher & Distributor, New Delhi.

Choudhury, B.P. 1998. Biodiversity conservation of Bhitarkanika Wildlife sanctuary Vis-a vis sustainable development. In: Biodiversity conservation: Problems and Prospects (ed. H.K. Patra): 59–67. Orissa Environmental Society, Bhubaneswar.

Choudhury, B.P. 1999. Mangroves of Bhitarkanika. In: Reference Orissa (ed. A.N. Tiwari) Raj Press, R–3, Inderpuri, New Delhi.

Choudhury, B.P. 2001. Role of Forest types in the environment of Orissa With special reference to mangrove vegetation. In: Mangrove Conservation and restoration (eds. S.K. Patanaik and H.K. Thatoi), **14**:14–21. Reproprit (P) Ltd. N-5/44.IRC Village.

Choudhury, B.P. 2001a. Biodiversity of Orissa Mangroves: Mangroves: Management and Conservational strategies for threatened taxa. *Nat. Bot. Soc.*, **55**:63–72.

Choudhury, B.P. 2002. Probable Solutions of environmental problems arising After Super Cyclone. *Orissa Review*, **56(6)**:32–34.

Choudhury, B.P., A.K. Biswal and H.N. Subudhi. 1991. Mangroves of Orissa and aspect of conservation. Rheedea, **1(1&2)**:62–67.

Choudhury, B.P., A.K. Biswal and S.P. Rath. 1995. Biodiversity in the Bhitarkanika wild life sanctuary in the state of Orissa. In: Environment Change and Management (ed. R.C. Monahty) 53–60. Kamal Raj Enterprises, New Delhi.

Choudhury, B.P., P.K. Nayak and J. Das. 2004. Mangroves of Chilika: Past and Present. *J. Econ. Tax. Bot.*, **28(1)**:18–24.

Choudhury, B.P., H.N. Subudhi and A.K. Biswal. 1999. The Mangrove forest of Bhitarkanika National Park (Proposed) Kendrapara district, Orissa In. Bhintarkanika: The wonderland of Orissa (ed. B.K. Behura): 19–29. Nature and Wildlife Con. Soc. of Orissa, Bhubaneswar.

Gupta, H.P. and A. Kandelwal. 1990. Mangroves of India: History and Palynostratigraph of Chilika lake, Orissa Proc. Sym. Vistas in India Palaeobotany. *Palaeobatanist*, **38**:379–393.

Haines, H.H. 1921–1925. The Botany of Bihar and Orissa, 6 parts, London.

Kar, C.S. and G.R. Satapathy. 1996. Mangroves ecosystem and its biodiversity in Orissa coast with respect of Bhitarkanika. Advances in Ecology and Env. Sci., Ashish Pub., New Delhi, pp. 419–438.

Majumdar, N.C. and L.K. Banerjee. 1985. A new species of *Heritiera* Dryand (Sterculiaceae) from Orissa. *Bull. Bot. Surv. Ind.*, **27(1-4)**:150–151.

Mishra, C.G. 1999. Green Forests of Orissa in Reference Orissa (ed. A.N. Tiwari): 16–20. Enterprising Publishers, New Delhi.

Mooney H.F. 1950. Supplement to the Botany of Bihar and Orissa, Catholic Press, Ranchi.

Naskar, K. and D.N. Guhabaxi. 1987. Mangroves of Sunderban (An Ecological Perspective) Nayaprakash, Calcutta.

Narayanswami, V. and H.G. Carter. 1922. Systematic list of plants of barkuda. *Mem. Asiat. Soc.*, Bengal, **7(4)**:289–319.

Nayak, R.K. and B.P. Choudhury. 2001. Mangroves of Mahanadi delta and their conservation. *Plant Sci. Res.*, **23(1&2)**:21–24.

Panda, P.C. and S.N. Patnaik. 1998. Contribution to the flora and vegetation of Chilika Wildlife Sanctuary Orissa in. Chilika (ed. S.N. Patro 81–85. Orissa Environmental Society.

Patnaik, S.K., C.S. Kar and S.K. Kar. 2000. Biodiversity in tidal swamps with special referneces to Bhitarkanika in Mahanadi delta (Geology) Resources and Biodiversity) 139–153. (ed. N.K. Mahalik) Ait Alumina Association (India chapter) New Delhi.

Patnaik, S.N. and B.P. Choudhury. 1989. Present status and future Development of mangrove vegetation in Bhitarkanika in INDO-US Workshop on wetland, mangrove and biosphere reserve: 89–95. Govt. of India, Ministry of Env. & Forest, New Delhi.

Raju, K.S.R. Krishna and U.N. Dev. 1982. Some ecological observation on Bhitarkanika Sanctuary in Orissa. Proc. Seminar Resources Dev. Env. Eastern Ghat: 117–118. Andhra University, Waltair, Visakhapatanam.

Rao, T.A. and R.K. Sastry. 1974. An Ecological approach towards Classification of coastal vegetation of India-11 Estuarine Border Vegetation. *Ind. For.*, **100**:438–452.

Saxena, H.O. and M. Brahmam 1994–1996. The Flora of Orissa, 4 vol. Regional Research Laboratory and Orissa Forest Development Corporation, Bhubaneswar.

Untawale, A.G. 1986. How to grow mangrove, National Institute of Oceanography, Donapaula, Goa.

CHAPTER 7

MEDICINAL PLANTS OF SOUTH WESTERN MAHARASHTRA

*Priti D. Mahekar** and *S.R. Yadav**

Introduction

Man has always turned to plants for help and utilized them for food, fodder, shelter, medicine and every requirement of his life. Plants have played major role in nearly all ancient and recent civilizations and cultures. Medicinal plants have been playing and will play important role in health care of human beings. Ancients discovered medicinal properties of plants, developed folk medicines and finally gave rise to traditional systems of herbal medicines such as Ayurveda, Siddha and Unani.

Primitive and ethnic populations have their own medicinal lore and some of their therapeutic practices have found place in today's medical science. Even today, the therapeutic hints from remote past and the primitive as well as folk medical lore form an important source of medical knowledge. Man has gathered this knowledge from time immemorial through human instincts, trial and error methods and keen observations on animal utility of plants. Large arrays of hints and claims, principles and practices have been handed over from generation to generation for welfare of human beings.

During dark period of Ayurvedic medicines, many quacks came into picture, which harmed greatly the development of herbal medicines in India shadowing the importance of the indigenous systems of medicine. Moreover, the indigenous systems of medicine were disregarded by practitioners of allopathy and homeopathy adding further to the deterioration of the systems. The clinical efficacy of the herbal medicines was usually not evaluated and the composition of the complex mixtures was only crudely analysed. Thus, herbal medicines became the domain of 'old wives tales', 'quacks medicines',

* Department of Botany, Shivaji University, Kolhapur-416 004.

exploitation of the sick, the disparity, gullibility and open to enormous and uncontrolled variation in composition and efficiency. In the process, we have lost some bits of precious information about our herbal medicines, however, after realizing the side effects of allopathic medicines, man has once again realized the importance of herbal medicine in health care.

Traditional systems of medicine have once again become a central theme of healthcare of modern society of modern age. There has been increasing demand for herbal medicines and the indigenous systems have now regained respectability among the scientific community all over the world. Ayurveda is gaining popularity and is appreciated not only in India but also abroad which is of great importance to all of us. The ancient knowledge of herbal medicine is a great source of information for scientific community, researchers and medical practitioners. In last couple of decades, a new trend in the preparation and marketing of drugs, based on medicinal plants has become increasingly apparent (Bisset, 1994; Rawls, 1996; Kinghorn and Seo, 1996; Tyler, 1997; Dev, 1997; Dev, 1999). These preparation labeled 'herbal drugs' or 'phytomedicines' are invariably single plant extracts or fractions thereof, as distinct from pure chemical entities which may be called 'molecular drugs' (Dev, 1997). This new breed of plant derived medicines are gaining importance because there have been a number of spectacular successes with herbal medicines. Majority of synthetic drugs have very unacceptable side-effects and diseases like cancer, cardiovascular disorder, chronic rheumatism, AIDS etc. all require new effective drugs. Similarly, known drugs are becoming ineffective in major tropical diseases and the very high cost of allopathic drugs is beyond reach of poor people.

Most of the populations of developing countries have relied and will continue to rely on indigenous medicines. Therefore, the indigenous systems of our medicines must be respected, restudied and evaluated in light of modern developments in science. India can well afford to develop new exciting and challenging plant-based drugs to treat common diseases and the diseases which have no satisfactory treatments. According to WHO (World Health Organization) herbal medicines serves the health need of about 80% of the world population. Thus, plant based therapeutic agents continue to have scientific, social and commercial significance and appear to be gathering momentum in certain health relevant areas (Sukhdev, 2001). But due to over exploitation and unrestricted and illegal cutting or deforestation and habitat fragmentation, forests are declining at alarming rate of 16.8 hectares per annum affecting tropical rain forests which harbor largest number of known medicinal plants. According to IUCN, by the end of this century about 20,000 plants used in traditional medicines as healing agents may have become extinct and by the end of century some 60,000 higher plant species may become extinct.

India is one of the twelve megacenters of biodiversity. The country is in fact an epitome of almost all climates, seasons and soils. The country abounds in all kinds of food, aromatic, timber, fiber, fodder, gum, spice, and medicinal plants. It is estimated that about 2000 species of medicinal plants grow in India. India is considered as a veritable emporium of medicinal plants. Owing to availability of land, variety of climatic, edaphic conditions and cheap labour, there is a great scope for cultivation of indigenous as well as exotic medicinal plants.

The Indian systems of medicine use plant-based raw materials in most of their preparations and formulations. There are some 1,250 Indian medicinal plants which go into formulating therapeutic preparations as per Ayurvedic or folklore traditions (Chatterjee and Pakrashi, 1991). The efficacy of these systems mainly depends upon the use of pure and genuine raw materials in the manufacture of drugs of these systems. Substitution of one plant species for another can have serious implication for human health.

In India, natural habitats especially forests have been hunting grounds for medicinal plants for last several centuries. As a result of continuous and unscrupulous exploitation of medicinal plants, a number of species have become vulnerable to extinction in natural habitat. In order to meet the above situation, measures are required to promote the cultivation of at least important and rare medicinal plants. There is a need to survey medicinal plants of different regions, their availability and present status in natural habitats. Similarly, it is equally important to gather information on local uses of medicinal plants, methods of administration and correct identification of folk medicines available in bazaar.

According to an all India co-ordinated project on Ethnobiology in India commissioned by the Ministry of Environment and Forestry, tribal people are reported to utilize around 7,500 species of plants for medicinal purpose (Pushpangadan undated report). Indian system of medicine use plant raw materials in most of their preparation and formulations. About 95% of raw materials for medicinal purpose are exploited from forest areas (Shiva, 1994). As a result of continuous and unscrupulous exploitation of medicinal plants, a number of plant species of medicinal potential have become vulnerable to extinction in natural habitat. Therefore, there is urgent need of studies on medicinal plants of different regions of the country.

In addition to these traditional systems, there is still parallel stream of tribal and folk medicines all over the world. Over 53 million tribal people belonging to 550 tribal communities are reported to inhabit the Indian subcontinent (Natesh, 2001). A rich, but still recent, neglected areas of Man's relation with biodiversity relates to small herbal vendors or tribal vendors. These tribal vendors move from place to place, selling medicinal herbs, herbal, animal and mineral extracts. These herbal drugs are given mostly in the form of barks, roots, twigs, leaves, flowers, fruits and seeds on the roadsides dispersing advice on the treatment of diseases and disorders. It is estimated that there could be 8–9 million vendors across the country selling about 100 herbs (Natesh, 2001). These people are possessing sound knowledge of folk medicines and this will provide an excellent clue for modern drug development programme. A case study dealing with only about 85 drugs dispensed by vendors in North India shows many less known recipes claimed by the vendors (Singh, 1996; Jain, 2000). Similarly, the correct identity of these folk medicines is little known to scientific community. Therefore, there is urgent need to survey folk medicine of our country and determine their correct botanical identity for further scientific investigation.

In recent years considerable attention has been paid to study medicinal plant species of our country. Although many attempts have been made to study medico or ethnomedico flora of India, no serious efforts have been made to study availability of medicinal plants in natural habitats, their status, local use and method of administration. Similarly, very little attention has been paid on correct botanical identity of folk medicines given by tribals, traditional practitioners; local people in various parts of our country. Identify of these folk medicines is little known to scientific community. There is need for correct botanical identity of folk medicines which will help in undertaking further research on these traditional medicines to confirm their effectiveness and utilization for health care of human society. For correct botanical identity, there is need to survey and to study external and internal morphology for identification of plants used as folk medicines.

Therefore, in present investigations attempts have been made to study medicinal plants and folk medicines of South Western Maharashtra with special emphasis on survey, distribution, documentation, present status of medicinal plants and correct botanical identity, pharmacognosy and method of administration of folk medicines.

With above things in mind, the present investigation was undertaken in 1999 with following objectives on medicinal plants of South Western Maharashtra.

1. To survey and prepare an account of medicinal plant wealth of the region.
2. To determine the availability and present status of medicinal plants.
3. To document medicinal plants in the form of herbarium specimens.
4. To gather information on local uses and methods of administration.
5. To determine correct identity of folk and bazaar medicines on the basis of external and internal morphology (Pharmacognosy) and chemical test etc.

Area of Study

South Western Maharashtra comprises five districts viz. Ratnagiri, Sindhudurg, Satara, Sangli and Kolhapur approximately between the latitude 15°28′ and 18°5′ N and longitude 73°22′ and 75°40′ E. The total area of about 40,157 sq. km. is bounded by Raigad and Pune districts on the

North, by Solapur district on the North and North-East, by Karnataka State on the East, by Goa state on the South by the Arabian sea on the West (Map-1 and Map-2). Geographically the study area falls under 3-regions viz. Konkan, Sahyadris and Desh.

Konkan region

Ratnagiri [15°36′ to 18°5′ N and 73°5′ to 74°30′ E] and Sindhudurg [17°3′ to 18°28′ N and 73°22′ to 74°14′ E] district falls under Konkan region. The geological formations in the area are coastal sands, lateritic Deccan

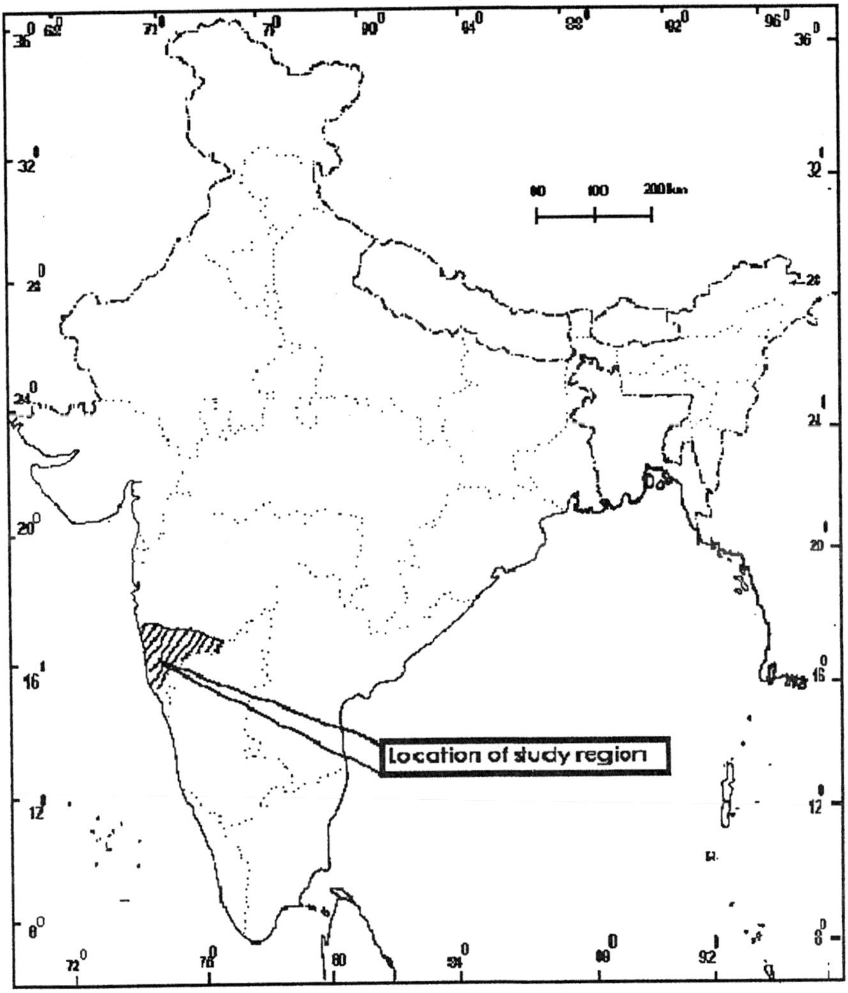

Map 1: Showing Study Area in India

traps, basic and altramatic intrusions, granite gneiss, acid dharwas, alluvium. Laterite soils are of common occurrence in the region. The loamy soils of the region are acidic in nature. Mainly four types of soil viz. Rice soils, Warkas soils, Salty soil and Red soils are found in the region. Both the districts receive heavy rainfall (250–300 mm/annum) mainly during June to August. Konkan region enjoys moderate humid equable climate.

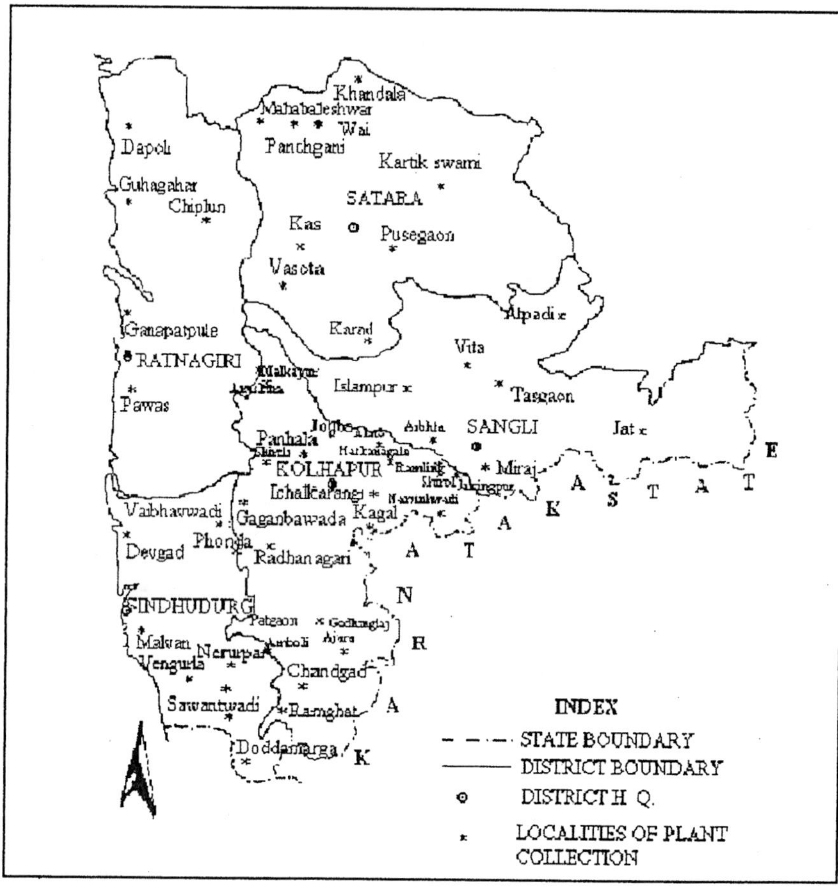

Map 2: Showing Localities of Plant Collection in South Western Maharashtra

Sahyadris

These are flat-topped mountains extend from Daman-Ganga to Baba Budhan hills in Mysore through the state of Maharashtra, Goa and Karnataka. They form a part of Western Ghats varying in height from 200–2000 meters above the sea level. Sahyadri Mountains play important role in hydrological cycles of the area, possess high biodiversity and harbors many medicinal plants. Valleys along Sahyadris provide unique climate and locations for plant growth. High flat tops and spurs, table lands and strong hill forts on the peaks and spurs of Sahyadris provide peculiar and protected habitat for plant vegetation. It is mainly dominated by semi evergreen forests, moist deciduous forests and dry deciduous forests. All the five districts under study are traversed by Sahyadri ranges. The forest areas are inhabited by tribal communities notably the Dhangars. Sahyadris form a part of every district under study area. Eastern parts of Ratnagiri and Sindhudurg and Western part of Satara [17°5′ to 18°11′ N and 73°33′ to 74°54′E], Sangali [16°45′ to 18°11′ N and 73°42′ to 75°40′ E] and Kolhapur [15°43′ to 17° N and 73°40′ to 74°72′ E] are occupied by main ranges of Sahyadris. The soils are of mixed type. The soils are mainly of lateritic origin and poor in nutrients. Soils in valleys are rich in nutrients due to addition of leaf-litter by vegetation. The region receives maximum rainfall (500–700) during rainy season (June–August). The region enjoys healthy climate with no extremes of temperature and possesses luxuriant vegetation.

Desh

The study area to the East of main ranges of Sahyadris fall under Desh region. It comprises eastern parts of Kolhapur, Sangli and Satara districts. Eastern parts of these districts fall under rain shadow areas and the rainfall decreases from main ranges of Sahyadris to easternmost parts of these districts. From main North-South crest of Sahyadris, ranges run eastward and southward in these districts forming river basins. Number of rivers with their tributaries originates in main ranges of Sahyadris and run eastward and southwards. Among these, river Krishna is the largest one, which originate at Mahabaleshwar. The other important rivers are Yenna, Koyana, Warna, Yerala, Panchganaga and Ghatprabha. The region possesses moist deciduous and thorny forests. Most of the land is under cultivation.

Desh region is characterized by light brown to deep black soils. Deep medium black cotton soils derived from Deccan traps are found mainly in easternmost parts of Satara and Sangli district. The soils are of mixed types. They are more or less neutral or alkaline in nature.

The climate of Desh region viz. Satara, Sangli and Kolhapur districts is hot and dry with low precipitation, low humidity and higher temperature fluctuations. The narrow strip along the western border of all the three districts receives heavy rainfall. The rainfall goes on decreasing as one moves from west to east and the eastern region of the study area lies in the rain shadow area. Significant variations in annual rainfall are observed at different places. For example, Mahabaleshwar, in the Ghats, receives over 633 cm rainfall and it suddenly drops down to 76 cm at Wai which is only 32 km east to Mahabaleshwar. Eastern part of Sangli and Satara districts fall under rain shadow area. The annual mean rainfall of the region varies between 60–65 cm, the temperature rises from middle of February to May and hottest month is May. Humidity is quite high for four months of rainy season (June–September) and low for remaining eight months (October–May). The relative mean humidity is high during July–August (80%) and lowest in March–April (25–30%).

Review of Literature

Studies on plants start with the history of man. The use of plants as a source of medicine is as old as humanity. There is a vast literature on medicinal plants as it is the subject directly concerned with human health. Plants have played and will play significant role in health care. Rural communities in most of the world continue to rely on locally produced, plant based remedies and 80–85% of the worlds population rely on these plant based medicines (Susan *et al.,* 2001). The international trade in medicinal plants is a major force in the global economy and demand is increasing in both developing and industrialized countries. The estimated value of the sale of plant based medicines exceeds 12 billion dollar with sale spread throughout the world (Susan *et al.,* 2001). Nowadays, the traditional medicines and medicinal plants are gaining importance throughout world.

Presently, the need for study of especially medicinal plants is well recognized. Number of authors have emphasized the need for detailed and systematic study of medicinal plants (Farnsworth and Bingel, 1977; von Reis Altschult, 1973; Balandran *et al.,* 1985, Duke, 1985; Naranjio, 1995 and Leaman, 2001).

Loss of Biodiversity all over the world, especially in tropics is an important issue of discussion and action today. We have entered in the age of mass extinction and it is estimated that with present rate of extinction, we will lose about 50% of our total biodiversity by the end of 21st century. It is alarming to note that roughly 10,000 species of medicinal plants are threatened with extinction worldwide (Leaman, 1996). Therefore, there is an urgent need to conserve atleast plant species potentials in human welfare. Most of the medicinal plants exploited from wild and are under threat of extinction. Medicinal plants species are under great pressure due to their increased demand and global trade. The story of Silphion is well known. By the first centuries A.D., Silphion was gone from markets, and all that remains is the image on the coin (Sckippmann, 1995). Similar story with some of the medicinal plants may repeat if, due care is not taken by human beings.

Traditional knowledge on medicinal plants is of great significance and forms basis for development of plant based drugs. Majority of the hints to plant based drugs come from ethnobotanical studies and traditional knowledge. In the process of globalization and modernization, we may loose traditional wisdom gathered for years together. Therefore, there is need for documentation of traditional knowledge of tribes and communities in different parts of the world. Similarly concentrated studies on various aspects of medicinal plants including systematics, cultivation, conservation and sustainable utilization is a need of a day.

India is one of the 12 centers of mega biodiversity (Nayar, 1987, 88, 89). It has significant wealth of medicinal plants. Indian systems of medicines are mainly based on medicines of plant and animal origin. Ayurveda is a very ancient science and has been practiced for several years. Indian history of medicinal plants is dated back to 3000 B.C. (Pal, 1998). India inherited a rich herbal heritage.

The history of medicines in India can be traced back to the oldest repository of human knowledge. 'Lord Dhanvantari', one of the disciplines of the emergence of 'Lord Ahwini' is the originator of Ayurveda (2500 B.C.–900 B.C.) who established foundation of the earliest science where the properties of various drugs have been given in detail of the eight divisions of Ayurveda. The origin of Ayurveda is lost in prehistoric antiquity, but its characteristic concepts appear to have matured between 2500–500 B.C. (Sharma, 1979) and it has vast literature (Majumdar, 1971; Sharma, 1979; Dev, 1997) in Sanskrit and several other Indian languages, covering various aspects of diseases, therapeutics and pharmacy.

Medicinal preparations in Ayurveda are invariably complex mixtures derived from plant and animal products, as well as minerals and metals. Plants form a dominant part of Ayurvedic pharmacopoeia. Earliest references are to be found in the Rig Veda where mention has been made for 67 medicinal plants and Atharveda in which mention has been made for about 290 plants used as charms for curing of diseases dating back to the second millennium B.C.

Important landmarks in Ayurvedic literature are Charak Samhita (900 B.C.) which mainly focuses on therapeutics; Sushruta Samhita (600 B.C.) dealing with surgery; Ashtang Haridaya (A.D. 700) which deals with principles and practices of medicines and Madhava Nidana (A.D. 800–900) which deals with diagnosis. The last celebrated writer on Hindu medicine was Bhava Mishra of Magadha and his treaties Bhava Prakash (A.D. 1550) held in high esteem by modern Ayurvedic practitioners. Besides this, a rather large number (770) of Nighantu Granthas were written, mostly between 6th and 7th centuries.

According to Chopra *et al.* (1958) at the end of compilation of Tantras and Siddhas period the knowledge of Ayurvedic medicines spread throughout western world. Greece and Rome took substantial materials from India to enrich their medical knowledge. The studies on medicinal plants spread rapidly all over the world. Some of important earlier literature includes that of Lindley (1838); Clutes (1919); Anonymous (1923); Button (1933); Ichimura (1932); Moanchino (1956); Holdsworth (1974, 1977); Anonymous (1975); Cantoria (1976); Anonymous (1976) and Abdallah (1977).

In recent years, studies on medicinal plants in western countries traveled long distance covering all the aspects related to medicinal plants such as documentation, taxonomical, anatomical, pharmacological and phytochemical screening of plants (Morton, 1981; Sofowora, 1982; Lassak, 1983; Aynehchi *et al.,* 1985; Ghazali, 1986; Li *et al.,* 1991; Pereira and Sheteolu, 1992; Willard, 1992; Iwu, 1993; Kim, 1994; Chevallier, 1996; Bocco, 1997; Van wyk *et al.,* 1997 and Chukul, 1997).

The systematic study of Indian medicinal plants started in early nineteenth century. The earliest contributions were from the writings of Sir William Jones (1799) entitled '*Botanical observations on selected plants*'. Garica da Orta, a physician and pharmacist, collected and studied plants from Sawantwadi and Goa region and published a book entitled, "*Coloquious dos simples Drogas da India*" in 1563. A detailed account of 57 more commonly used Indian Medicinal plants is given in the book "*Acostas Tractudo de las drogas medicines de las Indias orietales*", published at Burgos in 1578. It gives 69 Indian plants and other sources of drugs with good original full-page illustrations to 46 plants. With the help of native physicians and people, van Rheede studied plants of Malabar coast. This great collaborative work was published under the title, "*Hortus Indicus Malabricus*" in 12 volumes. This was followed by John Fleming's "*Catalogue of medicinal plants*" in 1810, Ainslie's "*Materia Medica of Hindustan*" in 1813 and Roxburgh's, "*Flora of Indica*" 1820. O. Shaughenssys, in 1844 published first book entitled "*Bengal Pharmacopoeia*" dealing with the properties and uses of the medicinal plants used in Bengal. Waring's '*Pharmacopoeia of India*' (1868) started a new epoch in establishing and recording the value of indigenous medicinal products on modern lines.

The translation of Sanskrit Materia Medica by Dutta (1877) has brought good account of drugs used by the Hindu physicians. Fluckiger and Handury's, (1883) published "*Pharmacographia and Materia Medica of western India*". These were followed by the publication of very comprehensive account on the medicinal plants by Warden and Hooper (1885), in book entitled "*Pharmocographia Indica*". This contains a mass of information regarding the uses of the Indigenous Materia Medica in Eastern and Western Himalaya. Dymock, Warden and Hooper (1890–1893) added valuable work on medicinal plants which deals with all the aspects related to herbal drugs. Sir George Watt (1889–96) published a book entitled "*Dictionary of the Economic products of India*". It is most elaborate work of all compiled by a large team of workers and refers to all the pervious work on medicinal plants. Kirtikar and Basu (1918) compiled and summarized literature on Indian Medicinal plants in more systemized and elaborated form. It contains not only information from Ayurvedic, Unani and Tibbi sources but also gives the results of personnel observations and experience of some of the writers. Chopra (1956) compiled information based on a critical study of the literature. Nadkarni, (1954) in his book entitled "*K. M. Nadkarni's Indian Materia Medica*" gives detailed account on drugs of plant, animal and Mineral origin and also incorporates drug preparations, pharmacological reports, methods of administrations and local uses if any.

Significant contributions to the studies on medicinal plants of India includes work of Dey (1896), Anonymous (1948–74, 1994), Dastur (1951), Jain (1968), Kurup, Ramdas and Joshi (1979), Kapoor and Mitra (1979), Satyavati (1976–1987), Sivrajan and Balachandran (1994). Agarwal and Ghosh (1985) compiled work on Indian root drugs and the work gives their trade name, distribution, medicinal uses, identifying characters and phytochemcial reports. Warrier (1996) compiled information on 500 medicinal plants of South India. Some of the other noteworthy publications on Indian medicinal plants include Thakur, Puri and Husain (1989), Jain and De Filips (1991), Rastogi (1991), Anonymous (1992), Subudhi, Choudhury and Acharya (1992), Hausain (1993), Chatterjee and Pakrashi (1994), Rai and Sharma (1994), Zafar (1994), Sanyal and Namhata (1995), Yoganarasimhan (1996), Kameshwar Rao (2000) and Saxena (2001).

Plant exploration on scientific basis started in Maharashtra only in early 18th century. Law, Graham, Nimmo, Sykes, Jaquemont, Gibson, Dalzell, Birdwood, Nairene, Hallberg, Acland, Cooke, Woodrow, Gammie and Ryne are the pioneer plant explorers of the region. Graham (1837) was the first to publish a catalogue of Bombay Plants. This initial publication was followed by other publications by Dalzell and Gibson (1861), Nairne (1894), Birdwood (1886, 1887, 1896, 1897) and Cooke (1901–1908). Nairne (1894) published an account of flowering plants of western India which covers Maharashtra. Cooke based on previous works (Dalzell and Gibson, 1861; Naire 1894; Talbot, 1894 etc.) and his own contribution published a comprehensive account of flowering plants, '*The flora of Presidency of Bombay*' covering Sind, Gujarat, Maharashtra and North Kanara. Even today, it is indispensable flora for the state of Maharashtra. Similarly, Talbot (1894), Graham (1839), Haines (1912–1914, 1916) and Gammie (1905–1912) made valuable contributions to the botany of Maharashtra during initial period of plant exploration.

The second period of botanical exploration in Maharashtra include that of Blatter (1926–1935), Santapau (1952, 1953, 1958), Santapau and Irani (1962) and Santapau and Henry (1973). Several significant contributions to the Botany of Maharashtra have been made during last 4–5 decades. (Vartak,

1966; Patel, 1968; Kamble and Pradhan, 1988; Kulkarni, 1988; Lakshminarasimhan and Sharma, 1991; Kothari and Moorthy, 1993; Deshpande *et al.*, 1993–95; Almeida, 1996–1998, Sharma *et al.*, 1996; Naik, 1998a; Singh and Karthikeyan, 2000; Singh *et al.*, 2001;Yadav and Sardesai, 2002). In addition to these important floristic works, several research papers in the forms of checklist, distributional records and new species have been made for the region, however, their review is out of scope in the present work. Although every medicinal plant that occur in Maharashtra is referred and recorded in different taxonomic accounts and floras of the region, there are only few publications exclusively devoted to the medicinal plants of the region.

Initial significant work on medicinal plants of Maharashtra include that of Agharkar (1953). He compiled a detailed account of medicinal plants of Bombay in, "*Bombay State Gazetteer*" Vol. III which includes about 350 flowering plants giving details on their medicinal uses, habit, habitat, localities, parts used, local uses and chemical constituents whenever available. Vartak (1957, 1959a, b, c, 1979, 1981a, b, c, 1982, 1989) made valuable contribution to medicinal plants of Pune and surrounding area. Similarly, several workers have enumerated medicinal plants of different regions viz. Khed taluka (Janardhanan, 1964–1965); Satpura mountains (Karnik, 1966); Chandrapur district (Kulkarni, 1968; Malhotra and Moorthy, 1976); Karnala district (Vartak and Mandvgane, 1981) and Dahanu region (Shaha, 1983; Ved Prakash and Mehrotra, 1987). Desai (1927) gave an exhaustive account of medicinal plants of Maharashtra, their uses and methods of administration. Recently, Naik (1998b) published an account of 350 medicinal plant species of Marathwada region in Marathi. While Singh and Sharma (2000) in "*Flora of Maharashtra State*" reported 443 species of medicinal value from Maharashtra, Upadhye and Kumbhojkar (1999) have given a list of 424 medicinal plants for the state of Maharashtra.

All the previous studies on medicinal plants from various regions are related to the various systems of medicines such as Ayurveda, Siddha, Unani etc. However, in 19th century, studies on medicinal plants started under various disciplines such as ethnobotany, ethnomedicines, ethnopharmacognosy, pharmacognosy, phyochemistry, pharmacology and ethnopharmacology.

According to Kamat (1972), the wandering tribes of the forest or aborigins and vaidyas or Hakims have the knowledge of various materials used as drugs and their dosages. Usually these nomadic forest tribes have their own local names to these drug plants. The branch of this knowledge goes under the title folk medicines. Folk medicines have no scientific basis but form important source for further scientific studies.

Robbins gave a broad definition of ethnobotany as the discovery and evaluation of the knowledge of all phases of life amongst aboriginal and the influence of the plant environment upon life, customs, beliefs and history of the tribal people. Vestal and Schultes (1939) stated that the ethnobotany is part of economic botany. Jones (1941) restricted ethnobotany as the study of interaction of the primitive man and plants. Schultes (1962) gave the definition of ethnobotany as " the study of relationships which exists between people of primitive societies and their plant environment". Castetter (1944) confined ethnobotany as a study of the botany of man in primitive state of culture.

The pioneer worker of ethnobotany is Hernandery (1570–75) who studied flora and fauna of Mexico in relation to human beings and published his work in 16 folio volumes. Some other well known workers in this field are Vestal and Schultes, 1939; Schultes, 1941, 1956, 1960, 1962, 1963, 1967, 1963a, b, 1990, 1992; Jones, 1941; Lipp, 1971; Vidal, 1971; Weiner, 1971; Doleres and Latorre, 1977 and Lischihichen, 1990. Traditional medicinal plants of the Central Provinces of Papua, New Guniea were studied by Holdsworth and Lacunineta (1981).

In India, S. K. Jain is a pioneer worker in the field of ethnobotany, who has made significant contributions on Indian medicinal plants and ethnobotany (Jain, 1963a, 1981, 1987, 1991a, b, 1997, 2001; Jain and De Filips, 1991). Very few studies on folk medicines and ethnobotany have been made in Maharashtra (Kamble and Pradhan, 1980; Upadhye *et al.*, 1995; Kulkarni and Kumbhojkar, 1992). Kulkarni and Kumbhojkar, (1992) published 40 Kitchen garden plants of Mahadeo Koli tribes in Maharashtra. Tiwari and Padhye (1993) in their ethnobotanical survey of Gond tribe of Chandrapur and Gadchiroli districts of Maharashtra reported 26 plants used in ethnomedicine.

Physical, sensory and botanical characteristic of drugs formed the main features of the study in earlier period (Wallis, 1967), but in the later half of the 19th century, the emphasis was laid on correct identity of crude drugs, their substitutes and adulterants of the commercial practices. This basis developed a new approach making studies for correct identity of crude drugs, their standardization and evaluation.

Seydler a German scientist who coined the term "*Pharmacognosy*" in his work entitled "*Analecta Pharmacognostica*" from combination of two Greek words viz. pharmakon-drug and ginghosco to acquire knowledge of further. Broadly it is defined as the scientific and systematic study of structural, physical, chemical and biological characters of crude drugs along with their history, methods of cultivation, collection and preparation for the market. The pioneer workers of pharmacognosy are Youngken (1950), Tyler (1958); Trease (1962) Trease and Evans (1962); Wallis (1967) and Tyler (1976). They established a new concept of studies on crude drugs of vegetable, animal and mineral origin. Studies generally deal with the source of drug, collection of drug, macro and microscopical characters, chemical constituents, uses and history.

In western countries, pharmacognosy is well developed and is an advanced science. Almost entire flora has been subjected to pharmacognostic investigations. Some of the important works include that of Youngken (1948); Berger (1949–1967); Rastad (1959); Anonymous (1962); Wallis (1967); Jackson and Snowdon (1969) and Pongprayoon (1991).

In India, some efforts have been made to provide proper place for this discipline among the sister sciences. In 1946, O.P. Basu established a Pharmacognosy laboratory at Lucknow. Subsequently, significant contributions in the field of pharmacognosy have been made by several workers (Bal, 1940; Datta, 1950; Datta and Mukerji, 1952; Aiyer and Kolammal, 1951–53; Iyengar, 1976; Mitra, 1969, 1979, 1985; Raghunathan and Roma Mitra, 1999).

Aiyer and Kolammal (1951–53) surveyed the Indian medicinal plants with respect to their pharmacognostical characters. Mukerji (1953) worked on Indian medicinal plants and it was first time he has given correct standard methods of preparations, dosage, specification of drugs, taxonomic characters of plants and parts, microscopic characters, action and uses. Pharmacognostic studies of Indian root and rhizome drugs were carried out by Datta (1950) and on leaf drugs by Datta and Mukerji (1952). Mehra *et al.* (1969) in their review on researches in Pharmacognosy described quite a number of drugs along with their proven substitutes and adulterants. Iyengar and Nayak (1984) have given an account of anatomy of crude drugs including tubers, leaf, stem, roots, seeds, rhizomes, barks etc with full anatomical descriptions and illustrations. Iyengar (1984) in his book '*Anatomy of crude drugs*' have given detailed account of about 84 drugs, their synonyms, regional names, active constituents, therapeutically and pharmaceutical uses and chemical tests. In 1986, Iyengar highlighted the major identifying characters of the powdered drugs. Roma Mitra (1985) published bibliography on Pharmacognosy of Indian medicinal plants containing all the available literature on medicinal plants till 1980. Recently, Raghunathan and Roma Mitra (1999) has complied pharmacognostic studies on 112 Indian medicinal plants giving detailed account of origin of drugs, synonyms, regional names, properties botanical description, histological characters and review of important previous works on each drug plant included in the book.

Recently Sarin, (1996) has published book "*Illustrated manual of Herbal Drugs used in Ayurveda*" in which he has provided detailed account of 118 crude botanical drug. Text of book contains detailed description on distinctive morphological and organoleptic characters of each drug. He incorporated 206 colour photographs of crude drugs to facilitate easy identifications. Sharma *et al.* (2001) has published Database on about 80 Indian Medicinal plants. In Maharashtra significant work on pharmacognosy includes that of Bhide (1941); Deshmukh *et al.* (1957), Surange and Pendse (1972) and Deokule (1995).

A need for detailed studies on wild medicinal plants has been emphasized by number of workers (Sangai, 2000, Leaman, 2001). Taking into account the need for study on correct identification of plant drug and documentation of traditional knowledge, the present studies have been undertaken. During present investigations, efforts have been made to survey and document medicinal plants of south western Maharashtra. Similarly, attempts have been made to gather information on their local uses and to determine correct botanical identity of folk medicines.

Material and Methods

Taxonomic Account of Medicinal Plants of South Western Maharashtra

Initially a primary list of medicinal plants recorded for south western Maharashtra (i.e. Kolhapur, Ratnagiri, Sangli, Satara and Sindhudurg districts) was prepared by referring relevant books on medicinal plants, floras, manuals, journals, reports and herbarium specimens deposited in herbarium of Shivaji University, Kolhapur. The list of medicinal plants was updated from time to time by procuring information from medicinal plants practitioners, vaidyas, tribal people, dhangars, medicinal shops, teachers and students of Ayurvedic colleges, information available on bottles of Ayurvedic drugs and preparations, internet and other possible available sources of information on medicinal plants.

After preparation of preliminary tentative list of medicinal plants of the region a detailed field survey on medicinal plants was undertaken in south western Maharashtra. The present account of medicinal plants is an outcome of 4 years (1998–2002) intensive and extensive field studies in the region. During the period of 4 years exploration, several tours of 1–3 days duration were arranged in different months to study and collect medicinal plant species from selected localities in the region (Map 2). Special attention was paid to under or unexplored remote areas in all the five districts of the region. Attempts were made to visit selected localities during all the seasons of a year. Field data on occurrence, habit, habitat, fragrance and colour of flowers, phenology, status, distribution, local names, local uses and methods of administration were recorded during field trips. Special attempts were paid to collect information on local names of plants, their local uses and methods of administration by consulting local people, tribal people of the region.

Plant specimens collected during survey were processed for herbarium by usual technique (Santapau, 1955; Jain & Rao, 1960; Rao & Sharma, 1990). At least three specimens of each medicinal plants except for cultivated plants have been deposited in herbarium of Botany Department Shivaji University, Kolhapur after proper processing of specimens were poisoned with saturated solution of Mercuric Chloride in rectified spirit.

Most of the specimens and field identifications were confirmed satisfactorily with the help of regional and national floras and available literature such as revision, monographs, research articles etc. Identification of doubtful specimens was confirmed by matching them with authentically identified species in the herbarium of Botany Department, Shivaji University, Kolhapur.

In the enumeration, the sequence of families has been followed after Bentham and Hooker's '*Genera Plantarum*' (1883–1962) with modifications in subscription of families according to recent workers such as J. Hutchinson (1973) and A. Cronquist (1988). The nomenclature has been adapted as far as practicable based on latest taxonomic literature and in accordance with the recommendations made by ICBN. The correct name is followed by citation of important pertinent literature. This is followed by basionyms and synonyms used in common earlier work on Maharashtra and India. Local name has been given after citation. The nomenclature of the species has been followed by full description, flowering and fruiting months, representative herbarium specimen (Exsiccata), distribution and local medicinal uses.

The references quoted in the prologue of the species are not cited in the bibliography in order to avoid repetition. Other relevant references on floristic are incorporated in review of literature. Attempts have been made to include photographs of maximum medicinal plants to facilitate easy identification. Some of the medicinal plants have been illustrated.

Folk Medicines and their Botanical Identity

Folk medicines sold in local bazaars by tribal and local people in various parts of south western Maharashtra were collected by weekly visits to different local bazaars of the region. Over 100 folk medicines have been collected from various bazaars of the region. All the bazaars have been visited during all seasons. The data such as local names and local uses, methods of administration on folk medicines was gathered by communicating with concerned people and noted in notebook. Collected samples were further processed for drying by using different methods. Some of the tuberous and rhizomatous specimens were preserved in FAA [Formaldehyde solution 5 ml + Glacial acetic acid 5 ml + 70% Ethyl alcohol 90 ml] or in 30% alcohol. The folk samples collected during present study are deposited in the Botany Department, Shivaji University, Kolhapur.

During investigation on correct identity of folk medicines, attempts were made to decide tentative botanical identity of plant species by knowing local names and by actual visits to field with dhangars and vaidyas. Finally the correct botanical identity was confirmed by comparison of morphological, anatomical and pharmacognostic characters of plant samples sold in markets with that of well identified plant specimens.

Pharmacognostic Studies

During present investigation over 100 folk plant samples were studied pharmacognostically. Pharmacognostic studies mainly dealt with macroscopic and microscopic characters of folk samples. Macroscopic characters such as form, size, shape, odor, color and taste of all folk specimens were studied. Study of microscopic characters of folk specimens mainly dealt with transverse section; studies of vessels, trachieds, phloem fibers, wood fibers, wood parenchyma, stone cells, starch gains, raphides, crystals of Calcium oxalate and Calcium carbonate.

For pharmacognostic study of woody folk specimens such as pieces of root, stem and bark the specimens were soaked in water for 4–8 hours and later on fixed in FAA. Fixed specimens were used for transverse sectioning. Transverse sections were taken by hand cut or by wood microtome. The transverse sections were stained by usual double staining technique given by Johanson (1940). Sections were mounted in DPX. Saturated solution of phosphoglucinol in 1N HCl was used to study xylem in tuberous specimens. To study wood elements such as vessels, tracheids, fibers etc. usual technique of maceration by Jeffery's fluid (Johanson, 1940) was employed.

For maceration, pieces of 2–3 × 0.5–1 cm first were soaked in water (for hard specimens) and then transferred to maceration fluid (10% nitric acid and 10% chromic acid in equal quantity) and kept in incubator at 50°C temperature for varying period and duration of incubation depend upon hardness of material. Macerated samples were washed thoroughly with water and then stained with Safranine O (M.S.), mounted in 10% glycerin and studied with the help of light microscope.

Anatomical characters of leaf samples were studied by taking hand cut sections. Cuticular studies of leaf samples were carried out by taking peels from both surfaces of leaf. The peels were stained with 2% aqueous Safranine O (M.S.) and mounted in 10% glycerin and studied for their characters by using microscope.

All permanent slides of folk samples of medicinal plants are deposited in Botany Department, Shivaji University, Kolhapur.

To study the powder characters of folk specimens, powders were prepared from folk samples with the help of grinder **Willey Type Grinding Mill Multiplux**. Materials for that were sliced into pieces of 5–6 × 1–2 cm size and dried in shade for 1–4 days or for weeks. Duration of drying varied for different samples depends upon nature of sample. Finally pieces were grinded in grinding machine into fine powder. Powders of folk medicines were studied for their color and contents such as, starch grains, stone cells, raphides, crystals of calcium oxalate, fiber pieces, vessel pieces etc.

For the microscopic examination of powder three types of slides were prepared for different purposes. These are as follows:

1. A judicious quantity of powder was taken on a glass slide to which few drops of chloral hydrate were added and after placing a coverslip, slide was heated for 1–2 minutes. Excess of chloral hydrate was allowed to drain. This was done to clear the tissues and to bring clarity of the view for better observation on contents of powder.
2. For the study of starch grains, powder was mounted in water with 1–2 drops of dilute Potassium iodide (KI).
3. For the study of mechanical tissue powder was treated with 1:1 phloroglucinol and conc. HCl. After 3–4 minutes powder was mounted in glycerin and observed under microscope.

Camera Lucida drawing at suitable magnifications of various anatomical structures was made by using light microscope and Prism type of Camera Lucida. Permanent preparations were deposited in Department of Botany, Shivaji University, Kolhapur.

All the field photographs of plants and plant parts were taken by Nikon Camera using microlens. Microphotographs were taken with the help of Carl Zeiss stereomicroscope with autographic unit (KL 1500 LCD).

During present investigation following chemical tests were used for the identification of folk medicines:

Qualitative tests for the identification of, tannin, starch, alkaloids and oil

Preparation of the extract: (Water extract, alcoholic extract, Chloroform extract). Take 0.2 gm of the powdered drug in test tube, add 2 ml of the solvent and shake it for 1 minute. Filter through cotton wool plug over a glass funnel. If required the solvent with the powder may be heated on a water bath for 2 min before filtration.

Tannins: Ferric chloride test: To the drug extract add 1% ferric chloride solution. Blue, green or brownish color indicates the presence of tannins.

Iodine: Add a few drops of iodine solution to about 1ml of the test solution. Appearance of blue color is due to the formation of starch iodine complex.

Glycosides: Take 0.5 ml of alcoholic extract of the drug add 1 ml of water and 1 ml of sodium hydroxide solution. A yellow color indicates the presence of glycosides.

Saponins: Weigh 0.1 gm of the drug and place it in a clean dry test tube. Add 10 ml of D/W put a stopper and vigorously shake for 30 seconds allow the tube to stand in a vertical position and observe for a period of 30 min if a honeycomb froth more than 3 cm above the surface of the liquid persists for 30 minutes, the drug is presumed to contain saponins.

Oil: To the thin section of the drug add alcoholic solution of Sudan III; Red color indicates presence of volatile oil.

Abbreviations used during present study are as follows:

Abbreviations

c	-	Collenchyma	phel	-	Phellogen
c. c.	-	Cell content	phl	-	Phloem
ck	-	Cork	phl. pa	-	Phloem parenchyma
cr	-	Crystals	pi	-	Pith
cu	-	Cuticle	ps	-	Palisade tissue
el	-	Lower epidermis	psl	-	Lower palisade
eu	-	Upper epidermis	psu	-	Upper palisade
fib	-	Fibers	rh	-	Rhytidoma
h	-	Hairs	s	-	Sclerenchyma
hyp	-	Hypodermis	s. c.	-	Secretory canal
ic	-	Inner cortex	scl	-	Sclerenchyma
idb	-	Idioblast	scl. sh	-	Sclerenchymatous sheath
lt	-	Laticifer	sec. phl	-	Secondary phloem
m	-	mesophyll	st	-	Stomata
mc	-	Mucilage cells	st. c.	-	Stone cells
mr	-	Medullary rays	st. h.	-	Stellate hairs
oc	-	Outer cortex	t	-	Tannin cells
og	-	Oil gland	trg	-	Glandular trichome
pa	-	Parenchyma	v	-	Vessel
per. fib	-	Pericyclic fibers	vb	-	Vascular bundle
phd	-	Phelloderm	xy	-	Xylem

Taxonomic Account of Medicinal Plants of South Western Maharashtra

RANUNCULACEAE Juss.

Clematis gouriana Roxb. ex DC. Syst. Nat. 1: 138. 1818; Hook. f. & Thoms. in Hook f. Fl. Brit. India 1: 4. 1872; Cooke, Fl. Pres. Bombay 1: 2. 1958 (Repr. ed.); Kamble in Singh *et al.,* Fl. Maharashtra State Dicot. 1: 151. 2000. MORIYEL, MORVEL, SHENDVEL.

Extensive climbers, stem and branches furrowed, glabrous, sparsely hairy when young. Leaves uni-bi or tripinnate; leaflets ovate, 2–10 × 1–4.5 cm, toothed, rounded or cordate at base, acute or acuminate at apex, 3–5 nerved;

upper surface shining. Inflorescence axillary or terminal, crowded, dense, up to 15 cm long. Flowers 1–1.2 cm in diameter, yellowish-white, pubescent. Sepals 4, oblong-obovate, pubescent. Stamens free, connectives scarcely produced above anthers. Achenes narrowly ovoid-oblong; more or less compressed with persistent feathery styles.

Fls. & Frts.: November–February. *Exsiccata*: SUK-238.
Frequent, along the Ghats.
Localities: Ambaghat, Amboli, Kas, Mahabaleshwar, Phonda, Radhanagari, Wai.

Local uses: Paste of leaves is applied on wound. Small pieces of roots are given once in a day with the roots of *Atlantia racemosa* orally with water as a blood purifier. Root paste is externally applied with ½ tea powder on wounds. Root paste is mixed with boiled rice water and decoction obtained is given thrice a day in dysentery. Roots are applied externally in leucoderma.

Clematis heynei M.A. Rau in Sharma *et al.* Fl. India 1: 67. 1993; Kamble in Singh *et al.* Fl Maharashtra State Dicot. 1: 152. 2000. *C. triloba* Heyne ex Roxb. Nov. Pl. Sp. 251. 1821 non Thunb. 1794; Hook. f. & Thoms. in Hook. f. Fl. Brit. India 1: 3. 1872; Cooke, Fl. Pres. Bombay 1: 2. 1958 (Repr. ed.). MOGRA-BEL, MORVEL, RANJAI.

Climbers with tomentose branches, stems sulcate. Leaves simple, or ternate, entire or 1–7 lobed; lobes ovate orbicular, 3–10 × 1.5–8 cm, rounded cuneate or cordate at base, mucronate at apex. Inflorescence in axillary corymbose panicles; bracts foliaceous, ovate, acute. Flowers ca 4.5 cm across, white. Sepals 4–6, spreading, pubescent on the back, glabrous within, mucronate. Stamens many, 1–1.2 cm long; without prolonged connective, filaments glabrous. Achenes ovoid, compressed, silky villous, with long feathery styles.

Fls. & Frts.: September–November. *Exsiccata*: SUK-345.
Common in deciduous forests.
Localities: Amba, Girgaon, Kas, Kolhapur, Ramghat.

Local uses: Roots are given orally for secretion of bile. Leaf paste applied externally in itch and in wounds, boils, used in leprosy, blood diseases and fever. Root decoction is given orally or small pieces of roots are placed in mouth in bilious vomiting. Decoction of root is given every night with boiled rice water to the children as anthelmintic.

DILLENIACEAE R. A. Salisbury

Dillenia indica L. Sp. Pl. 535. 1773; Hook. f. & Thoms. in Hook. f. Fl. Brit. India 1: 36 1872 p. p.; Cooke, Fl. Pres. Bombay 1: 6. 1958 (Repr. ed.); Venkanna

& Prasanna in Singh *et al.* Fl. Maharashtra State Dicot. 1: 160. 2000. MOTA-KARMAL.

Medium sized deciduous trees with straggling branches, 15–20 m high; young parts shining and silky. Leaves fascicled at the ends of branches, 70–80 × 15–20 cm, obovate, obtuse, margins dentate, nerves close, not forking at the margins. Flowers axillary, solitary ca 15 cm across, white fragrant. Sepals orbicular, concave, thick and fleshy. Petals oblong. Stamens many, inner larger and arching over the shorter ones. Fruit 8–10 cm across, hard; seeds reniform, black.

Fls. & Frts.: June–September. *Exsiccata*: SUK- 216
Localities: Cultivated in Public Garden at Kolhapur.

Local uses: One teaspoon powder of root bark is given in abdominal pains. One tablespoon of fruit juice is given twice a day in low fever. The ripe fruit remove dispels fatigue.

MAGNOLIACEAE Juss.

Michelia champaca L. Sp. Pl. 536. 1753; Hook. f. & Thoms. in Hook. f. Fl. Brit. India 1: 42. 1872; Cooke, Fl. Pres. Bombay 1: 8. 1958 (Repr. ed.); Kothari in Singh *et al.* Fl. Maharashtra State Dicot. 1: 162. 2000. CHAMPA, KUD-CHAMPA, PIVALA-CHAMPA, SON-CHAMPA.

Evergreen trees. Leaves 6.5–20 × 3.2–7.0 cm, elliptic oblong or lanceolate, coriaceous, glabrescent; petioles long, thickened below. Flowers bright yellow, fragrant, solitary, axillary, ca 5 cm across. Sepals oblong, acute. Petals linear lanceolate. Carpels subsessile. Follicles ca 2 cm long, oblong, gray with brownish spots. Seeds orbicular, reddish brown.

Fls. & Frts.: April–September. *Exsiccata*: SUK-614.
Usually cultivated as an ornamental tree for its flowers.
Localities: Kolhapur, Jaisingpur, Narsinghwadi.

Local uses: Paste of bark is applied to heal cracks on the feet's. Oil extracted from fruits and flowers is used for similar purpose. Flowers are kept in water for one night next day water is given early in the morning as diuretic. Flower paste applied externally in leprosy, skin diseases and ulcers.

ANNONACEAE Juss.

Annona reticulata L. Sp. Pl. 537. 1753; Hook. f. & Thoms. in Hook. f. Fl. Brit. India 1:78. 1872; Cooke, Fl. Pres. Bombay 1:15. 1958 (Repr. ed.); Mudaliar & Prasad in Singh *et al.* Fl. Maharashtra State Dicot. 1:163. 2000. RAMPHAL.

A tree, 8–12 m tall. Leaves oblong-lanceolate, 6–8 × 2.5–4.5 cm, glaucous beneath, pellucid punctate, apex acute. Flowers 2–4 on short peduncles, pedicels 2–3 cm long, with triangular bracts below the middle. Sepals triangular, 2–3 mm long, pubescent. Outer petals lanceolate ca 20 × 7 mm, tomentose on both surfaces; inner petals minute. Fruit syncarpous, yellowish red. Seeds smooth, black, ca 3 cm long.

Fls. & Frts.: July–November. *Exsiccata*: SUK-691.

A native of tropical America, cultivated throughout the region, also occur as an escape.

Localities: Kolhapur, Miraj.

Local uses: Leaves are made into paste with water and applied on wounds of animals mostly the goats and sheep. The bark is used as a tonic. The fruit is useful in blood complaints. It alleviates biliousness and thirst. Ripe fruit mixed with salt is used as a maturant. The seeds, leaves and immature fruit contain acrid principle, which is destructive to insect life.

Annona squamosa L. Sp. Pl. 537. 1753; Hook. f. & Thoms. in Hook. f. Fl. Brit. India 1: 78. 1872; Cooke, Fl. Pres. Bombay 1: 15. 1958 (Repr. ed.); Mudaliar & Prasad in Singh *et al.* Fl. Maharashtra State Dicot. 1: 165. 2000. SITAPHAL.

Shrubs or small trees. Leaves oblong-lanceolate, 5–7 × 2–4 cm, dark above, glaucous pubescent beneath when young. Flowers ca 2 cm, solitary or 2–4 on short axillary peduncles, leaf opposed. Sepals minute, pubescent, connate at base. Petals pubescent. Syncarp depressed globose, 5–10 cm in diam., glaucous when young, yellowish green when ripe; areoles 5–6 gonous, tuberculate. Seeds brown black, smooth, ca 1.5–2 cm long.

Fls. & Frts.: March–December *Exsiccata*: SUK-328.

Native of West Indies naturalized throughout Asia and also cultivated for its edible fruits.

Localities: Kolhapur, Satara, Sangli, Wai.

Local uses: The bruised leaves are applied as a poultice over boils, ulcers. The leaves are also used for destroying worms on the wounds of cattle.

Artabotrys hexapetalus (L. f.) Bhandari in Baileya 12: 147. 1964; Mudaliar & Prasad in Singh *et al.* Fl. Maharashtra State Dicot. 1: 175. 2000. *Artabotrys odoratissimus* R. Br. in Bot. Reg. 5: t. 423. 1820; Hook. f. & Thoms. in Hook. f. Fl. Brit. India 1: 54. 1872. HIRVA-CHAMPA, MADANMAST.

Scandent, evergreen shrubs climbing by means of hooks. Leaves oblanceolate, oblong or elliptic, 6–20 × 2–6 cm, acute or acuminate, shining green or punctate. Flowers solitary or paired on hooked peduncles, 5–6 mm long, greenish yellow, fragrant. Sepals ovate, 5–6 mm long, connate at base. Petals lanceolate, saccate at base, constricted above the stamens. Ripe carpels 6–10 or more, obovoid, glabrous, 2–3 cm long, yellow. Seeds oblong, ca 1.5 cm long, compressed and deeply grooved on one side.

Fls. & Frts.: Throughout year. *Exsiccata*: SUK-677

Native of Tropical Asia. Grown in gardens for the fragrant flowers.

Local uses: The flower powder is administered orally once in a day with water in excess sweating. Paste of flower is applied on forehead in headache. The flowers are smelled in headache. Paste of flower with the coconut oil is externally applied in itch.

Polyalthia longifolia (Sonner.) Thw. Enum. Pl. Zeyl. 398. 1864; Hook. f. & Thoms. in Hook. f. Fl. Brit. India 1: 62. 1872; Cooke, Fl. Pres. Bombay 1: 13. 1958 (Repr. ed.); Mudaliar & Prasad in Singh *et al.* Fl. Maharashtra State Dicot. 1: 175. 2000. ASHOK, DEDBARI.

Middle sized trees with spreading or pendulous branches. Leaves membranous, lanceolate, 15–20 × 3–5 cm, shining green above, paler beneath, gradually long acuminate at apex, with undulate margins. Flowers 2–3 cm across, yellowish green in axillary fascicles or shortly peduncled umbels. Sepals triangular, 4–5 mm long, pubescent, tips reflexed. Petals linear-lanceolate, ca 2 cm long, spreading, tapering to a point. Berries black, ovoid, 2–2.5 cm long. 1-seeded. Seeds smooth, pale brown, grooved in the center.

Fls. & Frts.: March July. *Exsiccata*: SUK-634.

Native of Tropical Asia planted in gardens and avenues.

Local uses: One teaspoon of bark powder twice in a day is administered in leucorrhoea and in menorrhoea.

Note: Stem bark of *Polyalthia longifolia* is the most common adulterant of the Ashoka bark i.e. *Saraca asoca.* It is readily distinguished from bark of *Polyalthia longifolia* (by absence of rough and warty protuberances and easily peeling off outer bark. The colour of inner bark in *Polyalthia longifolia* is dark brown in contrast with brick red colour of the bark of *Saraca asoca*).

Sageraea laurifolia (Grah.) Blatt. in J. Bombay nat. Hist. Soc. 34: 294. 1931; Mudaliar & Prasad in Singh *et al.* F1. Maharashtra State Dicot. 1: 173. 2000. *Bocagea dalzellii* Hook. f. & Thoms. in Hook. f. Fl. Brit. India 1: 92. 1872; Cooke, Fl. Pres. Bombay 1: 18. 1958 (Repr. ed.) p. p. ANDI, HAR-KINGAL, KILAND, SAGER, UNDI.

Trees, 6–12 m tall, evergreen. Leaves elliptic or elliptic-lanceolate or oblong lanceolate, coriaceous, 8–17.5 × 4.5–6.2 cm, shining, apex subacute, base rounded. Flowers small, white, solitary or in fascicles of 2–10, on small woody tubercles on the branches below the leaves, pedicels 1.5–2 cm long, with numerous scaly bracts at the base. Sepals orbicular, connate at the base, glabrous, ciliate. Petals orbicular, concave, the interior little shorter, ciliate. Stamens 12–18, outer row sometimes imperfect; connective broad, projecting above the dorsal anthers. Ovaries 3–5, oblong, hairy; style short, lateral glabrous; stigma small. Fruits ca 2.5 cm across, globular, wrinkled when dry.

Fls. & Frts.: October–April. *Exsiccata*: SUK-635.
Rare, in semi evergreen forests.
Localities: Ambaghat, Gaganbawada.

Local uses: The leaves have a pungent, astringent, and bitter taste. In the Konkan they are used for fomentation.

Uvaria narum (Dunal) Bl. Fl. Java, 5. 1828; Hook. f. & Thoms. in Hook. f. Fl. Brit. India 1: 50. 1872; Cooke, Fl. Pres. Bombay 1: 9. 1958. (Repr. ed.); Mudaliar & Prasad in Singh *et al.* Fl. Maharashtra State Dicot. 1: 174. 2000. *Unona narum* Dunal, Monogr. Farm. Anon. 99. 1817. SAPLIEVEL.

Scandent shrub, stellatly pubescent. Leaves oblong lanceolate, 12–24 × 3.0–6.5 cm, apex acute or acuminate. Flowers solitary, terminal, reddish, 2.7–3 cm in diam. Sepals orbicular-ovate, connate at the base, reflexed, minutely stellatly tomentose. Petals 6–8, ovate or oblong, connate at base. Stamens many, connective of anthers produced at the apex, broad, truncate. Carpel ovoid-oblong, green when young, 2–3 × 1.5–2.0 cm, shallowly constricted between seeds, ripens red. Seeds chestnut-brown, suborbicular, in a single row, usually 4–5.

Fls. & Frts.: May–October. *Exsiccata*: SUK-692.
Frequent, along Ghats.
Localities: Barki, Gaganbawada, Radhanagari, Ramghat.

Local uses: Bruised leaves are applied on forehead in hemicrania. Tincture of root powder is given after each 6 hrs in cough and in bronchitis. The oil obtained from roots is used to massage the body.

MENISPERMACEAE Juss.

Anamirta cocculus (L.) Wight & Arn. Prodr. 1: 446. 1834; Hook. f. Thoms. in Hook f. Fl. Brit. India 1: 98. 1872; Kulkarni in Singh *et al.* Fl. Maharashtra State Dicot. 1: 177. 2000. *Menispermum cocculus* L. Sp. Pl. 340. 1753. *Anamirta paniculata* Colebr. in Trans. Linn. Soc. 13: 66. 1822; Cooke, Fl. Pres. Bombay 1: 20. 1958 (Repr. ed.). KAKAPHAL, KAKMARI, VATOLI.

Dioecious stragglers, bark glaucous. Leaves broadly ovate, 15–25 × 10–15 cm, cordate to truncate at base, bluntly subacute to acuminate at apex, thinly coriaceous, glabrous above, pale beneath with tuft of hairs in the axils of the nerves except the basal ones. Inflorescence on old branches, paniculate, drooping, 16–40 cm long lateral branches, glabrous. Flowers green, scented, 4–5 mm across. Sepals subequal, deciduous, ovate oblong, concave, quite glabrous. Anthers: Male flowers forming globular head on short stout column of adherent filaments. Ovaries 3, on very short smooth gynophore surrounded at base by a ring of minute, fleshy, bifid staminodes; stigmas reflexed. Drupes 0.8–1.2 cm across, white, turning black.

Fls. & Frts.: December–February. *Exsiccata*: SUK-602.
Found in moist deciduous forest.
Localities: Amboli, Gaganbawada, Radhanagari, Ramghat, Vaibhavwadi.

Local uses: Fruits are externally applied in rheumatism and also in inflammation. Bruised leaves are immediately applied after insect bite. The berry is given orally in gas trouble. Dried fruits are pounded and powder is mixed with castor oil and applied externally in chronic skin diseases.

Cissampelos pareira L. var. **hirsuta** (Buch.-Ham. ex DC.) Forman in Kew Bull. 22: 356 1968; Kulkarni in Singh *et al.* Fl. Maharashtra State Dicot. 1: 178. 2000 *C. hirsuta* Buch.- Ham ex DC. Syst. Nat. 1: 535. 1817. *C. pareira* L. Sp. Pl. 1031. 1753; p. p.; Hook. f. & Thoms. in Hook. f. Fl. Brit. India 1: 103. 1872; Cooke, Fl. Pres. Bombay 1: 24. 1958 (Repr. ed.). PAHADMUL, TAHADVEL, TANNIVEL.

Climbing undershrubs, softly tomentose. Leaves peltate, broadly ovate to reniform, 3–5 × 2–4 cm, glabrous or sparsely pubescent above, pale and pubescent on nerves below, apex rounded, base truncate or subcordate. Male flowers in corymbose cymes, 4–6 cm long, bracts small subulate. Sepals 4, obovate-oblong, 2–3 mm long pubescent. Petals united to form a cup half as long as the sepals, pubescent outside, glabrous within. Female flowers in solitary or 2-nate axillary racemes. Bracts sessile, orbicular, foliaceous, persistent, more or less villous. Sepals one, ovate-oblong, ca 2 mm long. Petal one, half the length of the sepal. Drupe subglobose, 2–4 mm across, compressed, scarlet.

Fls. & Frts.: August–October. *Exsiccata*: SUK-162.
Common, in deciduous forest in plains.
Localities: Girgaon, Jotiba, Panhala, Ramling.

Local uses: Half teaspoon of root powder mixed in water and given orally in stomachache. In similar way it is given in urinary troubles. Decoction of root is given in fever, dysentery. Leaf powder mixed with Ghee and applied externally in skin eruption. Leaf powder is applied externally in chest pain. Leaf decoction is given in burning sensation in urinary passage.

Cocculus hirsutus (L.) Theob. in Mason, Burma ed. Teob. 2: 657. 1883; Kulkarni in Singh *et al.* Fl. Maharashtra State Dicot. 1: 179. 2000. *Menispermum hirsutum* L. Sp. Pl. 341. 1753. *Cocculus villosus* (Lam.) DC. Syst. Nat. 1: 525. 1817; Hook. f. & Thoms. in Hook. f. Fl. Brit. India 1: 101. 1872; Cooke, Fl. Pres. Bombay 1: 22. 1958 (Repr. ed.). JALJAMNI, KARROM, PARA-VEL, VASANVEL.

Woody twiners, young parts densely gray hairy. Leaves variable in size and shape, usually ovate, 2–12 × 1.5–12 cm, densely villous, cuneate to truncate or hastate at base, obtuse or mucronate at apex. Flowers small, dull green; Males in short axillary racemes or panicles. Sepals ca 1.5 mm long, outer ones oblanceolate, inner broad, obovate, densely hirsute on back, glabrous or hirsute within. Petals 6, membranous, deeply 2-lobed, embracing the filaments at base. Stamens 6. Female flowers 2–3, in axillary fascicles, sepals as in males. Petals thick and fleshy. Ovaries 3, smooth. Drupes purple, compressed globose, 4 mm in diam., transversely rugose.

Fls. & Frts.: September–March. *Exsiccata*: SUK-21.
Common, in hedges around fields, on bushes in hill forest and in fields.
Localities: Jaisingpur, Jat, Kolhapur, Miraj, Wai.

Local uses: Paste of leaves used externally to stop bleeding from the cut wounds. Bruised leaves are applied on forehead to stop headache. Leaf powder is boiled in water with coriander powder for 2–3 hours and dilute solution of this mixture is given thrice a day in burning sensation and in urinal discharges.

Cyclea peltata (Lam.) Hook. f. & Thoms. Fl. Ind. 201. 1855 & in Hook. f. Fl. Brit. India 1: 104. 1872 p.p.; Cooke, Fl. Pres. Bombay 1: 25. 1958 (Repr. ed.); Kulkarni in Singh *et al.* Fl. Maharashtra State Dicot. 1: 180. 2000. *Menispermum peltatum* Lam. Encycl. 4: 96. 1797. *Cyclea burmanni* (D.C.) Hook. f. & Thoms. op. cit. & in Hook. f. op. cit.; Cooke, op. cit. PAKAR.

Herbaceous twiners, young parts pubescent. Leaves deltoid or ovate, 4–9 × 3.5–6.0 cm, apex acuminate or shortly mucronate, base subcordate or hastate, glabrous or shining above, pubescent beneath. Flowers in axillary panicles. Male flowers: Calyx campanulate, divided nearly to the base into 4–5 segments, pilose. Corolla 4–5 segments. Stamens 4–6, connate. Female flowers: Sepals oblong glabrous. Petal one, orbicular, much shorter than the sepal. Ovary pilose. Drupes whitish green, obovoid, 3–5 mm.

Fls. & Frts.: September–February. *Exsiccata*: SUK-693.
Frequent in moist deciduous forests.
Localities: Chandgad, Gaganbawada, Halkarni, Radhanagari, Amboli, Vaibhavwadi.

Local uses: Stem paste is applied externally in combination with the bark of *Pittosporum floribundum* ('Vikari') in itching caused due to the biting of insects. Stem powder is also given orally in urinary problems.

Diploclisia glaucescens (Bl.) Diels in Engl. Pflanzenr. 46: 225, t. 77. 1910; Kulkarni in Singh *et al.* Fl. Maharashtra State Dicot. 1: 181. 2000. *Cocculus glaucescens* Bl. Bijdr. 25. 1825. *C. macrocarpus* Wight & Arn. Prodr. 13. 1834; Hook. f. & Thoms. in Hook. f. Fl. Brit. India 1: 101. 1872; Cooke, Fl. Pres. Bombay 1: 22. 1958 (Repr. ed.). RAMRIK, SAPTKAPI, WATENVEL, CHAKRVMBI, WARDHARA.

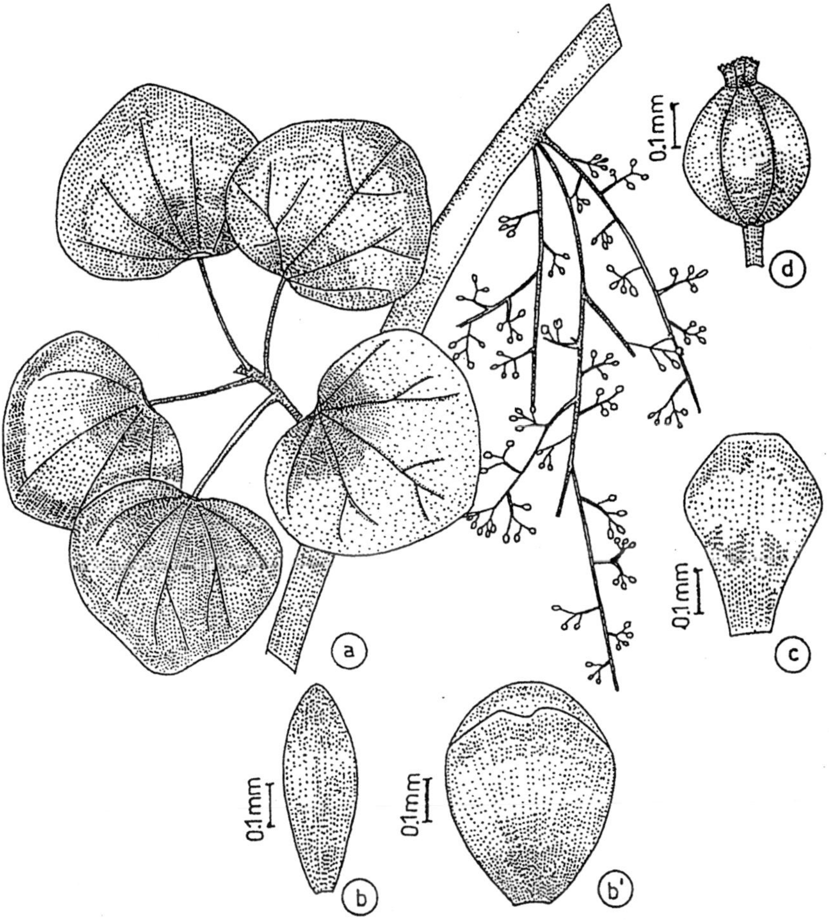

Text fig. 1: Diploclisia glaucescens a-twig; b & b'-sepals; c-petal; d-carpel.

Extensive climbers. Leaves broadly ovate to suborbicular, 4–10 × 4–12 cm, glaucous beneath, apex obtuse, base truncate or subcordate. Flowers in large pendulous panicles in divergent fascicles from the old wood. Flowers yellowish-green. Male flowers: Sepals 6, in two series, inner larger. Petals 6, shorter than the sepals. Stamens embraced by the petals; anthers subglobose. Female flowers: Staminodes 6 or 0. Ovaries oblong, incurved. Drupes obovoid, compressed, 2.0–2.5 cm across. Seeds kidney shaped.

Fls. & Frts.: January–May. *Exsiccata*: SUK-299.
Frequent, in moist deciduous forests.
Localities: Ambaghat, Gaganbawada, Radhanagari, Ramghat.

Local uses: Stem powder is given orally thrice a day to diabetic patient. Stem powder in combination with *Gnetum scandens* (Umbelvel) given orally with water every night in piles. Stem in combination with root of *Cinnamomum nitidum* (Mawarung) and root of *Salacia brunoniana* (Saptrangi) is given in asthma. Leaf powder is taken with milk as a cure for biliousness, gonorrhoea and syphilis.

Stephania japonica (Thunb.) Miers in Ann. Mag. Nat. Hist. 3, 18: 14. 1866; Kulkarni in Singh *et al.* Fl Maharashtra State I': 181. 2000. *Menispermum japonicum* Thunb. Fl. Jap. 193. 1784. *Stephania hernandifolia* (Willd.) Walp. Rep. 1: 96. 1842; Hook. f. & Thoms. in Hook. f. Fl. Brit. India 1: 103. 1872; Cooke, Fl. Pres. Bombay 1: 23. 1958 (Repr. ed.). PAHADVEL.

Slender twiners with tuberous roots. Leaves cordate or ovate, 5–8 × 3.5–6 cm, apex shortly acuminate, base rounded or shallowly cordate. Flowers in umbellate heads, greenish-yellow. Male flower: Sepal and petals 4–6; stamens 6–8, in peltate synandrium. Female flower: Sepals and petals 3–5, obovate-cuneate half the length of sepals. Stigma 3 lobed. Drupes solitary, subsessile, obovoid, compressed, glabrous, red, 4–8 mm across. Seeds curved.

Fls. & Frts.: January–February. *Exsiccata*: SUK-218.
Infrequent, in moist deciduous forests.
Localities: Ajara, Chandgad, Gaganbawada, Kas, Katyayani, Panhala, Vaibhavwadi.

Local uses: The root is given in urinary problems. The root paste is given in low fever. The root paste is also used in stomach complaint, one teaspoon of root powder is given with buttermilk in gas trouble. Root enters into the composition of many Ayurvedic preparations as a substitute for *Cissampelos pareira* L.

Tinospora cordifolia (Willd.) Miers ex Hook. f. & Thoms. Fl. India 184. 1855 & Hook. f. Fl. Brit. India 1: 97. 1872; Cooke, Fl. Pres. Bombay 1: 20. 1958 (Repr. ed.); Kulkarni in Singh *et al.* Fl. Maharashtra State Dicot. 1: 183. 2000. *Menispermum cordifolium* Willd. Sp. Pl. 4: 826. 1806. GULVEL.

Woody twiners with long pendulous aerial roots and striate or lenticulate glabrous branches. Leaves broadly ovate, 5–9 × 4–8.5 cm, acuminate, cordate. Flowers yellowish green in axillary racemes, males fascicled, females usually solitary; bracts lanceolate or foliaceous. Male sepals unequal, outer ones ovate oblong, ca 1 mm long, inner suborbicular, 2–3 mm across, concave. Petals clawed, limb subtrilobed, shorter than the inner sepals. Female sepals similar to those of males; petals oblong entire. Drupes 1–3, subglobose, 6–10 mm in diam., red. Seeds white, bean shaped, ca 7 × 4 mm, warty.

Fls. & Frts.: January–August. *Exsiccata*: SUK-20.
Common, in deciduous forests and also planted.
Localities: Kolhapur, Sangli, Wai.

Local uses: Small pieces of stem are tied around the neck in jaundice and also in low fever. Stem powder is given orally in urinary problems amd to diabetic patient.

NYMPHAEACEAE Salisb.

Nymphaea nouchali Burm. f. Fl. Ind. 120. 1768; Moorthy in Singh *et al.* Fl. Maharashtra State Dicot. 1: 185. 2000. *N. stellata* Willd. Sp. Pl. 2: 1153. 1799; Hook. f. & Thoms. in Hook. f. Fl. Brit. India 1: 114. 1872; Cooke, Fl. Pres. Bombay, 1: 27. 1958 (Repr. ed.). UPLIA KAMAL.

Rhizomatous perennial herbs. Leaves orbicular to elliptic, 12–20 cm across, often blotched with purple beneath, petioles long, slender. Flowers solitary, 6–15 cm across, peduncles long. Sepals oblong-lanceolate. Filaments 10–50, dilated at the base, anther appendage ligulate or subulate. Carpels 10–30. Fruits globose, spongy. Seeds ellipsoid.

Fls. & Frts.: Almost throughout the year. *Exsiccata*: SUK- 346.
Frequent, in shallow tanks and ponds.
Localities: Chandgad, Kagal, Kalamba, Kolhapur, Panhala, Patgaon, Shiroli.

Local uses: Root is orally given in piles. Flower is given orally in blood impurities and also given as aphrodisiac.

Nymphaea pubescens Willd. Sp. Pl. 2: 1154. 1799; Moorthy in Singh *et al.* Fl. Maharashtra State Dicot. 1: 185. 2000. *N. lotus* auct. non. L. 1753; Hook. f. & Thoms. Fl. Ind. 241. 1855 & in Fl. Brit. India 1: 114. 1872 p. p. *N. lotus* var. *pubescens* (Willd.) Hook. f. & Thoms. op. cit. & Hook. f. op. cit.; Cooke, Fl. Pres. Bombay 1: 27. 1958 (Repr. ed.). KAMAL.

Rhizomatous perennial herbs. Leaves orbicular to reniform, 18–25 cm across, deeply cordate at base with 6–8 cm deep sinus, glabrous and often nigropunctate above, velvety pubescent and prominently veined beneath, petioles long. Flowers solitary, 8–20 cm across, peduncles long, pubescent.

Sepals oblong-obtuse, 5–10 nerved. Petals ca 12, oblong, broad. Stamens ca 40; filaments much dilated at base. Fruits globose, many-seeded, spongy. Seeds ellipsoid.

Fls. & Frts.: Throughout year. *Exsiccata*: SUK-206
Frequent, in ponds and tanks.
Localities: Kagal, Kalamba, Kolhapur, Shiroli.

Local uses: Powdered roots are used for piles as a demulcent, also for dysentery, dyspepsia and cardiac tonic.

Nymphaea rubra Roxb. ex Salisb. Parad. Lond. 1, t. 14. 1805; Moorthy in Singh *et al.* Fl. Maharashtra State Dicot. 1: 186. 2000. *N. lotus* auct. non L. 1753; Hook. f. & Thoms. in Hook. f. Fl. Brit. India 1: 114. 1872, p. p. KUMUD, SALUK.

Aquatic herbs, rootstock short, ovoid, acute. Leaves peltate, 15–25 cm in diam., glabrous and often nigropunctate above, velvety pubescent and prominently veined beneath. Flowers solitary, 8–20 cm across, red, pale rose, or white; peduncles very long, pubescent. Sepals oblong obtuse, 5–10 ribbed. Petals oblong, obtuse, about thrice as long as broad. Anthers without appendages. Fruits globose, green, fleshy, ca 3 cm across. Seeds ovoid, rough, arillate.

Fls. & Frts.: Throughout year. *Exsiccata*: SUK-1777.
Cultivated in gardens.
Localities: Nerurpar in Sindhudurg district.

Local uses: Flowers are given in low fever. Root decoction given in diarrhoea and in piles. Flowers are crushed in water and water is given orally to the patient suffering from pimples. Flower decoction is given early in the morning to cure blood impurities.

NELUMBONACEAE Dumort.

Nelumbo nucifera Gaertn. Fruct. 1: 73, t. 19, f. 2. 1788; Moorthy in Singh *et al.* Fl. Maharashtra State Dicot. 1: 187. 2000. *Nelumbium speciosum* Willd. Sp. Pl. 2: 1258. 1799; Hook. f. & Thoms. in Hook. f. Fl. Brit. India 1: 116. 1872; Cooke, Fl. Pres. Bombay 1: 28. 1958 (Repr. ed.). KAMAL.

Perennial aquatic herbs with rhizomatous rootstock, slender, elongate, creeping, branched. Leaves peltate, orbicular, 30–90 cm across, concave or cupped, entire, radiately nerved, glaucous; petioles long, distinctly prickly. Flower solitary, red, pink, white or dull yellow. Torus 2 cm high, spongy, the top flat, 2.5 cm across. Sepals small. Petals elliptic oblong, 5–10 cm long, obtuse, spreading or suberect. Anther with clavate appendages. Carpels oblong, 1–1.5 cm long, glabrous. Seed with spongy testa.

Fls. & Frts.: Throughout year. *Exsiccata*: SUK-694.
Frequently, found in tanks and ponds, often Cultivated.
Localities: Shiroli.

Local uses: Flower powder is given as cardiac tonic. The seeds are given as tonic to the women in leucorrhoea. Petals are crushed in rose water, paste is applied on face to improve complexion. Flower powder is given twice a day with half cup of hot water in dysentery.

PAPAVERACEAE Juss.

Argemone mexicana L. Sp. Pl. 508. 1753; Hook. f. & Thoms. in Hook. f. Fl. Brit. India 1: 117. 1872; Cooke, Fl. Pres. Bombay 1: 29. 1958 (Repr. ed.); Madhusudan Rao in Singh *et al.* Fl. Maharashtra State Dicot. 1: 187. 2000. PIVLA DHOTRA.

Erect, glaucous annual herbs, 30–90 cm tall. Stem prickly, with yellow latex. Leaves sessile, oblanceolate, 12–20 × 2–4 cm, variously lobed and spinous dentate. Flowers solitary, terminal, 5–6 cm across, surrounded by small leafy bracts. Sepals 3, spinous, caducous, green. Petals 6, in two series of 3 each, broadly obovoid, bright golden yellow. Stigma red or purple. Capsules 3–4 cm long, prickly. Seeds globose, brown, ca 1 mm in diam.

Fls. & Frts.: Throughout year. *Exsiccata*: SUK-695.
Common, weed in waste places. Native of tropical America.
Localities: Kolhapur, Ratnagiri, Sangli, Satara.

Local uses: Leaf juice is mixed with *Tridax procumbens* and applied externally in leprosy. Seeds are smoked in asthma. The roots are applied externally in itch and inflammation. The yellow juice is rubbed on the body in rheumatic pain.

BRASSICACEAE Burm (*nom. alt.*)
[**CRUCIFERAE** Juss. (*nom. cons.*)]

Brassica juncea (L.) Czern. & Coss. in Consp. Fl. Chark. 8. 1859; Hook. f. & T. And. in Hook. f. Fl. Brit. India 1: 157. 1872; Cooke, Fl. Pres. Bombay 1: 37. 1958 (Repr. ed.); Pradhan in Singh *et al.* Fl. Maharashtra State Dicot. 1: 192. 2000. *Sinapis juncea* L. Sp. Pl. 668. 1753. MOHRI, RAI.

Erect, annual herb, 30–90 cm tall. Lower leaves lyrate-pinnatifid, obovate, 4–30 × 1.5–6 cm, the upper ones gradually smaller, lanceolate, dentate. Flowers in terminal, elongating racemes. Petals bright yellow. Siliqua erect-patent, 3–6 cm long. Seeds many, brown.

Fls. & Frts.: Cold Season.
Cultivated and escaped weed of wet lands.

Localities: Kolhapur, Miraj, Shiroli.

Local uses: Leaves are crushed and applied on swellings and in inflammatory diseases. Seed paste is applied in itching and other skin diseases. The seeds are taken moderately as a digestive condiment. Leaves are used to increase bile secretion.

Brassica nigra (L.) Koch in Roehl. Deutschl. Fl . ed. 3, 4: 713. 1833; Hook. f. & T. And. in Hook. f. Fl. Brit. India 1: 156. 1872; Pradhan in Singh *et al.* Fl. Maharashtra State Dicot. 1: 192. 2000. *Sinapis nigra* L. Sp. Pl. 668. 1753. KALI RAI.

Much branched, erect annual, 0.3–1.5 m or more tall. Leaves variable, radical pinnatifid or lobed, 10–12 × 2–4 cm, terminal lobe very large, stem leaves ovate, entire or pinnatifid, all petiolate, glabrous or slightly hairy. Flowers in terminal, usually much branched racemes. Petals bright yellow. Siliqua subtetragonous, 1.5–2.5 cm long, on short pedicels which are close to the rachis. Seeds ovoid, dark brown or black.

Fls. & Frts.: September–February.

Cultivated and escaped weed of wet lands.

Localities: Kolhapur, Miraj, Shiroli.

Local uses: Leaves are crushed and applied on swellings and in inflammatory diseases. Leaf juice warmed over flame and dropped in the ear to stop earache.

Raphanus sativus L. var. **sativus** Sp. Pl. 669. 1753; Hook. f. & T. And. in Hook. f. Fl. Brit. India 1: 166. 1872; Cooke, Fl. Pres. Bombay 1: 137. 1958 (Repr. ed.); Pradhan in Singh *et al.* Fl. Maharashtra State Dicot. 1: 204. 2000. MULA.

Annual or biennial herbs with thick, fleshy root. Stems 60–90 cm tall, much branched, more or less glaucous. Leaves lyrate-divided, lower ones long petioled, upper ones short petioled, entire or dentate. Flowers in terminal elongating racemes. Petals white or pink purple. Siliqua erect or ercto-patent, 3–9 cm long, spongy, 1–6 seeded, beaked, more or less constricted between the seeds. Seeds globose.

Fls. & Frts.: November–February.

Cultivated throughout the region for fleshy roots and fruits used as vegetable.

Localities: Kolhapur.

Local uses: The juice of fresh leaves is given in urinary complaint. Roots are given as diuretic and also in piles. It is given as a blood purifier. The seeds are used in menstrual problems.

CLEOMACEAE (Pax) Airy Shaw

Cleome chelidonii L. f. Suppl. Pl. 300. 1781; Hook. f. & Thoms. in Hook. f. Brit. India 1: 170. 1872; Cooke, Fl. Pres. Bombay 1: 42. 1958 (Repr. ed.); Londhe in Singh *et al.* Fl. Maharashtra State Dicot. 1: 219. 2000.

Erect, glaucous green, annual herbs, 30–90 cm tall. Stems often prickly. Leaves digitate, 3–9 foliolate, leaflets elliptic-obovate, 0.8–4.5 × 0.3–1.5 cm, mucronate at apex. Flowers 1.5–2 cm across, in terminal, corymbose, 20–60 cm long racemes. Sepals ovate, 2–3 mm long. Petals pink or dark-rosy, obovate, 0.7–2 cm long. Stamens many. Siliqua erecto-patent, cylindrical, 5.5–8.5 cm long, longitudinally striated. Seeds black, globose tubercled.

Fls. & Frts.: July–January *Exsiccata*: SUK-189.
Common, in wet paddy fields, ditches, shallow ponds etc.
Localities: Kagal, Kalamba, Kolhapur, Panhala,

Local uses: The roots are considered to be vermifuge. An infusion of the plant is commonly used in gingivitis and in the treatment of skin diseases.

Cleome gynandra L. Sp. Pl. ed. 1. 671. 1753; Londhe in Singh *et al.* Fl. Maharashtra State Dicot. 1: 221. 2000. *Gynandropsis pentaphylla* (L.) DC. Prodr. 1: 238. 1824; Hook. f. & Thoms. in Hook. f. Fl. Brit. India 1: 171. 1872; Cooke, Fl. Pres. Bombay 1: 42. 1958 (Repr. ed.). PANDHRI TILWAN.

Glandular pubescent herb, 30–70 cm tall. Leaves digitate, 3–9 foliate; leaflets subsessile, obovate-oblanceolate, 2.5 × 1.5–3 cm, terminal leaflet largest, all cuneate at the base, hairy. Flowers 1.5–2 cm across, in terminal corymbose elongating racemes. Sepals lanceolate. Petals white, spathulate. Stamen 6, inserted about half way up the gynophore. Ovary linear-oblong, on the top of gynophore. Siliqua linear, cylindric, 5–7 cm long. Seeds dark brown, muricate.

Fls. & Frts.: June–December. *Exsiccata*: SUK-696
Common, on wastelands and on old walls and roadsides.
Localities: Kagal, Kolhapur, Radhanagari.

Local uses: The leaves in combination with cow urine is poured in ear to stop earache. The leaves are applied on boils. Similarly leaf paste is applied on forehead to stop headache. Seeds decoction is given to expel stomach worms.

Cleome viscosa L. Sp. Pl. 672. 1753; Hook. f. & Thoms. in Hook. f. Brit. Fl. India 1: 170. 1872; Cooke, Fl. Pres. Bombay 1: 41. 1958 (Repr. ed.); Londhe in Singh *et al.* Maharashtra State Dicot. 1: 223. 2000. PIVALI TILWAN.

Erect, glandular pubescent herbs, 15–20 cm tall. Leaves digitate, 3–5 foliate; leaflets subsessile, elliptic-obovate, 1–4 × 0.5–2 cm, acute or obtuse at apex, cuneate at base. Flowers axillary, solitary, passing in terminal lax racemes. Sepals linear-lanceolate, 5–8 mm long. Petals yellow, oblanceolate. Stamens many, filaments unequal; anthers bluish. Siliqua cylindrical, 4–12 cm long, glandular pubescent, longitudinally striate. Seeds reniform, globose, ribbed.

Fls. & Frts.: September–May. *Exsiccata*: SUK-197.
Common, on wastelands and on old walls and roadsides.
Localities: Amba, Gaganbawada, Kagal, Kolhapur, Shirol.

Local uses: Leaf juice is given to favor digestion. Similarly poured in ear to cure earache, leaves are applied externally in headache. Roots are given to expel worms from stomach. Leaves are externally applied in toothache. Plant is used as cooling, stomachic, laxative, diuretic, reduces tumors and inflammations, useful in skin diseases, itching, ulcers, leprosy, useful in blood diseases and uterine complaints etc.

CAPPARACEAE Juss.

Cadaba fruticosa (L.) Druce in Rep. Bot. Exch. Cl. Soc. Brit. Isles 3: 415. 1914; Londhe in Singh *et al.* Fl. Maharashtra State Dicot. 1: 206. 2000. *Cleome fruticosa* L. Sp. Pl. 671. 1753. *Cadaba indica* Lam. Encycl. 1: 544. 1785; Hook. f. & Thoms. in Hook. f. Fl. Brit. India 1: 172. 1872; Cooke, Fl. Pres. Bombay 1: 45. 1958 (Repr. ed.) *C. farinosa* Forssk. Fl. Aeg.-Arab. 68. 1775; Hook. f. & Thoms. in Hook. f. op. cit. 173. KALASAKAL, KALITAKA.

Much branched, woody, erect or straggling, glandular pubescent shrubs, 2–3 m tall. Leaves elliptic-oblong, 1–6 × 0.5–2.5 cm, glabrous or nearly so, obtuse, retuse or mucronate at apex. Flowers in terminal, few-flowered corymbs, 2–2.5 cm across. Sepals ovate-oblong, 1.5 cm long, acute, pubescent. Petals white, oblanceolate-spathulate as long as the sepals. Disc prolonged into a funnel-shaped process, expanded and toothed at apex. Stamens 4–6, much exerted. Gynophore 2–2.5 cm long. Fruits cylindric, torulose, pubescent. Seeds many, brown, surrounded by orange-red aril.

Fls. & Frts.: November–March. *Exsiccata*: SUK-318.
Frequent, in dry deciduous forest.
Localities: Danoli, Jaisingpur, Kari, Tasgaon, Vita, Wai.

Local uses: The roots are given orally with tincture of boiled *Ricinus communis* flower powder early in the morning in intestinal worms. Root powder is also given in urinary problems. Leaves are given in small quantity to the children in case of indigestion. The leaves and roots are considered deobstrutant and anthelmintic and are prescribed in uterine obstructions.

Capparis decidua (Forssk.) Edgew. in J. Linn. Soc. Lond. 6: 184. 1862; Londhe in Singh *et al.* Fl. Maharashtra State Dicot. 1: 209. 2000. *Sodada decidua* Forssk., Fl. Aeg.-Arb. 81. 1775. *Capparis aphylla* Roth, Nov. Sp. Pl. Ind. Or. 238. 1821; Hook. f. & Thoms. in Hook. f. Fl. Brit. India 1: 174. 1872; Cooke, Fl. Pres. Bombay 1: 49. 1958 (Repr. ed.). NEPTI.

Low shrubs or small trees up to 6 m tall, stipular spines often straight, 3–7 mm long. Leaves early deciduous, linear lanceolate, 0.3–2 cm long. Flowers in 10–20 flowered corymbs on short branchlets. Sepals linear-lanceolate, 8–13 mm long, tomentose on the back, white ciliate along margins, mucronulate, the adaxial sepals saccate, the inner ones petaloid. Petals as long as the sepals, red to pink. Stamens 8–18. Gynophore 1–2 cm long, ovary globose. Fruits globose, 0.7–1.5 cm in diam., deep red. Seeds ovoid, 2–4 × 2 mm.

Fls. & Frts.: March–November. *Exsiccata*: SUK-159.
Common, in dry deciduous forests.
Localities: Aurwad, Narsimhwadi.

Local uses: Bark is used externally in case of rheumatism, it is also applied in any kind of inflammation. Fruits are given orally in toothache. Leaves are crushed and tied around the feet as a plaster in case of internal injury. The bark is given as laxative, anthelmintic, in cough, asthma, ulcers, boils, vomiting and piles.

Capparis grandis L. f. Suppl. Pl. 263. 1781; Hook. f. & Thoms. in Hook. f. Fl. Brit. India 1: 176. 1872; Cooke, Fl. Pres. Bombay 1: 50. 1958 (Repr. ed.); Londhe in Singh *et al.* Fl. Maharashtra State Dicot. 1: 210. 2000. KAUTEL, PACHARA, PACHUNDI.

Medium sized trees, 5–7 m tall. Leaves ovate or suborbicular, 3–7.5 × 2–2.5 cm, acuminate, velvety tomentose on both surface when young. Flowers in terminal corymbose racemes, 3–3.5 cm long. Sepals green, lanceolate to obovate, 5–10 mm long, puberulous, or glabrous on the back, ciliate along margins. Petals white, obovate, 8–12 × 3–4 mm, glabrous or puberulous. Stamens many. Gynophore 1–3 cm long; ovary ovoid, glabrous. Fruits globose, 2.5–3 cm in diam., red, glabrous. Seeds ca 1 cm across, compressed.

Fls. & Frts.: April–September. *Exsiccata*: SUK-362.
Localities: Hatkanagale, Jaisingpur, Kolhapur.

Local uses: The infusion of the bark and leaves is used internally for swellings and eruptions.

Capparis moonii Wight, Illustr. 1: 35. 1840; Hook. f. & Thoms. in Hook. f. Fl. Brit. India 1: 175. 1872; Cooke, Fl. Pres. Bombay 1: 49. 1958 (Repr. ed.); Londhe in Singh *et al.* Fl. Maharashtra State Dicot. 1: 210. 2000. GARABDUDI, POORWI, WAGATI, WAGHATI.

Climbing shrubs. Leaves elliptic-oblong with callous tip, 5–12 × 2–5 cm, stipular spines short, stout, hooked. Flowers white in terminal corymbs, 7–12 cm in diam. Sepals 2.5 cm in diam., orbicular, deeply cupped, much spreading, pubescent on the upper surface. Stamens many, 5–7 cm long. Gynophore 5–7 cm long; ovary ovoid, glabrous. Fruit subglobose, 5–10 cm in diam., pointed, seated on much thickened woody stalk. Seeds numerous, 1.8 cm in diameter.

Fls. & Frts.: January–June. *Exsiccata*: SUK-247.
Rare, in moist deciduous forests along Ghats.
Localities: Ajara, Borbet, Kolhapur, Narande.

Local uses: Paste of root bark applied to boils and swellings of testicles. Decoction of bark is given in syphylis. Fruit is used in cardiac troubles.

Capparis sepiaria L. Syst. Nat. ed. 10, 2: 1071. 1759; Hook. f . & Thoms. in Hook. f. Fl. Brit. India 1: 177. 1872; Cooke, Fl. Pres. Bombay 1: 51. 1958 (Repr. ed.); Londhe in Singh *et al.* Fl. Maharashtra State Dicot. 1: 214. 2000. KANTHTARA, PACHRA, PACHUDA.

Much branched straggling shrubs, 2–3 m tall, branchlets densely gray tomentose. Stipular spines recurved, 3–5 mm long. Leaves ovate-lanceolate to elliptic-oblong, 3.5–8.5 × 3–5 cm, acute or rounded at base, acute, obtuse or emarginate at apex, velvety gray tomentose on both surface when young. Flowers in axillary and terminal umbels, 6–9 mm across. Sepals ovate, 3–6 × 3–5 mm. Petals white lanceolate 3–7 × 1–3 mm. Stamens many. Gynophore 6–10 mm long. Ovary ovoid, 2 × 1 mm, glabrous. Fruit globose, 1–1.5 cm in diameter. Seeds 1–2, ovoid, 5–8 mm long.

Fls. & Frts.: March–July. *Exsiccata*: SUK-640.
Frequent, in hill forest, around field and along stream banks.
Localities: Ajara, Narande, Girgaon.

Local uses: Roots are given as a tonic. Decoction is given with concentrated juice of *Ocimum americanum* after every 2 hours in low fever, in cold and cough. Plant is used as a tonic, stomachic, improves appetite, blood troubles, tumors, inflammations and diseases of the muscles.

Capparis spinosa L. Sp. Pl. 503. 1753; Hook. f. & Thoms. in Hook. f. Fl. Brit. India 1: 173. 1872; Cooke, Fl. Pres. Bombay 1: 47. 1958 (Repr. ed.); Londhe in Singh *et al.* Fl. Maharashtra State Dicot. 1: 214. 2000. KABRA, KALAVRI.

A diffuse shrub, covered with soft yellowish tomentum. Leaves broadly ovate-orbicular, 2.2–4.5 × 1.5–3.0 cm, entire acute at apex, glabrous or pubescent. Flowers white, solitary, axillary. Sepals subequal, pubescent. Petals white, 2.7–3.5 cm long. Stamens many; filaments purple, longer than the petals. Gynophore 2.5–4 cm long. Fruit ellipsoid, 2.7–5 cm long, red when ripe. Seeds globose, embedded in scarlet pulp.

Fls. & Frts.: February–May. *Exsiccata*: SUK-641.
Occasional, in dry stream beds.
Localities: Ganpatipule, Jat, Khed, Vaibhavwadi.

Local uses: The root bark is applied on paralyzed part of body and massaged everyday for ½ hr. Root bark is also given orally as a diuretic. Fruit juice applied externally on wounds to kill worms in cattles. The root bark is given as a tonic, expectorant, anthelmintic, toothache emmenagogue, analgesic, in rheumatism, paralysis, toothache and tubercular glands.

Capparis zeylanica L. Sp. Pl. ed. 2, 720. 1762; Londhe in Singh *et al.* Fl. Maharashtra State Dicot. 1: 215. 2000. *C. horrida* L. f. Suppl. Pl. 264. 1781; Hook. f. & Thoms. in Hook. f. Fl. Brit. India 1: 178. 1872; Cooke, Fl. Pres. Bombay 1: 47. 1958 (Repr. ed.). GOVIND-PHAL, TORATTI, WAGATI.

Straggling shrubs, 2–5 m tall, with short, hooked, stipular spines. Branchlets densely covered with gray to brown indumentum. Leaves ovate to elliptic, 4–6 × 3–4 cm, emarginate rounded or mucronate at apex, Flowers 2–3 together in a supra axillary row, 2.5–3 cm across. Sepals obtuse, 1–1.5 × 1 cm, concave, densely brown tomentose on the back, ciliate on margins. Petals greenish-white, oblong-oblanceolate, 8–10 × 3–5 mm. Stamens many; filaments white turning purple. Gynophore 3–5 cm long; ovary ellipsoid; fruit subglobose, 3 cm in diam., reddish purple. Seeds globose, 5–6 mm in diameter.

Fls. & Frts.: January–June. *Exsiccata*: SUK-639.
Common, in dry deciduous forests.
Localities: Aundh, Debewadi, Kagal, Kolhapur, Vita.

Local uses: Roots are applied as a poultice on inflammation and also in muscular pains. Plant is used as a tonic, stomachache, cures fever, decoction of root is given in blood troubles.

Crateva adansonii DC. ssp. **odora** (Buch.-Ham.) Jacobs in Blumea 12 (2): 198. 1964; Londhe in Singh *et al.* Fl. Maharashtra State Dicot. 1: 223. 2000. *C. odora* Buch.-Ham. in Trans. Linn. Soc. 15: 188. 1827. *C. religiosa* var. *roxburghii* (R. Br.) Hook. f. & Thoms. in Hook. f. Fl. Brit. India 1: 172. 1872; Cooke, Fl. Pres. Bombay 1: 45. 1958 (Repr. ed.). VAIVARANA, WAIRODA.

Unarmed, deciduous, small trees, 3–5 m tall. Leaves digitate, leaflets 3, elliptic-lanceolate, 3–12.5 × 2–5 cm, acuminate, glabrous. Flowers 2–3.5 cm across, fragrant in terminal corymbs. Sepals ovate, 4.5 × 3mm, acute. Petals white or pale yellow, 2.5–3 cm long, clawed. Stamens many, filaments slender, 3–4 cm long turning purple. Gynophore slightly exceeding the stamens, stout. Ovary ovoid, 2 × 1.5 mm, pilose. Berry globose, smooth, many seeded.

Fls. & Frts.: March–June. *Exsiccata*: SUK-80.

Infrequent, along stream banks.
Localities: Barki, Kolhapur.

Local uses: The stem bark is given in excess obesity with buttermilk everyday for 3 months. One teaspoon of bark powder is given to reduce burning sensation in soles. The bark is stomachic, laxative, vesicant, anthelmintic, expectorant, good in strangury, diseases of the blood and chest, tuberculous glands, causes biliousness.

Crateva magna (Lour.) DC. Prodr. 1: 243. 1824; Londhe in Singh *et al.* Fl. Maharashtra State Dicot. 1: 225. 2000. *Capparis magna* Lour. Fl. Cochinch. 1: 331. 1790. *Crateva religiosa* var. *nurvala* (Buch.-Ham.) Hook. f. & Thoms. in Hook. f. Fl. Brit. India 1: 172. 1872; Cooke, Fl. Pres. Bombay 1: 44. 1958 (Repr. ed.). KATAR LINGAD, VARUN.

Trees, ca 5 m tall. Leaves 3-foliolate; leaflets ovate lanceolate, 8–15 × 5–6 cm, base attenuate, margins entire. Flowers greenish-white in corymbs. Sepals petaloid, small, distant, ovate, acute. Petals 2.5 × 0.8 cm; claw ca 0.6 cm long, very narrow. Stamens longer than the petals, spreading. Gynophore 2 cm long, terete, smooth. Ovary ellipsoid; stigma flat. Fruit globose or ovoid, woody, smooth, or scurfy berry, on the thickened gynophore. Seeds embedded in pulp, nearly smooth brown.

Fls. & Frts.: March–June. *Exsiccata*: SUK-79.
Occasional, along the steams.
Localities: Kolhapur, Malakapur, Shirol.

Local uses: The stem bark is given in obesity with buttermilk everyday for 3 month. It is also given in swelling of feet, and burning sensation in the soles of the feet. In Konkan it is given in rheumatism for both oral and in external applications. The bark is used as stomachic, laxative, anthelmintic, expectorant, good in strangury, diseases of the blood and tuberculous glands.

VIOLACEAE Batsch.

Hybanthus enneaspermus (L.) F. Muell. Fragm. Phyt. Austr. 10: 81. 1876; Godbole & Das Das in Singh *et al.* Fl. Maharashtra State Dicot. 1: 227. 2000. *Viola enneasperma* L. Sp. Pl. 937. 1753. *Ionidium suffruiticosum* (L.) R. & S. Syst. Veg. 5: 394. 1819; Hook. f. & Thoms. in Hook. f. Fl. Brit. India 1: 185. 1872. *I. heterophyllum* Vent. Jard. de la Malm fol. 27. 1803; Cooke, Fl. Pres. Bombay 1: 55. 1958 (Repr. ed.). RATANPARAS.

Erect or diffuse herbs, 10–30 cm tall, pubescent. Leaves sessile, linear-lanceolate, 1.5–5 × 0.2–1.5 cm, entire or obscurely serrate. Stipules subulate, ca 1mm long, gland tipped. Flowers axillary, solitary. Bracts subulate, attached above the middle of the pedicel. Sepals subulate, 2 mm long, green with hyaline margins. Petals deep pink, unequal, the two upper ones oblong,

2–3 mm long, lateral longer, falcate, the lowest much larger with orbicular or obovate limb, long claw and short spur. Stamens 5. Ovary ovoid. Capsules ovate, or ovate-oblong, 4–5 mm long, glabrous, 3-valved. Seeds ovoid, pale yellow, 2 mm long, longitudinally striate.

Fls. & Frts.: August–December. *Exsiccata*: SUK-71.
Frequent, on gravelly slopes of hills and amidst grasses.
Localities: Bahubali, Nesari, Ramling, Wai.

Local uses: Plant juice with coconut oil applied externally in headache. Decoction of plant with *Glycyrrhiza glabra* is given in gonorrhea. Leaf powder is given in bronchitis. Plant is used in urinary calculi, strangury, pain, dysentery, vomiting, burning, urethra discharges, blood troubles, asthma, epileptic fits, cures cough, gives tone to breast, alexeteric.

BIXACEAE Endl.

Bixa orellana L. Sp. Pl. 512. 1753; Hook. f . & Thoms. in Hook. f. Fl. Brit. India 1: 190. 1872; Cooke, Fl. Pres. Bombay 1: 57. 1958 (Repr. ed.); Shirodkar in Singh *et al.* Fl. Maharashtra State Dicot. 1: 231. 2000. SHENDRI.

Small, evergreen trees, 4–6 m tall. Leaves ovate, 7.5–18 × 5–12 cm, subcordate at base, acute or acuminate, glabrous or nearly so on both surface. Flowers in terminal, short panicles, 3–4 cm across. Sepals broadly ovate or orbicular, 8–10 mm across. Petals white or pink, obovate, 1.2–1.5 cm long. Stamens many, anthers dehiscing by apical pores. Capsule ovoid, clothed with long soft prickles, reddish-brown, 2-valved. Seeds red, trigonous.

Fls. & Frts.: August–November. *Exsiccata*: SUK-642.
Cultivated, native of Tropical America.
Localities: Ichalkaranji, Jaisingpur, Kolhapur.

Local uses: The plant is used in headache, leprosy, blood diseases, biliousness, vomiting, allays thirst, good remedy for dysentery and kidney diseases.

COCHLOSPERMACEAE Planch.

Cochlospermum religiosum (L.) Alst. in Trimen. Handb. Fl. Ceylon 6: 14. 1931; Kothari in Singh *et al.* Fl. Maharashtra State Dicot. 1: 231. 2000. *Bombax religiosum* L. Sp. Pl. 552. 1753. *Cochlospermum gossypium* (Cav.) DC. Prodr. 1: 527. 1824; Hook. f. & Thoms. in Hook. f. Fl. Brit. India 1: 190. 1872; Cooke in Fl. Pres. Bombay 1: 56. 1958 (Repr. ed.). GANERI.

Small, deciduous trees, 3–5 m tall. Leaves clustered near the ends of branchlets, 8–17 cm across, palmately 3–5-lobed, glabrous above, white tomentose beneath, lobes entire, acute. Petioles red. Flowers in terminal short, corymbose racemes, 8–12 cm across. Sepals unequal, oblong, reddish-brown, silky. Petals golden yellow, obovate, notched. Stamens many, inserted on a disk. Capsule ob-pyriform, 6–10 × 4–8 cm, reddish-brown, striate. Seeds 4–6 mm long, reniform, covered with white, silky wool.

Fls. & Frts.: February–May. *Exsiccata*: SUK-643.
Common in hilly deciduous forests or cultivated.
Localities: Kolhapur, Majale, Ramling.

Local uses: Gum extracted from the tree is used to soften the skin. The gum is given in eye disease such as inflamed granulation on inner surface of lids. The gum is used in gonorrhea, syphylis, asthma, eye troubles and cough.

FLACOURTIACEAE DC.

Casearia graveolens Dalz. in Hook. J. Bot. & Kew Gard. Misc. 4: 107. 1852; C. B. Cl. in Hook. f. Fl. Brit. India 2: 593. 1879; Cooke, Fl. Pres. Bombay 1: 552. 1958 (Repr. ed.); Jadhav in Singh *et al.* Fl. Maharashtra State Dicot. 1: 233. 2000. BOKHADA, KIRMIRA, PANDHRI-KARAI, RIVIL, SAPTRANGI.

Shrubs or small trees, 5–8 m tall. Leaves broadly elliptic, 9–10 × 3–10 cm, coriaceous, hard, obtuse or acuminate, crenate, serrate, glabrous. Flowers green, numerous in axillary clusters. Calyx 5-lobed; lobes oblong, 1–2 mm long, obtuse, concave, green, pubescent outside, persistent. Petals absent. Stamens 8; filaments glabrous. Stigma large, discoid. Fruit deep yellow, ellipsoid or subglobose, 1–2.5 cm long, 3-valved, yellow. Seeds ca 12, ovoid compressed, with scarlet aril.

Fls. & Frts.: August–January. *Exsiccata*: SUK-304, 364.
Frequent, in moist deciduous forest.
Localities: Amba, Amboli, Gaganbawada.

Local uses: Roots are given in diabetes. Leaves are fried on Ghee and given in stomach complaints and also in urinary complaints.

Flacourtia montana Grah. Cat. Bombay Pl. 10. 1839; Hook. f. & Thoms. in Hook. f. Fl. Brit. India 1: 192 1872; Cooke, Fl. Pres. Bombay 1: 58. 1958 (Repr. ed.); Jadhav in Singh *et al.* Fl. Maharashtra State Dicot. 1: 237. 2000. ATTAK, CHAMPARI, CHAPAR, CHAPRIN, TAR-BOR.

Trees, 5–15 m tall; stem thorny. Leaves ovate or broadly elliptic to oblong elliptic, 6–22 × 3–9 cm, base acute or rounded, margins crenate, tomentose along midrib and nerves beneath. Flowers in densely pubescent, bracteate, fascicled racemes, unisexual. Sepals 4–5, small imbricate. Petals absent. Stamens many, 2 mm long. Disk 2–5-celled. Fruit globose, 6–8 mm in diam., obtusely ribbed, size of a cherry, scarlet when ripe edible.

Fls. & Frts.: November–April. *Exsiccata*: SUK-37.

Frequent, in semi evergreen, moist deciduous forests along Ghats.
Localities: Tillarinagar.

Local uses: The fruits are used to relive nausea. Leaves are used as stomachic and given in diarrhoea and in weakness. Similarly, leaves are chewed in toothache. Fruits helps in digestion, allays thirst, useful in fever, in diarrhoea, piles, weakness of limbs, bleeding gums.

Hydnocarpus pentandra (Buch.-Ham.) Oken. Allg. Naturgesh. 3, (2): 1381. 1841; Jadhav in Singh *et al.* Fl. Maharashtra State Dicot 1: 239. 2000. *Chilmoria pentandra* Buch.-Ham. in Trans. Linn. Soc. London 13: 501. 1822; *Hydnocarpus wightiana* Bl. Rumph. 4: 22. 1848; Hook. f. & Thoms. in Hook. f. Fl. Brit. India 1: 196. 1872; Cooke, Fl. Pres. Bombay 1: 60. 1958 (Repr. ed.). KADU-KAVATH, KANTEL. KASTEL, KETI, KOBASED, KOWTI.

Trees, 15–20 m tall. Leaves oblong or oblong ovate, 8–20 × 3–7 cm, coriaceous, margins serrate, glabrous, base rounded or acute. Flowers solitary or in small racemes. Sepals 5, outer ovate, inner 3 larger, concave. Petals 5, broadly oblong orbicular, rounded, fringed with soft hairs. Male flowers: Stamens 5, filaments hairy, subulate. Female flowers: Stamens 5 as in the male, with pollen. Ovary globose, hairy; stigmas flat, bilobed. Berries 5–7 cm across, globose ovoid, tomentose. Seeds obtusely angular, numerous, yellowish.

Fls. & Frts.: March–August. *Exsiccata*: SUK-695.
Frequent, in deciduous forests.
Localities: Dajipur, Patgaon, Tillari, Udgiri.

Local uses: The seeds are used in several skin diseases, mainly it is used in leprosy. The oil is used in scaby eruptions with on equal portion of *Jatropha curcas* oil, sulphur, camphor and limejuice. Seeds are externally applied in dressing for wounds and ulcers. Seeds are used in skin disease and ophthalmia. The fatty oil from the seeds very closely resembles chaulmoogra oil in both the physical and chemical characters. It is used as an external application in certain cutaneous diseases, and has a special reputation in leprosy.

PITTOSPORACEAE R. Br.

Pittosporum dasycaulon Miq. in Anal. Bot. Ind. 3: 5. 1850; Hook. f. & Thoms. in Hook. f. Fl. Brit. India 1: 199. 1872; Cooke, Fl. Pres. Bombay 1: 62. 1958 (Repr. ed.); Kothari in Singh *et al.* Fl. Maharashtra State Dicot. 1: 243. 2000. GAPSUNDI, IKALI.

Large shrub or small trees, 4–8 m high. Leaves lanceolate, 1.7–8 × 0.8–2–5 cm, acute or acuminate, glabrous or the midrib beneath downy. Flowers in dense terminal corymbs, clothed with yellowish tomentum, fragrant. Sepals ovate, ca 2.5 cm long, very acute, densely pubescent outside, ciliate. Petals yellow, narrow-oblong, 0.6 × 0.8 cm long, capsules compressed, 0.6 cm in diam., densely tomentose, blackish, 2-valved, 5–6 seeded. Seeds obtusely angular, smooth, black.

Fls. & Frts.: November–January. *Exsiccata*: SUK-41
Rare, along Western Ghats.
Localities: Amboli, Phonda, Sawantwadi.

Local uses: Bark possesses narcotic properties. It is used as a febrifuge, in chronic bronchitis, it gave rise to dysenteric diarrhoea and used to cure leprosy, and itch. Oil is alternative tonic and a local stimulant.

Pittosporum wightii A. K. Mukherjee in J. Econ. Taxo. Bot. 5: 50. 1984; Kothari in Singh *et al.* Fl. Maharashtra State Dicot. 1: 244. 2000. *P. floribundum* Wight & Arn. Prodr. 154. Oct. 1834 (non Wight & Arn. ex Royle June, 1834); Hook. f. & Thoms. in Hook. f. Fl. Brit. India 1: 199. 1872 p. p.; Cooke, Fl. Pres. Bombay 1: 61. 1958 (Repr. ed.); p. p. VIKARI, VIKHARA, YEKADIA, YEKDI.

Large shrubs or small trees, 3–5 m tall. Leaves lanceolate or oblong-lanceolate, 7.5–12 × 2.5–6 cm, apex acute or acuminate, margins undulate. Flowers numerous in much branched, terminal, compound corymbs. Sepals ovate, ca 2.5 mm, acute, pubescent. Petals yellow, narrow, oblong, 0.9–1.2 cm long. Ovary hairy in lower part. Style simple, glabrous, stigma 2-lobed. Capsules pisiform, 1–2.5 cm in diam., 2-valved; 6-seeded. Seeds obtusely angular, smooth, black.

Fls. & Frts.: April–September. *Exsiccata*: SUK-644.
Common, along the W. Ghats.
Localities: Amba, Dajipur, Gaganbawada, Radhanagari, Tillari.

Local uses: The bark in combination with castor oil is externally applied in itch. The bark is externally applied over a leech and other insect bite, which causes itching in affected area. Bark in combination with fruits of *Terminalia chebula* and roots of *Salacia brunoniana* is given for external application in piles. Bark possesses narcotic properties. It is used as a febrifuge, in chronic bronchitis, it gave rise to dysenteric diarrhoea, leprosy and itch. Oil is alternative tonic and a local stimulant.

POLYGALACEAE R. Br.

Polygala arvensis Willd. Sp. Pl. 3 (2): 876. 1802; Lakshminarasimhan & Venkanna in Singh *et al.* Fl. Maharashtra State Dicot. 1: 246. 2000. *P. chinensis* auct. non L. 1753; Bennet in Hook. f. Fl. Brit. India 1: 204. 1872; Cooke, Fl. Pres. Bombay 1: 64. 1958 (Repr. ed.). PHUTANI.

Erect or diffuse, annual herbs, 3–25 cm tall, pubescent or glabrous. Leaves variable, orbicular, obovate, oblong or linear oblong, 2–4 × 0.3–2 cm. Flowers in axillary or extra axillary, short, capitate, few-flowered racemes, bracts oblong-ovate, acuminate, broad membranous, ciliate margins. Three outer sepals ovate, 1–1.5 mm long, acute, green margins ciliate. Wings herbaceous, ovate-oblong, acute. Petals yellow, larger one boat-shaped, crested on back, lateral ones oblanceolate, shorter. Stamens included in the larger petal. Capsule didymous, orbicular-oblong, 5 × 4 mm, obliquely obcordate, strongly ciliate. Seeds ovate, 3 mm long, brown hairy, strophiole 3-lobed, hairy at base.

Fls. & Frts.: June–September. *Exsiccata*: SUK-14.

Very common, in grasslands, in cultivated fields, in hill forests on slopes etc.

Localities: Jat, Kagal, Kolhapur, Ramling.

Local uses: Decoction of root is dropped into the nostrils to stop headache. Root paste is applied in muscular pains. Root decoction is given in urinary complaints.

Polygala persicariifolia DC. Prodr. 1: 326. 1824; Bennet. in Hook. f. Fl. Brit. India 1: 202. 1872 (as *persicariaefolia*); Cooke, Fl. Pres. Bombay 1: 63. 1958 (Repr. ed.); Lakshminarasimhan & Vekanna in Singh *et al.* Fl. Maharashtra State Dicot. 1: 249. 2000.

Erect, appressed hairy, annual herbs, 15–45 m tall. Leaves elliptic-lanceolate, 2.5 × 0.5–1.5 cm, mucronate, ciliate along margins. Flowers in terminal and extra axillary racemes, 1–2.5 cm long racemes; bracts minute, subulate, with broad membranous margins. Outer sepals unequal, ovate, 2–3 mm long, wings broadly ovate to orbicular, 5–7 mm across, thin, conspicuously nerved, subpetaloid. Petals pink or rose colored, as long as the wings. Capsules obcordate, 2–3 mm long, strongly ciliate on margins. Seeds oblong-elliptic, black, densely covered with white hairs, strophiole shortly 3-lobed, glabrous, obtuse.

Fls. & Frts.: August–September. *Exsiccata*: SUK-69

Very common, in grasslands, in cultivated fields, in hill forests on slopes etc.

Localities: Kagal, Ramling, Wai.

Local uses: Roots are used as diuretic, root paste is applied on forehead with salt in low fever.

CARYOPHYLLACEAE Juss.

Polycarpaea corymbosa (L.) Lam. Encycl. 2: 129. 1797; Edg. in Hook. f. Fl. Brit. India 1: 245. 1874; Cooke, Fl. Pres. Bombay 1: 70. 1958 (Repr. ed.);

Shirodkar in Singh *et al.* Fl. Maharashtra State Dicot. 1: 253. 2000. *Achyranthes corymbosa* L. Sp. Pl. 205. 1753.

Erect, dichotomously branched, pubescent, annual herb, 10–30 cm tall. Leaves sessile, linear lanceolate, 10–30 × 1–2 mm, mucronate. Flowers small numerous, in dense, terminal, corymbose, dichotomous cymes, bracts scarious or membranous, bifid at apex. Sepals like the bracts, 3–4 mm long. Petals ovate, orbicular, reddish-brown, less than half the length of the sepals, reddish brown. Capsules globose, 1 mm in diam., brown. Seeds obovoid, minute, brown.

Fls. & Frts.: August–November. *Exsiccata*: SUK-171.
Common, on rocky soil.
Localities: Kagal, Ramling, Wai.

Local uses: Leaves are bruised and applied as poultice in inflammatory swellings. Roots are given in urinary calculi. It is mixed with leaves of *Tridax procumbens* and applied externally on wounds. Plant is used in strangury, urinary calculi, boils, inflammatory swellings and ulcers.

PORTULACACEAE Juss.

Portulaca oleracea L. Sp. Pl. 445. 1753; T. Dyer in Hook. f. Fl. Brit. India 1: 246. 1874; Cooke, Fl. Pres. Bombay 1: 72. 1958 (Repr. ed.); Mudhusudan Rao in Singh *et al.* Fl. Maharashtra State Dicot. 1: 261. 2000. BHUIGOLI, KURFAH, MHOTIGHOL.

Succulent, prostrate herbs; stem reddish, 10–40 cm long. Leaves ovate spathulate, 0.5–5 × 3–1.5 cm. Flowers in terminal clusters, subtended by 2–8 involucral leaves; bracteoles ovate-acuminate, 2–4 mm long. Sepals ovate-triangular, 2.5–4 mm long, keeled or slightly winged, united below into a tube. Petals yellow, obovate-oblong, 4–5 mm long, connate at base. Capsule circumscisile, obovoid or ovoid, 6–9 mm long, green with reddish base. Seeds dull black, reniform, concentrically striate.

Fls. & Frts.: Throughout year. *Exsiccata*: SUK-180.
Very common, in moist locations, along gutters, nallas etc.
Localities: Kolhapur, Narsinhmwadi, Shirol.

Local uses: The juice of plant is applied on hands and feet when a burning sensation is felt. Whole plant is orally given in urinary diseases and in disease of kidney. It is given in bleeding gums, cough, urinary discharges. The leaves are used as stomachic, alexipharmic, laxative, reduces small tumors, inflammations, good for ulcers, asthma, urinary discharges, cure diarrhoea, dysentery, leprosy and piles.

Portulaca quadrifida L. Mant. Pl. 1: 73. 1767; T. Dyer in Hook. f. Fl. Brit. India 1: 247. 1874; Cooke, Fl. Pres. Bombay 1: 72. 1958 (Repr. ed.); Madhusudan

Rao in Singh *et al.* Fl. Maharashtra State Dicot. 1: 263. 2000. BHUICHAULI, KHATECHANVAL, RANGHOL.

Prostrate, succulent, glabrous or pilose, annual herbs. Stem often reddish. Leaves ovate-lanceolate, 3–10 × 2–4 mm, acute, nodes with whorl of hairs. Flowers often solitary, terminal, surrounded by 4 or rarely more leaves and long hairs. Sepals oblong, 3–4 mm long, membranous. Petals yellow, obovate, 5–7 × 4 mm. Stamens 8–12. Capsules obovate, 2–3 mm long, obscurely tetragonous, circumscissile, pale or deep red. Seeds dull black, orbicular renifrom, ca 1 mm in diam., concentrically muricate.

Fls. & Frts.: September–February. *Exsiccata*: SUK-696.
Common, on waste land, on rocky soil, in barren fields, black soil etc.
Localities: Ichalkaranji, Sangli, Shirol.

Local uses: Decoction of leaves is given twice a day in dysuria and also as a diuretic. Other uses are similar to the *Portulaca oleracea.*

CLUSIACEAE Lindl. (*nom. alt.*)
[**GUTTIFERAE** (*nom. cons.*)]

Calophyllum inophyllum L. Sp. Pl. 513. 1753; T. And. in Hook. f. Fl. Brit. India 1: 273. 1874; Cooke, Fl. Pres. Bombay 1: 84. 1958 (Repr. ed.); Godbole & Das Das in Singh *et al.* Fl. Maharashtra State Dicot. 1: 274. 2000. SURANGI, UNDI, WUMAKA.

Small or middle-sized trees. Leaves oblong elliptic, 15 × 10 cm, obtuse, or retuse at apex, rounded at base. Flowers 1.9–2.5 cm in diam., pure white, fragrant, in lax few flowered racemes, up to 8–15 cm long. Sepals 4, ovate-orbicular, concave, reflexed, ciliolate. Petals 4, oblong, obtuse, spreading. Stamens numerous, filaments united into 4–6 bundles. Style long, twisted. Stigma large mushroom shaped. Fruit globose, 3 cm across, smooth, yellow on maturity.

Fls. & Frts.: December–January. *Exsiccata*: SUK-131.
Common along sandy shores.
Localities: Ratnagiri, Ganapatpule, Pawas.

Local uses: Seeds are crushed, heated over flame again crushed in mottle and pistil and then paste is applied in rheumatism. Bark is used in dressing ulcers. Gum obtained from the branches is soaked in water for 24 hrs. and used to wash inflamed eyes. Bark is used to heal ulcers, and in inflammation of the eye, lessens appetite, astringent, improves the complexion. Fruit are emetic and purgative. The fixed oil, expressed from the kernels of the seeds, is said to cure scabies.

Garcinia indica (du Petit-Thou.) Choisy in DC. Prodr. 1: 561. 1824; T. And. in Hook. f. Fl. Brit. India 1: 261. 1874; Cooke, Fl. Pres. Bombay 1: 80. 1958

(Repr. ed.); Godbole & Das Das in Singh *et al.* Fl. Maharashtra State Dicot. 1: 276. 2000; *Brindonia indica* Du Petit-Thou. in Dict. Sci. Nat. 5: 339. 1804. AMSOL, BHINDA, BHIRAN, BHIRAND, KATAMBI, KOKAM, KOKAMBI, RATAMBA.

Trees, 10–20 m tall, branches drooping. Leaves oblong-lanceolate ca 10 × 5 cm, red when young, acute or abruptly acuminate at apex, narrowed at base. Flowers axillary or terminal, solitary or fascicled. Male flowers 4–8, outer sepals smaller than inner, stamens numerous, on a short column, anthers 2-celled. Female flowers solitary or 2–3 together, 4–8 rayed. Berries globose, 3 cm across, orange pink when ripe. Seeds 5–8, compressed.

Fls. & Frts.: November–February. *Exsiccata*: SUK-645.
Frequent, in evergreen and deciduous forest along coast.
Localities: Amba, Radhanagari, Tillari.

Local uses: Purple syrup from the juice of the fruit is given twice in a day in acidity, swellings occurred due to the acidity. Oil of the seeds is used in fissures of the lips, hand and on feet, by partly melting it and rubbing on the affected area. Syrup prepared from fruits also given in dysentery. Ripe fruit is used as appetizer, allays thirst, anthelmintic, cardiotonic, useful in bleeding piles, dysentery, tumors, pains and heart diseases.

Garcinia spicata (Wight & Arn.) Hook. f. in J. Linn Soc. 14: 486. 1875; Cooke, Fl. Pres. Bombay 1: 83. 1958 (Repr. ed.); Godbole & Das Das in Singh *et al.* Fl. Maharashtra State Dicot. 1: 276. 2000. *Xanthochymus spicatus* Wight & Arn. Prodr. 1: 102. 1834. *Garcinia ovalifolius* var. *spicata* (Wight & Arn.) T. And. in Hook. f. Fl. Brit. India 1: 269. 1874 non Oliver, 1868. HALDI, JANGLI-RAMPAHL, TAVIR.

Trees, 8–15 m tall, branches angular. Leaves broadly elliptic or orbicular, ca 6–8 × 3–6 cm, coriaceous, apex obtuse or retuse, rarely subacute. Flowers ca 6 cm in diam., in the axils of fallen leaves, the males numerous, sometimes mixed with the females in the same fascicles, but more crowded in a short spiciform racemes, the females in same fascicle, but more commonly crowded in a short spiciform racemes. Male flowers in axils of fallen leaves, sepals 5, orbicular, ciliate. Stamens opposite to petals. Female flowers fascicled. Staminodes in 5 trifid bundles. Ovary ovoid, stigma thick, fleshy, 3-lobed. Fruit oblong, 2.5–3 cm in diam., smooth, dark green, 1–3 seeded.

Fls. & Frts.: March–May. *Exsiccata*: SUK-697.
Rare, in semi evergreen forests.
Localities: Amba, Ambaghat, Ajra, Chandgad, Ratnagiri, Sawantwadi.

Local uses: Purple syrup from the juice of the fruit is given twice in a day in acidity, swellings occurred due to the acidity. Ripe fruit is used as appetizer.

Garcinia xanthochymus Hook. f. ex T. And. in Hook. f. Fl. Brit. India 1: 269. 1874; Cooke, Fl. Pres. Bombay 1: 82. 1958 (Repr. ed.); Godbole & Das Das in Singh *et al.* Fl. Maharashtra State Dicot. 1: 277. 2000. DHARAMBO, JHRAMBI, OTA.

Trees, up to 10 m high, branchlets angular. Leaves oblong lanceolate, ca 16–36 × 3–12, acute or acuminate, smooth, shining. Male flowers fascicled on old wood, ca 2 cm long, sepals 5, ciliate, fleshy, orbicular, concave, persistent; petals 5, orbicular; stamens many in 5 bundles. Hermaphrodite flowers similar to male. Ovary ovoid, pointed; stigma 5-lobed, oblong, spreading. Fruit large, of a middle sized apple, smooth, deep yellow when ripe. Seeds 1–4, oblong.

Fls. & Frts.: February–May. *Exsiccata*: SUK-698.
Occasional, near streams in semi evergreen to wet deciduous forests.
Localities: Gaganbawada, Radhanagari, Ramghat.

Local uses: Similar to *Garcinia indica.* The ripe fruit is used as a tonic, invigorating, alexipharmic, good in heart complaints, and biliousness. Fruits are given to reduce excess body weight.

Mammea suriga (Buch.-Ham. ex Roxb.) Kosterm. in Comm. For. Res. Inst. Indonesia Bogor. 72: 23, f. 19. 1961; Godbole & Das Das in Singh *et al.* Fl. Maharashtra State Dicot. 1: 279. 2000. *Calophyllum suriga* Buch.-Ham. ex Roxb. Fl. India 2: 608. 1832. T. And. in Hook. f. Fl. Brit. India 1: 276. 1874. *Ochrocarpus longifolius* Benth. & Hook. f. ex T. And. in Hook. f. op. cit. 270; Cooke, Fl. Pres. Bombay 1: 84. 1958 (Repr. ed.). NAGKESAR, PUNNAG, SURANG, SURINGI, TAMBRA.

Trees, 10–20 m tall. Leaves 18–24 × 6–6.8 cm, oblong, obtuse, coriaceous, subcordate at base. Flowers in fascicles in axils of fallen leaves. Calyx bursting into 2 valves, inflexed during flowering. Petals white, ovate-oblong, acute, thin. Stamens many, sterile in the female flowers. Style short, stout, stigma broad, peltate. Berries obliquely ovoid, ca 3.5 cm long, one seeded.

Fls. & Frts.: February–March. *Exsiccata*: SUK-699.
Frequent, in evergreen Ghats forests.
Localities: Amba, Amboli, Gaganbawada, Vaibhavwadi.

Local uses: The flowers are used as stomachic, aphrodisiac, in blood disease and leprosy. It is sometimes substituted under the name of *Mesua ferrea.*

Mesua ferrea L. Sp. Pl. 515. 1753; T. And. in Hook. f. Fl. Brit. India 1: 277. 1874; Cooke, Fl. Pres. Bombay 1: 86. 1958 (Repr. ed.); Godbole & Das Das in Singh *et al.* Fl. Maharashtra State Dicot. 1: 279. 2000. NAGCHAMPA, NAGCHPAHA, THORA-CHMPA.

Trees, 5--20 m tall. Leaves variable, elliptic-oblonglanceolate or lanceolate, apex acuminate or cuspidate, base acute, pale green above, glaucous beneath. Flowers white, 8–10 cm across, fragrant, axillary or terminal solitary or in pairs, subsessile. Sepals orbicular, densely pilose. Petals 4, obovate or obcordate, stamens numerous, yellow, forming globose mass. Style as long as stamens. Stigma peltate. Drupe woody, ovoid. 2.5–3 cm long, with a conical point, surrounded by enlarged sepals. Pericarp bivalved. Seeds 1–4, smooth, dark brown testa.

Fls. & Frts.: March–May. *Exsiccata*: SUK-30
Frequent, in semi evergreen forests.
Localities: Amboli, Gaganbawada, Patgaon, Sawantwadi.

Local uses: Paste of flower is applied in burning of the feet. Bark in combination with sugar is given to stimulate sweats and decoction of flower is given to reduce the sweat. Flower buds are given in dysentery. Paste of flower is used in bleeding piles, it is also used to stop bleeding from wounds. It is used in fevers, biliousness, foul breath, scabies, skin eruptions, itching, small tumors, headache, blood and heart trouble, sore throat, cough, hiccough, vomiting, thirst, dysentery and bleeding piles.

DIPTEROCARPACEAE Gaertn.

Vateria indica L. Sp. Pl. 515. 1753; T. Dyer in Hook. f. Fl. Brit. India 1: 313. 1874; Cooke, Fl. Pres. Bombay 1: 92. 1958 (Repr. ed.); Lakshminarasimhan & Venkanna in Singh *et al.* Fl. Maharashtra State Dicot. 1: 286. 2000. DHUP.

Large trees, 20–30 m high, bark white, branches clothed with stellate hairs. Leaves elliptic-oblong, ca 15–27 × 12–15 cm, shortly pointed or obtuse, cordate or rounded at the base, glabrous. Flowers remote, in large terminal panicles, 15–20 cm long. Calyx divided nearly to the base, lobes oblong-lanceolate, hoary puberulous on both surfaces, not enlarged in fruit, deflexed. Petals obtuse, white. Stamens numerous, anther connective adnate to apex, produced into subulate point; ovary pubescent, stigma small. Capsule oblong, 2–4 cm long, fleshy, 3-valved, valves 1-seeded.

Fls. & Frts.: April–May.
Rare, in semi evergreen forest.
Localities: Tillari.

Local uses: The resin is applied in hemicrania and tuberculous glands, boils. Smoke is given to the pregnant women and also in rheumatism. The bark is used as alexipharmic, cures cough, anemia, ear diseases, urinary discharges, skin eruptions, ulcers and wounds, useful in dysentery, leprosy and itch.

ANCISTROCLADACEAE Wall. (*nom. cons.*)

Ancistrocladus heyneanus Wall. ex Grah. Cat. Pl. Bombay 28. 1839; T. Dyer in Hook. f. Brit. India 1: 299. 1874; Cooke, Fl. Pres. Bombay 1: 93. 1958 (Repr. ed.); Venkanna & Prasanna in Singh *et al.* Fl. Maharashtra State Dicot. 1: 287. 2000. KARDAL, KAROTAL.

Large climbing shrubs, 6–8 m long, with crooked branches. Leaves elliptic-lanceolate or oblanceolate, ca 6–15 × 2–8 cm, Coriaceous crowed at ends of branches. Flowers creamy yellow, in lax terminal panicles. Calyx, lobes obovate cuneate, enlarged in fruit. Petals small, white, ovate oblong. Stamens 10, alternately shorter. Fruit small, surrounded by 5 wings.

Fls. & Frts.: January–May. *Exsiccata*: SUK-637.
Localities: Amba, Amboli, Ramghat, Tillari.

Local uses: Plant is used to treat AIDS patients.

MALVACEAE Juss.

Abelmoschus moschatus Medik. Malv. 46. 1781, Ampl. Hochr. in Candollea 2: 86. 1924; Venkanna & Das Das in Singh *et al.* Fl. Maharashtra State Dicot. 1: 293. 2000. *Hibiscus abelmoschus* L. Sp. Pl. 696. 1753; Mast. in Hook. f. Fl. Brit. India 1: 342. 1874; Cooke, Fl. Pres. Bombay 1: 118. 1958 (Repr. ed.). KASTURI-BHENDI.

Erect, tomentose herbs, 1–2.5 m tall. Leaves ovate or orbicular, 3–8 × 0.5–4.5 cm, 3–5 lobed, lobes triangular-acute, serrate. Flowers axillary, solitary, passing upwards into a short raceme, 5–8 cm across. Pedicels 0.5–1 cm long, hispid, bract filiform, ca 1 cm long, hispid, ciliate calyx slightly longer than the bracteoles, densely hispid on the back. Petals yellow with dark purple base, oblanceolate, 3–5 cm long. Capsules oval, 1–1.5 × 1–1.2 cm, glabrous or villous, acuminate. Seeds reniform, grayish-brown, sparsely golden, muricate, fragrant when bruised.

Fls. & Frts.: September–December. *Exsiccata*: SUK-363.
Cultivated for its fragrant seeds and occasionally escaped.
Localities: Kolhapur.

Local uses: Decoction of seeds is useful in nervous debility, hysteria and other nervous disorders. Juice of fresh plant is used as poultice and applied to chest in bronchitis. Mucilage made from root and leaves given twice a day in urinary discharges. Seeds are used in perfumery and as a substitute for musk.

Abutilon indicum (L.) Sweet, Hort. Brit. 54. 1826; Mast. in Hook. f. Fl. Brit. India 1: 326. 1874; Cooke, Fl. Pres. Bombay 1: 102. 1958 (Repr. ed.); Venkanna & Das Das in Singh *et al.* Fl. Maharashtra State Dicot. 1: 297. 2000; *Sida* *indica* L. in Torner, Cent. Pl. 2: 26. 1756. & Amoen. Acad. 4: 324. 1759. MUDRA, PETARI.

Erect, hoary tomentose, branched undershrubs, 1–2 m tall. Leaves broadly ovate, 3.5–11 × 3–10 cm, cordate, acuminate, toothed, rarely subtrilobed; stipules linear, 1 cm long, deflexed. Flowers 1.5–3.5 cm across, solitary, axillary. Sepals united below the middle, lobes ovate, acuminate, 4–6 mm long. Petals orange yellow without purple base, obovate, 1–3 cm long, toothed at apex. Carpels 15–20, with a distinct acute point, 8–13 × 3.5–9 mm, hairy but ultimately glabrous, shining brown. Seeds reniform, 3 mm long and as much broad, brown, dotted with minute scales, with a tuft of hairs near the hilum.

Fls. & Frts.: September–January. *Exsiccata*: SUK-12.
Common, on wasteland around fields.
Localities: Kolhapur, Sangli and Satara.

Local uses: Leaf paste is applied externally on ulcers and wounds of human and cattle. Leaves are used as a gargle to stop gumache and toothache. Root decoction is given as a diuretic. Leaves are eaten in bleeding piles. Leaf decoction is given in bronchitis, gonorrhea and in cough. Leaves are tied around the neck or hand in mental disorder. Bark is used as febrifuge, anthelmintic, alextetric, allays thirst, vomiting, lessens perspiration. Root cures uterine haemorrhagic discharges.

Hibiscus rosa-sinensis L. Sp. Pl. 694. 1753; Mast. in Hook. f. Fl. Brit. India 1: 344. 1874; Cooke, Fl. Pres. Bombay 1: 120. 1958 (Repr. ed.); Venkanna & Das Das in Singh *et al.* Fl. Maharashtra State Dicot. 1: 340. 2000. JASWAND.

Evergreen, nearly glabrous shrubs, 3–4 m tall. Leaves ovate, 6–10 × 3.5–7 cm, irregularly serrate dentate, shining green above, stipules lanceolate-subulate, 1–1.5 cm long, glabrous. Flowers 10–15 cm across, axillary, solitary, bracteoles 5–7, linear, lanceolate, 1–1.5 cm long. Calyx 2–3 cm long, divided almost to the middle, lobes lanceolate. Petals red, oblanceolate, 5–10 cm, long, entire or dentate at apex. Staminal tube exserted beyond the petals.

Fls: Throughout year, less in winter. *Exsiccata*: SUK-344.
Cultivated.

Local uses: Fresh root juice of the wild flower variety is given for gonorrhoea and the powdered root for menorrhoea. Flower juice is given as a demulcent and refrigerant drink in fevers. Leaves and stem bark is used for abortion. Staminal column is soaked in water for 24 hours and water is given in kidney trouble. Decoction of flower powder is given in bronchial catarrh. Flowers are fried in Ghee and given twice a day with betel leaves in menorrhagia. Flowers have been found to be effective in treating arterial hypertension and have significant antifertility effect. Flower powder is externally applied to stop graying of hair.

Kydia calycina Roxb. Pl. Cor. 3: 12, t. 215. 1811; Mast. in Hook. f. Fl. Brit. India 1: 348. 1874; Cooke, Fl. Pres. Bombay 1: 100. 1958 (Repr. ed.); Venkanna & Das Das in Singh *et al.* Fl. Maharashtra State Dicot. 1: 317. 2000. BHOTI, VARAS, WARUNG.

Moderate sized trees, 5–10 m tall, young parts densely white tomentose. Leaves ovate-orbicular, 7.5–15 × 6–19 cm, cordate at base, 5–7-lobed, lobes angular, green and stellate pubescent on the upper surface, densely white tomentose beneath. Flowers Polygamous, axillary or in terminal panicles. Bracteoles 4–6, 6–8 mm long, accrescent in fruit. Calyx 2–3 mm long, densely villous, lobes ovate-acute. Corolla white or pinkish, petals clawed, 4–5 mm long, emarginate. Style branches 3, stigmas peltate. Capsules 3-valved, globose, 2–4 mm in diam., covered with mealy tomentum. Seeds reniform, striate, dark brown.

Fls. & Frts.: August–October. *Exsiccata*: SUK-4

Common in moist deciduous forests.

Localities: Gaganbawada, Radhanagari, Amboli, Sawantwadi, Vaibhavwadi.

Local uses: Bark decoction is given in uterine complaints. Paste of flowers is externally applied on boils and ulcers.

Malachra capitata (L.). L. Syst. Nat. ed. 12, 2: 458. 1767; Mast. in Hook. f. Fl. Brit. Ind. 1: 329. 1874; Cooke, Fl. Pres. Bombay 1: 105. 1958 (Repr. ed.); Venkanna & Das Das in Singh *et al.* Fl. Maharashtra State Dicot. 1: 317. 2000. *Sida capitata* L. Sp. Pl. 685. 1753. RANAMBADI, RANBHENDI.

Erect, woody, hispid herbs, up to 1 m tall. Leaves orbicular or broadly ovate, 3–14.5 × 3.5–16.5 cm, cordate, crenate, obtuse, glabrous or glabrescent stipules linear or 2–3 branched, 1–2 cm long. Flowers subsessile, 4–6 together within a leafy involucre, pedicels with 4 leaves involucre at apex. Involucral leaves hispid, ovate-cordate, 8–10 mm long, on short stalk below the blade, with a white spot at base and with two opposite, ciliate appendages. Calyx membranous, lobes ovate, acute, the midrib produced into long bristly point. Petals yellow, obovate, 1.5–2 cm long. Ovary 5-locular; locules one ovuled, styles 10; stigmas capitate. Ripe carpels rounded on the back, wedge shaped on the inner side, white with brown reticulate veins. Seeds smooth, brown, black.

Fls. & Frts.: September–December. *Exsiccata*: SUK-647.

Occasional, weed in waste places.

Localities: Karad, Ratnagiri.

Local uses: Leaves are used in gonorrhoea. Half tablespoon of leaf powder is given after every meal to the diabetic patients. Seeds are used as a demulcent in cough, and ulcers in the bladder. It is given as a nervine tonic. Leaf paste is wrapped with leaves of betel. After warming paste is removed from the betel leaves and then applied externally as poultice to wounds and swellings.

Malvastrum coromandelianum (L.) Garcke in Boplandia 5: 295. 1857; Venkanna & Das Das in Singh *et al.* Fl. Maharashtra State Dicot. 1: 320. 2000. *Malva coromandeliana* L. Sp. Pl. 687. 1753. *Malvastrum tricuspidatum* (R. Br.) A Gray. Pl. Wright 1: 16. 1852; Mast. in Hook. f. Fl. Brit. Ind. 1: 321. 1874.

Erect or procumbent, woody, perennial herbs. Stems 15–50 cm long, stellate hairy. Leaves ovate-lanceolate, 2–10 × 1.5–5 cm, subcordate, truncate or cuneate at base, crenate, serrate, stipules lanceolate, 2–3 mm long. Flowers axillary, solitary or clustered, 1–1.5 cm across. Calyx campanulate, 8–10 mm long, lobes triangular, acute, hispid. Petals yellow or orange-yellow, obliquely obcordate, longer than the calyx. Fruits depressed globose, carpels 5–12, rugose, 4 awned at the top. Seeds black.

Fls. & Frts.: Throughout year. *Exsiccata*: SUK-648

A native of S. America, a naturalized along roadsides, gardens and in barren fields.

Localities: Shivaji University campus, Kolhapur.

Local uses: Daily consumption of 2–3 leaves twice a day is prescribed to diabetic patients. Leaves are applied to inflamed sores and wounds as a cooling and healing salve or as ointment. Flowers are given as a pectoral and diaphoretic.

Sida acuta Burm. f. Fl. Ind. 147. 1768; Cooke, Fl. Pres. Bombay 1: 98. 1958 (Repr. ed.); Venkanna & Das Das in Singh *et al.* Fl. Maharashtra State Dicot. 1: 325. 2000. *S. carpinifolia* Mast. in Hook. f. Fl. Brit. India 1: 323. 1874 non L. f. 1781. BALA, JANGLI-METHI, LANAN-CHIKANA, TUPKADI.

Erect, much-branched, woody annuals, 0.3–1 m tall. Leaves lanceolate, 2–10 × 0.6–3 cm, subcordate or rounded at base, serrate, acute, glabrescent on both surface. Stipules linear, lanceolate, longer than the petioles. Flowers 10–12 mm across. Calyx 6–7 mm long; lobes ovate-triangular. Corolla yellow; petals obovate-cuneate, longer than the sepals, obliquely truncate at apex. Fruits 4–5 mm across, carpels 5–9, reticulate-rugose, 2-awned, awns about half as long as the body of the carpels. Seeds 2 × 1 mm, brown, obtusely triangular.

Fls. & Frts.: September–January. *Exsiccata*: SUK-18.

Common, weed of gardens and barren fields.

Localities: Kolhapur, Sangli, Satara, Karad.

Local uses: Root powder is used with milk in gonorrhea. Leaf juice is given in chronic abdominal pain and in colic. Leaves are used with seeds of sesamum and paste is prepared, warmed on flame and applied on boils. Root is boiled with sesamum oil and named as a 'Balá Tel' and applied externally on weak patients. Leaves are made warm and moistened with gingelli oil and used to hasten suppuration. They are also used as abortifacient.

Sida cordata (Burm. f.) Borss. in Blumea 14: 182. 1966; Venkanna & Das Das in Singh *et al.* Fl. Maharashtra State Dicot. 1: 325. 2000. *Melochia cordata* Burm. f. Fl. Ind. 143. 1768. *Sida humilis* Cav. Diss. 5: 277. t. 134, f. 2. 1788; Mast. in Hook. f. Fl. Brit. Ind. 1: 322. 1874. *S. humilis* var. *veronicifolia* (Lam.) Mast. in Hook. f. op. cit. 322. *S. veronicifolia* Lam. Encyl. 1: 5. 1783; Cooke, Fl. Pres. Bombay 1: 97. 1958 (Repr. ed.). BHUIBALA.

Much branched, prostrate or trailing, perennial, woody herbs; branches hairy, 15–60 cm long. Leaves broadly ovate-cordate, 1.5–3 × 1–2.5 cm, acute or acuminate, serrate, sparsely stellate hairy; stipules filiform, 3–5 mm long. Flowers 1–1.5 cm across, 1–2 in leaf axils. Calyx 5–6 mm long, divided half way down from the top, lobes triangular-ovate, acute. Corolla yellow, slightly exceeding the calyx, petals obovate. Fruits 5–6 mm across, enclosed in persistent calyx, carpels often 5, smooth, awns 2, short. Seeds brown glabrous.

Fls. & Frts.: July–January. *Exsiccata*: SUK-19.

Common, on rocky soil under shade of trees.

Localities: Gaganbawada, Ichalkaranji, Kagal, Radhanagari, Shirol, Tasgaon, Wai.

Local uses: Leaf juice is given in dysentery during pregnancy. Flowers and young fruits are given in dysuria. Leaf paste is applied externally on wounds.

Sida cordifolia L. Sp. Pl. 684. 1753; Mast. in Hook. f. Fl. Brit. Ind. 1: 324. 1874; Cooke, Fl. Pres. Bombay 1: 99. 1958 (Repr. ed.); Venkanna & Das Das in Singh *et al.* Fl. Maharashtra State Dicot. 1: 326. 2000. BALA.

Erect, grey-tomentose undershrubs, 10–25 cm tall. Leaves ovate 0.1–5 × 0.5–4 cm, cordate, crenate, acute, softly hairy above, densely tomentose beneath. Flowers axillary, solitary or clustered at the ends of branches. Calyx divided half way down from the top, lobes ovate, acute. Corolla pale yellow; petals obovate. Fruits carpels 7–10, strongly reticulated, ciliate on the upper margins, awns 2, long retrosely scabrid-hairy. Seeds brown black.

Fls. & Frts.: August–November. *Exsiccata*: SUK-8.

Frequent, in open Situations.

Localities: Nesari.

Local uses: Root bark with sesamum oil and milk is effective in curing facial paralysis and sciatica. Powder of root given with milk and sugar for relief of frequent, micturition and leucorrhoea. Paste of leaves is used externally on wounds and itch.

Sida rhombifolia L. ssp. **rhombifolia** Sp. Pl. 684. 1753 emend. Mast. in Hook. f. Fl. Brit. Ind. 1: 323. 1874; Borss. in Blumea 14: 193. 1966; Venkanna & Das Das in Singh *et al.* Fl. Maharashtra State Dicot. 1: 327. 2000. *S. rhombifolia* var. *rhomboidea* (Roxb. ex Fleming) Mast. in Hook. f. op. cit. 324. *S. rhombifolia* var. *microphylla* (Cav.) Mast. in Hook. f. op. cit. 324. BALA.

Erect, much branched, stellate-hairy undershrubs, 30–90 cm tall, young parts gray tomentose. Leaves rhomboid, lanceolate, obovate or orbicular, 2.5–5 × 1.2–4.5 cm, crenate-serrate, acute. Flowers axillary solitary or 1–2 together, 1–1.5 cm across. Calyx angular, 4–6 mm long, lobes triangular, acute. Corolla yellow, 10–12 mm long; petals obovate. Fruits, 5–6 mm across; carpels 5–10, reticulated, with two short awns. Seeds smooth, black, obtusely trigonous, 2 mm long.

Fls. & Frts.: September–December. *Exsiccata*: SUK-19, 191.
Common, on wasteland along roadsides and in forest clearings.
Localities: Gaganbawada, Karad, Radhanagari.

Local uses: Root paste is given orally and applied externally in rheumatism. Leaf paste is applied externally on wounds. Whole plant decoction is used to wash wounds. Bark is used with milk is given in gonorrhea. Stem and leaf is used externally as a bandage in bone fracture and also given orally in similar case.

Sida spinosa L. Sp. Pl. 683. 1753; Mast. in Hook. f. Fl. Brit. Ind. 1: 323. 1874; Cooke, Fl. Pres. Bombay 1: 98. 1958 (Repr. ed.); Venkanna & Das Das in Singh *et al.* Fl. Maharashtra State Dicot. 1: 328. 2000. KATEBALA.

Erect, gray tomentose, woody herbs, 30–90 cm tall. Leaves ovate to obovate-oblong, 1–5 × 0.5–4 cm, cordate at base, crenate-serrate, grayish-green above, dense tomentose beneath, petioles with 1–3 small recurved spines near the base. Flowers 1–1.5 cm across, solitary or paired in leaf axils. Calyx gray tomentose, campanulate, 5–8 mm long, lobes triangular-ovate. Corolla white, slightly exceeding the calyx, petals obovate. Fruits 6–8 mm across, enclosed in the persistent calyx carpels often 5–2 awned, awned about half as long as the body of the carpels. Seeds smooth, black-brown.

Fls. & Frts.: September–January. *Exsiccata*: SUK-17.
Common weed of waste lands.
Localities: Shivaji University campus, Kolhapur.

Local uses: Leaves are given orally in gonorrhoea, gleet and sculding urine. Decoction of root is given as a tonic. The root, leaf and fruit is used as a tonic in wasting disease, cures ulcers and biliousness, useful in urinary discharges, scalding urine and skin infections.

Thespesia populnea (L.) Soland. ex Corr. Ann. Mus. Paris 9: 290, t. 8, f. 1. 1807; Mast in Hook. f. Fl. Brit. Ind. 1: 345. 1874; Cooke, Fl. Pres. Bombay 1: 121. 1958 (Repr. ed.); Venkanna & Das Das in Singh *et al.* Fl. Maharashtra State Dicot. 1: 330. 2000. *Hibiscus populneus* L. Sp. Pl. 694. 1753. BHEND-KAZED, PARAS-PIMPAL, PARAVHAJHADA.

Large, evergreen trees, 5–6 m tall. Leaves broadly ovate, 5–15 × 4–16 cm, cordate at base, entire, acuminate. Flowers axillary, solitary, 5–6 cm across. Calyx cupular, 1–1.5 cm long, slightly enlarging in fruit, truncate. Corolla yellow with pink purple center, petals oblanceolate. Capsule pyriform, depressed at the top, 2–3 cm in diam. Seeds ovoid, black pubescent.

Fls. & Frts.: September–January. *Exsiccata*: SUK-118
Occasionally, planted along roadsides.
Localities: Kolhapur.

Local uses: Fruits, leaves and roots soaked in turmeric water whole night then again soaked in cows urine for 12 hrs and then washed thoroughly with water and applied externally in scabies, psoriasis and other skin diseases. Root decoction is used as tonic. Poultice of leaves is used in painful joints. Seeds are given to the animals as aphrodisiac.

BOMBACACEAE Kunth.

Adansonia digitata L. Sp. Pl. 1190. 1753; Mast. in Hook. f. Fl. Brit. India 1: 348. 1874; Cooke, Fl. Pres. Bombay 1: 126. 1958 (Repr. ed.); Moorthy in Singh *et al.* Fl. Maharashtra State Dicot. 1: 345. 2000. GORAKH-AMLA, GORAKH-CHINCH.

Huge, deciduous trees, 10–20 m tall. Leaves digitate, leaflets 3–9, obovate-oblong, 4–12 cm long, acute. Flowers 10–12 cm across, axillary, solitary; pedicels stout, 12–20 cm long, softly hairy. Calyx soft tomentose outside, white silky-villous within, divided below the middle; lobes oblong-lanceolate. Petals white, obovate. Ovary densely villous, style 8–10 cm long, villous at base. Fruits oblong, 20–25 cm long, pale greenish brown, densely tomentose, 5-locular; seeds reniform embedded in brownish pulp.

Fls. & Frts.: May–July. *Exsiccata*: SUK-66.
Native of Africa, cultivated in gardens.
Localities: Kagal, Kolhapur, Panhala, Shirol.

Local uses: Pulp mixed with buttermilk is used as an astringent in diarrhoea and dysentery. Pulp mixed with sugar is given to reduce heat from the body and quenching thirst. Bark is given as a febrifuge. Fruit is used as antipyretic, useful in biliousness, vomiting, dysentery and excessive perspiration.

Bombax ceiba L. Sp. Pl. 511. 1753; Moorthy in Singh *et al.* Fl. Maharashtra State Dicot. 1: 343. 2000. *Bombax malabaricum* DC. Prodr. 1: 479. 1824; Mast in Hook. f. Fl. Brit. India 1: 349. 1874; Cooke, Fl. Pres. Bombay 1: 127. 1958 (Repr. ed.). KATESAVERI, SAVAR.

Deciduous trees, 10–30 m tall, prickles conical, black. Leaves crowded at the ends of branches, digitate, leaflets 5–7, ovate, lanceolate or elliptic-lanceolate, 5–23 × 1.5–9 cm, glabrous. Flowers sessile, crowded at the ends of branches. Calyx 3-lobed, lobes rounded, 1.5–2 cm long, glabrous outside, densely silky within. Corolla bright red tomentose outside, petals elliptic oblong, 5–7 cm long. Stamens about half as long as the corolla; filaments flattened. Staminal column divided into 5 bundles.

Localities: Hatkanangle, Ichalkaranji, Kolhapur, Karad.

Local uses: The roots are given as diuretic, in menorrhagia, useful in gonorrhoea and painful micturition. Cures diseases of the liver, spleen, blood, removes tumours, fat pain, alexiteric. Roots of one-year plant is sold under the name of "saphed musali" in markets (Naik 1998).

Ceiba pentandra (L.) Gaertn. Fruit. 2: 244, t. 133. 1791; Moorthy in Singh *et al.* Fl. Maharashtra State Dicot. 1: 345. 2000. *Bombax pentandrum* L. Sp. Pl. 511. 1753. *Eriodendron anfractuosum* DC. Prodr. 1: 479. 1824; Mast in Hook. f. Fl. Brit. India 1: 350. 1874; Cooke, Fl. Pres. Bombay 1: 128. 1958 (Repr. ed.). PANDHRI-SAVAR, SAFED SAVAR.

Tall trees 15–20 m tall. Leaves digitate, leaflets 5–9, lanceolate, 5–12 cm long, cuspidate, green above, glaucous beneath. Flowers 3–5 cm across, clustered in the leaf axils. Calyx 1.2–2 cm long, 5-lobed, glabrous outside, hairy within, lobes triangular-ovate. Corolla dull white or pinkish, petals obovate-oblong, 3.5–4 cm long, woolly outside. Anthers sinuous. Ovary conical, glabrous. Capsule fusiform, 3.5–4 cm long, hairy. Seeds pyriform, 5–6 mm long, black, embedded in silky wool. Seeds ovoid, 10–12 mm long, embedded in white silky cotton, inner three border, margins ciliate, enlarged in fruit.

Fls. & Frts.: March–June. *Exsiccata*: SUK-646.
Common, in valleys, on hills slopes and cultivated in gardens.
Localities: In all districts.

Local uses: Roots are commonly given in menorrhoea, leucorrhoea. Roots are used in dressing the wounds. Bark is similarly used. The gum or dried juice called as '*Mocharas*' is used as an aphrodisiac, in diarrhoea, dysentery and menorrhoea.

STERCULIACEAE Bartl.

Guazuma ulmifolia Lamk. Encycl. 3: 52. 1789; Moorthy in Singh *et al.* Fl. Maharashtra State Dicot. 1: 365. 2000. *Theobroma guazuma* L. Sp. Pl. 782. 1753. *Guazuma tomentosa* H. B. & K. Nov. Gen. Pl. Sp. 5: 320. 1826; Mast. in Hook. f. Fl. Brit. India 1: 375. 1874; Cooke, Fl. Pres. Bombay 1: 144. 1958 (Repr. ed.). RUDRAKSH.

Medium sized trees, 5–10 m tall, young parts tomentose. Leaves ovate, lanceolate, 8–20 × 4–7 cm, slightly oblique, base cordate, serrate acuminate, acute or obtuse, glabrescent above, hoary tomentose beneath. Flowers in axillary and terminal short cymes. Sepals distinct, ovate-oblong, 2–3 mm long, tomentose on the back. Petals pale yellow, hooded, clawed, longer than the sepals, with two long threadlike appendages at the bifid apex. Staminal columns with 5 lanceolate staminodes alternating with 5 groups, fertile anther 3. Ovary 5-lobed. Fruit globose, woody, tubercled, black. Seeds many, ovoid, 1 mm across, gray.

Fls. & Frts.: March–August. *Exsiccata*: SUK-638.
Planted in gardens and along roadsides.
Localities: Shivaji University campus, Kolhapur.

Local uses: The fruit is emollient and astringent. Bark is used as tonic and demulcent. It is externally applied in elephantiasis. Paste of fruit is given with paste of ginger twice a day in dysentery. Paste of fruit is applied externally on skin diseases.

Helicteres isora L. Sp. Pl. 963. 1753; Mast. in Hook. f. Fl. Brit. India 1: 365. 1874; Cooke, Fl. Pres. Bombay 1: 136. 1958 (Repr. ed.); Moorhty in Singh *et al.* Fl. Maharashtra State Dicot. 1: 355. 2000. ATI, KEWAN, MURUDSHEG.

Erect shrubs, 1–2.5 m tall, young parts stellate hairy. Leaves bifarious, subsessile, broadly ovate-oblong or orbicular, 8–13 × 5–10 cm, irregularly crenate-serrate pubescent on both surfaces, lobed. Flowers in axillary few flowered cymes. Calyx tubular, 1.5–2 cm long, 2-lobed, stellate pubescent outside, lobes triangular, unequal. Petals red, turning bluish-black, very unequal, closely reflexed on the calyx. Staminal column fused with the gynophore, much exserted, anther 10. Ovary conical on a curved 2–3 cm long gynophore. Follicles 4–6 cm long, linear twisted together, stellate tomentose. Seeds many angular black.

Fls. & Frts.: September–January. *Exsiccata*: SUK-117.
Common, on hill slopes.
Localities: Gaganbawada, Radhanagari.

Local uses: Fruits are used as antidysenteric and also given in colic. Paste of fruit is applied on itch in children. Juice of the root is given in diabetes. Root bark powder is used with seed powder of *Trigonella foenu-graceum* given in diabetes. The root and bark is reported to be expectorant, demulcent, astringent to the bowels, antigalactagogue and cures scabies.

Pterospermum acerifolium (L.) Willd. Sp. Pl. 3: 729. 1800; Mast. in Hook. f. Fl. Brit. India 1: 368. 1874; Cooke, Fl. Pres. Bombay 1: 137. 1958 (Repr. ed.); Moorthy in Singh *et al.* Fl. Maharashtra State Dicot. 1: 371. 2000. KANAKCHAMPA.

Evergreen trees up to 10 m tall, young parts and inflorescence densely rusty, stellate tomentose. Leaves somewhat obliquely oblong or obovate, 20–40 × 15–30 cm, deeply cordate or peltate, remotely toothed, glabrous and deep green above, gray to brownish tomentose beneath. Flowers solitary or in 2–3-flowered cymes, ca 15 cm long, fragrant. Sepals distinct, 10–15 × 1–1.5 cm, linear fleshy brown tomentose out side, tufted villous along the middle. Petals white, oblanceolate, shorter than the sepals. Staminal column with 5 staminodes alternating with 5 groups of 2–3 stamens. Ovary oblong, brown, hairy. Stigma club shaped.

Fls. & Frts.: March–April. *Exsiccata*: SUK-393.
Native of Eastern India. Planted in garden.
Localities: Kolhapur.

Local uses: Flower is used as tonic, laxative, anthelmintic, inflammation, blood troubles, abdominal pain, ascites, cures ulcers, leprosy, urinary discharges and tumors.

Sterculia foetida L. Sp. Pl. 1008. 1753; Mast in Hook. f. Fl. Brit. Ind. 1: 354. 1874; Cooke, Fl. Pres. Bombay 1: 131. 1958 (Repr. ed.); Moorthy in Singh *et al.* Fl. Maharashtra State Dicot. 1: 375. 2000. JANGLI BADAM.

Glabrous trees, 10–15 m tall; branches whorled, horizontal. Leaflets 5–9, subsessile, oblong-lanceolate, 8–18 × 3–5 cm, tapering to the base, acute or acuminate. Flowers in erect, axillary, 15–20 cm long panicles. Calyx deeply divided, lobes linear-oblong, 1–1.5 cm long, pubescent outside, woolly inside. Staminal column in male flowers curved; anthers 10–15. Carpels 5 in female flowers, pubescent, on stout long gynophore. Staminodes forming a ring beneath the carpels. Styles deflexed, hairy. Follicles 1–5, 10–12 cm, boat-shaped, thick, bright red when ripe. Seeds ovoid-oblong, black.

Fls. & Frts.: March–November. *Exsiccata*: SUK-703.
Planted along roadsides.
Localities: Ichalkaranji, Kolhapur.

Local uses: Oil from the seeds is a mild laxative and is used as a carminative. Seeds when swallowed incautiously bring nausea and vertigo. The leaves are repellent and operient.

Sterculia urens Roxb. Pl. Corom. 1: 25, t. 24. 1795 & Fl. Ind. 3:145. 1832; Mast. Hook. f. Fl. Brit. India 1: 35. 1874; Cooke, Fl. Pres. Bombay 1: 131. 1958 (Repr. ed.); Moorthy in Singh *et al.* Fl. Maharashtra State Dicot. 1: 359. 2000. KAD, PANDRUK.

Medium sized trees, 6–10 m tall, young parts pubescent. Leaves crowded at the ends of branches, palmately lobed, 15–30 cm across, cordate, glabrous above, velvety pubescent beneath, lobes 3–5, caudate acuminate. Flowers polygamous, in glandular pubescent terminal panicles appearing before the leaves. Calyx campanulate, viscous hairy; lobes oblong with small hairy gland at the base. Staminal column in male flowers short, stamens 10. Carpels 5 in hermaphrodite flowers, gynophore short; style short, stigmas 5. Follicles 5–6, ovoid oblong, 2–3 cm long, densely covered with dark purple, shinning, stinging hairs. Seeds oblong, black.

Fls. & Frts.: March–May. *Exsiccata*: SUK-704.
Common, in hill slopes in forests.
Localities: Amboli, Panhala, Phonda, Sawantwadi.

Local uses: The gum is useful in throat infection. Stem bark is made into poultice and applied on wounds. Roots used in gonorrhoea. The gum obtained from plant is called as a carraya gum used on various diseases.

TILIACEAE Juss.

Corchorus capsularis L. Sp. Pl. 529. 1753; Mast in Hook. f. Fl. Brit. India 1: 397. 1874; Cooke, Fl. Pres. Bombay 1: 157. 1958 (Repr. ed.); Pradhan in Singh *et al.* Fl. Maharashtra State Dicot. 1: 379. 2000.

Erect, annual; woody herbs, 1–1.5 m tall. Leaves lanceolate, 6–10 × 2–3 cm, rounded or acute base, serrate, the lower serrature on each side prolonged into a filiform appendage, acute or acuminate. Flowers 10–12 mm across, solitary or 2-nate on short axillary peduncles. Calyx oblong, 4–5 mm long. Petals yellow, oblanceolate, 8–10 mm long. Ovary subglobose, style short. Capsule subglobose, 10–12 mm in diam., depressed at apex, warted and ridged, 5-valved. Seeds wedge-shaped, smooth brown.

Fls. & Frts.: August–December. *Exsiccata*: SUK-1617.
Common, near wet places and seashore.
Localities: Dapoli, Shiroda, Devgad.

Local uses: A half glass of decoction of plant is given thrice a day with the water of boiled rice in diarrhoea. It is used as a stomachic, laxative, carminative, stimulant to increase appetite and flow of saliva and gastric juice, given as anthelmintic and antiperiodic.

Corchorus fascicularis Lam. Encycl. 2. 104. 1786; Mast in Hook. f. Fl. Brit. India 1: 398. 1874; Cooke, Fl. Pres. Bombay 1: 159. 1958 (Repr. ed.); Pradhan in Singh *et al.* Fl. Maharashtra State Dicot. 1: 380. 2000. BAHUPHALLI, HARANKHURI, MOTIBAHUTHALI.

Diffuse, annual, glabrous herbs, 30–60 cm tall. Leaves elliptic-oblong or lanceolate, 2.5–4 × 0.6–1.2 cm, rounded or cuneate at base, lower serratures prolonged on either side into filiform appendage, serrate, obtuse. Flowers 2–8 flowered, 2–3 mm across, in leaf opposed cymes. Sepals linear, 2–3 mm long, apicluate. Petals yellow, oblong-obovate, slightly exceeding the sepals. Capsule cylindric, 1–1.5 cm long, straight beaked. Seeds minute, black.

Fls. & Frts.: September–December. *Exsiccata*: SUK-705.
Common, in drying ditches, along roadsides, in most of fields etc.
Localities: Kolhapur, Ratnagiri.

Local uses: Plant is used to remove tumors, piles, cure dysentery. Seeds are used in stomach troubles, skin diseases and scabies.

Corchorus olitorius L. Sp. Pl. 529. 1753; Mast. in Hook. f. Fl. Brit. India 1: 397. 1874; Cooke, Fl. Pres. Bombay 1: 158. 1958 (Repr. ed.); Pradhan in Singh

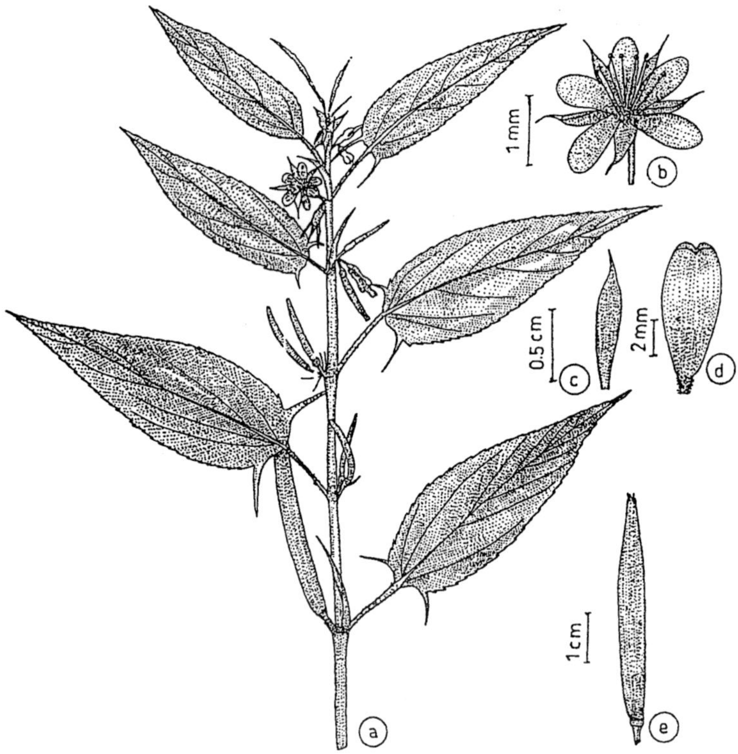

Text fig. 2: Corchorus olitorius a-twig; b-flower; c-sepal; d-petal; e-fruit.

et al. Fl. Maharashtra State Dicot. 1: 380. 2000. BANPAT, CHICHURIA, CHOOCTLA, MOTICHUNCH.

Stout, erect, much branched, woody, glabrous herbs, 45–120 cm tall. Leaves ovate lanceolate, 2–11 × 1.5–5.5 cm, rounded at base, lower serrature prolonged on either side into filiform appendages, serrate, acute or acuminate. Flowers 1–1.2 cm across, on 2–3-flowered short axillary peduncles. Sepals oblong, 3–4 mm long, apiculate. Petals yellow, oblong-lanceolate, 5–6 mm long. Capsule cylindric, 3–7 cm long, 10-ribbed, beaked. Seeds trigonous, 2–1.5 mm in diameter, black.

Fls. & Frts.: September–December. *Exsiccata*: SUK-246.
Common, throughout as a weed in cultivated fields and gardens.
Localities: Shivaji University campus, Kolhapur.

Local uses: Leaves are given in low fever, dried leaves with the *Curcuma longa* are given in diarrhoea. The leaves are given to patients recovering from acute dysentery to restore the appetite and improve the strength. Leaves are used to remove tumors, pain, ascites, abdominal tumors and piles.

Corchorus trilocularis L. Syst. Nat. 12: 369. 1764 & Mant. Pl. 77. 1767; Mast in Hook. f. Fl. Brit. India 1: 397. 1874; Cooke, Fl. Pres. Bombay 1: 158. 1958 (Repr. ed.); Pradhan in Singh *et al.* Fl. Maharashtra State Dicot. 1: 381. 2000. KADUCHINCH.

Erect, pubescent, annual, woody herbs, 20–120 cm tall. Leaves ovate to oblong-lanceolate, 2.5–9.5 × 1–5 cm, rounded or narrowed at base, serrate, base serratures prolonged or not obtuse at apex, pubescent. Flowers 1–1.5 cm across, in leaf opposed cymes. Sepals linear-oblong, 4–4.5 mm long. Petals yellow, oblong, 6–6.5 mm long. Capsules obtusely 3-angled, linear, 3.5–7 cm long, 3–4-valved, shortly beaked, pilose. Seeds trigonous and truncate at both ends, black.

Fls. & Frts.: August–March. *Exsiccata*: SUK-706.
Very Common, weed of wet locations.
Localities: Kolhapur, Panhala.

Local uses: Same as *C. fascicularis*.

Grewia asiatica L. Mant. Pl. 122. 1767; Mast. in Hook. f. Fl. Brit. India 1: 386. 1874 *excl.* var. *vestita*; Cooke, Fl. Pres. Bombay 1: 150. 1958 (Repr. ed.); Pradhan in Singh *et al.* Fl. Maharashtra State Dicot. 1: 385. 2000. PHALSI.

Small trees. Leaves ovate or suborbicular, 8–16 × 6–12 cm, acute or subacuminate, obliquely cordate or rounded at base, coarsely toothed, sub-glabrous above, hoary tomentose beneath, 5–6-nerved. Flowers in umbellate cymes. Sepals linear-oblong, 6 mm long, acute, pubescent outside. Petals yellow, oblong or ovate oblong, subacute, notched, smaller than sepals, glands white villous, less than 1/3 length of petals. Ovary villous. Drupes globose, 1–2-lobed, 1-celled, pilose.

Fls. & Frts.: April–June. *Exsiccata*: SUK-203.
Common, on hill slopes.
Localities: Ajara, Amba, Toseghar.

Local uses: Bark is given orally in urinary troubles and as a cardiac tonic. Fruits are chewed in throat complaints. Fruits are used to cure inflammation, heart and blood disorders and fevers. Fruit helps in removal of dead fetus, cures burning sensation in the vagina.

Grewia tiliifolia Vahl, Symb. Bot.1: 35. 1790; Mast. in Hook. f. Fl. Brit. India 1: 386. 1874, (*tiliaefolia*); Cooke, Fl. Pres. Bombay 1: 150. 1958 (Repr. ed.); Pradhan in Singh *et al.* Fl. Maharashtra State Dicot. 1: 393. 2000. *G. tiliifolia* Vahl. var. *leptopetala* (Brandis); Cooke, Fl. Pres. Bombay 1: 142. 1901 [1: 150. 1958 (Repr. ed.)]. DHAMAN, RODGI, SONSAMAN, TADSAL.

Trees 6–10 m tall, young parts densely hoary tomentose. Leaves broadly ovate or orbicular, 6–13 × 5–11 cm, cordate at base, crenate-dentate, acute or acuminate, tomentose on both surfaces, hoary beneath. Stipules foliaceous, falcate, auricled, tomentose. Flowers in axillary, peduncled cymes. Sepals oblong, sub-acute, gray-tomentose on the back, glabrous within. Petals yellow, oblong spathulate; glands densely white villous on margins. Ovary villous, style long, stigma 5-lobed. Drupes globose, 6–8 mm in diam., somewhat bilobed.

Fls. & Frts.: May–September. *Exsiccata*: SUK-707.
Common, throughout region in deciduous forests and also in semi evergreen forests.
Localities: Gaganbawada, Kagal, Kolhapur.

Local uses: Bark is rubbed with water and mucilage obtained is given in dysentery. Leaves are used to heal wounds. Bark is recorded to be used as aphrodisiac, heats the body and heals wounds.

Triumfetta rhomboidea Jacq. Enum. Syst. Pl. 22. 1760; Mast. in Hook. f. Fl. Brit. India 1: 395. 1874; Cooke, Fl. Pres. Bombay 1: 156. 1958 (Repr. ed.); Pradhan in Singh *et al.* Fl. Maharashtra State Dicot. 1: 399. 2000. THINJHIRA.

Much branched undershrubs, 30–120 cm tall. Leaves variable, rhomboid-ovate to lanceolate, lower ones 4–7 cm across, often 3-lobed, irregularly serrate, pubescent on both surfaces, upper ones simple. Flowers in dense, terminal and leaf opposed cymes arranged in racemes. Sepals oblong, 5–6 mm long, hooded and apiculate at apex. Petals obovate-oblong shorter than the sepals, long clawed, ciliate at base. Stamens many. Ovary globose, hairy. Fruits globose, 4 mm in diameter, spines hooked, glabrous. Seeds trigonous, smooth.

Fls. & Frts.: July–December. *Exsiccata*: SUK-710.
Common, in open situations.
Localities: Amba, Kolhapur, Radhanagari, Ratnagiri, Sawantwadi, Wai.

Local uses: Pounded leaves are applied over sprain, bruises and on tumors, as wound healing and in bone fractures. Root is given as milk inducing, as aphrodisiac, for cooling and in dysentery.

ELAEOCARPACEAE DC.

Elaeocarpus serratus L. Sp. Pl. 515. 1753; Mast in Hook. f. Fl. India 1: 401. 1874; Cooke, Fl. Pres. Bombay 1: 161. 1958 (Repr. ed.); Kothari in Singh *et al.* Fl. Maharashtra State Dicot. 1: 405. 2000. PERINKARA.

Trees, 10–15 m high. Leaves oblong or obovate, 5.0–11.2 × 2.5–5 cm, apex obtuse, acute or shortly acuminate, irregularly crenate, glabrous, often with glandular thickenings. Flowers in drooping racemes, 5–8 cm long. Sepals lanceolate, ca 1 cm long, acute. Petals white, cuneate-obovate, 7–8 mm long, laciniate half way down. Stamens many, anthers linear, one valve tipped at the apex with a small tuft of glistening hairs. Ovary pilose. 3-celled; style slender, longer than the stamens, the basal part hairy. Drupes oblong, 2.5–3.5 cm long, bluntly pointed at apex, greenish-yellow, 1–2-seeded.

Fls. & Frts.: January–March. *Exsiccata*: SUK-28.
Frequent, in semi evergreen forests.
Localities: Ajara, Amba, Lingmala falls, Patgaon.

Local uses: Leaf paste applied externally in rheumatism. The decoction of bark is given in biliousness. The nuts are used in rheumatism, typhoid fever and epilepsy.

Elaeocarpus tuberculatus Roxb. Fl. India 2: 594. 1832; Mast in Hook. f. Fl. Brit. India 1: 404. 1874; Cooke, Fl. Pres. Bombay 1: 162. 1958 (Repr. ed.); Kothari in Singh *et al.* Fl. Maharashtra State Dicot. 1: 409. 2000. RUDRAKSHA.

Trees, 80–90 m high. Leaves crowded at the ends of branches, obovate, 6–20 × 4–12 cm, apex rounded, glabrous hairy on midrib. Flowers in axillary, pubescent, racemes. Sepals oblong-lanceolate or linear-oblong, 1.3–1.5 cm long, acute, fulvous-pubescent. Petals wedge-shaped, hairy at base inside, silky villous outside, lacinate. Stamens many, anthers slender, linear, each with an erect awn, slightly curled or twisted. Ovary conical, silky villous; style silky near the base. Drupes 3–5 cm; stone compressed, tubercled on the flattened sides and with thickened margins.

Fls. & Frts.: December–October. *Exsiccata*: SUK-711.
Rare, in evergreen forests.

Localities: Ajara.

Local uses: Leaf paste applied externally in rheumatism. The decoction of bark is given in biliousness. The nuts are used in rheumatism, typhoid fever and epilepsy.

LINACEAE S.F. Gray

Linum usitatisimum L. Sp. Pl. 277. 1753; Hook. f. Fl. Brit. India 1: 410. 1874; Cooke, Fl. Pres. Bombay 1: 164. 1958 (Repr. ed.); Kothari in Singh *et al.* Fl. Maharashtra State Dicot. 1: 415. 2000. ALSI, JAVAS.

Erect, glabrous, annual herbs, 30–60 cm, tall. Leaves linear-lanceolate, 2–3 × 0.2–0.3 cm, narrowed at both ends, acute. Flowers 2–3 cm across, in terminal, corymbose panicles. Sepals unequal, outer two elliptic, 3–4 mm long, margins membranous, acute with 9–15 anthers. Ovary conical, glabrous; stigma linear. Capsules woody, ellipsoid, 10–12 cm long, 5-valved, thinly white silky. Petals blue, obovate, 6–8 mm long, crenate at apex. Ovary globose, apiculate. Capsules globose, 10–12 mm in diam., surrounded by persistent sepals. Seeds ellipsoid, compressed, smooth.

Fls. & Frts.: November–January.
Cultivated on a small scale.

Local uses: Linseed poultice is applied externally in gouty and rheumatic swellings. Oil mixed with lime water applied to burns. Seeds are soaked in over night and one teaspoon of paste is given orally in gonorrhoea for 5 days.

MALPIGHIACEAE Juss.

Hiptage benghalensis (L.) Kurz. in J. Asiat. Soc. Bengal 43: (2): 136. 1874; Kothari in Singh *et al.* Fl. Maharashtra State Dicot. 1: 419. 2000. *Banisteria benghalensis* L. Sp. Pl. 427. 1753. *Hiptage madablota* Gaertn. Fruct. 2: 169, t. 116, f. 4. 1791 *nom illeg.*; Hook. f. Fl. Brit. India 1: 418. 1874; Cooke, Fl. Pres. Bombay 1: 167. 1958 (Repr. ed.). HALADVEL, MADHUMALATI, MADHVILATA.

Large, woody twiners. Leaves elliptic-lanceolate, 8–15 × 3–6 cm, acute or rounded and with two glands at base, entire with glandular dots along margin, acuminate or acute, glabrous or nearly so. Flowers fragrant, 1.5–2.5 cm across, in terminal racemes. Gland on the calyx solitary, conspicuous, 2–6 mm long. Sepals ovate-elliptic, 2–5 mm long, rounded at apex. Petals pale yellow, orbicular to elliptic, 6–12 mm long, clawed; limb mostly fimbriate. Ovary angular globose. Fruits samaroid, brown, with 3 unequal wings.

Fls. & Frts.: March–June. *Exsiccata*: SUK-1621.

Rare, occasionally grown in gardens.

Localities: Amboli, Dajipur, Khindwadi, Radhanagari, Sawantwadi, Tillari.

Local uses: Leaves are applied in scabies; in chronic fever decoction of bark is given with the Neem leaf powder. Leaf paste applied externally on chronic wounds. Bark powder applied in similar case.

ZYGOPHYLLACEAE R.Br.

Fagonia schweinfurthii (Hadidi) Hadidi ex Ghafoor in Jafri & El-Gadi, Fl. Libya 38: 31. 1977; Kothari in Singh *et al.* Fl. Maharashtra State Dicot. 1: 423. 2000. *F. indica* Burm. f. var. *schweinfurthii* Hadidi in Reich. f. Fl. Iran. 98: 6. t. 1972. *F. cretica* auct. non L. 1753; Cooke, Fl. Pres. Bombay 1: 173. 1958 (Repr. ed.). *F. arabica* auct non L. 1753; Edgew. & Hook. f. in Hook. f. Fl. Brit. India 1: 425. 1874 p. p. DHAMASA.

Woody glaucous green, glabrous undershrubs, 60–90 cm tall; young parts glandular. Leaves pinnately 3-foliolate, 1–5 cm long; leaflets elliptic, 5–15 × 1–2 mm, cuspidate; stipular spines sharp, 5–12 mm long. Flowers 10–12 mm across, axillary solitary. Sepals ovate, 2–3 mm long, aristate. Petals spathulate, twice as long as the sepals, long-clawed. Filaments filiform. Ovary hairy. Fruits ovoid, 5 × 4 mm, acutely 5-angled, pubescent, splitting into 5-cocci. Seeds ovoid, acute, compressed, smooth.

Fls. & Frts.: October–December. *Exsiccata*: SUK-1622.

Rare, in dry places.

Localities: Jat, Sangali, Satara, Tasgaon.

Local uses: It is used in cases of abscesses from thorns. Decoction of plant is used in case of fever.

Tribulus terrestris L. Sp. Pl. 387. 1753; Edgew. & Hook. f. in Hook. f. Fl. Brit. India 1: 423. 1874 p.p.; Cooke, Fl. Pres. Bombay 1: 170. 1958 (Repr. ed.); Kothari in Singh *et al.* Fl. Maharashtra State Dicot. 1: 427. 2000. SARATA.

Prostrate annual herbs. Leaves in unequal pairs 2.5–5 cm long; leaflets 3–6 pairs, 5–10 mm long, cuneate or rounded at base, mucronate, pubescent. Flowers 7–15 mm across, axillary or leaf-opposed. Sepals lanceolate, 5–6 mm long, acute, pubescent outside. Petals yellow, obovate, twice as long as the sepals. Ovary pubescent; style very short. Fruits globose 6–10 mm in diam., with 5, woody mericarps, lateral spines much longer than the basal ones. Seeds many.

Fls. & Frts.: Throughout year. *Exsiccata*: SUK-1623.

Common, in dry regions.

Localities: Jat, Kolhapur, Sangali, Tasgaon.

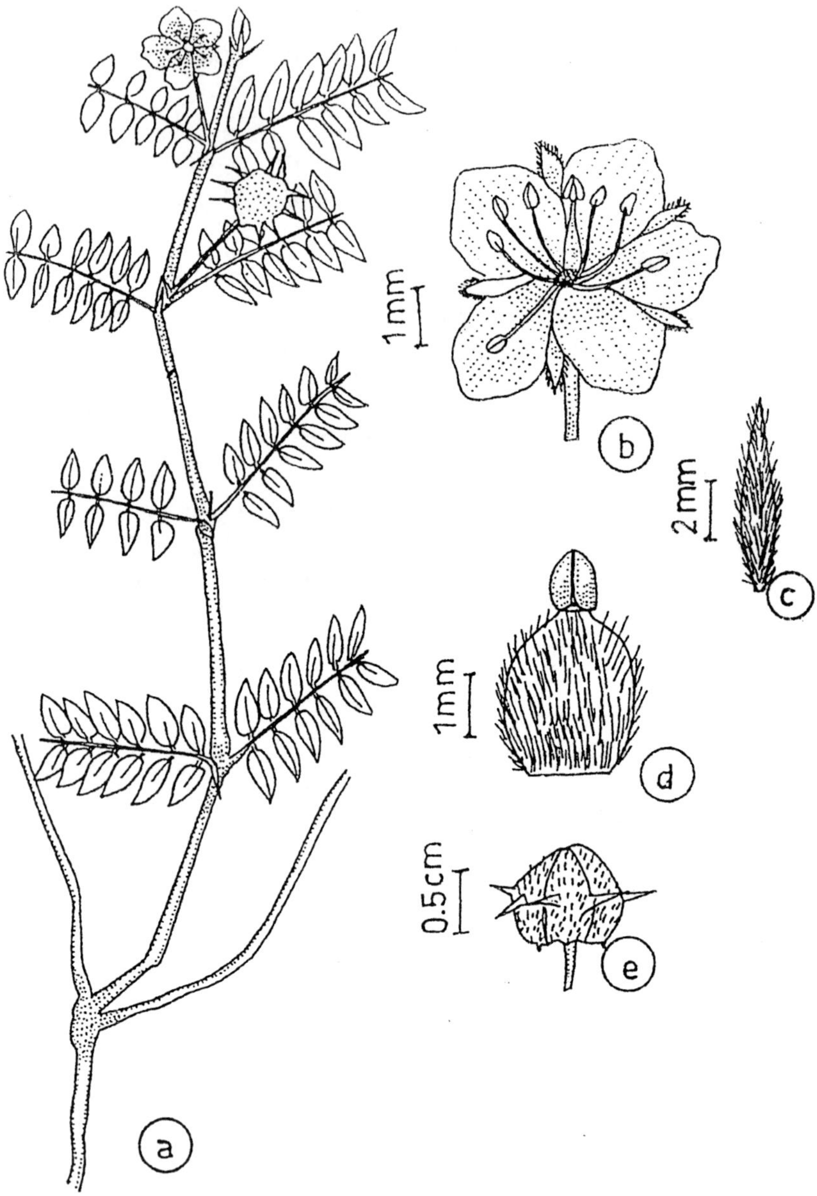

Text fig. 3: Tribulus terrestris a-twig; b-flower; c-calyx; d-carpel; e-fruit.

Local uses: The fruit is used as diuretic, removes gravel from the uterine and stone in bladder. Leaves are sometimes locally used to enrich the blood and also used to increase the menstrual flow. Root and fruits are used in urinary discharges, vesicualr calculi, reduces inflammation, cure skin and heart diseases, piles, leprosy, remove urinary stones in the bladder.

AVERRHOACEAE Hutch.

Averrhoa bilimbii L. Sp. Pl. 428. 1753; Edgew. & Hook. f. in Hook. f. Fl. Brit. India 1: 439. 1874; Cooke, Fl. Pres. Bombay 1: 178. 1958 (Repr. ed.); Kulkarni in Singh *et al.* Fl. Maharashtra State Dicot. 1: 434. 2000. BILIMBI.

Small trees, 5–8 m tall. Leaves odd pinnate, leaflets 5–21 pairs, oblong or elliptic oblong, 3–5 × 2 cm, base rounded, apex acuminate. Flowers from the trunk and branches, 5–8 mm across; panicles villous. Sepals ovate-lanceolate, 3–4 mm long, acute, pubescent. Petals oblong-spathulate, 5–8 mm long. Fruits oblong, 5–8 × 4–6 cm, obtusely 5-angled, seeds arillate, edible.

Fls. & Frts.: February–May. *Exsiccata*: SUK-1601.
Cultivated, Native of tropical America.
Localities: Kolhapur.

Local uses: Fruits are given orally to stop hair loss and applied externally in case of dandruff. The syrup of fruit is useful in relieving thirst, febrile excitement and also piles and scurvy. 1/2 cup of fruit juice diluted in same amount of water is given in scurvy.

OXALIDACEAE R. Br.

Biophytum sensitivum (L.) DC. Prodr. 1: 690. 1824; Edgew. & Hook. f. in Hook. f. Fl. Brit. India 1: 436. 1874 p. p; Cooke, Fl. Pres. Bombay 1: 177. 1958 (Repr. ed.); Kulkarni in Singh *et al.* Fl. Maharashtra State Dicot. 1: 436. 2000. *Oxalis sensitiva* L. Sp. Pl. 434. 1753. LAJALU.

Erect, annual, pubescent herbs, 2–20 cm tall. Leaves abruptly pinnate; leaflets 4–15 pairs, oblong, 5–10 mm long, obliquely rounded at base, entire, apiculate at apex, glabrous. Flowers 6–8 mm across, in terminal or axillary peduncled cymes. Sepals distinct, ovate lanceolate, 3–4.5 mm long. Petals yellow, obovate-oblanceolate, shortly clawed. Filaments in 2 whorls of 5 each, connate at base. Ovary oblong, styles distinct and often equal in length. Capsules slightly longer than the sepals. Seeds globose, irregularly tubercled.

Fls. & Frts.: July–December. *Exsiccata*: SUK-113.
Common, in wet fields and in grasslands.
Localities: Kagal, Kolhapur.

Local uses: Roots are given in cough and bronchitis. Seeds are applied to wound.

Oxalis corniculata L. Sp. Pl. 435. 1753; Edgew. & Hook. f. in Hook. f. Fl. Brit. India 1: 436. 1874; Cooke, Fl. Pres. Bombay 1: 177. 1958 (Repr. ed.); Kulkarni in Singh *et al.* Fl. Maharashtra State Dicot. 1: 437. 2000. AMBUSHI.

Small, appressed hairy, perennial herbs; stem creeping rooting at the nodes. Leaves digitately 3-foliolate; leaflets subsessile, obovate, 0.5–2 × 1.2–2.5 cm, cuneate at base, hairy. Flowers 6–8 mm across, in terminal or axillary peduncled cymes. Sepals lanceolate-oblong, 3–4.5 mm long, obtuse, hairy outside. Petals yellow, oblanceolate, 3–5 mm long, shortly clawed. Filaments in two whorls of 5 each, long and short alternating, connate at base. Ovary oblong; style distinct, hairy. Capsules oblong, narrowed at apex, pubescent. Seeds transversely ribbed.

Fls. & Frts.: Throughout year. *Exsiccata*: SUK-148.
Common, in shaded and wet localities.
Localities: Shivaji University campus, Kolhapur.

Local uses: The plant is rubbed down with water, boiled and the 1–3 tablespoon of white onion's juice added and this mixture is applied to the head in bilious headache. Infusion of leaves is given as a cooling medicine in fevers. Plant is used in dysentery and in fever.

BALSAMINACEAE Rich.

Impatiens balsamina L. Sp. Pl. 938. 1753; Edgew. & Hook. f. in Hook. f. Fl. Brit. India 1: 453. 1874; Cooke, Fl. Pres. Bombay 1: 184. 1958 (Repr. ed.); Mudaliar & Prasad in Singh *et al.* Fl. Maharashtra State Dicot. 1: 445. 2000. GAURI, TERDA.

Erect, flaccid, glabrous or sparsely hairy, annual herbs, 30–90 cm tall. Leaves alternate, elliptic lanceolate, 6–12 × 1.5–3 cm, acutely serrate, subacuminate. Flowers 1–3 together, in leaf axils, 2.5–3 cm across. Sepals ovate-oblong, 3–4 mm long, spurred sepals ovate-oblong, petaloid, 1–1.5 cm long; spur 2–3 mm long, curved. Petals often pink unequal, standard notched, 8–10 mm long, mucronate or awned, wings obovate, 2–2.5 cm long, auricled, deeply notched at apex. Ovary ovoid, stigmas 5. Capsule ellipsoid, 1.5–2.5 cm long, acuminate, densely silky-hairy. Seeds black, shining.

Fls. & Frts.: August–November. *Exsiccata*: SUK-1625.
Common, in open situations in plains.
Localities: Shivaji University campus, Kolhapur.

Local uses: Flower is cooling and tonic. Useful when applied to burns and scalds, emetic and diuretic.

RUTACEAE Juss.

Acronychia pedunculata (L.) Miq. Fl. India. Bat. Suppl. 532. 1861; Jadhav in Singh *et al.* Fl. Maharashtra State Dicot. 1: 464. 2000. *Jambolifera pedunculata* L. Sp. Pl. 349. 1753. *Acronychia laurifolia* Hook. f. Fl. Brit. India 1: 498. 1875 non Bl. 1823; Cooke, Fl. Pres. Bombay 1: 191. 1958 (Repr. ed.).

Trees or shrubs, 5–15 m high. Leaves unifoliolate, leaflets oblong-lanceolate or obovate, 6–15 × 4–8 cm, shortly or obtusely acuminate, entire, glabrous and shining above, base acute. Flowers fragrant, in paniculately, corymbose cymes. Calyx small; lobes ovate, ciliolate. Petals 6 mm long, yellowish-white, strap shaped, cucullate, inflexed at apex, hairy at base. Stamens 8; filaments subulate, villous towards base. Disk 8-angled, pubescent or villous. Ovary conical, softly hairy; stigma capitate, 4-lobed. Drupes subglobose, or ovoid or ellipsoid, 6–12 mm long, glabrous, rough with immersed glands, more or less lobed, tip angled or grooved.

Fls. & Frts.: September–November. *Exsiccata*: SUK-1602.
Rare, in semi evergreen forests.
Localities: Amboli.

Local uses: Bark is used as a tonic and also prescribed in scabies. Leaves are burnt near small pox patients.

Aegle marmelos (L.) Corr. in Trans. Linn Soc. London 5: 223. 1800; Hook. f. Fl. Brit. India 1: 516. 1875; Cooke, Fl. Pres. Bombay 1: 204. 1958 (Repr. ed.); Jadhav in Singh *et al.* Fl. Maharashtra State Dicot. 1: 464. 2000. *Crateva marmelos* L. Sp. Pl. 444. 1753. BEL.

Trees, 10–15 m tall. Leaves usually 3 foliate; leaflets lanceolate to elliptic-lanceolate, 4–5 × 2–3 cm, crenate, acuminate, glabrous, lateral leaflets subsessile, terminal one long-petioled. Flowers bisexual, greenish-white, scented, in axillary, short panicles. Sepals united to form short, 4–5-lobed calyx, lobes rounded, ciliolate. Petals 4, distinct, greenish white, oblong, 1.2 cm long. Stamens numerous, filaments distinct or sub fascicled. Ovary ovoid-oblong, disk obovate. Fruits globose, 5–10 cm in diam., rind yellowish-green, pulp sweet, orange coloured.

Fls. & Frts.: April–September. *Exsiccata*: SUK-332.
Common, often cultivated, also wild.
Localities: Kolhapur, Sangali.

Local uses: Leaves are used in high blood pressure. Leaf juice is applied externally in abscess. Root is used in Dashmula. The root bark is sometimes made into a decoction and used to the cure intermittent fever. Root is used to cure fevers, pain in the abdomen, palpitations of the heart, urinary troubles and hypochondriasis.

Atalantia monophylla (L.) Corr. Serr. in Ann. Mus. Natl. Hist. Nat. 6: 383. 1805; Hook. f. Fl. Brit. India 1: 511. 1875; Cooke, Fl. Pres. Bombay 1: 198. 1958 (Repr. ed.); Jadhav in Singh *et al.* Fl. Maharashtra State Dicot. 1: 465. 2000. *Limonia monophylla* L. Mant. Pl. 237. 1771. MAKAD–LIMBU.

Shrubs or trees. Leaves 1-foliolate; leaflets ovate oblong or elliptic, 2–8 × 1–3 cm, coriaceous, emarginate, bright green, entire, glabrous. Flowers crowded in axillary fascicles or in racemose much-contracted cymes. Calyx irregularly splited at the base. Petals white, cuneate-obovate, 9 mm long, rounded or subacute, clawed adnate to staminal tube. Stamens 8; filaments united into a tube almost throughout their entire length. Disk inconspicuous. Ovary conical. Berry globular, as large as nutmeg, 4-celled, each cell with 1 seed.

Fls. & Frts.: October–May. *Exsiccata*: SUK-1603.
Rare, in semi evergreen forest.
Localities: Gajapur.

Local uses: Leaves and roots are used as a blood purifier. Roots are given in cough and dysentery.

Atalantia racemosa Wight in Hooker's J. Bot. Kew, Gard. Misc.1: 64, t. 122 (Jan.). 1834; Hook. f. Fl. Brit. India 1: 512. 1875; Cooke, Fl. Pres. Bombay 1: 199. 1958 (Repr. ed.); Jadhav in Singh *et al.* Fl. Maharashtra State Dicot. 1: 465. 2000. MAKADLIMBU, RAN-LIMBU.

Trees, 8–10 m high. Leaves ovate or elliptic oblong, 4–5–11 × 2.5–6.0 cm, leathery, apex obtuse or retuse. Flowers white in racemes. Sepals usually 4, lobes rounded or subacute, ciliolate. Petals white, broadly elliptic or obovate, 9 mm long, concave in bud, glandular. Stamens 8; filaments united into a tube. Disk cupular. Ovary ovoid, oblong or obovoid, Berry globular, 9 mm in diam., creamy when ripe.

Fls. & Frts.: September–June. *Exsiccata*: SUK-398.
Common, in semi evergreen forest.
Localities: Amba, Barki, Gaganbawada, Panhala, Tillari, Amboli, Sawantwadi.

Local uses: Decoction of leaf given in bronchitis, asthma and cough. The leaves are chewed in bronchitis. Leaf powder used as blood purifier. Decoction of leaf is given every night as a blood purifier. One teaspoon of root powder is given in cough and sore throat after every 6 hrs. A teaspoon of root powder is given every night as blood purifier.

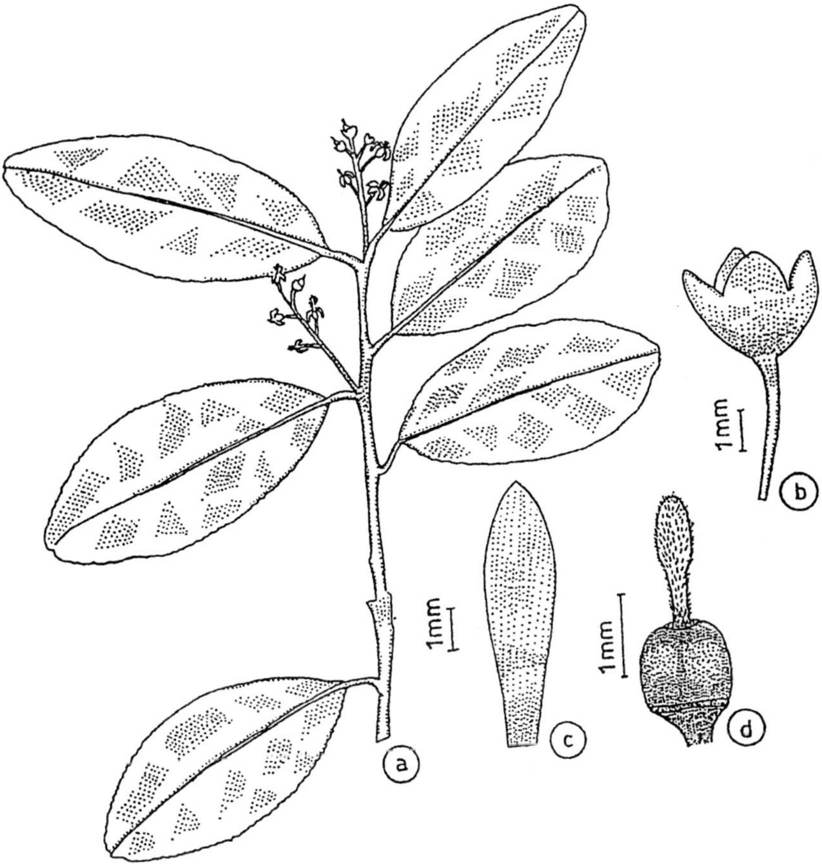

Text fig. 4: Atalantia racemosa a-twig; b-calyx; c-petal; d-carpel.

Citrus aurantifolia (Christm. & Panz.) Swingle in J. Wash. Acad. Sci. 3: 465. 1913; Jadhav in Singh *et al.* Fl. Maharashtra State Dicot.1: 483. 2000. *Limonia aurantifolia* Christm. & Panz. in Linn. Pflanzen. Syst. 1: 618. 1777. *C. medica* var. *acida* (Roxb.) Hook. f. Fl. Brit. India 1: 515. 1875; Cooke, Fl. Pres. Bombay 1: 201. 1958 (Repr. ed.). LIMBU, NIMBU.

Small glabrous trees, 3–6 m tall, with stout, stiff thorns. Leaves oblong to elliptic-ovate, 4–10 × 2–4 cm, rounded at base, crenate, acute or subacute; petioles 2–3 cm long, broadly winged. Flower 3–4 cm across, solitary or clustered in the leaf axils. Calyx cupular, 2–3 mm long, 3–5 fid. Petals white or tinged with red outside, oblong, 1.5–2 cm long. Ovary globose; style short; stigma capitate. Fruits globose 3–5 cm in diameter with thin, yellowish rind. Seeds many, ovoid, white smooth.

Fls. & Frts.: Throughout year. *Exsiccata*: SUK-331.
Cultivated throughout region.
Localities: Hatkanagale, Hupari, Kande.

Local uses: Fruit juice is given dysentery. Root paste is given orally in urinary calculi.

Citrus limon (L.) Burm. f. Fl. India 173. 1768; Jadhav in Singh *et al.* Fl. Maharashtra State Dicot. 1:484. 2000. *C. medica* var. *limon* L. Sp. Pl. 782. 1753. *C. medica* var. *limonum* (Wight & Arn.) Hook. f. Fl. Brit. India 1: 515. 1875; Cooke, Fl. Pres. Bombay 1: 201. 1958 (Repr. ed.). GADHA-LIMBU, IDLIMBU.

Shrubs or small trees, 3–5 m tall; branches with sharp 2–4 cm long thorns. Leaves elliptic-ovate or ovate lanceolate, 6–15 × 2.5–5 cm, crenate-serrate, obtuse; petioles 1–1.5 cm long narrowly winged or margined. Flowers solitary or few in the leaf axils; lobes obtuse. Petals white, usually tinged with red outside, oblong. Fruits globose or oblong, 5–10 cm in diameter, yellow when ripe; rind thick; juice plenty. Seeds ovoid, smooth rather compressed.

Fls. & Frts.: April–September. *Exsiccata*: SUK-1618.
Grown in garden.
Localities: Kolhapur, Satara, Sangali, Sindhudurg, Ratnagiri.

Local uses: The rind of the fruit is stomachic, carminative. The oil mixed with glycerin is applied to the eruption of acne.

Citrus medica L. var. **medica** Sp. Pl. 782. 1753; Hook. f. Fl. Brit. India 1: 514. 1875; Cooke, Fl. Pres. Bombay 1: 201. 1958 (Repr. ed.); Jadhav in Singh *et al.* Fl. Maharashtra State Dicot 1: 485. 2000. BIJORA, MAHLUNGI.

Shrubs or small trees, 5–7 meter tall; young parts glabrous, purple. Leaves elliptic, ovate, lanceolate, 6–15 × 2–5 cm, crenate, serrate, obtuse; petioles 1.5–2 cm long, necked or narrowly margined. Flowers 1–1.5 cm long, in 5–10-flowered short axillary racemes. Calyx cupular, minute, 4–5 lobed. Petals white or with reddish tinge outside, oblong. Stamens 20–40. Fruits oblong-ellipsoid, 10–20 cm long, yellowish; rind mammillate, very thick. Juice water colored. Seeds ovoid, 10–12 mm long, white.

Fls. & Frts.: May–October *Exsiccata*: SUK-1604.
Commonly cultivated.
Localities: Kolhapur, Miraj, Sangali.

Local uses: Fruit juice is given in hypotension. Fruit in combination with the powder of Asafetida is given in dysentery. Tea prepared from the fruit juice is given in diarrhoea and dysentery. Root and stem used in cough, hiccough, asthma; root is also given as blood purifier. Rind of fruit is used externally on skin for complexion. Root is anthelmintic, used in constipation and in tumors; removes colic; useful in vomiting, urinary calculus and caries of teeth.

Glycosmis pentaphylla (Retz.) DC. Prodr. 1: 538. 1824, quoad basionym; Cooke, Fl. Pres. Bombay 1: 192. 1958 (Repr. ed.); Jadhav in Singh *et al.* Fl. Maharashtra State Dicot. 1: 469. 2000. *Limonia pentaphylla* Retz. Obs. Bot. 5: 24. 1788. *G. pentaphylla* auct. non DC. 1824; Hook. f. Fl. Brit. India 1: 499. 1875.

Shrubs or small trees, 4–6 m high. Leaves 3–5 foliate; leaflets 3–9, 1–5.5 × 1–2 cm, cuneate or obovate, tip crenulate. Flowers small, 5–10 mm across, crowded in small clusters in erect axillary panicles, 5–10 cm long. Calyx small, lobes ovate-orbicular, margins membranous, ciliolate, glands on outside of each lobe a little below its apex. Petals white, broadly obovate or suborbicular, margins membranous. Filaments flattened, anthers with gland at the apex and sometimes with another small gland at the back of the anther. Ovary on short thick disk. Berry globose, smooth pinkish white or cream colored.

Fls. & Frts.: February–March. *Exsiccata*: SUK-1605.

Common, in moist evergreen forests.

Localities: Amba, Barki, Chiplun, Dajipur, Vaibhavwadi.

Local uses: Roots are given in low fever and in bronchitis. 4–5 leaves are crushed with 1–2 gm of ginger and two drops of honey and tincture of curcuma powder is added and pills prepared from the mixture are given thrice a day in bronchitis. Decoction of roots is useful in facial inflammation. Twigs are used as toothbrush. Leaf juice mixed with leaves of Neem is massaged on body in fever. Tender leaves are orally eaten in dysentery.

Limonia acidissima L. Sp. Pl. ed. 2. 1: 554. 1762; Cooke, Fl. Pres. Bombay 1: 196. 1958 (Repr. ed.); Jadhav in Singh *et al.* Maharashtra State Dicot: 1: 471. 2000. *Schinus limonia* L. Sp. Pl. 389. 1753 p. p. *Feronia elephantum* Corr. in Trans. Linn. Soc. London 5: 225. 1800; Hook. f. Fl. Brit. India 1: 516. 1875; Cooke, *loc. cit.* Bombay 1: 203. 1958 (Repr. ed.). KAVATH.

Moderate sized, glabrous trees, 10–15 m tall; branchlets with sharp, straight, 1–3.5 cm long spines. Leaves imparipinnate, 6–15 cm long; petioles and rachis flat, narrowly winged; leaflets 3–9, opposite, obovate, 2–5 × 1.2–2.5 cm, cuneate, obscurely crenulate at apex. Flowers in lateral or terminal lax panicles shorter than the leaves. Calyx small, lobes 5–6, triangular lobes. Petals 5–6, elliptic-oblong, 4–5 mm long, greenish or dull red. Ovary sessile. Fruits globose, 5–8 cm in diameter; pericarp woody, grayish-green. Seeds ovoid, compressed, embedded in brownish pulp.

Fls. & Frts.: March–September. *Exsiccata*: SUK-1607.

Cultivated, occasionally wild.
Localities: Kolhapur, Satara, Sindhudurg, Ratnagiri

Local uses: Powder of fruit is mixed with juice of *Phyllanthus acidus* fruits and given in hiccough. Swelling produced due to the acidity are cured by use of leaves. Leaves are soaked in water whole night and that water is given in hypertension. Leaf paste is applied externally in rheumatism. The fruit is used as cardiotonic, tonic to liver and lungs, astringent and binding, diuretic and used in strengthening of the gums.

Murraya koenigii (L.) Spr. Syst. 2: 315. 1825; Hook. f. Fl. Brit. India 1: 503. 1875; Cooke, Fl. Pres. Bombay 1: 193. 1958 (Repr. ed.); Jadhav in Singh *et al.* Fl. Maharashtra State Dicot. 1: 473. 2000. *Bergera koenigii* L. Mant. 1: 563. 1771. KADHI-LIMB. KARIPATTA.

Shrubs or small trees, 5–7 m tall. Leaves imparipinnate, 15–30 cm long; leaflets 11–25, alternate, obliquely ovate or somewhat rhomboid, 2–5 × 1–2 cm, acuminate, minutely crenate. Flowers greenish-white, in terminal corymbose cymes. Calyx cupular; lobe 5, triangular, acute, pubescent outside. Petals white, linear oblong, 5–6 mm long, rounded at apex. Filaments filiform, dilated at base. Ovary 2-locular. Fruits ovoid, 8–10 mm long, apiculate, rough with glands, purplish-black, seeds 2.

Fls. & Frts.: February–June. *Exsiccata*: SUK-1606.
Cultivated for aromatic leaves, also in semi evergreen forest.
Localities: Commonly cultivated in all districts for the leaves.

Local uses: Decoction of leaves is given in dysentery. Externally it is used to cure certain skin diseases and in itching. Roots are used to cure piles and itching.

Murraya paniculata (L.) Jack, in Mal. Misc. 1: 31. 1820; Jadhav in Singh *et al.* Fl. Maharashtra State Dicot.1: 475. 2000. *Chalcas paniculata* L. Mant. Pl. Alt. 68. 1767. *Murraya exotica* L. Mant. 2: 563. 1771 ('*Murraea*'); Hook. f. Fl. Brit. India 1: 502. 1875; Cooke, Fl. Pres. Bombay 1: 93. 1958 (Repr. ed.). KUNTI, PANDHRI.

Small trees, 4–10 m tall. Leaves imparipinnate, 10–15 cm long; leaflets 3–7, alternate, ovate-elliptic, obovate or rhomboid, 3–7 × 2–3 cm, acute and oblique at base, entire, acuminate with notched tip, glabrous and shining green. Flowers very fragrant, white, solitary or in terminal axillary corymbose cymes. Calyx cupular with small, glandular, oblong-obtuse lobes. Petals white, oblong-lanceolate, 5–6 mm long, subobtuse. Ovary 2-locular. Fruits ovoid, 1.5–2 cm long, pointed, reddish-purple. Seeds 2.

Fls. & Frts.: June–October. *Exsiccata*: SUK-225.
Cultivated. Common, in semi evergreen forests.

Localities: Gaganbawada, Kolhapur, Shirol.

Local uses: Leaf paste is used to cure itching. Leaf paste mixed with jaggry and made into pills and given in dysentery.

Paramignya monophylla Wight, Illus. Ind. Bot. 1: 108, t. 42. 1840; Hook. f. Fl. Brit. India 1: 510. 1875; Cooke, Fl. Pres. Bombay 1: 197. 1958 (Repr. ed.); Jadhav in Singh *et al.* Fl. Maharashtra State Dicot. 1: 479. 2000. KARVI-VAGETI, NARAYAN MAKADI

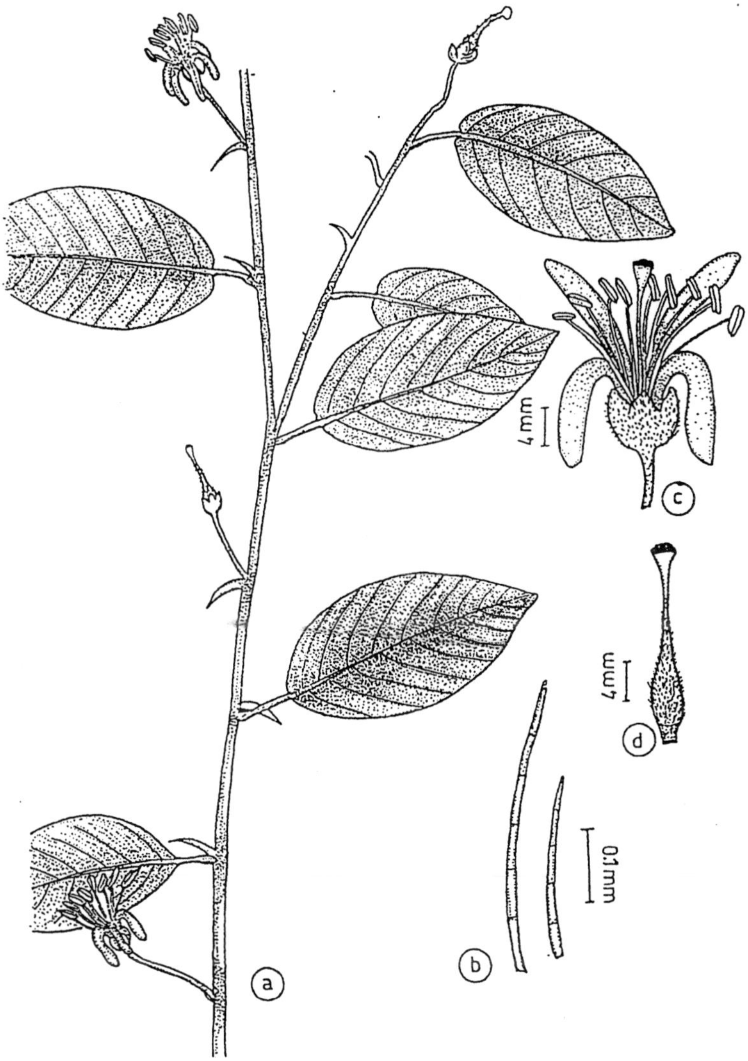

Text fig. 5: Paramignya monophylla a-twig; b-hair; c-flower; d-carpel.

Evergreen woody climbers, older branches armed with recurved axillary spines. Leaves unifoliolate, gland dotted; leaflets ovate-oblong, 5–12 × 3–7 cm, pale green, acute, obtuse or acuminate, base rounded. Flowers in cymes. Calyx softly woolly; lobes 5, shallow rounded. Petals linear-oblong, 1.9–2 cm long, recurved. Ovary hairy, 3–5-celled; style long, stout, lower part hairy. Berries 4–3 cm, obovoid, brownish-gray, pubescent.

Fls. & Frts.: December–April. *Exsiccata*: SUK-280.
Frequent, in moist deciduous and semi evergreen forests.
Localities: Amba, Amboli, Barki, Gaganbawada, Sawantwadi, Tillari.

Local uses: Roots are used as an alterative tonic, roots are given to the cattle in urinary infection. Root decoction is given as blood purifier. In skin diseases, paste of leaves applied externally and given orally with sugar candy every morning for 6 alternate days.

Ruta chalepensis L. Mant. Pl. 69. 1767; Jadhav in Singh *et al.* Fl. Maharashtra State Dicot. 1: 488. 2000. *R. graveolens* var. *angustifolia* (Pres.). Hook. f. Fl. Brit. India 1: 485. 1875 non. *R. angustifolia* Pers. 1805; Cooke, Fl. Pres. Bombay 1: 187. .1958 (Repr. ed.). SATAP.

Herbs, 30–60 cm high; strong scented, erect, glabrous and glaucous. Leaves 2–3 pinnate, 5–10 cm long; segments cuneate, spathulate oblong or linear oblong 2–10 × 1–4 mm. Flowers in spreading corymbs. Sepals distinct, ovate, 1.5–2 mm long, punctate. Petals yellow, oblong-obovate, slightly larger than sepals; limb pactinate, claw short. Ovary 3–5 lobed, ovoid, style short, stigma capitate. Fruits ovoid, 6–10 mm long, separating into 3–5, apically dehiscent mericarps, gland dotted. Seeds angled.

Fls. & Frts.: February–May.
Cultivated as potherb or in gardens.
Localities: Kolhapur.

Local uses: Decoction of leaf is used in fever. It is given in very small quantity. It may cause abortion. It is given in asthma which recurs in every rainy season associated with much nasal discharge. It is given with juice of ginger and some amount of jaggery only for 3 consecutive days. Leaf paste is applied in pains in bones, joints and cartilages. Leaf decoction is used in rheumatism.

Toddalia asiatica (L.) Lam. Tab. Encycl. Meth. 2: 116. 1797; Jadhav in Singh *et al.* Fl. Maharashtra State Dicot. 1: 479. 2000. *Paullinia asiatica* L. Sp. Pl. 365. 1753. *Toddalia aculeata* (Smith) Pers. Syn. Pl. 1: 249. 1805 nom. Superfl.; Hook. f. Fl. Brit. India 1: 497. 1875; Cooke, Fl. Pres. Bombay 1: 190. 1958 (Repr. ed.) JANGALI KALI MIRCHI.

Large scandent shrubs, prickles curved. Leaves 3-foliolate; leaflets coriaceous, sessile, oblong lanceolate or obovate, 4.5–9 × 2–4.5 cm, crenulate or subentire, midrib prickly. Flowers yellowish green in axillary cymes. Calyx lobes triangular. Petals 5, oblong, thickened at apex. Male flowers: Petals shorter sessile, 5–7 lobed. Fruit globose, 6–10 mm across, glabrous, depressed, pitted, yellow when ripe.

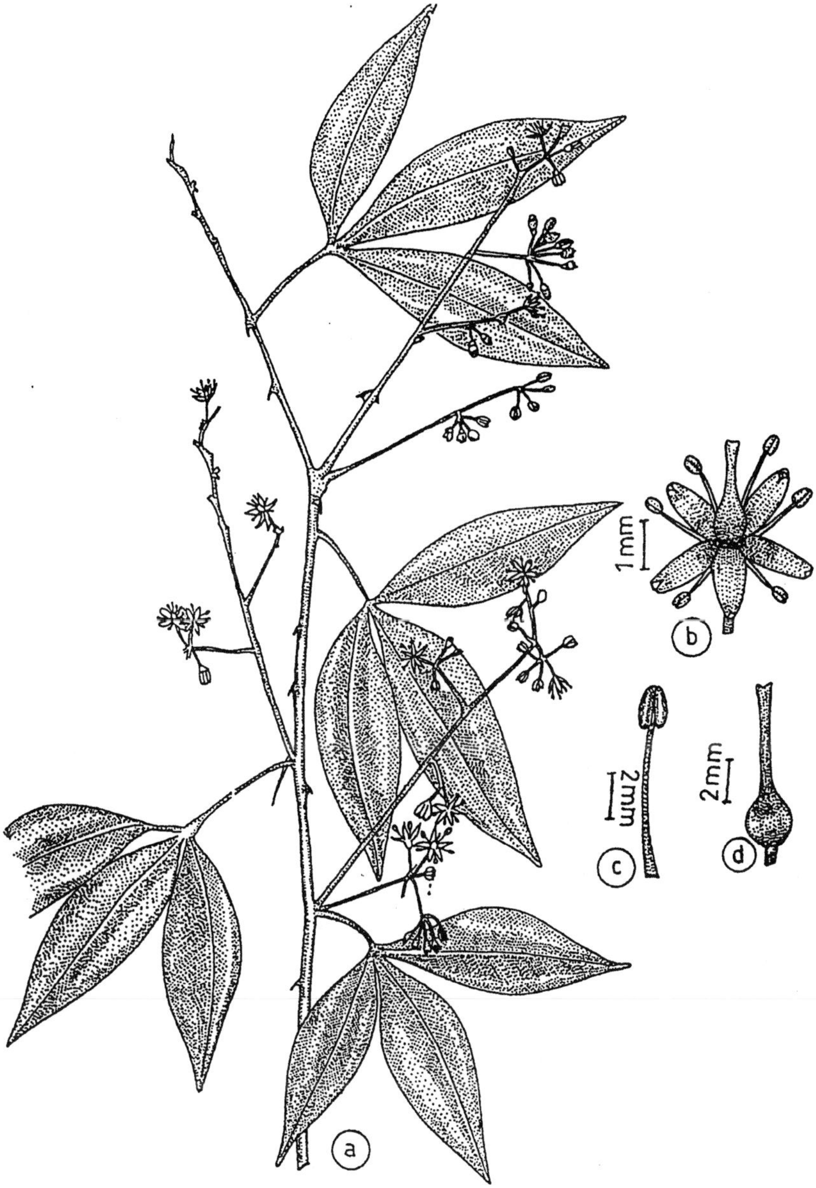

Text fig. 6: Toddalia asiatica a-twig; b-flower; c-stamen; d-carpel.

Fls. & Frts.: December–April. *Exsiccata*: SUK-52.
Common, in semi evergreen forests.
Localities: Amba, Barki, Gaganbawada, Radhanagari, Tillari.

Local uses: Root bark is used as tonic, blood purifier. Infusion of fresh root bark is stimulating tonic and carminative. Decoction of root is given as an antiperiodic or febrifuge. Roots are used to treat HIV patients and for urinary problems.

Zanthoxylum rhetsa (Roxb.) DC. Prodr. 1: 728. 1824; Hook. f. Fl. Brit. India 1: 495. 1875; Cooke, Fl. Pres. Bombay 1: 189. 1958 (Repr. ed.); Jadhav in Singh *et al.* Fl. Maharashtra State Dicot. 1: 481. 2000. KOKLI, TIRPHAL, TISAL.

Trees, 8–10 m tall; stem and branches prickly. Leaves crowded at the ends of the branches, equally or unequally pinnate; leaflets elliptic oblong or ovate, 3.5–12.0 × 3.0–5.0 cm, apex acuminate. Flowers greenish-yellow, in terminal cymes, 15–30 cm long. Calyx lobes triangular. Petals 4, elliptic, 2–3 mm long, yellow. Cocci globose, 8 × 6 mm, 1-seeded.

Fls. & Frts.: June–March. *Exsiccata*: SUK-105.
Common, in dry deciduous and semi evergreen forests.
Localities: Amboli, Dajipur, Gaganbawada Radhanagari, Tillari, Vaibhavwadi.

Local uses: Thorn is used to stop the flow of blood from wound. It is also used to wash the eyes. Fruits are given in pain; useful in diseases of the heart, mouth, teeth and throat; relieves hiccough and good for piles.

SIMAROUBACEAE DC.

Ailanthus excelsa Roxb. Pl. Cor. 1: 24, t. 23. 1795; Bennett in Hook. f. Fl. Brit. India 1: 518. 1875; Cooke, Fl. Pres. Bombay 1: 205. 1958 (Repr. ed.); Moorthy in Singh *et al.* Fl. Maharashtra State Dicot. 1: 489. 2000.GODLIMB, MAHARUK.

Tall trees, 8–20 m. Leaves odd pinnate, 20–100 cm long; leaflets 5–14 pairs, alternate or subopposite, falcate-lanceolate, 5–15 × 2–6 cm, unequal at base, coarsely and irregularly toothed or sublobate, glabrous or nearly so above, tomentose beneath. Flowers polygamous or unisexual, in large much branched, axillary panicles shorter than the leaves. Calyx short; lobes triangular. Petals greenish-yellow, ovate, 3–4 mm long, glabrous. Disk 10-lobed, stamens 10–12 in male flower, 2–3 in hermaphrodite and absent in female flowers. Ovary 2–6 partite; styles connate. Samaras 1–6, lanceolate, 3–6 × 0.5–1.2 cm, acute at both ends, reddish-brown, twisted near the base, many nerved. Seeds solitary, ovoid, orbicular.

Fls. & Frts.: February–May. *Exsiccata*: SUK-1626.

Common, along roadsides and in villages.

Localities: Sangli, Kolhapur, Satara districts.

Local uses: Leaf powder is given in debility. Similarly it is given in debility after childbirth. The juice of leaves is taken before the meal as an appetizer. Similarly it is given in dysentery. The bark is used as appetizer, anthelmintic, febrifuge, diarrhoea, dysentery, earache, cures skin diseases, troubles of the rectum, fevers, removes bad taste in the mouth due to the acidity.

BALANITACEAE Endl.

Balanites aegyptiaca (L.) Del. in Descr. Egypt. Hist. Nat. 2: 221, t. 28, f. 1. 1813; Moorthy in Singh *et al.* Fl. Maharashtra State Dicot.1: 491. 2000. *Ximenia*

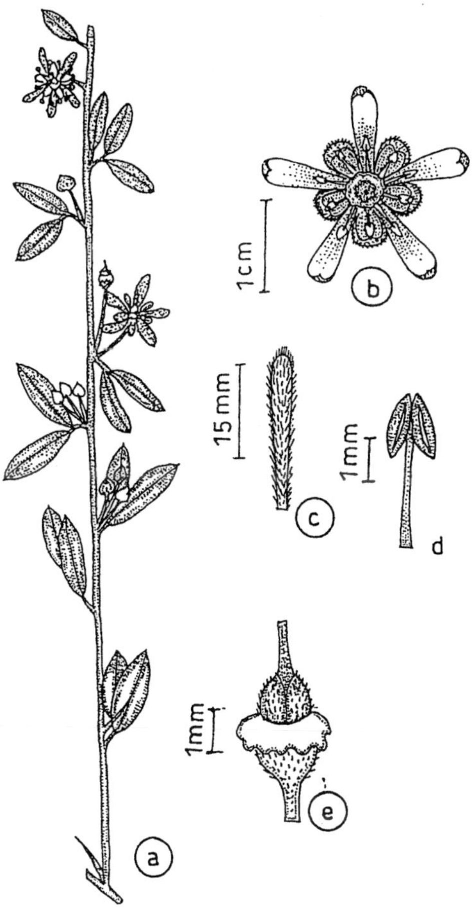

Text fig. 7: Balanites aegyptica a-twig; b-flower; c-petal; d-stamen; e-carpel.

aegyptiaca L. Sp. Pl. 1194. 1753. *Balanites roxburghii* Planch. in Ann. Sci. Nat. 4, 2: 258. 1854; Bennett in Hook. f. Fl. Brit. India 1: 522. 1875; Cooke, Fl. Pres. Bombay 1: 207. 1958 (Repr. ed.). HINGAN BET, HINGU.

Small trees reaching 10 m tall; young parts pubescent or tomentose; branches armed with straight, sharp, 3–5 cm long thorns. Leaves 2–foliolate; leaflets elliptic to obovate, 1.5–3 × 1–2.5 cm, acute at base, mucronate, obtuse or subacute at apex. Flowers in short, axillary, pubescent cymes, 1–1.5 cm across, fragrant. Sepals distinct, ovate, 3 mm long, pubescent outside, silky within. Petals dull white or greenish, oblong, obovate, 4–5 mm long, silky villous; style short. Drupes ovoid or ellipsoid, 4–5 × 3.3 cm, 5-angled, yellowish. Seed ovoid.

Fls. & Frts.: February–November. *Exsiccata*: SUK-315.
Common, in scrub and dry deciduous forests.
Localities: Alate, Atpadi, Bahubali, Jat.

Local uses: Oil extracted from the seeds is used to remove marks remained on body due to wounds or burned part. The fruit is used for boils. Diluted fruit pulp is externally applied on cut and wounds. The fruit is used as anthelmintic, in leucoderma and skin diseases.

OCHNACEAE DC.

Ochna obtusata DC. in Ann. Mus. Natl. Hist. Nat. 17: 411, t. 11. 1811; Moorthy in Singh *et al.* Fl. Maharashtra State Dicot. 1: 492. 2000. *O. squarrosa* auct. non. L. 1762; Bennett in Hook. f. Fl. Brit. India 1: 523. 1875 (excl. Syn. *O. lucida* Griff. & *O. rufescens* Thw.); Cooke, Fl. Pres. Bombay 1: 208. 1958 (Repr. ed.) KANAKCHAMPA.

Shrubs, 8–10 m tall. Leaves oblong-lanceolate, elliptic or obovate, 2–6 × 1–3 cm, narrowed at base into short petiole, finely serrulate, acute, glabrous. Flowers 8–10 mm across, fragrant, in umbellate panicles on old wood, or short branches. Sepals elliptic-oblong with close parallel veins, 1.2–1.3 cm long. Petals yellow, 5–12, oblanceolate, little longer than the sepals. Stamens many; filaments much shorter than the anthers. Styles connate. Fruits of 3–6 drupes, oblong-ovoid, 6–7 mm long, sessile, black, surrounded by the persistent calyx. Seeds oblong.

Fls. & Frts.: February–April. *Exsiccata*: SUK-1619.
Infrequent, in deciduous forests.
Localities: Kudal, Patgaon, Vaibhavwadi.

Local uses: The decoction of root is said to be useful in menstrual complaints and asthma.

BURSERACEAE Kunth.

Boswellia serrata Roxb. ex Colebr. in Asiat. Res. 9: 379, t. 5. 1807; Bennett in Hook. f. Fl. Brit. India 1: 528. 1875; Cooke, Fl. Pres. Bombay 1: 210. 1958 (Repr. ed.); Moorthy in Singh *et al.* Fl. Maharashtra State Dicot. 1: 494. 2000. SALAI, SALPHALI, SARPAL.

Trees, 4–6 m tall; young shoots pubescent. Leaves odd pinnate, 10–30 cm long; leaflets 8–15 pairs; opposite, subsessile, ovate-oblong, 2–6 × 1–3 cm, oblique at base, crenate, serrate, obtuse, pubescent. Flowers in axillary racemes shorter than the leaves. Calyx pubescent outside; lobes triangular-ovate, 2–3 mm long. Petals dull or greenish-white, ovate, pubescent outside. Stamens inserted at base of red, annular crenate disk. Ovary ovoid, sunk into the disk; style long. Drupe trigonous, ovoid, 1.5–2 cm long, pale green, pyrenes compressed, heart shaped.

Fls. & Frts.: February–June. *Exsiccata*: SUK-98.
Common, in deciduous forests.
Localities: Appachiwadi, Babujamal, Kartikswami.

Local uses: Gum in combination with Sulphur and bark of *Acacia catechu* is used in old wounds. Gum is used on scrofula (dishel), in rheumatism. In inflammation it is externally applied. Leaf powder is used as toothpaste. The gum is useful in skin and blood diseases, fevers, mouth sores, vaginal discharges and diseases of the testis.

Canarium strictum Roxb. Fl. India 3: 138. 1832; Bennett in Hook. f. Fl. Brit. India 1: 534. 1875; Cooke, Fl. Pres. Bombay 1: 214. 1958 (Repr. ed.); Moorthy in Singh *et al.* Fl. Maharashtra State Dicot. 1: 494. 2000. RANDHUP.

Trees, 20 m tall; young parts rufous-tomentose beneath. Leaves imparipinnate, coriaceous up to 60 cm long; leaflets 3–7 pairs, 10–20 × 4.5–8.7 cm, ovate-oblong or ovate to ovate-lanceolate, apex acuminate, base obtuse or semicordate, finely serrate along margins. Flowers in short-branched axillary panicles shorter than the leaves. Calyx usually 3-lobed, cupular, tube 3 mm long, rusty tomentose outside; lobes triangular. Petals oblong, obovate, ca 6 mm long, slightly hairy on the outside. Ovary ovoid, attenuated into style as long as the ovary, glabrous, rudimentary in male flowers. Drupes 3.7–5.0 × 1.5 cm, ellipsoid or ovoid, tapering at both ends; stone hard, bony.

Fls. & Frts.: February–May.
Rare, in evergreen forests.
Localities: Dicholi, Manoli, Takavali.

Local uses: The gum is used with gingelly oil in rheumatic pains.

Garuga pinnata Roxb. Pl. Cor. 3: 5, t. 208. 1819; Bennett in Hook. f. Fl. Brit. India 1: 528. 1875; Cooke, Fl. Pres. Bombay 1: 211. 1958 (Repr. ed.); Moorthy in Singh *et al.* Fl. Maharashtra State Dicot. 1: 495. 2000. KAKKAD.

Trees, ca 10 m high. Leaves odd pinnate; leaflets 6–10 pairs and odd one, opposite, subsessile, ovate-lanceolate, 5–12 × 2.5–5 cm, oblique, crenate, acuminate, pubescent, glabrescent. Flowers 6–10 mm across, in axillary panicles, which are clustered at the ends of branchlets. Calyx campanulate, 5–6 mm long, divided above the middle; lobes ovate-oblong, subobtuse. Petals yellowish-white, linear-oblong, shorter than the calyx, tomentose outside, pubescent within; tips inflexed. Disk thin, crenate filaments hairy. Drupes ovoid, pale yellow, up to 2 cm long, rugose. Seeds with membranous wings.

Fls. & Frts.: February–May. *Exsiccata*: SUK-63.
Occasionally, on hill forests.
Localities: Alate, Bahubali, Gadhinglaj, Katyayani, Radhanagari, Tillarinagar.

Local uses: The juice of leaves with leaves of *Justicia adhatoda* and *Vitex trifolia* mixed with honey is given in asthma. Fruits are eaten as a cooling and stomachic remedy.

MELIACEAE Juss.

Aphanamixis polystachya (Wall.) Parker in Indian Forester 57: 486. 1931; Moorthy in Singh *et al.* Fl. Maharashtra State Dicot. 1: 501. 2000. *Aglaia polystachya* Wall. in Roxb. Fl. Ind. 2: 429. 1824. *Andersonia rohituka* Roxb. Fl. Ind., 2: 213. 1832. *Amoora rohituka* (Roxb.) Wight & Arn. Prodr. 119. 1834; Hiern in Hook. f. Fl. Brit. India 1: 559. 1875; Cooke, Fl. Pres. Bombay 1: 224. 1958 (Repr. ed.). ROHITAK, SOHAGA.

Trees, 9–18 m tall; young parts finely silky. Leaves alternate, crowded at the ends of branches, 20–70 cm long; leaflets 9–19, opposite, oblong ovate or oblanceolate, 8–25 × 4–10 cm obliquely acute or caudate at base, margins entire, acuminate. Male flowers 4 mm across, in axillary or supra-axillary, solitary spikes. Calyx 5-lobed, ciliolate. Petals 3, broadly elliptic to orbicular, concave ca 5 × 4 mm, glabrous or puberulous outside, staminal tube as long as petals; anthers 6 subsessile. Disk broadly conical tawny-pubescent. Ovary subglobose or ellipsoid; stigma 3-lobed. Capsule obovoid, 2.5–4 cm across, 3-valved, coriaceous, yellow seeds oblong with scarlet aril.

Fls. & Frts.: January–December. *Exsiccata*: SUK-1620.
Infrequent, in evergreen forest.
Localities: Barki, Radhanagari.

Local uses: Seeds are used as laxative, anthelmintic; cure ulcers, diseases of the blood, eyes, ear; lessen muscular pain, ripe seeds yield an oil which is used as a stimulating liniment in rheumatism.

Azadirachta indica A. Juss. in Mem. Mus. Hist. Nat. Paris 19: 220, t. 12, f. 5. 1830; Cooke, Fl. Pres. Bombay 1: 220. 1958 (Repr. ed.). *Melia azadirachta* L. Sp. Pl. 385. 1753; Hiern in Hook. f. Fl. Brit. India 1: 544. 1875. NEEM, NIM, NIMB.

Trees, ca 20 m high. Leaves simply pinnate, 10–45 cm long, crowded near the ends of the branches; leaflets 9–13, sub–opposite, very oblique, lanceolate, 2–7 × 1–3 cm, serrate, glabrous. Flowers mildly fragrant, in axillary panicles shorter than leaves. Sepals round-ovate, minute, ciliolate. Petals white, obovate-oblong, 6–8 mm long, ciliolate. Staminal tube glabrous, shorter than the petals, toothed at apex; anther 10, opposite to the teeth. Disk absent. Ovary glabrous, 3–5 locular; ovules 2-in each locule; stigma 3-lobed. Drupes ellipsoid, 1–2 cm long, yellow, 1-seeded.

Fls. & Frts.: January–May.
Common, in drier parts.
Localities: Kolhapur, Sangli, Satara.

Local uses: Concentrated juice of leaves is used in washing wounds, ulcers and in many skin diseases. Leaves are boiled in water and bath is given to patients suffering from low fever, similarly, the leaf paste is applied externally in rheumatism. Leaf juice daily in early mornings' given in diabetes.

Cipadessa baccifera (Roth) Miq. Ann. Mus. Lugd.–Bat. 4: 6. 1968–69; Moorthy in Singh *et al.* Fl. Maharashtra State Dicot. 1: 503. 2000. *Melia baccifera* Roth, Nov. Pl. Sp. 215. 1821. *Cipadessa fruiticosa* Bl. Bijdr. 162. 1825; Hiern in Hook. f. Fl. Brit. India 1: 545. 1875; Cooke, Fl. Pres. Bombay 1: 219. 1958 (Repr. ed.). GUDMEI.

Shrubs 3–4 m high; branchlets tomentose. Leaves odd pinnate, 12 × 8 cm; leaflets 3–6 pairs, opposite, oblong, lanceolate, 3–7 × 1.5–2.5 cm, thin, coriaceous, glabrous, pilose below, base cuneate, margin distally serrate, apex acuminate. Inflorescence axillary corymbose panicles. Flowers bisexual, 5-merous. Calyx tube deeply lobed pubescent. Petals 5, white tomentose, white, lanceolate, pubescent. Stamens 10; anthers inserted between forks. Ovary obovoid, 5-celled, glabrous; style short; stigma turbinate. Drupe globose, red, 5-lobed. Seeds 5, angular.

Fls. & Frts.: September–December. *Exsiccata*: SUK-1690.
Occasional, on hill slopes of the ghats.
Localities: Kolhapur, Patgaon.

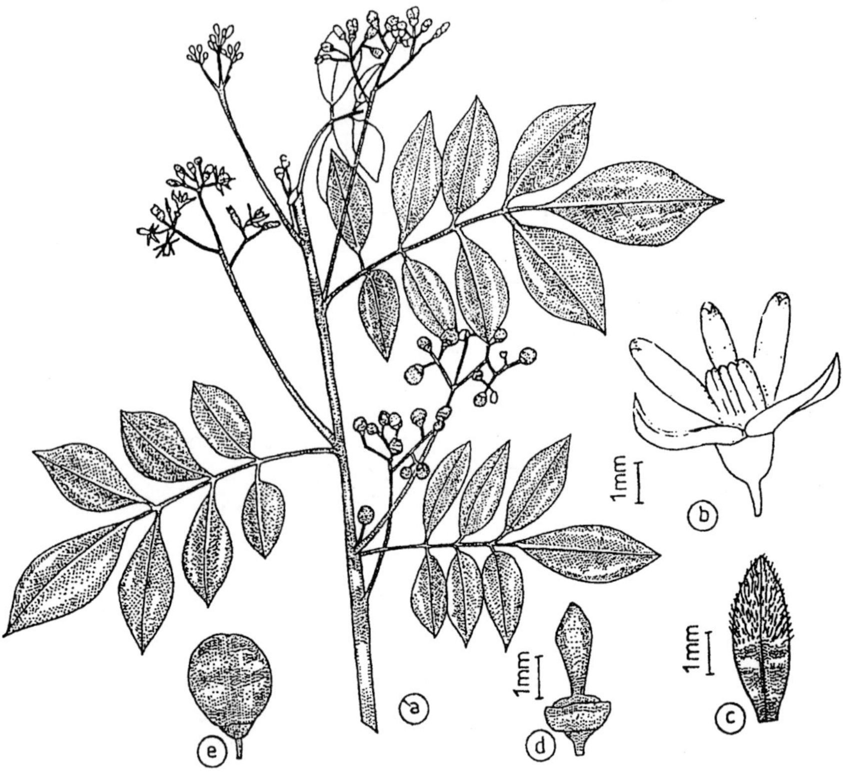

Text fig. 8: Cipadessa baccifera a-twig; b-flower; c-petal; d-carpel; e-fruit.

Local uses: Decoction of bark and leaves is given in scrofula. Root is given as blood purifier, heals ulcers, headache, uterine pains after delivery; cures fever, burning sensations and urinary discharges.

Melia azedarach L. Sp. Pl. 384. 1753; Hiern in Hook. f. Fl. Brit. India 1: 544. 1875; Cooke, Fl. Pres. Bombay 1: 218. 1958 (Repr. ed.); Moorthy in Singh *et al.* Fl. Maharashtra State Dicot. 1: 505. 2000. BAKAIN, BAKAN NIMB, BAKAVAN, PEJRI.

Moderate sized trees, 7–10 m tall. Leaves 2–3 pinnate, ca 50 cm long; pinnae alternate or opposite; leaflets opposite or subopposite, ovate-lanceolate, 1.5–5 × 0.5–3 cm, inequilateral at base, glabrous. Flowers fragrant in axillary panicles shorter than the leaves. Sepals ovate-oblong, minute, ciliolate. Petals obovate-oblong, 10–12 mm long, white with purple tips. Staminal tube dark purple, 6–7 mm long, 20-toothed, longitudinally ribbed; anthers sessile. Ovary globose, 5–6-locular; style stout; stigma capitate. Drupes elliptic-ovate, ca 2 × 1 cm, green, muricate. Seeds 4.

Fls. & Frts.: July–September. *Exsiccata*: SUK-1615.
Cultivated.
Localities: Kolhapur, Satara.

Local uses: Flower powder is applied on forehead in headache. Poultice of leaves is applied externally in various skin diseases. Decoction of bark and leaves is given in scrofula. Root is given as anthelmintic, removes pain in the heart; useful in vomiting, leucoderma; belching, blood impurities, heals ulcers, headache, uterine pains after delivery; cures fever, burning sensations, urinary discharges, lung complaints and rat-bite poisoning.

Naregamia alata Wight & Arn. Prodr. 117. 1834; Hiern in Hook. f. Fl. Brit. India 1: 542. 1875; Cooke, Fl. Pres. Bombay 1: 217. 1958 (Repr. ed.); Moorthy in Singh *et al.* Fl. Maharashtra State Dicot. 1: 507. 2000. KAPURBHENDI.

Shrubs, 10–25 cm tall, erect or decumbent, woody. Leaves trifoliolate; 3–9 cm long; leaflets sessile, cuneate-obovate, 2.5–3 × 3 cm, sometimes obtusely lobed entire, terminal slightly large. Flowers white, axillary solitary. Calyx hairy outside; lobes oblong-lanceolate. Petals linear-spathulate, 2.2 × 0.8 cm. Staminal tube 1.8–2 cm long, expanded at apex, 10-toothed, anthers 10, sessile. Ovary 3-locular, stigma discoid. Capsule membranous, 0.6–0.8 cm, 3-lobed, 3-valved; valves 2-seeded; seeds terete, tubercled, brownish-black, truncate at both ends.

Fls. & Frts.: August–December. *Exsiccata*: SUK-153.
Common, in open areas of deciduous forests.
Localities: Doddamarg.

Local uses: Decoction of leaves is given in biliousness; roots are given in acute dysentery. Oil extracted from the leaves is used in rheumatism. Root is used in asthma, bronchitis, biliousness and ulcers.

Soymida febrifuga (Roxb.) A. Juss. in Mus. Hist. Nat. Paris 19: 251, t. 22, f. 26. 1830; Hiern in Hook. f. Fl. Brit. India 1:567. 1875; Cooke, Fl. Pres. Bombay 1: 228. 1958 (Repr. ed.); Moorthy in Singh *et al.* Fl. Maharashtra State. Dicot. 1: 508. 2000. *Swietenia febrifuga* Roxb. Bot. Descr. Swietenia 1. 1793 & in Med. Facts Obs. 6: 129. 1795 & Pl. Cor. 1: 18, t. 17. 1795. ROHAN, RUHIN.

Trees, 10–12 m tall. Leaves unipinnate, 20–40 cm long; leaflets 3–6 pairs, elliptic oblong, 5–15 × 3.5–9 cm, rounded and unequilateral at base, entire, obtuse, glabrous. Flowers in large, axillary and terminal divaricate panicles. Often equaling the leaves. Sepals distinct, broadly ovate rounded with membranous margins. Petals white, obovate, 5–6 mm long, clawed, often notched at apex. Staminal tube urceolate, 2–3 mm long. Ovary globose, glabrous, style short, stigma 5-angled. Capsules obovoid, 5–7.5 cm long, 5-valved, woody glabrous, septifragal. Seeds compressed, winged at both ends, brown.

Fls. & Frts.: February–July. *Exsiccata*: SUK-1658.
Common, in deciduous forests.
Localities: Gajapur, Radhanagari.

Local uses: The bark is used in fever, in general debility, in advanced stage of dysentery and in diarrhoea. Leaf decoction is externally applied in ulcers and wounds, leaf decoction is also used in sore throat, cough and in asthma.

Toona hexandra (Wall. ex Roxb.) Roem. Syn. Hesp. 139. 1849; Moorthy in Singh *et al.* Fl. Maharashtra State Dicot. 1: 508. 2000. *Cedrela hexandra* Wall. ex Roxb. Fl. Ind. 2: 425. 1824. *Toona ciliata* Roem. Syst. Hesp. 136. 1846; Jain & Bennet in Hajra *et al.* Fl. India 4: 504 . 1997. *Cedrela toona* Roxb. ex Rottl. & Willd. in Ges. Naturf. Fr. Neue Schr. 2: 198. 1803; Bedd. Fl. Sylv. T. 10. 1869; Hiern in Hook. f. Fl. Brit. India 1: 568. 1875; Cooke, Fl. Pres. Bombay 1: 230. 1958 (Repr. ed.). KUDAK, MAHANIM, TOONA, TUNI

Trees, 20–27 m tall. Leaves unipinnate 30–60 cm long; leaflets opposite or alternate, 4–15 pairs, ovate lanceolate or lanceolate, 4.5–15 × 2.6 cm, uniequilateral at base, minutely serrate, acuminate, glabrous. Flowers white, honey scented, in large drooping terminal more or less pubescent panicles. Calyx lobes orbicular-ovate, ciliate. Petals broadly elliptic, ca 5 mm long, obscurely veined, ciliate, keeled at base inside. Stamens 5, each inserted on one of the orange colored hairy lobes of the disk; anthers oblong, cordate, rounded and apiculate at the apex. Ovary hairy, ovoid or subglobose. Capsules oblong, 1.5–3 cm long. Seeds with membranous wings at each end.

Fls. & Frts.: June–August. *Exsiccata*: SUK-1646.
Common, in moist deciduous forests.
Localities: Barki, Dajipur, Tillari.

Local uses: Bark is powerful astringent, used in chronic infantile dysentery. Bark is applied externally in various forms of ulceration. Bark is also given as cardiac tonic. Bark is useful in burning sensation; cures fevers, itching, headache and blood complaints.

Trichilia connaroides (Wight & Arn.) Bentv. in Acta Bot. Neerl. 11: 13. 1962; Moorthy in Singh *et al.* Fl. Maharashtra State Dicot. 1: 509. 2000. *Zanthoxylum connarioides* Wight & Arn. Prodr. 148. 1834. *Heynea trijuga* Roxb. ex Sims. in Bot. Mag. 41: t. 1738. 1875; Hiern in Hook. f. Fl. Brit. India 1: 565. 1875; Cooke, Fl. Pres. Bombay 1: 227. 1958 (Repr. ed.). LIMBARA.

Trees, 8–10 m tall; branches spreading. Leaves 15–35 cm long; leaflets ovate oblong or oblong-lanceolate, 7–9, 5–10 × 2.5–4.0 cm, acuminate, base rounded or acute, glabrous above, pubescent beneath. Flowers white in corymbosely branched panicles. Calyx lobes ovate, subacute. Petals much longer than the calyx, 5–6 mm long, midrib thickened, membranous, margins ciliolate. Staminal tube shorter than the petals, much longer than the calyx; lobes linear, bidentate, densely hairy within; anther between the teeth of the lobes. Ovary glabrous; style obconical, stigma 2-dentate with thickened ring at the base. Capsules obovoid, 1.2 × 1.4 cm, reddish-pink when ripe. Seeds 2–5.

Fls. & Frts.: March–September. *Exsiccata*: SUK-29.
Infrequent, in moist deciduous forests.
Localities: Adhoshi, Karanjphen, Kusawade, Radhanagari, Tudye.

Local uses: It is given same way as that of *Azadirachta indica.*

Turraea villosa Benn. Pl. Jav. Rar. 182. 1840; Hiern in Hook. f. Fl. Brit. India 1: 542. 1875; Cooke, Fl. Pres. Bombay 1: 216. 1958 (Repr. ed.); Moorthy in Singh *et al.* Fl. Maharashtra State Dicot. 1: 511. 2000.

Shrubs, 3–5 m tall. Leaves membranous, appearing after the flowers, elliptic ovate, 5–10 × 3.5–6 cm, acuminate, entire, softly villous when young, glabrous at maturity, base acute or rounded. Flowers pale yellow, in axillary solitary or in fascicles or in short peduncled umbels. Calyx campanulate, pubescent, 5-toothed; teeth triangular. Petals linear spathulate, 2–3 cm long. Staminal tube pale yellow, 3–3.5 cm in length, glabrous, 10-toothed; anthers longer than the teeth of staminal tube. Ovary subglobose, 5-celled; style 3.5–4 cm in length, far exserted; stigma large, ovoid. Capsule subglobose, 1–2 cm across, glabrous. Seeds oblong.

Fls. & Frts.: April–August. *Exsiccata*: SUK-89.
Common, in moist deciduous forest.
Localities: Amba, Barki, Gaganbawada, Radhanagari, Tillari.

Local uses: Roots are used as on application to fistulas and is administered internally in leprosy. In market it is given under the name of *Naregamia alata.*

FLINDERSIACEAE (Engl.) C. T. White & Airy Shaw

Chloroxylon swietenia DC. Prodr. 1: 625. 1824; Hiern in Hook. f. Fl. Brit. India 1: 569. 1874; Cooke, Fl. Pres. Bombay 1: 231. 1958 (Repr. ed.); Moorthy in Singh *et al.* Fl. Maharashtra State Dicot. 1: 514. 2000. BEHRU, BHIRRA, BILLU, GIRYA, HALDA

Deciduous trees, 5–10 m tall. Leaves unipinnate, 15–30 cm long; leaflets 10–20 pairs, oblong, 2–3.5 × 0.5–1.8 cm, unequilateral at base, glabrous; glaucous, obtuse. Flowers in axillary and terminal, many-flowered, branched panicles shorter than leaves. Sepals distinct, ovate, minute. Petals white, ovate, 4–5 mm long, sagittate and clawed at base. Disk fleshy, stamens inserted between the lobes of the disk, shorter than petals. Ovary sunk into the disk, 3-lobed, pubescent. Capsules oblong, acute. Seeds compressed, winged, 1.5–2 cm long.

Fls. & Frts.: March–May. *Exsiccata*: SUK-84.
Occasional, in open forests.
Localities: Kowad, Nesari, Tarewadi.

Local uses: The leaves are applied to wounds and also prescribed in rheumatism. It is a powerful irritant, causing dermatitis when applied to the skin.

ICACINACEAE Miers.

Nothapodytes nimmoniana (J. Grah.) Mabberley in Manilal, Bot. & Hist. Hort. Malab. 88. 1980. Moorthy in Singh *et al.* Fl. Maharashtra State Dicot.

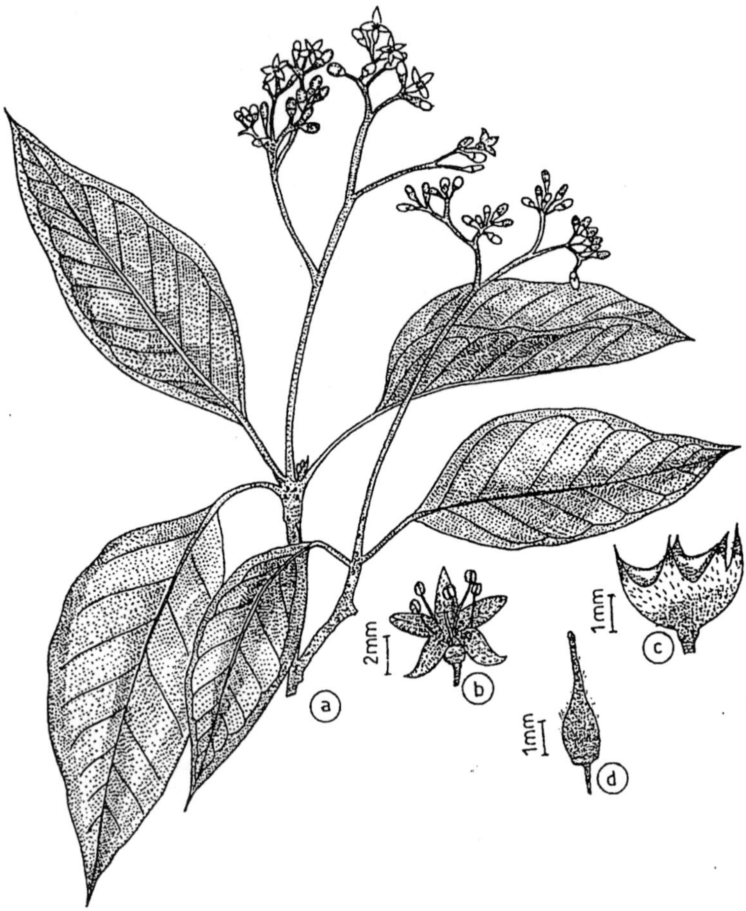

Text fig. 9: Nothapodytes nimmoniana a-twig; b-flower; c-calyx; d-carpel.

1: 517. 2000. *Premna nimmoniana* J. Grah. Cat. Pl. Bombay 155. 1839. *Mappia foetida* (Wight) Miers, Contrib. 1: 64. 1851; Mast. in Hook. f. Fl. Brit. India 1: 589. 1875; Cooke, Fl. Pres. Bombay 1: 239. 1958 (Repr. ed.). AMRUT, GHANERA, KALGUR.

Trees, ca 3 m high. Leaves crowded towards the ends of the branches, ovate oblong, 9–18 × 4.5–12 cm, acuminate, dark green, glabrous above, paler and pubescent beneath. Flowers in terminal panicles. Calyx small, pubescent externally, 5-toothed; teeth triangular, very acute. Petals cream yellow, linear-oblong, acute, 5–8 mm long, densely sericeo-villous on both surfaces, tip inflexed. Stamens 5; filaments dilated, attached to the edges of the petals at the base. Disk shallow, cup-shaped, villous within. Ovary ovoid, densely clothed with silky hairs, free, attenuated into style. Drupes ellipsoid or obovoid, 9–15 mm long, glabrous and purple when ripe.

Fls. & Frts.: October–November. *Exsiccata*: SUK-3.
Common, along the Ghats.
Localities: Amboli, Gaganbawada, Mahabaleshwar, Panchgani, Tillari, Vaibhavwadi.

Local uses: Decoction of plant is given in fever. Flowers are inhaled in headache. Stem is used in treatment of cancer.

CELASTRACEAE R. Br.

Cassine glauca (Rottb.) O. Ktze. Rev. Gen. Pl. 1: 114. 1891; Londhe in Singh *et al.* Fl. Maharashtra State Dicot. 1: 520. 2000. *Mangifera glauca* Rottb. Nye. Samml. Kongel. Norske Vid. Selsk. Skr. 2: 534, t. 4, f. 1. 1783. *Elaeodendron glaucum* (Rottb.) Pers. Syn. Pl. 1: 241. 1805; Law. in Hook f. Fl. Brit. India 1: 623. 1875, p. p. Cooke, Fl. Pres. Bombay 1: 248. 1958 (Repr. ed.). BHUTKAS, TAMRUJ.

Medium sized or fairly large trees. Leaves ovate, lanceolate, 7–8 × 2.5–4 cm, acute or rounded at base, crenate-serrate or subentire acute or acuminate. Flowers numerous in divaricate, axillary or extra axillary dichotomously branched cymes, 1–1.5 cm across. Calyx deeply divided; lobes unequal, orbicular with membranous margins. Petals greenish-white, oblong, 4–5 mm long, obtuse. Stamens shorter than the petals. Disk thick fleshy. Drupes ovoid, 8–10 mm long, apiculate. Stone 1–3 locular, 3–6 seeded.

Fls. & Frts.: February–December. *Exsiccata*: SUK-690.
Common, in dry deciduous forest.
Localities: Kagal, Ramling.

Local uses: Bark is given internally in extreme cases of anemia, as a blood purifier, given as cardiac tonic and as a wound healing.

Cassine paniculata (Wight & Arn.) Lobreau-Collen in Adansonia 2, 15: 220. 1976; Londhe in Singh *et al.* Fl. Maharashtra State Dicot. 1: 521. 2000. *Elaeodendron paniculatum* Wight & Arn. Prodr. 157. 1834. *E. glaucum* Sensu Laws. in Hook. f. Fl. Brit. India 1: 623. 1875, p. p. non (Rottb.) Pers.1805. RAGATWAD.

Medium sized or fairly large trees. Leaves ovate, lanceolate, 7–8 × 2.5–4 cm, acute or shortly acuminate. Flowers axillary, 1–1.5 cm across, in ca 10 cm long cymes. Drupes fleshy, elliptic. Seeds 2.

Fls. & Frts.: April–July. *Exsiccata*: SUK-1610.
Rare, in semi evergreen forests.
Localities: Barki, Borbet, Dhangarmola, Radhanagari, Tillari.

Local uses: Bark is given internally in extreme cases of anemia, as a blood purifier, given as cardiac tonic and as a wound healing.

Celastrus paniculatus Willd. Sp. Pl. 1: 1125. 1798; Laws. in Hook. f. Fl. Brit. India 1: 617. 1875; Cooke, Fl. Pres. Bombay 1: 245. 1958 (Repr. ed.); Londhe in Singh *et al.* Fl. Maharashtra State Dicot. 1: 521. 2000. DHIMAR-BEL, JYOTISHMATI, KANGLI, KANGUN, KARKANGANVEL, KAZAD-KUNGONI, KIRI, MALKANGONI, PIWALVEL.

Scandent or twining, glabrous shrubs. Leaves oblong to orbicular, 3–10 × 2.5–7 cm, entire and rounded or acute at base, crenate-serrate in the upper half, shortly acuminate. Flowers often unisexual, 3–4 mm across, in terminal panicles. Calyx pubescent; lobes semiorbicular, ciliate. Petals creamy-white, oblong, 3–4 mm long, rounded at apex. Ovary in female flowers globose; style short; stigma 3-lobed. Staminodes with ovate empty anthers. Disk concave, 5–10-lobed. Capsule ovoid or globose, 6–10 mm in diam., pale greenish-yellow, 3-valved. Seeds ovoid, dark brown, striate, completely enveloped in scarlet fleshy aril.

Fls. & Frts.: March–June. *Exsiccata*: SUK-68.
Common, in all forests.
Localities: Gaganbawada, Kagal, Radhanagari.

Local uses: Roots are given as an external and internal remedy in rheumatism. Root powder is used as a poultice, they are good stimulant; application to foul, unhealthy and indolent ulcers. The seeds are given as brain and the liver tonic, cure joint pain, paralysis and weakness; oil is stomachic, tonic, good for cough and asthma; used in leprosy; cures headache and leucoderma.

Salacia brunoniana Wight & Arn. Prodr. 105. 1834; Laws. in Hook. f. Fl. Brit. India 1: 626. 1875; Cooke, Fl. Pres. Bombay 1: 251. 1958 (Repr. ed.); Londhe in Singh *et al.* Fl. Maharashtra State Dicot. 1: 527. 2000.

A scandent shrub. Leaves oblong or elliptic, 4–8 × 1.5–3 cm, base acute, apex acuminate, minutely serrate, glabrous, coriaceous. Flowers few, greenish-yellow, 1–2 from axillary tubercles. Calyx lobes 5, triangular. Petals 5, ovate. Disk cup-shaped. Stamens 3, inserted on disk. Ovary ovoid, 3-celled, sunk in the disk; style short; stigma simple. Berries globose; rind woody.

Fls. & Frts.: February–April. *Exsiccata*: SUK-662.
Rare in semi evergreen forests.
Localities: Gaganbawada, Juna Vasot.

Local uses: Roots are extensively used in diabetes. The roots are given in urinary calculi. Externally applied in rheumatism, gonorrhoea and skin diseases.

Salacia chinensis L. Mant. Pl. 2: 293. 1771; Londhe in Singh *et al.* Fl. Maharashtra State Dicot. 1: 527. 2000. *S. prionoides* DC. Prodr. 1: 571. 1824; Laws. in Hook. f. Fl. Brit. India 1: 626. 1875; Cooke, Fl. Pres. Bombay 1: 250. 1958 (Repr. ed.); NISUL-BONDI, SAPTARANGI.

A large straggling shrub. Leaves oblong-ovate or lanceolate, 4–7.5 × 1.5–2.5 cm, thin, coriaceous, base subacute to obtuse, margin serrate-crenate, apex acute. Flowers ca 6 mm in diam., greenish-yellow in axillary umbels. Calyx lobes 5, broadly ovate. Petals 5, oblong-obovate, clawed, rounded at apex and with membranous margins. Disk intrastaminal, urceolate, thick, minutely papillose. Stamens 3; filaments linear. Ovary ovoid, 3-celled; style short; stigma simple. Berries globose; 1 cm in diam., minutely apiculate, fleshy, red when ripe, 1-celled, 1-seeded.

Fls. & Frts.: December–February. *Exsiccata*: SUK-362.
Common, in plains and on Ghats.
Localities: Amboli, Barki, Karanjphen, Patgaon.

Local uses: Same as *S. brunoniana*.

Salacia macrosperma Wight, Ic. t. 962. 1840; Laws. in Hook. f. Fl. Brit. India 1: 628. 1875; Cooke, Fl. Pres. Bombay 1: 251. 1958 (Repr. ed.); Londhe in Singh *et al.* Fl. Maharashtra State Dicot. 1: 529. 2000. SAPTARANGI.

Scandent shrubs. Leaves elliptic-oblong, 7.5–10.5 × 3.2–4.5 cm, base cuneate, margin entire, apex acuminate, coriaceous, glabrous. Flowers small; yellowish-green in axillary or extra axillary in dense fascicles. Calyx lobes 5, oblong. Petals 5, elliptic oblong. Disk convex, slightly depressed in the center. Stamens 3, inserted close to ovary, filaments very short. Ovary 3-celled, sunk in the disk, style very short. Berries globose, 3–5 cm in diam., verrucose, orange when ripe.

Fls. & Frts.: January–March. *Exsiccata*: SUK-363.

Common, shrub in forests, and sacred groves.
Localities: Amba, Amboli, Tillari, Vasota.

Local uses: Similar to as that of *S. brunoniana*.

RHAMNACEAE Juss.

Ventilago maderaspatana Gaertn. var. **maderaspatana** Fruct. 1: 223. t. 49, f. 2, 1788; Laws. in Hook. f. Fl. Brit. India 1: 631. 1875; Cooke, Fl. Pres. Bombay 1: 253. 1958 (Repr. ed.); Pradhan in Singh *et al.* Fl. Maharashtra State Dicot. 1: 542. 2000. LOKANDE.

Woody climber. Leaves elliptic, ovate or lanceolate, 3–6 × 2–3 cm, thin, coriaceous, pubescent, obtuse or subacute at base, margin entire, apex acute-acuminate. Flowers 3 mm in diam., greenish with an offensive odor, in large spreading panicles. Calyx pubescent or puberulous outside, glabrescent within; lobes triangular erect, very acute, keeled on inner face and with a hard incurved point at apex within. Petals much smaller than the calyx lobes. Disk 5-lobed, pubescent, lining the calyx. Stamens 5, adnate to petals. Ovary embedded in disk, 2-celled, pubescent; style 2-fid; stigma simple. Nut globular, 5 mm in diam., supported by persistent calyx, wings oblong, 3.5–5 × 0.8–0.5 cm, 1-nerved.

Fls. & Frts.: December–February. *Exsiccata*: SUK-1614.
Frequent, in moist deciduous forests.
Localities: Barki, Borbet, Gajapur, Sawantwadi, Tillari.

Local uses: Leaves are crushed and applied for itch. Powdered root bark is carminative, stomachic, tonic and stimulant; useful as tonic dyspepsia; debility and in slight cases of fever. Stem bark powdered and mixed with gingelly oil and used for the external application on skin diseases and itch.

Ziziphus oenoplia (L.) Mill. Gard. Dict. ed. 8, 3. 1768; Laws. in Hook. f. Brit. India 1: 634. 1875; Cooke, Fl. Pres. Bombay 1: 257. 1958 (Repr. ed.); Pradhan in Singh *et al.* Fl. Maharashtra State Dicot. 1: 545. 2000. *Rhamnus oenoplia* L. Sp. Pl. 194. 1753. BURGI, YERUNI

Straggling shrubs. Leaves distichous, obliquely rhomboid or ovate, 2–6 × 1.5–2.5 cm, 3-nerved from the oblique base, entire, acute, glabrous or pubescent above, rusty tomentose, especially on nerves beneath; stipular spines solitary, hooked. Flowers minute, in subsessile, pubescent, paniculate cymes. Calyx hairy outside; lobes ovate acute. Petals greenish, obovate, 1.5–2 mm long, cuneate, shorter than the calyx. Stamens shorter than the petals. Disk 10-lobed, glabrous. Styles united almost to the apex. Drupes globose, ca 5 mm in diam., black; pulp scanty; stone woody, 2-celled.

Fls. & Frts.: December–February. *Exsiccata*: SUK-1611.

Common, throughout the region.
Localities: Kolhapur, Sangali, Satara.

Local uses: Roots are prescribed in itching; leaf decoction is given in mild cases of dysentery and stomachache. Decoction of root is used in fever and powder is applied to old wounds and ulcers. Seed powder is given to reduce obesity.

Ziziphus rugosa Lam. Encyl. 3: 319. 1789; Laws. in Hook. f. Fl. Brit. India 1: 636. 1875; Cooke, Fl. Pres. Bombay 1: 258. 1958 (Repr. ed.); Pradhan in Singh *et al.* Fl. Maharashtra State Dicot. 1: 545. 2000. CHURAN, TAURN.

Straggling armed shrubs. Leaves ovate or oblong-elliptic 4–8 × 2.2–6.2 cm, 5-nerved, base cordate, margin serrate, glabrous above, fulvous tomentose beneath; prickles from a broad base, solitary, short, recurved. Flowers yellowish-green on spinous, tomentose panicles. Calyx pubescent outside; lobes ovate, acute, keeled on inner surface. Petals greenish obovate, 1.5–2 mm long, shorter than the calyx. Disk 5-lobed. Stamens 5, enclosed in and shorter than the petals. Ovary ovoid, tomentose, 2-celled; style 2, connate below the middle. Drupes globose pyriform, stone thin, 1-celled, 1-seeded.

Fls. & Frts.: June–August. *Exsiccata*: SUK-292.
Common, in moist deciduous forests.
Localities: Gaganbawada, Radhanagari.

Local uses: Roots mixed with jaggery and small pills prepared from it is given in leucorrhoea and also as a tonic. The roots are dipped in hot water for 24 hrs and then used to stop toothache.

VITACEAE Juss.

Ampelocissus latifolia (Roxb.) Planch. in J. Vigne, Amer. 374. 1884 & in DC. Monogr. Phan. 5(2): 370. 1887; Kulkarni in Singh *et al.* Fl. Maharashtra State Dicot. 1: 549. 2000. *Vitis latifolia* (Lam.) Roxb. Fl. India. 1: 661. 1820; Laws. in Hook. f. Fl. Brit. India 1: 652. 1875; Cooke, Fl. Pres. Bombay 1: 269. 1958 (Repr. ed.). DUKELA, NADENA.

Climbers; stem striate; glabrous, tendrils forked. Leaves orbicular,cordate, 10–17 cm long; 3–7-lobbed or palmate, sharply and irregularly toothed, reticulately veined, tomentose beneath when young; 5–7 nerved, lowest pair almost horizontal, branched on lower side. Flowers 5-merous, 2–3 mm across, deep reddish-brown, in compact thyrsoid cymes. Calyx small, truncate. Petals oblong, 2–2.5 mm long. Stigma sessile, foveolate. Berries globose, 6–9 mm in diam., glabrous 2-seeded, black. Seeds ellipsoid rounded, 6–7 mm in diam., transversely grooved on the back, flattened, smooth, polished.

Fls. & Frts.: June–August. *Exsiccata*: SUK-101.
Common, in hill forests along gullies.
Localities: Amboli, Vaibhavwadi.

Local uses: The roots are externally applied in wounds, itching and elephantiasis.

Cayratia trifolia (L.) Domin. in Biblioth. Bot. 89. 371. 1927; Kulkarni in Singh *et al.* Fl. Maharashtra State Dicot. 1: 552. 2000. *Vitis trifolia* L. Sp. Pl. 203. 1753; Cooke, Fl. Pres. Bombay 1: 271. 1958 (Repr. ed.). *V. carnosa* Wight & Arn. Prodr. 127. 1834; Laws. in Hook. f. Fl. Brit. India 1: 654. 1875. AMBAT-VEL, DHAVRI.

Herbaceous climbers; stem pubescent; tendril slender, forked. Leaves subsucculent, 3-foliolate, leaflets ovate-lanceolate, 3–5 × 2–3 cm, crenate serrate, acute or obtuse, more or less pubescent on both the surfaces. Flowers in long-peduncled, cymes. Calyx pubescent outside, truncate or minutely 4 toothed. Petals greenish-white, oblong, rounded at apex, hooded, pubescent outside. Disk cup-shaped. Berries depressed, globose, 7–8 mm in diam., fleshy, 2–4-seeded, black. Seeds trigonous, transversely rugose.

Fls. & Frts.: August–October. *Exsiccata*: SUK-175.
Common, on fence around fields and creeping on bare rocks in hill forests.
Localities: Halkarni, Patgaon, Wai.

Local uses: Roots in combination with piper fruits are given in abscess.

Cissus elongata Roxb. Fl. India 1: 429. 1820; Kulkarni in Singh *et al.* Fl. Maharashtra State Dicot. 1: 553. 2000. *Vitis elongata* (Roxb.) Wall. Num List. no. 6016. 1831–32; Laws. in Hook. f. Fl. Brit. India 1: 658. 1875; Cooke, Fl. Pres. Bombay 1: 273. 1958; (Repr. ed.).

Large climbers; stem compressed, succulent smooth, tendrils long simple, leaf-opposed. Leaves usually digitately 5–3-foliolate; leaflets oblong, lanceolate or elliptic-lanceolate; 10–15 × 2–6 cm, acuminate, serrate, narrow into very short petiole. Flowers greenish purple, 4-merous, in di or di-trichotmous leaf-opposed cymes. Calyx membranous, cup-shaped, truncate. Petals 4, oblong, subacute, 2.5–3 mm long. Disk 4-lobed. Ovary ovoid; style slender; stigma simple. Berries 1–2 cm in diam., ellipsoid, 1-seeded, black. Seeds ca 1.2 mm in diam., ellipsoid, pointed at one end, rugose on the back, with large elliptic pit on the face.

Fls. & Frts.: May–October. *Exsiccata*: SUK-220.
Common, in moist deciduous forests.
Localities: Amba, Barki, Panhala.

Local uses: Stem is given in certain skin disease for external application such as in elephantiasis, scabies and in itching. Stem is also used as wound healing. Fresh stem is applied on insect bite to stop itching and inflammation.

Cissus pallida (Wight & Arn.) Planch. in DC. Monogr. Phan. 5: 477. 1887; Kulkarni in Singh *et al.* Fl. Maharashtra State Dicot. 1: 557. 2000. *Vitis pallida* auct. non. Wight & Arn. in Wight. Cat. India Pl. 1833; Laws. in Hook. f. Fl. Brit. India 1: 647. 1875; Cooke, Fl. Pres. Bombay 1: 265. 1958 (Repr. ed.).

Climbing shrubs; stem woody below, terete; tendrils simple. Leaves ovate or orbicular; 7–15 cm in diam., pale green, shortly acuminate, glabrous, margins with shallow, distant glandular incurved serratures. Flowers in compound umbellate cymes. Calyx saucer-shaped, margins membranous. Petals 4, calyptrately deciduous, triangular ovate, free at apex. Disk 4-lobed. Berries 1–2-seeded, pyriform or subglobose, size of pea, mucronate, supported on a incurved pedicels.

Fls. & Frts.: June–July. *Exsiccata*: SUK-1616.
Common, in dry regions.
Localities: Kartikswamighat, Shingnapur.

Local uses: Stem is externally applied in rheumatism. It is given with leaves of *Azadirachta indica* in chronic or old wounds for 15 days.

Cissus quadrangularis L. Mant. 1: 39. 1767; Kulkarni in Singh *et al.* Fl. Maharashtra State Dicot. 1: 557. 2000. *Vitis quadrangularis* Wall. ex Wight & Arn. Prodr. 125. 1834; Laws. in Hook. f. Fl. Brit. India 1: 645. 1875; Cooke, Fl. Pres. Bombay 1: 266. 1958 (Repr. ed.). CHAUDARI, HARSANKAR, KANDVEL.

Extensive climbers with long aerial roots; stem leafless when old, very long, fleshy, glabrous; angles of the young branches winged; tendrils long, slender, simple. Leaves broadly ovate or reniform, 3–5 cm long, sometimes obtusely 3–7-lobed, denticulate, glabrous. Flowers in shortly peduncled, umbellate, axillary cymes. Calyx cup-shaped, truncate or obscurely 4 toothed. Petals greenish-white, ovate-oblong, 2–3 mm long, acute and hooded at apex. Disk erect, 4-lobed. Berries obovoid, 4–6 mm long, red, apiculate, 1–2 seeded.

Fls. & Frts.: March–May. *Exsiccata*: SUK-613.
Occasionally, planted and also running wild.
Localities: Ichalkaranji, Kolhapur, Shirol.

Local uses: The juice of slightly burned stem with honey is given in irregular menstruation. The fresh stem is beaten and paste is given in asthma. In backache, crushed stem is tied around the back. Decoction of stems is given in bone fractures. Fresh stem pounded and applied for burns and wounds. In indigestion, the leaves and stems are used. Stem is used as stomachic, tonic, piles, constipation, chronic ulcers and in bone fractures.

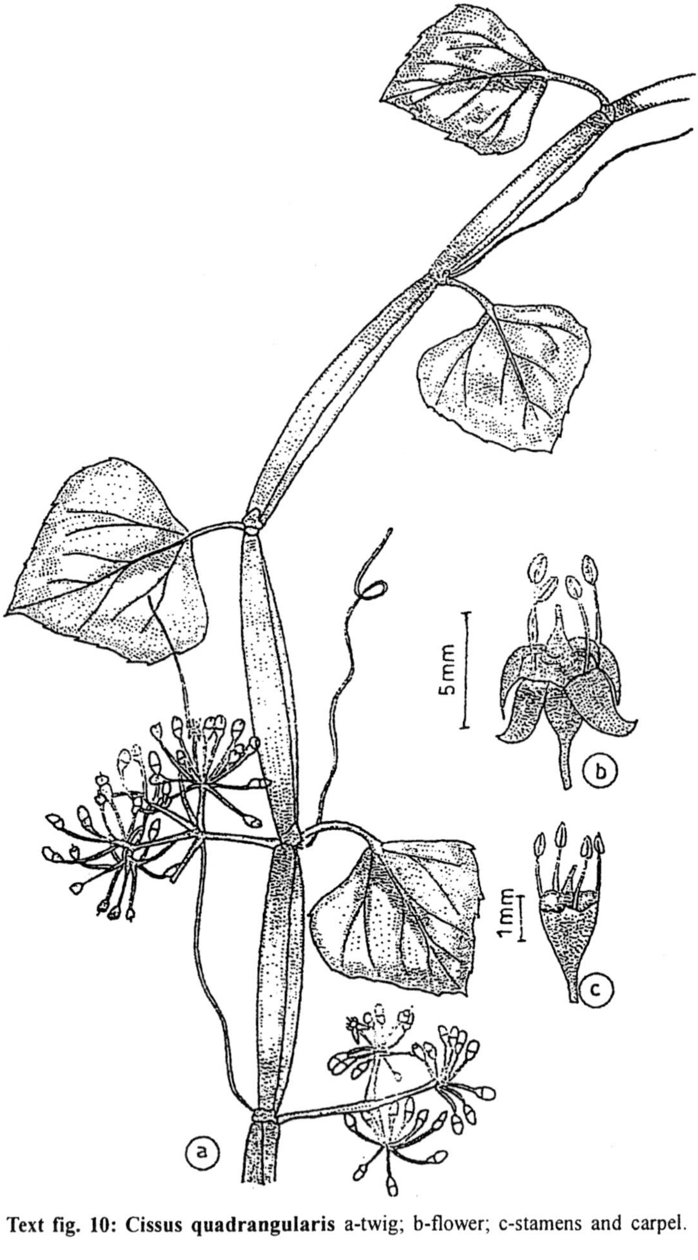

Text fig. 10: Cissus quadrangularis a-twig; b-flower; c-stamens and carpel.

Cissus repens Lam. Encycl. 1: 31. 1789; Kulkarni in Singh *et al.* Fl. Maharashtra State Dicot. 1: 559. 2000. *Vitis repens* (Lam.) Wight & Arn. Prodr. 125. 1834; Laws. in Hook. f. Fl. Brit. India 1: 646. 1875; Cooke, Fl. Pres. Bombay 1: 265. 1958 (Repr. ed.).

Slender, scandent, glabrous shrubs, 60–100 cm tall; tendrils forked. Leaves membranous, broadly ovate, 5–15 × 2–7 cm, cordate, distantly spinous toothed, acuminate. Flowers in compound umbellate, extra-axillary cymes. Calyx membranous, truncate. Petals greenish-white, triangular-ovate, 2–2.5 mm long. Berries pyriform, 3–4 mm long, 1-seeded.

Fls. & Frts.: February–April. *Exsiccata*: SUK-91.

Common, in semi evergreen forests.

Localities: Amboli, Panhala, Phonda, Sawantwadi, Shiroda, Vaibhavwadi.

Local uses: Plant is applied to sloughing and fetid ulceration, also to boils, and small abscesses as a maturant. Root powder is heated and applied to cuts and fracture.

Cissus setosus (Roxb.) Alston in Trimen, Handb. Fl. Ceylon 6: Suppl. 53. 1931; Kulkarni in Singh *et al.* Fl. Maharashtra State Dicot. 1: 563. 2000. *Vitis setosa* (Roxb.) Wall. ex Wight & Arn. in Wight, Cat. Ind. Pl. 26. 1833 & Prodr. 127. 1834; Laws. in Hook. f. Fl. Brit. India 1: 654. 1875; Cooke, Fl. Pres. Bombay 1: 270. 1958 (Repr. ed.). KHAJAGOLICHA-VEL. TRTPHAN-KHAND.

Succulent, herbaceous climbers; stem and leaves covered with scattered glandular bristles. Leaves sessile; leaflets shortly stalked, elliptic-obovate-oblong, 3–10 × 2–6 cm, irregularly toothed, obtuse, glabrous above, glandular hispid beneath. Flowers small, in leaf opposed, or apparently terminal, divaricate cymes. Calyx cup-shaped, truncate. Petals creamy white, oblong, 2 mm long, reflexed. Berries ovoid or globose, 1–1.5 cm in diam., scarlet, glandular hispid. Seeds solitary, subglobose, smooth.

Fls. & Frts.: August–November. *Exsiccata*: SUK-184.

Common, on rocky slopes in hill forests.

Localities: Alate, Babujamal, Bahubali, Ramling.

Local uses: Root is given in heart troubles. It is externally applied in small body tumors; orally with the honey or with sugar cane juice it is given in biliousness. Root cures tumors, pains and spleen complaints; purifies blood and given for inflammation of the spleen.

Vitis vinifera L. Sp. Pl. 202. 1753; Laws. in Hook. f. Fl. Brit. India 1: 652. 1875; Cooke, Fl. Pres. Bombay 1: 275. 1958 (Repr. ed.); Kulkarni in Singh *et al.* Fl. Maharashtra State Dicot. 1: 565. 2000. DRAKSHA, ANGOOR.

Extensive, perennial climbers; stem fistular, striate; young parts tomentose. Leaves rotund-cordate, 10–12 cm across, dentate and 3–7-lobed, glabrous above, tomentose beneath when young. Flowers in long peduncled, pubescent cymes. Calyx cup-shaped, truncate, pubescent outside. Petals white, oblong, connate, hooded at apex, calyptrately deciduous. Ovary globose. Berries globose or oblong, varying in size, pale green or purple, 2–4-seeded. Seeds oblong-ovoid, brown.

Fls. & Frts.: November–May.
Cultivated.
Localities: Kolhapur, Sangli, Satara.

Local uses: The leaves are externally applied in skin disease. The leaves are useful in piles, juice cures headache, piles, inflammation of piles, diuretic, stops bleeding from the mouth, fruits are used as tonic to the liver and in chronic bronchitis.

LEEACEAE (DC.) Dumort.

Leea indica (Burm. f.) Merr. in Philipp. J. Sci. 14: 245. 1919; Moorthy in Singh *et al.* Fl. Maharashtra State Dicot. 1: 567. 2000. *Staphylea indica* Burm. f. Fl. Ind. 75. t. 24. f. 2. 1768. *Leea sambucina* (L.) Willd. Sp. Pl. 1: 1177. 1797 *nom. illeg.*; Laws. in Hook. f. Fl. Brit. India 1: 666. 1875 p. p.; Cooke, Fl. Pres. Bombay 1: 277. 1958 (Repr. ed.). DINDA, KARKANI.

Erect shrubs, 2–3 m tall. Leaves 2–3 pinnate, 30–60 cm long; leaflets variable in shape, usually elliptic, 7–20 × 3–8 cm, acute, or crenate at base, irregularly serrate, acuminate. Flowers in branched, corymbose, paniculate cymes. Calyx shortly divided; lobes subacute or rounded with membranous margins. Staminal tube with lobes rounded and notched at apex; anthers united in bud. Berries globose, 6–8 mm in diam., 2–6 lobed, smooth, purple black.

Fls. & Frts.: August–March. *Exsiccata*: SUK-158.
Common, throughout the region.
Localities: Amba, Anuskura, Chandgad, Tillari, Sawantwadi.

Local uses: Roots are given with fruits of *Tamarindus indica* in chronic dysentery. Leaves and roots are given in abdominal pain.

Leea macrophylla Roxb. ex Horn. Hort. Hafn. 1: 231. 1813; Laws. in Hook. f. Fl. Brit. India 1: 664. 1875; Cooke, Fl. Pres. Bombay 1: 276. 1958 (Repr. ed.); Moorthy in Singh *et al.* Fl. Maharashtra State Dicot. 1: 568. 2000. ANDERPHOD, BENDAR, DINDA, DINDO, SUPAD, VHODLO.

Stout, erect herbs, 1.5 m tall; root tuberous, perennial. Leaves broadly ovate, 15–90 cm long, cordate at base, shallowly lobed, serrate, acute, glabrous above, pubescent beneath. Flowers in terminal, much branched, 20–30 cm long, puberulous, corymbose cymes. Peduncles stout, deeply grooved. Calyx divided about 1/3 way down, lobes triangular-ovate, tipped with a hard point. Petals greenish-white, oblong, 2–3 mm long. Staminal tube deeply divided; lobes oblong, entire emarginate. Berries globose, 6–8 mm in diam., black, 3–6 lobed.

Fls. & Frts.: July–October. *Exsiccata*: SUK-112.
Occasional, in hill forests.
Localities: Ganpatpule, Pawas, Sawantwadi.

Local uses: Roots are pasted and applied on wounds and ulcers. Roots are also used in itching and elephantiasis.

Leea setuligera C. B. Cl. in Trim. J. Bot. 19: 105. 1881; Cooke, Fl. Pres. Bombay 1: 279. 1958 (Repr. ed.); Moorthy in Singh *et al.* Fl. Maharashtra State Dicot. 1: 569. 2000. LAHAN-DINDA, TAMBDA-DINDA.

Perennial herbs; root tuberous. Leaves compound 2–3-pinnate; leaflets elliptic or elliptic lanceolate, 4–5 × 2–3 cm, acuminate, sharply and closely serrate, clothed with stiff bristles between nerves above and beneath, base rounded or acute. Flowers in small compact umbellate cymes, reddish-white. Petals 5, oblong. Stamens 5, united at base; lobes of the staminal tube bifid; anthers united in bud. Berries irregularly lobed.

Fls. & Frts.: June–October. *Exsiccata*: SUK-1609.
Common, in deciduous and semi evergreen forests.
Localities: Amba, Amboli, Gaganbawada.

Local uses: Tubers are used to cure skin diseases paste of tuber is applied externally in elephantiasis and itch. Tubers are also applied to heal wounds. Tubers are also used on affected areas of skin due to biting of leech to stop blood and used in fissures

SAPINDACEAE Juss.

Allophylus cobbe (L.) Raeusch. Nom. Bot. ed. 3, 108. 1797; Hiern in Hook. f. Fl. Brit. India 1: 673. 1875; Cooke, Fl. Pres. Bombay 1: 282. 1958 (Repr. ed.); Kulkarni in Singh *et al.* Fl. Maharashtra State Dicot. 1: 570. 2000. *Rhus cobbe* L. Sp. Pl. 267. 1753. TIPAN.

Small trees or shrubs. Leaves 3-foliolate; leaflets 6–12.5 × 3.5–5.6 cm, ovate elliptic, acute or acuminate, sharply and distantly serrate, dentate, more or less hairy pubescent, terminal leaflet usually acute at the base, lateral leaflets rounded and more or less oblique. Flowers small, white, shortly pedicelled, in fascicles along the branches of spikes, axillary 2–4 branched racemes. Sepals glabrous. Petals cuneate, with fringed scale above the claw. Stamens 8. Ovary 2-lobed hairy. Fruit globose, smooth size of a pea, red when ripe.

Fls. & Frts.: May–August. *Exsiccata*: SUK-67.
Common, in moist deciduous, semi evergreen forests.
Localities: Gaganbawada, Radhanagari, Panhala, Mahabaleshwar.

Local uses: Leaf paste is applied externally as a bandage in bone fracture, similarly, stem is also used for the same purpose.

Cardiospermum halicacabum L. Sp. Pl. 366. 1753; Hiern in Hook. f. Fl. Brit. India 1: 670. 1875; Cooke, Fl. Pres. Bombay 1: 280. 1958 (Repr. ed.); Kulkarni in Singh *et al.* Fl. Maharashtra State Dicot. 1: 571. 2000. GHANPHODI, KAPALPHODI, TEJOVATI.

Annual, climbing, glabrous or pubescent herbs. Leaves biternate; uppermost often trifoliate; leaflets lanceolate-oblong to ovate, 1–6 × 0.5–3 cm, incised serrate to pinnatifid, ciliate, acute or obtuse and mucronate, glabrous above, pubescent beneath. Flowers in umbellate cymes on axillary peduncles; peduncles with 2 tendrils at the top. Sepals distinct, suborbicular, unequal. Petals white, obovate-spathulate, 2.5–3 mm long. Stamens 8, filaments hairy. Pistillode in male flowers minute. Capsule 3-gonous, bladdary brown, 1.5–2 × 2–2.5 cm, pubescent. Seeds glabrous, smooth.

Fls. & Frts.: July–December. *Exsiccata*: SUK-72.
Common, in wet fields, on waste lands and in hedges.
Localities: Ichalkarangi, Kolhapur, Panhala.

Local uses: Leaves are bruised in castor oil and applied externally in inflammations, leaf decoction is dropped in ear to stop earache. It is also used to cure rheumatism.

Dimocarpus longan Lour. Fl. Cochinch. 233. 1079; Kulkarni in Singh *et al.* Fl. Maharashtra State Dicot. 1: 572. 2000. *Nephelium longana* (Lam.) Camb. in Mem. Mus. Hist. Nat. 18: 30. 1829; Hiern in Hook. f. Fl. Brit. India 1: 688. 1875; Cooke, Fl. Pres. Bombay 1: 285. 1958 (Repr. ed.). UMB, VUMB.

Trees, 13–15 m high. Leaves 8–36 cm long, equally or unequally pinnate; leaflets 2–5 pairs, oblong, ovate-lanceolate, 5–10 × 3–5 cm, oblique, shortly acuminate, glabrous, reticulately veined, base cuneate. Flowers small, white in clusters on the branches of terminal, axillary large much branched puberulous panicles. Calyx 5–6 lobed, lobes obtuse, densely pubescent. Petals 5–6, linear-spathulate, pubescent. Disk pubescent. Stamens 6–10, included in female, exerted in male. Ovary 2–3 lobed; style long; stigmas 2, recurved. Fruit supported on persistent calyx, globose, 5–8 cm, reddish, tubercled. Seeds solitary, enveloped in fleshy aril, black, shining.

Fls. & Frts.: February–April. *Exsiccata*: SUK-360.
Common, in semi evergreen forests.
Localities: Amba, Barki, Borbet, Gajapur, Tillari.

Local uses: Fruits are used like that of *Sapindus laurifolius*. Fruits are internally given in small children to stop stomachache. It is used as a tonic.

Dodonea viscosa (L.) Jacq. Enum. Syst. Pl. 19. 1760; Hiern in Hook. f. Fl. Brit. India 1: 697. 1875; Cooke, Fl. Pres. Bombay 1: 287. 1958 (Repr. ed.); Kulkarni in Singh *et al.* Fl. Maharashtra State Dicot. 1: 572. 2000. *Ptelea viscosa* L. Sp. Pl. 118. 1753. BAAD, JAKHMI.

Shrubs or small trees; young parts glandular pubescent. Leaves, oblanceolate, 3–10 × 0.6–3 cm, subsessile, entire, acute; margins often revolute. Flowers unisexual, inconspicuous, in axillary or terminal cymes. Sepals oblong. Petals absent. Stamens 5–10, usually 8. Disk minute in male flowers, small in female flowers. Ovary 3-angled. Capsules orbicular, 1–2 cm across, with wide marginal wings, glabrous. Seeds compressed, brown.

Fls. & Frts.: June–January. *Exsiccata*: SUK-63.
Common, in dry deciduous forests.
Localities: Alate, Babujamal.

Local uses: Leaves are applied on burned part. Leaf paste is applied externally in rheumatism.

Lepisanthes tetraphylla (Vahl) Radlk. Sap. Holl. Ind. 35 (105) 1877; Kulkarni in Singh *et al.* Fl. Maharashtra State Dicot. 1: 573. 2000. *Hemigyrosa canescens* Bl. in Rumphia 3: 166. in Obs. 1847; Hiren in Hook. f. Fl. Brit. India 1: 671. 1875; Cooke, Fl. Pres. Bombay 1: 281. 1958 (Repr. ed.). KARPA, LOKHANDI.

Middle sized trees. Leaves variable in length, rachis smooth; leaflets 2–3 pairs, opposite or subopposite, oblong lanceolate, 6–12 × 3–5 cm, obtuse, acute or emarginate, glabrous, shining ashy green. Flowers in simple or panicled velvety-pubescent racemes which are axillary or scattered. Sepals unequal, the outer ovate orbicular, ciliate, inner larger, orbicular, concave, oblong spathulate, shortly clawed. Scale hairy at base, 2-lobed. Stamens slightly longer than petals, filaments hairy. Fruits subglobose trigonous, 1.5–2.5 cm in diam. Velvety tomentose, apiculate yellow. Single seeded.

Fls. & Frts.: February–May. *Exsiccata*: SUK-293.
Rare, in semi evergreen and moist deciduous forests.
Localities: Barki, Borbet, Chandgad, Dajipur, Gajapur, Panhala.

Local uses: Leaf ash is mixed with castor oil and applied externally in itch and elephantiasis.

Text fig. 11: Lepisanthes tetraphylla a-twig; b-flower; c-bract; d-calyx; e-petal; f-carpel.

Sapindus laurifolius Vahl, Symb. Bot. 3: 54. 1794; Cooke, Fl. Pres. Bombay 1: 284. 1958 (Repr. ed.); Kulkarni in Singh *et al.* Fl. Maharashtra State Dicot. 1: 576. 2000. *S. trifoliatus* auct. non L. 1753; Hiern in Hook. f. Fl. Brit. India 1: 682. 1875 p. p. ARITHA, RINGI, RITHA.

Trees; young parts rusty tomentose. Leaves abruptly pinnate, 10–25 cm long; leaflets subopposite, 2–3 pairs, 5–18 × 2–10 cm, acute or obtuse at base, glabrous above. Flowers in terminal, rusty tomentose panicles shorter than the leaves, males numerous, bisexual few. Calyx 5-cleft; sepals rotund-ovate, fulvous tomentose outside, glabrous within. Petals white, lanceolate, slightly longer than the sepals, clawed and with lateral scales above the claw. Disk concave, margin hairy. Stamens 8; filaments villous. Ovary densely hairy. Drupes 2–3-lobed, clothed with fulvous tomentum when young, glabrous and wrinkled when ripe, dark brown, 2–3 seeded. Seeds blackish, smooth, 6–8 mm in diameter.

Fls. & Frts.: October–February. *Exsiccata*: SUK-616.
Usually cultivated.
Localities: Ichalkaranji, Gadhinglaj, Kolhapur, Panhala.

Local uses: Fruits are given internally as an emetic in cough and asthma. 2–3 drops of the fruit juice are inhaled in nose to stop hemicrania. Fruits in combination with the fruits of *Acacia sinuata* is applied on hairs to stop dandruff and graying of hairs. Fruits are used in chronic dysentery, diarrhoea, cholera, hemicrania, tubercular glands, paralysis of the limbs and in lumbago.

Schleichera oleosa (Lour.) Oken in Allg. Naturgesch. 3(2): 1341. 1841; Kulkarni in Singh *et al.* Fl. Maharashtra State Dicot. 1: 577. 2000. *Pistacia oleosa* Lour. Fl. Cochinch. 2: 615. 1790. *Schleichera trijuga* Willd. Sp. Pl. 4 (2): 1096. 1805; Hiern in Hook. f. Fl. Brit. India 1: 681. 1875; Cooke Fl. Pres. Bombay 1: 283. 1958 (Repr. ed.). KOSAB, KOSHIMB, KUSUM.

Large trees, 12–15 m tall. Leaves abruptly pinnate, 15–30 cm long; petiole grooved. Leaflets 3 pairs, opposite, elliptic-oblong to obovate-oblong, 6–12 × 2–10 cm, unequal rounded, subacute, or subcordate base, entire, obtuse, glabrous above, densely brown tomentose beneath. Flowers minute, in axillary, lax panicles shorter than the leaves. Calyx pubescent outside, lobes acute. Petals absent. Disk with wavy margin. Stamens 5–8. Ovary ovoid, pointed. Drupes ovoid globose, 1.5–2 cm long, pointed, echinate, with small blunt prickles. Seeds 1 or 2, enclosed, in pulpy, edible aril.

Fls. & Frts.: March–June. *Exsiccata*: SUK-1612.
Occasional, in hill forests.
Localities: Barki, Chandgad, Panhala, Tillari.

Local uses: Leaf powder is given in leprosy with the oil of *Azadirachta indica*. Leaves are also used for the cure of itch and acne (an inflammatory diseases of the skin). Stem bark is mixed with castor oil and applied externally on skin diseases. Oil is used as tonic, stomachic, anthelmintic, purgative; cure skin disease and heal ulcers.

ANACARDIACEAE Lindl.

Anacardium occidentale L. Sp. Pl. 383. 1753; Hook. f. Fl. Brit. India 2: 20. 1876; Cooke, Fl. Pres. Bombay 1: 292. 1958 (Repr. ed.); Prasanna in Singh *et al.* Fl. Maharashtra State Dicot. 1: 581. 2000. KAAJU.

Small trees 3 m tall. Leaves obovate or elliptic, 8–12.5 × 5–8 cm, coriaceous, rounded at apex, glabrous, cuneate. Flowers in terminal panicles. Sepals lanceolate, puberulous outside. Petals linear lanceolate, deflexed from others. Ovary glabrescent, attenuated, subulate. Fruit 2–3 cm long; pedicel large, fleshy, dark coloured.

Fls. & Frts.: January–March. *Exsiccata*: SUK-300.
Cultivated.
Localities: Amba, Ratnagiri, Sawantwadi.

Local uses: Extract of bark is given to lower the blood sugar, Tar or acrid oil from the bark is externally applied in leprosy, corns, ulcers and as counter irritant and rubefacient. Shell oil is used in treatment of hook-worms; also for cracks on soles of feet, warts. The resinous juice from the cashewnut is useful in mental derangement, sexual debility.

Buchanania cochinchinensis (Lour.) Almeida, Fl. Mah. 1: 287. 1996; Prasanna in Singh *et al.* Fl. Maharashtra State Dicot. 1: 582. 2000. *Toluifera cochinchinensis* Lour. Fl. Cochinch. 262. 1790. *Buchanania lanzan* Spreng. in J. Bot. (Schrad.) 2: 234. 1800; Cooke, Fl. Pres. Bombay 1: 293. 1958 (Repr. ed.). *B. latifolia* Roxb. Fl. Ind. 2: 385. 1832; Hook. f. Fl. Brit. India 2: 23. 1876. CHAR, CHAROLI, PIYAL.

Large shrubs or small trees; young parts silky pubescent. Leaves thickly coriaceous, oblong ovate to obovate, 9–20 × 5–10 cm, rounded at base, entire, obtuse or emarginate, glabrous above, densely pubescent beneath. Flowers small, sessile in terminal and axillary, brown tomentose panicles. Calyx short, 3–5 lobed, lobes broadly ovate, ciliate. Petals greenish-white, ovate-oblong, 2–2.5 mm long. Disk fleshy. Stamens 10, shorter than the petals, flattened. Ovaries 5, one perfect, conical, villous the other 4 reduced to filaments. Drupes lenticular, 10–12 mm long, black; stone hard, 2-valved.

Fls. & Frts.: February–May. *Exsiccata*: SUK-260, 264.
Common in hill forests.
Localities: Achirne, Babu Jamal, Nesari.

Local uses: Oil extracted from the seeds is used on skin diseases. It is also employed to remove spots and blemishes from the face. Oil is applied in glandular swellings of neck. Leaf powder is applied externally in inflammation of gum. Plant extract is used as anticancer. Leaf juice used as blood purifier, in biliousness and burning sensation.

Holigarna arnottiana Hook. f. Fl. Brit. India 2: 36. 1876; Cooke, Fl. Pres. Bombay 1: 298. 1958 (Repr. ed.); Prasanna in Singh *et al.* Fl. Maharashtra State Dicot. 1: 583. 2000. BIBBA, HULGERI.

Trees, 10–25 m tall, branches densely leafy at apex. Leaves obovate-lanceolate, 12–25 × 6–12 cm, acute or rounded at apex, glabrous, paler beneath; 2 petiolar spurs or tubercles at the top of petiole, early deciduous. Flowers minute, crowded, in rufous-tomentose panicles. Calyx-teeth very short. Petals 5, oblong, 2–2.5 mm long, acute. Drupes obliquely ovoid, ca 2.5 cm long, rounded at top, glabrous, long-pedicelled, entirely included in the torus.

Fls. & Frts.: January–February. *Exsiccata*: SUK-366, 660.
Common, in evergreen forests.
Localities: Amboli, Doddamarg, Sawantwadi.

Local uses: Plant is poisonous, juice of the fruit is used in old incurable wounds.

Holigarna grahamii (Wight) Kurz. in Asiat. Soc. Beng. 41 (2): 305. 1872; Hook. f. Fl. Brit. India 2: 37. 1876; Cooke, Fl. Pres. Bombay 1: 298. 1958 (Repr. ed.); Prasanna in Singh *et al.* Fl. Maharashtra State Dicot. 1: 583. 2000. *Semecarpus grahamii* Wight, Ic, t. 235. 1839. BIBBA, HULGERI.

Trees, 3–4 m tall; young branches ferrugineo-tomentose. Leaves oblanceolate, 30–50 × 10–15 cm, rigidly coriaceous, triangular above the middle, acute or acuminate, glabrous or shining above, ferrugineo-pubescent beneath; petioles with 1–2 spurs on each side. Flowers in rusty tomentose terminal panicles, clustered, the male panicle about one feet long, the female shorter. Calyx cup shaped; lobes minute. Petals 5, 2.5 mm long, densely villous. Drupes ca 2 cm long, enclosed in the cup-shaped torus, except 1/6 or 1/4 of the upper part which is exserted.

Fls. & Frts.: November–May. *Exsiccata*: SUK-372.
Common, in semi evergreen forests.
Localities: Amboli, Barki, Borbet, Dajipur, Tillari, Udegiri.

Local uses: Plant is poisonous, juice of the fruit is used in old incurable wounds.

Lannea coromandelica (Houtt.) Merr. in J. Arnold Arbor. 19: 353. 1939; Prasanna in Singh *et al.* Fl. Maharashtra State Dicot. 1: 584. 2000. *Dialium coromandelicum* Houtt. Nat. Hist. 2, 2: 39, t. 5, f. 2. 1774. *Odina wodier* Roxb. Fl. India 2: 293. 1832; Hook. f. Fl. Brit India 2: 29. 1876; Cooke, Fl. Pres. Bombay 1: 296. 1958 (Repr. ed.). MOYA, SHIMTI.

Small deciduous trees; bark ash colored. Leaves imparipinnate, 20–45 cm long; leaflets 3–5 pairs and an odd one ovate, elliptic, 7.5–11.5 × 4–6 cm, oblique at base, entire, acuminate, glabrous, flower appearing when tree is bare of leaves, crowded in cymose fascicles. Male racemes compound, pendulous, female ones simple, erect; pedicels very short, pubescent. Calyx divided to the base; lobes ovate orbicular, ca 1.5 mm long, ciliate. Petals greenish-white or tinged with purple, ovate-oblong, 3–4 mm long. Ovary ovoid-oblong, in the males rudimentary. Drupes oblong, reniform, compressed, reddish-brown.

Fls. & Frts.: February–April. *Exsiccata*: SUK-661.
Common, on slopes in hill forest.
Localities: Amba, Barki, Borbet, Dajipur, Kagal, Panhala.

Local uses: Bark is mixed with coconut oil and applied to sprain and bruises. Bark is used in toothache and sore throat. Bark is used in skin eruptions, heart diseases and in rheumatism. It is considered as a valuable application to old and obstinate ulcers as also to leprous ulcers.

Mangifera indica L. Sp. Pl. 200. 1753; Hook, f. Fl. Brit. India 2: 13. 1876; Cooke, Fl. Pres. Bombay 1: 291. 1958 (Repr. ed.); Prasanna in Singh *et al.* Fl. Maharashtra State Dicot. 1: 584. 2000. AMBA, AAM.

Tall evergreen trees, with dense crown of spreading branches. Leaves oblong-lanceolate, 8–20 × 4–6 cm, rounded or narrow at base, entire, acute or acuminate, glabrous. Flowers monoecious, 5 mm long, odor disagreeable, in large terminal panicles. Calyx deeply lobed; sepals ovate, pubescent outside. Petals dull white, oblong, 4–5 mm long, deflexed, with 3 strong orange colored ridges on the inner face. Disk fleshy, 5-lobed. Stamens solitary. Ovary glabrous. Drupes often obliquely pyriform, 6–20 cm long; stone compressed, fibrous. Seed oblong.

Fls. & Frts.: January–June. *Exsiccata*: SUK-338.
Cultivated throughout region.

Local uses: Fruit cures blood impurities, astringent to bowels; cures throat troubles, ulcers, dysentery, urinary discharge, vaginal troubles, useful in bleeding piles and menorrhagia.

Rhus mysorensis G. Don, Gen, Syst. 2: 74. 1832; Hook. f. Fl. Brit. India 2: 9. 1876; Cooke, Fl. Pres. Bombay 1: 290. 1958 (Repr. ed.); Prasanna in Singh *et al.* Fl. Maharashtra State Dicot. 1: 585. 2000. AMANI, AMBONI.

Shrubs with thorny branches; young parts densely pubescent. Leaves 3 or rarely 5 foliolate; leaflets ovate, 1–3 × 0.7–2.2 cm, cuneate at base, broadly crenate, obtuse, often emarginate, pubescent on both surfaces. Flowers minute in small axillary panicles. Calyx pubescent outside with 4–6 ovate lobes. Petals greenish-white, oblong, 2–2.5 mm long, subacute. Disk 5-lobed, lobes often notched. Ovary globose, 1-locualr. Drupes globose, 4–5 mm in diam., brownish-black, glabrous, shining.

Fls. & Frts.: August–December. *Exsiccata*: SUK-44.

Common, in scrub jungles.

Localities: Alate, Ashta, Babujamal, Tasgaon, Vita, Wai.

Local uses: Fruits are mixed with tincture of salt and paste is given orally in dysentery. Leaf decoction is given in itching.

Semecarpus anacardium L. f. Suppl. 182. 1781; Hook. f. Fl. Brit. India 2: 30. 1876; Cooke, Fl. Pres. Bombay 1: 296. 1958 (Repr. ed.); Prasanna in Singh *et al.* Fl. Maharashtra State Dicot. 1: 586. 2000. BIBBA.

Small dioecious trees, 3–4 m tall; young parts grey pubescent. Leaves oblong to oblanceolate, 10–40 × 5–20 cm, rounded or subcordate at base, with entire cartilaginous margins, rounded or subacute at apex, glabrous above, ashy gray or buff and pubescent beneath. Flowers subsessile, clustered in pubescent panicles as long as or shorter than the leaves. Calyx 5–6 lobed. Petals creamy white, ovate, 4–5 mm long, acute. Ovary in male flowers minute, in females subglobose, densely pilose. Drupes shining black, obliquely ovoid, 2–2.5 cm long, seated on a fleshy, edible, orange-colored thalamus.

Fls. & Frts.: January–September. *Exsiccata*: SUK-326.

Common, throughout in moist and dry deciduous forests.

Localities: Amba, Panhala.

Local uses: Fruit is punctured with the needle one drop of oil is applied on the upper part of throat to stop throat troubles, sore throat and bronchitis. Nut bruised and applied to Os Uteri to procure abortion. Oil is applied externally in rheumatism and leprous nodules. Fruit is given in dysentery; cures tumors, fevers, loss of appetite, urinary discharge; heals ulcers; strengthens the teeth. Fruit cause rash or redness of the skin or an itchy or uneasy sensation in any part of the body or to skin.

Spondias pinnata (L. f.) Kurz. in Prelim. Rep. For. & Veg. Pegu Append. A. 44. & B. 42. 1875; Prasanna in Singh *et al.* Fl. Maharashtra State Dicot. 1: 587. 2000. *Mangifera pinnata* L. f. Suppl. 156. 1781. *Spondias mangifera* Willd. Sp. Pl. 2: 751. 1799 *nom. illeg.*; Hook. f. Fl. Brit. India 2: 42. 1876; Cooke, Fl. Pres. Bombay 1: 299. 1958 (Repr. ed.). AMBADA.

Medium sized glabrous trees, 10–12 m tall. Leaves pinnate, 40–50 cm long; leaflets 3–5 pairs, opposite, oblong or elliptic oblong, 5–12 × 2–7 cm, slightly oblique at base, entire, acuminate. Flowers bisexual or polygamous, sessile, clustered in terminal panicles. Calyx teeth minute, triangular. Petals pinkish-green, ovate-oblong, 2–3 mm long, acute. Disk 10-lobed. Stamens 10, half as long as the petals. Drupes obovoid, ca 1.5–2 cm, pale yellowish-green; stone woody. Seed one or more.

Fls. & Frts.: February–June. *Exsiccata*: SUK-673.

Occasional, in hill slopes and around fields.

Localities: Amboli, Kusumbi, Patgaon, Radhanagari, Tillarinagar, Vaibhavwadi.

Local Uses: Unripe fruit juice with sugar and black pepper seeds are given in acidity. Leaf juice is dropped in earache. Leaves are used as a appetizer. Roots are used in chronic dysentery. 2–5 leaves are crushed with tincture of turmeric powder and pills prepared are given twice a day as blood purifier. Similarly it is given in sore throat. Decoction of wood is used in gonorrhoea and leucorrhoea. Paste of bark is rubbed in articular and muscular rheumatism.

MORINGACEAE Dumort.

Moringa concanensis Nimmo in Grah. Cat. Bombay Pl. 43: 1839; Hook. f. Fl. Brit. India 2: 45. 1876; Cooke, Fl. Pres. Bombay 1: 301. 1958 (Repr. ed.); Prasanna in Singh *et al.* Fl. Maharashtra State Dicot. 1: 589. 2000. RAN SHEVGA.

Trees, 5–8 m tall. Leaves 2 pinnate, ca 15–20 cm long; pairs 5–6, distant, 12–20 cm long, the primary rachis thickened at the base; leaflets 4–6 pairs and an odd one, broadly elliptic or suborbicular, obtuse 2–3 × 1–2 cm, pale beneath. Flowers in lax, divaricate, thinly pubescent panicles reaching ca 45 cm long. Calyx thinly tomentose; segments white, oblong, reflexed, 12–15 mm long. Petals yellow, veined with red, oblong or oblong-spathulate, the lower ca 15 mm long. Stamens 5 fertile; 4–5 staminodes; filaments hairy at the base. Capsules straight, acutely triquetrous, slightly constricted between the seeds, 30–45 × 1.5 cm; valves hard. Seeds white or pale yellow, three-angled, 6–8 mm in diameter, three winged; wings very thin, hyaline.

Fls. & Frts.: January–June. *Exsiccata*: SUK-22.
Rare, in deciduous forests.
Localities: Chandoli.

Local uses: Used in chest complaints.

Moringa oleifera Lam. Encycl. 1: 398. 1785; Prasanna in Singh *et al.* Fl. Maharashtra State Dicot. 1: 590. 2000. *M. pterygosperma* Gaertn. Fruct. 1: 314. 1791; Hook. f. Fl. Brit. India 2: 45. 1876; Cooke, Fl. Pres. Bombay 1: 301. 1958 (Repr. ed.). SHEVGA.

Middle sized trees. Leaves 3–4 pinnate, up to 50 cm long; rachis thickened and articulated at base; pinnae and pinnules opposite; leaflets orbicular to obovate, 1–2 cm long, mucronate, glabrous. Flowers in large, lax, terminal, puberulous panicles. Calyx cup-shaped, 5-lobed, segments, unequal petaloid, linear-lanceolate, 10–12 mm long, reflexed. Petals white, unequal, spathulate, 12–15 mm long. Perfect stamens 5, alternating with

5–7 staminodes. Ovary oblong, style cylindric. Pod linear, up to 50 cm long, obtusely triangular, 9-ribbed. Seeds 3 angled and winged on angles.

Fls. & Frts.: January–June. *Exsiccata*: SUK-253.
Cultivated, Native of South India.

Local uses: Root bark is mixed with salt and applied as a poultice to tumors. The bark decoction dropped in nose to stop headache. Seed powder is given in mild dysentery. Poultice of leaves is applied in glandular swellings. Leaves are used as poultice in inflammation. Root in combination with root of *Ricinus communis* is used externally in paralysis. Oil from the seeds is used for external application in rheumatism. Root bark is anthelmintic, improves appetite; cures tuberculous glands in the neck.

CONNARACEAE R. Br.

Connarus monocarpus L. Sp. Pl. 675. 1753; Hook. f. Fl. Brit. India 2: 50. 1876; Cooke; Fl. Pres. Bombay 1: 303. 1958 (Repr. ed.); Prasanna in Singh *et al.* Fl. Maharashtra State Dicot. 1: 590. 2000. *C. wightii* Hook. f. op. cit. 51.; Cooke, op. cit. 304. *C. ritchiei* Hook. f. op. cit.; Cooke, op. cit. KOKANI, KUTKUTAR, NAGOTRI.

Small trees or shrubs. Leaves imparipinnate; leaflets 1–2 pairs, elliptic lanceolate, 7–10 × 3–4.5 cm, acuminate, rounded or acute at the base. Flowers crowded in erect pyramidal, densely pubescent terminal panicles. Sepals ovate-oblong, subacute, densely pubescent. Petals longer than the sepals, linear-oblong, more or less pubescent. Stamens 10. Ovaries 5, densely pubescent; style slender; stigma capitellate. Follicles bright scarlet, fusiform, ca 3–5 cm long, slightly falcate, tapering into narrow stalk, surrounded by accrescent calyx, smooth. Seeds ovoid, slightly compressed, 2.5–3 cm long, aril yellow; testa black, shining.

Fls. & Frts.: February–March. *Exsiccata*: SUK-45.
Common, in semi evergreen forests.
Localities: Amba, Barki, Borbet, Gaganbawada, Phonda, Radhanagari.

Local uses: Pulp of the fruit is used in diseases of the eye. Decoction of the root is given in syphilis.

Tali minor (Gaertn.) Almeida, Fl. Mah. 1: 283. 1996; Prasanna in Singh *et al.* Fl. Maharashtra State Dicot. 1: 591. 2000. *Aegiceras minor* Gaertn. Fruct. 1: 216, t. 46. 1788. *Rourea santaloides* (Vahl) Wight & Arn. Prodr. 144. 1844; Hook. f. Fl. Brit. India 2: 47. 1876; Cooke, Fl. Pres. Bombay 1: 303. 1958 (Repr. ed.). VERDHARA, WAKERI.

Climbing or sarmentose shrubs. Leaves imparipinnate; leaflets 2–4 pairs, opposite or alternate, elliptic or lanceolate, 2–7 × 2–3.5 cm, coriaceous, caudate-acuminate, shining above. Flowers small in glabrous racemose panicles. Sepals broadly ovate or suborbicular, more or less minutely ciliolate. Petals spathulate-oblong, 4 mm long. Ovary ovoid, glabrous; stigma bilobed. Follicles 2 cm long, sessile conical ovoid, tapering to a point, falcately curved, chestnut brown, finely striate, supported on large calyx, dehiscing laterally.

Fls. & Frts.: October–December. *Exsiccata*: SUK-368.
Rare, in semi evergreen forests.
Localities: Amboli, Dajipur, Gajapur, Kasal, Patgaon, Sawantwadi, Shiroda, Vasota.

Local uses: Root is used in diabetes, similarly in urinary calculi. It is externally applied in inflammations. Crushed leaves are warmed on flame and applied externally in joint pains. Bark is boiled with water and salt and again washed in water crushed with turmeric powder and given in case of anemia. Bark is also used in tuberculosis and urinary discharges.

FABACEAE Lindl. (*nom. alt.*)
[**PAPILIONACEAE** Giseke (*nom. cons.*)]

Abrus precatorius L. Syst. Nat. ed. 2, 2: 472. 1767; Baker in Hook. f. Fl. Brit. India 2: 175. 1876; Cooke, Fl. Pres. Bombay 1: 382. 1958 (Repr. ed.); Kothari in Singh *et al.* Fl. Maharashtra State Dicot. 1: 597. 2000. GUNJ.

Glabrous, perennial twiners. Leaves pinnate, 5–10 cm long; leaflets 10–20 pairs, opposite, oblong, 6–15 × 4–6 mm, rounded at both ends, apiculate. Flowers crowded in many flowered racemes shorter than leaves. Calyx shortly toothed, 2–3 mm long. Petals pink or white with pinkish tinge, 8–12 mm long. Pods oblong, 3–4.5 × 1–1.5 cm, truncate with sharp, deflexed beak, silky pubescent. Seeds globose, 3–5 mm in diam.; red or white with or without black spot, shining.

Fls. & Frts.: September–January. *Exsiccata*: SUK-145.
Common, on bushes and in dry deciduous forests.
Localities: Alate, Ashta, Kagal, Narande, Ramling, Tasgaon, Vita.

Local uses: Leaves are given in hoarseness. Root and leaf of white seeded variety is used in urinary complaints. Leaves are applied on wounds, ulcers and in cough. Fruits are used as brain tonic, as aphrodisiac (harmful to old men). The roots are used as substitute for liquorice. Seeds are boiled and taken as tonic. Paste of seeds applied locally in sciatica, stiffness of the shoulder and paralysis. Poultice of seeds used as suppository to bring about abortion. Cures fevers, stomatitis, head complaints, asthma, thirst, tuberculous gland and caries of teeth.

Butea monosperma (Lam.) Taub. in Engl. & Prantl. Pflanzenfam. 3(3): 365. 1894; Kothari in Singh *et al.* Fl. Maharashtra State Dicot. 1: 610. 2000. *Erythrina* *monosperma* Lam. Encycl. 2: 391. 1785. *Butea frondosa* Koen. ex Roxb. in Asiat. Res. 3: 469. 1792; Baker in Hook. f. Fl. Brit. India 2: 194. 1876; Cooke, Fl. Pres. Bombay 1: 395. 1958 (Repr. ed.). PALAS.

Deciduous trees, 12–15 m tall. Leaves 3-foliolate; petioles swollen at base; leaflets obovate-rhomboid, 8–15 × 5–8 cm, oblique, cuneate at base, obtuse, glabrous above, densely gray silky beneath. Flowers large in dense rigid racemes, 15–30 cm long, 2–3 together on tumid nodes. Calyx 10–15 mm long dark olive green, densely velvety outside, clothed with silky hairs within, the 3 lower deltoid. Corolla salmon orange coloured, 3–6 cm long, clothed outside with slivery hairs. Pods stalked, oblong, 10–20 × 2–5 cm, 1-seeded, reticulately veined, silvery tomentose.

Fls. & Frts.: January–June. *Exsiccata*: SUK-309.
Common, throughout in deciduous forest of the state.
Localities: Kolhapur, Sangali, Shirol.

Local uses: Seeds are given as aphrodisiac. Paste of seed is also given in irregular menstrual flow. One teaspoon of flower paste is given to pregnant women in case of diarrhoea. Flower and gum decoction is given in bone fractures. Roots are applied externally to cure wound. In bone fractures roots are crushed and tied around fractured part as a poultice and also given orally. 1–2 and 5–6 cm long roots are soaked in water and that water is given in irregular menstrual flow. Gum is used in corneal opacities, cures excessive perspiration.

Cajanus lineatus (Wight & Arn.) van der Maesen in Agric. Univ. Wageningen Papers 85 (4): 143, f. 18, t. 10 (1985) 1986; Kothari in Singh *et al.* Fl. Maharashtra State Dicot. 1: 613. 2001. *Atylosia lineata* Wight & Arn. Prodr. 258. 1834; Baker in Hook. f. Fl. Brit. India 2: 213. 1876; Cooke, Fl. Pres. Bombay 1: 408. 1958 (Repr. ed.). JANGLITUR, RANTUR.

Shrubs, 1–2 m tall; stem and branches sulcate, densely clothed with soft hairs. Leaves 3-foliolate; leaflets obovate, 1.5–3 × 0.6–1 cm, acute, faintly mucronate, densely silky-villous when young on both surfaces, glabrous at maturity, cuneate at base, 3-nerved. Flowers axillary, solitary or in pairs, without a common peduncle. Calyx ca 5 mm long, hairy; teeth shorter than the tube, deltoid, cuspidate. Corolla ca 1 cm long, yellow. Pods linear, 1–1.5 × 6.6 cm, turgid, depressed, clothed with long whitish hairs. Seeds 2 with a conspicuous divided strophiole, black.

Fls. & Frts.: October–April. *Exsiccata*: SUK-38.
Common in semi evergreen forests.
Localities: Amba, Barki, Borbet, Lingmala, Phonda, Radhanagari, Wai.

Local uses: Leaf powder or leaf juice is applied externally in toothache and in gum complaints. Leaves are given orally for diarrhoea in cattle's.

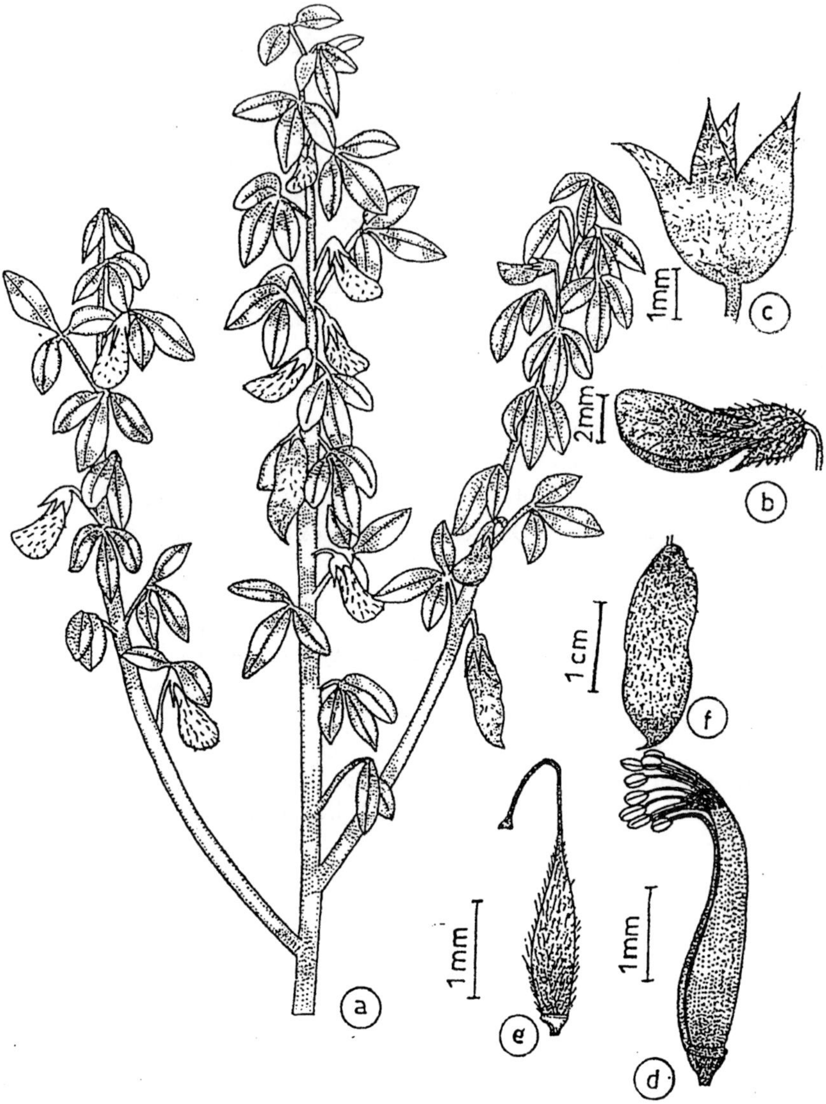

Text fig. 12: Cajanus lineatus a-twig; b-flower; c-calyx; d-stamens; e-carpel; f-fruit.

Castanospemum australe A. Cunn. & Fraser in Hook. Bot. Misc. 1: 241, t. 51, 1830; Kothari in Singh *et al.* Fl. Maharashtra State Dicot. 1: 768. 2000. BLACK-BEAN.

Trees, ca 1.8 m high. Leaves imparipinnate; leaflets 11–15, elliptic or ovate-oblong. Flowers yellow-orange reddish, in racemes. Pods oblong or terete, woody. Seeds chestnut like.

Fls. & Frts.: September–December.
Cultivated in Town Hall garden Kolhapur.

Local uses: Fruits used in AIDS.

Clitoria ternatea L. Sp. Pl. 753. 1753; Baker in Hook. f. Fl. Brit. India 2: 208. 1876; Cooke, Fl. Pres. Bombay 1: 405. 1958 (Repr. ed.); Kothari in Singh *et al.* Fl. Maharashtra State Dicot. 1: 619. 2000. GOKARN.

Twining, rather woody annual or perennial herbs. Leaves imparipinnate, 5–9 foliolate, 5–15 cm long; leaflets opposite, elliptic oblong or elliptic, 2–5 × 1.5–3 cm, acute or obtuse at base, rounded or often notched at apex, glabrescent above, hairy beneath. Flowers large, axillary solitary. Calyx tubular, 1.5–2 cm long; teeth lanceolate, shorter than the tube. Corolla light or dark blue or white, 3–5 cm long. Pods oblong, 5–12 × 0.8–1.2 cm, flattened, beaked, appressed hairy. Seeds 6–10, quadrate, brown, smooth.

Fls. & Frts.: May–December. *Exsiccata*: SUK-666.
Common, throughout the region, often cultivated.
Localities: Shivaji University campus, Kolhapur, Hatkanangle.

Local uses: Root decoction is given in urinary problems and in cough. Leaf juice with ginger paste is given to stop sweating. Decoction of leaf and root is given in hemicrania. Root is used in ulcers of cornea, tuberculous glands, elephantiasis, headache; cures leucoderma, burning sensation, pain, biliousness, inflammation, ulcers and abdominal enlargement.

Crotalaria retusa L. Sp. Pl. 715. 1753; Baker in Hook. f. Fl. Brit. India 2: 75. 1876; Cooke, Fl. Pres. Bombay 1: 318. 1958 (Repr. ed.); Kothari in Singh *et al.* Fl. Maharashtra State Dicot. 1: 645. 2000. KHULKHULA.

Woody, much branched, erect annual herbs, up to 1 m tall; branches grooved, clothed with appressed white hairs. Leaves oblong to oblanceolate, 2.5–8 × 1–2.5 cm, narrowed at base, retuse at apex, glabrous and dark green above, paler and appressed hairy beneath. Flowers large and showy, in terminal-many flowered racemes, 15–30 cm long. Calyx 10–12 mm long, glabrous or nearly so; upper teeth ovate, acute, connate at base; 3 lower teeth narrower, triangular, acuminate, connate below the middle. Corolla golden yellow, exserted. Pods oblanceolate, ca 4 × 1 cm, beaked, black when ripe, glabrous, shortly stalked. Seeds many, black, reniform.

Fls. & Frts.: October–January. *Exsiccata*: SUK-259.
Common, weed of cultivated field, riverbanks.
Localities: Amba, Panchgani, Radhanagari.

Local uses: Paste of leaves applied on skin diseases. Seeds are given to reduce fats from the body.

Cullen corylifolia (L.) Medik., Vorles. Churpf. Phy.-Oek. Ges. 2: 380. 1787; Kothari in Singh *et al.* Fl. Maharashtra State Dicot. 1: 652. 2000. *Psoralea corylifolia* L. Sp. Pl. 764. 1753; Baker in Hook. f. Fl. Brit. India 2: 103. 1876; Cooke, Fl. Pres. Bombay 1: 341. 1958 (Repr. ed.); BAVCHI.

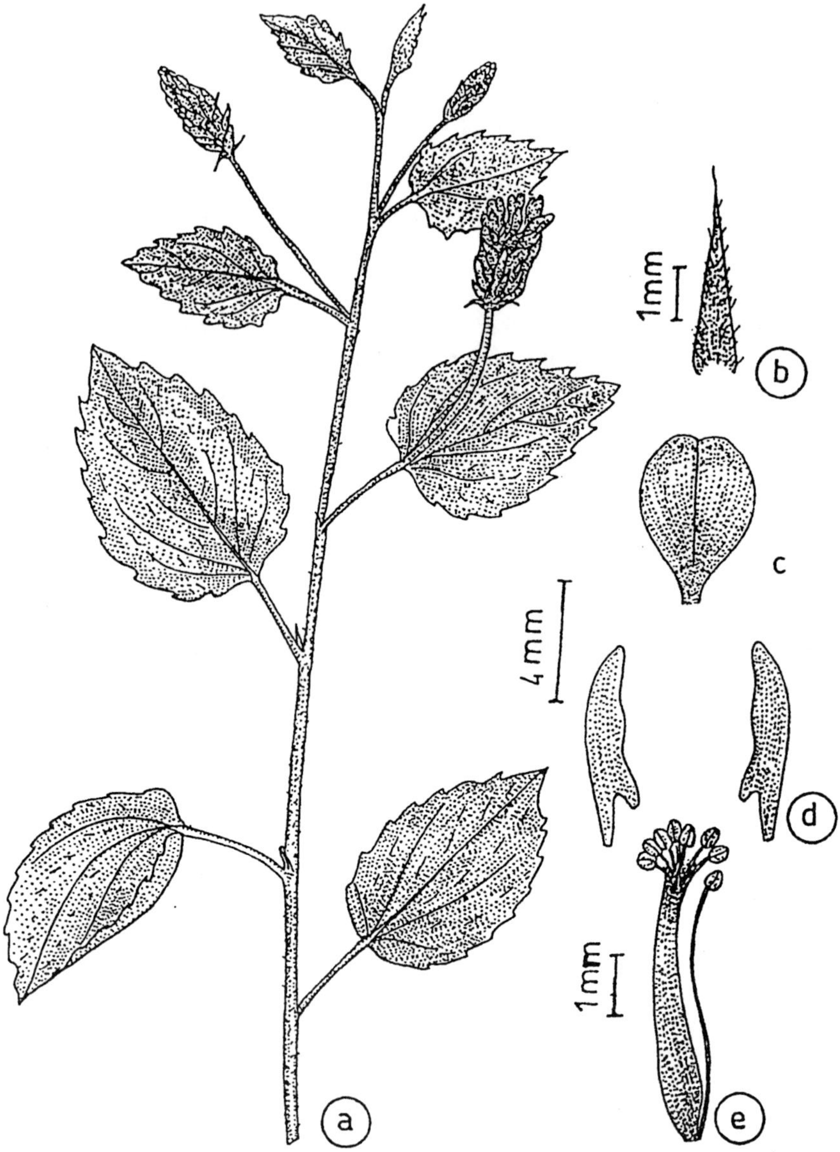

Text fig. 13: Cullen corylifolia a-twig; b-bract; c-wing; d-keel; e-stamens.

Herbs, 60–80 cm high; branchlets glandular pubescent, warty. Leaves simple ovate, 3.5–5.7 × 2.8–4.2 cm, punctate, pubescent, base obtuse, margin dentate, apex acute apiculate. Flowers purplish-violet in axillary racemes. Calyx tube campanulate, lobes 5, lanceolate, upper ones connate. Corolla 8 mm long, exserted, standard oblong-obovate; wings oblong, keels connate, obtuse. Stamens diadelphous; anthers slightly dimorphous. Ovary sessile; style apically dilated, incurved; stigma simple. Pods ovoid oblong. Seeds solitary reniform.

Fls. & Frts.: September–February. *Exsiccata*: SUK-13.
Common, weed of black cotton soil.
Localities: Kolhapur, Uchgaon.

Local uses: The seeds are commonly used in treatment of leucoderma, leprosy, psoriasis and inflammatory diseases of the skin.

Cyamopsis tetragonolobus (L.) Taub. in Engl. & Prantl. Nat. Pflanzenf. 3(3): 259. 1894; Kothari in Singh *et al.* Fl. Maharashtra State Dicot. 1: 652. 2000. *Psoralea tetragonoloba* L. Mant. Pl. 104. 1767. *Cyamopsis psoralioides* (Lam.) DC. Mem. Leg. 231. 1825 and Prodr. 2: 216. 1825; Baker in Hook. f. Fl. Brit. India 2: 92. 1876; Cooke, Fl. Pres. Bombay 1: 328. 1958 (Repr. ed.). GAWAR.

Erect, annual herbs, 60–90 cm tall; stem grooved, clothed with appressed white hairs. Leaves 3-foliolate; leaflets elliptic, 5–7 × 2–4 cm, acute at base and apex, sharply dentate, clothed on both surfaces with mixed white hairs. Flowers small, in axillary dense racemes, bracts linear-subulate, 4–5 mm long, persistent. Calyx hairy outside, 5–6 mm long; Calyx teeth unequal, the 3 upper triangular-acute, shorter than the 2 linear-subulate teeth. Corolla pale pink, bluish-purple or white, slightly longer than the calyx. Pods thick, fleshy, subtetragonous, 3–15 cm long, acuminate. Seeds 5–6, compressed, reniform.

Fls. & Frts.: April–May. *Exsiccata*: SUK-664.
Cultivated, throughout the region for pods as vegetables.
Localities: In all districts.

Local uses: Paste of leaves in combination with juice of *Allium sativum* is used in elephantiasis. Boiled fruits are given in early stages of leucoderma orally daily for the period of 5–6 months. Fruits are laxative, improve appetite; removes biliousness and night blindness.

Dalbergia sissoo Roxb. ex DC. Prodr. 2: 416. 1825 and Roxb. in Fl. Ind. 3: 223. 1832; Baker in Hook. f. Fl. Brit. India 2: 231. 1876; Cooke, Fl. Pres. Bombay 1: 421. 1958 (Repr. ed.); Kothari in Singh *et al.* Fl. Maharashtra State Dicot. 1: 662. 2000. SHISHAM, SISVI.

Evergreen trees, 10–15 m tall; young parts pubescent; branches drooping. Leaves 3–5-foliolate; leaflets alternate, broadly ovate, 5–7 × 3.5–5 cm, cuneate at base, acuminate, glabrous on both surfaces. Flowers sessile, in axillary panicles shorter than the leaves, branches densely hairy, bracts linear-subulate, hairy. Calyx 4–5 mm long, hairy; teeth short, ciliate, the 2 upper connate except at the tip, the lateral linear, obtuse, the lowest the longest, acute, corolla creamy white, 6–8 mm long. Pods strap shaped, 5–3 × 0.9–1.2 cm, subacute or obtuse, glabrous, narrowed into a long stalk, often only 1 seeded, rarely up to 3 seeded.

Fls. & Frts.: February–October. *Exsiccata*: SUK-667.
Common, in deciduous forests, often planted.
Localities: Jat, Kartikswami.

Local uses: Leaf juice is used to wash the eyes, oil is applied externally in the cutaneous affections, in joint pains; used in scabies. Bark and wood are aphrodisiac, abortifacient, expectorant, anthelmintic, appetizer, allays thirst, vomiting, burning sensations, cures skin diseases, troubles of the anus, ulcers, diseases of the blood, leucoderma, dyspepsia and dysentery.

Desmodium gangeticum (L.) DC. Prodr. 2: 327 1825; Baker in Hóok. f. Fl. Brit. India 2: 168. 1876 (incl. var. *maculatum* (L.) Baker.); Cooke, Fl. pres. Bombay 1: 379. 1958 (Repr. ed.); Kothari in Singh *et al.* Fl. Maharashtra State Dicot. 1: 670. 2000. *Hedysarum gangeticum* L. Sp. Pl. 746. 1753. SALPARNI, SALVAN.

Erect undershrubs, ca 1 m, tall; branches angled, clothed with appressed white hairs. Leaves 1-foliolate; leaflets ovate-oblong, 7–13 × 5–7.5 cm rounded, truncate, or subcordate at base, acute to acuminate, glabrous and green above, paler and clothed with dense soft white hairs beneath. Flowers in few-flowered clusters along axillary terminal racemes, 15–30 cm long; rachis pubescent. Calyx 2 mm long, hairy; teeth triangular, longer than the tube. Corolla pink or white, 4 mm long. Pods subfalcate, 10–20 × 2 mm, 6–8 jointed, sparsely clothed with hooked hairs.

Fls. & Frts.: March–November. *Exsiccata*: SUK-675.
Occasional, along the stream in Konkan.
Localities: Dapoli, Ratnagiri, Salva Donger.

Local uses: It is one of the species used in Dashmula. Decoction of root is given in chronic fever and in diarrhoea. Leaves are applied in inflammations. Decoction of root with the powder of pepper, salt and ginger is given in fever after each 4 hrs. Roots are externally applied in hemicrania. Root is also used to prevent abortion.

Erythrina variegata L. in Stickm. Herb. Amboin. 101. 1754 & Amoen. Aca. 4: 122. 1759; Kothari in Singh *et al.* Fl. Maharashtra State Dicot. 1: 683. 2000. *E. indica* Lam. Encycl. 2: 391. 1876; Baker in Hook. f. Fl. Brit. India 2: 188. 1876; Cooke, Fl. Pres. Bombay 1: 390. 1958 (Repr ed.). PANGARA.

Medium sized trees, up to 15 m tall; branches smooth or prickly, glabrous. Leaves 3-foliolate; leaflets broadly rhomboid-ovate, 8–15 × 7–12 cm, truncate or narrowed at base, acute or acuminate, glabrescent. Flowers in terminal, horizontal spreading, spicate racemes, appearing before leaves. Calyx 2–3 cm long, clothed with deciduous tomentum, oblique, spathaceous. Corolla bright red, 4–5 cm long, with very unequal petals. Pods subcylindric, 10–30 cm long, distinctly tortulose. Seeds 6–8, subreniform, brown.

Fls. & Frts.: January–June. *Exsiccata*: SUK-674.
Common, in deciduous forests and also planted.
Localities: Kolhapur, Shirol.

Local uses: Bark is applied in joint pains. After delivery it is given with coconut milk as galactogogue. Roots are given with cold milk as an aphrodisiac. Root is an emmenagogue. Bark is used in dysentery. Leaves are given in urinary discharges, inflammation and cures ulcers.

Flemingia tuberosa Dalz. in Hook. Kew. J. Bot. 2: 34. 1850; Baker in Hook. f. Fl. Brit. India 2: 230. 1876; Cooke, Fl. Pres. Bombay 1: 420. 1958 (Repr. ed.); Kothari in Singh *et al.* Fl. Maharashtra State Dicot. 1: 689. 2000.

Perennial, prostrate herbs; root tuberous, 5 cm long; stem 60–70 cm long, copiously branched; branches glabrous or with a few spreading hairs, faintly striate. Leaves 3-foliolate; leaflets oblong-lanceolate, 3–5 × 0.9–1.1 cm, acute, sparsely hairy above, more strongly on the nerves beneath, not or obscurely gland-dotted. Flowers in few flowered lax axillary racemose cymes, longer than leaves, the branches very slender. Calyx 6–9 mm long, densely hairy; teeth subequal, longer than the tube, linear acute, 3-nerved corolla lilac, slightly longer than the calyx; standard hairy on the back. Pods oblong, slightly longer than the calyx, glabrous, transversely reticulate towards the apex. Seeds 1–2.

Fls. & Frts.: September–November. *Exsiccata*: SUK-670.
Apparently endemic, rare on plateaus.
Localities: Kas, Panchgani.

Local uses: Tuber and roots are astringent; useful in dysentery and vaginal discharges.

Indigofera cordifolia Heyne ex Roth, Nov. Pl. Sp. 357. 1821; Baker in Hook. f. Fl. Brit. India 2: 93. 1876; Cooke, Fl. Pres. Bombay 1: 331. 1958 (Repr. ed.); Kothari in Singh *et al.* Fl. Maharashtra State Dicot. 1: 701. 2000. BECHKA, GODADI.

Erect or procumbent herbs, 10–30 cm tall; branches clothed with white hairs. Leaves simple, ovate-cordate, 8–10mm, subsessile, hairy on both surfaces, stipules minute, setaceous. Flowers in dense, subsessile, 4–8 flowered axillary clusters. Calyx 3 mm long, hairy outside; tube very short; teeth linear, acute, hairy. Corolla red, hardly exserted. Pods oblong, 3–4 mm long, straight, densely pubescent. Seeds 1–2, small, brown.

Fls. & Frts.: August–January. *Exsiccata*: SUK-622.
Common, throughout in open situations.
Localities: Kolhapur, Jaisingpur, Wai.

Local uses: Plant decoction is given in dysentery twice a day. Seeds are given as anthelmintic.

Indigofera linifolia (L.) Retz. Obs. Bot. 4: 29. 1786; Baker in Hook. f. Fl. Brit. India 2: 92. 1876; Cooke, Fl. Pres. Bombay 1: 330. 1958 (Repr. ed.); Kothari in Singh *et al.* Fl. Maharashtra State Dicot. 1: 707. 2000. *Hedysarum linifolium* L. f. Suppl. 331. 1781. PANDARPHALI.

Prostrate or erect, grayish-green, woody herbs; stems 10–40 cm long. Leaves simple, linear-lanceolate, 10–20 × 2 mm, acute at both ends. Flowers in dense, axillary racemes, 5–15 mm long. Calyx 3 mm long, deeply divided, clothed with white silvery hairs; teeth linear subulate, longer than the tube. Corolla bright red, 2–3 times as long as the calyx. Pods globose, 1.5–2 mm diam., apiculate, shining white with appressed silky hairs, 1-seeded.

Fls. & Frts.: August–January. *Exsiccata*: SUK-677.
Common throughout in open grasslands and rocky soils.
Localities: Jat, Kagal, Ramling, Wai.

Local uses: It is given in urinary troubles. It reduces heat from the body. Decoction is given in acidity.

Indigofera tinctoria L. Sp. Pl. 751. 1753; Baker in Hook. f. Fl. Brit. India 2: 99. 1876 p. p.; Cooke, Fl. Pres. Bombay 1: 339. 1958 (Repr. ed.); Kothari in Singh *et al.* Fl. Maharashtra State Dicot. 1: 714. 2000. NEEL.

Shrubs or undershrubs, 30–100 cm tall; branches with appressed silvery hairs. Leaves pinnate, 5–10 cm long; leaflets 7–13, opposite, elliptic to obovate or oblanceolate, 1–2 × 0.4–1 cm, acute at base, obtuse or emarginate at apex, glabrous or nearly so above, appressed hairy beneath. Flowers many, in axillary, subsessile, dense racemes, 5–10 cm long. Calyx 1–1.5 cm long, hairy outside; teeth as long as the tube, acute. Corolla pink, 4 mm long. Pods linear, cylindric, 2–3.5 cm long, straight or slightly curved, deflexed, apiculate. Seeds 8–12, rarely up to 20.

Fls. & Frts.: August–February. *Exsiccata*: SUK-678.
Common, throughout in plains, in open situations and also cultivated.
Localities: Kondave, Radhanagari, Wai.

Local uses: Leaves and seeds are used in burns. It is externally employed on skin diseases and on insect bites. It is used in abdominal complaints, heart diseases, insanity; cure fever, leucoderma, enlarged spleen, cephalagia, injuries etc.

Mucuna monosperma DC. ex Wight in Hook. Bot. Misc. 2: 346. 1831; Baker in Hook. f. Fl. Brit. India 2: 185. 1876; Cooke, Fl. Pres. Bombay 1: 388. 1958 (Repr. ed.); Kothari in Singh *et al.* Fl. Maharashtra State Dicot. 1: 724. 2000.

Large woody twiners; young branches clothed with rusty tomentum. Leaves 15–22 cm long; leaflets 5.5–10 × 5–7 cm, ovate-oblong or elliptic, shortly acuminate, glabrous above, pubescent beneath, base rounded. Flowers in 6–12 flowered corymbose axillary racemes, shorter than the leaves. Calyx 9 mm long, clothed with irritant bristles; teeth about half as long as the tube, the upper truncate, the lateral deltoid, the lower linear. Corolla 3 cm long, purple; keel abruptly inflexed at the tip. Pods 5–7 × 5 cm, winged on both sutures and obliquely plaited on the faces, covered with brown deciduous irritant bristles. Seeds solitary, compressed, 3–5 cm across, nearly circular in horizontal cross section, slightly dark brown, smooth shining.

Fls. & Frts.: November–March. *Exsiccata*: SUK-671.
Frequent, along the Ghats and along banks.
Localities: Barki, Chandgad, Tillari, Vaibhavwadi.

Local uses: Seed is used as one of the ingredient in children's 'Bal Ghooti'. Seeds are externally applied as a sedative. Stem is used in old wounds. Seeds are used as cardiac tonic in very small quantity.

Mucuna pruriens (L.) DC. Prodr. 2: 405. 1825; Baker in Hook. f. Fl. Brit. India 2: 187. 1876; Cooke, Fl. Pres. Bombay 1: 389. 1958 (Repr. ed.); Kothari in Singh *et al.* Fl. Maharashtra State Dicot. 1: 724. 2000. *Dolichos pruriens* L. Stickman, Diss. Herb. Amb. 23. 1754. KHAJ-KUILEE, KAVACH.

Large, gray, pubescent, twining annual herbs; branches slender. Leaves 3-foliolate; leaflets membranous, rhomboid-ovate, 5.5–15.5 × 5–10.5 cm, lateral ones inequilateral and rounded at base, the terminal one cuneate at base, all subacute, mucronate at apex, pubescent above, densely gray silky beneath. Flowers in many-flowered, 10–30 cm long, dense, pendulous racemes, solitary or 2–3 together. Calyx 1 cm long, gray tomentose outside; teeth lanceolate, as long as tube, the upper two connate. Corolla dark purple, 2.5–3.5 cm long. Pods 's' shaped, turgid, 5–8 cm long, densely clothed with brown shining bristles. Seeds 5–6.

Fls. & Frts.: August–January. *Exsiccata*: SUK-672.
Common, on fences around fields, along hedges.
Localities: Chandgad, Nesari, Wai.

Local uses: Hairs of the fruits are mixed with Jaggery, pills prepared and given to children to expel worms/ascarids. Roots are prescribed in blood diseases. Bark powder is applied externally in wounds and in bone fracture. Decoction of bark with the little salt and bark powder of *Bombax ceiba* is given in urinary troubles. The root is given in dysentery and in uterine troubles, biliousness.

Paracalyx scariosus (Roxb.) Ali in Univ. Stud. Karachi 5(3): 95. 1968; Kothari in Singh *et al.* Fl. Maharashtra State Dicot. 1: 727. 2000. *Cylista scariosa* Roxb. Pl. Cor. 1: 64, t. 92. 1798; Baker in Hook. f. Fl. Brit. India 2: 219. 1876; Cooke, Fl. Pres. Bombay 1: 412. 1958 (Repr. ed.). RANGHEVDA.

Twining, pubescent, undershrubs; stems and branches finely downy. Leaves 3- foliolate; leaflets ovate-rhomboid, 3–5 × 2–4.5 cm, subcordate at base, acute, softly velvety pubescent above, densely so and prominently reticulately veined beneath. Flowers in axillary racemes or panicles. Calyx 4–5 mm long, enlarging to 2–2.5 cm in fruit, downy outside, 2-lipped, lower lip boat shaped, longer than the upper lip. Corolla yellow, concealed in the calyx. Pods ovoid, 10 × 6 mm, oblique, short-hairy. Seed solitary, oblong, 5 mm long, dull-black, polished.

Fls. & Frts.: November–April. *Exsiccata*: SUK-263.
Common, in dry deciduous forests.
Localities: Amba, Dajipur, Gaganbawada.

Local uses: Roots are prescribed in leucorrhoea. Leaf decoction is given in dysentery. Leaves are externally applied in tumors, ulcers and in boils. Fruits enrich the blood, cures biliousness, good for throat troubles.

Pongamia pinnata (L.) Pierre, For. Fl. Cochinch. t. 385. 1899; Kothari in Singh *et al.* Fl. Maharashtra State Dicot. 1: 728. 2000. *Cytisus pinnatus* L. Sp. Pl. 741. 1753. *Pongamia glabra* Vent. Jard. Mal. t. 28. 1803; Baker in Hook. f. Fl. Brit. India 2: 240. 1876; Cooke, Fl. Pres. Bombay 1: 429. 1958 (Repr. ed.). KARANJ.

Small or middle sized trees. Leaves pinnate, 10–20 cm long; leaflets opposite, 5–7, ovate-oblong or elliptic, 4–9 × 3–5 cm, acute or rounded at base, shortly acuminate. Flowers in axillary racemes shorter than the leaves. Calyx truncate or obscurely toothed, 3 mm long, brown pubescent outside. Corolla pinkish-white, 10–12 mm long. Pods woody, oblong-ellipsoid, 4–5 × 2–2.5 cm, compressed, recurved, mucronate. Seeds-solitary, reniform, brown.

Fls. & Frts.: February–May. *Exsiccata*: SUK-374.
Common, along the riverbanks, often cultivated.
Localities: Kolhapur, Satara.

Local uses: Leaves in form of poultice applied to ulcers infected with worms. Leaves are externally applied in joint pains. Fresh bark used internally in bleeding piles. Flowers soaked in ½ glass of water for 12 hours and decoction is given in diabetes after meal. Seed oil is used as antiseptic and stimulant healing properties, useful in cutaneous affections, herpes and scabies, itching. Bark is used in teeth complaints.

Pterocarpus marsupium Roxb. Pl. Cor. 2: 9, t. 116. 1799; Baker in Hook. f. Fl. Brit. India 2: 239. 1876; Cooke, Fl. Pres. Bombay 1: 427. 1958 (Repr. ed.), Kothari in Singh *et al.* Fl. Maharashtra State Dicot. 1: 729. 2000. BIBLA, BIWALA.

Large, deciduous, glabrous trees, 10–15 m tall. Leaves imparipinnate, 10–20 cm long; leaflets 5–9, subopposite, ovate-oblong, 4.5–13 × 4–7 cm, subacute at base, obtuse, rounded, truncate or retuse at apex, shining above, paler beneath, glabrous. Flowers in short, axillary and terminal panicles shorter than the leaves. Calyx 5–6 mm long, brown pubescent outside; teeth broadly triangular, much shorter than the tube. Corolla yellow, 10–12 mm long; petals with crisped margins. Pods orbicular, 4–5 cm across, 1–2 seeded, with reticulately veined, broad, glabrous wing. Seeds reniform, 10 mm across, pale brown.

Fls. & Frts.: December–March. *Exsiccata*: SUK-679.
Occasional, in semi evergreen forests.
Localities: Amba, Amboli, Panhala, Radhanagari, Tillari.

Local uses: Bark powder is used for toothache; decoction of bark is used in diseases of the blood, eruptions on the body. Bark is dipped whole night in water and one full glass of water is given in diabetes early in the morning. The plant gum is used as laxative, anthelmintic, alterative, cures diseases of the blood, eruptions on the body, leucoderma, erysipelas, urinary discharges and anal troubles.

Pueraria tuberosa (Roxb. ex Willd.) DC. in Ann. Sci. Nat. 4: 97. 1825; Baker in Hook. f. Fl. Brit. India 2: 197. 1876; Cooke, Fl. Pres. Bombay 1: 399. 1958 (Repr. ed.); Kothari in Singh *et al.* Fl. Maharashtra State Dicot. 1: 731. 2000. *Hedysarum tuberosum* Roxb. ex. Willd. Sp. Pl. 3. 1197. 1802. BHUIKOHLA, KHALVEL, PITHANA, VIDARI KAND.

Large, perennial, twining shrubs; root tuberous. Leaves 3-foliolate; leaflets subcoriaceous, broadly ovate, 8–15 × 6–11 cm, truncate or cuneate at base, acuminate, glabrescent above, silky beneath. Flowers in many flowered, lax, axillary racemes, 15–30 cm long. Calyx 6–8 mm long, densely silky outside; teeth shorter than the tube, oblong, obtuse, ciliate. Corolla bluish, exerted. Pods oblong, 6–9 cm long, constricted between the seeds, clothed with long, silky, brown hairs. Seeds 3–10, compressed, brown.

Fls. & Frts.: February–April. *Exsiccata*: SUK-378.
Frequent, in deciduous forests.
Localities: Bhudargad, Ganpatpule, Pawas, Ratnagiri.

Local uses: Tuber is given as a tonic. It is used to cure rheumatic pains. Tuber is given orally in epilepsy. Tuber is given as aphrodisiac, tonic, glactogogue, diuretic, clears the voice; cures blood diseases.

Sesbania grandiflora (L.) Poir. in Lam. Encycl. 7: 127. 1806; Baker in Hook. f. Fl. Brit. India 2: 115. 1876; Cooke, Fl. Pres. Bombay 1: 350. 1958 (Repr. ed.); Kothari in Singh *et al.* Fl. Maharashtra State Dicot. 1: 774. 2000. *Robinia grandiflora* L. Sp. Pl. 722. 1753. HADGA.

Soft wooded trees; up to 10 m tall; branches thinly pubescent. Leaves abruptly pinnate, 15–30 cm long; leaflets 10–30 pairs, oblong-elliptic, 1.5–3 × 0.5–1.5 cm, rounded or subcrodate at base, rounded at apex, glabrous above, thinly pubescent beneath. Flowers in few flowered, axillary, short racemes. Calyx campanulate, 1.5–2 cm long, glabrous, obscurely toothed. Corolla glabrous. Pods green, glabrous. Seeds many.

Fls. & Frts.: February–April. *Exsiccata*: SUK-633.
Cultivated, around villages and in betle yards.

Local uses: Leaf is used as tonic and antipyretic, cure night blindness and biliousnes. Paste of leaf is externally applied in joint pains, juice of leaves or flower is a popular remedy in nasal catarrh and headache, it blows up the nostrils and causes a very copious discharge of fluid, relieving the pain and sense of weight in the frontal sinuses. It cures itching, leprosy, night blindness, epilepsy, gout; flowers are useful in night blindness.

Tephrosia purpurea (L.) Pres. Syn. Pl. 2: 329. 1807; Baker in Hook. f. Fl. Brit. India 2: 112. 1876. p. p.; Cooke, Fl. Pres. Bombay 1: 346. 1958 (Repr. ed.); Kothari in Singh *et al.* Fl. Maharashtra State Dicot. 1: 753. 2000. *Cracca purpurea* L. Sp. Pl. 752. 1753. SARPUNKA, UNHALI.

Much branched, perennial herbs or undershrubs, 30–80 cm tall; stem woody. Leaves pinnate, 4.5–10 cm long; leaflets 9–21, oblong-oblanceolate, 0.8–3 × 0.3–1 cm, acute or cuneate at base, emarginate and apiculate at apex; glabrous above, sparsely appressed hairy beneath. Flowers 2–4 together in extra axillary racemes, 5.5–12 cm long. Calyx 4 mm long, thinly silky outside; teeth triangular, subulate, as long as the tube. Corolla pink or pinkish purple, twice as long as the calyx. Pods oblong-linear, 2.5–4.5 cm long, curved at the tip, glabrescent. Seeds 5–6, oblong, 2–3 mm long, pale brown, mottled with black spots.

Fls. & Frts.: Throughout year. *Exsiccata*: SUK-680.
Common, in waste places and in dry regions.

Localities: Babu Jamal, Sangli, Wai.

Local uses: Decoction of root is given in cough, fever and as a diuretic. It is also given to cure acidity. Fresh root or leaves are given with a black pepper in obstinate colic. The root is used as diuretic, useful in enlargement of the spleen, cures tumors, ulcers, leprosy, asthma and bronchitis.

Teramnus labialis (L. f.) Spreng. Syst. Veg. 3: 235. 1826; Baker in Hook. f. Fl. Brit. India 2: 184. 1876; Cooke, Fl. Pres. Bombay 1: 387. 1958 (Repr. ed.); Kothari in Singh *et al.* Fl. Maharashtra State Dicot. 1: 755. 2000. *Glycine labialis* L. f. Suppl. 325. 1781. RANUDID.

Widely spreading, twining herbs; root tuberous; stems slender, appressed hairy. Leaves 3-foliolate; leaflets 2–6 × 1–4.5 cm, acute or rounded at base, acute, apiculate, appressed hairy beneath. Flowers in axillary, few-flowered, lax racemes, 2–15 cm long, solitary or fascicled along the slender hairy rachis. Calyx 3–4 mm long, silky outside; teeth lanceolate, as long as the tube. Corolla white turning reddish, 4–6 mm long. Pods narrowly linear, 3–5 × 0.3–0.4 cm, slightly curved, apiculate. Seeds 8–12, oblong, 3–4 mm long, brown, polished.

Fls. & Frts.: September–December. *Exsiccata*: SUK-262.

Common, in hedges around fields and under the shade of tree.

Localities: Amboli, Dindewadi, Mahabaleshwar, Phonda, Radhanagari, Vaibhavwadi.

Local uses: Root is tuberous given in rheumatism. Roots are given as a tonic and galactagogue. Fruits are used as antipyretic, tonic, inflammation; cure blood diseases, gout, bronchitis, burning sensation; useful in paralysis, rheumatism, and affections of the nervous system.

Trigonella foenum-graecum L. Sp. Pl. 777. 1775; Baker in Hook. f. Fl. Brit. India 2: 87. 1876; Kothari in Singh *et al.* Fl. Maharashtra State Dicot. 1: 775. 2000. METHI.

Erect simple or branched, annual herbs, 20–40 cm tall; stem and branches patently hairy. Leaves 3-foliolate, leaflets obovate, 0.6–4 × 0.4–1.5 cm, cuneate at base, rounded or truncate or emarginate at apex, glabrous, glaucous beneath. Flowers sessile, 1–1.5 cm long, axillary, solitary or in pairs. Calyx 3 mm long, hairy; teeth subulate, ciliate, as long as the tube. Corolla white or pale yellow, 1–1.5 cm long. Pods linear, 10–25 cm long, straight or slightly curved, long beaked. Seeds oblong, 2–3 mm long, angular, brownish-yellow.

Fls. & Frts.: February–April.

Cultivated, throughout the region as a leafy vegetable and for seeds used in cooking.

Local uses: Seeds are soaked in water for 2 hours and the paste prepared from the soaked seeds is applied on hairs to stop dandruff and graying of hairs. Similarly seeds are soaked in oil and used for the same purpose. Powder of seeds in given orally in diabetes. Seeds are used as tonic, bronchitis, piles, useful in heart diseases; removes bad taste of mouth.

Uraria picta (Jacq.) Desv. ex DC. Prodr. 2: 324. 1825; Baker in Hook. f. Fl. Brit. India 2: 155. 1876; Cooke, Fl. Pres. Bombay 1: 367. 1958 (Repr. ed.); Kothari in Singh *et al.* Fl. Maharashtra State Dicot. 1: 758. 2000. *Hedysarum pictum* Jacq. Coll. Bot. 2: 262. 1788. NATYA, PITHVAN, PRISHNAPARNI.

Suffruticose under shrubs, 1–1.5 cm tall; stems stout, downy with hooked hairs. Leaves imparipinnate, 15–30 cm long; leaflets subcoriaceous, 6–20 × 1–2 cm, rounded at base, acute, green blotched with white glabrous above, minutely pubescent beneath. Lower leaves 1–3 foliolate; leaflet sub-orbicular or oblong. Flowers in close fascicles along the rachis of spicate, cylindric racemes, 10–30 cm long. Calyx 4 mm long; teeth plumose, much longer than the short tube. Corolla 5–6 jointed, joints smooth, polished. Pods glabrous pale lead color; joints 3–6, smooth, polished, folded on one another. Seeds ovate, 2 mm long, light brown.

Fls. & Frts.: August–September. *Exsiccata*: SUK-681.
Infrequent, in deciduous forests and islands.
Localities: Amba Ghat, Salava Donger, Panhala.

Local uses: Root is one of the component of Dashmula. Root in combination with *Justicia adhatoda*, *Sida acuta* and *Desmodium gangeticum* is given in Jaundice, fever and in acidity.

CAESALPINIACEAE R. Br.

Bauhinia purpurea L. Sp. Pl. 375. 1753; Baker in Hook. f. Fl. Brit. India 2: 284. 1878; Cooke, Fl. Pres. Bombay 1: 461. 1958 (Repr. ed.); Kothari in Singh *et al.* Fl. Maharashtra State Dicot. 1: 780. 2000. KURUL, RAKTA-KANCHAN.

Glabrous trees, 12–15 cm tall; branches stout. Leaves roundish, as broad as long, 8–12 cm across, cordate at base, divided nearly ½ way down into 2 obtuse lobes, glabrous above, thinly puberulous beneath. Flowers fragrant, in terminal and axillary, few-flowered corymbose racemes; bracts triangulate-acute, pubescent. Calyx fulvous-pubescent, 2.5 cm long; tube 6–10 mm long; limb twice as long as the tube, splitting into two coriaceous segments. Corolla purple, 3.5–4 cm long; petals oblanceolate, acute, long-clawed. Fertile stamens 3–4; stigma oblique, peltate. Pods linear, 10–25 × 2–2.5 cm, flat, pointed at both ends, glabrous. Seeds 10–15, oblong-ellipsoid, 12–15 × 8–10 mm, brown, polished.

Fls. & Frts.: October–January.

Common, in deciduous forests and often cultivated.

Localities: Cultivated in Kolhapur, Satara, Sangli.

Local uses: Bark is used as wound healing, useful in ulcers. It is given as emetic.

Bauhinia racemosa Lam. Encycl. 1: 390. 1785; Baker in Hook. f. Fl. Brit. India 2: 276. 1878; Cooke, Fl. Pres. Bombay 1: 459. 1958 (Repr. ed.); Kothari in Singh *et al.* Fl. Maharashtra State Dicot. 1: 780. 2000. APTA, SHEETHSONA, VANU-RAJA.

Small trees, 5–8 m tall. Leaves broader than long, 1.8–4 × 2.5–5.5 cm, cordate at base, divided a little less than ½ way down into 2 rounded lobes, glabrous above, more or less tomentose beneath. Flowers in terminal or leaf opposed racemes, 4–12 cm long; rachis densely pubescent. Calyx pubescent; tube very short, limb 6–8 mm long spathaceous, at length reflexed. Corolla dull white or pale yellow, 8–10 mm long; petals oblanceolate. Stamens 10, all fertile, stalked; filaments densely hairy at the base. Pods stalked, linear, 10–25 × 1–2 cm, flat, glabrous, tapering at base. Seeds 10–15, oblong, 6–10 × 5–6 mm, compressed, black.

Fls. & Frts.: April–October. *Exsiccata*: SUK-83.

Common, in deciduous forests.

Localities: Shivaji University campus, Kolhapur.

Local uses: Root bark is given orally in diarrhoea and dysentery. Decoction of root bark is useful in inflammation of liver and for gargle during aphthae. Juice of young flower along with pepper and onion is used for diarrhoea and dysentery. Decoction of bark is used in tuberculous glands, cough, blood diseases. Bark is used with ginger as an internal remedy for scrofula. Other uses are similar to *B. purpurea*.

Bauhinia variegata L. Sp. Pl. 375. 1753; Baker in Hook. f. Fl. Brit. India 2: 284. 1878; Cooke, Fl. Pres. Bombay 1: 462. 1958 (Repr. ed.); Moorthy in Singh *et al.* Fl. Maharashtra State Dicot. 1: 795. 2000. CHAMOL, KACHNAR, KANCHAN.

Middle-sized trees, 4–8 m tall. Leaves broadly ovate, 6–10 × 8–12 cm, cordate at base, divided 1/3 way down into two rounded lobes, glabrous above, pubescent especially on the nerves beneath. Flowers in axillary, short, few-flowered, corymbose racemes. Calyx 1.5–2 cm long; spathaceous, tube very short or none, limb broadly ovate, thinly pubescent. Corolla pink purple; petals oblanceolate, 3–4.5 cm long the upper variegated with white and broader than the rest. Stamens 5, shorter than the petals. Pods linear, 15–25 × 1.5–2 cm, flat, glabrous-black. Seeds 10–15, oblong black.

Fls. & Frts.: Throughout the year.

Cultivated.

Local uses: Decoction of root is given twice a day for lessening fatness and against tumor.

Caesalpinia bonduc (L.) Roxb. Fl. India 2: 362. 1832 p. p. excl. in J. Bot. 76: 179. 1938; Moorthy in Singh *et al.* Fl. Maharashtra State Dicot. 1: 782. 2000. *Guilandina bonduc* L. Sp. Pl. 381. 1753. *Caesalpinia bonducella* (L.) Fleming in Asiat. Res. 11: 159. 1810 *nom. illeg.*; Baker in Hook. f. Fl. Brit. India 2: 254. 1878; Cooke, Fl. Pres. Bombay 1: 437. 1958 (Repr. ed.). GAJAGA, SAGARGOTA.

Extensive, scandent shrubs; branches finely gray downy, armed with hooked and straight, hard yellow prickles. Leaves 30–60 cm long, bipinnate; pinnae 6–8 pairs, 5–7 cm long. Leaflets 5–9 pairs, oblong, 1.53 × 1–1.5 cm, obtuse, strongly mucronate, glabrous and shining above, paler and pubescent beneath. Flowers in dense, long peduncled, terminal and supra axillary racemes, 10–25 cm long. Calyx 6–8 cm long, fulvous hairy; lobes ovate-oblong, obtuse. Corolla greenish-yellow; petals oblanceolate, scarcely longer than the calyx. Filaments declinate, flattened at base, clothed with long white silky hairs. Pods oblong ellipsoid, 5–7 × 3 cm, armed with prickles. Seeds 1–3, ovoid globose, ash colored.

Fls. & Frts.: August–February. *Exsiccata*: SUK-682.
Common, in deciduous forests and on hedges.
Localities: Atpadi, Miraj.

Local uses: Seed is used with castor oil in inflammation; oil is employed in earache, on pimples paste of seeds applied externally. Powder of seeds is mixed with clove and given in stomachache. Seeds are useful in colic. Seed powder is given in diabetes twice a day. Seeds are also given in menstrual problems. The seeds are very useful in cheap and antiperiodic, antipyretic and tonic, valuable in all ordinary cases of simple and in continued and intermittent fevers.

Cassia absus L. Sp. Pl. 376. 1753; Baker in Hook. f. Fl. Brit. India 2: 265. 1878; Cooke, Fl. Pres. Bombay 1: 451. 1958 (Repr. ed.); Moorthy in Singh *et al.* Fl. Maharashtra State Dicot. 1: 785. 2000. CHANKAN, CHIMAR, SUPALYEL.

Erect, glandular-pubescent undershrubs, 20–45 cm tall. Leaves 3–5 cm long; leaflets 2 pairs, very oblique, elliptic-oblong to elliptic obovate, 2–2.4 × 1.3–2.5 cm, obtuse or subcordate, mucronate, glabrous or nearly so above, pubescent beneath. Flowers in terminal or leaf opposed erect few flowered racemes. Calyx 4 mm long, hairy; teeth oblong obtuse, subequal. Corolla yellow, turning brick red; petals obovate, cuneate, 5–6 mm long. Stamens 5, all perfect. Pod linear, 2–3.5 × 0.6–0.7 cm, flat, slightly oblique, clothed with gray hairs. Seeds 4–6, obovoid, black polished.

Fls. & Frts.: August–November. *Exsiccata*: SUK-683.
Common, in open situations.
Localities: Amba, Gaganbawada, Kolhapur.

Local uses: The seeds are used in ringworms. It is externally applied in leucoderma. Seeds are used to heal ulcers, well in eye disease, pile; cure bronchitis, pains, itching, hiccough.

Cassia alata L. Sp. Pl. 378. 1753; Baker in Hook. f. Fl. Brit. India 2: 264. 1878; Cooke, Fl. Pres. Bombay 1: 454. 1958 (Repr. ed.); Moorthy in Singh *et al.* Fl. Maharashtra State Dicot. 1: 797. 2000.

Much branched shrubs, 1–2 m tall. Leaves 30–60 cm long; leaflets 8–14 pairs, 5–12 × 2.5–8.0 cm, lower oblong-elliptic, upper broadly obovate. Flowers 2.5 cm across, yellow or orange. Pods 10–20 cm long, 1.5 cm broad with a wing down the middle of each valve, oblong, thin, shortly stipitate. Seeds many, transverse.

Fls. & Frts.: September–December. *Exsiccata*: SUK-369.
Cultivated.
Localities: Karad, Kolhapur, Sangli.

Local uses: Bruised fresh leaves are mixed with limejuice, the paste is spread upon the affected part in eczema. Decoction of leaves and flowers is used in eczema. Leaf decoction is considered as a cure for herpes

Cassia auriculata L. Sp. Pl. 1753; Baker in Hook. f. Fl. Brit. India 2: 264. 1878; Cooke, Fl. Pres. Bombay 1: 448. 1958 (Repr. ed.); Moorthy in Singh *et al.* Fl. Maharashtra State Dicot. 1: 785. 2000. ARAT, AWAL, TARVAD.

Much branched shrubs, 1–2 m tall. Leaves 7.5–9.5 cm long, rachis densely fulvous pubescent, with erect linear gland between each pair of leaflets; stipules large, auriculate and leaflets 7–12 pairs, oblong-obovate, 1.5–2.5 × 0.7–1.5 cm, rounded at both ends, mucronate, glabrous. Flowers ca 5 cm across, in terminal and axillary racemes. Calyx ca 1 cm long, glabrous; tube very short; lobes unequal, 2 larger than the other 3, ovate-orbicular. Corolla bright yellow; petals long clawed and limbs with crisped margins perfect stamens 7, of which 3 are longer than the 4 lateral ones; staminodes 3. Pods linear-oblong, 7.5–15 × 1–1.5 cm flat, thin, pale brown depressed between the seeds, pubescent. Seeds 10–20, ovoid, brown, polished.

Fls. & Frts.: January–July. *Exsiccata*: SUK-237.
Common, along the drier tracts in open situations.
Localities: Dry region of Kolhapur, Sangli and Satara districts.

Local uses: Stem is used in tooth complaints. Seed powder is given in diabetes. Flowers are used in urinary discharges. Tea prepared from the leaves is used in chronic fever. Leaves are used for turbid urine or chylos urine. Flowers are used for spermatorrhoea.

Cassia fistula L. Sp. Pl. 377. 1753; Baker in Hook. f. Fl. Brit. India 2: 261. 1878; Cooke, Fl. Pres. Bombay 1: 444. 1958 (Repr. ed.); Moorthy in Singh *et al.* Fl. Maharashtra State Dicot. 1: 786. 2000. AMALTAS, BAHAVA.

Deciduous, moderate sized, glabrous trees, 5–10 m tall. Leaves 15–30 cm long, rachis pubescent; leaflets 4–8 pairs, opposite, ovate, 5–12.5 × 2.5–7.5 cm, cuneate at base, acute, bright green and glabrous above, paler and pubescent especially on nerves beneath. Flowers in lax, drooping racemes, 20–40 cm long. Calyx ca 1 cm long, divided to base, pubescent; teeth oblong, obtuse. Corolla bright yellow, 2.5–3.5 cm across; petals subequal, obovate, shortly clawed. Stamens 10, all fertile, the 3 longest with curved filaments, their anthers dehiscing by longitudinal slits; the four lateral with straight short filaments and versatile, anthers opening by pores at base, the remaining 3 much smaller with indehiscent anthers. Pods cylindric, 30–60 cm long, straight or slightly curved, black. Seeds many, oblong, dark brown, polished.

Fls. & Frts.: April–October. *Exsiccata*: SUK-303.
Common, in deciduous forests, also cultivated.
Localities: Shivaji University campus, Kolhapur, Shirol.

Local uses: Fruit pulp is given as purgative in piles. Root is given as a tonic and febrifuge. Juice of young leaves is used to cure ringworm. Leaf decoction is given orally in paralysis, similarly applied externally. Fruits lessen inflammation and heat of the body, useful in chest complaints, throat troubles, liver complaints, diseases of the eye and griping rheumatism.

Cassia obtusifolia L. Sp. Pl. 377. 1753; Cooke, Fl. Pres. Bombay 1: 447. 1958 (Repr. ed.); Moorthy in Singh *et al.* Fl. Maharashtra State Dicot. 1: 789. 2000. *C. tora sensu* Baker in Hook. f. Fl. Brit. India 2: 263. 1878 p. p. non L. 1753. TAKALA, TARWAD.

Erect, undershrubs, 40–150 cm tall; stem and branches often angular. Leaves 6–10 cm long, rachis grooved, pubescent, with a conical gland between the lowest pair of leaflets; leaflets 3 pairs, opposite, obovate-oblong, 2–5 × 0.8–2 cm, somewhat oblique at base, subacute or obtuse, pale green, pubescent on both surfaces. Flowers in subsessile pairs in the axil of upper leaves. Calyx pubescent divided to the base; segments ovate, 5–8 mm long obtuse, pubescent on the back. Perfect stamens 7, staminodes 3, anthers opening by pores. Pods linear, obtusely 4-gonous, 15–25 cm long, obliquely septate between the seeds. Seeds 30–40, rhomboidal, 3–5 mm long, areole 3–5 mm broad.

Fls. & Frts.: September–October. *Exsiccata*: SUK-611.
Common, throughout in waste place.
Localities: Karad road, Uchgaon.

Local uses: Leaf paste is warmed on flame and applied on any tuberculous glands. Leaves are pounded and applied externally in eczema.

Cassia occidentalis L. Sp. Pl. 377. 1753; Baker in Hook. f. Fl. Brit. India 2: 262. 1878; Cooke, Fl. Pres. Bombay 1: 445. 1958 (Repr. ed.); Moorthy in Singh *et al.* Fl. Maharashtra State Dicot. 1: 789. 2000. KASIVDA, RAN-TAKLA, TAKLA, TARWAD.

Glabrous, annual, undershrubs, 30–150 cm tall; branches furrowed, often purplish. Leaves fetid when bruised, 15–20 cm long; rachis glabrous with a single sessile gland near its base. Leaflets 3–5 pairs, opposite, ovate-lanceolate, 2.5–4.5 × 1.5–3.5 cm, rounded and some what oblique at base, gradually narrowed to acute apex, glabrous above, thinly pubescent beneath. Flowers in short peduncled, few flowered, corymbose racemes. Calyx ca 1 cm long, divided to the base, glabrous; segments membranous, white tinged with pink, oblong-obtuse. Petals yellow, veined with red, ovate-oblong, 10–12 mm long. Perfect stamens 7; staminodes 3. Pods linear-oblong, 10–12.5 × 0.6–0.8 cm compressed, glabrous, transversely septate between the seeds. Seeds 20–30, ovoid, 6 × 4 mm, black green, shining.

Fls. & Frts.: August–December. *Exsiccata*: SUK-684.
Common, in waste lands.
Localities: Jaisingpur, Miraj, Satara.

Local uses: Seeds are used as a cure for the convulsions. Roots are prescribed as a diuretic. Leaves are externally applied in various skin disease. Fresh leaves are grounded and applied to wounds and swellings. Seeds are used as diuretic; root is useful in elephantiasis; cures cough and hiccough, asthma, fever and sore throat.

Cassia sophera L. Sp. Pl. 379. 1753; Baker in Hook. f. Fl. Brit. India. 2: 262. 1878; Cooke, Fl. Pres. Bombay 1: 446. 1958 (Repr. ed.); Moorthy in Singh *et al.* Fl. Maharashtra State Dicot. 1: 790. 2000. JANGLI-TAKLA, RAN-TAKLA.

Glabrous undershrubs, 1–2 m tall. Leaves 15–20 cm long; rachis grooved, glabrous, with a solitary conical gland near the base. Leaflets 5–10 pairs, elliptic-lanceolate, 2.5–7.5 × 0.8–2 cm, rounded or oblique at base, acute or acuminate, glabrous. Flowers in axillary, short, few-flowered corymbose racemes. Calyx 5–6 mm long, divided to the base; segments ovate-oblong, obtuse. Petals yellow, subequal, ovate, 10–12 mm long, obtuse. Stamens 10 of which 3 upper reduced to staminodes, remaining perfect. Pods linear, 6–9 × 0.6–0.7 cm, slightly turgid, straight or curved, thick margined. Seeds 30–40, broadly ovoid, 5–6 × 4 mm, compressed, dark brown.

Fls. & Frts.: November–March. *Exsiccata*: SUK-685.
Common, in waste places.
Localities: Hatkanagale, Kolhapur.

Local uses: A decoction of whole plant is given in diminishing urine and as expectorant. It gives relief in case of acute bronchitis. Infusion of fresh leaves is useful in gonorrhea in subacute stage. Leaf and seed paste is externally applied in eczema and itch. Bark infusion or powdered seeds with honey is given in diabetes.

Cassia surattensis Burm. f. ssp. **glauca** (Lam.) K. & S. Larsen in Fl. C. L. V. 18: 102. 1980; Moorthy in Singh *et al.* Fl. Maharashtra State Dicot. 1: 790. 2000. *C. glauca* Lam. Encycl. 1: 647. 1784; Baker in Hook. f. Fl. Brit. India 2: 265. 1878; Cooke, Fl. Pres. Bombay 1: 453. 1958 (Repr. ed.). MOTHA-TARVAD.

Glabrous, deciduous trees, 5–10 mm tall. Leaves 10–25 cm long; rachis thinly pubescent with a clavate gland between each of the 2–3 lower pairs of leaflets; leaflets 4–9 pairs, ovate-oblong or obovate, 1.5–6.5 × 1.2–3.0 cm, rounded at base, subacute, pale green and glabrous above, glaucous beneath. Flowers in axillary, corymbose racemes shorter than the leaves. Calyx yellow, glabrous, 10–12 mm long, divided to the base; segments broadly ovate-orbicular, subequal. Petals bright yellow, broadly oblong, 1.8–2 cm, obtuse, shortly clawed. Stamens 10, all prefect, subequal. Pods linear-oblong, 15–20 × 1–2 cm flat, tapering to both ends, glabrous. Seeds 15–25, oblong, 5–6 × 3 cm, compressed, smooth, dark brown.

Fls. & Frts.: September–May. *Exsiccata*: SUK-610.
Infrequent, along streams and Nalas.
Localities: Kolhapur.

Local uses: Bark and leaves are prescribed in diabetes.

Cassia tora L. Sp. Pl. 376. 1753; Baker in Hook. f. Fl. Brit. India 2: 263. 1878 p. p.; Cooke, Fl. Pres. Bombay 1: 447. 1958 (Repr. ed.); Moorthy in Singh *et al.* Fl. Maharashtra State Dicot. 1: 791. 2000. TAKLA, TAROTA.

Annual, fetid, undershrubs, 30–90 cm tall. Leaves 4–10 cm long; rachis grooved with conical gland between each of the 2 lowest pairs of leaflets; leaflets 3 pairs, obovate, 1.5–7.5 × 0.8–4 cm, oblique at base, obtuse, glabrous or thinly pubescent, dark green. Flowers in subsessile pairs in the axils of leaves, crowded upwards. Calyx glabrous, divided to the base, 5 mm long; segments subequal, ovate, obtuse. Petals yellow, subequal, oblong, 6–8 mm long, the upper often emarginate. Stamens 10 of which 3 upper reduced to staminodes, three longest with quadrate-oblong anthers and four lateral ones with similar but smaller anthers, all dehiscing by apical pores. Pods subterete, linear, 9–20 × 0.3 cm, glabrous, often spreading. Seeds oblong, 2–3 mm long, obliquely truncate at ends, brown, smooth, areole 1.5–2 mm broad.

Fls. & Frts.: August–February. *Exsiccata*: SUK-123.
Abundant, in waste places.
Localities: Kagal, Kolhapur, Radhanagari road.

Local uses: Leaf paste is warmed on flame and applied on any tuberculous glands. Leaves are pounded and applied externally in eczema. Seeds are mixed with oil and paste is prepared and applied on elephantiasis. Root paste is mixed with 1/2 cup of diluted limejuice and given orally in ringworm and itch.

Hardwickia binata Roxb. Pl. Cor. 3: 6, t. 209. 1819; Baker in Hook. f. Fl. Brit. India 2: 270. 1878; Cooke, Fl. Pres. Bombay 1: 456. 1958 (Repr. ed.); Moorthy in Singh *et al.* Fl. Maharashtra State Dicot. 1: 792. 2000. ANJAN.

Glabrous, deciduous trees, 15–20 m tall; branches slender, drooping. Leaves 2–5 cm long; leaflets 2, subsessile, ovate-oblong, 2–7 × 1–3 cm, oblique at base, obtuse. Flowers in axillary and terminal panicles. Calyx 3–5 mm long, glabrous; tube short; segments oblong, obtuse, membranous. Petals absent. Stamens 10; filaments slender, subequal. Pods 4–7 × 1.5–2 cm, strap-shaped, narrowed at both ends, thin. Seeds solitary near the tip of the pod.

Fls. & Frts.: August–January. *Exsiccata*: SUK-686.
Common, in deciduous forests, often cultivated.
Localities: Jaisingpur, Kolhapur.

Local uses: Latex is boiled in water and oil obtained from it is used in gonorrhoea. Plant is used in treatment of cancer.

Moullava spicata (Dalz.) Nicols. in Manilal, Bot. Hist. Hort. Malab. 184. 1980; Moorthy in Singh *et al.* Fl. Maharashtra State Dicot. 1: 793. 2000. *Caesalpinia spicata* Dalz. in Hook. Kew Gard. Misc. 3: 90. 1851. *Wagatea spicata* (Dalz.) Wight, Ic. t. 1995. 1853; Baker in Hook. f. Fl. Brit. India. 2: 261. 1878; Cooke, Fl. pres. Bombay 1: 443. 1958 (Repr. ed.) VAKERI, WAGHTI.

Robust, woody, scandent undershrubs; branches armed with numerous recurved prickles. Leaves 8–30 cm long; rachis armed with prickles; pinnae 4–6 pairs, 5–12 cm long; leaflets 5–7 pairs, coriaceous, oblong, 2–4 × 1–2 cm, rounded or cordate at base, obtuse or sub acute, dark green and shining above, paler beneath. Flowers nearly sessile in dense spicate racemes, reaching 60 cm long; rachis stout, grooved, brown hairy, armed with scattered prickles. Calyx red, densely puberulous, 10–12 mm long; segments oblong, obtuse, nearly equaling the campanulate tube. Petals orange yellow, 10–12 mm long, inserted on the top of the calyx tube. Stamens 10, alternately long and short; filaments dilated and pilose at base. Pods linear-oblong, 3–6 × 1–1.5 cm, torulose with thickened sutures. Seeds 3–4, obovate-oblong, brown polished.

Fls. & Frts.: November–April. *Exsiccata*: SUK-687.
Frequent, in moist deciduous forests.
Localities: Amba, Amboli, Dajipur, Gaganbawada, Tillari.

Local uses: Stem is applied in any kind of tuberculous glands or exo or endo cortosis. Stem is given in premature or night ejaculation. It is also used as wound healing, reduces heat from the body.

Saraca asoca (Roxb.) de Wilde in Blumea 15: 393. 1968; Moorthy in Singh *et al.* Fl. Maharashtra State Dicot. 1: 793. 2000. *Jonesia asoca* Roxb. in Asiat. Res. 4: 365. 1795; Wight, Ic. T. 206. 1839. *Saraca indica* auct. non L. (1753); Baker in Hook. f. Fl. Brit. India 2: 271. 1878; Cooke, Fl. Pres. Bombay 1: 456. 1958 (Repr. ed.). ASHOK, JASVANT.

Trees, 10–15 m tall. Leaves abruptly pinnate, 15–25 cm long, rachis glabrous; stipules intrapetiolar, completely united, scarious, ovate-oblong; leaflets 10–15 × 2–5.5 cm, oblong, lanceolate, obtuse or acute, quite glabrous, base rounded or cuneate, slightly oblique. Flowers fragrant, numerous, in dense axillary corymbs, 7–10 cm across. Calyx passing from yellow to orange and finally red, tube, 1.2–1.9 cm long, cylindric, solid at base; segments 4, oblong or obovate or obovate-oblong, 1 cm long. Petals 0. Stamens 7 or 8, much exerted, filaments filiform, thrice as long as the calyx-segments; anthers purple. Ovary pubescent, especially on the sutures; styles curved into a ring. Pods black, 10–25 × 3–5 cm, linear oblong, tapering to both ends, compressed, glabrous, veined. Seeds 4–8, ellipsoid-oblong, 2.5 cm long, slightly compressed.

Fls. & Frts.: January–June. *Exsiccata*: SUK-64.
Infrequent, in semi evergreen and moist deciduous forest.
Localities: Amba, Tillari.

Local uses: Bark is orally given in leucorrhoea for 15 days after menstrual flow. Liquid extract is given in menorrhagia. Decoction of bark is used in burning sensation, diseases of the blood, biliousness, effects of fatigue, tumors, enlargement of the abdomen, colic, piles, bloody discharges from the uterus and in menorrhagia.

Tamarindus indica L. Sp. Pl. 34. 1753; Baker in Hook. f. Fl. Brit. India 2: 273. 1878; Cooke, Fl. Pres. Bombay 1: 457. 1958 (Repr. ed.); Moorthy in Singh *et al.* Fl. Maharashtra state Dicot. 1: 794. 2000. AMLI, CHINCH.

Large evergreen trees, 15–20 m tall. Leaves 1–4 cm long; rachis minutely pubescent; leaflets 10–20 pairs, oblong, 8–20 × 5–8 mm, subcoriaceous, glabrous. Flowers in few flowered lax racemes at the ends of the branchlets; rachis slender, pedicels jointed below the calyx; bracts boat shaped, enclosing bud, caducous. Calyx 4–5 mm long, minutely pubescent outside; tube short; teeth lanceolate, two lowest connate. Petals yellow, obovate, 10–12 mm long, striped with red. Stamens 3, monadelphous. Pods oblong, 3–15 × 2 cm, rather turgid, with thin brown brittle epicarp and brown pulpy mesocarp. Seeds 1–10, ovate-quadrate, ca 10 × 8 mm, dark brown, polished.

Fls. & Frts.: February–June. *Exsiccata*: SUK-321.
Common, around villages and fields.
Localities: Kolhapur, Sangli, Satara.

Local uses: A gargle of tamarind water is recommended in sore throat. Boiled seeds are applied on boils, juice of fruit is given in constipation and in inflammation. Ash of fruit rind is given in excessive menstrual flow. Bark is used to heal ulcers, fruits useful on liver complaints, scabies, sore throat and stomatitis.

MIMOSACEAE R. Br.

Acacia catechu (L. f.) Willd. Sp. Pl. 4: 1079. 1806; Baker in Hook. f. Fl. Brit. India 2: 295. 1878; Cooke, Fl. Pres. Bombay 1: 476. 1958 (Repr. ed.); Londhe in Singh *et al.* Fl. Maharashtra State Dicot. 1: 808. 2000. *Mimosa catechu* L. f. Suppl. Pl. 439. 1781. BABHAL GHUBATA, KHIR, KUMTIA, SON. KHAIRI.

Small trees, 5–7 m tall. Leaves 10–15 cm long; rachis white-villous with glands between many of the pairs of pinnae and a large gland near the middle of the petiole; stipular spines hooked; pinnae 20–30 pairs, 3–5 cm long, subsessile. Leaflets 30–50 pairs, linear 4–6 × 1 mm, sessile, subacute, often ciliate. Flowers sessile, in 1–4 nate axillary spikes, 5–10 cm long; rachis white pubescent. Calyx campanulate, 1–1.5 mm long, hairy outside; teeth deltoid, ciliate. Corolla 4–5 mm long, dull white; lobes oblong, obtuse. Stamens much exserted. Pods stalked, 6–10 × 1.5 cm, flat, thin, brown, tapering at both ends, glabrous. Seeds 3–10, compressed, brown, polished.

Fls. & Frts.: July–February. *Exsiccata*: SUK-688.
Common, in deciduous forests.
Localities: Panhala, Kartik Swami Ghat, Radhanagari

Local uses: Bark is used in bleeding gums. Decoction of bark is used to wash ulcers. Kathbol is a mixture of catechu and myrrh, given to women after confinement as a tonic, and is useful in cough. Bark is given as antidysentric, antipyretic, cures sore throat, bronchitis, indigestion, heaviness, ulcers, boils, psoriasis, inflammations, leprosy, anemia, leucoderma; given in elephantiasis and strengthen the teeth.

Acacia concinna (Willd.) DC. Prodr. 2: 464. 1825; Baker in Hook. f. Fl. Brit. India 2: 296. 1878; Cooke, Fl. Pres. Bombay 1: 479. 1958 (Repr. ed.); Londhe in Singh *et al.* Fl. Maharashtra State Dicot. 1: 809. 2000. *Acacia sinuata* (Lour.) Merr. Trans. Amer. Philos. Soc. 24 (2) 186. 1935. *Mimosa sinuata* Lour. Fl. Cocninch. (Willd..) DC. Prodr. 2: 464. 1825. SIKAKAI.

Large, scandent shrubs armed with numerous small hooked prickles; branches brown, dotted white. Leaves 5–10 cm long; main rachis prickly, with a large gland about the middle of the petiole and one between the upper most pair of pinnae; stipules ovate, cordate; pinnae 4–8 pairs, 2–6 cm long, their rachis slender, grooved, ending in a weak spine. Leaflets subsessile, 15–25 pairs, linear, 6–12 × 1.5–2.5 mm, unequal sided, acute or mucronate, glabrous. Flowers in globose heads, on fascicled peduncles, 2–3 cm long arranged in terminal panicles, bracts oblique, membranous. Calyx funnel shaped, 2–3 mm long, divided nearly half way down; teeth lanceolate. Corolla dull white, 3–4 mm long; lobes lanceolate. Pods linear-oblong, 10–15 × 1.2–1.5 cm, flat, thin, rugose, rusty brown, tomentose, depressed between the seeds. Seeds 6–10, ovoid, black polished.

Fls. & Frts.: Throughout year. *Exsiccata*: SUK-689.
Occasional, on hill slopes.
Localities: Ajara, Amba, Barki, Dajipur, Mahabaleshwar.

Local uses: Leaves are given in chronic cough; leaves are used with black pepper in jaundice. Fruits are used to remove dandruff. Pod is used as emetic and laxative. Externally used in eczema, erysipelas and in burning sensation.

Acacia farnesiana (L.) Willd. Sp. Pl. 4: 1083. 1806; Baker in Hook. f. Fl. Brit. India 2: 292. 1878; Cooke, Fl. Pres. Bombay 1: 473. 1958 (Repr. ed.); Londhe in Singh *et al.* Fl. Maharashtra State Dicot. 1: 809. 2000. *Mimosa farnesiana* L. Sp. Pl. 521. 1753. DEV-BABHAL, KANKRI, SHENGATI.

Shrubs or low trees, 2–5 m tall; branchlets zig-zag, glabrous. Leaves 2–6 cm long; rachis slender, pubescent with a minute gland about the middle of the petiole, stipular spines 4–8 mm long, sharp, sometimes wanting. Pinnae 4–8 pairs, 2–3 cm long; leaflets 10–20 pairs, linear-oblong, 3–6 × 1–2 mm, unequal sided, glaucous above, paler beneath. Flowers sessile in globose heads, 6–8 mm diam., fragrant; peduncles 2–3 cm long, crowded on axillary solitary, deltoid, on long slender stalk, ciliolate. Calyx 1–5 mm long, membranous; teeth short, triangular, acute. Corolla yellow, 2.5 mm long; lobes very short, obtuse. Pods subcylindric, 3–8 × 1 cm, turgid, dark brown, glabrous. Seeds oval, 5–6 mm long, dark brown, compressed.

Fls. & Frts.: August–May. *Exsiccata: SUK-174.*
Occasional, in dry deciduous forests.
Localities: Ichalkaranji, Kolhapur, Tasgaon, Vita.

Local uses: Bark is used as antidysenteric, cures stomatitis, caries of the eeth, diseases of the blood, itching, bronchitis, leucoderma, ulcers, nflammations and erysipelas.

Acacia ferruginea DC. Prodr. 2: 458. 1825; Baker in Hook. f. Fl. Brit. India 2: 295. 1878; Cooke, Fl. Pres. Bombay 1: 477. 1958 (Repr. ed.); Londhe in Singh *et al.* Fl. Maharashtra State Dicot. 1: 811. 2000. ICHAN, KANTI, PANDHARA-KHAIR.

Moderate sized trees, 6–10 m tall. Leaves 5–10 cm long; rachis slender, glabrous, with a large gland on the petiole, another between uppermost pair of pinnae; stipular spines short, hooked; pinnae 3–6 pairs, 4–7 cm long. Leaflets 10–20 pairs, linear-oblong, 6–8 × 2–3 mm, unequal sided, obtuse, glaucous above, paler beneath, glabrous. Flowers sessile in slender axillary spikes, 5–7.5 cm long; peduncles many, glabrous. Calyx campanulate, 1.5 mm long, glabrous; teeth deltoid. Corolla pale yellow or dull white, 3–4 mm long; lobes oblong-lanceolate, as long as the tube. Pods oblong, 10–15 × 2–2.5 cm, straight, flat, thin, brown, reticulately veined; the upper sutures narrowly winged. Seeds 4–8, ovoid-orbicular, 10–12 mm across, pale brown.

Fls. & Frts.: May–October. *Exsiccata:* SUK- 1500.
Planted along roadsides.
Localities: Ramling, Ratnagiri, Satara, Sindhudurg.

Local uses: Bark is useful in diseases of the blood. Leaves are useful in liver complaints, diseases of the eye, diarrhoea, dysentery, gonorrhoea, gleet, burns and scalds.

Acacia leucophloea (Roxb.) Willd. Sp. Pl. 4: 1083. 1806; Baker in Hook. f. Fl. Brit. India 2: 294. 1878; Cooke, Fl. Pres. Bombay 1: 475. 1958 (Repr. ed.); Londhe in Singh *et al.* Fl. Maharashtra State Dicot. 1: 811. 2000. *Mimosa leucophloea* Roxb. Pl. Cor. 2: 27, t. 150. 1800. HIVAR.

Small trees, rarely reaching 8 m tall. Leaves 6–8 cm long; rachis pubescent, with 1–3 glands near the base and a cup-shaped gland between each pair of pinnae; stipular spines short, straight, variable in length; pinnae 4–15 pairs, subsessile. Calyx ca 1 mm long; teeth short. Corolla dull white, twice as long as the calyx; lobes oblong, subacute. Pods oblong, 10–20 × 0.6–0.8 cm, flat, brown tomentose, thickened on sutures. Seeds 10–20, obovoid, 5–7 mm long, glabrous, brown.

Fls. & Frts.: August–February. *Exsiccata*: SUK-156.
Common, along roadsides and hill slopes.
Localities: Babu Jamal, Ramling, Uttur.

Local uses: Bark is used as anthelmintic, antipyretic, cures inflammation, bronchitis, leprosy. Bark is used for the same purpose as *A. catechu.*

Acacia nilotica (L.) Willd. ex Del. ssp. **indica** (Benth.) Brenan, Kew Bull. 12 84. 1957; Londhe in Singh *et al.* Fl. Maharashtra State Dicot. 1: 813. 2000 *Mimosa nilotica* L. Sp. Pl. 521. 1753. *A. arabica* (Lam.) Willd. 1806; Baker in Hook. f. Fl. Brit. India 2: 293. 1878; Cooke, Fl. Pres. Bombay 1: 472. 1958 (Repr. ed.). BABHUL.

Trees, 3–8 m tall. Leaves 5–8 cm long; rachis downy, with gland at the base, stipular spines variable, 0.5–6 cm long, straight, sharp, ash-white, smooth, pinnae 4–8 pairs, 2–4 cm long. Leaflets subsessile, 10–25 pairs, linear-oblong, 3–6 × 1–2 mm, subobtuse, glabrous or nearly so. Flowers in globose heads; peduncle broadly ovate, acute, pubescent. Calyx campanulate, 1–1.5 mm long; teeth very short. Corolla bright yellow, 3 mm long; lobes triangular acute. Pods stalked, 6–15 cm long, densely fine gray tomentose. Seeds 6–10, discoid, 6–8 mm across, margined, brown, polished.

Fls. & Frts.: Throughout year. *Exsiccata*: SUK-316.
Very common, around fields, along river banks, along roadsides.
Localities: Kagal.

Local uses: Bark is used to strengthen the teeth. Bark decoction is used to wash ulcers. Used in colic.

Acacia pennata (L.) Willd. Sp. Pl. 4: 1090. 1806; Baker in Hook. f. Fl. Brit. India 2: 297. 1878; Cooke, Fl. Pres. Bombay 1: 480. 1958 (Repr. ed.); Londhe in Singh *et al.* Fl. Maharashtra State Dicot. 1: 813. 2000. *Mimosa pennata* L. Sp. Pl. 507. 1753. CHILAR, GHATTAD-KHAIR, SHEMBI.

Scandent shrubs armed with numerous stout hooked or straight prickles; young parts pubescent. Leaves 8–15 cm long; main rachis densely pubescent, prickly, with a large gland above the middle of the petiole and one between each of the 2–4 upper pairs of pinnae, 6–18 pairs, 3–6 cm long, subsessile, their rachis densely pubescent. Leaflets 10–50 pairs, closely set, linear oblong, 3–5 × 1–1.5 mm, obtuse or acute, ciliate on margins, other wise glabrous. Flowers in globose heads arranged in large terminal panicles; heads 10–12 mm in diam., peduncles pubescent, 2–4 together, bracts linear. Calyx glabrous, 2–2.5 mm long, divided 1/3 way down; teeth linear-lanceolate, acute. Corolla dull yellow or white, 4–5 mm long, lobes linear acute. Pods stalked, strap shaped, 10–15 × 2–3 cm, flat, thin, apiculate, brown, glabrous. Seeds 7–14, ovoid-oblong, compressed, dark-brown.

Fls. & Frts.: August–March. *Exsiccata:* SUK-169.
Common, in dry deciduous forests.
Localities: Kagal, Babujamal, Jaisingpur, Ramling.

Local uses: Leaf juice is used in indigestion of infants. Leaves are chewed in bleeding gums.

Albizia amara (Roxb.) Boiv. Encycl. 19, Siecle 2: 34. 1834; Baker in Hook. f. Fl. Brit. India 2: 301. 1878; Cooke, Fl. Pres. Bombay 1: 483. 1958 (Repr. ed.); Londhe in Singh *et al.* Fl. Maharashtra State Dicot. 1: 815. 2000. *Mimosa amara* Roxb. Pl. Cor. 1: 40, t. 122. 1799. KANSAR, LALAI, TUGLI.

Small much branched, unarmed trees, 5–8 cm tall; young parts, clothed with yellowish grey pubescent. Leaves 4–10 cm long; main rachis densely pubescent with a small gland about the middle of the petiole and another between the upper most pair of pinnae; leaflets 10–30 pairs, sessile, close, linear-oblong, 4–5 × 2.5–3 mm, obtuse glaucous above, pubescent beneath. Flowers subsessile, fragrant, on 2–4-nate, peduncled, globose heads. Calyx 1.5–2.5 mm long, pubescent outside; teeth short, deltoid. Corolla pale yellow, 4–6 mm long; lobes linear-lanceolate, pubescent on the back of the tips. Pods linear-oblong, 10–15 × 1.5–2 cm, flat, thin, dark brown. Seeds 6–8, ovoid, brown.

Fls. & Frts.: April–August. *Exsiccata*: SUK-383.
Frequent, in hill forests.
Localities: Narande, Ramling.

Local uses: Seeds are given in piles, oil extracted from them is said to cure white leprosy.

Albizia lebbeck (L.) Willd. Sp. Pl. 4: 1066. 1806; Baker in Hook. f. Fl. Brit. India 2: 298. 1878; Cooke, Fl. Pres. Bombay 1: 481. 1958 (Repr. ed.); Londhe in Singh *et al.* Fl. Maharashtra State Dicot. 1: 816. 2000. *Mimosa lebbeck* L. Sp. Pl. 576. 1753. KALASIRIS, SIRAS.

Unarmed deciduous trees, 10–20 m tall. Leaves 8–15 cm long, main rachis glabrescent, with a large gland above the base of the petiole and another below the upper most pair of pinnae; pinnae 2–4 pairs, 8–12 cm long, rachises glabrous. Leaflets 5–9 pairs, elliptic-oblong, 3–5 × 2–2.5 cm, cuneate at base, obtuse, glabrous above, glabrescent beneath. Flowers fragrant, in globose umbellate heads, 2–3.5 cm broad, solitary or 2–4 together. Calyx 4 mm long, pubescent; teeth short, deltoid. Corolla dull white, 8–10 mm long; tube glabrous; lobes 2–2.5 mm long; pubescent. Stamens much longer than the corolla; filaments connate at base. Pods oblong, 10–30 × 2.5–5 cm, flat, thin, straw coloured, obtuse or subacute at apex. Seeds 5–10 ovoid oblong, 7–8 mm long, pale brown, smooth.

Fls. & Frts.: March–December.
Planted along roadsides, and also wild.
Localities: Gaganbawada, Kolhapur, Shirol.

Local uses: Root bark or stem bark is used in toothache, strength the gums and teeth. Seeds are used in tuberculous glands. Root is used in hemicrania, cures diseases of the blood, leucoderma, itching and skin diseases.

Albizia odoratissima (L. f.) Benth. in Hook. f. London. J. Bot. 3: 88. 1844; Baker in Hook. f. Fl. Brit. India 2: 299. 1878; Cooke, Fl. Pres. Bombay 1: 482. 1958 (Repr. ed.); Londhe in Singh *et al.* Fl. Maharashtra State Dicot. 1: 817. 2000. *Mimosa odoratissima* L. f. Suppl. Pl. 437. 1781. KALA SIRAS.

Tall, unarmed trees, up to 15 m tall. Leaves 15–30 cm long; main rachis downy, with a large sessile gland on the petiole a little above the base and also at the bases or upper 1–2 pinnae; pinnae often 4-pairs, 6–10 cm long, their rachis pubescent. Leaflets 8–15 pairs, sessile, obliquely oblong, 2–3 × 1–1.5 cm, rounded at apex, dark green above, densely clothed with spreading gray hairs, ultimately dull brown beneath. Flowers fragrant, sessile, in numerous small heads, 2–2.5 cm across. Calyx 1 mm long, pubescent; teeth obsolete. Corolla 3–3.5 mm long, dull white, divided ¼ way down; lobes subulate villous at the tips. Stamens 8–10 mm long; filaments pale yellow, connate at base. Pods shortly stalked, oblong, 10–20 × 2–3 cm, thin, downy often with 1 cm long beak. Seeds 8–12, broadly ovate, 8–6 mm, yellowish-brown.

Fls. & Frts.: May–December. *Exsiccata*: SUK-1624.
Common, in plains and deciduous forests.
Localities: Alate, Amba, Narande, Panhala;

Local uses: Uses are same as *A. lebbeck.*

Dichrostachys cinerea Brenen & Brummit in Bot. Soc. Borter 39: 111. 1935; Londhe in Singh *et al.* Fl. Maharashtra State Dicot. 1: 819. 2000. *D. cinerea* (L.) Wight & Arn. Prodr. 271. 1834; Baker in Hook. f. Fl. Brit India 2: 288. 1878; Cooke, Fl. Pres. Bombay 1: 468. 1958 (Repr. ed.). *Mimosa cinerea* L. Sp. Pl. 577. 1753. DURANGI BABHUL, SIGAM-KATI, VELATUR.

Large shrubs or small trees, 2–3 m tall, armed with sharp, straight thorns, 2–4 cm long. Leaves 2–4.5 cm long; pinnae 3–12 pairs, 1–2 cm long; leaflets 7–12 pairs, linear-oblong, 1–2.5 mm long, rounded at apex, glabrous or nearly so. Flowers dimorphic in elongate spikes; the upper bisexual, the lower neuter. Flowers bisexual: Calyx 1–1.5 mm long, 4–5 toothed. Corolla yellow, twice as long as the calyx; petals linear-oblong. Stamens 8–10; filaments distinct; anthers gland-crested. Ovary many ovuled. Neuter flowers. Calyx and corolla pale pink, longer than those of bisexual flowers. Staminodes 10 with long, pink filaments. Pistillode minute. Pods linear-oblong, 6–10 cm long, curled or twisted, pubescent. Seeds 2–10, orbicular, compressed, brown smooth.

Fls. & Frts.: August–December. *Exsiccata*: SUK-317.
Common, throughout in scrub forests.
Localities: Babu Jamal, Katavadi, Ramling.

Local uses: Root paste is externally applied in rheumatism. Decoction of bark is given as an appetizer. Decoction is used to wash ulcers. Root is used in urinary calculi and renal troubles, diseases of vagina and uterus and pain in joints.

Entada rheedei Spreng. Syst. Veg. 2: 325. 1825; Londhe in Singh *et al.* Fl. Maharashtra State Dicot. 1: 819. 2000. *E. pusaetha* DC., Mem. Legum. 12. 1825 and Prodr. 2: 425. 1825. *E. scandens* auct. non Benth. 1841; Baker in Hook. f. Fl. Brit. India 2: 287. 1878 p. p.; Cooke, Fl. Pres. Bombay 1: 465. 1958 (Repr. ed.). GARAMBI, GARBI, GARDUL.

Large woody climbers with thick trunks. Leaves 2-pinnate; main rachis glabrous, grooved, usually ending in a bifid tendril; pinnae 2–3 pairs, stalked;

Text fig. 14: Entada rheedei a-twig; b-bract; c-flower; e-carpel.

leaflets oblong or obovate, 2.5–7 × 0.6–3 cm, obtuse, emarginate, rigidly coriaceous, dark green, glabrous, reticulately veined, base acute. Flowers in peduncled panicle, or simple spikes, 15–25 cm long, axillary or from the nodes of old leaflets branches. Calyx 1–2 mm long; minutely toothed. Corolla 2–3 mm long, yellow. Pods 30–90 × 7–10 cm, and 2.5 m thick, slightly curved, woody. Seeds woody, the sutures very thick, indented between the seeds. Seeds 6–15, orbicular or orbicular-oblong, 2.5–3 cm in diam., compressed, smooth and shining, brown. The joints of the pod break away from the thickened sutures which remain as an empty frame.

Fls. & Frts.: March–May. *Exsiccata*: SUK-301.
Infrequent, in semi evergreen forests.
Localities: Amba, Mahabaleshwar.

Local uses: Seeds are immediately applied after scorpion sting. Stem is used as wound healing. Seed paste is applied externally in hemicrania and in joint pains.

Mimosa pudica L. Sp. Pl. 518. 1753; Baker in Hook .f. Fl. Brit. India 2: 291. 1878; Cooke, Fl. Pres. Bombay 1: 470. 1958 (Repr. ed.); Londhe in Singh *et al.* Fl. Maharashtra State Dicot. 1: 822. 2000. LAJALU, LAJARI.

Prickly, prostrate or suberect undershrubs; stem and branches softly hairy. Leaves 4–8 cm long; pinnae 1–2 pairs, 4–7 cm long; their rachises clothed with bristles. Leaflets 12–20 pairs, sessile, linear-oblong, 6–8 × 3–4 mm, acute, glabrous above pubescent beneath. Flowers in globose heads, 6–8 mm in diam.; peduncles 2–25 cm long, prickly, often in axillary pairs; calyx minute. Corolla pink, 2–2.5 cm long; lobes 4, ovate-oblong, obtuse. Stamens much exserted. Pods 1.5–2.5 × 0.3–0.4 cm, flat, slightly curved; joints clothed with bristles on sutures. Seeds compressed, brown.

Fls. & Frts.: September–January. *Exsiccata*: SUK-1627.
Common, in semi evergreen and moist deciduous forests in moist places.
Localities: Kudal, Malwan, Sawantwadi.

Local uses: Leaf decoction is used with coconut oil as ear drops. The leaf paste is used to wash and remove films of the conjunctiva. Root is used in vaginal and uterine complaints, inflammations and burning sensation.

Prosopis cineraria (L.) Druce, Rep. Bot. Soc. Exch. Cl. Brit. Is 1913, 3: 422. 1914; Londhe in Singh *et al.* Fl. Maharashtra State Dicot. 1: 823. 2000. *Mimosa cineraria* L. Sp. Pl. ed. 2, 2: 1500. 1763 ('*cinerea*'). *Prosopis spicigera* L. Mant. 68. 1767; Baker in Hook. f. Fl. Brit. India 2: 288. 1878; Cooke, Fl. Pres. Bombay 1: 467. 1958 (Repr. ed.). SHAMI.

Trees, 8–12 m tall; branches slender, glabrous armed with straight prickles, 3–6 mm long. Leaves 3–3.5 cm long; rachis glabrous; stipular spines 3–6 mm long. Pinnae two pairs, 2–7 cm long; leaflets 7–12 pairs, oblong, 1–1.5 × 0.3–0.4 cm, subsessile, rounded and very oblique at base, rounded and mucronate at apex, glabrous. Flowers sessile. Calyx 1–1.5 mm long, membranous, obscurely 5-toothed. Corolla pale yellow or creamy white, 3 mm long. Stamens shortly exerted; anthers crested with a deciduous gland. Pods 5–15 cm long, glabrous. Seeds 10–15, ovoid-oblong, compressed, shining brown.

Fls. & Frts.: November–April. *Exsiccata*: SUK-176.

Infrequent, in dry regions.

Localities: Babu Jamal, Devale, Gadhinglaj, Ichalkaranji, Kolhapur, Nesari, Tasgaon.

Local uses: Leaf decoction is given as a brain tonic.

Xylia xylocarpa (Roxb.) Taub. Bot. Centralbl. 67: 395. 1891; Londhe in Singh *et al.* Fl. Maharashtra State Dicot. 1: 824. 2000. *Mimosa xylocarpa* Roxb. Pl. Cor. t. 100. 1798. *Xylia dolabriformis* Benth. in Hook. London J. Bot. 4: 417. 1845; Baker in Hook. f. Fl. Brit. India 2: 286. 1878; Cooke, Fl. Pres. Bombay 1: 465. 1958 (Repr. ed.). JABLA, JAMBA, YORUL.

Tall tree; young parts tomentose. Leaves 2-pinnate; pinnae 2, terminal on a petiole 2–3 cm long which is furnished with a gland at the apex between the pinnae. Leaflets 2–4 pairs, diminishing downwards, with often a small unopposed leaflets on the outside below the lowest pair of leaflets, the terminal leaflets 7–15 × 2–6 cm. Flowers sessile, in dense globose heads, 1.2–1.9 cm in diam. Calyx tubular 5 mm long; teeth 5, valvate. Corolla yellowish white, 6 mm long; petals lanceolate, valvate. Stamens 10; filaments free, 1.2 cm long, exerted; anthers crested when young with a stalked deciduous gland. Ovary sessile; ovules many. Pods 12–15 × 2–5 cm, large, woody, oblong, falcate, flat, rusty-tomentose, septate between the seeds, finally dehiscent. Seeds 6–10, oblong-ellipsoid, compressed, 1.2–1.5 × 0.6–0.9 cm, brown, smooth, polished.

Fls. & Frts.: March–December. *Exsiccata*: SUK-302.

Occasional, in moist deciduous or semi evergreen forests.

Localities: Amboli, Barki, Doddamarg, Phonda, Sawantwadi, Radhanagari.

Local uses: Oil from the seeds is given in rheumatism. Bark is used in washing of ulcers and applied externally in joint pains.

ROSACEAE Juss.

Rosa damascena Mill., Gard. ed. 8: 15. 1768; Hook. f. Fl. Brit. India 2: 364. 1878; Bailey, Man. Cult. Pl. 535. 1949 (Repr. ed.); Londhe in Singh *et al.* Fl. Maharashtra State Dicot.1: 836. 2000. DAMASK, GULAB.

Scandent shrubs, ca 1–2 m in height; prickles hooked. Leaflets mostly 5, ovate-oblong, up to 6 cm long. Flowers 6–12 in a corymbose inflorescence 4–6 cm across. Calyx campanulate, glandular hispid; lobes often pinnate like the leaves. Corolla pale or dark red, fragrant.

Fls. & Frts.: March–April and September–October.
Cultivated, native of Middle East Asia.

Local uses: Paste of petals is applied externally as astringent. Flower buds are considered aperient, cardiac, tonic, cephalic, removing bile and cold humours.

Rosa indica L. Sp. Pl. 492. 1753; Hook. f. Fl. Brit. India 2: 364. 1878; Londhe in Singh *et al.* Fl. Maharashtra State Dicot.1: 836. 2000. DESHIGULAB.

Evergreen, glabrous shrubs; prickles stout, hooked. Leaves 5–15 cm long; stipules narrow adnate to the top. Leaflets 3–7, elliptic lanceolate, 3–7 cm long, acuminate, glabrous. Flowers 3–5 cm across, solitary or in short panicles, scentless. Calyx globose; lobes often reflexed. Corolla red; petals 5-many.

Fls. & Frts.: Throughout year.
Cultivated.

Local uses: Paste of petals is applied externally as astringent. Flower buds are considered aperient, cardiac, tonic, cephalic, removing bile and cold humours.

Rosa multiflora Thunb. Fl. Jap. 214. 1784; Hook. f. Fl. Brit. India 2: 364. 1878; Londhe in Singh *et al.* Fl. Maharashtra State Dicot. 1: 837. 2000. PANDHRA GULAB.

Erect scandent shrubs; prickles stout, hooked. Leaves 2–5 cm long; stipules pectinate; leaflets 5–11, ovate to obovate, 2–3 cm long. Flowers numerous in terminal panicles, fragrant, ca 2 cm across. Calyx globose; lobes ovate acute. Corolla white.

Fls. & Frts.: Throughout year.
Native of Japan, cultivated.

Local uses: Same as *R. damascena.*

CRASSULACEAE DC.

Kalanchoe pinnata (Lam.) Pers. Syn. 446. 1805; Londhe in Singh *et al.* Fl. Maharashtra State Dicot.1: 842. 2000. *Cotyledon pinnata* Lam. Encycl. 2: 141. 1786. *Bryophyllum calycinum* Salisb., arad. Lond. t. 3. 1805; C. B. Cl. in Hook. f. Fl. Brit. India 2: 413. 1878; Cooke, Fl. Pres. Bombay 1: 494. 1958 (Repr. ed.). PANPHUTI

Succulent, glabrous, erect herbs, 30–120 cm tall. Leaves usually variable, the lower usually simple, the upper often 3–7 foliolate, long petioled. Leaflets ovate-elliptic, 8–20 × 5–13 cm, crenate and often producing plantlets from the crenatures. Flowers pendent, from terminal corymbose cymes. Calyx 2–3 cm long, green with red lines; teeth triangular, acute. Corolla reddish, swollen at base, constricted at the middle, twice as long as the calyx; lobes triangular. Hypogynous scales subquadrate. Follicles enclosed in persistent papery calyx and corolla. Seeds oblong ellipsoid, smooth.

Fls. & Frts.: December–February. *Exsiccata*: SUK-281.
Cultivated, native of Tropical Africa.
Localities: Jasingpur, Kolhapur, Shiroli.

Local uses: Leaves are chewed with salt in cough and in bronchitis. Leaves are warmed on the flame and applied externally to cure wounds. Leaf decoction is given twice a day in urinary calculi. Leaves are eaten to control diabetes. Juice of leaves is mixed with leaf juice of *Aegle marmelos* given in amoebic dysentery..

DROSERACEAE Salisb.

Drosera burmanni Vahl, Symb. Bot. 3: 50. 1794; C. B. Cl. in Hook. f. Fl. Brit. India 2: 424. 1878; Cooke, Fl. Pres. Bombay 1: 499. 1958 (Repr. ed.); Godbole & Das Das in Singh *et al.* Fl. Maharashtra State Dicot.1: 844. 2000.

Acaulescent herbs, erect, glandular. Leaves rosulate, obovate-orbicular, 5–8 × 3.5–6 mm, reddish-green, upper surface with gland tipped hairs, base attenuate or subtruncate-obtuse. Flowers pale pink in 15–20 flowered secund racemes. Calyx lobes 5, oblong, 4–5 mm long. Petals 5, oblong-oblanceolate, stamens 5. Ovary ovoid, 3-celled, styles 5, apically shortly lobed. Capsule subglobose. Seeds numerous, minute, black, reticulate.

Fls. & Frts.: December–February. *Exsiccata*: SUK-1675.
Occasional, on lateritic plateaus.
Localities: Ajara, Amba, Amboli, Sawantwadi, Vaibhavwadi,

Local uses: Plant is given as cardiac tonic. Paste of plant is applied externally in boils.

Drosera indica L. Sp. Pl. 282. 1753; C. B. Cl. in Hook. f. Fl. Brit. India 2: 424. 1878; Cooke, Fl. Pres. Bombay 1: 499. 1958 (Repr. ed.); Godbole & Das Das in Singh *et al.* Fl. Maharashtra State Dicot.1: 844. 2000.

Erect or decumbent, glandular-pubescent, annual herbs, 10–25 cm tall. Leaves cauline, alternate, narrow, linear, 2–5 cm long, circinate in venation, upper part copiously fringed with fine gland tipped tentacles; flowers in leaf opposed, 1-few flowered racemes, glandular hairy. Calyx 4–5 mm long;

segments oblong, lanceolate, acute, more or less glandular pubescent. Petals deep pink, narrowly spathulate, slightly exceeding the calyx. Capsules 3-valved, ovoid, 5–7 mm long. Seeds many, obovoid, pale brown, reticulate.

Fls. & Frts.: September–December. *Exsiccata*: SUK-139.
Common, on gravelly soil and around ponds and in pastures.
Localities: Amba, Amboli, Dajipur, Kas, Thoseghar, Pawas.

Local uses: Same as *D. burmanii.*

COMBRETACEAE R. Br.

Anogeissus latifolia (Roxb. ex DC.) Wall. ex. Guill. & Perr. Fl. Seneg. Tent. 1: 280. 1832; C. B. Cl. in Hook. f. Fl. Brit. India 2: 450. 1878; Cooke, Fl. Pres. Bombay 1: 512. 1958 (Repr. ed.); Scott in Kew Bull. 33: 560. 1979; Diwakar in Singh *et al.* Fl. Maharashtra State Dicot: 2: 1. 2001. *Conocarpus latifolia* Roxb. ex DC. Prodr. 3: 17. 1828. DHAWDA.

Deciduous, glabrescent trees, 5–10 m tall. Leaves opposite or subopposite, elliptic-oblong, 3–8.5 × 2.5–6 cm, rounded at base, obtuse, emarginate or shortly cuspidate at apex, glabrous. Flowers sessile, in small globose heads. Calyx tube short, pubescent; teeth short, broadly triangular. Stamens 10, exserted. Style thickened at base, villous. Fruits crowed in globose heads, orbicular, 4–8 mm in diam., compressed, winged with a long beak, pale yellow or brown. Seeds ovoid, solitary.

Fls. & Frts.: September–December. *Exsiccata*: SUK-97.
Common, in deciduous forests.
Localities: Babu Jamal, Kagal.

Local uses: Bark is used in bad taste due to acidity, useful in anemia, urinary discharges. Bark is externally applied in bone fractures.

Calycopteris floribunda (Roxb.) Poir. in Lam. Encyl. Suppl. 2: 41. 1811; C. B. Cl. in Hook. f. Fl. Brit. India 2: 449. 1878; Cooke, Fl. Pres. Bombay 1: 512. 1958 (Repr. ed.); Diwakar in Singh *et al.* Fl. Maharashtra State Dicot. 2: 2. 2001. *Getonia floribunda* Roxb. Pl. Cor. t. 87. 1798. BAGULI, UKSHI.

Scandent shrubs. Leaves opposite, ovate-lanceolate or elliptic-oblong, 6–12 × 3–6 cm, acute, base rounded, both surfaces tomentose. Flowers in fulvous-pubescent terminal panicles, sessile, yellowish-green. Calyx densely hairy within, at the mouth of the tube; tube produced above the ovary; lobes accrescent, 1.8–4.5 cm long in fruit, 3-nerved, tomentose. Petals absent. Ovary densely villous. Fruit ovoid, 5-ribbed, pubescent, crowned by the persistent calyx lobes, one seeded.

Fls. & Frts.: March–May. *Exsiccata*: SUK-659.
Frequent, in moist deciduous forests.

Localities: Amba, Barki, Gaganbawada.

Local uses: Leaf paste is applied externally in fresh wounds and cuts. Leaves are grounded and given with Ghee in dysentery. Flowers are used as a poultice for headache. Fruits are carminative and used in jaundice.

Terminalia bellirica (Gaertn.) Roxb. Pl. Cor. 2: 54, t. 198. 1805 ('*bellerica*'); C. B. Cl. in Hook. f. Fl. Brit. India 2: 445. 1878; Cooke, Fl. Pres. Bombay 1: 508. 1958; (Repr. ed.); Diwakar in Singh *et al.* Fl. Maharashtra State Dicot. 2: 4. 2001. *Myrobalanus bellirica* Gaertn. Fruct. 2: 90, t. 97. 1791. BEHDA, BEILIRINA, HELA.

Large, deciduous trees, 8–12 m tall. Leaves collected at the ends of branches, coriaceous, 10–20 × 7–15 cm, obtuse or rounded at apex, glabrous. Flowers in axillary terminal, simple or branched spikes, upper ones male, the lower hermaphrodite. Calyx pubescent outside, 3–4 mm long; teeth recurved, deltoid, 1.5 mm long. Disk brown hairy. Drupes broadly ellipsoid or subglobose, 2–2.5 × 1.5–2 cm, densely finely velutinous, brown, faintly 5-ribbed.

Fls. & Frts.: May–September. *Exsiccata*: SUK-96.

Common, in moist deciduous forests.

Localities: Ajara, Amboli, Barki, Gaganbawada, Mahabaleshwar, Tillari, Vaibhavwadi.

Local uses: Fruits used with seeds of *Butea monosperma* is given as anthelmintic. Fruit decoction is used to wash eyes. Fruits are prescribed in piles. Oil is applied externally on swellings. Fruit is used as laxative, anthelmintic, useful in bronchitis, sore throat, inflammation, asthma and bronchitis. It is one of the ingredient in 'Triphla Churna'

Terminalia catappa L. Syst. Nat. ed. 2: 674. 1767; C. B. Cl. in Hook. f. Fl. Brit. India 2: 444. 1878; Cooke, Fl. Pres. Bombay 1: 511. 1958 (Repr. ed.); Diwakar in Singh *et al.* Fl. Maharashtra State Dicot. 2: 7. 2001. BADAM, DESHI-BADAM, JANGLI-BADAM.

Deciduous trees, 8–12 cm, tall. Leaves chartaceous or papyraceous, crowded at the ends of branches, 15–20 × 8–12 cm, rounded or shortly acuminate at apex, glabrous above, pubescent especially on the nerves beneath; 1–2 glands at the tip on the lower surface. Flowers sessile in axillary spikes, mostly male flowers, a few bisexual flowers present at the base. Calyx tube 1–3 mm long, above the ovary; lobes ovate-triangular, 1–1.5 mm long. Disk hairy. Fruits usually green, ovoid or ellipsoid, 3–7 × 2–5 cm, glabrous.

Fls. & Frts.: April–October. *Exsiccata*: SUK-663.

Native of tropical Asia, planted in gardens and along roadsides.

Local uses: Decoction of bark is given in leucorrhoea and gonorrhea Oil is applied externally in headache caused by acidity.

Terminalia chebula Retz. Obs. Bot. 5: 31. 1788; C. B. Cl. in Hook. f. Fl. Brit. India 2: 446. 1878; Cooke, Fl. Pres. Bombay 1: 509. 1958 (Repr. ed.); Diwakar in Singh *et al.* Fl. Maharashtra State Dicot. 2: 4. 2001. HIRADA.

Trees up to 20 m tall. Leaves elliptic-oblong, 6–15 × 4–8 cm, rounded or cordate at base, acute or acuminate, glabrous above, pubescent beneath, usually 2 glands near the summit. Flowers all hermaphrodite, in terminal, often panicled spikes. Calyx campanulate, 3 mm long, glabrous outside, hairy within; lobes short obscure. Drupes pendulous, ellipsoid, 2–3 × 1.5–2 cm, brown, glabrous, obscurely 5-ribbed.

Fls. & Frts.: April–October. *Exsiccata*: SUK-382.
Frequent, in moist deciduous or dry deciduous forests.
Localities: Amboli, Borbet, Mahabaleshwar, Phonda, Tillari, Vaibhavwadi.

Local uses: Ripe fruits are given as a purgative and as a tonic. Fruits are used in sore throat, bleeding and ulceration of the gums. One gram of fruit powder is wetted with Ghee and taken daily to avoid diseases. Fruit powder with sugar is given in piles. Powder is mixed with coconut oil and applied on joint pains and swellings. It is one of the ingredients of *'Triphala churna'*.

Terminalia cuneata Roth, Nov. Pl. Sp. 379. 1821; Diwakar in Singh *et al.* Fl. Maharashtra State Dicot. 2: 5. 2001. *Pentaptera arjuna* Roxb. In Hort. Beng. 34. 1814 and Fl. Ind. 2: 438. 1832. *Terminalia arjuna* (Roxb. ex DC.) Wight & Arn. Prodr. 314. 1834; C. B. Cl. in Hook. f. Fl. Brit. India 2: 447. 1878; Cooke, Fl. Pres. Bombay 1: 509. 1958 (Repr. ed.). ARJUN, SADADA, KAHU.

Large trees reaching 12 m in height. Leaves often subopposite, oblong or elliptic-oblong, rounded or subcordate at base, obtuse or subacute shallowly crenate-serrate or subentire, glabrous, pale green above, brownish beneath. Flowers sessile, in short axillary and terminal panicles. Calyx glabrous, 4 mm long; teeth triangular, 1.5 mm long. Disk clothed with yellowish or reddish hairs. Drupes ovoid oblong, 3–5 cm long, dark brown, with 5 hard wings.

Fls. & Frts.: April–August. *Exsiccata*: SUK-168.
Common, throughout semi evergreen forests.
Localities: Hatkanagale, Kolhapur.

Local uses: Leaves are used to cover for sores and ulcers. Juice of fresh leaves used to stop earache. In bone fractures bark is given orally and applied externally. Bark decoction is used to wash ulcers.

Terminalia elliptica Willd. Sp. Pl. 4: 969. 1806; Diwakar in Singh *et al.* Fl. Maharashtra State Dicot. 2: 5. 2001. *T. crenulata* Roth, Nov. Sp. Pl. 380. 1821. *Pentaptera crenulata* Roxb. Fl. Ind. 2: 438. 1832. *T. tomentosa* Wight & Arn. var. *crenulata* Cl. In Hook f. Fl. Brit. India 2: 447. 1878. *T. tomentosa* Wight & Arn., Prodr. 314. 1834 non Mart; Cooke, Fl. Pres. Bombay 1: 510. 1958 (Repr. ed.). AIN.

Trees, 10–15 m tall, young parts yellowish-brown tomentose. Leaves alternate or subopposite, elliptic-oblong, 5–17 × 2.5–8 cm, coriaceous, often cordate at base, obtuse or subacute, softly tomentose when young, becoming glabrous above when mature, with one to two glands near the base of midrib. Flowers bisexual, in axillary, fulvous pubescent, spikes or terminal panicles. Calyx 3–4 mm long, hairy outside; teeth 1.5 mm long, triangular. Disk clothed with long fulvous hairs. Drupes 3–5 cm long, with 5 broad, coriaceous, brown glabrous wings.

Fls. & Frts.: April–November. *Exsiccata*: SUK-94.
Frequent, in deciduous forests.
Localities: Gaganbawada, Kartikswami Ghat, Kolhapur, Vaibhavwadi.

Local uses: Same as *Terminalia cuneata*.

Terminalia paniculata Roth, Nov. Pl. Sp. 383. 1821; C. B. Cl. in Hook. f. Fl. Brit. India 2: 448. 1878; Cooke, Fl. Pres. Bombay 1: 510. 1958 (Repr. ed.); Diwakar in Singh *et al.* Fl. Maharashtra State Dicot. 2: 6. 2001. KINJAL, HONAL.

Large trees. Leaves coriaceous, alternate or subopposite, oblong, 5–15 × 3–6 cm, rounded at base, acute, glabrous, or few scattered hairs, 2 glands near the base of midrib below. Flowers sessile, in compound panicles. Calyx cup-shaped; teeth triangular, 1.5 mm long, densely villous within. Disk densely clothed with long, yellowish-white hairs. Fruit 6–12 mm long, sessile, closely set in densely spreading panicles, rusty tomentose, 3 winged, 2 of which smaller than other.

Fls. & Frts.: August–September. *Exsiccata*: SUK-666.
Frequent, in moist deciduous forests.
Localities: Amba, Amboli.

Local uses: Bark is given as cardiotonic.

MYRTACEAE Juss.

Eucalyptus globulus Labill. Rel. Voy. Rech. Perouse 1: 153, t. 13. 1799; Kulkarni in Singh *et al.* Fl. Maharashtra State Dicot. 2: 18. 2001. NILGIRI.

Glabrous trees up to 20 m tall. Leaves lanceolate, opposite sessile, cordate, those on the old shoots 15–20 cm long, acuminate, glaucous. Flowers large, solitary, or 2–3 together. Calyx tube hard, bluish-white. Lid of the calyx and corolla conical. Fruit angular 2 cm across.

Fls. & Frts.: February–May. *Exsiccata*: SUK-319.
Cultivated.

Local uses: Leaves yield strong aromatic oil, valued in medicine as febrifuge. Leaves are inhaled in conyza. Oil is applied on forehead to stop headache due to conyza. Infusion of leaves given in apthous ulcerations of mouth and applied locally in erysipelas. Lotion is useful in foul ulcers.

Psidium guajava L. Sp. Pl. 470. 1753; Duthie in Hook. f. Fl. Brit. India 2: 468. 1878; Cooke, Fl. Pres. Bombay 1: 529. 1958 (Repr. ed.); Kulkarni in Singh *et al.* Fl. Maharashtra State Dicot. 2: 8. 2001. PERU.

Small, glabrous trees. Leaves sessile, opposite, elliptic-oblong, 3–12 × 2–6 cm, cordate or rounded at base, obtuse, entire, softly hairy beneath when young. Flowers 2–3 cm across, solitary or 2–3 in axillary clusters. Calyx-lobes 4, ovate, 8–10 mm long, acute or obtuse. Petals 4, white, ovate-oblong or orbicular, exceeding the calyx. Fruits pyriform, 5–10 cm across, fleshy, green with white or rosy pulp. Seeds globose, hard, smooth, pale brown.

Fls. & Frts.: April–July. *Exsiccata*: SUK-1637.
Cultivated. A native of tropical America.
Localities: Kolhapur.

Local uses: Decoction of leaves in combination with clove is used in toothache and sore throat. Bark is employed in the diarrhoea of children; young leaves are used as a tonic in the diseases of the digestive functions. Extract of leaves used in epilepsy and chorea. Tincture is rubbed over the spine of children suffering from convulsions.

Syzygium cumini (L.) Skeels in U. S. Dept. Agric. Bur. Pl. Indrustr. Bull. 248: 25. 1912; Kulkarni in Singh *et al.* Fl. Maharashtra State Dicot. 2: 10. 2001. *Myrtus cumini* L. Sp. Pl. 471. 1753. *Eugenia jambolana* Lam. Encycl. 3: 198. 1789; Duthie in Hook. f. Fl. Brit. India 2: 499. 1879; Cooke, Fl. Pres. Bombay 1: 523. 1958 (Repr. ed.) JAMBHUL, JAMBHOOL.

Trees reaching 20 m in height. Leaves opposite, oblong-elliptic, 8–16 × 4–6 cm, rounded or acute at base, acuminate, glabrous on both surfaces. Flowers in paniculate cymes arising from old leaf scars. Calyx 4–5 mm long, with cup-shaped limb, truncate or obscurely 4 toothed. Petals white, calyptrate, ovate-orbicular, 2–3 mm long. Stamens exserted. Fruits dark violet, globose or ellipsoid, smooth variable in size, 1–3 cm long, crowned with truncate calyx limb. Seeds solitary, globose or oblong, grayish-brown.

Fls. & Frts.: March–July. *Exsiccata*: SUK-291.
Common, throughout the region.
Localities: Barki, Borbet, Gaganbawada, Kolhapur.

Local uses: Seeds are used in diabetes. Bark and root is also given in diabetes. Leaves are given with honey in diarrhoea. Decoction of leaves is used in gargle to cure mouth ulcers. Seed paste is externally employed on pimples and acne.

Syzygium hemisphericum (Wight.) Alst. in Trim. Handb. Fl. Ceylon 6: 115. 1931; Kulkarni in Singh *et al.* Fl. Maharashtra State Dicot. 2: 10. 2001. *Eugenia hemispherica* Wight, Ic. t. 525. 1842; Duthie in Hook. f. Fl. Brit. India 2: 477. 1878; Cooke, Fl. Pres. Bombay 1: 519. 1958 (Repr. ed.).

Moderate sized trees. Leaves ovate-lanceolate or oblanceolate, 6–15 × 3–6.5 cm, acuminate, dotted, glabrous and shining, base tapering. Flowers in axillary and terminal paniculate cymes, fragrant. Calyx rugose outside, 6–9 mm long; tube hemispheric; lobes rounded, reflexed. Disk thick. Petals broadly ovate or suborbicular, ca 9 mm long, concave, gland dotted, reflexed, cream coloured. Fruits globose, 2–3 cm in diam., crowned with persistent calyx, one seeded.

Fls. & Frts.: March–April. *Exsiccata*: SUK-1628.
Infrequent, in moist deciduous and semi evergreen forests.
Localities: Amboli, Barki, Dajipur, Vaibhavwadi.

Local uses: Same as *Syzygium cumini.*

Syzygium zeylanicum (L.) DC. Prodr. 3: 260. 1828; Diwakar in Singh *et al.* Fl. Maharashtra State Dicot. 2: 16. 2001. *Myrtus zeylanica* L. Sp. Pl. 675. 1753. *Eugenia spicata* Lam. Encycl. 3: 201. 1789; Cooke, Fl. Pres. Bombay 1: 521. 1958 (Repr. ed.). *E. zeylanica* (L.) Wight, Illus. 2: 15. 1850; Duthie in Hook. f. Fl. Brit. India 2: 485. 1878. p. p. PITKULI.

Large shrubs. Leaves ovate elliptic to linear lanceolate, ca 3.5–10 × 2–3.5 cm, coriaceous, caudate-acuminate, margins recurved, aromatically fragrant, smooth shining above, paler and glandular beneath. Flowers in axillary and terminal many-flowered paniculate cymes. Calyx funnel shaped, 5 mm long, tubercled with glands on the outside; lobes 4–5, broadly ovate, concave, erect. Petals 4–5, orbicular. Fruits pisifrom, pure white, 1-seeded.

Fls. & Frts.: February–April. *Exsiccata*: SUK-319.
Infrequent, in semi evergreen forests along Western Ghats.
Localities: Amboli, Sawantwadi.

Local uses: Leaves are applied externally in joint pains. Oil obtained from the leaves is used in rheumatism. Bark is used in sore throat. Leaf decoction is given in fever. Leaf paste, with camphor, is applied on body in cold, fever and hemicrania.

LECYTHIDACEAE Poir. (*nom. cons.*)

Barringtonia acutangula (L.) Gaertn. Fruct. 2: 97, t. 101. 1791; C. B. Cl. in Hook. f. Fl. Brit. India 2: 508. 1879; Cooke, Fl. Pres. Bombay 1: 528. 1958 (Repr. ed.); Pradhan in Singh *et al.* Fl. Maharashtra State Dicot. 2: 23. 2001. *Eugenia acutangula* L. Sp. Pl. 421. 1753. INGLI, TIVAR.

Glabrous trees, 8–10 m tall. Leaves obovate-oblong or elliptic, 7.5–15 × 2.2–6.5 cm, narrowed at base, rounded or subacute at apex. Flowers fragrant in long pendent racemes. Calyx 5 mm long, tube hairy outside, as long as the teeth; teeth 4, oblong rounded, ciliolate. Petals red, ovate-oblong, 5 mm long. Stamens many; filaments 2 cm long. Fruits bluntly quadrangular, 2–3 × 1–2 cm, truncate at both ends, crowned by persistent calyx. Seeds ovoid black.

Fls. & Frts.: August–October. *Exsiccata*: SUK-1613.
Common, near the coast, but also occurs along streams and rivers.
Localities: Nerurpar, Sakharppa.

Local uses: Leaf decoction is given in diarrhoea. Juice of fruit used as anthelmintic, bronchitis, hallucinations, useful in abdominal colic, lumbar pain. Alcoholic extract of root is hypoglycemic. Fruit is used as emetic and expectorant.

Careya arborea Roxb. Pl. Cor. 3: 14, t. 218. 1819; C. B. Cl. in Hook. f. Fl. Brit. India 2: 511. 1879; Cooke, Fl. Pres. Bombay 1: 528. 1958 (Repr. ed.); Pradhan in Singh *et al.* Fl. Maharashtra State Dicot. 2: 22. 2001. KUMBHA, KUMBOHYA.

Glabrous, deciduous trees, 5–10 m tall. Leaves sessile, obovate-oblong, 10.5–30 × 6–15 cm, tapering at base, rounded or shortly acuminate, crenate denticulate, glabrous. Flowers sessile, 3–6 cm across, on thick rachis, ovate, obtuse, erect. Calyx 2–2.5 cm long, obtuse 8–10 mm long erect. Petals dull white, elliptic oblong, 3–4 cm long, obtuse or subacute. Fruits 6–8 cm in diam., green.

Fls. & Frts.: March–June. *Exsiccata*: SUK-351.
Common, in semi evergreen or moist deciduous forests.
Localities: Amba, Amboli, Borbet, Gaganbawada, Phonda, Sawantwadi, Vaibhavwadi.

Local uses: Bark with honey is given demulcent in cough and colds. Bark is applied externally in tumors. Crushed bark is applied on inflammation.

MELASTOMACEAE Juss.

Memecylon umbellatum var. **umbellatum** Burm f. Fl. Ind. 87. 1768; Cooke,. Fl. Pres. Bombay 1: 535. 1958 (Repr. ed.). *M. edule* Roxb. Pl. Cor. 1: 59, t. 82. 1798; C. B. Cl. in Hook. f. Fl. Brit. India 2: 563. 1879 p. p. ANJAN.

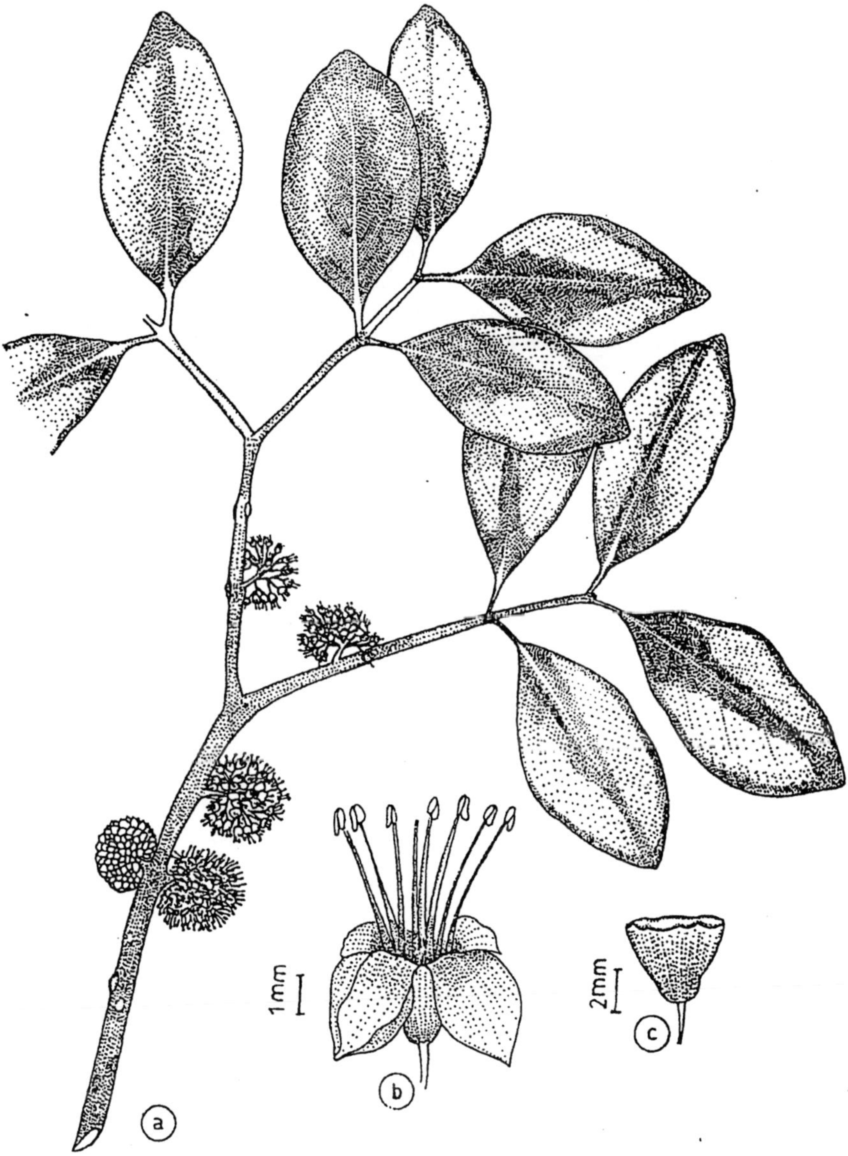

Text fig. 15: Memecylon umbellatum a-twig; b-flower; c-calyx.

Trees, 5–8 m high. Leaves ovate-elliptic, 5–8 × 2.5–4.5 cm, coriaceous, glabrous, base cuneate, margins entire, apex obtuse. Flowers dark blue in pedunculate cymes, from the axils of fallen leaves, on old wood. Calyx campanulate, becomes saucer shaped after expansion; tube 1 mm long; lobes 4. Petals 4, orbicular. Stamens 8. Ovary globose, 1-celled; style 1 cm long. Berries globose, 6 × 4–5 mm, purple when mature, shining black when dried.

Fls. & Frts.: January–April. *Exsiccata*: SUK-1629.
Common, in moist evergreen forests.
Localities: Barki, Gaganbawada, Kas, Thoseghar.

Local uses: Root decoction is given in excess menstrual discharge and in cough.

LYTHRACEAE J. St. Hil.

Ammannia baccifera ssp. **baccifera** L. Sp. Pl. 120. 1753 & ed. 2, 1:175. 1762; C. B. Cl. in Hook. f. Fl. Brit. India 2: 569. 1879; Cooke, Fl. Pres. Bombay 1: 541. 1958 (Repr. ed.); Diwakar in Singh *et al.* Fl. Maharashtra State Dicot. 2: 30. 2001. BHARJAMBHUL.

Glabrous, erect, annual, herbs, 15–60 cm tall; stem and branches ascending. Leaves sessile, linear-oblong or oblanceolate, 1.5–4 × 0.3–1 cm, acute. Flowers in dense axillary clusters. Calyx red, 1–1.5 mm long, hemispheric; marked with 8 faint vertical lines; teeth 4, triangular acute. Petals 0. Capsules globose, irregularly circumscissile above the middle not whole covered by the calyx. Seeds hemispheric, pitted on plane face, yellowish-brown.

Fls. & Frts.: October–December. *Exsiccata*: SUK-1630.
Common, as weed of harvested rice fields, wet soil along stream banks.
Localities: Amboli, Kagal, Sawantwadi, Vaibhavwadi.

Local uses: Whole plant is used with ginger, roots of *Cyperus rotundus,* leaves of *Vernonia cinerea* in herpetic eruptions. Ash of plant is used with castor oil in skin diseases. Leaves mixed with oil and applied externally in herpetic eruptions.

Lagerstroemia indica L. Syst. Nat. ed. 10: 1076. 1759; Clarke in Hook. f. Fl. Brit. India 2: 575. 1879; Cooke, Fl. Pres. Bombay 1: 546. 1958 (Repr. ed.); Diwakar in Singh *et al.* Fl. Maharashtra State Dicot. 2: 39. 2001. CHINAI–MENDHI.

Glabrous shrubs, 2–3 m tall. Leaves opposite, subsessile, 2–5 × 1.5–2.5 cm, rounded or subcordate at base, obtuse. Flowers in short axillary and terminal, 2–3 flowered paniculate cymes. Calyx campanulate, 8–10 mm long, reddish-green glabrous; teeth triangular, 6, acute 3–4 mm long, erect in fruit. Petals 6, white or purple, broadly orbicular with crisped margins, 1.5 × 2 cm; claw slender. Stamens many, half as long as the petals. Capsules globose, 1–1.5 cm in diam., acute at apex. Seeds compressed, winged at one end, brown.

Fls. & Frts.: May–August. *Exsiccata*: SUK-1631.
Native of China, cultivated for ornamental purpose.
Localities: Kolhapur, Sangli, Satara.

Local uses: Decoction of bark is used in fever. Fruit is used as local application for apthae of mouth.

Lawsonia inermis L. Sp. Pl. 349. 1753; Cooke, Fl. Pres. Bombay 1: 544. 1958 (Repr. ed.); Diwakar in Singh *et al.* Fl. Maharashtra State Dicot. 2: 40. 2001. *L. alba* Lam. Encycl. 3: 106. 1789; Clarke in Hook. f. Fl. Brit. India 2: 573. 1879. MHENDI.

Shrubs or small trees, 3–5 m tall; branches thorny. Leaves opposite, broadly elliptic lanceolate. ca 2 × 1 cm, acute at both ends, glabrous. Flowers in axillary panicles. Calyx campanulate, 3–5 mm long; teeth 4, ovate, 2–3 mm long, acute. Petals pale yellow or creamy white, suborbicular, 3 mm long, and as much broad, undulate. Stamens 8 inserted in pairs on the calyx tube. Capsules globose, supported by persistent calyx and tipped with a style. Seeds many, trigonous, tuberculate, brown.

Fls. & Frts.: April–August. *Exsiccata:* SUK-1632.
Commonly cultivated as hedge plant.
Localities: In all districts.

Local uses: Leaves are applied externally in burning sensation. Leaf paste is applied externally after abortion. The oil and essence are rubbed over the body to keep the body cool. Flower paste is applied in headache. It is given as a tonic in general debility. Leaves are used in bronchitis.

Woodfordia fruticosa (L.) Kurz. in J. Asiat. Soc. Bengal 40: 56. 1871; Diwakar in Singh *et al.* Fl. Maharashtra State Dicot. 2: 39. 2001. *Lythrum fruiticosum* L. Sp. Pl. ed. 2, 641. 1762. *Woodfordia floribunda* Salisb. Prasad. Lond. t. 42. 1806; C. B. Cl. in Hook. f. Fl. Brit. India 2: 572. 1879; Cooke, Fl. Pres. Bombay 1: 543. 1958 (Repr. ed.). DHAYATI.

Straggling shrubs, 1–4 m tall. Leaves opposite, bifarious, ovate-lanceolate, 3.5–9.5 × 1.0–3.5 cm, sessile, rounded or cordate at base, acute, green and softly pubescent above, hoary and nigro-punctate beneath. Flowers in short, 2–15 flowered cymes from axils of leaves. Calyx bright red, campanulate at base, tubular above, 1.2–1.5 cm long, striate, covered with glandular dots, oblique at mouth; teeth triangular, acute. Petals red, slightly longer than the calyx teeth, narrowly linear. Stamens 12. Capsules ellipsoid, 1–1.5 cm long, irregularly bursting. Seeds numerous, obovoid, brown smooth.

Fls. & Frts.: January–May. *Exsiccata:* SUK-53.

Common, in crevices of rocks, walls of old forts, along streams, road cutting etc.

Localities: Borbet, Kagal, Radhanagari.

Local uses: Dried flowers are given internally in leucorrhoea, menorrhoea. Flower decoction is given in a safe stimulant in pregnancy. Flower powder is given with honey in dysentery. Roots are given in diabetes and also applied externally in wounds.

PUNICACEAE Horan.

Punica granatum L. Sp. Pl. 472. 1753; Clarke in Hook. f. Fl. Brit. India 2: 581. 1879; Cooke, Fl. Pres. Bombay 1: 548. 1958 (Repr. ed.); Venkanna & Prasanna in Singh *et al.* Fl. Maharashtra State Dicot. 2: 42. 2001. DALIMB, ANAR.

Deciduous shrubs or small trees, 5–6 m tall, branches thorny, angular. Leaves opposite or subsessile, ovate-oblong or elliptic-oblong, 2–5 × 1–1.5 cm, rounded or cordate at base, obtuse or subacute, glabrous. Flowers showy, extra-axillary, solitary on short pedicels. Calyx tube campanulate, 1.5–2 cm long, red; sepals 5–7, thick, leathery, persistent in fruit. Petals 5–7, bright red, inserted on the edge of the hypanthium, orbicular, short clawed; limb wrinkled. Stamens many, clothing the inside of the hypanthium; filaments slender. Ovary inferior, 3–7 locular. Seeds surrounded by juicy pulp.

Fls. & Frts.: August–November.

Native of south Asia. Planted in orchards for edible fruits.

Local uses: Bark decoction is given in bronchitis; bark is used to strength the gums and used in piles.

TRAPACEAE

Trapa natans L. var. **bispinosa** (Roxb.) Makino in Iinuma, Somoku. Dzusetzii ed. 3. 1: 137. 1907; Lakshminarasimhan & Venkanna in Singh *et al.* Fl. Maharashtra state Dicot. 2: 46. 2001. *T. bispinosa* Roxb. Pl. Cor. 3: 29, t. 234. 1815; C. B. Cl. in Hook. f. Fl. Brit. India 2: 590. 1879; Cooke, Fl. Pres. Bombay 1: 551. 1958 (Repr. ed.). SHINGADA.

Aquatic herbs. Leaves crowded, submerged, rhomboid, toothed, 3–5 × 5.5–7.0 cm, densely pubescent below. Flowers axillary, solitary. Calyx tube short, adnate to the lower part of ovary; lobes 5, persistent. Petals 4, sessile, oblong, margins crenulate, pure white. Fruit obovoid, angular, bihorned, ca 2–3.8 cm, across.

Fls. & Frts.: July–November. *Exsiccata:* SUK-1635.
Occasional, often cultivated.
Localities: Shiroli, Shivaji University campus, Kolhapur.

Local uses: Fruits are used in leprosy, burning sensation, fatigue, inflammations, diseases of the blood and urinary discharges.

PASSIFLORACEAE Juss.

Passiflora edulis Sims Curtis Bot. Mag. t. 1989. 1818; Cooke, Fl. Pres. Bombay 1: 557. 1958 (Repr. ed.); Venkanna & Prasanna in Singh *et al.* Fl. Maharashtra State Dicot. 2: 48. 2001. KRISHNA KAMAL.

Large herbaceous climbers; tendrils simple. Leaves palmately 3-lobed, 5–12 cm long, rounded or subcordate at base, serrate; lobes lanceolate, acuminate, glabrous, shining above, paler beneath. Flowers solitary, axillary, on long peduncles. Sepals lanceolate, 2–2.5 × 0.5–0.8 cm, green outside, white within, horned on the back. Petals white, lanceolate. Corona double ring of bluish-purple threads. Stamens 5, united into a tube around gynophore. Styles 3. Berries ovoid oblong, purple dotted.

Fls. & Frts.: September. *Exsiccata*: SUK-1642.
Native of Brazil, planted in gardens for showy flowers.

Local uses: Fruit pulp is used as tonic.

Passiflora foetida L. Sp. Pl. 959. 1753; Cooke, Fl. Pres. Bombay 1: 557. 1958 (Repr. ed.); Venkanna & Prasanna in Singh *et al.* Fl. Maharashtra State Dicot. 2: 48. 2001. VELI–GHANI.

Climbing fetid, annual herbs. Leaves ovate, 3–8 × 2–6 cm, deeply 3-lobed, cordate at base, glandular hairy; lobes lanceolate, acute or acuminate, denticulate. Flowers solitary or 2-nate. Involucral bracts bipinnatifid, with filiform, glandular-hairy segments. Calyx 1.2–1.5 cm long, divided half way down; segments lanceolate, acuminate. Petals white, oblong, 1–5 × 0.5–0.6 cm. Corona double, with white purple threads, 8–10 mm long. Stamens 5; filaments forming a tube around a short gynophore. Styles 3. Fruits globose, 3–4 cm in diam. Seeds ovate, 8 × 3 mm, reticulate, brown.

Fls. & Frts.: July–December. *Exsiccata*: SUK-54.
A native of South America cultivated in gardens and occasionally escaped.
Localities: Ratnagiri, Sawantwadi.

Local uses: Leaves are applied on the head for giddiness and headache, a decoction is given in biliousness and asthma.

CARICACEAE Dumort.

Carica papaya L. Sp. Pl. 1036. 1753; Masters in Hook. f. Fl. Brit. India 2: 599. 1879; Cooke, Fl. Pres. Bombay 1: 557. 1958 (Repr. ed.); Venkanna & Prasanna in Singh *et al.* Fl. Maharashtra State Dicot. 2: 50. 2001. PAPAYA, POPAI.

Rapidly growing, soft wooded, dioecious, or monoecious trees, 2–6 m tall. Leaves palminerved, palmatifid, 20–60 cm wide, 7–9 lobed, glabrous. Male flower in drooping lax panicles from the axils of upper leaves. Corolla polypetalous, white. Stamens 10 in two series. Female flowers subsessile, solitary in few-flowered corymbs in leaf axils. Corolla yellow. Ovary globose; stigmas 5. Fruits oblong-ovoid to obovoid, 30 × 15 cm or rarely larger, yellow and orange yellow. Seeds many, black, globose, 3–4 mm in diam.

Fls. & Frts.: Throughout year.

A native of tropical America, grown for edible fruits.

Local uses: Leaves are used as diuretic. They are used to expel worm. Milky juice of unripe fruit is used as cosmetic to remove freckles and other blemishes from the skin.

CUCURBITACEAE Juss.

Benincasa hispida (Thunb.) Cogn. in A. & C. DC. Monogr. Phan. 3: 513. 1881; Madhusudan Rao in Singh *et al.* Fl. Maharashtra State Dicot. 2: 71. 2001. *Cucurbita hispida* Thunb. Fl. Jap. 322. 1784. *Benincasa cerifera* Savi, Bibl. Iatl 9: 158. 1818; C. B. Cl. in Hook. f. Fl. Brit. India 2: 616. 1879; Cooke, Fl. Pres. Bombay 1: 582. 1958 (Repr. ed.). KAWALU, KOHLA.

Large annual climbers; tendrils slender. Leaves reniform-rounded, 10–25 cm across, deeply cordate, 5–7 lobed, sinuate-dentate along margins, scabrous above, hispid beneath. Flowers large; male 5–15 cm long, female 2–4 cm long. Probracts foliaceous, ovate, 6–15 × 5–10 mm. Calyx tube densely villous; 12–15 mm broad; teeth lanceolate, 8–12 × 3–5 mm. Petals yellow, obtuse, 3–5 cm long. Fruits 25–60 × 10–25 cm, densely hairy when young, thickly deposited with waxy bloom in mature. Seeds compressed, ovoid, 10–11 × 5–7 mm, white or pale yellowish, distinctly marginate.

Fls. & Frts.: December–April. *Exsiccata*: SUK-1633.

Cultivated, for fruits used as vegetable and various other edible purposes.

Local uses: Fruit juice is given in diabetes fresh juice of fruit is used for internal hemorrhage. Seed powder is given as a purgative and anthelmintic. Fruit cures strangury, urinary discharges, and urinary calculi, given as heart tonic.

Bryonia dioica Jacq. Fl. Aust. 2: 59, t. 199. 1773; Chakr. in Fasc. Fl. India 11: 14 1982; Madhusudhan Rao in Singh *et al.* Fl. Maharashtra State Dicot. 2: 52. 2001. CHAMBHAR GONDA.

Dioecious, climbing annual herbs; root tuberous. Tendril simple. Leaves broadly ovate or sagittate in outline 6–12 cm long and as broad; 3–5 lobed, lobes acute, margin undulate denticulate. Flowers white or yellow. Male flowers in 10–15 flowered racemes. Calyx tube campanulate; lobes 5, short, triangular. Corolla yellow, subcordate, 5-partite, lobes ovate. Stamens 3, free; anthers broadly ovate, one unilocular, others bilocular, connective broad. Female flowers in 4–7 flowered corymb. Fruits globose, greenish-red, wrinkled when dry, 8–10 mm in diam. Seeds black, finely rugulose, thinly marginate.

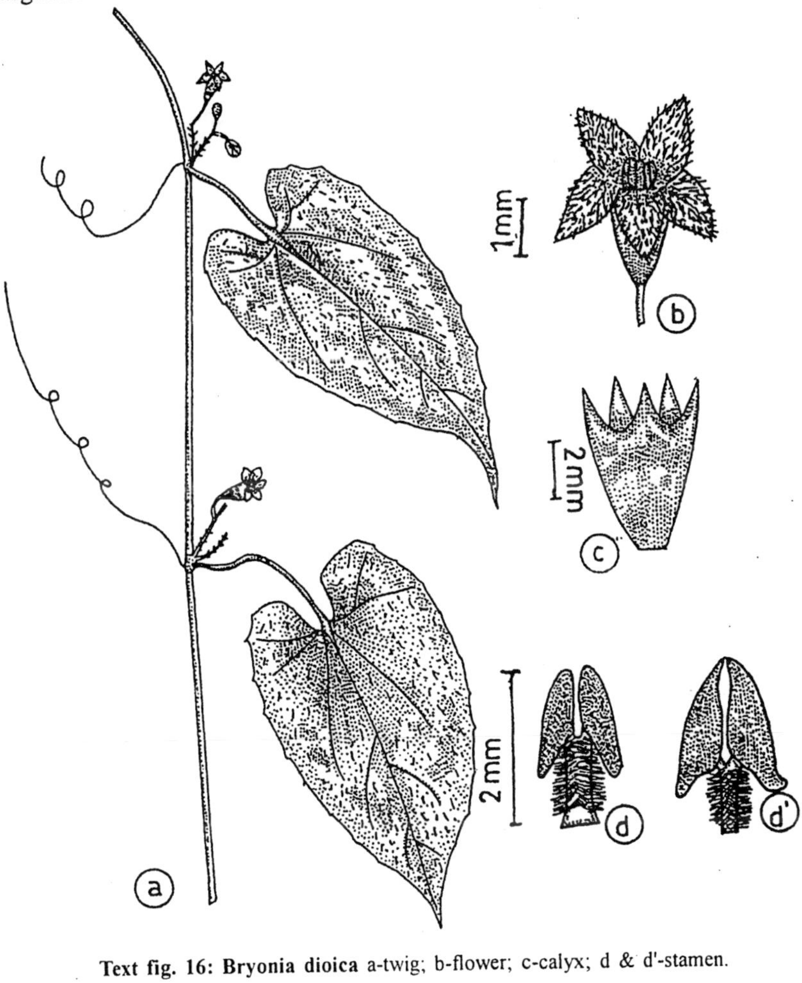

Text fig. 16: Bryonia dioica a-twig; b-flower; c-calyx; d & d'-stamen.

Fls. & Frts.: May–August. *Exsiccata*: SUK-399.
Rare, in semi evergreen forests.
Localities: Amboli, Sawantwadi, Tillarinagar.

Local uses: Roots are given in stomachache. It is not given to pregnant women as it may cause abortion.

Citrullus colocynthis (L.) Schrad. in Linnaea 12: 414. 1838; C. B. Cl. in Hook. f. Fl. Brit. India 2: 620. 1879; Cooke, Fl. Pres. Bombay 1: 571. 1958 (Repr. ed.); Madhusudan Rao in Singh *et al.* Fl. Maharashtra State Dicot. 2: 53. 2001. *Cucumis colocynthis* L. Sp. Pl. 1011. 1753. INDRAYAN, KADUINDRAYAN.

Monoecious, perennial herbs; stem angular, hirsute; tendrils slender, short. Leaves ovate in outline, 5–12 × 3–8 cm, pinnately 5–7 lobed, each lobe again lobulate, cordate at base, hirsute. Male flowers solitary or 2–4 together in leaf axils. Calyx tube campanulate, 4–8 mm long, hispid; teeth subulate, 2–3 mm long. Corolla pale yellow; petals ovate, 4–6 × 2–3 mm, acute. Fruit globose, depressed at apex, 5–7 cm in diam., variegated green and white, glabrous; pulp spongy, bitter. Seeds ovate-oblong, yellowish-brown.

Fls. & Frts.: July–September. *Exsiccata*: SUK-1634.
Localities: Atpadi.

Local uses: Root is externally applied in swellings. Paste of root is applied to the enlarged abdomen of children. Paste of fruit and root is applied to boils and pimples. Roots are soaked in castor oil and paste is mixed with *Celastrus* oil and applied externally in rheumatism. Fruits are used in leucoderma, ulcers, asthma, bronchitis, tuberculous glands of the neck, anemia, throat diseases, elephantiasis. Fruit is externally applied in inflammation of the breasts and in joint pains.

Coccinia grandis (L.) Voigt, Hort. Suburb. Calc. 59. 1845; Madhusudan Rao in Singh *et al.* Fl. Maharashtra State Dicot. 2: 53. 2001. *Bryonia grandis* L. Mant. Pl. 1: 126. 1767. *Coccinia indica* Wight & Arn. Prodr. 347. 1834; Cooke, Fl. Pres. Bombay 1: 572. 1958 (Repr. ed.). *Cephalandra indica* (Wight & Arn.) Naud. Ann. Sci. Nat. Bot. Ser. 5, 5: 16. 1859; C. B. Cl. in Hook. f. Fl. Brit. India 2: 621. 1879. TONDLI

Perennial, glabrous, climbing herbs; nodose below; tendrils slender. Leaves broadly ovate, 5–10 cm across, 5-angled or 3–5 lobed; cordate at base, minutely denticulate along margins. Flowers axillary, solitary. Male flowers: Peduncle striate. Calyx tube campanulate, 4–5 mm long, teeth ovate, acute. Corolla white, 2.5–3.5 cm long; lobes shorter than the tube, staminal column 2–3 mm long; anthers 6–7 mm thick. Female flowers: Calyx and corolla as in male. Staminodes 3, subulate, 2–3 mm long. Ovary 12–15 mm long; style slender; stigma papillose. Fruits ellipsoid, rounded at both ends, red when ripe. Seeds oblong yellowish.

Fls. & Frts.: March–December. *Exsiccata*: SUK-155.
Frequent, on hedges.
Localities: Kagal, Kolhapur, Narisimhawadi, Sangali.

Local uses: Leaf juice is applied to the whole body and also given internally to induce perspiration in fever. Leaf juice is given to pregnant women suffering from bleeding. Leaf juice is given orally with honey and applied externally in tuberculous glands. Leaves applied externally in skin eruption. Fruit, leaf, root juice is given in diabetes. Root stops vomiting, fruit is emetic, used in bronchitis, diseases of the blood.

Corallocarpus epigaeus (Rottl.) Hook. f. in Oliv. Fl. Trop. Afr. 2: 558. 1871; C. B. Cl. in Hook. f. Fl. Brit. India 2: 628. 1879; Cooke, Fl. Pres. Bombay 1: 578. 1958 (Repr. ed.); Madhusudan Rao in Singh *et al.* Fl. Maharashtra State Dicot. 2: 54. 2001. *Bryonia epigaea* Rottl. in Neue. Schr. Ges. Naturf. Fr. Berl. 4: 212. 1803. KARVINA, KARMUGLIU, MIRCHIKAND.

Climbing monoecious herbs; root tuberous; branches zig-zag. Tendrils filiform. Leaves broadly ovate to orbicular in outline, 2–8 × 2.5–9 cm, 3–5-lobed, lobes often lobulate, glabrous above, pubescent beneath. Male flowers 5–15 at the apex of a glabrous peduncle. Calyx tube 1–1.5 mm long; teeth minute, subulate. Petals greenish-yellow, ovate, 1 mm long. Anthers yellow; connective produced beyond the cells, bifid. Female flowers solitary, on short axillary peduncles. Fruits ellipsoid, 1–3 cm long, scarlct in the middle, base and beak green, circumscissilely dehiscent in middle. Seeds 6–9, pyriform turgid, brown.

Fls. & Frts.: June–August. *Exsiccata*: SUK-324.
Localities: Kartikswami Ghat, Babu Jamal.

Local uses: Roots are applied externally in rheumatism. Plant is used as emetic, cures inflammations, cures chronic rheumatism, given in dysentery.

Cucumis melo L. Sp. Pl. 1011. 1753; Madhusudan Rao in Singh *et al.* Fl. Maharashtra State Dicot. 2: 56. 2001. *C. pubescens* Willd. Sp. Pl. 4: 614. 1805; C. B. Cl. in Hook. f. Fl. Brit. India 2: 619. 1879. *C. trigonous* Roxb. Fl. Ind. 3: 722. 1832; C. B. Cl. in Hook. f. op. cit.; Cooke, Fl. Pres. Bombay 1: 569. 1958 (Repr. ed.). *C. melo*. var. *agrestis* Naud. in Ann Sci. Nat. 4. 11: 73. 1859; Cooke, op. cit. MEKI, SHENDAD.

Perennial herbs; stem rough with short rigid hairs. Leaves suborbicular in outline, 3–6 cm across, deeply 5–7 lobed, cordate at base, lobes ovate-oblong, narrowed at base, rounded at apex, dentate, hispid. Male flowers often solitary. Calyx tube campanulate; 6–8 mm long, teeth subulate. Corolla yellow, ca 2 cm long; lobes ovate-oblong, acute. Filaments very short; anthers 3–4 mm long, appendage of the connective shorter than the anthers. Female flowers densely hairy. Fruits obovoid, ca 3 cm long, round obscurely trigonous, longitudinally variegated with 10 green stripes, pale yellow when ripe. Seeds ovate-oblong, compressed, smooth, pale yellow.

Fls. & Frts.: June–October. *Exsiccata*: SUK-1641.
Common, on hill slopes.
Localities: Kolhapur, Ramling, Wai.

Local uses: Pulp of fruit is drastic purgative. Seeds are given in very mild proportions in stomach complaints as a purgative. Tincture of powder is given in diabetes.

Cucurbita maxima Duch. ex Lam. Encycl. 2: 151. 1786; C. B. Cl. in Hook. f. Fl. Brit. India 2: 622. 1879; Madhusudan Rao in Singh *et al.* Fl. Maharashtra State Dicot. 2: 73. 2001. TAMBADA BHOPLA, KADDU.

Prostrate or climbing annual herbs, stem grooved, hispid. Leaves reniform with 5, shallow rounded lobes, 6–19 × 7–30 cm, minutely dentate, hispid. Male flowers solitary, axillary. Calyx tube 5–10 mm long; teeth linear, hairy. Corolla orange-yellow, campanulate, 5–10 cm long; lobes reflexed, ovate, acute. Filaments thick, glanduliform, 5–8 mm long; anthers 12–15 mm long. Female flowers 5–7 cm long. Fruits large, orange-yellow, 20–50 cm in diam. Seeds obovoid, narrowly margined, rounded at apex, narrowed at base, white or pale yellow.

Fls. & Frts.: September–January. *Exsiccata*: SUK-1636.
Cultivated, in most of the parts as rainy season vegetable.

Local uses: Fruit pulp is used as poultice, applied to burns, inflammation and boils. Fresh seeds are given with milk and honey as anthelmintic.

Diplocyclos palmatus (L.) C. Jeffrey in Kew Bull. 15: 352. 1962 & Ibid. 34: 800. 1980; Madhusudan Rao in Singh *et al.* Fl. Maharashtra State Dicot. 2: 58. 2001. *Bryonia palmata* L. Sp. Pl. 1012. 1753. *Bryonopsis laciniosa* (L.) Naud. in Ann. Sci. Nat. Bot. Ser. 4, Bot. 12: 141. 1859; Cooke, Fl. Pres. Bombay 1: 568. 1958 (Repr. ed.). *Bryonia laciniosa* L. Sp. Pl. 1013. 1753. C. B. Cl. In Hook. f. Fl. Brit. India 2: 622. 1879. SHIVLINGI.

Glabrous, annual, climbing herbs; stem slender, grooved. Tendrils slender bifid. Leaves membranous, broadly ovate 10–15 cm across, deeply cordate at base, 5-lobed oblong lanceolate, sinuate, distantly dentate along margins, shining green above, paler beneath. Male flowers in fascicles. Calyx 2–2.5 mm long, glabrous; teeth subulate, shorter than tube. Corolla pale yellow, lobes ovate oblong, acute. Female flowers solitary or few together on axillary short peduncles. Fruits subsessile, globose, bluish-red when ripe. Seeds ovoid, yellowish-brown.

Fls. & Frts.: August–December. *Exsiccata*: SUK-16.
Frequent, on hill slopes and valleys.
Localities: Shivaji University campus, Kolhapur.

Local uses: Leaves are externally applied in inflammation. Seed and root powder is given to help conception in women.

Kedrostis foetidissima (Jacq.) Cogn. in A & C. Mon. Phan. 3: 634. 1881; Madhusudan Rao in Singh *et al.* Fl. Maharashtra State Dicot. 2: 59. 2001.

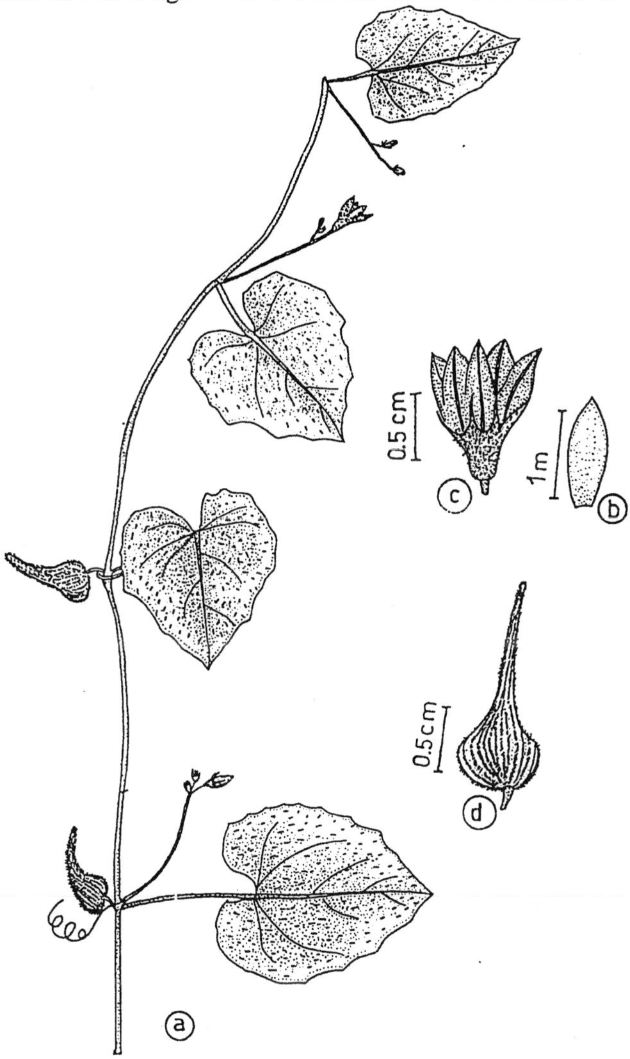

Text fig. 17: Kedrostis foetidissima a-twig; b-bract; c-flower; d-fruit.

Trichosanthes foetidissima Jacq. Collet2: 341. 1789. *Kedrostis rostrata* (Rottl.) Cogn. in DC. Mongr. Phan. 3: 636. 1881; Cooke, Fl. Pres. Bombay 1: 577. 1958 (Repr. ed.). *Rhynchospora foetida* sensu C. B. Cl. in Hook. f. Fl. Brit. India 2: 627. 1879 non Schard. 1838. NURAKVEL.

Monoecious succulent, viscous, climbing herbs; stem 30–90 cm long, sparsely pilose. Tendrils filiform, leaves broadly ovate, 1.5–4.5 × 2–5 cm, cordate at base, entire or shallowly lobed, denticulate; petioles slender, densely villous. Male flowers on 2-nate axillary peduncles. Calyx tube hemispherical, ca 2 × 1.5 mm; teeth subulate. Petals yellow, oblong lanceolate, 3–4 mm long, acute. Anthers suborbicular; connectives shortly produced. Female flowers solitary, axillary. Fruits deep red, ovoid, 1.5–2 × 0.8–1.2 cm, tapering into long beak. Seeds obovoid, turgid, finely granular marginate.

Fls. & Frts.: October–January. *Exsiccata*: SUK-135.
Localities: Atpadi, Dhondewadi.

Local uses: Root decoction with the bark of *Alstonia scholaris* is given in dysentery. Fruits are given internally in case of colic and abdominal pain. Roots externally applied with bark of *Pittosporum floribundum* on piles.

Lagenaria siceraria (Mol.) Standl. in Publ. Field Mus. Nat. Hist. Chicago Bot. Ser. 3: 435. 1930; Madhusudan Rao in Singh *et al.* Fl. Maharashtra State Dicot. 2: 74. 2001. *Cucurbita siceraria* Mol. Sagg. Chile, 133: 335. 1782. 6: 43. 1896. *Lagenaria vulgaris*. Seringe, Mem. Soc. Phys. Hist. Nat. Geneve. 3(1): 25, t. 2. 1825; C. B. Cl. in Hook. f. Fl. Brit. India 2: 613. 1879. Cooke, Fl. Pres. Bombay 1: 581. 1958 (Repr. ed.). DUDHYA BHOPALA.

Large, climbing, annual herbs; stem thick, softly pubescent; tendrils bifid. Leaves ovate-orbicular, 10–40 cm across, cordate at base, angular or shortly trilobed, acute or shortly acuminate, white pubescent. Male flowers on long axillary peduncles. Calyx tube 2–3 cm long; sepals narrow. Petals white, obovate, 3–4 × 2–3 cm, emarginate at apex, tomentose. Female peduncles shorter than male. Ovary villous. Fruits clavate, 30–60 × 6–15 cm, greenish-yellow, turning brown. Seeds obovate, oblong white, truncate, bidentate at apex.

Fls. & Frts.: September–January. *Exsiccata*: SUK-1649.
Cultivated, in all regions. Fruits are used as vegetable.

Local uses: Pulp is applied externally to the soles in burning of feet. Decoction of leaves mixed with sugar given in jaundice.

Luffa acutangula (L.) Roxb. Fl. Ind. 3: 713. 1832; C. B. Cl. in Hook. f. Fl. Brit. India 2: 615. 1879; Cooke, Fl. Pres. Bombay 1: 556. 1958 (Repr. ed.); Madhusudhan Rao in Singh *et al.* Fl. Maharashtra State Dicot. 2: 59. 2001. *Cucumis acutangula* L. Sp. Pl. 1: 1011. 1753. *Luffa acutangula* var. *amara*

Roxb.) C. B. Cl. in Hook. f. op. cit.; Cooke, op. cit. 567. KADUDODKA, RAN-DODKA, RANTURAI.

Extensive, annual, climbers; stem 5-angular, glabrous, often with scabrous angles. Tendrils 3-fid. Leaves broadly ovate to roundish, 15–20 cm across, 5–7 angled or lobed, glabrous. Male flowers on 10–15 cm long, 15–20 flowered peduncles; probract ovate fleshy green with 3–10 glistening glands on the upper surface. Calyx tube campanulate; teeth lanceolate, winged. Petals bright yellow, obcordate. Stamens 3, 1 unilocular, 2 bilocular. Fruits 10-angled, constricted at apex. Seeds ovate, compressed, black.

Fls. & Frts.: August–December. *Exsiccata*: SUK-199.
Cultivated, in all districts. Fruits used as vegetables.

Local uses: Two parts of leaf juice, one part Ghee and one part of wax warmed and ointment prepared is used on all kinds of ulcers and wounds.

Luffa cylindrica (L.) M. S. Roem. Syn. Mon. 2: 63. 1846; Madhusudan Rao in Singh *et al.* Fl. Maharashtra State Dicot. 2: 74. 2001. *Momordica cylindrica* L. Sp. Pl. 1009. 1753. *Luffa aegyptiaca* Mill. Gard. Dict. ed. 8. 1768; C. B. Cl. in Hook. f. Fl. Brit. India 2: 614. 1879; Cooke, Fl. Pres. Bombay 1: 565. 1958 (Repr. ed.). GHOSALE.

Extensive, annual, climbers; stem 5-angular, glabrous. Tendrils trifid. Leaves broadly ovate, 15–25 cm across, 5 lobed, often gland dotted. Male flowers on 15–20 flowered axillary peduncles. Calyx tube short, broadly campanulate; teeth oblong-cuneiform, 2–3 cm long, rounded at apex. Petals bright yellow, orbicular, 2–3 cm across, rounded or emarginate at apex. Stamens 3 with 1 unilocular and 2 bilocular anthers or 5 with 5 unilocular anthers. Female flowers on 5–10 cm long, axillary peduncles. Fruits fusiform, 15–40 × 6–10 cm, strongly fibrous when ripe. Seeds ovate, 10–12 × 8–9 mm, smooth.

Fls. & Frts.: August–December. *Exsiccata*: SUK-1648.
Cultivated, for its tender fruits used as vegetable.

Local uses: Two parts of leaf juice, one part Ghee and one part of wax warmed and ointment prepared is used on all kinds of ulcers and wounds. Fruit is used as tonic, in spleen diseases, leprosy, piles and fevers.

Momordica charantia L. Sp. Pl. 1009. 1753; C. B. Cl. in Hook. f. Fl. Brit. India 2: 616. 1879; Cooke, Fl. Pres. Bombay 1: 562. 1958 (Repr. ed.); Madhusudan Rao in Singh *et al.* Fl. Maharashtra State Dicot. 2: 74. 2001. KARALE.

Extensive, annual climbers. Tendrils simple, slender, pubescent. Leaves suborbicular reniform, 5–12 cm across, 5–7 lobed, cordate at base, prominently nerved on both surfaces; glabrous; lobes ovate oblong, dentate, acute. Male flowers solitary, bracts foliaceous, reniform, entire. Calyx tub campanulate, 4–5 mm long; teeth subulate. Corolla yellow, 12–20 mm long lobes obovate obtuse. Female flowers solitary; bracteate. Fruits oblon ellipsoid, 8–25 cm long, muricate, tuberculate, dehiscing at apex. Seeds dul white, quadrate, subtridentate at apex, sculptured on face.

Fls. & Frts.: June–October. *Exsiccata*: SUK-1663
Cultivated, fruits are used as vegetables.

Local uses: Leaf decoction is given as purgative to children. Leaf juice is dropped in nostrils in hemicrania. Fruits are given to diabetic patient Fruit is digestible, laxative, antipyretic, anthelmintic, appetizer, given in anemia, urinary discharges, asthma, ulcers, bronchitis, juice is useful in cholera.

Momordica cymbalaria Hook. f. in Oliv. Fl. Trop. Afr. 2: 540. 1871; C. B. Cl. in Hook. f. Fl. Brit. India 2: 618. 1879; Cooke, Fl. Pres. Bombay 1: 564. 1958 (Repr. ed.); Madhusudan Rao in Singh *et al.* Fl. Maharashtra State Dicot. 2: 63. 2001. KADAVACHI.

Prostrate or climbing, perennial herbs; roots tuberous. Stem slender, striate; tendrils simple filiform. Leaves nearly orbicular, 1.5–3 × 2.5–3 cm, 5–7 lobed, entire. Male flowers in axillary, 2–5 flowered racemes. Calyx broadly campanulate; tube lanceolate, 6–7 mm long. Stamens 2; filaments 0.5–2 mm long; anthers 2 mm long, one bipartite, the other 3 partite. Female flowers on slender, ebracteate, 2–8 cm long peduncles. Fruits angular, pyriform 2–4 × 1.5 cm, with 8 sharp ridges, narrowed at base and apex. Seeds broadly ovoid, 6–7 mm long, smooth, blackish-gray.

Fls. & Frts.: June–October. *Exsiccata*: SUK-1643.
Occasional, in dry region.
Localities: Donewadi, Tasgaon, Vita.

Local uses: Tubers are used to procure abortion.

Momordica dioica Roxb. ex Willd. Sp. Pl. 4: 605. 1805; C. B. Cl. in Hook. f. Fl. Brit. India 2: 617. 1879 (excl. syn.); Cooke, Fl. Pres. Bombay 1: 563. 1958 (Repr. ed.); Madhusudan Rao in Singh *et al.* Fl. Maharashtra State Dicot. 2: 63. 2001. KARTOLI.

Perennial climbers; root tuberous. Leaves 4–10 cm across, cordate at base, entire or 3–5 lobed, remotely denticulate. Male flowers solitary, bracts sessile at apex of peduncle, cucullate. Calyx tube short; teeth linear lanceolate, 5–7 mm long, acute densely villous. Corolla yellow, 2–3.5 cm long; lobes oblong-obtuse, prominently black-veined. Female flowers ebracteate, on short peduncles. Fruits ovoid, 3–5 cm long, green-echinate, red when ripe. Seeds pale yellow, broadly ovoid, emarginate.

Fls. & Frts.: July–October. *Exsiccata*: SUK-130.

Localities: Bhudergad, Gaganbawada, Patgaon, Sawantwadi, Vaibhavwadi.

Local uses: Root powder is given orally in diabetes. Roots are given in urinary calculi and also as diuretic. Root is used to stop bleeding from piles. Roots are used with benefit in head troubles, urinary calculi, as an errhine in jaundice, cures elephantiasis.

Mukia maderaspatana (L.) Roem. Syn. Monogr. 2: 47. 1846; Madhusudhan Rao in Singh *et al.* Fl. Maharashtra State Dicot. 2: 64. 2001. *Cucumis maderaspatanus* L. Sp. Pl. 1012. 1753. *Mukia scabrella* (L. f.) Arn. in Hook. J. Bot. 3: 376. 1841; C. B. Cl. in Hook. f. Fl. Brit. India 2: 623. 1879. *Melothria maderaspatana* (L.) Cong. in A. & C. DC. Mon. Phan. 3: 623. 1881; Cooke, Fl. Pres. Bombay 1: 573. 1958 (Repr. ed.). CHIRATI, CHIRBUT, GHUGRI.

Climbing, annual, herbs; stem hispid-scabrid. Tendril simple. Leaves ovate or subdeltoid, 4–15 × 3–13 cm, 3–5 lobed, sinuate cordate at base, minutely denticulate, scabrous above, pubescent beneath. Male flowers in axillary fascicles. Calyx tube ca 2 × 1.5 mm, villous, hirsute; teeth erect, subulate, 1–1.5 mm long. Corolla yellow, 2–4 mm long; lobes ovate oblong, rounded at apex. Filaments short, pilose; anthers oblong ciliate, distinctly appendiculate at base. Female flowers solitary or sub-fasiculate. Fruits red, smooth shining. Seeds ovoid-oblong, 3–4 × 2–2.5 mm, gray, prominently scrobiculate.

Fls. & Frts.: July–September. *Exsiccata*: SUK-5.

Common, on hedges around fields.

Localities: Gaganbawada, Shivaji University campus, Kolhapur, Wai.

Local uses: Roots are chewed in toothache. Tender leaves are used as a gentle aperient and recommended in vertigo and biliousness. Roots are used as diuretic.

Solena amplexicaulis (Lam.) Gandhi in Sald. & Nicols. Fl. Ḥassan Dt. 1.79. 1976; Madhusudhan Rao in Singh *et al.* Fl. Maharashtra State Dicot. 2: 65. 2001. *Bryonia amplexicaulis* Lam. Encycl. 1: 496. 1785. *Zehneria umbellata* Thw. Enum. Pl. Zeyl. 125. 1859; C. B. Cl. in Hook f. Fl. Brit. India 2: 625. 1879 p. p. *Melothria heterophylla* (Lour.) Cogn. in A. & C. DC. Mon. Phan. 3: 618. 1881; Cooke, Fl. Pres. Bombay 1: 575. 1958 (Repr. ed.). GOMATI, GOMETTA.

Glabrous, perennial climbers; root tuberous; stem slender, furrowed. Tendril simple. Leaves ovate, lobed, hastate, 5–15 cm long, entire or palmately 3–5 lobed, cordate at base, scabrid and pale green above, paler beneath, remotely denticulate along margins; lobes acute. Male flower subumbellate; umbels 15–20 flowered. Calyx tube campanulate, 4–6 mm long; teeth minute subulate. Corolla small; lobes triangular, 1–1.5 mm long, acute. Filaments slender, 2–3 mm long. Female flowers solitary on axillary peduncles. Ovary oblong, 10 ribbed. Fruits oblong, tapering towards the apex, bright red when ripe. Seeds obovoid or subglobose, smooth white.

Fls. & Frts.: July–October. *Exsiccata*: SUK-221.
Frequent, in moist deciduous and semi evergreen forests.
Localities: Amba, Ambaghat, Bhuibawada.

Local uses: Roots are given as nutritive food, juice of root is used with milk and given in menorrhoea. Roots are orally given in rheumatism.

Trichosanthes cucumerina L. Sp. Pl. 1008. 1753; C. B. Cl. in Hook. f. Fl. Brit. India 2: 609. 1879; Cooke, Fl. Pres. Bombay 1: 560. 1958 (Repr. ed.); Madhusudhan Rao in Singh *et al.* Fl. Maharashtra State Dicot. 2: 66. 2001. RANPADWAL.

Extensive, climbing, annual monoecious herbs; Tendrils 2–3 fid. Leaves broadly ovate to orbicular-reniform, 7–10 × 8–12 cm, cordate at base, 5–7 lobed or angled, distantly denticulate, glabrous above, pubescent beneath. Male flowers 5–15 cm long, 8–15 flowered axillary racemes. Calyx tube 2–2.5 cm long; teeth short, acutely triangular. Petals white, lanceolate oblong, 8–10 mm long. Female flowers solitary. Fruits 3–8 cm long; tapering at both ends, with a long sharp beak, scarlet when ripe. Seeds semi ellipsoid, 10–12 × 5–6 mm, compressed, truncate at apex, narrowed at base.

Fls. & Frts.: August–November. *Exsiccata*: SUK-103.
Frequent, in Konkan region.
Localities: Amboli, Sawantwadi, Tillari.

Local uses: Plant juice is given 2 to 3 times in a day as a febrifuge; leaf juice is rubbed over the whole body in remittent fevers and in rheumatism. Oil from the seeds is given in bronchitis. Plant is used in bronchitis; root is used to cure headache, boils; fruits are used in itching, leucoderma, diseases of the blood, burning sensation, leprosy and ulcers.

Trichosanthes tricuspidata Lour. Fl. Cochinch. 589. 1790; Madhusudhan Rao in Singh *et al.* Fl. Maharashtra State Dicot. 2: 67. 2001. *T. palmata* Roxb. Fl. Ind. 3: 704. 1832; C. B. Cl. in Hook. f. Fl. Brit. India 2: 606. 1879; Cooke, Fl. Pres. Bombay 1: 560. 1958 (Repr. ed.). KAUNDAL.

Robust, climbing herbs; stem woody. Leaves ovate, 5–12 cm long and as much broad, shallowly cordate at base, deeply or shallowly 3–5 lobed, dark green above, paler and with black glands beneath; lobes ovate oblong, dentate-serrate, acute. Male flowers 15–20 cm long, 5–10 flowered axillary racemes. Bracts broadly ovate, fringed, dotted with dark green glandular spots. Calyx tube 2.5–3 cm long, pubescent; lanceolate; laciniate. Petals white, obovate-cuneate. Filaments villous. Female flowers solitary, on short axillary peduncles. Fruits globose, 3–5 cm in diam., red with 10 orange coloured stripes. Seeds obovoid, smooth, brown or black.

Fls. & Frts.: August–November. *Exsiccata*: SUK-212.

Common, in moist deciduous and semi evergreen forests.

Localities: Amba, Amboli, Barki, Borbet, Gaganbawada, Sawantwadi, Vaibhavwadi.

Local uses: Fruit is smoked as a remedy for asthma. Fruit pounded and mixed with warm coconut oil, is used to sores under the ears and nostrils. Fruit is used in fever. Stem is externally applied in backache. 10–20 ml root decoction is given in fever. Fruit is purgative, abortifacient, lessens inflammation, cures hemicrania, weakness of limbs, heat of brains, leprosy, used in epilepsy and rheumatism.

AIZOACEAE Rudolphi

Trianthema portulacastrum L. Sp. Pl. 223. 1753; Kamble and Lakshminarasimhan in Singh *et al.* Fl. Maharashtra State Dicot. 2: 89. 2001.

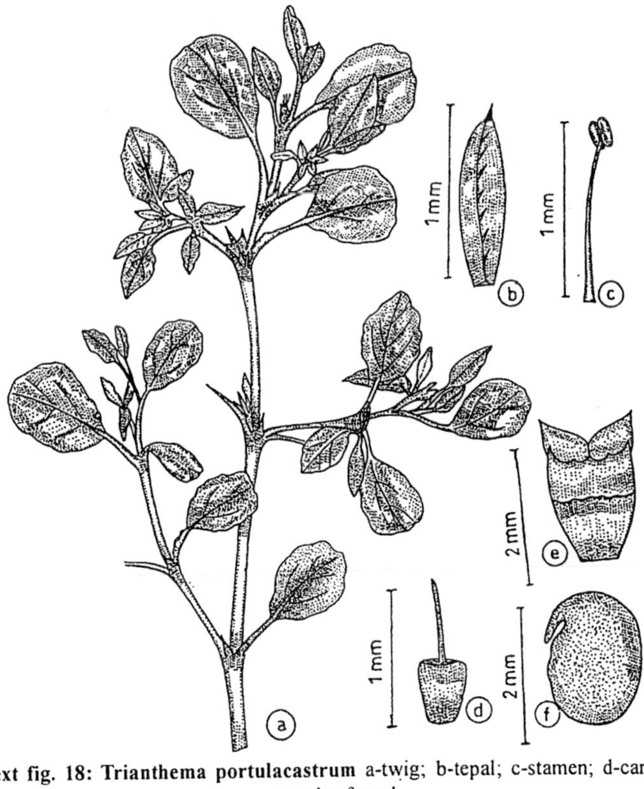

Text fig. 18: Trianthema portulacastrum a-twig; b-tepal; c-stamen; d-carpel; e-capsule; f-seed.

T. monogyna L. Mant. 1: 69. 1767; C. B. Cl. in Hook. f. Fl. Brit. India 2: 660. 1879; Cooke, Fl. Pres. Bombay 1: 589. 1958 (Repr. ed.). GHOL, PUNDHARIGHENTUT, WASU.

Prostrate, succulent herbs; stem slightly angular, glabrous or minutely pubescent. Leaves subfleshy, obliquely opposite, unequal, the upper one of the pair larger, obovate or orbicular, 3–4.5 × 2–3 cm, the lower broadly ovate or rounded, 1–1.2 × 0.6–1 cm, apiculate at apex. Flowers solitary, sessile, almost concealed by the pouch of the petiole. Calyx lobes ovate, green or reddish outside, pink within, with membranous margin. Stamens many. Ovary truncate; style one. Capsule concealed in the pouch of the petiole, lid truncate with two spreading teeth. Seeds 5–8, reniform, muriculate, dull black.

Fls. & Frts.: July–January. *Exsiccata*: SUK-95.
Common, on wasteland along roadsides and in fields etc.
Localities: Jat, Hatkanagale, Ramling, Shivaji University campus, Kolhapur.

Local uses: Leaf decoction is given in cough; leaf in combination with betel leaves is given in anemia. Leaves are given as a diuretic. Roots are given in hepatitis. Leaf juice is dropped into nostrils in hemicrania.

MOLLUGINACEAE Hutch.

Glinus oppositifolius (L.) A. DC. in Bull. Herb. Boiss. 2, 1: 559. 1992; Kamble & Lakshminarasimhan in Singh *et al.* Fl. Maharashtra State Dicot. 2: 93. 2001. *Mollugo oppositifolia* L. Sp. Pl. 89. 1753; Cooke, Fl. Pres. Bombay 1: 593. 1958 (Repr. ed.). *M. spergula* L. Syst. Nat. ed. 10. 881. 1759; C. B. Cl. in Hook. f. Fl. Brit. India 2: 662. 1879.

Prostrate or procumbent, glabrous herbs. Leaves whorled, oblanceolate, linear, lanceolate or sometimes spathulate, tapering into an obscure petiole, rounded, acute or apiculate at apex, glabrous or pubescent when young. Flowers 2-many in axillary fascicles. Sepals oblong, 4–5 mm long, green and glabrous on the back, white within and with broad scarious margins. Stamens only 3, no staminodes. Ovary glabrous, trilocular; styles 3, short; stigma spreading. Capsules ellipsoid, shorter than the sepals. Seeds numerous, subreniform, tubercled, dark brown, appendaged with a small, bristle like strophiole.

Fls. & Frts.: July–November. *Exsiccata*: SUK-1645.
Common, on wet mud along nallas, around tanks
Localities: Kolhapur.

Local uses: Leaf powder is given as stomachic. Leaf decoction is applied externally in skin diseases. Plant is warmed with castor oil and poured in ear to stop earache.

Mollugo pentahylla L. Sp. Pl. 89. 1753; Cooke, Fl. Pres. Bombay 1: 594. 1958 (Repr. ed.) Kamble & Lakshminarasimhan in Singh *et al.* Fl. Maharashtra State Dicot. 2: 94. 2001. *M. stricta* sensu C. B. Cl. in Hook. f. Fl. Brit. India 2: 663. 1879 non L. 1762.

Erect or diffuse, glabrous, annual herbs, 10–30 cm tall; branches dichotomous, 4-angled. Leaves in whorls of 2–9; basal one obovate-obtuse or oblanceolate, cauline, linear-lanceolate, 1–3 × 0.3–0.6 cm, acute. Flowers in terminal cymes or pseudoracemses. Sepals elliptic-oblong, 1.5–2.5 mm long, obtuse, with narrow green keel and broad white margins. Stamens 3. Styles 3. Capsule subglobose, 2–2.5 mm in diam. Seeds numerous; reniform, compressed, dark brown.

Fls. & Frts.: July–December. *Exsiccata*: SUK-177.

Common, in grasslands, cultivated fields, along stream banks and on wet rocks.

Localities: Kagal, Kolhapur, Panhala.

Local uses: An infusion of the plant is given to women to promote menstrual discharge other uses are same as *G. oppositifolius*.

APIACEAE Lindl.

Anethum graveolens L. Sp. Pl. 263. 1753; Pradhan in Singh *et al.* Fl. Maharashtra State Dicot. 2: 108. 2001. *Peucedanum graveolens* (L.) Hiern, Fl. Trop. Afr. 3: 19. 1877; C. B. Cl. in Hook. f. Fl. Brit. India 2: 709. 1879 non S. Wats. 1871. SHEPU, BALANT SHEPU.

Erect, annual herbs, 60–90 cm tall; stem smooth, glabrous. Leaves 10–20 cm long, pinnately divided into narrow linear segments; the ultimate segments 1–2 cm long. Flowers in large terminal, compound umbels, ca 15 cm across; rays rather unequal. Calyx obscurely toothed, 2 mm long, glabrous. Corolla yellow; petals ovate, oblong, 2.5 mm long, deciduous. Disk yellow thin. Styles divaricate. Fruits ellipsoid, 4 mm; dorsal ribs thin, sharp; lateral ribs with distinct narrow or broad wings

Native of Europe, cultivated.

Localities: Kolhapur, Satara.

Local uses: Fruits are given as anthelmintic in children. It is given to women's during pregnancy. Decoction of fruits is given in stomach complaints. Seeds are anthelmintic, abortifacient, useful in bronchitis, hiccough and in abdominal pains. An infusion is given as cordial drink to women after confinement.

Centella asiatica (L.) Urban in Mart. Fl. Bras. 11: 287, t. 78, f. 1. 1879; Pradhan in Singh *et al.* Fl. Maharashtra State Dicot. 2: 97. 2001. *Hydrocotyle asiatica* L. Sp. Pl. 234. 1753; C. B. Cl. in Hook. f. Fl. Brit. India

2: 668. 1879; Cooke, Fl. Pres. Bombay 1: 592. 1958 (Repr. ed.). BRAHMI, MUNDUKPARNI.

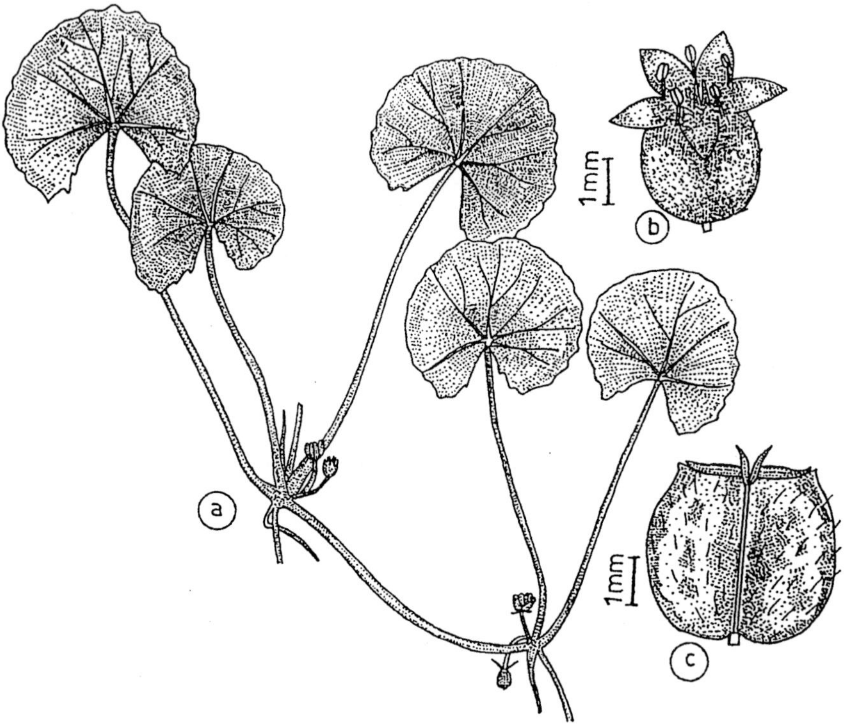

Text fig. 19: Centella asiatica a-twig; b-flower; c-fruit.

Prostrate or creeping annual herbs; stems 30–40 cm long, rooting at nodes. Leaves broadly suborbicular to reniform, 2–4 cm across, deeply cordate at base, entire or shallowly dentate, glabrous on both surfaces, prominently veined. Flowers in axillary, fascicled umbels, peduncle bearing 3–4 sessile flowers; bracts 2 beneath each umbel, ovate. Calyx campanulate; teeth absent. Petals minute, white or pink, ovate. Fruits ovoid, 3–4 mm long, hard, greenish-brown, often crowded with the persistent petals.

Fls. & Frts.: April–December. *Exsiccata*: SUK-283.

Common, in wet shaded situations.

Localities: Barki, Dajipur, Kolhapur, Radhanagari, Sawantwadi, Tillari, Vaibhavwadi.

Local uses: Leaf decoction is given as appetizer, similarly used as brain tonic. Leaves are used in hair loss and graying of hairs. It is given as cardiac tonic.

Coriandrum sativum L. Sp. Pl. 256. 1753; C. B. Cl. in Hook. f. Fl. Brit. India 2: 717. 1879; Cooke, Fl. Pres. Bombay 1: 609. 1958 (Repr. ed.); Pradhan in Singh *et al.* Fl. Maharashtra State Dicot. 2: 108. 2001. KOTHIMBIR, KOTHMIR, DHANE, DHANIYA.

Strong smelling, annual herbs, 15–45 cm tall. Leaves 2–6 cm long, pinnately or ternately decompound, glabrous; segments 8–15 × 2–8 mm, variously lobed; petioles variable in length, with thin, membranous sheathing base. Flowers in terminal compound umbels; rays few; calyx tube campanulate, 2–2.5 mm long; teeth minute. Petals white or tinged with pink, unequal, obovate-emarginate, those of the outer flowers in the umbels spreading like rays and much larger. Disk often yellowish, flat. Styles spreading. Fruits globose, 4–5 mm in diam., strongly ribbed, brown.

Fls. & Frts.: July–September.

Cultivated, Native of southern Europe.

Local uses: Fruits are soaked in water for 12 hrs and that water is taken early in the morning in dysuria and in gonorrhea. Leaf decoction is given early in the morning in burning sensation. The seeds are chewed to correct foul breath.

Cuminum cyminum L. Sp. Pl. 254. 1753; C. B. Cl. in Hook. f. Fl. Brit. India 2: 218. 1879; Cooke, Fl. Pres. Bombay 1: 609. 1958 (Repr. ed.); Pradhan in Singh *et al.* Fl. Maharashtra State Dicot. 2: 109. 2001. JIRE.

Glabrous or pubescent, annual herbs, 10–30 cm tall. Leaves 3–5 cm long, with few, filiform acerose-tipped divisions, ultimate segments filiform, glabrous; petioles short, dilated into membranous sheath. Flowers in few flowered terminal umbels; primary and secondary rays short. Calyx 1.5–2 mm long; teeth minute. Corolla white or pink; petals ovate-orbicular, minute. Disk white, thin. Fruits ellipsoid, 5–6 mm long, brown, bristly.

Fls. & Frts.: February–March.

Cultivated, native of the Mediterranean region.

Localities: Alate, Alas, Chokak, Talsande.

Local uses: Fruits are used in stomach complaints; fruits decoction is given in gonorrhoea. Other properties are same as *Coriandrum sativum*.

Foeniculum vulgare Mill. Gard. Dist. ed. 8, no. 1. 1768; Cl. in Hook, f. Fl. Brit. India 2 : 695. 1879; Cooke, Fl. Pres. Bombay 1: 609. 1958 (Repr. ed.); Pradhan in Singh *et al.* Fl. Maharashtra State Dicot. 2: 109. 2001. BADISHEP.

Erect annual, glaucous herbs, 60–90 cm tall; stem terete striate. Leaves 5–15 cm long, much dissected into linear segments; petioles dilated at base. Flowers small in large terminal compound umbels; involucral bracts and bracteoles absent; primary and secondary rays many, 1–2 cm long. Calyx minute; teeth obscure. Petals yellow, ovate, ca 1 mm long. Fruits ovate-elliptic, 5–8 mm long, green, 10-ribbed, beaked.

Fls. & Frts.: December–April.
Localities: Hatkanagale, Kolhapur, Sangavade.

Local uses: Aqueous extract is used as enema for expulsion of foetus. Hot infusion helps to increase lacteal secretion and stimulates sweating.

Trachyspermum ammi (L.) Sprague in Kew Bull. 1929: 228. 1929; Pradhan in Singh *et al.* Fl. Maharashtra State Dicot. 2: 107. 2001. *Sison ammi* L. Sp. Pl. 252. 1753. *Carum copticum* Benth. ex C. B. Cl. in Hook. f. Fl. Brit. India 2: 682. 1879; Cooke, Fl. Pres. Bombay 1: 600. 1958 (Repr. ed.). ONWA, AJWAYAN.

Erect annual herbs, 30–90 cm tall; branches slender, ascending, glabrous. Leaves 2–5 cm long, pinnate decompound; segments narrowly linear. Flowers in axillary and terminal, compound umbels; rays 4–12. Calyx 1–1.5 mm long, hairy outside, teeth minute. Petals white, ovoid, ca 1 mm long, retuse or emarginate. Fruits ellipsoid-oblong, 2–2.5 mm long, distinctly ribbed, muricate, hispid.

Fls. & Frts.: January–November. *Exsiccata*: SUK-1650.
Cultivated, native of Southern Europe.
Localities: Alate, Hatkanagale.

Local uses: The roots are used as a diuretic. Fruits are given in stomach complaints with buttermilk. Root extract is used in ear boils, paste is externally applied in inflammation.

Trachyspermum roxburghianum (DC.) Craib. Fl. Siam. Enum. 1: 788. 1931; Pradhan in Singh *et al.* Fl. Maharashtra State Dicot.2: 107. 2001. *Ptychotis roxburghiana* DC. Prodr. 4: 109. 1830. *Carum roxburghianum* (DC.) Kurz. In J. Asiat Soc. Bengal 46 (2): 114. 1877; C. B. Cl. In Hook. f. Fl. Brit. India 2: 682. 1879. *Carum stictocarpum* C. B. Cl. in Hook. f. Fl. Brit. India 2: 681. 1879; Cooke, Fl. Pres. Bombay 1: 600. 1958 (Repr. ed.). RANOWA.

Erect, glabrous, much branched herbs, 15–90 cm tall; root fusifrom, perennial. Leaves 2-pinnate; leaflets deeply cut into linear-lanceolate segments, glabrous or minutely pubescent. Flowers in axillary and terminal compound umbels; primary rays 5–10. Calyx hispid outside; teeth obscure. Petals suborbicular, rounded at apex, white or pink. Fruits ovoid, 1–1.5 mm long, glabrous or hispid brown, compressed.

Fls. & Frts.: December–February. *Exsiccata*: SUK-1659.
Rare, in moist deciduous forest.
Localities: Amba.

Local uses: The roots are used as a diuretic. Fruits are given in stomach complaints with buttermilk.

ALANGIACEAE DC.

Alangium salvifolium (L. f.) Wanger in Engl. Pflanzenr. 41: 9, f. 2 A-E. 1910; Londhe in Singh *et al.* Fl. Maharashtra State Dicot. 1: 114. 2001. *Grewia salvifolia* L. f. Suppl. 409. 1781. *Alangium lamarkii* Thw. Enum. 133. 1859; C. B. Cl. in Hook. f. Fl. Brit. India 2: 741. 1879; Cooke, Fl. Pres. Bombay 2: 1. 1958 (Repr. ed.). ANKUL, ANKWAL, ANSAROLI.

Shrubs or small trees with more or less thorny branches. Leaves narrowly oblong to ovate lanceolate, 6–12 × 2–6 cm, rounded or acute at base, acuminate, entire, glabrous above, pubescent beneath. Flowers few in axillary umbels. Calyx turbinate, 3 mm long, densely silky pubescent outside; teeth triangular. Petals dull white, 5–6, narrowly linear, 15–20 mm long and 4–5 mm wide, reflexed. Stamens about 20; filaments as long as petals. Style long; stigma capitate. Fruit subglobose, 2–2.5 × 1.5–2 cm, purplish-red, crowned with calyx limb. Seed oblong, brown.

Fls. & Frts.: February–April. *Exsiccata*: SUK-143.
Occasional, in deciduous forest.
Localities: Jotiba, Nesari, Panhala.

Local uses: It is reputed single drug for the treatment of rabies. Root paste is used to cure skin diseases. Bark extract is taken orally for lowering blood pressure. It is useful in diarrhoea, simple continued fevers. Oil of the root bark is useful in external application in acute rheumatism. Extracts of flowers and fruits are used to cure eyesores.

RUBIACEAE Juss.

Catunaregam spinosa (Thunb.) Tirveng. in Bull. Mus. Hist. Nat. Paris 3, 35: 13. 1978 & in Taxon 27: 515. 1978; Mudaliar & Prasad in Singh *et al.* Fl. Maharashtra State Dioct. 2: 121. 2001. *Gardenia spinosa* Thunb. Diss. Gard. no. 7. 1780. *Randia dumatorum* (Retz.) Poir. in Lam. Encycl. Suppl. 2: 829. 1811; Hook. f. Fl. Brit. India 3: 110. 1880; Cooke, Fl. Pres. Bombay 2: 27. 1958 (Repr. ed.). GEHELA, GEL, MADANPHAL.

Large shrubs or small trees, armed with strong, straight, nearly horizontal thorns, 1.5–3.5 cm long. Leaves usually fascicled from suppressed branches, obovate, 2.5–5 × 2–3 cm, cuneate at base, entire or wrinkled, obtuse, pubescent on both surfaces. Flowers at the ends of short branches, fragrant, solitary or 2–3 together. Calyx 10–12 mm long, densely hairy outside; teeth 5, foliaceous, ovate-oblong, 5 mm long. Corolla white, turning pale yellow, 2–2.5 cm long, tube densely hairy outside; lobes obovate-oblong, rounded at apex, pubescent outside. Fruits leathery, ovoid, 2.5–3.5 cm long; yellowish obscurely longitudinally ribbed, crowned with large calyx limb, 2-locular. Seeds many, flat, brown.

Fls. & Frts.: July–September. *Exsiccata*: SUK-1700.
Common, in deciduous and moist deciduous forests.
Localities: Ajara, Amba, Borbet, Dajipur, Kas, Pachgani, Vaibhavwadi.

Local uses: Fruit decoction is used as emetic, it is used in chronic bronchitis, pain in the muscles. Fruits are used to wash the boils; similarly it is applied externally to remove black marks from the skin due to excess secretion of melanin after pregnancy. Roots are used for the same purpose. Root decoction is used as emetic in cough. Paste is applied externally in skin diseases. Fruit is used as carminative; cures ulcers, inflammations, tumours and skin diseases.

Ceriscoides turgida (Roxb.) Tirveng. in Bull. Mus. Natn. Hist. Nat. Paris, 3e 521, Bot. 35: 15. 1978; Mudaliar & Prasad in Singh *et al.* Fl. Maharashtra State Dicot. 2: 123. 2001. *Gardenia turgida* Roxb. Fl. Ind. 2: 557. 1832; Hook. f. Fl. Brit. India 3: 118. 1880. *G. turgida* var. *montana* (Roxb.) Hook. f. Fl. op. cit.; Cooke, Fl. Pres. Bombay 2: 29. 1958 (Repr. ed.). PHETRA.

Small, deciduous trees, 3–4 m tall; bark smooth, yellowish; branches armed with strong, straight thorns. Leaves oblong or suborbicular, 3–7.5 × 2.5–3.5 cm, tapering at base, obtuse, glabrous above, velvety tomentose beneath. Flowers from young leafless shoots, dimorphic, males in fascicles, female solitary. Male flowers: Calyx turbinate, pubescent; mouth truncate or with minute teeth. Corolla white turning pale yellowish, 2–2.5 cm long, glabrous; lobes oblong, 4–5 mm long, obtuse. Female flowers: Calyx flask–shaped, 1–2 cm long; teeth foliaceous, linear, 3–4 mm long. Corolla as in the males. Fruits ovoid or subglobose, 1.5–2.5 cm in diam., glabrous bluntly pointed.

Fls. & Frts.: April–June. *Exsiccata*: SUK-1639
Infrequent, in dry deciduous forests.
Localities: Hatkanagale, Narande, Nerli, Sawantwadi, Shiroda.

Local uses: Roots are given as a remedy for indigestion in children, roots are applied externally on wounds and in cases of roundworm.

Haldina cordifolia (Roxb.) Ridsd. in Blumea 24: 361. 1978; Mudaliar & Prasad in Singh *et al.* Fl. Maharashtra State Dicot. 2: 127. 2001. *Nauclea cordifolia* Roxb. Pl. Cor. 1: 40, t. 53. 1798. *Adina cordifolia* (Roxb.) Hook. f. ex Brandis, For. Fl. 263, t. 33. 1874; Hook. f. Fl. Brit. India 3: 24. 1880; Cooke, Fl. Pres. Bombay 2: 7. 1958 (Repr. ed.). HALDU, HEDU.

Large, deciduous trees, 10–20 m tall. Leaves broadly ovate-orbicular, 7.5–15 × 9–16 cm, cordate at base, subacute, or acuminate, glabrous, or thinly pubescent above, densely so beneath; petioles 5–10 cm long. Flowers numerous in axillary 1–3 peduncled globose heads 2–2.5 cm in diam., peduncles stout, 2–5 cm long. Calyx 2–2.5 mm long, hairy; teeth club shaped. Corolla yellow, 5–6 mm long; tube slender, dilated above, pubescent beneath; lobes ovate-oblong, acute, recurved. Style much exserted above, pubescent. Capsules cuneate, 3–4 mm long, densely hairy. Seeds minute.

Fls. & Frts.: June–September. *Exsiccata*: SUK-286, 670.
Frequent, in deciduous forests.
Localities: Ambarde.

Local uses: Bark paste is applied to heal wounds. Bark paste mixed with black pepper and sniffed into the nose in headache. Bark is used in inflammations, blood diseases and skin diseases.

Hedyotis corymbosa (L.) Lam. Encycl. 1: 272. 1792; Mudaliar & Prasad in Singh *et al.* Fl. Maharashtra State Dicot. 2: 130. 2001. *Oldenlandia corymbosa* L. Sp. Pl. 119. 1753; Hook. f. Fl. Brit. India 3: 64. 1880 p. p.; Cooke, Fl. Pres. Bombay 2: 15. 1958 (Repr. ed.). PITAPAPDA.

Diffuse, slender, glabrous or puberulous annual herbs; stem 10–25 cm long, quadrangular, thinly hairy, rooting at lower nodes. Leaves sessile, linear-lanceolate; 1–3 × 0.2–0.4 cm, acute; stipules membranous, truncate. Flowers small, paired or 1–4 on axillary peduncles. Calyx 2–2.5 mm long, pubescent; teeth narrowly triangular, equaling the calyx tube. Corolla white, lobes as long as the tube, 2–2.5 mm long, acute. Capsule globose, broader than long, 2 × 2.5 mm, compressed, crowned with persistent calyx. Seeds 2, black, rugose.

Fls. & Frts.: July–December. *Exsiccata*: SUK-165, 671
Common, on old walls, in waste lands.
Localities: Kolhapur.

Local uses: Juice of leaves is applied in burning of palms of the hand and soles of the feet from fever. It is considered as a cooling in the treatment of fever and supposed to be caused by deranged air and bile, i.e. remittent fever, with gastric irritability and nervous depression.

Ixora coccinea L. Sp. Pl. 110. 1753; Hook. f. Fl. Brit. India 3: 145. 1880; Cooke, Fl. Pres. Bombay 2: 40. 1958 (Repr. ed.); Mudaliar & Prasad in Singh *et al.* Fl. Maharashtra State Dicot. 2: 140. 2001. BAKARA.

Glabrous shrubs, up to 1 m tall. Leaves sessile or nearly so, oblong, 4–10 × 3.6 cm, subcordate and amplexicaule at base, obtuse, apiculate;

stipules with long rigid, cuspidate point. Flowers numerous, in dense sessile or peduncled corymbiform cymes. Calyx 2 mm long; teeth triangular acute. Corolla red, tube slender; 2.5–3 cm long; lobes elliptic-oblong, 4–5 mm long, subacute. Fruits globose, 5–6 mm in diam., purple black. Seeds deeply ventrally concave, brown.

Fls. & Frts.: Throughout year. *Exsiccata*: SUK-1690.

Frequent, in open and deciduous forests.

Localities: Amba, Amboli, Barki, Gaganbawada, Mahabaleshwar, Radhanagari.

Local uses: Flowers are fried with seeds of Cumin and Nagkesar and made into bolus with butter and sugarcandy and given in dysentery. Leaves applied externally as a poultice on boils. Root is administered in gonorrhea. Bark paste is applied externally on boil and in headache.

Ixora nigricans R. Br. ex Wight & Arn. Prodr. 428. 1834; Hook. f. Fl. Brit. India 3: 148. 1880; Cooke, Fl. Pres. Bombay 2: 38. 1958 (Repr. ed.); Mudaliar & Prasad in Singh *et al.* Fl. Maharashtra State Dicot. 2: 141. 2001. KATKUDA.

Large shrubs or small trees, up to 5 m tall. Leaves opposite or in whorls of 3, oblanceolate to elliptic-oblong, 8–20 × 4–7 cm, coriaceous and hard, rounded or cordate at base, obtuse, glabrous and shining; stipules broadly ovate, cuspidate. Flowers numerous, in terminal and axillary, trichotomously branched, paniculate cymes shorter than the leaves. Calyx ovoid-oblong, 2–2.5 mm long; teeth minute, 1 mm long, triangular. Corolla tube 10–15 mm long; lobes 4, linear-oblong, obtuse. Style densely hairy, stigmas elliptic-lanceolate. Fruits didymous, 5–6 mm in diam., black.

Fls. & Frts.: December–April. *Exsiccata*: SUK-78.

Common, in hill forests, along banks of rivers, streams etc.

Localities: Amboli, Barki, Dajipur, Mahabaleshwar, Radhanagari, Tillari.

Local uses: Flowers are fried with seeds of Cumin and Nagkesar and made into bolus with butter and sugarcandy and given in dysentery. Root is administered in gonorrhea. Bark paste is applied externally on boils and in headache.

Ixora pavetta Andr. Bot. Rep. t. 78. 1799; Mudaliar & Prasad in Singh *et al.* Fl. Maharashtra State Dicot. 2: 141. 2001. *I. arborea* Roxb. ex J. E. Sm. In Rees, Cyclop. 19: 5. 1811. *I. parviflora* Vahl, Symb. 3: 11, t. 52. 1794 non Lam. 1791; Hook. f. Fl. Brit. India 3: 142. 1880; Cooke, Fl. Pres. Bombay 2: 39. 1958 (Repr. ed.). LOKHANDI, RAIKUDA.

Woody shrubs, 2.5–6 m tall. Leaves elliptic obovate or oblong, 8–12 × 3.0–8.5 cm, thickly coriaceous, upper surface dark green, dull beneath, base rounded or cordate; stipules broadly ovate, cuspidate. Flowers white, odorous, small and very numerous, in subglobose clusters, in sessile cymes. Calyx ovoid-oblong, 1.3–1.5 mm long; teeth 4, minute, triangular, subacute. Corolla tube 6–11 mm in length; lobes 4, linear-oblong, obtuse, style densely clothed with white hairs; branches of the stigma elliptic-lanceolate. Fruit 6 mm in diam., reddish-green, bilobed.

Fls. & Frts.: Throughout the year. *Exsiccata*: SUK-354.
Infrequent, in deciduous forests.
Localities: Amba, Dajipur, Gaganbawada, Ramling.

Local uses: Leaf paste is applied externally in rheumatism. Paste of stem is externally applied in bone fracture. Stem and leaf paste is applied in bone pain during fever.

Meyna laxiflora Robyns in Bull. Jard. Bot. Brux. 11: 228. 1928; Mudaliar & Prasad in Singh *et al.* Fl. Maharashtra State Dicot. 2: 145. 2001. *Vangueria spinosa* Roxb. Fl. Ind. 2: 172. 1824 p. p. excl. type; Hook. f. Fl. Brit. India 3: 136. 1880; Cooke, Fl. Pres. Bombay 2: 36. 1958 (Repr. ed.). ALU, HELU.

Shrubs or small trees, 3–10 m tall, armed with thorns, 2.0–2.5 cm long. Leaves elliptic oblong, ovate or ovate lanceolate, 5.0–12.5 × 3.7–8.5 cm, shining, acute, base cuneate tapering into 0.6–2.5 cm long petioles; stipules glabrous, broadly triangular at the base. Flowers greenish-white, in peduncled cymes from the old scars below the leaves. Calyx 2–2.5 mm long, glabrous, tube cup-shaped, ribbed, rugose; teeth 5, scarious, triangular, very acute. Corolla 8 mm in length, broad, glabrous outside, throat densely hairy within; lobes 5, ovate-lanceolate, acute, equaling the tube. Fruit globose, 2–5 cm across, flesh, chocolate colour when ripe. Seeds, 5–6.

Fls. & Frts.: March–June. *Exsiccata*: SUK-1673.
Frequent, in deciduous forests.
Localities: Amba, Dajipur, Gaganbawada, Mahabaleshwar, Radhanagari, Ramling.

Local uses: Fruit is used as cooling and expellant of phlegm and bile.

Mitragyna parvifolia (Roxb.) Korth. Obs. Naucl. Ind. 19. 1839; Cooke, Fl. Pres. Bombay 2: 8. 1958. (Repr. ed.); Mudaliar & Prasad in Singh *et al.* Fl. Maharashtra State Dicot. 2: 145. 2001. *Nauclea parvifolia* Roxb. Pl. Cor. 1: 40, t. 52. 1795. *Stephegyne parvifolia* Kunth. in Verh. Nat. Ges. Bot. 161. 1840; Hook. f. Fl. Brit. India 3: 25. 1880. KADAM.

Large, deciduous, glabrous or pubescent trees, 6–10 m tall. Leaves elliptic-obovate or suborbicular, 7.5–18.5 × 4–12 cm, acute, rounded or subcordate at base, acute or bluntly acuminate at apex, glabrous; stipules oblong spathulate, pink, obtuse, parallel-veined. Flowers in axillary solitary, peduncled, globose heads, 1.5–2.0 cm in diam. Calyx funnel shaped; 2 mm long; limb truncate. Corolla creamy-white, 6–8 mm long; tube narrowly funnel-shaped, lobes triangular-ovate, 2–2.5 mm long, acute, recurved. Style white much exerted; stigma mitriform. Capsule oblong, 3 mm long, forming a head, 12–15 mm in diam., seeds many, black, minute.

Fls. & Frts.: April–August. *Exsiccata*: SUK-161.
Common, in semi evergreen forests.
Localities: Anuskara, Mahabaleshwar, Sawantwadi, Shiroda, Vaibhavwadi.

Local uses: Leaf paste is applied in strains and bruises. Bark is orally given once in a day in general debility.

Morinda pubescens J. E. Sm. in Rees, Cyclop. 24, n. 3. 1813; Mudaliar & Prasad in Singh *et al.* Fl. Maharashtra State Dicot. 2: 146. 2001. *M. tinctoria* Roxb. var. *tomentosa* (Heyne ex Roth.) Hook. f. Fl. Brit. India 3: 156. 1880; Cooke, Fl. Pres. Bombay 2: 43. 1958 (Repr. ed.). AHJHL, BARTONDI.

Trees, 2–8 m tall; young branches quadrangular, tomentose. Leaves broadly elliptic or oblong lanceolate, 8.5–18 × 3.11–5 cm, acuminate, tomentose on both surfaces; stipules bifid; lobes triangular, acute. Flowers pure white, in globose heads. Calyx truncate; 1–1.5 mm long, limbs foliaceous, oblong obtuse or subacute, hairy outside. Corolla tube 2–2.5 cm long, hairy outside; lobes oblong, obtuse or subacute. Syncarps of pyrenes ellipsoid, fleshy, ca 3 cm across, not winged.

Fls. & Frts.: March–December. *Exsiccata*: SUK-100.
Common, throughout the region.
Localities: Alate, Babu Jamal, Borbet, Kagal, Kolhapur, Ramling, Tasgaon, Wai.

Local uses: Leaves are used as a healing application to wounds and ulcers. Juice of leaves is externally applied to gout to relieve pain.

Mussaenda glabrata (Hook .f.) Hutch. ex Gamble, Fl. Pres. Madras 610. 1921. [2: 430. 1957 (Repr. ed.)]; Mudaliar & Prasad in Singh *et al.* Fl. Maharashtra State Dicot. 2: 147. 2001. *M. frondosa* var. *glabrata* Hook. f. Fl. Brit. India 3: 90. 1880. *M. frondosa* L. Sp. Pl. 177. 1753; Hook. f. op. cit. 89; Cooke, Fl. Pres. Bombay 2: 24. 1958 (Repr. ed.). BHUTKES, SARVAD.

Shrubs. Leaves broadly elliptic, 7.5–13.7 × 4.5–8.0 cm, apex shortly acuminate; base rounded or tapering into a petiole of 2.0–2.5 cm length; stipules twin, linear, lanceolate, acute, hairy. Flowers in terminal and axillary cymes, one calyx lobe modified into a creamy-yellow, elliptic bract, acute at apex, base tapering into a claw. Calyx teeth linear-subulate, 1.5 cm long, hairy, the enlarged segments leaf like, creamy white, persistent, 6–11 × 3–3.5 cm, broadly ovate, elliptic, acuminate, 5–7 nerved. Corolla 3–4 cm long, deep golden yellow, tube slender, very hairy outside and with a dense lining of yellowish hairs completely blocking the mouth; lobes ca 6 mm long, broadly ovate or elliptic, apiculate, pubescent on the back. Berries obovoid, 1.0–1.3 cm long, sparsely hirsute.

Fls. & Frts.: April–November. *Exsiccata*: SUK-239.

Infrequent, in deciduous forests.

Localities: Ajara, Amba, Amboli, Dajipur, Radhanagari, Sawantwadi, Vaibhavwadi.

Local uses: Leaves are given in jaundice. Roots are externally applied in leucoderma. Decoction of flower is used to wash ulcers.

Neolamarckia cadamba (Roxb.) Bosser in Bull. Mus. Natn. Hist. Nat., Paris, 6. Sect. B, Adansonia 3: 247. 1984; Mudaliar & Prasad in Singh *et al.* Fl. Maharashtra State Dicot. 2: 155. 2001. *Nauclea cadamba* Roxb. Fl. Ind. 2: 121. 1824. *Anthocephalus cadamba* (Roxb.) Miq. Fl. Ind. Bat. 2: 135. 1856; Hook f. Fl. Brit. India 3: 23. 1880. *A. indicus* A. Rich. in Mem. Soc. Hist. Nat. Paris. 5: 238. 1834; Cooke. Fl. Pres. Bombay 2: 6. 1958 (Repr. ed.). KADAM, NIT.

Trees, 10–12 m tall. Leaves elliptic to elliptic-oblong, 10–23 × 5–15 cm, coriaceous, upper surface shining, lower pubescent, apex acute, base rounded or cordate; stipules lanceolate, acute, caducous. Flowers fragrant, in solitary, terminal, globular heads of ca 3 cm in diam. Calyx 5–6 mm long, 5-lobed; lobes linear, obtuse, hairy. Corolla orange or pale yellow, tube 0.7–12 mm in length, infundibuliform, slender; segments linear oblong, ca 2.5 mm long, erect, acute, imbricate, slightly pubescent outside. Stigma long. Capsules in globose receptacles, 0.3 × 0.5 cm.

Fls. & Frts.: February–November. *Exsiccata*: SUK-256.

Frequent, in moist deciduous forests.

Localities: Amba, Amboli, Gaganbawada, Mahabaleshwar, Sawantwadi, Vaibhavwadi.

Local uses: Bark is given as a tonic. Decoction of bark is used in stomach complaints, similarly decoction is warmed with equal parts of ginger and leaf of *Justicia adhatoda* is given in cough. Paste prepared from above 3 combination is applied externally in headache. Bark is used in uterine complaints. Fruit is used to cure leprosy, dysentery and fever.

Pavetta crassicaulis Bremek. in Feddes Report. 37: 112. 1934 & 47: 25. 1939; Mudaliar & Prasad in Singh *et al.* Fl. Maharashtra State Dicot. 2: 158. 2001. *P. indica* auct. non L. 1753; Hook. f. Fl. Brit. India 3: 150. 1880; Cooke, Fl. Pres. Bombay 2: 41. 1958 (Repr. ed.). ASAVAL, PAPAT.

Erect shrubs, 1–2 m tall; young branches sub quadrangular, glabrous. Leaves obovate-lanceolate, 4–12 × 4–6 cm, acute, base cuneate. Flowers white, odorous, in terminal sessile corymbose pubescent cymes. Calyx 6 mm long, densely pubescent, tube campanulate, teeth triangular, acute, reflexed. Corolla tube 1.2 cm in length; lobes linear oblong, subacute. Style white, glabrous; stigma green, narrowly clavate, puberulous. Berries globose, 1 cm in diam., black, smooth.

Fls. & Frts.: March–November. *Exsiccata*: SUK-354.
Frequent, in open areas, along Ghats and plains.
Localities: Amba, Barki, Gaganbawada, Radhanagari.

Local uses: Root and leaf decoction is given in rheumatism. Leaf paste is applied externally to heal wound and in the bone fracture. Leaf decoction is given in cough and bronchitis.

Rubia cordifolia L. Syst. Nat. ed. 12, 3: 229. 1768; Hook. f. Fl. Brit. India 3: 202. 1881; Cooke, Fl. Pres. Bombay 2: 54. 1958 (Repr. ed.); Mudaliar & Prasad in Singh *et al.* Fl. Maharashtra State Dicot. 2: 162. 2001. ILTA, MANJISTH.

Perennial, prostrate, decumbent or climbing herbs; stem sharply 4-angled, retrosely scabrid, hairy on angles, on leaf nerves and margins. Leaves whorled, 6–8 at each node, elliptic, cordifrom or ovate-lanceolate, 2.5–10 × 1.2–4 cm, acute, rounded or cordate at base; stipules 0. Flowers in terminal panicled glabrous cymes. Calyx 3 mm long, tube globose, glabrous. Corolla greenish, divided up to base; lobes 5, ovate, 3 mm long, acute. Styles 2; stigmas globose. Fruits bicarpellate, globose, 3–6 mm in diam., purplish-black when ripe.

Fls. & Frts.: January–December. *Exsiccata*: SUK-10.
Frequent, in shady places in deciduous and semi evergreen forests.
Localities: Ajara, Amboli, Dajipur, Panchgani.

Local uses: Roots are applied externally in leucoderma. Roots are used with other combination in uterine pains. Decoction of root is given every night in skin diseases caused by blood impurities. Root improves the voice and complexion; cures disease of the uterus, vagina, eye, ear, blood; cures leucoderma erysipelas and oleagenous; increase appetite.

Spermadictyon suaveolens Roxb. Pl. Cor. 3: 32, t. 236. 1815; Mudaliar & Prasad in Singh *et al.* Fl. Maharashtra State Dicot. 2: 167. 2001. *Hamiltonia suaveolens* Roxb. Fl. Ind. 1: 554. 1832; Hook. f. Fl. Brit. India 3: 197. 1880; Cooke, Fl. Pres. Bombay 2: 51. 1958. (Repr. ed.). GITSAYA, GIDAS.

Trichotomously branched small shrubs, 1–2 m tall. Leaves elliptic-lanceolate 7.5–2 × 3–6 cm, narrowed at base, acute, pubescent on both surfaces; stipules triangular hairy. Flowers numerous; sessile or nearly so, in subglobose heads in terminal trichotomous, pubescent. Calyx 3–4 mm long, hairy, tube narrow; teeth 4–5, linear-lanceolate, acute. Corolla white, pale blue or pale pink, 10–12 mm long; tube slender, slightly widened at the mouth; lobes 4–5, ovate, acute. Capsules ellipsoid, 3–4 mm long. Seeds few, triquetrous, surrounded by a loose lace like covering.

Fls. & Frts.: February–July. *Exsiccata*: SUK-34.

Common, on rocky soil of hill slopes.

Localities: Borbet, Panchgani, Satara, Wai Ghat.

Local uses: Root and leaf decoction is given in stomach complaints, root is boiled in rice, with dried ginger and given in peritionitis. It is externally applied on herpes virus, and in hypertension.

Tamilnadia uliginosa (Retz.) Tirveng. & Sastre in Mauritius Inst. Bull. 8(4): 85. 1979; Mudaliar & Prasad in Singh *et al.* Fl. Maharashtra State Dicot. 2: 167. 2001. *Gardenia uliginosa* Retz. Obs. Bot. 2: 14. 1781. *Randia uliginosa* (Retz.) DC. Prodr. 4: 386. 1830; Hook. f. Fl. Brit. India 3: 110. 1880; Cooke, Fl. Bombay 2: 27. 1958 (Repr. ed.). PANDHARA.

Small, glabrous trees, 3–5 m tall, thorn stout, 2–5 cm long, sharp. Leaves fascicled on suppressed branchlets, obovate or obovate-oblong, 7.5–15 × 3–6 cm, tapering to the base, glabrous above, pubescent beneath. Stipules triangular, acute. Calyx 1.2 cm long, tubular; teeth 5–7, rounded, ciliate. Corolla white, glabrous; tube short with a ring of hairs near the mouth; lobes 5–7, orbicular, much overlapping. Stamens 5–7, sessile. Stigma large; style ribbed. Fruits ovoid, 5–6 cm long, yellowish-brown, crowned with persistent calyx limb. Seeds numerous, compressed, smooth, brown.

Fls. & Frts.: April–June. *Exsiccata*: SUK-378.

Localities: Amba, Radhanagari.

Local uses: Root is applied externally on boils and seeds are mixed with fruits of *Garcinia indica* and given in dysuria and also as diuretic. Root is used as cooling and diuretic.

ASTERACEAE Dumort. (*nom. alt.*) [COMPOSITAE Juss. (*nom. cons.*)]

Ageratum conyzoides L. Sp. Pl. 839. 1753; Hook. f. Fl. Brit. India 3: 243. 1881; Cooke, Fl. Pres. Bombay 2: 70. 1958 (Repr. ed.); Shirodkar & Lakshminarasimhan in Singh *et al.* Fl. Maharashtra State Dicot. 2: 181. 2001. BURANDU, OSADI, SAHADEVI.

Erect, hispid, annual herbs, 15–80 cm tall; stem terete, clothed when young, with rough spreading hairs. Leaves opposite or upper alternate, broadly ovate, 3–7 × 1–5 cm, cuneate at base, crenate and with ciliate margins, acute, hairy. Heads in dense terminal corymbs; florets with bad smell. Involucral bracts linear, acute, ciliolate and with scarious margins. Pappus of 5 scales, aristate, dilated at base, serrualte. Corolla tubular, pale blue, pink or white, 2–3 mm long; lobes ovate, acute. Achenes linear, 1.5–2 mm long, 4-gonous, slightly curved black, short hairy.

Fls. & Frts.: July–February. *Exsiccata*: SUK-1672.

Common, weed in harvested and barren fields, along stream banks, on wastelands.

Localities: Kagal, Kolhapur, Sangali, Shivaji University campus, Wai.

Local uses: Leaves are externally applied to wounds, it acts as a styptic and heals them quickly. Leaves are applied externally in fever for quick perspiration. Leaf paste with 4 tender leaves of *Ocimum sanctum* mixed with 1/4 teaspoon powder of pepper given after every 4 hours in fever. A poultice of leaves is applied on boils; it is said to prevent tetanus. It is applied to wounds. Decoction of flower bud cures cancerous growth.

Blumea eriantha DC. in Wight, Contrib. Bot. Ind. 15. 1834; Hook. f. Fl. Brit. India 3: 266. 1881; Cooke, Fl. Pres. Bombay 2: 78. 1958 (Repr. ed.); Shirodkar & Lakshminarasimhan in Singh *et al.* Fl. Maharashtra State Dicot. 2: 188. 2001. BURADI, NIMURDI.

Erect, aromatic, viscid-pubescent herbs, 30–150 cm tall; stem striate. Leaves elliptic-oblong to ovate-oblong or obovate-oblong, 5–15 × 3–10 cm, lower ones shortly petioled, upper ones sessile, irregularly spinous dentate, pubescent to silky villous. Head globose or obovoid, 3–10 mm across, on long slender, hairy peduncles of terminal cymes. Receptacle convex, glabrous, pitted. Involucral bracts 3–4 seriate, outer ones linear acuminate, green, villous; the inner ones similar but less pubescent. Marginal floret 5–6 mm long, with 2–3-fid, filiform, yellow corollas; corolla tube glabrous; lobes pubescent. Central floret with 5-fid, tubular, yellow corollas. Achenes oblong 4-gonous, minute, often not well developed.

Fls. & Frts.: December–May. *Exsiccata*: SUK-1682

Endemic. Common, in fields and in open forests.

Localities: Amba, Borbet, Barki, Chandgad, Tillari, Sawantwadi.

Local uses: Leaves are used with leaves of *Vitex negundo* and mustard oil and applied externally in body pain and rheumatism. Leaves are externally applied in scabies.

Blumea lacera (Burm. f.) DC. in Wight, Contrib. Bot. Ind. 14. 1834; Hook. f. Fl. Brit. India 3: 263. 1881; Cooke, Fl. Pres. Bombay 2: 76. 1958 (Repr. ed.); Shirodkar & Lakshminarasimhan in Singh *et al.* Fl. Maharashtra State Dicot. 2: 189. 2001. *Conyza lacera* Burm. f. Fl. Ind. 180, t. 59, f. 1. 1768. BURANDO.

Erect or procumbent, glandular villous herbs, 25–120 cm tall; stem terete. Leaves elliptic-ovate or obovate, 2.5–20 × 1.5–6.5 cm, obtuse, crenate-lobulate, serrate dentate, lower shortly petiolate, upper sessile, all densely, villous. Heads 4–6 × 5–7 mm, in axillary or terminal, spiciform panicles; receptacle glabrous, pitted. Involucral bracts 4–5 seriate, linear; outer ones shorter, green with narrow scarious margins, densely hairy on the back, inner ones gradually longer, more scarious and less pubescent. Marginal florets with filiform, 2–3-fid, yellow corollas. Ovaries linear, minute, glabrous; style long; stigma papillate. Pappus bristle 5, scabrid, as long as the corolla. Achenes not well developed, glabrous, subtetragonous.

Fls. & Frts.: October–May. *Exsiccata*: SUK-626.
Occasional, on waste lands.
Localities: Kas, Kolhapur, Panchgani, Satara, Wai.

Local uses: Plant with black pepper is given in fever. Root is chewed to cure diseases of the mouth. Cures bronchitis.

Echinops echinatus Roxb. Fl. Ind. 3: 447. 1832; Hook. f. Fl. Brit. India 3: 358; Cooke, Fl. Pres. Bombay 2: 112. 1958 (Repr. ed.); Shirodkar & Lakshminarasimhan in Singh *et al.* Fl. Maharashtra State Dicot. 2: 207. 2001. KATECHUMBAK, UTKATKAR.

Much branched, rigid annual herbs, 30–60 cm tall. Leaves alternate, sessile, oblong, 5–12 × 2–5 cm, lyrate or deeply pinnatifid; lobes triangular-oblong, sinuate, spinous pointed. Compound heads globose, 2–4 cm in diam., on stout peduncles. Involucral bracts of individual simple heads scale-like; outer oblanceolate, glabrous, spinous tipped; intermediate bracts often turned into sharp spines, 2–4 cm long; inner most bracts connate or more than half their length enclosing a single floret; lobes lanceolate, acute, laciniate at apex, ciliate with scarious margins. Flowers bisexual with tubular, 5-lobed white corolla; lobes of the corolla linear, 4–5 mm long, acute. Achenes obconic, 3–4 mm long, brush-like, yellowish.

Fls. & Frts.: September–January. *Exsiccata*: SUK-173.
Common, in dry regions on wastelands.
Localities: Kagal, Kolhapur, Jaisingpur, Tasgaon, Satara, Wai.

Local uses: Seeds are used as aphrodisiac, leaf juice is given in chronic fever. Powder of root is applied to wounds in cattle to destroy maggots. Plant is used in urinary discharges, gleet, thirst and diseases of the heart. Root is abortifacient and aphrodisiac.

Eclipta prostrata (L.) L. Mant. 2: 286. 1771; Shirodkar & Lakshminarasimhan in Singh *et al.* Fl. Maharashtra State Dicot. 2: 207. 2001. *Verbesina prostrata* L. Sp. Pl. 902. 1753. *Eclipta alba* (L.) Hassk. Fl. Jav. Rar. 528. 1848; Hook. f. Fl. Brit. India 3: 304. 1881. *E. erecta* L. Mant. 2: 286. 1771; Cooke, Fl. Pres. Bombay 2: 112. 1958 (Repr. ed.). MAKA.

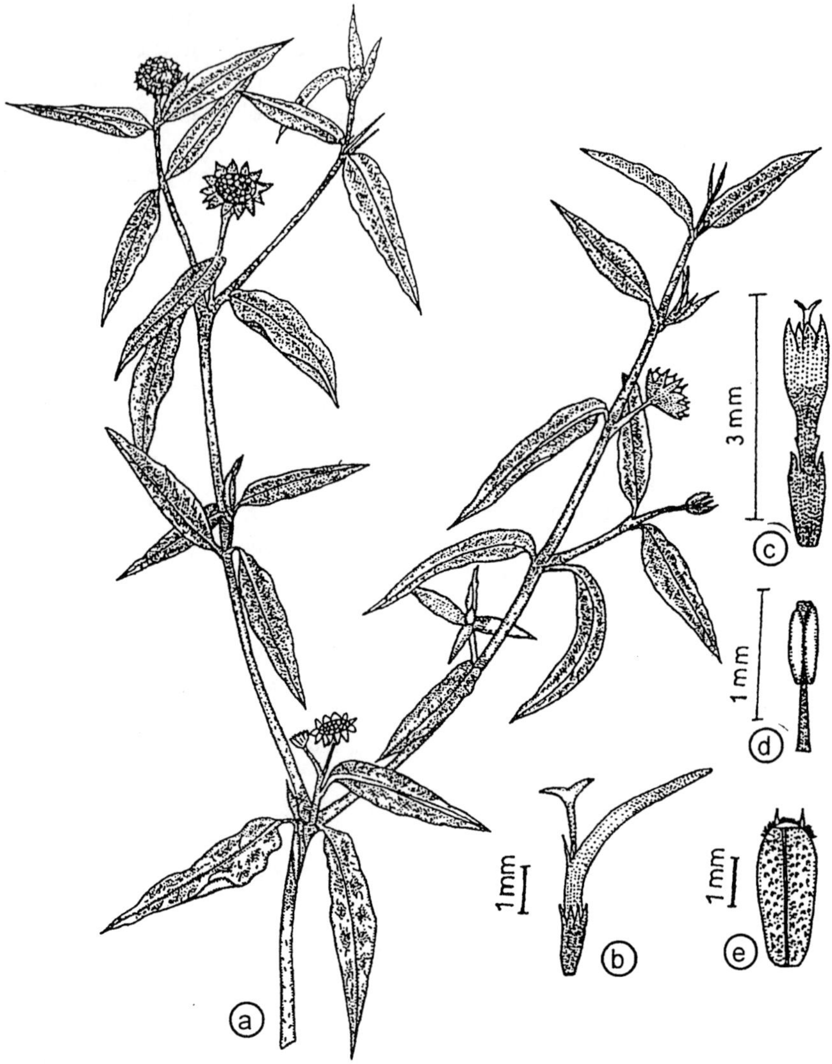

Text fig. 20: Eclipta prostrata a-twig; b-ray floret; c-disc floret; d-stamen; e-achene.

Erect or prostrate, much branched annual herbs; often rooting at lower nodes. Stems 15–60 cm long, strigose with appressed white hairs. Leaves opposite, sessile or subsessile, oblong-lanceolate or ovate-lanceolate, 2–15 × 0.5–2 cm, narrowed to the base, entire or faintly serrate, acute, hispid. Heads 5–10 mm across, solitary or 2–7 together, axillary and terminal, on 0.5–5 cm, long peduncles. Receptacle convex, paleate; palea narrow, enclosing florets. Involucral bracts 2-seriate, ovate-lanceolate, acute, appressed hairy. Marginal florets with white, 2-dentate, 3–6 mm long ligulate corollas. Central florets with tubular, 4–5 fid, white corollas, 4–5 mm long. Achenes oblong-turbinate or cuneate, 2 mm long, compressed, with a narrow wing, covered with warty projections, brown.

Fls. & Frts.: July–February. *Exsiccata*: SUK-54.

Common, in wet situations.

Localities: Kolhapur, Sangli, Miraj, Jaisingpur, Wai.

Local uses: Leaf powder is given in diabetes. It is used as a tonic, and deobstruent in hepatic and spleen enlargement. One full glass of leaf juice is given early in the morning with empty stomach in fever and jaundice for 3 continuous days. Leaves are used with *Lawsonia inermis* and applied externally to stop inflammation and wounds and ulcers in burns. Fresh leaves rubbed on gums to stop toothache.

Elephantopus scaber L. Sp. Pl. 814. 1753; Hook. f. Fl. Brit. India 2: 242. 1881; Cooke, Fl. Pres. Bombay 2: 68. 1958 (Repr. ed.); Shirodkar & Lakshminarasimhan in Singh *et al.* Fl. Maharashtra State Dicot. 2: 208. 2001. HASTIPATA, GOJEEBH.

Scapigerous, annual herbs, 15–60 cm tall; scape usually dichotomously branched near the top, strigose with appressed white hairs. Leaves mostly basal forming a rosette on the ground, obovate-oblong, 10–20 × 4–6 cm, tapering into an obscure petiole, coarsely serrate-dentate, rounded, or subacute at apex, hairy on both surfaces; cauline leaves few, smaller sessile. Heads numerous, sessile, closely packed, forming a large, terminal, compound head 2–4 cm across, subtended by 3, large ovate-cordate, leafy bracts. Involucral bracts 2-seriate; outer ones ovate-lanceolate, 1-nerved, half as long as the inner, the inner ones 3–5 nerved, linear, scarious, cuspidate. Florets 3–5, all similar with tubular, 5-fid, pink purple corollas; limb of the corolla ligule like. Achenes ca 5 mm long, truncate, 10-ribbed, pubescent.

Fls. & Frts.: August–January. *Exsiccata*: SUK-125.

Occasional, under shade of trees.

Localities: Ajara, Amboli, Borbet, Gaganbawada, Radhanagari, Vaibahvwadi, Vasota.

Local uses: Leaves in combination with *Sauromatum venosum* tuber, castor oil and fumes are given in piles. Decoction of roots and leaves are given in dysuria. Leaf paste is applied externally in inflammated swellings. Plant is used to remove foul taste from the mouth; useful in all poisoning from the bites of nails of animals; good in diseases of the blood and heart.

Emilia sonchifolia (L.) DC. in Wight Contrib. Bot. Ind. 24. 1834; Hook. f. Fl. Brit. India 3: 336. 1881; Cooke, Fl. Pres. Bombay 2: 107. 1958 (Repr. ed.); Shirodkar & Lakshminarasimhan in Singh *et al.* Fl. Maharashtra State Dicot. 2:

208. 2001. *Cacalia sonchifolia* L. Sp. Pl. 836. 1753. DHAMAPAN, DADAMANDI.

Erect or decumbent, more or less glabrous herbs, 12–40 cm tall. Leaves alternate, mostly near the base of stem, very variable; the lower petioled, lyrate, obovate or spathulate, 6–15 cm long, entire or toothed; the cauline more or less amplexicaule and auricled at base, acute. Heads 1–1.5 cm long, few in lax corymbs. Receptacle flat, naked. Involucral bracts 1-seriate, more or less connate, glabrous or nearly so, linear-oblong, acute, margins scarious. Florets all bisexual, fertile with tubular 5-fid, pink or purple, 6–8 mm long corollas. Achenes oblong, obscurely 5-gonous, 2.5–3 mm long, brown, scabrous on angels. Pappus hairs copious, white, soft as long as the corolla.

Fls. & Frts.: August–December. *Exsiccata*: SUK-1681.
Common, weed of cultivated fields and on wastelands.
Localities: Kagal, Satara, Shivaji University campus, Kolhapur, Wai.

Local uses: Leaves are applied externally on wounds. Leaf decoction is given to cure sore throat. Leaf decoction is given as febrifuge.

Glossocardia bosvallea (L. f.) DC. Wight Contrib. Bot. Ind. 19. 1834; Shirodkar & Lakshminarasimhan in Singh *et al.* Fl. Maharashtra State Dicot. 2: 213. 2001. *Verbasina bosvallea* L. f. Suppl. 379. 1781. *Glossocardia linearifolia* Cass. in Dict. Sci. Nat. 19: 62. 1821; Hook. f. Fl. Brit. India 3: 308. 1881; Cooke, Fl. Pres. Bombay 2: 100. 1958 (Repr. ed.). PITTAPAPDA.

Prostrate, suberect or erect herbs; stems numerous spreading from the central root stock, 5–20 cm long, grooved, glabrous or with few weak hairs. Leaves alternate, pinnatisect or decompound, 1–5 cm long; segments linear, apiculate. Heads axillary and terminal, 5 mm across, ca 1–1.5 cm long, on long or short peduncles. Receptacle flat with few linear paleae. Involucral bracts 1–2 seriate, outer ones ovate-lanceolate, acute, margin membranous. Marginal florets 1-seriate, 3–8 female, with ligulate, 2-fid, yellow corollas. Central florets few, bisexual with tubular 4–5-fid, yellowish corollas. Achenes narrowly oblong, 4–10 mm long, brown, flattened, densely hairy, ciliate. Pappus of 2 awns.

Fls. & Frts.: August–December. *Exsiccata*: SUK-138.
Common, in open places.
Localities: Babu Jamal, Hatkanagale, Kagal, Kolhapur, Ramling, Tasgaon, Wai.

Local uses: Roots and leaves are given in fever. Leaf paste is applied externally in fungal infection. It is given orally in menstrual and uterine complaints.

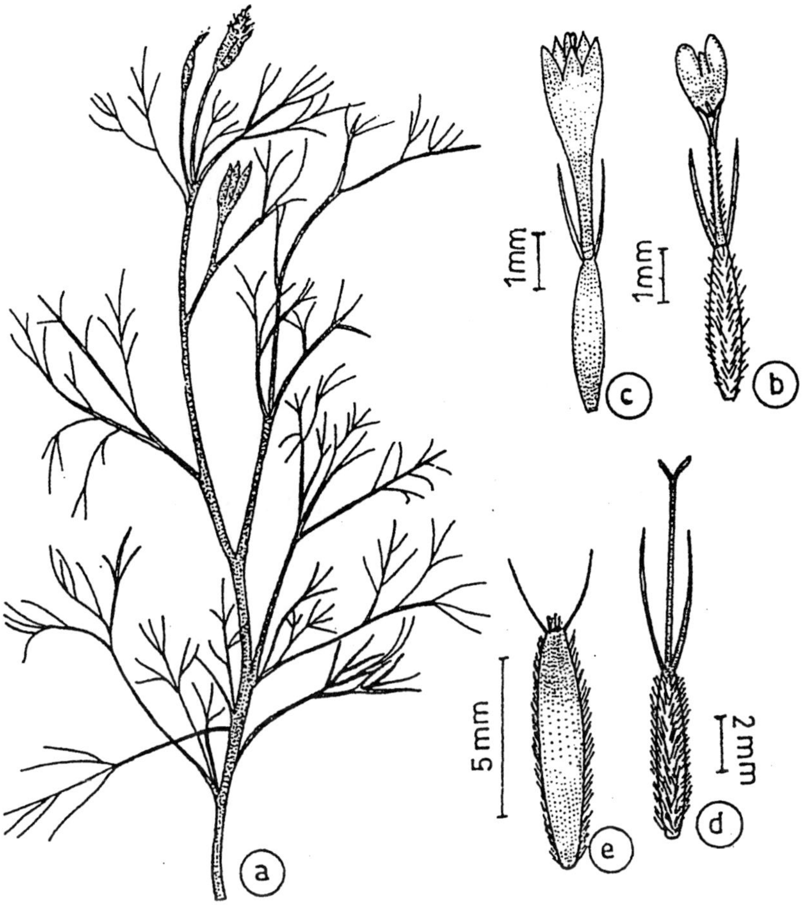

Text fig. 21: Glossocardia bosvallea a-twig; b-ray floret; c-disk floret; d-carpel; e-achene.

Grangea maderaspatana (L.) Poir. in Lam. Encycl. (Suppl. 2.) 825. 1812; Hook. f. Fl. Brit. India 3: 247. 1881; Cooke, Fl. Pres. Bombay 2: 72. 1958 (Repr. ed.); Shirodkar & Lakshminarasimhan in Singh *et al.* Fl. Maharashtra State Dicot. 2: 218. 2001. *Artemisa maderspatana* L. Sp. Pl. 849. 1753. MASHIPATRI.

Prostrate, annual herbs; stems many, spreading from the center, 10–30 cm long, pubescent. Leaves alternate, sessile, pinnatifid, 2–5 cm long, lobes opposite or suboppsite, the terminal lobe the largest, all coarsely serrate-dentate, pubescent. Heads globose, 4–8 mm in diam., solitary or 2-nate on short leaf opposed peduncles. Receptacle convex, naked. Involucral bracts few seriate, elliptic, obtuse, dense pubescent. Marginal florets female, with filiform, 2-fid yellow corollas. Central florets bisexual, with tubular 4–5 fid, yellow corollas. Anther bases obtuse. Style arms flattened, obtuse. Achenes subterete, 2–3 mm long, somewhat compressed, glandular. Pappus a short ring of scales.

Fls. & Frts.: December–May. *Exsiccata*: SUK-1680.
Frequent, in marshes around tanks, along riverbanks.
Localities: Ajara, Amba, Amboli, Chandgad, Mahabaleshwar, Radhanagari.

Local uses: Leaf juice is applied as an instillation for earache. Leaf juice is given in bronchitis and in cough. It is applied externally to reduce perspiration. Leaf paste is applied in rheumatism in the knee joints. Juice is used with *Caesalpinia bonduc* and seeds of *Holarrhena pubescens* in fever.

Lagascea mollis Cav. in Anal. Sci. Nat. 6: 332, t. 44. 1803; Hook. f. Fl. Brit. India 3: 302. 1881; Cooke, Fl. Pres. Bombay 2: 93. 1958 (Repr. ed.); Shirodkar & Lakshminarasimhan in Singh *et al.* Fl. Maharashtra State Dicot. 2: 222. 2001. THARVAD.

Much branched, annual herbs, 20–60 cm tall; stem weak, glabrescent; young parts densely gray hairy. Leaves alternate, rhomboid-ovate, 3–5 × 2.5–3.5 cm, crenate, acute, or acuminate, cuneate at base, coarsely hairy above, silky beneath. Heads compound, 2–2.5 cm across, subtended by an involucre of leafy bracts which are ovate-acute pubescent, containing many simple heads. Each simple head consisting of an involucre and a single floret. Involucral bracts 5, lanceolate, connate into a tube for about half their length enclosing the ovary, ciliate. Florets with tubular, 5-fid, white corollas. Anther bases sagittate. Style arms long hairy. Achenes cuneate, 2 mm long, smooth. Pappus a short fimbriate cup.

Fls. & Frts.: July–December. *Exsiccata*: SUK-1679.
Common, in waste lands as weed.
Localities: Hatkanagale, Karad road, Kolhapur, Sangli.

Local uses: Leaf juice is dropped into nostrils to relieve headache. Paste of leaf applied externally on wounds.

Lamprachaenium microcephalum (Dalz.) Benth. in Benth. & Hook. f. Gen. Pl. 2: 226. 1873; Hook. f. Fl. Brit. India 3: 229. 1881; Cooke, Fl. Pres. Bombay 2: 64. 1958 (Repr. ed.); Shirodkar & Lakshminarasimhan in Singh *et al.* Fl. Maharashtra State Dicot. 2: 225. 2001. *Decaneurum microcephalum* Dalz. in Hook. Kew. J. Bot. 3: 231. 1851. BRAMHDANDI.

Erect herbs 30–60 cm high; stem hairy glandular, often tinged with purple. Leaves alternate, elliptic, 5–7.5 × 2.5–3.5 cm, acuminate, sparsely clothed with short bristly hairs above, densely covered with white woolly tomentum beneath, distantly bristle serrate, tapering at the base. Heads small, 4–8 mm across, numerous on hairy slender peduncles. Involucral bracts many seriate, dry, white tomentose on the back. Receptacle flat, naked. Corollas purple, all tubular, equal, regular; tube slender, narrowly 5-fid. Anthers obtuse auricled. Style arm subulate hairy. Pappus reddish, equaling the corolla tube, rigid. Achenes obovoid, 2–2.5 mm long, turgid, slightly compressed, smooth shining.

Fls. & Frts.: October–November. *Exsiccata*: SUK-32.

Common, in waste places. Endemic to India.

Localities: Radhanagari.

Local uses: Root is externally used in various skin diseases, itching and in leucoderma. Leaf is applied externally in inflammation and insect bite.

Launaea procumbens (Roxb.) Ramayya & Rajagopal in Kew Bull. 23: 465. 1969; Shirodkar & Lakshminarasimhan in Singh *et al.* Fl. Maharashtra State Dicot. 2: 226. 2001. *Prenanthes procumbens* Roxb. Fl. Ind. 3: 404. 1832. *Launaea nudicaulis* sensu auct. plur. (India) non *L. nudicaulis* (L.). Hook. f. *Sensu. Stricto.* Fl. Brit. India 3: 416. 1881; Cooke, Fl. Pres. Bombay 2: 122. 1958 (Repr. ed.). PATHARI.

Procumbent herbs; stems much branched, 15–30 cm or more long, with yellow juice. Leaves in a basal rosette, obovate-oblong, 8–24 × 3.5–8 cm, narrowed at base, dentate, pinnately lobed; lobes obtuse, spinulose-dentate with cartilaginous teeth, glabrous. Heads 1.5–2 cm long on short bracteate peduncles, solitary or clustered, along terminal, subracemose inflorescence. Involucral bracts 3-seriate; outer ones ovate, inner ones lanceolate, all florets with pale yellow, 2–3 dentate, 1.5–2.5 cm long ligules. Achenes oblong, 3–4 mm long, rugulose, black, truncate. Pappus white.

Fls. & Frts.: August–October. *Exsiccata*: SUK-1678.

Common, in open situations, in waste places and weed in cultivated field.

Localities: Kolhapur, Panhala.

Local uses: It is given in chronic skin diseases. Leaf decoction is given as lactagogue.

Parthenium hysterophorus L. Sp. Pl. 988. 1753; Shirodkar & Lakshminarasimhan in Singh *et al.* Fl. Maharashtra State Dicot. 2: 227. 2001. GAJARGAVAT.

Erect, profusely branched, leafy herbs, 0.5–1.5 m tall; stem fistular, striate, pubescent. Leaves sessile, radical and alternate, pinnately divided, 5–20 cm long; lobes again lobulate, entire or broadly serrate, pubescent on both surfaces. Heads 4–5 mm across, peduncled in terminal, lax, corymbose cymes. Involucral bracts often 10, 2-seriate, outer ovate, inner ones suborbicular, larger than the outer, subtending a female floret with one male floret on either side. Marginal florets female, fertile with white, 3-dentate, spreading corolla. Central florets male, with tubular, white, corollas. Achenes compressed, obovate, 0.5–1 mm long, black. Pappus absent.

Fls. & Frts.: September–January. *Exsiccata*: SUK-625.

Native tropical America, naturalized and now has become troublesome weed throughout the region.

Localities: Kolhapur, Satara, Sangali, Wai.

Local uses: Root decoction is given in chronic fever. One to 3 drops are dropped in nose to stop inflammation of sinus.

Sonchus oleraceus L. Sp. Pl. 794. 1753; Hook. f. Fl. Brit. India 3: 414. 1881; Cooke, Fl. Pres. Bombay 2: 119. 1958 (Repr. ed.); Shirodkar & Lakshminarasimhan in Singh *et al.* Fl. Maharashtra State Dicot. 2: 242. 2001. MHATARA.

Erect, annual herbs, 20–90 cm tall; stem fistular, striate, branched upwards. Leaves oblong or obovate-oblong, 10–30 × 2–6 cm, narrowed at base and decurrent into short or long petioles, often runcinate-pinnatifid, upper sessile, ampelexicaule and with spreading acute auricles; lobes acutely dentate. Heads 1.5–2 cm long on 1–5 cm long peduncles, combined into subumbellate corymbs. Involucral bracts 2–3 seriate, lanceolate, acute, fluccose with few gland-bristles. All florets with yellow, 1.5–2 cm long, linear, obscurely dentate, ligular corollas. Achenes rugose, 3-ribbed on faces, up to 3 mm long. Pappus hairs white, 5–8 mm long.

Fls. & Frts.: September–May. *Exsiccata*: SUK-1674.

Common, on wet lands around fields.

Localities: Kolhapur.

Local uses: Gum is taken orally in a dose of 2–4 grains, behaves as an intensely powerful hydrogogue.

Sphaeranthus indicus L. Sp. Pl. 1314. 1753; Hook. f. Fl. Brit. India 3: 275. 1881 p. p.; Cooke, Fl. Pres. Bombay 2: 84. 1958 (Repr. ed.); Shirodkar & Lakshminarasimhan in Singh *et al.* Fl. Maharashtra State Dicot. 2: 242. 2001. GORAKHMUNDI.

Strongly aromatic, procumbent herbs; branches terete, 15–45 cm long, with green, toothed wings. Leaves sessile, obovate to oblanceolate, 1–6 × 0.4–2 cm, decurrent on the stems, serrate dentate, obtuse or subacute, glandular pubescent. Head compound, spherical, 1–1.5 cm in diam., containing several, small, sessile, few flowered simple heads. Simple heads with 10–15, oblong, scarious, glandular pubescent involucral bracts which are ciliate. Marginal florets 5–8, female, with filiform, 2–3-fid, purple corollas. Central florets 2–3, bisexual, with 5-lobed, tubular, dark purple corollas. Ovaries of both florets glandular pubescent. Achenes ovoid-oblong, glabrous or pubescent, minute, pappus absent.

Fls. & Frts.: September–January. *Exsiccata*: SUK-49.
Common, in harvested rice fields.
Localities: Amba, Borbet, Pawas, Ratnagiri.

Local uses: Paste of leaves is used with coconut oil and externally and orally given in skin diseases. Leaf decoction is used in sore throat, cough, bronchitis. Flowers are used as a tonic and cooling agent. Herb is used as laxative, tonic, cures tubercuolus glands, indigestion, bronchitis, diseases of the spleen, elephantiasis, anemia, pain in the uterus and vagina, piles, leucoderma, dysentery hemicrania and looseness of the breasts.

Spilanthus calva DC. in Wight, Contrib. Bot. Ind. 19. 1834; Shirodkar & Lakshminarasimhan in Singh *et al.* Fl. Maharashtra State Dicot. 2: 243. 2001. *S. acmella* auct. non Murr. 1774; Hook. f. Fl. Brit. India 3: 307. 1881; Cooke, Fl. Pres. Bombay 2: 99. 1958 (Repr. ed.). AKKALKARA.

Decumbent, perennial herbs, 15–40 cm tall; rooting from the lower nodes. Leaves opposite, ovate-oblong, 2–6 × 0.6–2.5 cm, narrowed at base into a short petiole, entire or serrate, acute, thinly hairy on both, surfaces. Heads discoid, 5–8 mm across, on slender, 3–12 mm long peduncles. Receptacle columnar with a narrowed top, paleaceous; palea conduplicate, keeled, embracing the central florets, with lobed top and membranous margins. Involucral bracts many seriate, broadly ovate or oblong, obtuse or acute, glabrous, bisexual, yellow, one-seriate; female with yellow, 3-dentate, ligulate corollas. Central florets numerous, bisexual, yellow 5-lobed corollas. Anthers base obtuse. Style arms truncate. Achenes with ciliate margins. Pappus absent.

Fls. & Frts.: September–January. *Exsiccata*: SUK-1699.
Occasional, weed in wet fields and in open lands.
Localities: Amboli, Kolhapur, Lingmalafalls, Radhanagari, Sawantwadi, Tillari.

Local uses: Flower and roots are chewed to induce or provoke salivation. Leaves are given in stomach pains. It is also given as diuretic, flower heads and root are chewed in toothache, paste of root is applied externally in paralysis and given internally as a stimulant and as a brain tonic.

Spilanthus paniculata Wall. ex DC. Prodr. 5: 625. 1836; Shirodkar & Lakshminarasimhan in Singh *et al.* Fl. Maharashtra State Dicot. 2: 244. 2001.

S. acmella var. *paniculata* (DC.) C. B. Cl. Comp. Ind. 139. 1876; Hook. f. Fl. Brit. India 3: 307. 1881. AKKALKHAR, AKKALKARA.

Erect, ascending, annual herbs, 20–46 cm tall; stem and branches striate, more or less pubescent. Leaves opposite, broadly ovate or ovate-lanceolate, 2–5 × 1.5–4 cm, narrowed at base, entire or remotely serrate, acute, glabrous. Heads radiate or discoid, ovoid, 8–12 mm long, on 2–15 cm long peduncles. Receptacle columnar, paleaceous; palea embracing the central florets. Involucral bracts many seriate, ovate-oblong, obtuse. Marginal florets 5–12, with yellow, broadly ovate, ligulate corollas. Central florets many, with yellow, tubular, 5-lobed corollas. Anther base obtuse. Style arms short truncate. Achenes often glandular-verrucose. Pappus bristles subequal.

Fls. & Frts.: September–January. *Exsiccata*: SUK-1698.
Frequent, in wet fields under shade of trees.
Localities: Ichalkaranji, Kolhapur.

Local uses: Flowers and roots are chewed to induce or provoke salivation. Leaves are given in stomach pains. It is also given as diuretic. Roots are chewed in toothache, paste of root is applied externally in paralysis and given internally as a stimulant and as a brain tonic.

Tridax procumbens L. Sp. Pl. 900. 1753; Hook. f. Fl. Brit. India 3: 311. 1881; Cooke, Fl. Pres. Bombay 2: 102. 1958 (Repr. ed.); Shirodkar & Lakshminarasimhan in Singh *et al.* Fl. Maharashtra State Dicot. 2: 247. 2001. DAGADI PALA.

Procumbent herbs; branches spreading, 30–60 cm long, hairy. Leaves opposite, ovate-elliptic, 2–5 × 8–2.5 cm, pinnatifid or deeply incisodentate, acute, clothed on both sides with hairs from bulbous bases. Heads 10–15 mm across, solitary on long slender peduncles. Receptacle flat, paleaceous; palea membranous, linear. Involucral bracts few-seriate; the outer ones ovate, densely hairy; the inner ones membranous, linear, oblong, pubescent. Marginal florets 4–8, with white or yellow, 3-dentate ligules. Central florets many, with tubular 5-lobed yellow corollas. Anther base sagittate with short, acute auricles at base. Achenes oblong, 2–2.5 mm long, densely silky black. Pappus of many, aristate, feathery bristles.

Fls. & Frts.: Throughout year. *Exsiccata*: SUK-81.
Common, weed in waste places and fields.
Localities: Hatkanagale, Ramling, Shivaji University campus.

Local uses: Leaf juice is used to reduce heat from the body. Leaf juice or paste is applied on cut wounds to stop swelling and bleeding from wounds. Leaf juice is given twice a day in urinary complaints.

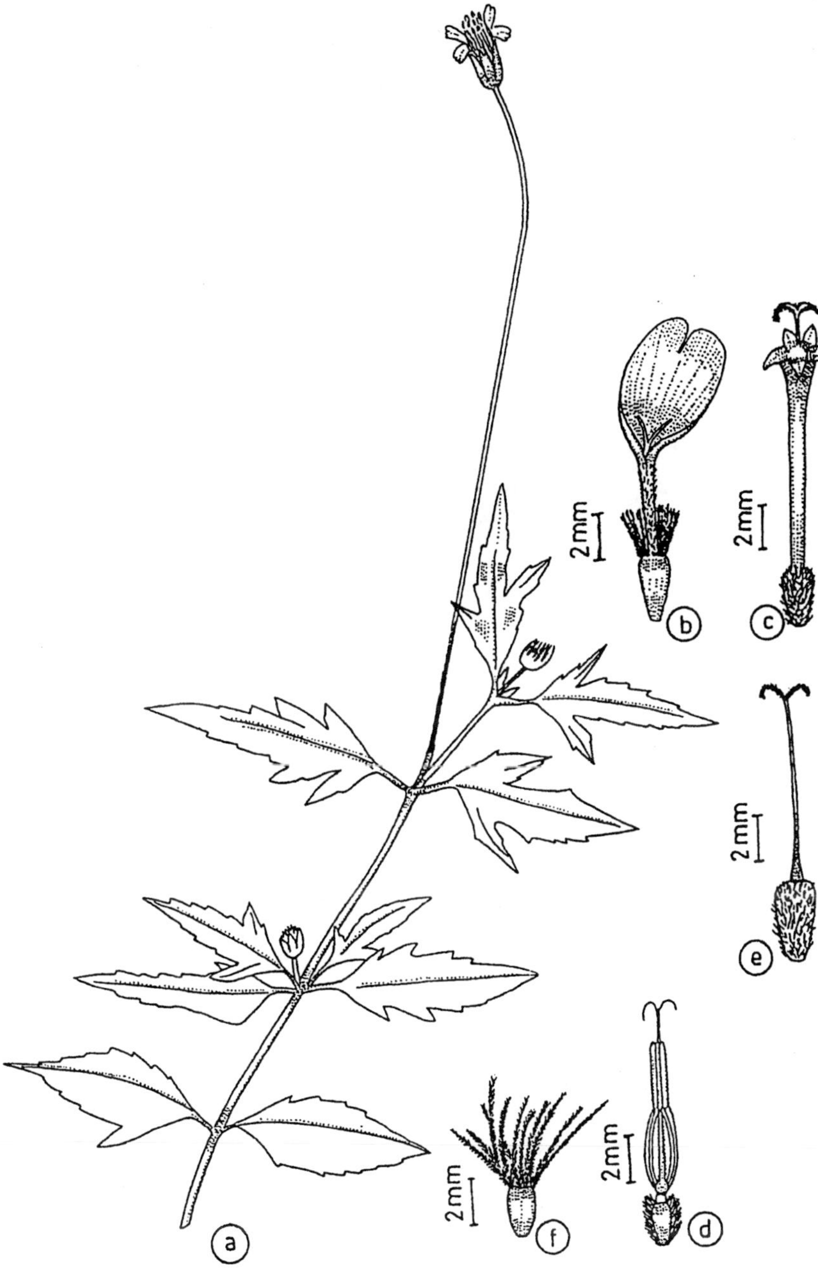

Text fig. 22: Tridax procumbens a-twig; b-ray floret; c-disk floret; d-stamen; e-carpel; f-achene.

Vernonia cinerea (L.) Less. in Linnea 4: 291. 1829; Hook. f. Fl. Brit. India 3: 233. 1881; Cooke, Fl. Pres. Bombay 2: 65. 1958 (Repr. ed.); Shirodkar & Lakshminarasimhan in Singh *et al.* Fl. Maharashtra State Dicot. 2: 248. 2001. *Conyza cinerea* L. Sp. Pl. 862. 1753. SAHADEVI.

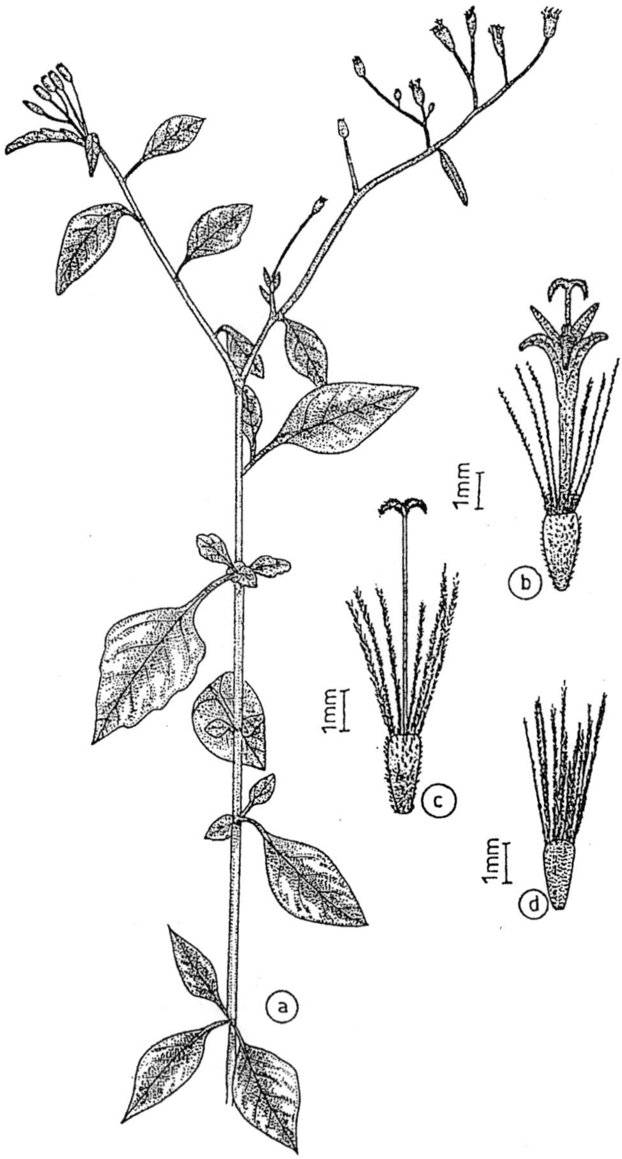

Text fig. 23: Vernonia cinerea a-twig; b-flower; c-carpel; d-achene.

Erect, annual herbs, 15–120 cm tall; stem striate, pubescent. Leaves alternate, variable in shape, ovate-lanceolate, orbicular spathulate, 2–5 × 2.5–3 cm, narrowed at base, entire or shallowly crenate-serrate, acute, shortly mucronate, pubescent. Heads 3–5 mm across, in lax, divaricate, terminal corymbose cymes. Receptacle convex, naked or with few bristles. Involucral bracts many striate, green, the outer ones short, ovate, pubescent, inner ones linear-lanceolate, awned, silky on the back. All florets with pink-purple, tubular, 5-lobed, 1.5–2 mm long corolla. Anther base obtuse. Style arms subulate. Achenes obovate-oblong, 1.5–2 mm long hairy. Pappus bristles in two series.

Fls. & Frts.: July–February. *Exsiccata*: SUK-1.
Common, weed along roadsides, fields and in gardens.
Localities: Shivaji University campus, Kolhapur, Panhala.

Local uses: Leaf juice is given in fever. It is useful in malarial fever. Juice is used to wash ulcers it is said to be very effective. It is given in cancer.

LOBELIACEAE R. Br.

Lobelia nicotianaefolia Roth ex R. & S. Syst. Veg. 5: 47. 1819 (*nicotianaefolia*); C. B. Cl. in Hook. f. Fl. Brit. India 3: 427. 1881; Cooke, Fl. Pres. Bombay 2: 131. 1958 (Repr. ed.); Londhe in Singh *et al.* Fl. Maharashtra State Dicot. 2: 277. 2001. DEVNALA.

Biennial or perennial herbs; stem stout hollow, 1–4 m tall. Leaves alternate, sessile, oblong-lanceolate, 10–25 × 4–6 cm, acute, finely serrulate, glabrous above, glabrous or pubescent beneath, tapering at the base. Flowers in terminal racemes sometimes 15–20 cm long. Calyx tube pubescent or glabrous; lobes linear lanceolate, acuminate, with toothed or ciliate margins. Corolla 2.5–3.5 cm long, much curved pubescent or glabrous, white; lobes 1.2 cm long, linear, 3 usually connate throughout. Filament connate into tube, 12 mm long, curved, dilated at base; anthers dark purple, linear, obtusely pointed, curved, glabrous or with few white hairs on the back, the 2 lower often with a bunch of long white glistening hairs at the tip. Capsule subglobose, 2 cm in diam., opening by 2 valves. Seeds very small, ellipsoid, compressed, yellowish-brown, extremely acrid.

Fls. & Frts.: October–May. *Exsiccata*: SUK-1694.
Common, along Ghats.
Localities: Amba, Gaganbawada, Radhanagari.

Local uses: Dried leaves are chewed and applied externally to remove leech and stop itching. Leaves in small quantity given as a stomachic, diuretic. It is used to remove cough. Leaves are aphrodisiac, stomachic, diuretic, cures diseases of blood, heart, uterus; cure burning sensation causes biliousness and erysipelas.

PLUMBAGINACEAE Juss.

Plumbago zeylanica L. Sp. Pl. 151. 1753; C. B. Cl. in Hook. f. Fl. Brit. India 3: 480. 1882; Cooke, Fl. Pres. Bombay 2: 136. 1958 (Repr. ed.); Londhe in Singh *et al.* Fl. Maharashtra State Dicot. 2: 282. 2001. CHITRAK.

Glabrous, undershrubs, 1–1.5 m tall. Leaves ovate-oblong, 2.5–7.5 × 2–3.5 cm, rounded at base, undulate along margins, acute, glabrous above, white punctate beneath. Flowers in axillary and terminal racemes combined into 20–30 cm long panicles. Calyx glandular-hairy outside; teeth linear subulate, very short. Corolla white with 2–2.5 cm long narrow tube, 1.5–2 cm across limb; lobes oblong, apiculate. Style glabrous, 5 fid. Capsules oblong, 5–6 mm long, acute, enclosed in calyx. Seed solitary, ellipsoid.

Fls. & Frts.: July–October. *Exsiccata*: SUK-1693.
Frequent, in hill forests along gullies, often cultivated.
Localities: Dondewadi, Kolhapur, Tasgaon, Wai.

Local uses: Root powder is mixed with alcohol for 8 days and given in Typhoid. Root is given in small quantity as appetizer. In large doses it is poisonous and may cause abortion. It is also given in menstrual complaints. Root and root bark is used to cure intestinal troubles, dysentery, leucoderma, inflammation, piles, bronchitis, itching, disease of the liver and good in anemia.

PRIMULACEAE Vent.

Anagallis arvensis L. Sp. Pl. 211. 1753; Hook. f. Fl. Brit. India 3: 506. 1882; Cooke, Fl. Pres. Bombay 2: 139. 1958 (Repr. ed.); Londhe in Singh *et al.* Fl. Maharashtra State Dicot. 2: 285. 2001.

Annual or perennial herbs, 10–30 cm tall; stem and branches tetragonous, glabrous. Leaves sessile, opposite, ovate-elliptic, 1–2 × 0.3–1 cm, acute, glabrous, entire, gland dotted. Flowers axillary, solitary; peduncles 1.5–3 cm long, slender, erect in flower, decurved in fruit. Calyx divided nearly to the base, 5-partite; segments narrow, lanceolate, acuminate, slightly shorter than the corolla. Corolla rotate, 5-partite, blue. Stamens 5, filaments filiform, villous. Ovary globose; ovules many; style simple; stigma obtuse. Capsule globose, 3 mm wide, operculately dehiscent. Seeds many, minute, plano-convex, trigonous.

Fls. & Frts.: August–May. *Exsiccata*: SUK-1683.
Common, in cultivated fields.
Localities: Amba, Amboli, Borbet, Gaganbawada, Panhala, Radhanagari.

Local uses: It causes vomiting and loose motion, reduces swelling of extremities in liver diseases. It is to be used only occasionally. If used properly it is very useful to modify the course of rheumatoid arthritis. Herb cures inflammations, sores, pain in the liver, kidney, opthalmia keratitis, leprosy, hydrophobia, improves eyesight, juice as an errhine removes bad smell and secretion.

MYRSINACEAE R. Br.

Ardisia solanacea Roxb. Pl. Cor. 1: 27, t. 27. 1795; Cooke, Fl. Pres. Bombay 2: 146. 1958 (Repr. ed.); Londhe in Singh *et al.* Fl. Maharashtra State Dicot. 2: 286. 2001. *A. humilis* auct. non Vahl, 1794; C. B. Cl. in Hook. f. Fl. Brit. India 3: 529. 1882. BUGDI, DIKNA.

Shrubs or small trees, 1–2 m tall. Leaves alternate, sessile or petiolate, obovate or lanceolate, 4–8 × 2.5–4 cm, obtuse or shortly acuminate, glabrous, punctate, base gradually tapering and decurrent on the petioles. Flowers pink in axillary corymbs. Calyx divided to the base, 5-lobed; lobes rotund-ovate, 3 mm long, densely gland dotted and with membranous ciliate margins. Corolla rose-coloured, free; lobes ovate-elliptic, 8–9 mm in length. Stamens 5, shorter than the petals; filaments very short, anthers large, lanceolate, acute. Ovary ovoid, pyramidal or subglobose; style short to elongate; stigma minute. Drupes depressed globose, 0.7–1.3 cm wide, purple-black.

Fls. & Frts.: March–January. *Exsiccata*: SUK-11.

Frequent, in semi evergreen forests.

Localities: Amboli, Dajipur, Mahabaleshwar, Radhanagari, Tillari.

Local uses: Root paste is applied externally in wounds and ulcers. Decoction is given in fever. It is given as a stimulant.

Embelia basaal (R. & S.) A. DC. Trans. Linn. Soc. 17: 131. 1837; Londhe in Singh *et al.* Fl. Maharashtra State Dicot. 2: 287. 2001. *Ardisia basaal* R. & S. Syst. Veg. 4: 517. 1819. *E. tsjerium-cottam* sensu Sant. in Rec. Bot. Surv. India. 16 (1): 141. 1967 (3rd Rev. ed.) p. p. quoad. Syn. *E. basaal* non (R. & S.) A. DC. 1837. *E. robusta.* auct. plur. non Roxb. 1820; C. B. Cl. in Hook. f. Fl. Brit. India 3: 515. 1882 p. p. quoad. *E. basaal*; Cooke, Fl. Pres. Bombay 2: 143. 1958 (Repr. ed.). AMBALDANGE, AMBATI, ELANG, VOWDING.

Rambling shrubs or small trees, up to 3 m high. Leaves broadly elliptic, 3–20 × 2–8 cm, shortly acuminate, base rounded or subacute, gland dotted, margins entire, or irregularly toothed, glabrous above, paler, reddish and rusty pubescent on nerves beneath. Flowers pale greenish-yellow, fragrant, in axillary-rusty puberulous racemes shorter than the leaves. Calyx 1.5 mm long, free, or slightly connate at base, 4–5 fid; lobes ovate, subacute, gland dotted. Petals 5, free, elliptic-oblong, 2–3 mm long, subacute, puberulous, densely papillose inside. Stamens 4–5; filaments adnate to petals; anthers not apiculate. Fruits globose, apiculate with style, striated, 4 mm across, ripens red.

Fls. & Frts.: Throughout the year. *Exsiccata*: SUK-88.
Common, in semi evergreen and deciduous forest.
Localities: Amba, Amboli, Dajipur, Gaganbawada, Mahabaleshwar, Radhanagari.

Local uses: Root paste is applied on any kind of ulcers, chronic wounds, decoction is given to reduce heat from the body. Fruits are placed on burning coal and smoke is given to women after pregnancy as a laxative, antiseptic and fruits are given twice a day for a child to expel worms.

Embelia drupacea (Dennst.) M.R. & S.M. Almeida in J. Bombay Nat. Hist. Soc. 90. 426. (1993) 1994; Londhe in Singh *et al.* Fl. Maharashtra State Dicot. 2: 287. 2001. *Pothos drupacea* Dennst. Schlussel 15, 24, 33. 1818. *Embelia viridiflora* Scheft. Myrs. Arch. Ind. 45. 1867; C. B. Cl. in Hook. f. Fl. Brit. India 3: 516. 1882; Cooke, Fl. Pres. Bombay 2: 145. 1958 (Repr. ed.). AMBAT.

Extensive climbing shrubs; branches lenticellate, warted. Leaves elliptic-oblong, coriaceous, 4–22 × 3–8 cm, punctate at both sides, obtuse or subacute, margins narrowly recurved. Flowers 4-merous, in short cylindric racemes from the axils of fallen leaves. Calyx free, 2.5–1.5 mm long; segments 4, triangular-ovate, acute, gland dotted, margins glandular-ciliate. Petals free, elliptic oblong, 3–4 mm in length, obtuse, gland dotted, papillose inside. Stamens 4, exserted; filaments connate with the petals; anthers oblong, rounded or slightly apiculate at apex, glandular at the back with the dark coloured glands. Fruit globose, 6–8 mm wide, tipped with persistent calyx, smooth, red when ripe.

Fls. & Frts.: January–June. *Exsiccata*: SUK-630.
Common, on margins of plateaus of higher altitude.
Localities: Kas, Lingmala falls, Thoseghar.

Local uses: Fruit are placed on burning coal and smoke is given to women after pregnancy as a laxative, antiseptic and fruits are given twice a day for a child to expel worms.

Embelia ribes Burm. f. Fl. Ind. 62, t. 23. 1768; C. B. Cl. in Hook. f. Fl. Brit. India 3: 513. 1882; Cooke, Fl. Pres. Bombay 2: 143. 1958 (Repr. ed.); Londhe in Singh *et al.* Fl. Maharashtra State Dicot. 2: 289. 2001. MURUDVEL, VAIVARANG, VAVDING.

Large scandent shrubs. Leaves elliptic lanceolate, 4–10 × 2–4 cm, shortly and obtusely acuminate, entire, glabrous, shining above, paler or silvery beneath, gland dotted, base rounded or acute. Flowers 5-merous, greenish-yellow, small, in lax terminal panicled racemes. Calyx connate;

segment 5, free, elliptic, 3–4 mm, subobtuse, pubescent on both sides. Petals 5, greenish-yellow, free, 3–4 mm long, elliptic subobtuse, pubescent on both sides. Stamens 5, shorter than petals, erect, filaments inserted little below the middle of the petals. Fruit globose, 5 mm wide, black when ripe, tipped with persistent style, warty, one seeded.

Fls. & Frts.: February–June. *Exsiccata*: SUK-322.
Very rare, in moist deciduous or semi evergreen forests.
Localities: Amba, Gajapur, Udgiri, Vasota.

Local uses: Root paste is applied on any kind of ulcers, chronic wounds, decoction is given to reduce heat from the body. Fruits are placed on burning coal and smoke is given to women after pregnancy as a laxative, antiseptic and fruits are given twice a day for a child to expel worms. Fruit is good appetizer, carminative, anthelmintic, alexiteric, laxative, alternative, cures tumors, ascites, bronchitis, mental diseases, diseases of the heart, urinary discharges, jaundice and hemicrania.

Maesa indica (Roxb.) A. DC. in Trans. Linn. Soc. London 17: 134. 1834; C. B. Cl. in Hook. f. Fl. Brit. India 3: 509. 1882; Cooke, Fl. Pres. Bombay 2: 141. 1958 (Repr. ed.). Londhe in Singh *et al.* Fl. Maharashtra State Dicot. 2: 289. 2001. *Baeobotrys indica* Roxb. Fl. Ind. 2: 230. 1824. ATKI.

Shrubs, 1.5–5 m tall. Leaves ovate oblong or elliptic lanceolate, 5–12 × 3–10 cm, acute or acuminate, base acute, margin serrate, dentate, thin, glabrous and shining above, glabrous or with few scattered hairs on the nerves beneath. Flowers very small, faintly fragrant, pale white, numerous in compound, panicled, glabrous racemes. Calyx 1.5 mm long, tube divided ½ way down; lobes 5; rotund-ovate, obtuse, glabrous on the back, marked with dark lines; tube enlarged in fruit, enclosing berry, rugose and obscurely ribbed outside. Corolla 2.5 mm long, white, divided to the middle; lobes-5, rotund-ovate, abruptly or shortly contracted at base, rotate. Berry globose, creamy white, 3 mm in diam., covered almost to the apex by the persistent calyx and tipped with short style. Seeds black.

Fls. & Frts.: Throughout the year. *Exsiccata*: SUK-47.
Common, in moist deciduous forests.
Localities: Amba, Amboli, Chandgad, Gaganbawada, Mahabaleshwar, Radhanagari.

Local uses: Same as *E. ribes*. Root is given in syphilis, fruit is considered anthelmintic.

SAPOTACEAE Juss.

Madhuca latifolia (Roxb.) Chev. in Rev. Bot. Appl. 23: 149. 1943. *Bassia latifolia* Roxb. Pl. Cor. 1: 20. 1795; C. B. Cl. in Hook. f. Fl. Brit. India 3: 544.

1882; Cooke, Fl. Pres. Bombay 2: 152. 1958 (Repr. ed.); Pradhan in Singh *et al.* Fl. Maharashtra State Dicot. 2: 293. 2001. MAHOA, MOHVA.

Large, deciduous trees, 15–30 m tall; sap milky. Leaves alternate or subopposite, often clustered at the ends of branchlets, elliptic to ovate, 7.5–15 × 4–7.5 cm, rounded or acute at base, entire and undulate, shortly acuminate, tomentose when young, glabrous at maturity. Flowers in dense fascicles, terminal or from axils of fallen leaves. Calyx divided nearly to the base; segments ovate, acute, rusty tomentose. Corolla cream coloured, 13–16 mm long; tube 7–8 mm long; lobes 7–14, ovate lanceolate, acute. Stamens 20–30; anthers in 3 series, acuminate, hairy on the back. Ovary hirsute. Fruits ovoid, 2–3.5 cm long, greenish tomentose, 1–3 seeded. Seeds ovoid-ellipsoid, polished brown.

Fls. & Frts.: February–May. *Exsiccata*: SUK-1640.
Infrequent, often cultivated in the regions.
Localities: Katyayani, Panhala.

Local uses: Spirit obtained from the flowers is used in rheumatism, used to wash wounds and ulcers, in sore throat and in mouth ulcers. Plant is used to cure biliousness, burning sensation, ulcers, fatigue, bark heals wounds and ulcers, cures leprosy; used in fracture, hastens suppuration, cures rheumatism, flower is used in diseases of heart, cures burning sensation, biliousness, ear complaints. Fruit is used in bronchitis, fever and burning sensation.

Manilkara hexandra (Roxb.) Dub. in Ann. Mus. Col. Mars. 3, 3: 9, f. 2. 1915; Pradhan in Singh *et al.* Fl. Maharashtra State Dicot. 2: 294. 2001. *Mimusops hexandra* Roxb. Pl. Cor. 1: 16, t. 15. 1798; C. B. Cl. in Hook. f. Fl. Brit. India 3: 549. 1882; Cooke, Fl. Pres. Bombay 2: 155. 1958 (Repr. ed.). KHIRNI, RAJANI.

Trees, evergreen, 15–18 m tall. Leaves obovate-oblong, 4.5–10 × 2.5–5 cm, rounded or emarginate at apex, acute at base, dark glossy green above. Flowers whitish, solitary, axillary, or in fascicles. Calyx 4 mm long, rusty tomentose outside; segments 6, reflexed, ovate, subacute. Corolla tube 7 mm long; lobes usually 18, in 3 series, the 6 inner oblanceolate, shortly clawed, outer 12 linear acute. Stamens 6; filaments as long as anthers; staminodes 6, alternate, bifid or denticulate at apex, glabrous. Ovary silky pubescent, 12-celled. Berries 1-seeded, reddish-yellow when ripe. Seeds ovoid, reddish-brown, shining.

Fls. & Frts.: September–November. *Exsiccata*: SUK-7.
Common, in dry deciduous and moist deciduous forests.
Localities: Amba, Satara.

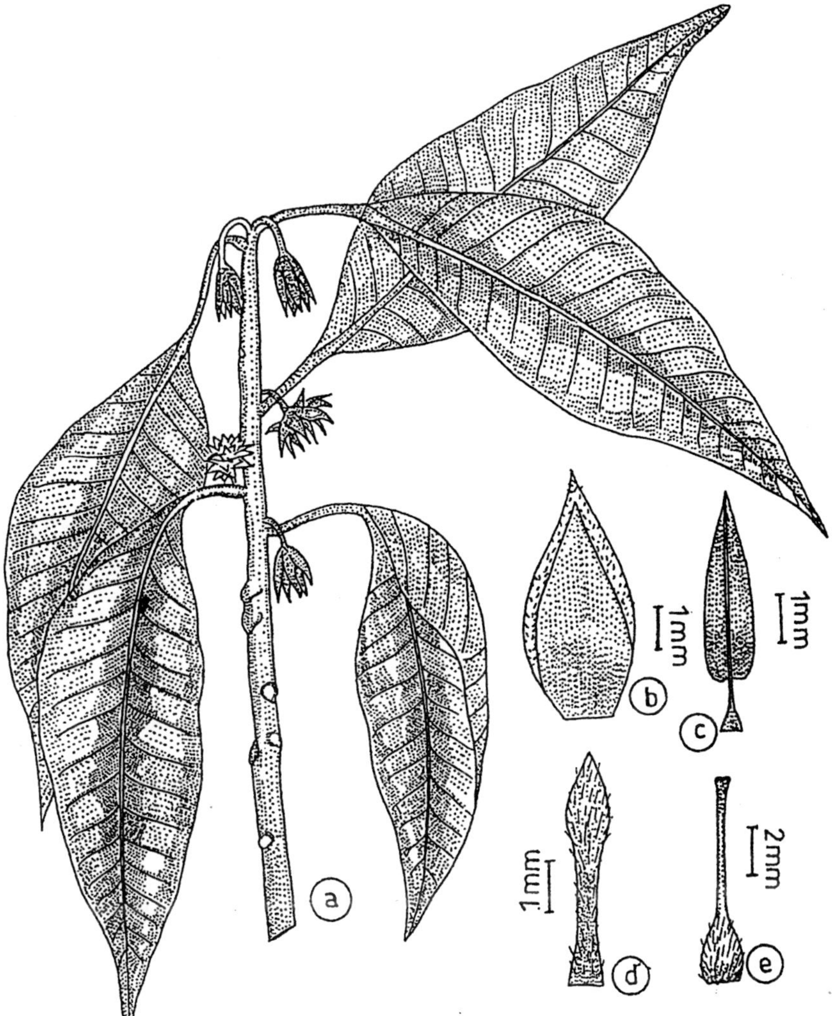

Text fig. 24: Manilkara hexandra a-twig; b-tepal; c-stamen; d-staminode; e-carpel.

Local uses: Bark decoction is given in chronic fever. Bark decoction is used as fomentation to fix loose teeth complaints. Unripe fruits are used to prepare a lotion for sores and wounds. Paste of flower is applied externally in headache. Flowers are smoked in asthma. Seed pulp and black pepper mixed together and powder taken orally with water early in the morning is useful in constipation. Bark is used as cardiotonic, stomachic, cures diseases of the gums and teeth. Flower is used to cure diseases of the blood, good for the teeth. Seeds fix loose teeth and an errhine cure trouble in head.

Manilkara zapota (L.) van Royen in Blumea 7: 410. 1953; Pradhan in Singh *et al.* Fl. Maharashtra State Dicot. 2: 296. 2001. *Achras zapota* L. Sp. Pl. 2: 1190. 1753, quoad typum synonymis sloqnei & Plukenetii exclusis. *A. sapota* L. Sp. Pl. ed. 2: 470. 1762; C. B. Cl. in Hook. f. Fl. Brit. India 3: 534. 1882. CHIKKU.

Evergreen trees, 3–10 m tall. Leaves alternate, elliptic-lanceolate, 8–12 × 2.5–3 cm, acute at base, entire, acute or shortly acuminate, shining green above, paler beneath. Flowers 8–10 mm across, solitary axillary. Sepals in two whorls of 3 each, broadly ovate to orbicular, 4–5 mm long, brown hairy outside. Corolla white, urceloate, 8–10 mm long; lobes ovate, half as long as the tube. Stamens alternating with petaloid staminodes. Ovary villous, 10–12-locular. Berries globose or ovoid-ellipsoid, 5–8 cm in diam., rusty brown. Seeds ellipsoid or obovoid, compressed, 1.5–2 cm long, black.

Fls. & Frts.: Throughout year.

Cultivated, native of West Indies and Tropical America.

Local uses: Seeds are given as a diuretic. Bark is used as a tonic and febrifuge.

Mimusops elengi L. Sp. Pl. 349. 1753; C. B. Cl. in Hook. f. Fl. Brit. India 3: 548. 1882; Cooke, Fl. Pres. Bombay 2: 155. 1958 (Repr. ed.); Pradhan in Singh *et al.* Fl. Maharashtra State Dicot. 2: 294. 2001. BAKUL.

Evergreen trees, 5–8 m tall. Leaves alternate, elliptic-lanceolate, 5–12 × 3–5 cm, acute or rounded at base, entire and undulate, shortly acuminate, glabrous. Flowers fragrant, 1.5–2 cm across, solitary or in fascicles of 2–6 from leaf axils. Calyx 8–10 mm long, brown-pubescent outside; segments 8 in 2 series; outer ones ovate-lanceolate; inner ones lanceolate. Corolla white, slightly exceeding the calyx; tube short; lobes linear-oblong, 8–10 mm long, about 24, in 2 series. Stamens 8, alternating with lanceolate-acuminate staminodes. Ovary silky pubescent. Berries ovoid, 2–2.5 cm long, yellow when ripe. Seeds solitary, ovoid, compressed brown.

Fls. & Frts.: January–March. *Exsiccata*: SUK-57.

Common, in semi evergreen forests, often cultivated.

Localities: Amboli, Chandgad, Devghar, Gaganbawada, Mahabaleshwar, Sawantwadi.

Local uses: Bark decoction in given in chronic fever. Bark decoction is used as fomentation to fix loose teeth. Other uses same as *Mimusops hexandra*.

EBENACEAE Guerke

Diospyros ebenum Koen. ex Retz. Phys. Salsk. Handl. 1: 176. 1780; C. B. Cl. in Hook. f. Fl. Brit. India. 3: 558. 1882; Cooke, Fl. Pres. Bombay 2: 163. 1958

(Repr. ed.); Pradhan in Singh *et al.* Fl. Maharashtra State Dicot. 2: 299. 2001. KARE, MALYA.

Trees. Leaves elliptic oblong-lanceolate, 5–15 × 1.8–3.5 cm, shortly and bluntly acuminate, bright green and shining above, paler beneath. Male flowers 3–12 together, subsessile, in shortly peduncled cymes. Calyx 2–4 mm long, infundibuliform, glabrous or hairy; lobes 4, very short, rounded, ciliate. Corolla pale greenish-yellow, tubular 9–10 mm long before expansion; lobes 4, ovate, equaling or slightly shorter than the tube. Stamens many, unequal; filaments pubescent; anthers linear lanceolate, acuminate. Female flowers solitary, larger than males, calyx much longer, deeply divided, lobes 4, ovate acute. Staminodes 8. Ovary glabrous, 8-celled; style 1; stigmas 4. Fruit subglobose, 9–10 mm in diam., calyx lobes persistent, spreading outwards. Seeds 3–8, black.

Fls. & Frts.: February–March. *Exsiccata*: SUK-1638.
Rare, in semi evergreen forests.
Localities: Dajipur, Udgiri.

Local uses: Bark and decoction of leaves is given in dysentery.

Diospyros melanoxylon Roxb. Pl. Cor. 1: 36, t. 46. 1795; C. B. Cl. in Hook. f. Fl. Brit. India 3: 564. 1882; Cooke, Fl. Pres. Bombay 2: 159. 1958 (Repr. ed.); Pradhan in Singh *et al.* Fl. Maharashtra State Dicot. 2: 301. 2001. TAMRAJ, TAMRANG, TEMRU.

Trees, ca 15 m tall. Leaves alternate, elliptic-oblong or elliptic, 5–15 × 2–5 cm, coriaceous, obtuse or subacute, softly tawny-tomentose on both sides when young, glabrous above, pubescent beneath, base acute. Male flowers, 4–6-merous, 3–12 in tawny-tomentose panicled drooping cymes longer than the petioles. Calyx 4–6 mm long, campanulate, hairy; lobes 4–6, shorter than the tube, subacute. Corolla tubular, twice long as the calyx, pubescent outside, glabrous inside; lobes 4–6, broadly ovate, acute, ca 3 mm long. Stamens many; anthers linear-oblong, acuminate. Female flowers longer than the male, solitary, subsessile, 4–5-merous, calyx and corolla as in male. Staminodes 8–10, ovary hairy, 4–8 celled; ovule solitary in each cell; styles 2-fid. Fruit yellow, ovoid globose, 2–3 cm long; fruiting calyx thickly coriaceous, nearly flat, the lobes undulate, margins reflexed. Seeds 2–8, compressed oblong.

Fls. & Frts.: February–April. *Exsiccata*: SUK-1644.
Common, in dry deciduous forests.
Localities: Amba, Kas, Panhala.

Local uses: Bark is used as laxative and given as a tonic. Decoction of bark is given in diarrhoea and dyspepsia as a tonic. Decoction of bark is given to wash the eyes. Fruit is used to cure biliousness and dysentery.

Diospyros montana Roxb. Pl. Cor. 1: 37, t. 48. 1795; C. B. Cl. in Hook. f. Fl. Brit. India 3: 555. 1882; Cooke, Fl. Pres. Bombay 2: 165. 1958 (Repr. ed.); Pradhan in Singh *et al.* Fl. Maharashtra State Dicot. 2: 301. 2001.

Small or medium sized trees. Leaves ovate-oblong or elliptic, 5–10 × 2.5 cm, obtusely acuminate, softly puberulous or tomentose when young, at length glabrous, base rounded or acute, truncate. Male flowers in short few flowered panicles. Calyx 4-lobed, glabrous; lobes broadly ovate, acute, strongly ciliate, much enlarged in fruit. Corolla white, urceolate, 3–4 times longer than the calyx; lobes little shorter then tube. Stamens many in two rows; anthers lanceolate, with long slender awn, glabrous. Female flowers, axillary,· solitary; calyx glabrous; lobes ovate-oblong, obtuse. Corolla as in male. Staminodes 4. Ovary globular, glabrous, 8-celled; ovules solitary in each cell; styles 4, bifid at the apex, supported by persistent calyx. Seeds 2–8.

Fls. & Frts.: January–April. *Exsiccata*: SUK-1665.
Common, in dry, moist deciduous as well as semi evergreen forests.
Localities: Amboli, Kagal, Barki.

Local uses: Fruit paste is applied externally in case of wounds and in boils.

Diospyros peregrina (Gaertn.) Guerke in Engl. & Prantl, Pflanzenf. 4(1): 164. 1891. *Embryopteris peregrina* Gaertn. Fruct. 1: 145, t. 29, f. 2. 1788. Pradhan in Singh *et al.* Fl. Maharashtra State Dicot. 2: 302. 2001. *Diospyros embryopteris* Pers. Syn. 2: 624. 1807; C. B. Cl. in Hook. f. Fl. Brit. India 3: 556. 1882; Cooke, Fl. Pres. Bombay 2: 167. 1958 (Repr. ed.). TEMBURNI.

Trees, 8–15 m tall. Leaves oblong-lanceolate to oblong, 10–25 × 5.5–7 cm, obtuse to subacute, glabrous and shining, base tapering. Male flowers in shortly peduncled fulvous pubescent, racemose axillary, drooping cymes containing 2–7 flowers. Calyx 6 mm long, hairy; lobes 4, ovate, pubescent, margins ciliate. Corolla campanulate, 9 mm long; lobes thick, rounded, pubescent. Stamens many in pairs; filaments hairy; anthers nearly equal, linear with a line of hairs up the middle. Female flowers: larger than the male, 2.5 cm across, solitary;. calyx deeply divided; lobes 4, cordate, obtuse, recurved. Staminodes 1–12, hairy. Ovary hairy, 8-celled; styles 4; stigma lobed and undulate. Fruit large 2.5–6.5 cm in diam., solitary subglobose, yellow, covered with a rusty, easily detachable scurfiness, 8-celled; calyx persistent, reflexed, spreading. Seeds 4–8, flattened, smooth reddish, brown.

Fls. & Frts.: March–May. *Exsiccata*: SUK-1664.
Often, cultivated along roadsides.
Localities: Kolhapur.

Local uses: Bark is used in dysentery and diarrhoea. Infusion of the fruit is used as a gargle in apthae and sore throat. Unripe fruit cures ulcers, ripe fruit cures urinary losses and stone in urinary tract. Flowers and fruits given in hiccough of children.

SYMPLOCACEAE Desf.

Symplocos racemosa Roxb. Fl. Ind. 2: 539. 1832; C. B. Cl. in Hook. f. Fl. Brit. India 3: 576. 1882; Pradhan in Singh *et al.* Fl. Maharashtra State Dicot. 2: 305. 2001. *S. beddomei* C. B. Cl. in Hook. f. op. cit. 3: 582. 1882; Cooke, Fl. Pres. Bombay 2: 169. 1958 (Repr. ed.). LENDA.

Moderate sized trees, 8–10 m tall. Leaves crowded at the ends of branches, elliptic-oblong, 5–15 × 3.5–5 cm, acute, narrowed at base, subcoriaceous, glandular-serrate, crenate or subentire, glabrous, polished and shining above. Flowers sessile or shortly pedicelled, white, fragrant, in pubescent spikes or racemes, 5–10 cm long. Calyx 4 mm long, glabrous; lobes 5, ovate, subacute. Corolla divided to the base, segments scarcely connate; Petals 6 mm long, oblong, obtuse, either 5 in one series or 6–8 in one or two series. Stamens many, in bundles, adnate to base of petals. Ovary inferior, 3-celled, style thick with a ring of hairs and orange colored gland at the base. Fruit ovoid-oblong 1.2 × 0.6 cm, dark blue, surrounded by distinct calyx lobes.

Fls. & Frts.: November–April. *Exsiccata*: SUK-61.
Infrequent, scattered in moist deciduous and semi evergreen forests.
Localities: Amba, Amboli, Gaganbawada, Mahabaleshwar, Radhanagari, Vasota.

Local uses: Bark paste is applied externally to promote the maturation or resolution of stagnant tumors. Decoction is used as a gargle for giving fitness to bleeding and spongy gums. Bark is useful in eye diseases, for spongy gums and bleeding, cures disease of the blood, dysentery, inflammation and good for ulcers in the vagina.

OLEACEAE Hoffm. ex Link

Jasminum auriculatum Vahl, Symb. Bot. 3: 1. 1794; C. B. Cl. in Hook. f. Fl. Brit. India 3: 600. 1882; Cooke, Fl. Pres. Bombay 2: 174. 1958 (Repr. ed.); Pradhan in Singh *et al.* Fl. Maharashtra State Dicot. 2: 307. 2001. JAI.

Scandent or twining shrubs. Leaves opposite, 3-foliolate, the two lateral leaflets very small or absent, the terminal leaflet ovate-elliptic, 2.5–3 × 0.5–2.2 cm, rounded at base, acute or shortly acuminate, glabrous above, pubescent beneath. Flowers in many flowered lax, corymbose cymes. Calyx 3–4 mm long, pubescent, lobes short, obtuse. Corolla white, glabrous, tube

10–12 mm long; lobes 5–7, elliptic-oblong, 6 mm long, acute. Fruits ellipsoid or globose, 4–5 mm in diam., black.

Fls. & Frts.: March–October. *Exsiccata*: SUK-605.

Common, in hill forests forming large bushes.

Localities: Babu Jamal, Kolhapur, Ramling.

Local uses: Leaves are chewed to cure mouth ulcers and to remove bad taste. Leaf decoction with decoction of *T. bellerica*, *T. chebula* and *Emblica officinalis* is used as a gargle for curing mouth ulcers. Flowers are inhaled to stop headache. Flower paste applied externally in skin diseases. Flower decoction is given after the menstrual cycle for continuous 5 days in menstrual irregularities.

Jasminum malabaricum Wight. Ic. t. 1250. 1848; C. B. Cl. in Hook. f. Fl. Brit. India 3: 594. 1882; Cooke, Fl. Pres. Bombay 2: 172. 1958 (Repr. ed.); Pradhan in Singh *et al.* Fl. Maharashtra State Dicot. 2: 309. 2001. KASUR.

Large climbing shrubs. Leaves opposite, broadly ovate, 7–4 × 5–6 cm, acuminate, entire, glabrous, base round or cordate. Flowers white, fragrant in lax trichotomous compound many-flowered terminal cymes. Calyx 12 mm long, pubescent; teeth 5–7, subulate from a slightly broader base, usually reflexed, hairy. Corolla glabrous; tube 10–22 mm long; lobes 6–10, oblong lanceolate, 6–10 mm long, very acute. Berries 1 × 0.6 cm, ovoid, purplish-black.

Fls. & Frts.: March–June. *Exsiccata*: SUK-1697.

Common, in moist deciduous forests.

Localities: Amba, Amboli, Barki, Dajipur, Gaganbawada, Tillari, Udgiri, Vaibhavwadi.

Local uses: Root is given for external application on wounds and ulcers. Root decoction is given in general debility and in menorrhoea. Fresh juice of leaves applied to corms.

Jasminum odoratissimum (L.) Bailey, Man. Cult. Pl. (ed. 2). 789. 1949; Pradhan in Singh *et al.* Fl. Maharashtra State Dicot. 2: 313. 2001. *J. grandiflorum* L. Sp. Pl. ed. 2, 9. 1762; C. B. Cl. in Hook. f. Fl. Brit. India. 3: 603. 1882. JAI.

Scandent or twining shrubs; stem and branches green, grooved, glabrous. Leaves opposite, pinnate; leaflets elliptic or ovate, 0.8–5 × 0.5–2.5 cm, acute at base, acute or acuminate, pubescent on both surfaces and often ciliate. Flowers fragrant in lax terminal cymes. Calyx cup 3 mm long; teeth linear reaching 12 mm long in fruit. Corolla white streaked with purple on the back, tube narrow, 15–18 mm long; lobes oblong-elliptic, acute, about as long as the tube. Fruits globose, 3–4 mm in diam., black.

Fls. & Frts.: March–September. *Exsiccata*: SUK-1675.

Cultivated, in gardens.

Local uses: Paste of flower is applied externally on skin diseases and in headache. Flowers are tied around the back in urinary complaints and menstrual problems. Leaf decoction is used as gargle in toothache. Flowers are used in diseases of head, teeth and eyes, good in toothaches and suppurations of eyes.

Nyctanthes arbor-tristis L. Sp. Pl. 6. 1753; C. B. Cl. in Hook. f. Fl. Brit. India 3: 603. 1882; Cooke, Fl. Pres. Bombay 2: 176. 1958 (Repr. ed.); Pradhan in Singh *et al.* Fl. Maharashtra State Dicot. 2: 311. 2001. PARIJATAK.

Small trees, 2–5 m tall; branches quadrangular; scabrid with stiff whitish hairs. Leaves opposite, ovate, 5–11 × 3–7.5 cm, rounded or cuneate at base, entire or with a few distant teeth in the lower half, acute or acuminate, rough, scabrous above with short bulbous based hairs, densely pubescent beneath with appressed hairs. Flowers fragrant at night, sessile, pedunculate, axillary and terminal paniculate cymes. Calyx narrowly campanulate, hairy, truncate or obscurely toothed; teeth ciliate. Corolla white with orange-red, 6–10 mm long tube; lobes white, unequally obcordate, cuneate as long as the tube. Anthers 2, sessile. Fruits obcordate or orbicular, 1.5–2 cm across, compressed, 2-locular, glabrous.

Fls. & Frts.: Almost throughout the year. *Exsiccata*: SUK-1670. Cultivated.

Local uses: Seeds are soaked in water for 12 hours and paste is prepared and applied on bald to cure skin infection and enhance new growth of hairs on bald area. Leaf decoction or leaf powder is used in ulcers or wounds in cattle. Warm leaves are applied to reduce inflammation. Fresh juice of leaves is given with honey in chronic fever. Juice of leaves is used to cure fevers, bark is used to cure bronchitis. A decoction of the bark, leaves, roots, and flowers is given in excessive diuresis and in enlargement of the spleen; seeds are useful in piles and skin diseases.

Olea dioica Roxb. Fl. Ind. 1: 106. 1832; C. B. Cl. in Hook. f. Fl. Brit. India 3: 612. 1882; Cooke, Fl. Pres. Bombay 2: 129. 1958. (Repr. ed.); Pradhan in Singh *et al.* Fl. Maharashtra State Dicot. 2: 311. 2001. HADAWYA, KARAMB, PARJAMBHUL.

Trees, 3–10 m tall. Leaves elliptic-lanceolate, 7–10 × 3–5 cm, coriaceous, acute or acuminate, entire or distantly and sharply serrate, glabrous. Flowers polygamous, in axillary or leaf opposed panicles. Male flowers; calyx 1 mm long, glabrous; 4-lobed; lobes triangular, acute. Corolla 3 mm long, tube short; lobes 4; 1 mm long, lobes triangular, ovate, subacute. Stamens 2, filaments short; anthers oblong. Hermaphrodite flowers: calyx and corolla as in male; stigma capitate or bifid. Drupes ellipsoid, 0.6–1.5 cm long, often slightly pointed, purple and covered with a whitish powdery bloom when ripe. Seeds ellipsoid, shortly pointed, rugose, glabrous, intensely bitter to the taste.

Fls. & Frts.: January–April. *Exsiccata*: SUK-284.

Common, in moist deciduous semi evergreen forests.

Localities: Amba, Barki, Borbet, Dajipur, Mahabaleshwar.

Local uses: Ash of fruit is mixed with roots of *Hemidesmus indicus* and used on skin diseases. Bark and fruit paste is applied in rheumatism, decoction is used to wash old wounds and given in fever.

SALVADORACEAE

Salvadora persica L. var. **wightiana** (Planch. ex. Thw.) Verdc. in Kew Bull. 19: 150. 1964; Pradhan in Singh *et al.* Fl. Maharashtra State Dicot. 2: 316. 2001. *S. wightiana* Planc. ex Thw. Enum. Pl. Zeyl. 190. 1860. *S. persica* sensu Roxb. Fl. Ind. 1: 389. 1837. non L. 1753; C. B. Cl. in Hook. f. Fl. Brit. India 3: 619. 1882; Cooke, Fl. Pres. Bombay 2: 182. 1958 (Repr. ed.). KHATAU, PILU, PIVLA.

Large shrubs or small trees; branches many, drooping, glabrous, terete, finely striate. Leaves opposite, somewhat fleshy, ovate, 2.5–4 × 1.5–3.5 cm, acute or rounded at base, entire, acute, or obtuse and mucronate, glabrous. Flowers small, in axillary and terminal panicles. Calyx ca 1 mm long, glabrous, cleft half way down; teeth rounded. Corolla 3 mm long, white, divided nearly to the base, lobes oblong 2–2.5 mm long, obtuse, much reflexed. Stamens 4, exerted, shorter than the reflexed lobes. Drupes globose, 3 mm in diam., red or purple, smooth. Seed one, globose.

Fls. & Frts.: November–February. *Exsiccata*: SUK-400.

Often found along sea shore.

Localities: Ratnagiri.

Local uses: Root is employed for the manufacture of toothbrush, it strengthens teeth, gums, keeps them from becoming spongy and improves digestion. Leaves are used as an external application in rheumatism. The fruits are useful in biliousness. Leaves are used as a tonic to the liver, diuretic, analgesic, anthelmintic, useful in nose troubles, in piles, scabies and leucoderma.

APOCYNACEAE Juss.

Aganosma cymosa (Roxb.) G. Don, Gen. Syst. 4: 1837–38; Hook. f. Fl. Brit. India 3: 665. 1882; Cooke, Fl. Pres. Bombay 2: 203. 1958. (Repr. ed.); Pradhan in Singh *et al.* Fl. Maharashtra State Dicot. 2: 317. 2001. *Echites cymosa* Roxb. Fl. Ind. 2: 16. 1832. MALATI.

Climbing shrubs, stem woody. Leaves elliptic or elliptic lanceolate, 5–8 × 2.5–3 cm, acute or acuminate, glabrous above, base acute. Flowers yellowish-white, in dense tomentose cymes. Calyx divided to the base; sepals linear lanceolate, acute, tomentose. Corolla tomentose outside; lower portion to the tube 1 mm long, oblong, tube inflated above, and with longitudinal hairy lines inside behind the anthers; lobes oblong lanceolate, acute. Anthers linear, acuminate, shortly spurred at the base. Carpels of the ovary distinct, hairy at the tip. Follicles 16–20 × 0.5–0.6 cm, linear, cylindric. Seeds elliptic oblong, compressed, coma 3 cm long.

Fls. & Frts.: April–June. *Exsiccata*: SUK-365.
Common, in semi evergreen forests.
Localities: Amboli, Dajipur, Phonda, Radhanagari, Sawantwadi, Tillari.

Local uses: Plant is emetic, anthelmintic. Leaf decoction is useful in bronchitis, leprosy, skin diseases, ulcers, inflammation, purulent discharges from the ear; diseases of mouth. Flower is good in treatment of diseases of the eye. Leaves cure biliousness.

Alstonia scholaris (L.) R. Br. Mem. Wern. Nat. Hist. Soc. 1: 76. 1811; Hook. f. Fl. Brit. India 8: 642. 1882; Cooke, Fl. Pres. Bombay 2: 194. 1958 (Repr. ed.); Pradhan in Singh *et al.* Fl. Maharashtra State Dicot. 2: 318. 2001. *Echites scholaris* L. Mant. Pl. 53. 1767. SATVIN.

Trees, 13–20 m tall. Leaves 4–7 in a whorl, oblong-lanceolate, 10–17 × 2.5–5 cm, obtuse, or shortly, bluntly acuminate, dark green above, pale, or whitish beneath, base tapering. Flowers greenish-white, in umbellately branched many flowered pubescent capitate cymes. Calyx 1–1.5 mm long, pubescent; lobes oblong, obtuse, ciliate. Corolla tube 5–6 mm long, villous; lobes cuneate oblong, rounded or sub-truncate at the apex, pubescent. Disk 0, carpels pubescent. Follicles cylindric, 30–60 × 0.3 cm, pendulous in clusters. Seeds linear oblong, 6 mm long, compressed, rounded with hairy fringed at both ends.

Fls. & Frts.: December–March. *Exsiccata*: SUK-110.
Common, in moist and semi evergreen tracts, often cultivated.
Localities: Amboli, Dajipur, Mahabaleshwar, Sawantwadi, Tillari, Vaibhavwadi.

Local uses: Bark is used to cure wounds and ulcers. Juice of leaves with fresh ginger is administered to women after confinement. Bark decoction is useful in malarial fever. In colic diseases and it is used with bark of *Holarrhena pubescens* and leaves of *Moringa olerifera*.

Anodendron manubriatum Merr. in Philipp. J. Sci. 7: 333. 1912; Pradhan in Singh *et al.* Fl. Maharashtra State Dicot. 2: 319. 2001. *A. paniculatum* (Roxb.)

A. DC. in DC. Prodr. 8: 444. 1844; Hook. f. Fl. Brit. India 3: 668. 1882; Cooke, Fl. Pres. Bombay 2: 204. 1958 (Repr. ed.). LAMTANI.

Climbing shrubs; stems woody. Leaves elliptic-oblong, 8–15.2 × 3–3.5 cm, shortly or abruptly acuminate, thinly coriaceous, glabrous and shining, base usually rounded. Flowers yellow, small, numerous, in large lax trichotomously branched terminal and axillary glabrous, paniculate cymes. Calyx glabrous; lobes ovate, 1.5 mm long, subacute. Corolla tube pubescent inside, very slightly enlarged, lobes linear, 2–3 mm long, obtuse, falcate, with a few scattered white hairs on upper surface. Stamens near the base of corolla tube; filaments short, broad; anthers sagittate, spurs short, divergent. Stigma subsessile, conical pointed. Follicles straight, 10–15 cm long, obtuse, terete, hard, woody. Seeds 8–12, elliptic-lanceolate, 2–3 cm long, compressed beaked, reddish-brown, coma white.

Fls. & Frts.: December–March. *Exsiccata*: SUK-31.

Common, in semi evergreen forests.

Localities: Gajapur, Patgaon, Tillari, Vaibhavwadi.

Local uses: Root paste is applied externally in case of scabby eruptions. Poultice of leaves is applied externally in case of rheumatic inflammation.

Carissa congesta Wight. Ic. t. 1289. 1848; Pradhan in Singh *et al.* Fl. Maharashtra State Dicot. 2: 320. 2001. *C. congesta* auct. non L. 1767 ; Hook. f. Fl. Brit. India 3: 630. 1882, p. p.; Cooke, Fl. Pres. Bombay 2: 186. 1958 (Repr. ed.). KARVAND.

Shrubs; stem with stout horizontal spines, 2.5–3 cm long. Leaves elliptic or obovate, 3–7 × 2.5–5.0 cm, coriaceous, obtuse and mucronate at apex, subacute at base. Flowers white, in terminal corymbose cymes. Calyx pubescent, lobes 5, lanceolate, 2–3 mm long, acute, ciliate. Corolla tube 1.5 cm long, swollen, pubescent at top, lobes 5; oblong-lanceolate, acute, pubescent and ciliolate. Stamens included; filaments very short, anthers linear oblong. Ovary glabrous, 2-celled; stigma penicillate. Berries ellipsoid, or subglobose, 1.5–2.5 cm long, purplish black, smooth, 4 or more seeded.

Fls. & Frts.: June–April. *Exsiccata*: SUK-35.

Common, throughout in dry deciduous forests.

Localities: Ajara, Barki, Mahabaleshwar, Ramling, Vaibhavwadi.

Local uses: Root is used with camphor and limejuice in itching due to insect bite. Fruit is used in bilious complaints, lessens thirst. Root is used as anthelmintic. Fruit is used in biliousness, used as appetizer; diminishes sexual power.

Catharanthus roseus (L.) G. Don, Gen. Syst. 4: 95. 1837; Pradhan in Singh *et al.* Fl. Maharashtra State Dicot. 2: 331. 2001. *Vinca rosea* L. Syst. ed. 10. 944. 1759; C. B. Cl. in Hook. f. Fl. Brit. India 3: 640. 1882. SADAPHULI.

Glabrous, bushy, perennial herbs, 30–90 cm tall. Leaves oblong, 3–6 × 1.5–2.5 cm, obtuse, or rounded at apex, acute at base, glabrous or finely pubescent. Flowers solitary or in pairs, axillary on short pedicels. Calyx 5–6 mm long, teeth lanceolate, acute, hairy. Corolla white, with yellow center or rose pink with dark red center; tube 2.5–3 cm long, hairy near the throat; lobes broadly obovate, cuneate, acute, 1.5–1.7 cm long. Filament short; anthers free from stigma. Carpels 2; ovaries distinct, united by the style, alternating with 2 linear glands. Follicles linear, 2–3 cm long, straight, pointed. Seeds oblong, black.

Fls. & Frts.: Throughout year. *Exsiccata*: SUK-187.
Cultivated, native of West Indies.

Local uses: Decoction of leaves given in diabetes. Decoction of flower is given in leucorrhoea. It is given in chronic febrile illness, hypertension and indigestion.

Holarrhena pubescens (Buch.-Ham.) Wall. ex G. Don, Gen. Syst. 4: 78. 1837; Pradhan in Singh *et al.* Fl. Maharashtra State Dicot. 2: 322. 2001. *Echites pubescens* Buch.-Ham. in Trans. linn. Soc. London 13: 521. 1821. *Holarrhena antidysenterica* (Roxb.) Wall. ex A. DC. in DC. Prodr. 8: 413. 1844; Hook. f. Fl. Brit. India 3: 644. 1882; Cooke, Fl. Pres. Bombay 2: 195. 1958 (Repr. ed.). PANDHRAKUDA

Trees or large shrubs, 3–4 m tall. Leaves broadly ovate to elliptic, 10–20 × 5–11 cm, obtuse, at base, obtusely acuminate. Flowers white, in terminal corymbose cymes. Calyx lobes oblong-lanceolate, acute, ciliate. Corolla puberulous outside; tube 8–12 mm long; lobes oblong, rounded at apex. Stamens included; anthers free. Ovaries distinct; style filiform. Follicles cylindric, ca 15–30 × 0.6–0.8 cm; seeds linear-oblong, 0.8 cm long, coma of brown hairs.

Fls. & Frts.: February–June. *Exsiccata*: SUK-1669.
Common, throughout the state in dry deciduous and semi evergreen forests.

Localities: Ajara, Amba, Atigre, Babu Jamal, Barki, Borbet, Dajipur.

Local uses: Bark is given in dysentery. Bark is used in amoebic dysentery, fever, piles and skin disease. Seeds are used in bleeding piles, fatigue and hallucinations.

Nerium indium Mill. Gard. Dict. ed. 8, n. 2. 1768; Pradhan in Singh *et al.* Fl. Maharashtra State Dicot. 2: 332. 2001. *N. odorum* Soland. in Ait. Hort. Kew 1: 297. 1789; C. B. Cl. in Hook. f. Fl. Brit. India 3: 655. 1882; Cooke, Fl. Pres. Bombay 2: 206. 1958 (Repr. ed.). KANER.

Shrubs reaching 4 m tall. Leaves whorled, elliptic lanceolate, 10–20 × 1–2 cm, narrowed at both ends, entire, shining green above, paler beneath.

Flowers 4–5 cm across in terminal, polychasial cymes. Calyx tubular; 5-partite; sepals lanceolate, 6–8 mm long, acuminate. Corolla white, pink, deep red; tube short, with fimbriate scales near the mouth; lobes ovate-oblong, 1.5–2.5 cm long, overlapping to the right, Stamens inserted near the throat; anther cells long spurred with twisted hairs. Ovaries distinct. Follicles lanceolate, 10–15 cm long, acutely angled. Seeds villous with long terminal coma.

Fls. & Frts.: Throughout the year.

Native of W. Himalaya and Nepal, grown in gardens.

Local uses: Roots are made into paste and applied on cancers and ulcers of penis. It is also used as abortifacient. Decoction of roots is used to reduce swellings. Oil prepared from root is used in skin diseases of a scaly nature and in leprosy.

Plumeria rubra L. Sp. Pl. 209. 1753; Pradhan in Singh *et al.* Fl. Maharashtra State Dicot. 2: 332. 2001. *P. acutifolia* Poir. in Lam. Encycl. Suppl. 2: 667. 1812; C. B. Cl. in Hook. f. Fl. Brit. India. 3: 641. 1882; Cooke, Fl. Pres. Bombay 2: 205. 1958 (Repr. ed.). LAL CHAPHA.

Small trees, 3–5 m tall; stem soft, hollow, leafless for major part of the year; latex ample. Leaves alternate, oblong-lanceolate, 15–30 × 5–12 cm, narrowed at base and passing into short petiole, acuminate, glabrous on both surfaces. Flowers in terminal, peduncled corymbs, fragrant. Calyx 3–8 mm long, deeply divided; sepals suborbicular. Corolla lobes 2–4 × 1.5–3 cm, overlapping to the left. Follicles not seen.

Fls. & Frts.: Throughout the year.

Cultivated.

Local uses: Root decoction is given as a purgative. Branches are tied around the back to induce abortion. Similarly leaf juice is given orally to induce abortion. Milky juice is applied externally in case of rheumatism.

Quirivelia frutescens (L.) M. R. & S. M. Almeida in J. Bombay Nat. Hist. Soc. 90: 427. (1993) 1994; Pradhan in Singh *et al.* Fl. Maharashtra State Dicot. 2: 325. 2001. *Apocynum frutescens* L. Sp. Pl. 213. 1753. *Ichnocarpus frutescens* (L.) R. Br. Mem. Wern. Nat. Hist. Soc. 1: 62. 1811; Hook. f. Fl. Brit. India 3: 669. 1882; Cooke, Fl. Pres. Bombay 2: 205. 1958 (Repr. ed.). KRISHNASARIVA.

Climbing shrubs. Leaves elliptic oblong, 4.5–7.5 × 1.8–3.5 cm, acute or acuminate, glabrous above, paler beneath, base rounded. Flowers greenish-white, numerous, in axillary cymes. Calyx fulvous-hairy, divided ½ way down; lobes ovate, acute, without glands inside. Corolla tube 0.1–0.3 long, middle portion of tube inflated over stamens, upper portion constricted below the lobes; lobes pubescent on the upper side, with white hairs, broad and oblong, slender with twisted acumen which is deflexed in bud and flower. Disk of 5 erect linear lobes. Ovary hairy. Follicles straight, 10–15 cm long. Seeds linear, 1.2–1.8 cm long; coma white, scanty.

Fls. & Frts.: November–December. *Exsiccata*: SUK-1666.
Frequent, in deciduous and semi evergreen forests.
Localities: Panhala, Jotiba, Gajapur, Tillari.

Local uses: Decoction of leaves is applied externally in case of skin eruptions. Roots are given as a blood purifier. Other uses are same as *Hemidesmus indicus*.

Rauvolfia serpentina (L.) Benth. ex Kurz, For. Fl. Brit. Burma 2: 171. 1877; Hook. f. Fl. Brit. India 3: 632. 1882; Cooke. Fl. Pres. Bombay 2: 188. 1958

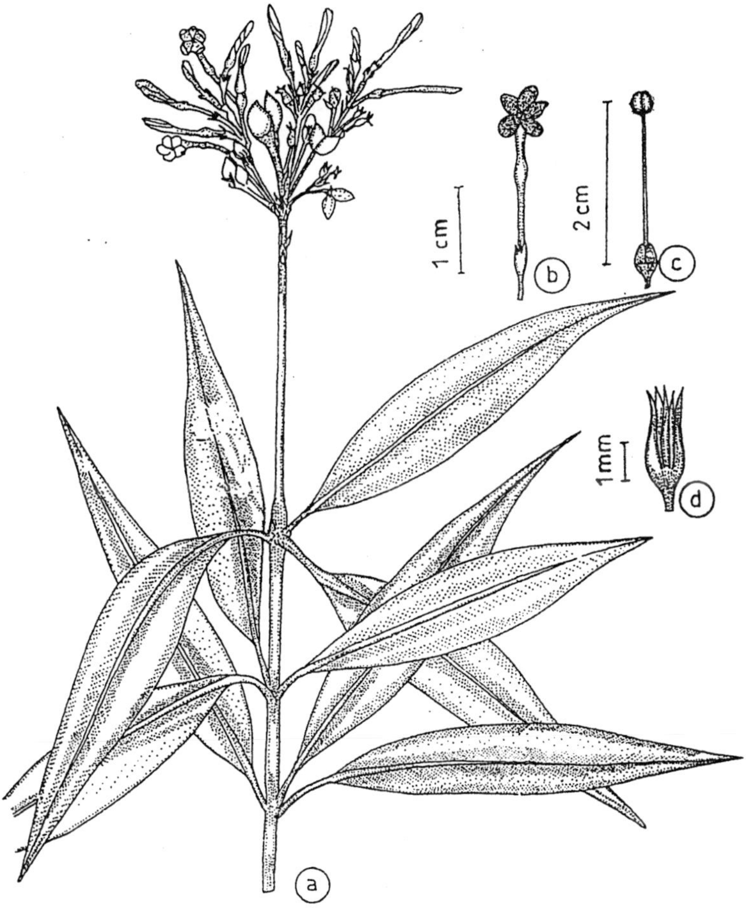

Text fig. 25: Rauvolfia serpentina a-twig; b-flower; c-carpel; d-calyx.

(Repr. ed.); Pradhan in Singh *et al.* Fl. Maharashtra State Dicot. 2: 325. 2001. *Ophioxylon serpentinum* L. Sp. Pl. 1043. 1753. HADKI, SARPAGANDHA.

Shrubs, 30–40 cm tall. Leaves whorled, oblanceolate or obovate, 8–15 × 2.5–5.5 cm, acute, light green above, paler beneath, base tapering. Flowers white or pink, appearing with the young leaves, in lax few-flowered corymbose cymes. Calyx glabrous; segments 5, lanceolate, 1–1.2 cm long, acute. Corolla tube 0.9 cm long; segments 5; elliptic oblong, rounded at the apex. Disk cup-shaped, membranous, obscurely lobed. Drupes of 2 pyrenes, connate at base, mature berries purplish black, pulp light purple.

Fls. & Frts.: March–May. *Exsiccata*: SUK-108.
Infrequent, on lateritic plateaus in Konkan.
Localities: Doddamarg, Pawas, Ratnagiri, Sawantwadi.

Local uses: Root in combination with *Cocculus hirsutus* or *Cissampelos pereira* is given in cholera. In colic it is given with fruits of *Helicters isora*. Processed products of this plant containing various alkaloids are useful in high blood pressure. Powder and decoction of whole plant is very useful in chronic fevers and debility following fever.

Rauvolfia verticillata Baillon in Bull. Soc. Linn. Paris 1: 768. 1888; Pradhan in Singh *et al.* Fl. Maharashtra State Dicot. 2: 326. 2001. *R. densiflora* (Wall.) Benth. ex Hook. f. Fl. Brit. India 3: 633. 1882; Cooke, Fl. Pres. Bombay 2: 189. 1958 (Repr. ed.).

Shrubs, 3.5 m tall. Leaves 3–4, whorled, obovate, 6–10 × 2.5–5 cm, acuminate, glabrous, light-green above, paler beneath, base tapering. Flowers white or pink, appearing with the young leaves, in lax few-flowered corymbose cymes. Calyx segments 4 mm long, lanceolate, acute, and glabrous. Corolla tube 0.9 cm long; tube stout, inflated at the top, throat hairy within; lobes elliptic-oblong; rounded at the apex. Drupes ellipsoid, 1–1.5 cm long, free, brownish purple when ripe.

Fls. & Frts.: March–April. *Exsiccata*: SUK-371.
Frequent, in moist and semi evergreen forests.
Localities: Patgaon, Ajara, Amboli, Dajipur, Sawantwadi, Tillari, Vaibhavwadi.

Local uses: Same as *R. serpentina* and used as adulterant to *R. serpentina*.

Tabernaemontana alternifolia (Roxb.) Nicols. & Suresh in Taxon 35: 354. 1986 & in Nicols. *et al.*, Interpret. van Rheede's Hortus Malab. 57. 1980; Pradhan in Singh *et al.* Fl. Maharashtra State Dicot. 2: 326. 2001. *Ervatamia heyneana* (Wall) Cooke, Fl. Pres. Bombay 2: 134. 1904 [2: 196. 1958 (Repr. ed.)]. *Tabernaemontana heyneana* Wall. Bot. Reg. 7: Sub. t. 1273. 1829; Hook f. Fl. Brit. India 3: 646. 1882. NAGKUDA.

Shrubs or small trees, 2–3 m high. Leaves oblong or oblong-lanceolate, 7–20 × 5–7 cm, coriaceous, shortly acuminate, glabrous, base acute. Flowers in pedunculate many flowered cymes. Calyx 5 mm long, coriaceous, glabrous; lobes 2 mm long, broadly oblong, obtuse with membranous margins. Corolla tube 1.5–2.5 cm long, inflated near the top; limb 2.5 cm across with obtuse crisped lobes overlapping to the right. Follicles 2.5–4.5 × 1.9–1.2 cm, yellow when ripe, subsessile curved, somewhat boat-shaped, usually with short beak which is often recurved and two sharp side ribs. Seeds 8 mm long, surrounded by red pulp.

Fls. & Frts.: March–April. *Exsiccata*: SUK-1676.
Apparently endemic, in moist deciduous forests.
Localities: Amba, Amboli, Gaganbawada, Mahabaleshwar, Phonda, Radhanagari.

Local uses: Stem bark and leaves applied as a poultice to the wounds. Seeds are used as purgative. Seeds are poisonous. Stem bark and leaves applied as a poultice to the wounds. Seeds are used as purgative.

Vallaris solanacea (Roth) O. Ktze. Rev. Gen. Pl. 2: 417. 1891; Pradhan in Singh *et al.* Fl. Maharashtra State Dicot. 2: 326. 2001. *Peltanthera solanacea* Roth, Nov. Pl. Sp. 132. 1821. *Vallaris heynei* Spr. Syst. Veg. 635. 1824; Hook. f. Fl. Brit. India 3: 650. 1882; Cooke, Fl. Pres. Bombay 2: 199. 1958 (Repr. ed.).

Climbing shrubs, stems woody. Leaves elliptic-lanceolate or oblong lanceolate, 5–11 × 2.5–3.5 cm, acuminate, pellucid-dotted, base acute. Flowers white, fragrant, in axillary 3–10 flowered lax cymes. Calyx 2.5–3 mm long, pubescent; lobes ovate-oblong, obtuse or subacute, pubescent. Corolla finely pubescent outside; limbs 1.9 cm across, spreading, divided ½ way down into 5 obovate, obtuse lobes. Anthers acute, forming a sharp cone, exerted; filaments broad linear, connivent, broad, linear, pubescent. Disk with 5 oblong obtuse, ciliate lobes. Ovary hairy; style hairy; stigma pointed, fibrously striate. Seeds ovoid, 1.5 cm long, beaked; coma silvery white.

Fls. & Frts.: December–April. *Exsiccata*: SUK-327.
Common, often cultivated.
Localities: Babu Jamal, Kolhapur.

Local uses: Milky juice is mild irritant, applied to old sores and sinus. It excites some degree of inflammation in them and thereby expedites the process of healing.

Wrightia arborea (Dennst.) Mabb. in Taxon 26: 533. 1977; Pradhan in Singh *et al.* Fl. Maharashtra State Dicot. 2: 327. 2001. *Periploca arborea* Dennst.

Schluss. 13, 23 & 25. 1818 in Forsts. Allg. Teutsch. Gaerten.-Mag. 3: 32, 41, 83. 1818. *Wrightia tomentosa* (Roxb.) R. & S. Veg. ed. Nov. 4: 414. 1819; Hook. f. Fl. Brit. India 3: 653. 1882; Cooke, Fl. Pres. Bombay 2: 199. 1958 (Repr. ed.). BURALICODE.

Trees, 8–9 m tall; young parts tomentose. Leaves elliptic-oblong, 7.5–15.0 × 3.5–7.0 cm, acuminate, tomentose, base acute. Flowers white malodorous, in short, dense, erect terminal corymbose, tomentose cymes. Calyx pubescent outside, glandular, broadly ovate obtuse; lobes oblong, 1.2–1.5 cm long, rounded at apex. Corolla white; tube 2.5–3 mm long; lobes oblong, 6–8 mm long, obtuse. Corona orange, with 5–10, laciniate scales. Fruit cylindric, 15–25 cm long, groove on each side at the junction of the carpels, rough, white tubercled. Seeds 1.5–2 cm long, slender, attenuated at the apex, coma white, deciduous.

Fls. & Frts.: April–June. *Exsiccata*: SUK-608.
Common, in moist forests.
Localities: Babu Jamal.

Local uses: Bark is given orally to the cattle to stop excess secretion of saliva.

Wrightia tinctoria ssp. **tinctoria** R. Br. in Mem. Wern. Nat. Soc. 1: 74. 1811; Pradhan in Singh *et al.* Fl. Maharashtra State Dicot. 2: 327. 2001; Cooke, Fl. Pres. Bombay 2: 200. 1958 (Repr. ed.). KALAKUDA.

Trees, 3–4 m tall. Leaves elliptic-lanceolate or oblong-lanceolate, 7.5–15 × 2.5–6 cm, acuminate, glabrous, base acute or rounded. Flowers white, fragrant, in lax terminal cymes. Calyx glabrous, glandular inside; segments 2.5 mm long, oblong, rounded, margins membranous. Corolla tube 3 mm long; lobes 9 mm long, oblong, obtuse, corona of numerous linear scales. Fruits of 2 distinct pendulous follicles, 25–45 × 0.6–1.2 cm, cylindric, tapering at both ends, glabrous, striate, cohering the tip only. Seeds 1.2–1.8 cm long, pointed; coma deciduous.

Fls. & Frts.: March–May. *Exsiccata*: SUK-608.
Common, in deciduous forests.
Localities: Amba, Amboli, Mahabaleshwar, Radhanagari, Sawantwadi, Vaibhavwadi.

Local uses: Bark is given orally to the cattle to stop excess secretion of saliva.

ASCLEPIADACEAE R. Br.

Asclepias curassavica L. Sp. Pl. 215. 1753; Hook, f. Fl. Brit. India 4: 18. 1883; Cooke, Fl. Pres. Bombay 2: 245. 1958 (Repr. ed.); Jagtap & Das Das in Singh *et al.* Fl. Maharashtra State Dicot. 2: 335. 2001. KURKI.

Erect undershrubs, up to 1 m high; stem terete. Leaves opposite, linear lanceolate, 8–12 × 1.2–2.2 cm, acute at both ends, glabrous. Flowers showy, in extra axillary umbellate cymes. Calyx divided to the base; lobes lanceolate, acute, pubescent outside, ciliolate. Corolla rotate, reflexed; lobes elliptic, 8–10 mm long, acute, pubescent outside. Corona saffron-yellow; staminal lobes 5, oblong, 3–5 mm long, hooded. Anther appendages subulate, 4–5 mm long. Pollinia obliquely ovoid. Ovaries globose. Follicles erect, lanceolate, 10–12 cm long, glabrous. Seeds broadly ovate, 5–6 × 4 mm; coma silky white, 2–3 cm long.

Fls.: Throughout years; *Frts.*: February–May. *Exsiccata*: SUK-296.
Native of West Indies, naturalized.
Localities: Kolhapur, Miraj, Narsingwadi, Tasgaon.

Local uses: It is used in acute asthmatic bronchitis associated pharyngitis and sinusitis with expectoration of sticky greenish sputum. It is also a remedy in piles and gonorrhea.

Calotropis gigantea (L.) Ait. Hort. Kew. ed. 2, 2: 78. 1811; Hook, f. Fl. Brit. India 4: 17. 1883; Cooke, Fl. Pres. Bombay 2: 214. 1958 (Repr. ed.); Jagtap & Das Das in Singh *et al.* Fl. Maharashtra State Dicot. 2: 341. 2001. *Asclepias gigantea* L. Sp. Pl. 214. 1753. AKRA, MANDAR, RUI.

Shrubs, 2–4 m tall, branches and leaves beneath clothed with white tomentum. Leaves opposite, sessile or shortly petioled, broadly obovate or elliptic-oblong, 10–17 × 3–10 cm, cordate and auricled at base, subacute or obtuse, fugaceously pubescent on both sides. Flowers in lateral sub-corymbose cymes. Calyx divided almost to the base, cottony pubescent outside, glandular within; teeth broadly ovate, 6–8 mm long, acute. Corolla 3–4 cm long, with revolute margins. Pollinia ovoid. Corona of 5, fleshy, laterally compressed, 1–1.2 cm long-lobes, with a curved spur at base and two rounded auricles just below the rounded apex. Follicles boat-shaped with hooked tips, 6–10 × 3–4 cm, cottony pubescent. Seeds broadly ovate, compressed, 6–7 × 4–5 mm, brown, coma silky white.

Fls. & Frts.: October–July. *Exsiccata*: SUK-296.
Common, throughout in dry regions.
Localities: Kolhapur, Miraj, Jaisingpur, Ramling, Sangali.

Local uses: Latex from the leaf is dropped in wounds or ulcer to expel thorn from the wounds. Leaves are warmed on flame and applied externally on swellings caused by wounds. Latex given in a very minute amount increases the body secretions. It is used for patients with strong constitution. It can be used for treatment of sinusitis, pneumonia, asthmatic bronchitis.

Calotropis procera (Ait.) R. Br. in Ait. Hort. Kew ed. 2, 2: 78. 1811; Hook, f. Fl. Brit. India 4: 18. 1883; Cooke, Fl. Pres. Bombay 2: 215. 1958 (Repr. ed.);

Jagtap & Das Das in Singh *et al.* Fl. Maharashtra State Dicot. 2: 342. 2001. PANDHARI MANDAR, RUI.

Erect shrubs, 1–2 m tall. Leaves opposite, sessile or shortly petioled, broadly ovate, obovate or ovate-oblong, 5–15 × 4–8 cm, cordate and auricled at base, cottony pubescent on both surfaces. Flowers in lateral umbellate cymes. Calyx 6–8 mm long, ovate, acute, pubescent. Corolla 2.5–3 cm across, lobes ovate, 9–10 mm long, acute. Corona lobes 0.5–0.6 cm long, truncate, with a recurved spur at base and without auricles at apex. Follicles boat shaped, recurved, 6–9 × 3.5 cm, cottony pubescent. Seeds ovoid, compressed, 6–7 mm long, brown; coma silky white, 2.5–3 cm long.

Fls.: Throughout the year; *Frts.*: January–June. *Exsiccata*: SUK-394.
Very Common, and abundant on wastelands.
Localities: Alate, Atigre, Babu Jamal, Hatkanagle, Miraj.

Local uses: Same as *Calotropis gigantea.*

Caralluma adscendens var. **attenuata** (Wight.) Gravely & Mayur in Bull. Madras Govt. Mus. n. s. Nat. Hist. Soc. 4(1): 13. 1931; Jagtap & Das Das in Singh *et al.* Fl. Maharashtra State Dicot. 2: 343. 2001. *C. attenuata* Wight Ic. t. 1268. 1850; Hook. f. Fl. Brit. India 4: 76. 1883. MAKADSHING.

Dwarf, succulent, perennial, glabrous herbs, 20–45 cm tall; stem and branches 4-angled and distally attenuated. Leaves opposite, sessile, deltoid-ovate, 1–2 mm long, acute, ciliate, caducous. Flowers solitary, from upper nodes, 1.2–1.5 cm across. Calyx 2–2.5 mm long, stellate, glabrous outside, glandular within; teeth lanceolate, acute. Corolla rotate; tube short; lobes valvate, ovate-lanceolate, 5–6 mm long, conspicuously purple-ciliate along margins. Outer corona of 5-deeply cleft lobes, each with two long filamentous teeth; inner corona 5-lobed, truncate. Pollinia ovoid. Follicles lanceolate-cylindric, 6–10 cm long, tapering to a sharp point. Seeds oblong, 5 mm long, brown; coma dull-white or brownish.

Fls. & Frts.: June–October. *Exsiccata*: SUK-99.
Common, in dry barren places in drier parts.
Localities: Alate, Babu Jamal, Kagal, Kartikswami Ghat.

Local uses: Whole plant is eaten to reduce acidity. Plant is anthelmintic, useful in leprosy, diseases of the blood, causes constipation. Plant is given orally in asthma and in acidity.

Ceropegia bulbosa var. **bulbosa** Roxb. Pl. Cor. 1: 11, t. 7. 1798; Hook. f. Fl. Brit. India 4: 67. 1883; Cooke, Fl. Pres. Bombay 2: 240. 1958 (Repr. ed.); Jagtap & Das Das in Singh *et al.* Fl. Maharashtra State Dicot. 2: 347. 2001. GALYA, KANDWEL, KANDIL WEL.

Twining perennial herbs; tuber globose, 2–3 cm in diam. Leaves opposite, petiolate, orbicular to ovate or elliptic, 2–3.5 × 1.5–2.5 cm, rounded or subrcodate at base, glabrous. Flowers 6–10, in lateral umbellate cymes. Calyx glabrous; teeth lanceolate, acute. Corolla often 1.5 cm long, inflated at base, narrow in the middle, funnel-shaped below the lobes, grayish outside, purplish and more or less hairy inside, lobes subulate-linear, 4–10 mm long, folded longitudinally on their back, connate at the tips, deep purple and villous within. Outer corona saucer shaped, inner one with 5, filiform, 2–3 mm long, gradually tapering to a point. Seeds oblong, 1–2 mm long, comose.

Fls. & Frts.: August–October. *Exsiccata*: SUK-669.
Frequent, in bushes on hill slopes in drier parts.
Localities: Alate, Atigre, Babu Jamal, Hatkanagale, Kagal.

Local uses: Tubers are used in inflammation of gums. Tubers are given as a tonic.

Cosmostigma racemosum (Roxb.) Wight, Contrib. 42. 1894 & Ic. t. 591. 1842; Hook. f. Fl. Brit. India 4: 46. 1836; Cooke, Fl. Pres. Bombay 2: 232. 1958 (Repr. ed.); Jagtap & Das Das in Singh *et al.* Fl. Maharashtra State Dicot. 2: 360. 2001. *Asclepias racemosa* Roxb. Fl. Ind. 2: 32. 1832. JATI, MARVEL, SHENDVEL.

Climbing shrubs. Leaves deciduous, ovate, 5.5–14 × 5–12 cm, acute, glabrous, base cordate. Flowers small, greenish in axillary or sublateral cymes. Calyx divided ¾ down, glabrous; segments broadly oblong, rounded at apex, margins membranous, ciliate. Corolla 3 mm long; lobes ovate-oblong, 3 mm long, obtuse, fleshy, speckled with brown dots; corona of 5 usually 2-fid broad membranous erect scales. Staminal-column arising from near the base of the corolla; pollinia clavate, slightly curved, erect, waxy. Follicles ovoid-oblong, 7–17 × 3 cm, bluntly pointed, smooth. Seeds ovate 1.2 cm long, margined; coma silky.

Fls. & Frts.: June–August. *Exsiccata*: SUK-341.
Occasional, in moist deciduous and semi evergreen forests.

Local uses: Leaves are used to cure ulcerous sores. Root bark is administered internally in vataka, a disease in which lumps of undigested food are passed. Useful in dyspepsia, accompanied by a febrile condition and absence of bile in the stools.

Gymnema sylvestre (Retz.) R. Br. ex Schultes in R. & S. Syst. Veg. 6: 57. 1819; Hook. f. Fl. Brit. India 4: 29. 1883; Cooke, Fl. Pres. Bombay 2: 224. 1958 (Repr. ed.); Jagtap & Das Das in Singh *et al.* Fl. Maharashtra State Dicot. 2: 362. 2001. *Periploca sylvestris* Retz. Obs. 2: 15. 1781. BEDKI, KAVALI, PTIANI.

Much branched, twining shrubs; young parts densely gray pubescent. Leaves opposite, ovate or elliptic-lanceolate, 3–5 × 1.5–3 cm, acute, pubescent, rounded at base. Flower minute, in lateral, subsessile cymes. Calyx cupular, pubescent outside, eglandular within; teeth ovate, 1–1.5 mm long, obtuse with membranous, ciliolate margins. Corolla campanulate, greenish-yellow, 3–4 mm across; tube 1–5 mm long, pubescent within, lobes deltoid, shorter than the tube, appearing as ridges. Pollinia minute, falcate. Follicles linear lanceolate, 6–8 × 0.6–0.8 mm, tapering, glabrous. Seeds ovoid-oblong, 10–12 × 3–4 mm, brown, with broad, thin margin, coma silky.

Fls. & Frts.: April–July. *Exsiccata*: SUK-84.

Very Common, in deciduous forests.

Localities: Amba, Borbet, Panhala.

Local uses: Leaves are given orally in diabetes. Decoction is used with leaves of *H. pubescens* in asthma. Useful in eye complaints, cures opacities of the lens, cornea; good in diseases of the heart, piles, leucoderma, inflammations; cures burning sensation. Powdered leaves are taken to reduce blood sugar but after continued use it loses efficiency.

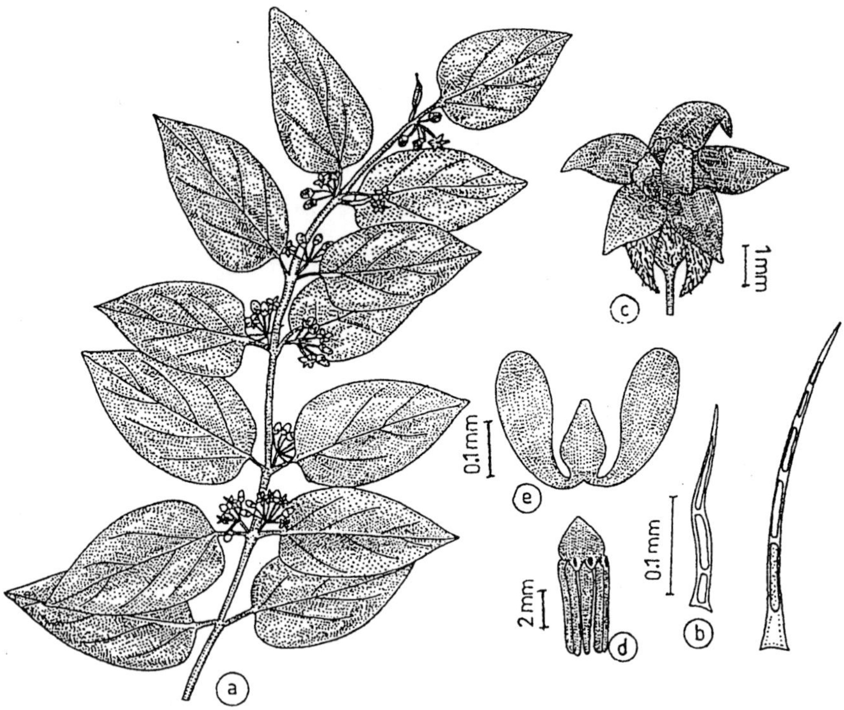

Text fig. 26: Gymnema sylvestre a-twig; b-hair; c-flower; d-gynostegium; e-pollinia.

Holostemma annulare (Roxb.) K. Schum. in Engl. & Prantl, Pflanzenf. 4(2): 250. 1895 (*annularis*); Jagtap & Das Das in Singh *et al.* Fl. Maharashtra State Dicot. 2: 367. 2001. *Asclepias annularia* Roxb. Fl. Ind. 2: 37. 1832. (*annularis*). *Holostemma ada–kodien* Sch. in L. Syst. Veg. 6: 95. 1820 nom. illeg. *H. rheedei* Wall. Pl. Asiat. Rar. 2: 51. 1831; Hook. f. Fl. Brit. India 4: 21. 1883. *H. rheedeianum* Spreng. Syst. 1: 851. 1825; Cooke, Fl. Pres. Bombay 2: 220. 1958 (Repr. ed.). SHIDODI, TULTULI.

Extensive, twining shrubs; stem terete. Leaves membranous when dry, broadly ovate, 5–15 × 3–10 cm, cordate at base, acuminate, glaucous, glabrous. Flowers in lateral short, fasciculate cymes. Calyx rotate, glabrous; sepals broadly ovate, 5–6 mm long, obtuse, with membranous margins. Corolla rotate, pinkish-white outside, deep red within, divided ¾ way down; lobes ovate, 10–12 cm long, obtuse, glabrous. Corona staminal, ring like. Pollinia linear-falcate. Follicles linear-oblong, 10–12 × 0.5–0.6 cm, tapering to a blunt point. Seeds ovoid, 6–8 mm long, compressed glabrous; coma 2–2.5 cm long.

Fls. & Frts.: July–October. *Exsiccata*: SUK-1753.
Common, along hilly forests.
Localities: Amba.

Local uses: Roots are given in diabetes with water. It is given in leucorrhoea with the roots of *Bombax ceiba*. Flowers are made into paste and applied externally an itching and in insect bite. Plant is used to cure ulcers, diseases of blood, leucoderma, useful in gonorrhea and for vesicular calculi.

Leptadenia reticulata (Retz.) Wight & Arn. in Wight, Contrib. Ind. Bot. 47. 1834; Hook. f. Fl. Brit. India 4: 63. 1883; Cooke, Fl. Pres. Bombay 2: 237. 1958 (Repr. ed.); Jagtap & Das Das in Singh *et al.* Fl. Maharashtra State Dicot. 2: 370. 2001. *Cynanchum reticulatum* Retz. Obs. 2; 15. 1781. HIRANDODI, KHARKHODA, SHINGUTI.

Twining shrubs. Leaves opposite, glaucous green, ovate to ovate-lanceolate, 5–10 × 3–8 cm, rounded at base, acute, gray tomentose especially beneath when young, glabrescent with age. Flowers in lateral, many-flowered, umbellate cymes. Calyx divided 2/3 way down; teeth ovate-lanceolate, 1–1.5 mm long. Corolla rotate, dull white, divided ¾ way down; lobes lanceolate, 2.5–3 mm long, acute, densely pubescent outside. Corona double, coralline of 5 small scales; staminal corona of fleshy, wavy ring. Pollinia globose, erect. Follicles often solitary, broadly lanceolate, 6–9 × 2.5–3 cm, with thick curved beak. Seeds narrowly ovoid, 8–10 mm long, compressed; coma 2–4 cm long.

Fls. & Frts.: September–January. *Exsiccata*: SUK-151.

Common, on hedges along riverbanks in drier parts of the region.
Localities: Ajara, Borbet, Chandgad, Kagal, Kolhapur.

Local uses: Decoction of root is used to wash ulcerated wounds. Latex from stem is applied externally in rheumatism. Poultice of leaves warmed with onion and applied externally in case of wounds.

Oxystelma esculentum (L. f.) R. Br. in Mem. Wern. Soc. 5: 40. 1809, printed in 1811 ex Schult. in L. Syst. Veg. 6: 89. 1820; Hook. f. Fl. Brit. India 4: 17. 1883; Cooke, Fl. Pres. Bombay 2:216. 1958 (Repr. ed.); Jagtap & Das Das in Singh *et al.* Fl. Maharashtra State Dicot. 2: 372. 2001. *Periploca esculenta* L. f. Suppl. Pl. 168. 1781. DUDHANI.

A twining undershrubs. Leaves opposite, linear or linear lanceolate, 4–8 × 0.5–2 cm apex mucronate, base truncate. Flowers large strikingly handsome, drooping, in pedunculate, lateral subumbellate or racemose, few-flowered cymes longer than the leaves. Calyx divided nearly to the base; lobes oblong lanceolate, 3 mm long, acute, glandular inside. Corolla pale rose colored, saucer shaped, 2.5 cm across, tube scarcely any, the mouth with a densely pubescent ring; lobes 8–9 mm long, deltoid, acute, conspicuously ciliate, veined with purple; corona staminal, the lobes ovate lanceolate, acuminate, with incurved entire subulate tips. Pollen masses 1 in each cell, pendulous. Follicles ovoid-lanceolate, tapering to a point, 8 × 1.5 cm, glabrous. Seeds numerous, ovate-flat, 3 mm long, margins thin, often minutely and obscurely denticulate.

Fls. & Frts.: May–February. *Exsiccata*: SUK-70.
Rare, growing along riverbanks, in moist deciduous forests.
Localities: Kolhapur, Ratnagiri, Shirgaon.

Local uses: Decoction of plant is used as gargle in apthous ulceration of mouth and sore throat. Roots are considered specific for jaundice. Milky sap is used as a wash for ulcers. Fruit juice is given in cough, in muscle pain and gonorrhea. It is also useful in leucoderma.

Pergularia daemia (Forssk.) Chiov. Result. Sc. Miss. Stefan. Paoli. Somal. Ital. 1: 115. 1916; Jagtap & Das Das in Singh *et al.* Fl. Maharashtra State Dicot. 2: 375. 2001. *Asclepias daemia* Forssk. Fl. Aeg.-Arb. 51: 1775. *Daemia extensa* R. Br. in Mem. Wern. Soc. 5 (1): 50. 1811; Hook. f. Fl. Brit. India 4: 20. 1883; Cooke, Fl. Pres. Bombay 2: 219. 1958 (Repr. ed.). UTARNI.

Extensive, hoary tomentose, twining undershrubs. Leaves opposite, reniform-ovate, 2–6 × 2.5–6.5 cm, deeply cordate at base with rounded lobes, acuminate, glabrous or nearly so above, pubescent beneath. Flowers in drooping, corymbose cymes becoming racemose with age. Calyx villous, divided almost to the base, sepals ovate-lanceolate, acute, ciliate. Corolla glabrous outside, hairy within, dull greenish-yellow or white, 10–12 mm long tube half as long as the elliptic oblong lobes. Corona staminal, outer of 5-lobed membrane, inner of laterally compressed lobes; pollinia oblong. Follicles strongly reflexed, lanceolate, 5–8 × 1–1.2 cm, attenuate into a long beak, echinate with soft spines, pubescent, ash-grey with purplish tinge. Seeds broadly ovate, 6–7 mm long, dentate along margin, densely pubescent; coma silky white, 2.5–3 cm long.

Fls. & Frts.: July–February. *Exsiccata*: SUK-150.

Frequent, on hedges around fields and near villages in drier parts of the region.

Localities: Hatkanagale, Ramling, Shiroli.

Local uses: Leaf juice is used in the treatment of asthma and applied to rheumatic swellings in combination with lime or ginger. Plant is used to cure ulcers, useful in urinary discharges, leucoderma. It is not safe medicine because of its purgative and emetic properties.

Sarcostemma viminale ssp. **viminale** (L.) R. Br. in Mem. Wern. Soc. 1: 51. 1809; Printed in 1811 and Prodr. Fl. Nov. Holl. 464. 1810 var. **viminale**; Jagtap & Das Das in Singh *et al.* Fl. Maharashtra State Dicot. 2: 376. 2001. *Euphorbia viminalis* L. Sp. Pl. 452. 1753. *Sarcostemma brevistigma* Wight & Arn. in Wight, Contrib. 59. 1834; Hook. f. Fl. Brit. India 4: 26. 1883; Cooke, Fl. Pres. Bombay 2: 222. 1958 (Repr. ed.). RANSHER, SOMA, SOMVEL.

Much branched, leafless shrubs. Flowers in terminal, sessile umbels. Calyx puberulous outside, glandular within, sepals ovate, 8–12 mm long, acute, ciliate. Corolla rotate, white, 15–20 mm across, glabrous; lobes ovate-oblong, 6–8 mm long, acute, with revolute margins. Corona double, staminal; outer cupular, crenately 10-lobed; inner one of 5, erect, fleshy lobes. Pollinia clavate, pendulous. Follicles usually solitary, lanceolate, 6–10 cm long, tapering at both ends. Seeds ovate, 4–5 mm long, flattened; coma silky white, 2 cm long.

Fls. & Frts.: February–May. *Exsiccata*: SUK-65, 320.

Frequent, in hilly areas in crevices of rocks, in drier parts of the region.

Localities: Babu Jamal, Jat, Ramling, Tasgaon.

Local uses: Stem latex is used to heal wound. In small quantity it is used in biliousness. The dried stem has emetic properties.

Tylophora indica (Burm. f.) Merr. in Phil. J. Sci. 19: 373. 1921; Jagtap & Das Das in Singh *et al.* Fl. Maharashtra State Dicot. 2: 379. 2001. *Cynanchum indicum* Burm. f. Fl. Ind. 70. 1768. *Tylophora asthamatica* Wight & Arn. in Wight, Contrib. 51. 1834; Hook. f. Fl. India 4: 44. 1833; Cooke, Fl. Pres. Bombay 2: 228. 1958 (Repr. ed.). ANANTIMUL, KHADKI, RASNA, PITMARI.

Twining undershrubs with thin creeping rootstock; sap watery. Leaves opposite, ovate-oblong or elliptic-oblong, 5–7 × 3–4.5 cm, rounded at base, acute or apiculate, glaucous and glabrous above, pale green and pubescent beneath. Flowers in lateral umbellate cymes. Calyx pubescent outside glandular within, divided almost to the base; teeth linear-lanceolate, 2.5–3 mm long, acute. Corolla dull greenish-white, 1–1.5 cm across; tube short, lobes elliptic, 5–6 mm long, acute. Corona single, staminal of 5 small, fleshy obovate lobes. Pollinia globose, horizontal. Follicles divaricate, lanceolate, 8–10 × 1 cm, tapering to a point. Seeds ovoid, 6–7 × 3–4 mm, flattened, strongly margined, reddish-brown, glabrous, coma silky white, 2–3 cm long.

Fls. & Frts.: July–November. *Exsiccata*: SUK-397.
Occasional, along streams in moist deciduous and evergreen forests.
Localities: Patgaon, Vaibhavwadi.

Local uses: Leaves are used with *Acorus calamus* in cough. Leaf juice is given as an emetic. It is also dried and made into pills which are administered in dysentery. Root juice is given in chronic bronchitis and early stages of whooping cough. It is locally used in rheumatism and as blood purifier.

Wattakaka volubilis (L. f.) Stapf in Bot. Mag. Sub. T. 897. 1923; Jagtap & Das Das in Singh *et al.* Fl. Maharashtra State Dicot. 2: 381. 2001. *Asclepias volubilis* L. f. Suppl. 170. 1781. *Dregea volubilis* (L. f.) Benth. ex Hook. f. Fl. Brit. India 4: 46. 1883. *Marsdenia volubilis* (L. f.) Cooke, Fl. Pres. Bombay 2: 166. 1904 [2: 230. 1958 (Repr. ed.)]. AMBARI, HIRANDODI.

Large, perennial, twining shrubs with watery sap. Leaves opposite, broadly ovate, 5–10 × 4–8 cm, shortly acuminate at apex, glabrescent, with numerous glands at the base of the midrib on the upper surface. Flowers in dense lateral, umbellate cymes. Calyx pubescent outside, glandular within; teeth ovate, 2.5 mm long, acute, ciliolate. Corolla green, rotate, glabrous, divided ¾ way down; lobes 4–5 mm long, broadly ovate; corona of 5-fleshy lobes. Pollinia obliquely oblong. Follicles lanceolate, 7–10 × 3–4 cm, divaricate, tapering to a blunt point, brown tomentose. Seeds broadly ovate, 10–12 × 7–8 mm, glabrous; coma 4–5 cm long.

Fls. & Frts.: July–October. *Exsiccata*: SUK-1754.
Common, in bushes and in hill forests.
Localities: Kagal, Kartikswami Ghat, Kolhapur.

Local uses: Leaves are used as tonic, cures tumors, piles, leucoderma, asthma, urinary discharges. Leaves are much employed as an application to boils and abscesses. Roots and tender stocks are considered emetic and expectorant.

PERIPLOCACEAE Schlecht

Cryptolepis buchanani R. Br. ex R. & S. Syst. 4: 409. 1819; Hook. f. Fl. Brit. India 4: 5. 1883; Cooke, Fl. Pres. Bombay 2: 210. 1958 (Repr. ed.); Jagtap & Das Das in Singh *et al.* Fl. Maharashtra State Dicot. 2: 384. 2001. KAWALE, KALI-TAKALI.

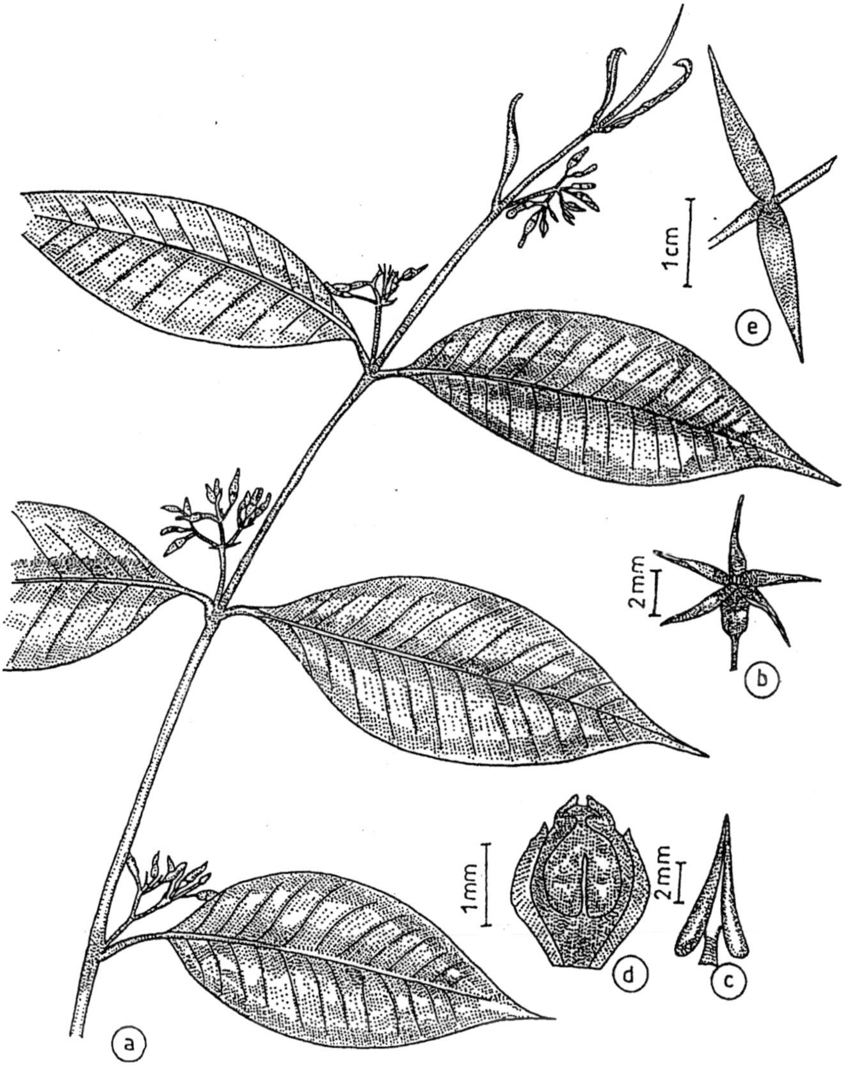

Text fig. 27: Cryptolepis buchanani a-twig; b-flower; c-stamens; d-gynostegium; e-fruit.

Extensive, glabrous, twining shrubs. Leaves opposite, elliptic oblong or oblong-lanceolate, 6–12 × 3–6 cm, acute, entire, retuse or apiculate, shining green above, whitish beneath, glabrous. Flowers in short, axillary, paniculate cymes. Calyx glabrous; tube very short; teeth ovate, 1.5–2 mm long, subacute. Corolla tube 2 mm long; lobes linear-lanceolate, 5–6 mm long, subacute; corona scales clavate. Filaments distinct; anthers with a connective produced into a fleshy apiculus, more or less connivent into a cone and united with the style. Follicles divaricate, cylindric, 5–10 × 1–2 cm, tapering to a blunt point. Seeds ovate-oblong, black, 6–8 cm long, with 2–2.5 cm long coma.

Fls. & Frts.: Throughout the year. *Exsiccata*: SUK-213.
Common, throughout the region.
Localities: Kagal, Kolhapur, Shiroli.

Local uses: Leaves are boiled in water and bath is given to small children to stop their stomach complaints. Stem and root powder is used to heal wounds and ulcers.

Hemidesmus indicus (L.) Schult. in Roem. & Schult. Syst. Veg. 6: 126. 1819; Hook. f. Fl. Brit. India 4: 5. 1883; Cooke, Fl. Pres. Bombay 2: 210. 1958 (Repr. ed.); Jagtap & Das Das in Singh *et al.* Fl. Maharashtra State Dicot. 2: 388. 2001. *Periploca indica* L. Sp. Pl. 211. 1753. ANANTVEL, UPARSUL.

Creeping or twining glabrous shrubs; rootstock woody; stem terete. Leaves opposite, very variable, elliptic-oblong to linear-lanceolate, 3–10 × 1.5–5 cm, rounded or subcordate at base, entire, acute or obtuse and apiculate at apex, glabrous, dark green or often variegated with white above, dull white beneath. Flowers in dense axillary sessile cymes. Calyx glabrous outside; teeth ovate, 1–1.5 mm long, with membranous ciliolate margins, acute. Corolla greenish, outside, purple within, tube very short; lobes ovate oblong, 4 mm long, fleshy, acute or acuminate. Filaments distinct; anthers cohering at apex, terminated by inflexed membrane. Follicles cylindric, linear, 10–15 × 0.5–0.6 cm, tapering to a point at apex, glabrous. Seeds ovate-oblong, 6–8 mm long, flattened, black; coma silvery white, 2–2.5 cm long.

Fls. & Frts.: August–December. *Exsiccata*: SUK-39.
Common, throughout region.
Localities: Amba, Amboli, Gaganbawada.

Local uses: Root powder is used with root powder of *Rubia cordifolia* to remove black marks remained due to pimples and also externally applied on boils. Root paste is given orally and applied in every skin disease. Root powder is given every night in blood impurities. It is helpful to maintain health of AIDS patient. Root is used to cure leucoderma, diarrhoea, blood diseases, leucorrhoea, burning sensation; useful in piles and eye troubles. Repeated abortions can be tackled successfully by using *Withania somnifera*,

Tectona grandis root, *Tribulus terristris* and Devdar with *Hemidesmus indicus*.

LOGANIACEAE Mart.

Strychnos dalzellii C. B. Cl. in Hook. f. Fl. Brit. India 4: 87. 1883; Cooke, Fl. Pres. Bombay 2: 250. 1958 (Repr. ed.); Jadhav in Singh *et al.* Fl. Maharashtra State Dicot. 2: 392. 2001. KAJARVEL.

Large climbers with clavate ferruginous pubescent tendrils. Leaves ovate, 7.5–12.5 × 3.5–5.6 cm, acute glabrous, 3-nerved, base rounded or cuneate. Flowers in short dense axillary and terminal subcapitate puberulous cymes. Calyx 2 mm long, puberulous, outside; segments 5, orbicular oblong, 1 mm long, hairy at base inside and with ciliate margins. Corolla 3 mm long; 5-lobed; lobes 2.5 mm long, thick, oblong, subacute, with a dense transverse fringe of white hairs about ½ way down the inner face. Stamens 5; filaments short; anthers deltoid, ovate, cordate, the lobes slightly beard at the base. Ovary ovoid, quite glabrous, tapering into the style. Berry 2.5–3.5 cm in diam., crustaceous, many seeded. Seeds 1 cm in diam., discoid.

Fls. & Frts.: February–November. *Exsiccata*: SUK-343.
Rare, scattered in forests.
Localities: Amboli, Mahabaleshwar, Patgaon.

Local uses: Leaf powder mixed with aromatic oils and applied externally as a poultice on wounds and ulcers.

Strychnos nux-vomica L. Sp. Pl. 189. 1753; C. B. Cl. in Hook. f. Fl. Brit. India 4: 90. 1883; Cooke, Fl. Pres. Bombay 2: 250. 1958 (Repr. ed.); Jadhav in Singh *et al.* Fl. Maharashtra State Dicot. 2: 392. 2001. KAJARA, KASARKAMORA.

Trees, 10–15 m tall, often with short sharp strong axillary spines. Leaves broadly elliptic or elliptic ovate, 6–12 × 4.5–6 cm, obtuse or shortly acuminate, glabrous and shining, 5-nerved, base rounded. Flowers numerous, greenish-white, in terminal pedunculate, pubescent compound cymes. Calyx 2.5 mm long, pubescent outside; segments 5, lanceolate, acute, 1.5 mm long. Corolla less than 1.5 cm long, 5-lobed, glabrous; tube cylindric, hairy below, the throat glabrous; lobes 4 mm long, narrowly oblong acute. Ovary and style glabrous. Fruits globose, 2.5–7.5 cm in diam., slightly rough, but shining, orange red when ripe. Seeds usually many, discoid, 8–10 mm, much compressed, concave on one side and convex on other, clothed with very fine, appressed grey silky hairs radiating from the center.

Fls. & Frts.: February–November. *Exsiccata*: SUK-1789.
Scattered in forests, along roadsides and coasts.

Localities: Malwan, Tillari, Vaibhavwadi.

Local uses: Same as *S. potatorum*. Small doses of seeds cure diabetes. Stem is also given to diabetic patients.

Strychnos potatorum L. f. Suppl. Pl. 148. 1782; C. B. Cl. Hook. f. Fl. Brit. India 4: 90. 1883; Cooke, Fl. Pres. Bombay 2: 251. 1958 (Repr. ed.); Jadhav in Singh *et al.* Fl. Maharashtra State Dicot 2: 392. 2001. NIRMALI, NIVALI.

Trees, ca 5 m tall. Leaves ovate or elliptic, 3–6 × 2.5–3.5 cm, acute, subacuminate, glabrous and shining, spurious, 3–5 nerved. Flowers in short, glabrous, nearly sessile axillary cymes. Calyx 2 mm long, glabrous; segments 5, ovate, acute, 2–5 mm long. Corolla 4–5 mm long, 5-lobed; lobes 2.5 mm long, oblong, acute, with a tuft of hairs inside towards the base of each lobe. Ovary ovoid, glabrous, tapering into a long glabrous style; stigma obscurely 2-lobed. Berry black when ripe, 1.8–2 cm across. Seeds 1–2, circular, 8 mm in diam., bluntly lenticular, greatly compressed, shining with short appressed silky yellow, hairs.

Fls. & Frts.: June–January. *Exsiccata*: SUK-1778.
Common, in hill forests.
Localities: Amboli, Sawantwadi, Tillari, Vaibhavwadi.

Local uses: Seeds are soaked in buttermilk for 3 to 4 days. Washed with water and small doses of the seeds are given with aromatic oil in colic. Decoction of seeds used to cure fever. Leaf paste is applied as a poultice in case of wounds and ulcers. Seeds are used in leucoderma, itching, piles, ulcers, anemia, jaundice and urinary discharges.

GENTIANACEAE Juss.

Canscora decurrens Dalz. in Hook. J. Bot. & Kew Gard. Misc. 2: 136. 1850; C. B. Cl. in Hook. f. Fl. Brit. India 4: 103. 1883; Cooke, Fl. Pres. Bombay 2: 257. 1958 (Repr. ed.); Jadhav in Singh *et al.* Fl. Maharashtra State Dicot. 2: 398. 2001.

Herbs, 40 cm high; stem 4-winged. Leaves opposite and alternate, sessile, broadly ovate, 0.5–2 × 0.5–1 cm, acute, 3-nerved, rounded at base; lower petiolate, elliptic. Flowers subtrichotomous cymes. Calyx not winged; lobes 4; teeth lanceolate, acute. Corolla pale rose colored or white; tube 5 mm long; teeth lanceolate, 5 mm long; lobes shorter than the tube, oblong, obtuse. Ovary 1-celled, ovules many; stigma bilobed. Capsules narrowly oblong, 5 mm long; membranous.

Fls. & Frts.: September–April. *Exsiccata*: SUK-1755.
Common, in moist shady places.
Localities: Amba, Amboli, Mahabaleshwar, Pawas, Ratnagiri, Wai.

Local uses: Plant juice is dropped in nose to cure headache due to fever. It is used with *Hoppea dichotoma* as a nervine tonic. Juice is orally given in tuberculosis.

Canscora diffusa (Vahl) R. Br. ex R. & S. Syst. Veg. 3: 301. 1818; C. B. Cl. in Hook. f. Fl. Brit. India 4: 103. 1883; Cooke, Fl. Pres. Bombay 2: 257. 1958 (Repr. ed.). *Gentiana diffusa* Vahl, Symb. Bot. 3: 47. 1794.

Herbs, 30–50 cm high; stem-obtusely quadrangular, obscurely winged. Leaves membranous, broadly ovate, 5–12.0 × 0.5–1 cm, upper sessile, acute, rounded at base, lower petiolate, elliptic. Flowers numerous in lax diffuse paniculate cymes. Calyx not winged, lobes 4, 6 mm long, lanceolate acute. Corolla 1.5 cm long; tube 0.5 cm long, green; limb rosy; lobes unequal, ovate, obtuse. Stamens 4, one longer than the other 3. Capsules ellipsoid, 0.5 cm long, membranous.

Fls. & Frts.: September–February. *Exsiccata*: SUK-230.
Common, weed in open, moist places.
Localities: In all districts.

Local uses: Plant juice is dropped in nose to cure headache due to fever. It is used with *Hoppea dichotoma* as a nervine tonic. Juice is orally given in tuberculosis.

Centaurium meyeri (Bunge) Druce in Rep. Bot. Exch. Club. Brit. Isles 4: 613. 1917; Jadhav in Singh *et al.* Fl. Maharashtra State Dicot. 2: 405. 2001. *Erythraea meyeri* Bunge in Ledb. Fl. Altai 1: 220 1829; C. B. Cl. in Hook. f. Fl. Brit. India. 4: 102. 1883. *Erythraea roxburghii* D. Don in Lond. & Edinb. Phill. Mag. & J. Sci. 8: 77. 1836; C. B. Cl. op. cit.; Cooke, Fl. Pres. Bombay 2: 255. 1958 (Repr. ed.). LUNTAK.

Herbs, 30–50 cm high, erect. Radical leaves rosulate, obovate-oblong, 2.5 × 0.2–1.0 cm, obtuse, cauline, smaller; linear, oblong, all sessile, glabrous. Flowers in dichotomous cymes with a pedicelled flower in the fork. Calyx shortly or deeply 5-cleft; lobes keeled; linear lanceolate. Corolla white or pink, 12–15 mm long; tube 7–8 mm long; lobes star-like, elliptic, acute. Stamens 5; anthers twisted at apex. Capsules oblong, 8–12 mm long, glabrous, enclosed in persistent calyx. Seeds numerous, ovoid, light brown.

Fls. & Frts.: November–March. *Exsiccata*: SUK-1756.
Common, in moist places and harvested fields.
Localities: Shivaji University campus, Kolhapur, Kagal.

Local uses: Decoction of plant is given after every 3 hours in fever. Plant is chewed in diabetes every morning.

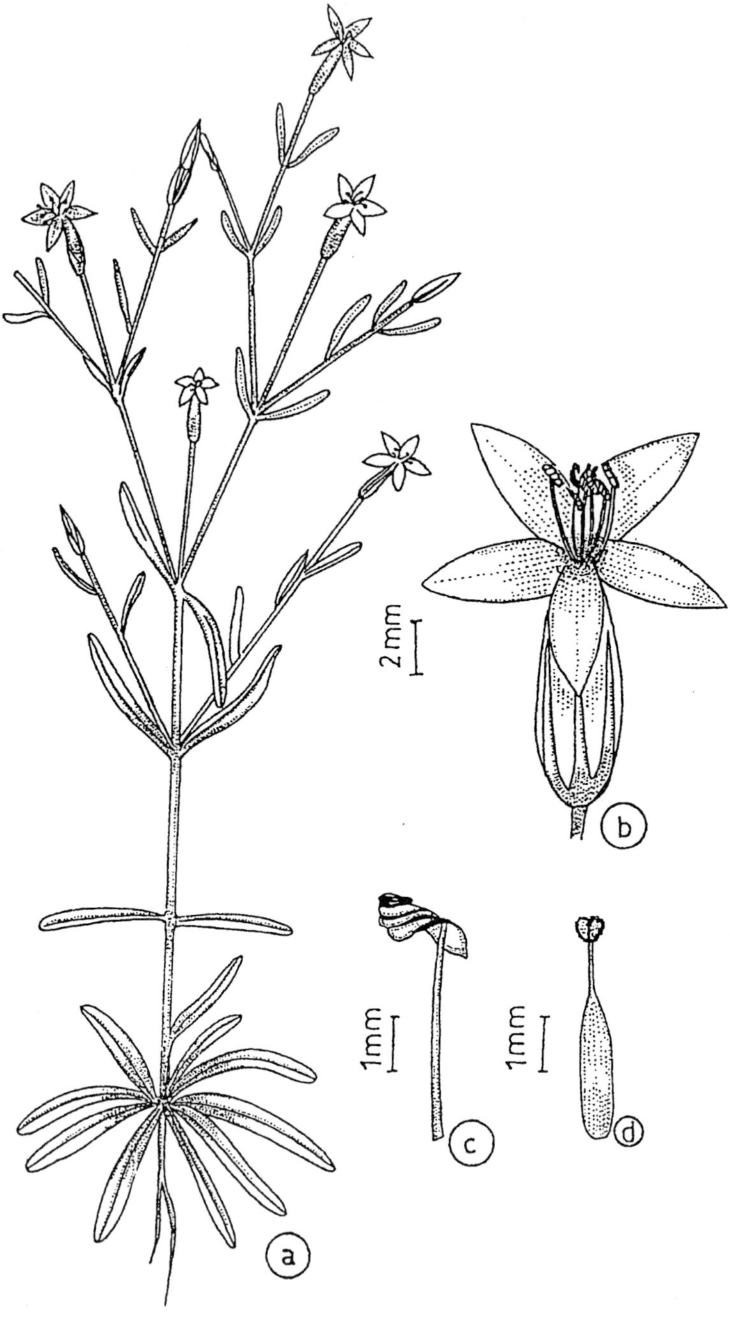

Text fig. 28: Centaurium meyeri a-habit; b-flower; c-stamens; d-carpel.

Exacum pedunculatum L. Sp. Pl. 112. 1753; C. B. Cl. in Hook. f. Fl. Brit. India 4: 97. 1883; Cooke, Fl. Pres. Bombay 2: 253. 1958 (Repr. ed.); Jadhav in Singh *et al.* Fl. Maharashtra State Dicot. 2: 407. 2001.

Herbs, 10–30 cm high, erect; stem quadrangular. Leaves elliptic oblong or oblong-lanceolate. Flowers in terminal much branched cymes. Calyx 4-lobed, thinly membranous; lobes lanceolate, furnished with a large almost semicircular strongly nerved wing, subauricled at base. Corolla blue, 0.6–1.5 cm long; lobes 4, lanceolate, acute, 0.4–0.8 cm long. Stamens on the corolla throat; anthers oblong, cordate at base. Capsule globose, 4 mm diam., smooth, shining.

Fls. & Frts.: August–December. *Exsiccata*: SUK-229.
Common, along ghats, open places and banks of nallas.
Localities: Bhudergad, Kas, Radhanagari.

Local uses: Decoction is used in fever.

Exacum tetragonum Roxb. Fl. Ind. ed. Carey & Wall. 1: 413. 1820; C. B. Cl. in Hook. f. Fl. Brit. India 4: 95. 1883; Jadhav in Singh *et al.* Fl. Maharashtra State Dicot. 2: 408. 2001. *E. tetragonum* var. *stylosa* (Wall. ex. D. Don) C. B. Cl. in Hook. f. op. cit. *E. bicolor* Roxb. op. cit. 413; C. B. Cl. in Hook. f. op. cit. 96; Cooke, Fl. Pres. Bombay 2: 252. 1958. UDICHIRAYET.

Herbs, 30–60 cm high, annual, erect; stem quadrangular. Leaves variable, ovate, elliptic or lanceolate, 1.5–17 × 0.5–5 cm, acute or obtuse, 5-nerved. Flowers showy, in copious terminal glabrous cymes. Calyx 12 mm long, glabrous, deeply 4-partite; segments ovate, aristate-acuminate, margins membranous, keel furnished with long membranous wing. Corolla 2–3 cm long; lobes 2.5 cm long, obovate, acute, the lower half white, the upper blue. Anthers 8–12 mm long, curved. Capsule ellipsoid, 12 mm long, smooth and shining tipped with the remains of the style, yellowish-brown.

Fls. & Frts.: August–February. *Exsiccata*: SUK-228.
Common, on plateaus.
Localities: Barki, Bhudergad, Mahabaleshwar, Panchgani, Radhanagari.

Local uses: Plant is used as a tonic in fever and as a stomachic.

Hoppea dichotoma Heyne ex Willd. in Ges. Naturf. Fr. Neue. Schr. 3: 435. 1801; C. B. Cl. in Hook. f. Fl. Brit. India 4: 100. 1883; Cooke, Fl. Pres. Bombay 2: 256. 1958 (Repr. ed.).

Herbs, 3–12 cm high, erect, dichotomously branched; quadrangular, more or less winged. Leaves ovate, 0.3–0.6 × 0.2–0.4 cm. Flowers pale yellow, in terminal dichasial cymes. Calyx campanulate, 2.5–3 mm long, membranous; lobes 4, lanceolate, overtopping the corolla; acute. Corolla tubular, 3 mm long, pale yellow; lobes ½ as long as the tube, ovate-oblong, obtuse. Stamens

1 fertile, 3 barren. Ovary 1-celled. Capsule ellipsoid, 0.3 cm long. Seeds many, minute, obscurely reticulate.

Fls. & Frts.: October–December. *Exsiccata*: SUK-1777.
Common, in open moist places amongst grasses.
Localities: Malwan, Pawas, Ratnagiri, Sawantwadi.

Local uses: Plant is used as stomachic. It is used to cure fever. Applied externally in itch and insect bite.

Swertia densifolia (Griseb.) Kashyapa in Kew Bull. 15: 42. 1961; Jadhav in Singh *et al.* Fl. Maharashtra State Dicot. 2: 410. 2001. *Ophelia densifolia* Griseb. in DC. Prodr. 9: 125. 1845. *Swertia decussata* Nimmo ex C. B. Cl. in Hook. f. Fl. Brit. India 4: 127. 1883; Cooke, Fl. Pres. Bombay 2: 259. 1958 (Repr. ed.). KADV, KAVDADI.

Herbs, 25–80 cm high, annual, stems erect, obtusely quadrangular, more or less 4-winged. Leaves sessile, broadly ovate, 1.9–3 × 1.2–2.5 cm, obtuse, 5-nerved. Flowers very numerous, in dense corymbose cymes. Calyx 9 mm long, tube 1 mm long; lobes lanceolate, acute. Corolla white with blue veins; tubes 8–10 mm long; lobes broadly ovate, acuminate, with a yellow circular pits at the base, fringed all round with rather long scaly fimbriae. Capsule ovoid, 6–10 mm long, beaked, 2-valved. Seeds many, dark brown.

Fls. & Frts.: December–January. *Exsiccata*: SUK-1776.
Common, on plateaus and on open slopes.
Localities: Barki, Borbet, Kas, Mahabaleshwar.

Local uses: It is useful as a tonic or mild febrifuge in fever. It is good for vomiting in pregnancy. Leaf juice is used to cure urinary discharges. Mild dose of leaf decoction is given in bronchitis.

Swertia minor (Griseb.) Knobl. in Bot. Centralbl. 60: 321. 1894; Cooke, Fl. Pres. Bombay 2: 259. 1958 (Repr. ed.); Jadhav in Singh *et al.* Fl. Maharashtra State Dicot. 2: 412. 2001. *Ophelia minor* Griseb. in DC. Prodr. 9: 126. 1845. *Pleurogyne minor* (Griseb.) Benth. in Gen. Pl. 2: 816. 1876; C. B. Cl. in Hook. f. Fl. Brit. India 4: 120. 1883.

Herbs, 15 cm high, annual; stem 4-winged. Leaves sessile, ovate 0.5–1 × 0.5 cm, acute, obscurely 3-nerved. Flowers 4-merous, in leafy paniculate cymes. Calyx divided almost to the base, somewhat enlarged in fruit; segments 4, ovate, 3–4 mm long, acute. Corolla 6–7 mm long, tube very short; lobes 4, ovate-oblong, acute, with 5-parallel nerves, each lobe with 2 triangular pockets near the base. Anthers versatile. Capsule ovoid, 5 mm long. Seeds suborbicular, compressed, not winged.

Fls. & Frts.: July–October. *Exsiccata*: SUK-1775.
Common, on plateaus of higher altitudes admist grasses

Localities: Gaganbawada, Kas, Panchgani, Barki, Panhala.

Local uses: Leaf decoction is given in fever. Other uses are similar as *S. densifolia*.

HYDROPHYLLACEAE R. Br.

Hydrolea zeylanica (L.) Vahl, Symb. Bot. 2: 46. 1791; C. B. Cl. in Hook, f. Fl. Brit. India 4: 133. 1883; Cooke, Fl. Pres. Bombay 2: 263. 1958 (Repr. ed.); Kulkarni in Singh *et al.* Fl. Maharashtra State Dicot. 2: 415. 2001. *Nama zeylanica* L. Sp. Pl 226. 1753. KERITI, POPTI.

Herbs, aquatic or semi-aquatic, procumbent-ascending, rooting at lower nodes. Leaves lanceolate or linear-lanceolate, 2–5 × 0.3–1.5 cm, membranous, acute, glabrous. Flowers numerous, in racemes on short lateral branches. Calyx glandular hairy, enlarged in fruit; segments 5, lanceolate, 6 mm long, acute. Corolla of a fine blue with darker veins, deeply divided; segment 5, 6–7 mm long; lobes ovate, acute. Stamens exerted, filaments dilated at base; anthers sagittate. Styles 2, distinct, spreading. Capsule ovoid-oblong, 4–6 mm, enclosed in enlarged persistent sepals. Seeds numerous, minute, oblong.

Fls. & Frts.: November–February. *Exsiccata*: SUK-1771.
Frequent, in wet situations and along banks of streams, in rice fields.
Localities: Amba, Patgaon.

Local uses: The leaves, beaten into a pulp and applied as poultice, are considered to have a cleaning and healing effect on neglected and callous ulcers.

BORAGINACEAE Juss.

Coldenia procumbens L. Sp. Pl. 125. 1753; C. B. Cl. in Hook. f. Fl. Brit. India 4: 144. 1883; Cooke, Fl. Pres. Bombay 2: 271. 1958 (Repr. ed.); Kulkarni in Singh *et al.* Fl. Maharashtra State Dicot. 2: 416. 2001. TRIPANKI.

Prostrate or procumbent herbs; branches spreading, 3–10 cm long; white pubescent. Leaves ovate-oblong, 1–2 × 0.5–1.5 cm, tapering at base, with crenate and crisped margin, rounded or obtuse at apex, hairy on both surfaces. Flowers solitary-axillary, nearly sessile. Calyx divided to the base, densely hairy; sepals 4, ovate, 2–2.5 mm long, acute, ciliate. Corolla white, fading pale yellow, slightly exerted; lobes 4, oblong, 1.5–2 mm long, rounded at apex. Stamens 4, little exerted from the corolla tube. Fruit dry, pyramidal, 4-lobed, 3–4 mm long, separating into 1-celled, beaked nutlets muriculate, hairy.

Fls. & Frts.: December–May. *Exsiccata*: SUK-1772.
Common, in dried pond and harvested rice fields.

Localities: Kagal, Ichalkaranji, Shirol.

Local uses: Fresh leaves are boiled and paste is applied on rheumatic swellings. Juice of plant is used to wash chronic and old wounds.

Cordia dichotoma Forst. f. Prodr. 18. 1786; Kulkarni in Singh *et al.* Fl. Maharashtra State Dicot. 2: 417. 2001. *C. obliqua* Willd. Phytogr. 4, t. 4. 1794; C. B. Cl. in Hook. f. Fl. Brit. India 4: 137. 1883, excl. var. *C. myxa* auct. plur. non L. 1753; C. B. Cl. in Hook. f. op. cit. 136; Cooke, Fl. Pres. Bombay 2: 265. 1958 (Repr. ed.). BHOKAR.

Moderate-sized, deciduous trees, up to 10 m tall. Leaves alternate, ovate, elliptic or orbicular, 4–9 × 3–5.5 cm, rounded, cordate or shortly cuneate at base, entire sinuate, dentate, obtuse, scabrous above, glabrous beneath. Flowers polygamous, in large, lax, terminal and axillary pedunculate cymose panicles. Calyx 4–6 mm long, glabrous outside, pubescent within; teeth 5, shallow. Corolla white; tube as long as the calyx; lobes oblong, obtuse or emarginate, recurved, as long as the tube. Drupes globose or ovoid, 1–2.5 cm diam., pale orange colored, supported by saucer-shaped, irregularly 5–10 lobed, longitudinally striate glabrous calyx; pulp sticky. Seeds 2–4, bony stone.

Fls. & Frts.: December–May. *Exsiccata*: SUK-1750.
Common, in moist deciduous or semi evergreen forests often cultivated.
Localities: Ajara, Gaganbawada, Kolhapur, Malakapur.

Local uses: Bark decoction is given to expel worms. Fruit syrup is given in dysuria, urinary inflammation. Bark is used to heal wounds. Fruit is used as an expectorant and astringent, valuable in all diseases of lungs. Fruits are purgative, useful in dry cough, diseases of the chest, chronic fever and burning of throat.

Cordia macleodii (Griff.) Hook. f. & Thoms. in J. Linn. Soc. 2: 128. 1858; C. B. Cl. in Hook. f. Fl. Brit. India 4: 139. 1883; Cooke, Fl. Pres. Bombay 2: 266. 1958 (Repr. ed.); Kulkarni in Singh *et al.* in Fl. Maharashtra State Dicot. 2: 419. 2001. *Hemigymnia macleodii* Griff. in Calc. J. Nat. Hist. 3: 363. 1843. DAHIVAN.

Deciduous trees, 8–10 m tall; bark smooth, ovate. Leaves alternate or sub opposite, broadly ovate or orbicular, 8–14 × 7–14 cm, truncate or cuneate at base, entire, obtuse, pubescent on the upper surface when young, rough with white scales when old, woolly tomentose beneath with stellate hairs. Flowers polygamous; subsessile, in dense paniculate; terminal and axillary, tomentose cymes. Calyx obconic, 6–8 mm long, densely tomentose, outside, ribbed, teeth short, obtuse. Corolla white, 12–15 mm long, lobes spathulate-oblong, 7–8 mm long, longer than the tube. Stamens 5–6 exerted; filaments hairy at base. Male flowers with rudimentary ovary but without style and stigma. Drupes ovoid, 15–20 mm long, acute, supported on the cup-shaped, ribbed, tomentose, persistent calyx.

Fls. & Frts.: February–August. *Exsiccata*: SUK-668.
Common, in deciduous forests.
Localities: Kalamba.

Local uses: Leaves are applied as a plaster in strain and bruises.

Ehretia laevis Roxb. Pl. Cor. 1: 42, t. 55. 1795 & Fl. Ind. 2: 341. 1824; C. B. Cl. in Hook. f. Fl. Brit. India 4: 141. 1883; Cooke, Fl. Pres. Bombay 2: 269. 1958 (Repr. ed.); Kulkarni in Singh *et al.* Fl. Maharashtra State Dicot. 2: 424. 2001. DATRNG.

Small trees, 4–10 m tall. Leaves 5–12 × 3–7.5 cm, variable, often unsymmetric, elliptic-oblong or obovate, subacute, acuminate, obtuse or emarginate at apex, entire, glabrous, paler beneath, unequally cuneate. Flowers

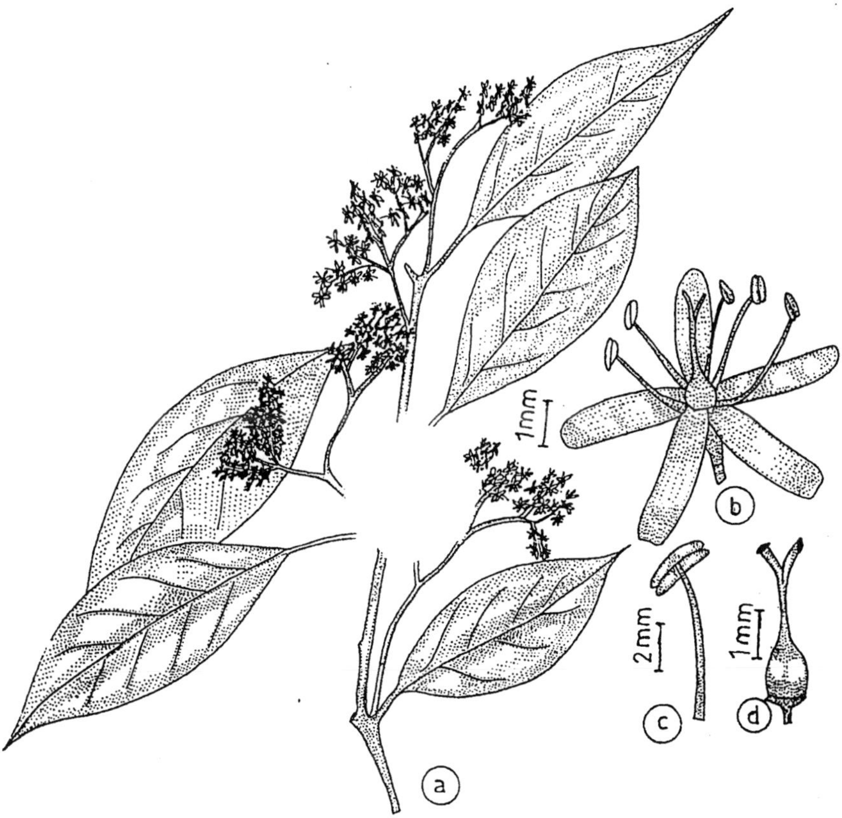

Text fig. 29: Ehretia laevis a-twig; b-flower; c-stamens; d-carpel.

white, sessile or nearly so, in much-branched, glabrous, or pubescent corymbose axillary and terminal cymes. Composed of unilateral spikes or racemes. Calyx 1–1.5 mm long, pubescent; segments ovate, acute. Corolla 5 mm long; lobes 3 mm long, obtuse, spreading. Filaments glabrous; anthers not apiculate. Style stout, more or less deeply bifid; stigma capitate. Drupes 2-lobed, rather broad than long, red at first, black and wrinkled when dry.

Fls. & Frts.: January–June. *Exsiccata*: SUK-1757.

Frequent, in deciduous forests.

Localities: Amboli, Sawantwadi, Tillari.

Local uses: Leaves are boiled in water with little salt. Juice prepared from boiled leaves is given in debility due to fever.

Heliotropium indicum L. Sp. Pl. 130. 1753; C. B. Cl. in Hook f. Fl. Brit. India 4: 152. 1883; Cooke, Fl. Pres. Bombay 2: 273. 1958 (Repr. ed.); Kulkarni in Singh *et al.* Fl Maharashtra State Dicot. 2: 424. 2001. BHURUNDI.

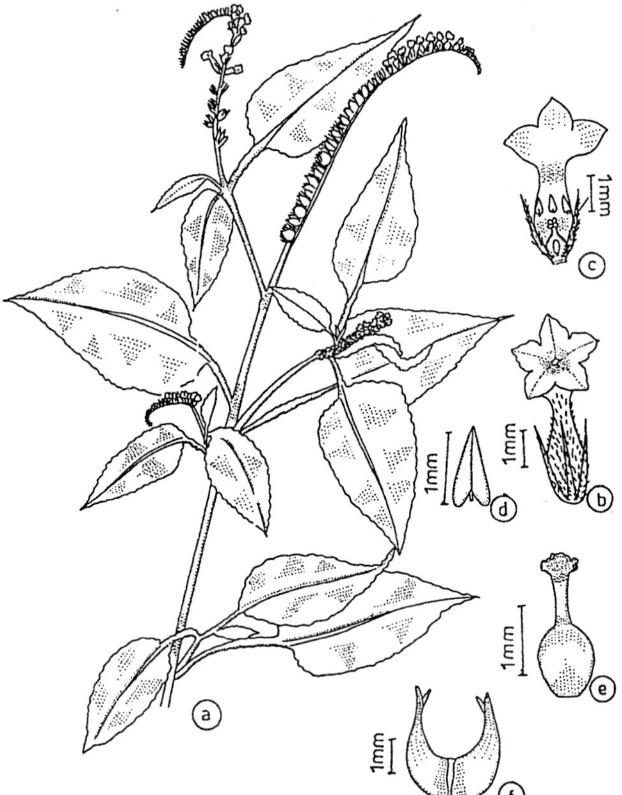

Text fig. 30: Heliotropium indicum a-twig; b-flower; c-l. s. of flower; d-stamen; e-carpel; f-fruit.

Erect or decumbent-ascending, annual herbs; branches hirsute. Leaves alternate, ovate-elliptic or ovate-oblong, 3–8 × 2–4 cm, sparsely hairy on both sides, apex acute, base, contracted into partially winged petiole. Flowers white or bluish-white, sessile in 2-ranked, dense, scorpoid cymes; calyx 2–2.5 mm long, bristly with few long hairs outside; segments 1–1.5 mm, linear-lanceolate, acute, unequal. Corolla 3–3.5 mm long, tube 3 mm long, hairy; limb infundibuliform; orbicular-oblong. Stamens inserted below the middle of the corolla tube; filaments very short. Stigma with annular frill at the base. Fruit 2-lobed, nutlets beaked, angled. One seeded.

Fls. & Frts.: July–September. *Exsiccata*: SUK-617.
Common, weed in open moist places.
Localities: Kas, Kolhapur, Satara, In all districts.

Local uses: Plant is externally applied to itch caused by Leech. Decoction of leaf is used for cleaning and healing ulcers. Fresh leaf is mixed with equal quantity of leaf of *Cryptolepis buchanani* paste externally in bee sting.

Trichodesma zeylanicum (Burm. f.) R. Br. Prodr. 496. 1810; C. B. Cl. in Hook. f. Fl. Brit. India 4: 154. 1883; Cooke, Fl. Pres. Bombay 2: 282. 1958 (Repr. ed.); Kulkarni in Singh *et al.* Fl. Maharashtra State Dicot. 2: 436. 2001. *Borago zeylanica* Burm. f. Fl. Ind. 41, t. 14, f. 2. 1768. JALASHIRASI.

Erect, annual herbs, 30–60 cm tall; stem clothed with long soft hairs. Leaves opposite, oblong-lanceolate, 2.5–12 × 1–3.5 cm, narrowed at base, entire, acute or subacute, densely hispid above, finely pubescent beneath. Flowers in terminal, elongate, bracteate, paniculate cymes or occasionally a few solitary flowers in the upper axils. Calyx divided almost to the base; sepals ovate-lanceolate, 6–10 mm long, rounded at base. Corolla pale blue, shorter than the calyx; lobes ovate or obovate rounded with sudden spirally twisted acumination. Staminal cone exerted; connectives long, twisted. Nutlets 4, broadly ovoid, 4–5 × 3–3.5 mm, subobtuse, slightly rugose on the inner face, enclosed in persistent calyx.

Fls. & Frts.: November–February. *Exsiccata*: SUK-1758.
Common, in dry regions.
Localities: Kagal, Kartikswami Ghat, Kolhapur, Ramling.

Local uses: Root is made into a paste and applied to reduce swellings. Infusion of leaves is given as diuretic.

CONVOLVULACEAE Juss.

Argyreia boseana Sant. & Patel in Trans. Bose Res. Inst. Calcutta 22: 35, t. 3. 1958; Venkanna & Das Das in Singh *et al.* Fl. Maharashtra State Dicot. 2: 440. 2001. *A. hookeri* sensu Cooke, 2: 325. 1958 (Repr. ed.). GAVELI, MARALMATHNGI.

Climbing shrubs; stems grooved. Leaves broadly ovate, 10–13 × 8–10 cm, acuminate, glabrous, base cordate. Flowers in long pedunculate few-flowered cymes; peduncles 15–20 cm long, grooved, hairy. Sepals 5, ovate, lanceolate, 1–1.5 cm long, sparsely hairy on back and with membranous margins, the outer more than 1.2 cm long, slightly narrower and longer than the inner. Corolla rose-colored, 5–6 cm long, tubular, campanulate; tube 2.5 cm long, glabrous; limb 5–6 cm wide at mouth, with 5 acute lobes, the bands sparsely hairy with appressed hairs. Stamens 5; filaments hairy at base; anthers oblong, obtuse. Berry 1–1.2 cm in diam., globose.

Fls. & Frts.: July–December. *Exsiccata*: SUK-278.
Rare, in hilly tracts and in plains.
Localities: Amba, Amboli, Barki, Borbet, Dajipur, Gaganbawada.

Local uses: Stem is used to heal wounds and ulcers. Paste of stem with bark of *Pittosporum floribundum* is applied externally in scabies and elephantiasis. Stem decoction is given to reduce obesity.

Argyreia nervosa (Burm. f.) Bojer, Hort. Maurit. 224. 1837; Venkanna & Das Das in Singh *et al.* Fl. Maharashtra State Dicot. 2: 445. 2001. *Convolvulus nervosus* Burm. f. Fl. Ind. 48, t. 20. F. 1. 1768. *Argyreia speciosa* (L. f.) Sweet, Hort. Brit. 1: 289. 1827; C. B. Cl. in Hook. f. Fl. Brit. India 4: 185. 1883; Cooke, Fl. Pres. Bombay 2: 324. 1958 (Repr. ed.). SAMUDRASHOK.

Large climbing shrubs; branches densely tomentose. Leaves ovate, 10–25 × 6–18 cm, apex acute, base cordate, glabrous above, persistently white tomentose beneath. Flowers in sub-capitate cymes; peduncles 7–15 cm long, white tomentose; calyx white tomentose outside; sepals oblong-elliptic, obtuse. Corolla 5–6 cm long, tubular-infundibuliform, band silky-pubescent outside, tube inflated, white pubescent outside, rose purple and glabrous inside. Stamens included; filaments pilose at base. Fruit globose, 2 cm across. Seeds 2–4.

Fls. & Frts.: July–December. *Exsiccata*: SUK-160.
Occasional, along hedges near villages.
Localities: Alate, Kagal, Kolhapur.

Local uses: Root is used in rheumatic affections and diseases of the nervous system. Leaves are maturative and absorptive and are used as emollient poultice for wounds and externally in skin diseases. Root is used in gleet, gonorrhoea, strangury and in chronic ulcers. Roots are mixed with vinegar the sap is rubbed over the body to reduce obesity.

Convolvulus arvensis L. Sp. Pl. 153. 1753; C. B. Cl. in Hook. f. Fl. Brit. India 4: 219. 1883; Cooke, Fl. Pres. Bombay 2: 303. 1958 (Repr. ed.); Venkanna & Das Das in Singh *et al.* Fl. Maharashtra State Dicot. 2: 449. 2001. CHANDVEL, HIRANPAG.

Perennial herbs; stems angular, glabrous or pubescent. Leaves ovate-oblong, 2.5–3 cm long, glabrous, obtuse and apiculate, hastate at the base, lower leaves often lobed. Flowers axillary, 1 or 2 or more flowered; pedicels much longer than the calyx 3.5–5 mm long, slightly unequal, oblong-elliptic; obtuse or truncate, glabrous or nearly so, unequal. Corolla 2 cm long, broadly infundibuliform, pink or nearly white, glabrous. Stamens unequal ovary glabrous. Capsules globose, 0.6–0.8 cm across. Seeds 4, subtrigonous, dark reddish-brown.

Fls. & Frts.: December–June. *Exsiccata*: SUK-1742.
Common, in cultivated lands, along road sides and railway lines.
Localities: Ashta, Atigre, Shiroli, Talsande, Tasgaon.

Local uses: Root is given orally in piles and as a purgative in constipation.

Evolvulus alsinoides (L.) L. Sp. Pl. ed. 2, 2: 392. 1762; C. B. Cl. in Hook. f. Fl. Brit. India 4: 220. 1883; Cooke, Fl. Pres. Bombay 2: 297. 1958 (Repr. ed.); Venkanna & Das Das in Singh *et al.* Fl. Maharashtra State Dicot. 2: 453. 2001. *Convolvulus alsinoides* L. Sp. Pl. 157. 1753. VISHNUKRANTA.

Herbs, prostrate or suberect, pubescent. Leaves elliptic-oblong, 0.5–2 × 0.3–0.5 cm, subsessile, appressed silky, acute to obtuse at base and also at apex. Flowers axillary, pedunculate in 1 to several flowered dichasia or solitary, or in 1–3 flowered cymes; light blue; peduncles very long, filiform. Calyx densely silky; sepals 5, 3–4 mm long, lanceolate, acute. Corolla 5 mm long, infundibuliform or subrotate; limb plicate, subentire. Capsules 4 mm in diam., globose, thin, 4-valved. Seeds 4, glabrous, black.

Fls. & Frts.: July–November. *Exsiccata*: SUK-121.
Common, in grasslands, waste places in dry areas.
Localities: Kagal, Shivaji University campus, Kolhapur.

Local uses: It is used in fever with dried ginger and milk. It is mixed with oil and applied externally to promote the growth of hair. Orally it is given to strengthen the brain and memory. Juice of plant is given to stop bleeding.

Ipomoea aquatica Forssk. Fl. Aeg.–Arab. 44. 1775; C. B. Cl. in Hook. f. Fl. Brit. India 4: 210 1883; Cooke, Fl. Pres. Bombay 2: 315. 1958 (Repr. ed.); Venkanna & Das Das in Singh *et al.* Fl. Maharashtra State Dicot. 2: 456. 2001. NELICHI-BHAJI.

Annual or biennial herbs; floating, rooting at nodes; stems thick, hollow. Leaves elliptic-oblong or subdeltoid, 4–8 × 3–6 cm, acute, glabrous, base usually dilated, cordate or hastate with rounded or acute lobes, less commonly truncate. Flowers solitary or in few-flowered, axillary cymes; peduncles 1.5–8 cm long. Calyx 6–8 mm long, glabrous; sepals subequal, ovate, oblong, obtuse and minutely mucronate, with thin pale margins. Corolla pale pink purple, 3–5 cm long, glabrous. Stamens included; filaments hairy at base. Capsules, ovoid, 7–8 mm long, glabrous. Seeds 2–4, broadly trigonous, grayish pubescent.

Fls. & Frts.: September–December; April–June. *Exsiccata*: SUK-120.
Common, on margins of ponds and lakes.
Localities: Kagal, Kolhapur, Shiroli.

Local uses: Leaves are externally applied on toes to lessen the inflammation. Leaf decoction is given as a tonic in general debility. Plant is used as anthelmintic, useful in leprosy, leucoderma, fever, jaundice, biliousness etc.

Ipomoea mauritiana Jacq. Collect. 4: 216. 1791; Venkanna & Das Das in Singh *et al.* Fl. Maharashtra State Dicot. 2: 465. 2001. *I. digitata* auct. non L. 1753; C. B. Cl. in Hook. f. Fl. Brit. India 4: 202. 1883; Cooke, Fl. Pres. Bombay 2: 318. 1958 (Repr. ed.). BHUIKOHLA.

Twiners, sometimes prostrate; roots tuberous; stems terete. Leaves palmately divided to or mostly beyond the middle, base cordate or truncate; 6–14 × 6–16 cm, segments 3–7 (9), lanceolate, apex mucornulate; petioles slightly muricated, 3–10 cm long. Flowers in many-flowered corymbosely paniculate cymes. Peduncles solitary, axillary, ca 15 cm long. Sepals 6–12 mm long, orbicular-oblong, subobtuse, concave, glabrous. Corolla purple, 2.5–5.5 cm long, campanulate, infundibuliform, lobes of the limb emarginate. Ovary 4-celled, glabrous. Capsules ovoid, 1.2–1.4 cm long, 4-celled, 4-valved, surrounded by the enlarged rather fleshy sepals. Seeds clothed with brownish cottony readily detachable hairs.

Fls. & Frts.: August–January. *Exsiccata*: SUK-119.
Occasional, in moist deciduous forests.
Localities: Dindewadi, Patgaon, Ratnagiri.

Local uses: Tuberous roots are used as a tonic. It is orally given in rheumatism and also applied externally with wine. Tubers are given with roots of *Asparagus racemosus* to increase secretion of milk. Root is used as tonic, aphrodisiac, galactogogue, diuretic, stimulant, useful in burning sensation, vomiting diseases of the blood, improves voice and the complexion.

Ipomoea nil (L.) Roth, Cat Bot. 1: 36. 1797; Venkanna & Das Das in Singh *et al.* Fl. Maharashtra State Dicot. 2: 465. 2001. *Convolvulus nil* L. Sp. Pl. ed. 2. 219. 1762. *Ipomoea hederacea* auct. non Jacq. 1786; C. B. Cl. in Hook. f. Fl. Brit. India 4: 199. 1883; Cooke, Fl. Pres. Bombay 2: 321. 1958 (Repr. ed.).

Annual or perennial, herbaceous twiners; stem hirsute, 1–3 m or more long. Leaves broadly ovate in outline, entire or 3-lobed, 3–10 × 4–12 cm, cordate at base, acuminate, appressed hairy on both surfaces. Flowers in axillary, 1-few flowered cymes; peduncles 2–10 cm long, hispid. Sepals herbaceous, narrowly lanceolate-acuminate, patently hispid on the back near the base. Corolla blue, funnel-shaped, 5–6 cm long, turning red or purple on fading; tube white; limb 3–4 cm across. Stamens and styles included; filaments hairy at base. Capsules ovoid-globose, 8–10 mm in diam., mucronate, glabrous, 3-valved. Seeds 6 or less, ovoid, 3-gonous, black, velvety or gray pubescent.

Fls. & Frts.: September–December. *Exsiccata*: SUK-1743.
Common, around fields, in hedges, on bushes in forests.
Localities: Kolhapur, Miraj, Sangli.

Local uses: The seeds are used externally to cure inflammations. Seed is used with dried ginger and rock salt and powder is given as anthelmintic, cure cough and acts as laxative. Seeds are externally applied on scabies. Seeds cure inflammations, abdominal diseases, fevers, headache, diseases of the head, bronchitis, leucoderma, pains in the joints and dry the phlegm.

Merremia gangetica (L.) Cufod. in Bull. Jard. Bot. Brux. Suppl. 31: 743. 1961; Venkanna & Das Das in Singh *et al.* Fl. Maharashtra State Dicot. 2: 475. 2001. *Merremia emarginata* (Burm. f.) Hall. f. in Bot Jahrb. 16: 552. 1893; Cooke, Fl. Pres. Bombay 2: 305. 1958 (Repr. ed.). *Convolvulus gangeticus* L. in Torner, Cent. Pl. 2: 9. 1756. *Ipomoea reniformis* Choisy in Mem. Soc. Phys. Geneve 6: 446. 1834; C. B. Cl. in Hook. f. Fl. Brit. India 4: 206. 1883.UNDIRKANI.

Perennial, prostrate herbs, rooting at nodes, sparsely hairy or glabrescent, 50–60 cm long. Leaves reniform or ovate-cordate, 1.2–2.5 cm broad, obtuse, crenate, glabrous. Flowers yellow, axillary solitary or few together on a very short peduncle. Sepals slightly hairy on the back, strongly ciliate on the margins, the 2 outer 4 mm long, ovate, mucronate, the 3 inner longer, reaching nearly 6 mm long, subquadrate, deeply divided at the apex into 2 truncate divaricate lobes. Corolla 6 mm long; lobes of the limb acute. Capsules subglobose, 4–5 mm in diam. smooth, surrounded by the slightly enlarged, ciliate sepals. Seeds glabrous, dark chestnut colored.

Fls. & Frts.: September–October. *Exsiccata*: SUK-268.
Common, in wastelands and water logged areas.
Localities: Kolhapur, Jaisingpur, Narsinghwadi.

Local uses: It is orally given in constipation. It is given with milk and sugar. Externally applied in certain skin diseases.

Merremia vitifolia (Burm. f.) Hall. f. in Bot. Jahrb. Syst. 16: 552. 1893; Cooke, Fl. Pres. Bombay 2: 307. 1958 (Repr. ed.); Venkanna & Das Das in Singh *et al.* Fl. Maharashtra State Dicot. 2: 479. 2001. *Convolvulus vitifolius*

Burm. f. Fl. Ind. 45, t. 18, f. 1. 1758. *Ipomoea vitifolia* (Burm. f.) Bl. Bijdr. 709. 1825; C. B. Cl. in Hook. f. Fl. Brit. India 4: 213. 1883. NAVIT.

Twiners; stems slender, terete, striate, patently hirsute with yellowish hairs, old stems woody, glabrescent and warted. Leaves palmately divided to base, with 5–7, lanceolate, mucronulate lobes, margins coarsely dentate to irregularly pinnately lobed segments, glabrous or hairy on veins beneath the middle segments, 2.5–10 × 0.5–3 cm, lateral ones smaller. Flowers 1–7 in pedunculate cymes; peduncles 6–1.2 cm long, patently hairy. Calyx more or less densely clothed outside with long spreading hairs; sepals 1–1.2 cm long, enlarged in fruit, broadly elliptic, obtuse; the outer slightly shorter, apiculate. Corolla yellow, 2.5–5 cm long, subanulate; bands distinct, strongly lineate. Capsules 1.2 cm in diam., subglobose. Seeds glabrous.

Fls. & Frts.: October–November. *Exsiccata*: SUK-1745.

Infrequent, along hedges in secondary forests, on river banks and waysides of deciduous forests.

Localities: Amba, Amboli, Dajipur, Radhanagari, Ramling, Sakharpa, Tillari.

Local uses: It is orally given in constipation. It is given with milk and sugar. Externally applied in certain skin diseases.

Operculina turpethum (L.) S. Manso, Enum. Subst. Bras. 16. 1836; Cooke, Fl. Pres. Bombay 2: 309. 1958 (Repr. ed.); Venkanna & Das Das in Singh *et al.* Fl. Maharashtra State Dicot 2: 479. 2001. *Convolvulus turpethum* L. Sp. Pl. 155. 1753. *Ipomoea turpethum* (L.) R. Br. Prodr. 485. 1810; C. B. Cl. in Hook. f. Fl. Brit. India 4: 212. 1883. DUDHKALMI, NISHOTTAR, NISOTHAR.

Perennial twiners with long fleshy much branched roots; stems narrowly or broadly 3–5 winged, glabrous or sparsely hairy, younger parts sometimes tomentose. Leaves variable in shape, mostly orbicular or ovate or ovate-lanceolate, 5–15 × 1–14 cm, base cordate or hastate, apex obtuse or acuminate, mostly entire, glabrous above, pubescent beneath; petioles terete or winged. Flowers in axillary 1-few flowered cymes; peduncles 2–18 cm long, terete or winged. Sepals 2.2 cm long, much enlarged in fruit, broadly ovate or suborbicular, obtuse, mucronate, concave, pubescent, 3 inner sepals smaller, 1.9 cm long. Corolla white, 3–5 cm long, subcampnulate. Anthers narrowly oblong, cordate. Capsules depressed globose, 1.5–1.7 cm in diam., seeds ovoid, 5–6 mm long black.

Fls. & Frts.: September–April. *Exsiccata*: SUK-1748.

Common, on bushes and in hedges.

Localities: Tillarinagar.

Local uses: Tuberous roots of white variety are used as a purgative. Black variety is powerful irritant and drastic purgative. Root is anthelmintic;

useful in ascites, itch, leucoderma, constipation, abdominal troubles, inflammation, fever, anemia, piles and erysipelas.

Xenostegia tridentata (L.) Austin & Staples in Brittonia 32: 533. 1980; Venkanna & Das Das in Singh *et al.* Fl. Maharashtra State Dicot 2: 483. 2001. *Convolvulus tridentatus* L. Sp. Pl. 157. 1753. *Ipomoea tridentata* (L.) Roth in Roem. Arch. Bot. 1, 2: 38. 1798; C. B. Cl. in Hook. f. Fl. Brit. India 4: 205. 1883. *Merremia tridentata* (L.) Hall. f. Bot. Jahrb. Syst. 16: 552. 1893; Cooke, Fl. Pres. Bombay 2: 306 (Repr. ed.). MORGA.

Herbs, prostrate with woody rootstock; branches 10–80 cm long, angular, glabrous. Leaves linear, hastate or oblong-obovate, 4–20 × 0.5–6 mm, apex acute, murcronulate. Flowers in 1–2 (-3) flowered racemes. Sepals 4–5 mm long, elliptic, subobtuse, strongly mucronate, glabrous, the outer rather shorter than the inner. Corolla pale yellow, less than 1 cm long. Bands not very strongly lineate. Capsules ovoid, 4 mm in diam., 4-valved, straw colored. Seeds 4 or less, dull black.

Fls. & Frts.: September–April. *Exsiccata*: SUK-1749.
Localities: Halkarni, Nesari, Hatkanangle.
Local uses: Juice of leaves is given with milk and sugar in urinary complaints. Leaf juice is applied externally in rheumatism and pile. It is used as tonic and laxative.

CUSCUTACEAE Dumort.

Cuscuta chinensis Lam., Encycl. 2: 229. 1786; C. B. Cl. in Hook. f. Fl. Brit. India 4: 226. 1883; Cooke, Fl. Pres. Bombay 2: 293. 1958 (Repr. ed.); Venkanna & Das Das in Singh *et al.* Fl. Maharashtra State Dicot. 2: 491. 2001. AMARVEL.

Leafless, parasitic annuals. Stems slender, filiform, much branched, often forming a tangled mass. Flowers solitary or in shortly pedunculate cymes. Calyx 1–5 mm long divided ½ way down; lobes deltoid-ovate, subobtuse, often with a tubercular keel on the back. Corolla campanulate, ovoid, globose, 2–5 mm long; lobes equaling the tube, ovate-oblong, subobtuse; scales at the base of the filaments fimbriate. Filaments 0.5 mm long. Styles 2 distinct. Capsules 1.5 mm in diam., globose, hyaline, divided into 2 lobes by a deep furrow on top. Seeds 1.0–1.5 mm long, brown, minutely foveolate.

Fls. & Frts.: August–October. *Exsiccata*: SUK-1764.
Common, parasite, mostly on *Lawsonia inermis*.
Localities: Kolhapur, Satara.

Local uses: It is used with black pepper and used as a purgative, reduces swellings. Juice is mixed with sugar and given in fever and cough.

Cuscuta hyalina Roth, Nov. Pl. Sp. 100. 1821; C. B. Cl. in Hook. f. Fl. Brit. India 4: 226. 1883; Cooke, Fl. Pres. Bombay 2: 292. 1958 (Repr. ed.); Venkanna & Das Das in Singh *et al.* Fl. Maharashtra State Dicot. 2: 492. 2001.

Leafless, parasites, stem slender, closely twining. Flowers pale yellow in fascicle or in shortly pedunculate cymes or racemes. Calyx 3 mm long, divided rather more than ½ way down; lobes 5, ovate, acute or acuminate, slightly unequal. Corolla 1.5 mm long, 5-lobed; lobes ovate-lanceolate, 2 mm long, very acute, or acuminate; scales 0. Filaments distinct; styles 2 slender, distinct. Capsules globose, 3–4 mm across, brown, 2-celled with a narrow interstylar apertures, surrounded by persistent withered corolla. Seeds oblong, 1.5 mm long, horizontal.

Fls. & Frts.: August–November. *Exsiccata*: SUK-1765.
Common, in drier parts of the region.
Localities: Ichalkaranji, Kolhapur, Satara.

Local uses: It is used with black pepper as a purgative, reduces swellings. Juice is mixed with sugar and given in fever and cough.

SOLANACEAE Juss.

Datura innoxia Mill. Gard. Dict. ed. 8, no. 5. 1768; Madhusudan Rao in Singh *et al.* Fl. Maharashtra State Dicot 2: 495. 2001. *D. metel* sensu Sims in Curits, Bot. Mag 35: t. 1440. 1812 non L. 1753; C. B. Cl. in Hook. f. Fl. Brit. India 4: 243. 1883; Cooke, Fl. Pres. Bombay 2: 344. 1958 (Repr. ed.). DHOTRA.

Erect, gray-tomentose, much-branched herbs, 60–90 cm tall; stems woody below. Leaves ovate; 7–15 × 5–9 cm, unequal sided and sub-cordate at base, obscurely sinuate-dentate, acute, densely soft, pubescent on both surfaces. Flowers solitary. Calyx 6–10 cm long, tubular, pubescent outside; teeth triangular, acute, 1–2 cm long. Corolla pure white, 10–12 cm across; lobes deltoid, acute, with 5–6 mm long teeth alternating with them. Filaments adnate to corolla tube for a considerable length. Capsules globose, 2.5–3.5 cm in diam., minutely gray-tomentose, clothed with long, slender, weak spines. Seeds subreniform, pale brown, smooth.

Fls. & Frts.: September–March. *Exsiccata*: SUK-1766.
Common, weed in waste places.
Localities: Hatkanagale, Karad, Shiroli.

Local uses: Smoke from the burning leaves is recommended for relieving attacks of asthma. Seeds are heated for few minutes and then applied them as a poultice over the spleen in intermittent fever. Root powder is applied to the gums in order to relieve the pain of toothache.

Datura metel L. Sp. Pl. 179. 1753; Madhusudan Rao in Singh *et al.* Fl. Maharashtra State Dicot. 2: 496. 2001. *D. fastuosa* L. Syst. Nat. ed. 10, 2: 932.

1759; C. B. Cl. in Hook. f. Fl. Brit. India 4: 243. 1883; Cooke, Fl. Pres. Bombay 2: 343. 1958 (Repr. ed.); *D. fastuosa* var. *alba* C. B. Cl. in Hook. f. op. cit. ; Cooke, op. cit. 344. KALA-DHOTRA.

Erect, glabrous, much-branched, annual herbs, 0.5–1 m tall; branches often tinged with purple. Leaves ovate, 4.5–20 × 3–15 cm, obliquely cuneate at base, sinuate-dentate, acute or acuminate. Flowers solitary. Calyx 8–10 cm long, tubular, pubescent outside; lobes lanceolate, acuminate. Corolla 15–18 cm long, purplish outside, white within; lobes triangular-acute. Filaments adnate to the corolla for half their length. Capsules globose, 4 × 3.5 cm, spinous with short tubercles.

Fls. & Frts.: November–February. *Exsiccata*: SUK-163.
Occasional, in wastelands.
Localities: Kagal, Kolhapur, Miraj, Wai.

Local uses: Whole plant is toxic, narcotic, removes the pain of tumors and piles; seeds are narcotic, febrifuge, anthelmintic, good in inflammations, cause headache. Other uses are same as *Datura innoxia.*

Datura stramonium L. Sp. Pl. 179. 1753; C. B. Cl. in Hook. f. Fl. Brit. India 4: 242. 1883; Madhusudan Rao in Singh *et al.* Fl. Maharashtra State Dicot. 2: 496. 2001. DHOTRA.

Erect, sparsely pubescent, annual herbs, 40–90 cm tall; branches green. Leaves ovate, 8–15 × 5–10 cm, obliquely cuneate at base, deeply sinuate, dentate, acute or acuminate. Calyx 3.5–5 cm long, tubular, 5-ribbed, teeth lanceolate; 5–6 mm long, hairy within. Corolla 6–8 cm long; lobes rounded, acuminate. Filaments adnate to the corolla tube for more than half their length. Capsules globose, 4–5 cm in diam., minutely tomentose, covered with numerous, sharp spines. Seeds subreniform, brown, smooth.

Fls. & Frts.: September–October. *Exsiccata*: SUK-609.
Infrequent in waste lands.
Localities: Kolhapur.

Local uses: Same as *Datura metel.*

Nicandra physalodes (L.) Gaertn. Fruct. 2: 237, t. 141, f. 2. 1791 ('*physaloides*'); C. B. Cl. in Hook. f. Fl. Brit. India 4: 240. 1883; Cooke, Fl. Pres. Bombay 2: 346. 1958 (Repr. ed.); Madhusudan Rao in Singh *et al.* Fl. Maharashtra State Dicot. 2: 497. 2001. *Atropa physalodes* L. Sp. Pl. 181. 1753. RANPOPATI.

Erect, much fistular, annual herbs, 15–150 cm tall; stem fistular, glabrous. Leaves ovate lanceolate, 6–15 cm long, acute and decurrent at base, lobed or coarsely toothed, acute, glabrous; petioles winged in the upper part, 2–10 cm long. Flowers 1.5–2.5 cm across, solitary, axillary, at first erect, nodding with age. Calyx 0.5–1 cm long, enlarging up to 3 cm long in fruit, divided to the base; sepals ovate-lanceolate, acute. Corolla pale blue with white short tube; lobes ovate-orbicular, 1.5–2 cm long. Stamens 5, included. Berries globose, 2–15 mm in diam., brown, enclosed in enlarged calyx. Seeds numerous, flat brown.

Fls. & Frts.: September–December. *Exsiccata*: SUK-1767.
Frequent, in on plains as a weed.
Localities: Islampur, Kolhapur, Satara.

Local uses: Leaf juice is given in urinary complaints.

Nicotiana glauca R. Grah. in Edinb. N. Phil. J. 5: 175. 1828 et in Bot. Mag. t. 2827. 1829; Madhusudan Rao in Singh *et al.* Fl. Maharashtra State Dicot. 2: 514. 2001.

Glaucous shrubs, 2.5–3 m tall; stem terete, fistular, glabrous. Leaves ovate-lanceolate, 5–15 × 2–7 cm, acute and decurrent on the petiole, entire, acute, glabrous, black on drying. Flowers in axillary and terminal, lax panicles. Calyx tubular, 10–12 mm long, pubescent outside; teeth triangular subulate, 1–1.5 mm long, acuminate. Corolla tube narrow, 3–3.5 cm long, pale yellow with bright yellow lobes, pubescent outside; lobes minute, less than 1 mm long. Stamens included. Capsules ovoid or ellipsoid, little shorter than the calyx, membranous. Seeds numerous, globose, minute, black.

Fls. & Frts.: February–May. *Exsiccata*: SUK-628.
Cultivated, maintained in botanical garden in Shivaji University, Kolhapur.
Localities: Narsihanwadi, Sangli.

Local uses: Leaf powder is chewed and applied to insect bites and to rid the skin of ticks and leech. It is apparently toxic to ticks and causes them to drop off. Dried leaves warmed and deeped in oil are applied to fresh wounds. Decoction is given as an enema. Dried leaves are chewed in caries of the teeth, bronchitis, tubercular glands of the neck, dimness of sight. Leaves are laxative, tonic, emetic, carminative, anthelmintic, useful in bronchitis, asthma, caries of the teeth, skin diseases, scorpion sting, inflammations; cause bad eyesight.

Nicotiana tabacum L. Sp. Pl. 180. 1753; C. B. Cl. in Hook. f. Fl. Brit. India 4: 245. 1883; Cooke, Fl. Pres. Bombay 2: 346. 1958 (Repr. ed.); Madhusudan Rao in Singh *et al.* Fl. Maharashtra State Dicot. 2: 515. 2001. TAMBAKHU.

Erect, glandular-pubescent, much-branched, annual or perennial herbs, 30–90 cm tall. Leaves sessile, elliptic-oblong to ovate-lanceolate, 20–45 × 10–20 cm, narrowed and semi-amplexicaule at base, sinuate-repend, acute or acuminate, glabrous or hairy, upper one smaller and passing into bracts flowers in terminal panicles. Calyx cupular, 1–1.5 cm long, slightly enlarging in fruit, glandular hairy outside; lobes ovate, 5–6 mm long, acuminate. Corolla greenish-white, 5–8 mm long, shortly acuminate. Filaments hairy at base. Capsules ovoid, 15–18 mm long, 2–4 valved, more or less enclosed by persistent calyx, glabrous. Seeds numerous, globose, black.

Fls. & Frts.: January–March. *Exsiccata*: SUK-629.

Cultivated, native of tropical America.

Local uses: Same as *Nicotiana glauca*.

Physalis angulata L. Sp. Pl. 183. 1753; Madhusudan Rao in Singh *et al.* Fl. Maharashtra State Dicot. 2: 516. 2001. POPATI.

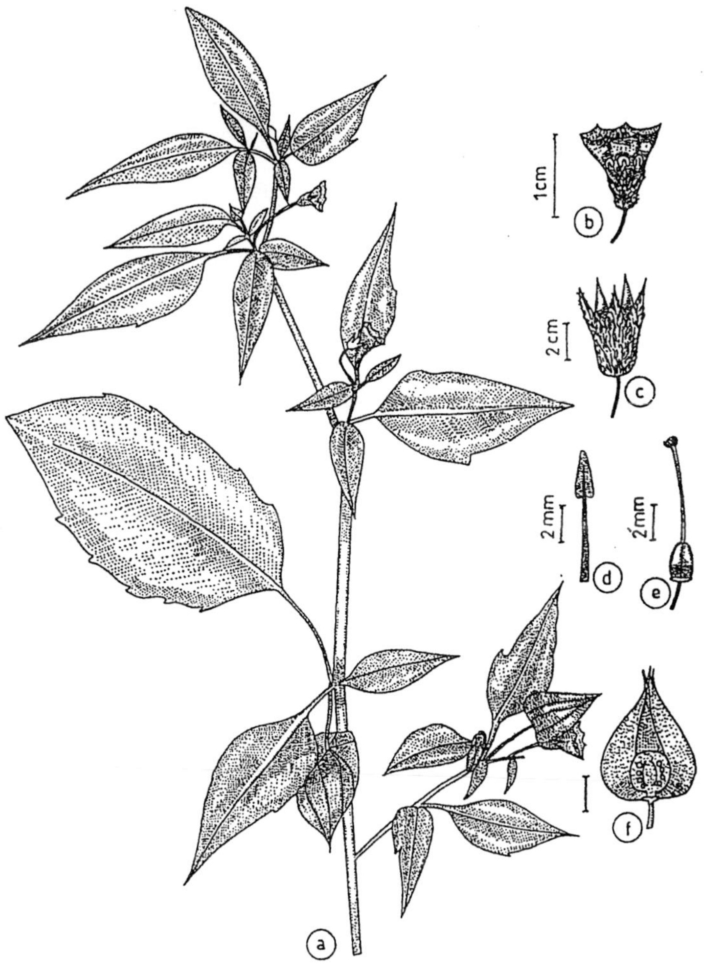

Text fig. 31: Physalis angulata a-twig; b-flower; c-calyx; d-stamen; e-carpel; f-fruit.

Soft, erect, much branched herbs, 20–45 cm, tall; stem prominently angular, glabrous, often rooting from lower nodes. Leaves broadly ovate or elliptic, 3–10 × 2–6 cm, cuneate and unequal sided at base, long acuminate, glabrous. Flowers solitary, axillary or terminal, nodding, 1–1.5 cm long. Calyx 4–5 mm long, divided half way down, glabrous; teeth triangular-ovate, acuminate, ciliate. Corolla pale yellow with 5, faint brown spots inside; lobes ovate, acute, ciliate. Anthers yellow, blue or violet. Berries ovoid-globose, 6–7 × 5–6 mm, orange yellow, enclosed in an enlarged, persistent calyx. Seeds many, compressed, granulate.

Fls. & Frts.: October–February. *Exsiccata*: SUK-137.

Common, weed of wet field.

Localities: Shiroli.

Local uses: Plant is made into paste with rice water and applied to restore flaccid breasts. Berries are given in abdominal troubles. Paste of plant is applied on palms and toes in burning sensation and inflammation.

Solanum anguivi Lam. Encycl. 2: 23. 1794; Madhusudan Rao in Singh *et al.* Fl. Maharashtra State Dicot. 2: 503. 2001. *S. indicum* auct. non L. 1753 p. p.; C. B. Cl. in Hook. f. Fl. Brit. India 4: 234. 1883; Cooke, Fl. Pres. Bombay 2: 336. 1958 (Repr. ed.). DORLI, MOTHIRINGNI, RINGNI.

Erect, much branched, prickly undershrubs, 30–120 cm tall; prickles stout, with broad base; branches covered with stellate hairs. Leaves ovate, 2–8 × 1.5–6 cm, cordate, cuneate or truncate at base, sinuately lobed into ovate acute lobes, acute, prickly on midnerve, pubescent above, densely so beneath; hairs simple mixed with stellate ones. Flowers in extra-axillary, racemose cymes. Calyx 2.5–3 mm long, stellate hairy outside; teeth triangular, 1.5 mm long, acute. Corolla purple, 6–8 mm long, clothed outside with purple, stellate hairs, lobes deltoid, ovate. 6.5 mm long, acute. Berries globose, yellow, glabrous. Seeds flat, 3–4 mm across, minutely pitted, brown.

Fls. & Frts.: August–December. *Exsiccata*: SUK-1798.

Common, in moist deciduous forests.

Localities: Dajipur, Gaganbawada, Kagal, Kolhapur, Mahabaleshwar, Radhanagari, Wai.

Local uses: Root is one of the component of Dashmula. Juice of the leaves with fresh juice of the ginger is administered to stop vomiting. Leaf paste is used on skin diseases. Decoction of root is given in bronchitis attended with fever. Root is used to remove foulness of the mouth, useful in leucoderma, cardiac troubles, fever, asthma, pain, bronchitis and vomiting.

Solanum nigrum L. Sp. Pl. 186. 1753; C. B. Cl. in Hook. f. Fl. Brit. India 4: 229. 1883; Cooke, Fl. Pres. Bombay 2: 332. 1958 (Repr. ed.); Madhusudan Rao in Singh *et al.* Fl. Maharashtra State Dicot. 2: 505. 2001. KANGONI.

Erect, glabrous or pubescent, annual herbs, 30–60 cm tall. Leaves ovate-lanceolate, 2–7 × 1.5–5 cm, cuneate at base, entire or remotely dentate, acute, glabrous. Flowers in extra-axillary, sub-umbellate, 3–8 flowered, pendulous cymes. Calyx 3–4 mm long; glabrous; teeth oblong, 1 mm long, obtuse. Corolla white, 4–8 mm long, divided more than half way down; lobes oblong, subacute. Anthers yellow, notched at apex. Berries globose, 7–8 mm in diam., seeds ovoid, yellowish, 1.5–2 mm long.

Fls. & Frts.: August–January. *Exsiccata*: SUK-1794.
Common, weed in open situations.
Localities: Ajara, Devale, Kolhapur.

Local uses: Young shoots are given in chronic skin diseases and used in psoriasis. Leaf juice is dropped in nose in headache and diseases of the nose. Fruit decoction is used to improve appetite. Berries are used as a tonic, diuretic; improves appetite and taste, useful in diseases of heart, liver, piles, inflammation and facilitate delivery.

Solanum trilobatum L. Sp. Pl. 188. 1753; C. B. Cl. in Hook. f. Fl. Brit. India 4: 236. 1883; Cooke, Fl. Pres. Bombay 2: 337. 1958 (Repr. ed.); Madhusudan Rao in Singh *et al.* Fl. Maharashtra State Dicot. 2: 506. 2001.

Scandent undershrubs, 1–3 m tall; branches densely prickly, divaricate, glabrescent; prickles sharp, hooked. Leaves ovate or rotund-ovate, 2–6 × 1.5–4 cm, acute or rounded at base, irregularly sinuate, prickly on the midrib, sparsely stellate hairy. Flowers in extra-axillary, racemose cymes. Calyx cyathiform, 3–4 mm long, stellate hairy outside; teeth triangular, 1–1.5 mm long, acute. Corolla violet purple, 12–15 mm long; lobes oblong-lanceolate, stellate hairy outside. Anthers yellow, oblong. Berries bright red, globose, 6–8 mm in diam., seeds ovoid-orbicular, 2.5–3 mm across, minutely pitted.

Fls. & Frts.: August–January. *Exsiccata*: SUK-1782.
Infrequent, in open situations in plains.
Localities: Ajara, Islampur, Wai.

Local uses: A decoction of the entire plant is administered in cases of acute and chronic bronchitis.

Solanum virginianum L. Sp. Pl. 187. 1753; Mudhusudan Rao in Singh *et al.* Fl. Maharashtra State Dicot. 2: 506. 2001. *S. xanthocarpum* Schard. & Wendl. Sert. Hanov. 1: 8, t. 2. 1795; C. B. Cl. in Hook. f. Fl. Brit. India 4: 236. 1883; Cooke, Fl. Pres. Bombay 2: 335. 1958 (Repr. ed.). BHUI-RINGANI, KATE-RINGNI.

Procumbent or trailing, herbs or undershrubs; stems woody at base; branches many, clothed with long, sharp, 10–15 mm long priekles; young parts stellate hairy. Leaves elliptic-oblanceolate, 5–10 × 2–6 cm, sinuate or subpinnatified, subacute, prickly on both surfaces, glabrescent. Flowers in extra-axillary, few-flowered cymes. Calyx 1–1.2 cm long, densely stellate hairy and prickly outside; teeth short; 7–10 mm long, linear-lanceolate, acute. Corolla violet-purple, 2–2.5 cm long, deltoid-acute, hairy outside. Anthers yellow, oblong. Berries globose, 1.5–2 cm in diam., or white with green veins. Seeds subreniform, 2–2.5 mm in diam., glabrous.

Fls. & Frts.: June–January. *Exsiccata*: SUK-266.

Common, on wastelands.

Localities: Alate, Atigre, Babu Jamal, Kagal, Kolhapur, Ramling, Tasgaon, Wai.

Local uses: Root is one of the component of Dashmula. Smoke of seeds is given in inflammatory piles. Similarly, seeds are smoked as a Tobacco to relive toothache. Decoction of root is given with addition of long pepper and honey in cough and catarrh, and with rock salt and asafetida in spasmodic cough. Seed smoke is inhaled in hemicrania. Fruits juice is given in sore throat. Root is useful in bronchitis, asthma, fever, thirst and urinary concretions. Fruit improve appetite.

Withania somnifera (L.) Dunal in DC. Prodr. 13: 453. 1852; C. B. Cl. in Hook. f. Fl. Brit. India 4: 239. 1883; Cooke, Fl. Pres. Bombay 2: 341. 1958 (Repr. ed.); Mudhusudan Rao in Singh *et al.* Fl. Maharashtra State Dicot. 2: 508. 2001. *Physalis somnifera* L. Sp. Pl. 182. 1753. ASHWAGANDHA.

Erect, undershrubs, 60–150 cm tall; branches terete, densely clothed with mealy stellate tomentum. Leaves ovate-ellipsoid, 4–8 × 3–5 cm, cuneate at base, entire, obtuse, pubescent above, densely tomentose beneath. Flowers in sessile or subsessile fascicles. Calyx 4–5 mm long, stellate tomentose outside; teeth linear, 2–2.5 mm long, acute. Corolla greenish-yellow, 7–8 mm long, divided more than half way down; lobes lanceolate, acute pubescent outside. Anthers elliptic-orbicular. Berries red, globose, 5–6 mm in diam., enclosed in enlarged persistent calyx. Seeds subreniform, 2 mm in diam., yellow.

Fls. & Frts.: September–December. *Exsiccata*: SUK-127.

Common, weed in waste places, roadsides and in cultivated fields.

Localities: Atpadi, Ichalkaranji, Kumbhoj, Tasgaon.

Local uses: Root powder is much useful in arthritis, which is secondary to immunological disturbance e.g. Rheumatoid arthritis. Root along with *Mucuna monosperma* seed in 1: 2 proportion is a sure remedy in Parkinson's diseases. Daily consumption of 5 gm powder with tincture of ginger along with ghee is used to induce natural sleep. 3 gm root powder is given to AIDS patient before weight loss and useful to control progression of disease.

SCROPHULARICEAE Juss.

Bacopa monnieri (L.) Penn. in Proc. Acad. Nat. Sci. Philadelphia 98: 94. 1946; Godbole & Prasad in Singh *et al.* Fl. Maharashtra State Dicot. 2: 524. 2001. *Lysimachia monnieri* L. in Torner, Cent. Pl. 2: 9. 1756. *Herpestis monniera* H. B. & K. Nov. Gen. & Sp. Pl. 2: 366. 1888; Hook. f. Fl. Brit. India 4: 272. 1884. *Moniera cuneifolia* Michx. Fl. Bor. Am. 2: 22. 1803; Cooke, Fl. Pres. Bombay 2: 356. 1958 (Repr. ed.). MUNDUKPARNI, NIRA-BRAHMI.

Prostrate, creeping or procumbent herbs; stems fleshy, glabrous. Leaves opposite or the upper alternate, sessile obovate to oblanceolate, 1–2.5 × 0.3–7 cm, fleshy, narrowed at base, obtuse or subacute, glabrous. Flowers on slender, 5–15 mm long pedicles. Calyx glabrous, divided to the base; sepals 5, upper sepals ovate, 3–4 mm long, acute; other sepals shorter, lanceolate, acute. Corolla blue or pale purple, 7–8 mm long; lobes subequal, rounded. Stamens 4; anthers in pairs, bluish. Capsules ovoid, oblong, scarcely 1 mm long, striate.

Fls. & Frts.: Throughout year. *Exsiccata*: SUK-40.
Common, in marshes along streams and river banks.
Localities: Hatkanagale, Kolhapur, Miraj, Sangli, Satara.

Local uses: Plant juice is given to children in cough, conyza and cold. A teaspoonful of the juice of the leaves given to infants suffering from catarrh, severe bronchitis gives relief by causing vomiting and purging. It is used as nervine tonic, useful in insanity, epilepsy and hoarseness and in hysteria.

Kickxia ramosissiama (Wall.) Janchen in Oest. Bot. Zeit. 82: 152. 1933; Godbole & Prasad in Singh *et al.* Fl. Maharashtra State Dicot. 2: 530. 2001. *Linaria ramosissima* Wall. Pl. As. Rar. 2: 43, t. 153. 1831; Hook. f. Fl. Brit. India 4: 251. 1883; Cooke, Fl. Pres. Bombay 2: 353. 1958 (Repr. ed.).

Diffuse or pendent, glabrous herbs, branching from the base; branches slender, 30–75 cm long. Leaves alternate or the lowest opposite, triangular hastate or 3–7 lobed, 1–5 × 0.7–2.5 cm, truncate or subcordate at base, acute at apex, glabrous or glabrescent, the upper ones lanceolate-sagitate, much smaller and passing into bracts. Flowers axillary solitary; pedicles slender, 10–30 mm long. Calyx divided almost to the base, 4–5 mm long; sepals linear-lanceolate, acute. Corolla yellow, 12–15 mm long, pubescent outside; spur 3–4 mm long, straight or curved; limb 2-lipped; lobes 1–1.5 mm long, obtuse. Stamens 4; filaments pubescent. Capsules subglobose, 2.5–3 mm in diam., pubescent. Seeds many, angled, echinulate.

Fls. & Frts.: October–December. *Exsiccata*: SUK-194.
Common, on old wall and rocks.
Localities: Kartikswami Ghat, Panhala, Ram Donger, Wai.

Local uses: Decoction of plant is given in chronic cough. It is highly valued as a remedy for diabetes.

Striga gesnerioides var. **gesnerioides** (Willd.) Vatke in Oest. Bot. Zeit. 2. 11. 1875; Godbole & Prasad in Singh *et al.* Fl. Maharashtra State Dicot. 2: 550. 2001. *Buchnera gesnerioides* Willd. Sp. Pl. 3: 338. 1800. *Striga orobanchioides* (R. Br. ex Endl.) Benth. Comp. Bot. Mag 1: 361, t. 19. 1836; Hook. f. Fl. Brit. India 4: 299. 1884; Cooke, Fl. Pres. Bombay 2: 374. 1958 (Repr. ed.).

Erect, much branched, glabrous or pubescent, annual herbs, 5–45 cm tall; stems simple or branched, dark reddish-purple. Leaves scale like, ovate-oblong to lanceolate, 0.5–1 × 0.2–0.3 cm, acute, passing upwards into bracts. Flowers many, sessile, in terminal spikes. Calyx lanceolate, 6–7 mm long, cupular, 4–5 ribbed; teeth 4–5, 2–2.5 mm long, acute, bristle tipped. Corolla white or pink; tube 8–12 mm long, incurved about the middle; limb 2-lipped; lobes orbicular, obtuse. Stamens 4; anthers stipitate. Capsules oblong-ellipsoid, 5–6 mm long, glabrous. Seeds oblong, truncate at both ends, dark brown.

Fls. & Frts,: September–December. *Exsiccata*: SUK-1780.
Common, parasitic on roots of Acanthaceae members.
Localities: Amboli, Bhudergad, Borbet, Dajipur, Kas, Mahabaleshwar, Vaibhavwadi.

Local uses: Plant is used in diabetes.

Verbascum chinese (L.) Sant. Fl. Purandhar 90. 1958; Godbole & Prasad in Singh *et al.* Fl. Maharashtra State Dicot. 2: 554. 2001. *Scrophularia chinensis* L. Mant. 2: 250. 1771. *Celsia coromandeliana* Vahl, Symb. 3: 79. 1794; Hook. f. Fl. Brit. India 4: 240. 1883; Cooke, Fl. Pres. Bombay 2: 352. 1958 (Repr. ed.). KUTAKI.

Scapigerous, annual herbs, 30–75 cm tall; scapes often branched near the top. Leaves basal ones oblanceolate, 3–20 × 2–8 cm, narrowed at base, lyrate or pinnatisect acute, crenate. Flowers in simple or branched, terminal racemes; rachis 30–60 cm long, glandular pubescent. Calyx deeply divided, 4–5 mm long, glandular pubescent outside; sepals 5, linear-oblong, 3–4 mm long, acute. Corolla rotate, yellow, 10–15 mm across, lobes 5, rounded, subequal. Stamens 4; filaments densely bearded. Capsules subglobose, 6–8 mm in diam., seeds many, oblong, truncate, warted, pale yellow.

Fls. & Frts.: September–May. *Exsiccata*: SUK-1762.
Common, on waste places.
Localities: Kolhapur, Sangli.

Local uses: Decoction of plant is given orally in skin diseases and applied externally. It reduces the inflammation of hands and feet.

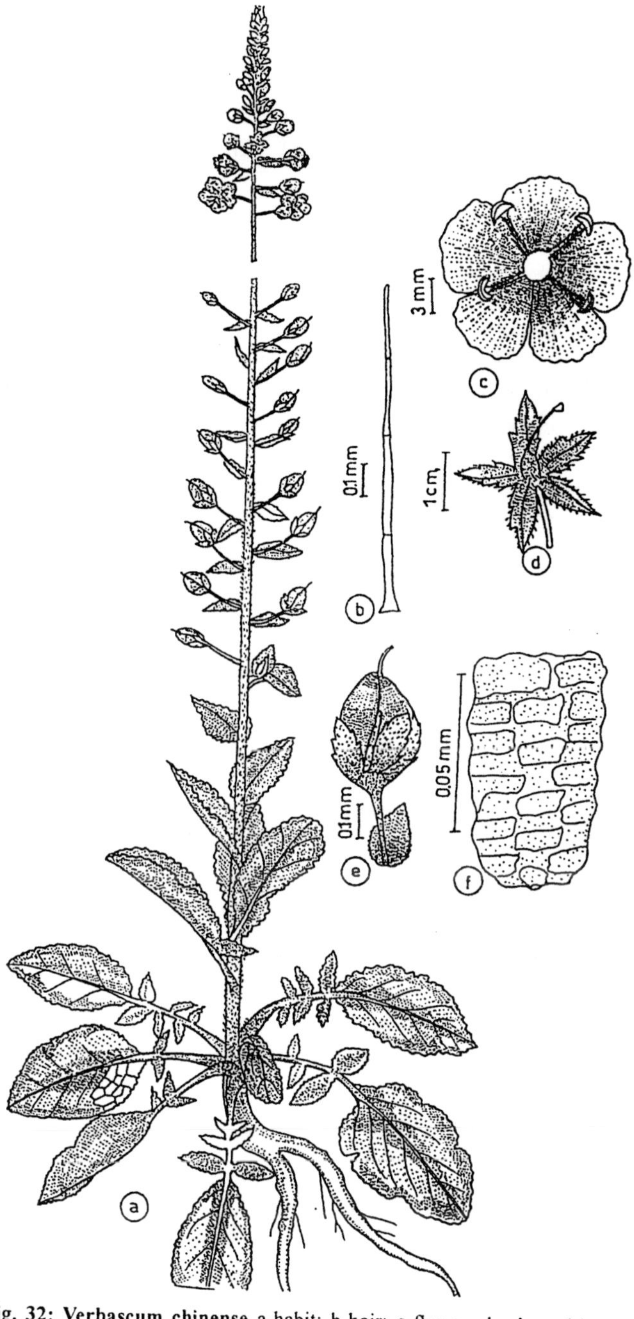

Text fig. 32: Verbascum chinense a-habit; b-hair; c-flower; d-calyx with carpel; e-fruit; f-seed.

BIGNONIACEAE Juss.

Dolichandrone falcata (Wall. ex DC.) Seem. in J. Bot. 8: 381. 1870; C. B. Cl. in Hook. f. Fl. Brit. India 4: 380. 1884; Cooke, Fl. Pres. Bombay 2: 404. 1958 (Repr. ed.); Godbole & Prasad in Singh *et al.* Fl. Maharashtra State Dicot. 2: 574. 2001. *Spathodea falcata* Wall. ex DC. Prodr. 9: 206. 1845. MEDH-SHINGI, MEADSHINGI.

Deciduous trees, 4–10 m tall. Leaves opposite, simply pinnate, 7–15 cm long; rachis pubescent; leaflets usually 5–9, opposite with an odd one, orbicular to obovate, 1–3.5 × 1.2–2.7 cm, cuneate or rounded at base, with rather undulate margins, emarginate or rounded with short flowered racemes. Calyx 1.5–2 cm long, pubescent, with a short, stout mucronate at apex. Corolla 2–2.5 cm long; tube very narrow at base, gradually expanding upwards; lobes obovate-oblong with crisped, undulate margins. Stamens 4, included; anther-cells oblong, parallel. Disk annular, capsules flat, falcately curved, 25–45 cm long, acute. Seeds numerous, 2–2.5 cm long, winged at two ends, rectangular, brown.

Fls. & Frts.: March–June. *Exsiccata*: SUK-1768.

Common, in hill forests on rocky soils.

Localities: Alate, Atigre, Babu Jamal, Hatkanagale, Ichalkaranji, Kagal, Tasgaon.

Local uses: Leaves are boiled in water and given in joint pains, backache and in muscular pain. Leaves are applied externally in swellings. Bark is used in abortion.

Heterophragma quadriloculare (Roxb.) K. Schum. in Engl. & Prantl, Pflanzenf. 4(3b): 243. 1895; Prasanna in Singh *et al.* Fl. Maharashtra State Dicot. 2: 575. 2001. *Bignonia quadrilocularis* Roxb. Pl. Cor. 2: 24, t. 145. 1798. *Heterophragma roxburghii* A. DC. in DC. Prodr. 9: 210. 1845; C. B. Cl. in Hook. f. Fl. Brit. India 4: 381. 1884; Cooke, Fl. Pres. Bombay 2: 404. 1958 (Repr. ed.). VARAS.

Trees, 5–15 m tall. Leaves simply pinnate, 30–45 cm long, tomentose when young, afterwards glabrous, crowded near the ends of the branches; leaflets 3–5 pairs, elliptic oblong, 5–12 × 2–6 cm, acuminate, base oblique. Flowers fragrant in large terminal, many-flowered, densely fulvous-tomentose panicles. Calyx 1.5–2 cm long, densely tomentose outside, clothed with long silky tawny hairs, irregularly lobed, obscurely 2 lipped. Corolla reaching 5 cm long, pale rose colored or white when young; tube pubescent outside when young, glabrous at maturity; mouth of limb 5 cm across; lobes 1.2 cm long with crisped margins. Filaments densely woolly at base. Capsules straight, 1.2 cm thick, 30 × 7 cm, linear, pointed, velvety when young, glabrous at maturity; dissepiment 4-winged, making the capsules 4-celled.

Fls. & Frts.: February–April. *Exsiccata*: SUK-43.
Frequent, in hilly forests.
Localities: Amba, Amboli, Gaganbawada, Radhanagari, Sawantwadi, Vaibhavwadi.

Local uses: Bark is applied externally on skin diseases. Root is given in menorrhoea and in premature ejaculation or night emission.

Oroxylum indicum (L.) Vent. Dec. Gen. Nov. 8. 1808; C. B. Cl. in Hook. f. Fl. Brit. India 4: 378. 1884; Cooke, Fl. Pres. Bombay 2: 401. 1958 (Repr. ed.); Prasanna in Singh *et al.* Fl. Maharashtra State Dicot. 2: 575. 2001. *Bignonia indica* L. Sp. Pl. 625. 1753. TETU.

Small deciduous trees, 6–10 m tall. Leaves opposite, 2–3 pinnate, 1–2 m long; pinnae opposite; leaflets 2–4 pairs, ovate or elliptic, 5–12 × 3–10 cm, rounded or cordate at base, entire, acuminate, glabrous. Flowers fleshy, fetid, in erect, 30–60 cm long, racemes. Calyx leathery, oblong-campanulate, 2–2.5 cm long, glabrous. Corolla purple, 8–10 cm long; lobes 3–3.5 cm long, with crisped margins. Stamens slightly exerted; filaments hairy at base. Capsules straight, elliptic-oblong, 30–60 × 5–8 cm tapering to both ends, with woody valves. Seeds numerous, 5–6 cm long, winged all around except at base.

Fls. & Frts.: May–July. *Exsiccata*: SUK-205.
Infrequent, in deciduous forests.
Localities: Dajipur, Phonda, Salva Donger, Radhanagari, Sawantwadi, Tillari.

Local uses: Root is one of the component of Dashmula. Root bark is useful in diarrhoea and dysentery. Decoction of root is given in fever and bronchitis. Root bark is astringent to the bowels, cooling, aphrodisiac, tonic, biliousness, fevers, bronchitis, intestinal worms, vomiting and dysentery.

Stereospermum chelenoides. (L.) DC. Bibl. Univ. Geneve II, 17: 125. 1838 et Prodr. 9: 210. 1845 p. p.; Prasanna in Singh *et al.* Fl. Maharashtra State Dicot. 2: 577. 2001. *Bignonia chelenoides* L. f. Suppl. 282. 1781, p.p. *Stereospermum suaveolens* (Roxb.) DC. Prodr. 9: 211. 1845; C. B. Cl. in Hook. f. Fl. Brit. India 4: 382. 1884; Cooke, Fl. Pres. Bombay 2: 406. 1958 (Repr. ed.). PARAL.

Deciduous trees, 10–15 m tall; young parts viscid-pubescent. Leaves opposite or rarely in whorls of 3, unipinnate, 30–50 cm long; leaflets 3–4 pairs with an odd one, broadly elliptic to ovate oblong, 3.5–7.5 × 3.5–5 cm, rounded and unequal sided at base, serrulate, acuminate, rough above, pubescent beneath. Flowers fragrant, in large, lax, trichotomous, viscid hairy panicles. Calyx campanulate, 6–8 mm long, viscid-pubescent outside, shortly acutely 5-toothed. Corolla dull, purple, 2–3 cm long, pubescent outside and in lower half inside; limb 2-lipped; lobes rounded, crisped-crenate. Stamens

4, filaments without a tuft of hairs at base. Capsules straight, cylindric, 30–45 × 2 cm, rough due to whitish dots. Seeds 3–3.5 cm, winged on both ends.

Fls. & Frts.: April–June. *Exsiccata*: SUK-85.
Occasional, in deciduous forests.
Localities: Barki, Borbet, Panhala, Radhanagari, Sawantwadi.

Local uses: Root is an ingredient in Dashmula. It is given in urinary complaints and used as a tonic. Decoction is given in fever. Root is useful in inflammations, eructations, vomiting, asthma, fevers, diseases of the blood, thirst and loss of taste.

Stereospermum colais var. **colais** (Buch.-Ham. ex Dillw) Mabb. in Taxon 27: 553. 1978; Prasanna in Singh *et al.* Fl. Maharashtra State Dicot. 2: 577. 2001. *Bignonia colais* Buch.-Ham. ex Dillw. Rev. Ref. Hort. Mal. 28. 1839. *Dipterospermum personatum* Hassk. Flora 25 (2): 28. 1842. *Stereospermum chelonoides* auct. non (L.f.) DC. 1845; C. B. Cl. in Hook. f. Fl. Brit. Ind. 4: 382. 1884; Cooke, Fl. Pres. Bombay 2: 405. 1958 (Repr. ed.). PARAL.

Deciduous trees, 10–15 m tall; branches spreading, more or less glabrous. Leaves opposite or in whorls of 3, unipinnate, 30–45 cm long; leaflets 3–5 pairs and an odd one, elliptic, 4–8 × 2–4 cm, rounded at base, subentire, caudate-acuminate, glabrous. Flowers fragrant, in lax, drooping, terminal panicles; branches of the panicle slender, glabrous, articulate. Calyx campanulate, 5–6 mm long, glabrous, shortly 3–5-toothed. Corolla 18–20 mm long, pubescent outside; limb 2-lipped; lobes rounded crisped. Stamens 4; filaments with a tuft of woolly hairs at base. Capsules linear, 30–40 cm long, obtusely 4-gonous. Seeds 2.5–3 cm long, winged at both ends.

Fls. & Frts.: February–September. *Exsiccata*: SUK-344.
Occasional, in hilly forests.
Localities: Amboli, Barki, Dajipur, Panhala, Patgaon.

Local uses: Same as *S. sauveolens*.

PEDALIACEAE R. Br.

Pedalium murex L. Syst. Nat. ed. 10. 1123. 1759; C. B. Cl. in Hook. f. Fl. Brit. India. 4: 386. 1884; Cooke, Fl. Pres. Bombay 2: 412. 1958 (Repr. ed.); Londhe in Singh *et al.* Fl. Maharashtra State Dicot. 2: 586. 2001. MOTHE-GOKHRU, SARATA.

Much branched annual glabrous herbs, 30–40 cm high; stem and branches often slightly rough with scaly glands. Leaves opposite, pale glaucous-green; broadly ovate-oblong, 2–5 × 1.9–3.5 cm, truncate, or obtuse, coarsely crenate-serrate, glabrous above, the lower side usually covered with minute scales, base acute. Flowers axillary, solitary. Calyx small, scarcely 3 mm long, minutely scaly outside, divided rather more than ½ way down; lobes 5, linear-triangular, acute. Corolla 2.5 cm long, bright yellow; tube 2 cm long, slender; lobes broad, rounded. Stamens 4, didynamous; filaments glandular hairy at the base. Fruit 1.2–2 cm long, narrowed at base, pyramidal-ovoid, above the spines, bluntly 4-angled, with stout sharp horizontal spines from the angles.

Fls. & Frts.: September–January. *Exsiccata*: SUK-107.
Occasional, on the sandy shore.
Localities: Ratnagiri.

Local uses: Infusion of stem and leaves is used in gonorrhea and dysuria. Decoction is given for incontinence of urine, gleet and nocturnal emissions. Decoction of root is given in urinary calculi.

Sesamum orientale L. Sp. Pl. 634. 1753; Londhe in Singh *et al.* Fl. Maharashtra State Dicot. 2: 587. 2001. *S. indicum* L. Sp. Pl. 634. 1753; C. B. Cl. in Hook. f. Fl. Brit. India 4: 387. 1884; Cooke, Fl. Pres. Bombay 2: 413. 1958 (Repr. ed.). RANTIL, TIL.

Erect, glandular-pubescent, annual herbs, 60–90 cm tall. Leaves alternate or the lower opposite and often deeply 3-lobed; lobes lanceolate, 3–15 × 1.5–6 cm, serrate, acute, glabrous above, puberulous beneath; upper leaves entire, lanceolate, much smaller, passing into bracts. Flowers ill-smelling, axillary, solitary, forming a false raceme at the ends of branches. Pedicels 1–2 cm long, with glands at the base. Calyx campanulate, 6–7 mm long, deeply divided, glandular-pubescent outside. Sepals lanceolate-subulate, obtuse. Corolla white, pink or pale purplish with yellow plate and deep purple lobes, 2–3 cm long, pubescent outside, limb 2-lipped; lobes rounded. Stamens 4, included. Capsules bluntly quadrangular, shortly beaked, pubescent. Seeds obovoid, compressed, glabrous, black or white.

Fls. & Frts.: August–October. *Exsiccata*: SUK-106.
Commonly cultivated.

Local uses: Paste of seeds is applied to burns. Seeds are also given as a blood purifier. Seed paste is used as a hair wash and is supposed to promote the growth of hair and make it black. Seeds are given every night with warm water in constipation. Seeds are given as a tonic, diuretic, promote the growth of hair; useful in diarrhoea, gouty joints, urinary concretions, eye disease, applied to ulcers and piles.

MARTYNIACEAE Stapf

Martynia annua L. Sp. Pl. 618. 1753; Londhe in Singh *et al.* Fl. Maharashtra State Dicot. 2: 589. 2001. *M. diandra* Glox. Obs. 14, t. 1. 1785; C. B. Cl. in Hook. f. Fl. Brit. India 4: 386. 1884; Cooke, Fl. Pres. Bombay 2: 414. 1958 (Repr. ed.). VINCHU.

Erect, glandular hairy, stout, annual herbs, 1–2 m tall; stem fistular. Leaves opposite, or the upper alternate, broadly ovate-orbicular, 10–30 cm across, cordate at base, dentate, sub obtuse at apex, densely glandular hairy; petioles stout, fistular, as long as the lamina. Flowers in dense axillary, 5–15 cm long racemes; bracts pink, spathulate-rounded, 1–1.5 cm across, caducous. Calyx yellowish-white, divided near to the base, 1.5–2 cm long, glandular pubescent; sepals unequal. Corolla 5–6 cm long, glandular pubescent, pale pink outside, yellow with stalked glands and purple dots on the petals; lobes rounded, pink, with dark purple blotch at base. Stamens 2; staminodes 2–3, clavate, included. Fruit green, ovoid, 2–3 × 1.5–2 cm; beak short, upturned. Pyrene black, hard with serrate-dentate ribs. Seeds solitary.

Fls. & Frts.: July–November. *Exsiccata*: SUK-149.

Native of Mexico, naturalized on wastelands.

Localities: Hatkanagale, Kolhapur, Shiroli.

Local uses: Leaves are applied externally in tuberculous glands of the neck. Juice of leaves used as a gargle for sore throat.

ACANTHACEAE Juss.

Acanthus ilicifolius L. Sp. Pl. 639. 1753; C. B. Cl. in Hook. f. Fl. Brit. India 4: 481; Cooke, Fl. Pres. Bombay 2: 427. 1958 (Repr. ed.); Moorthy in Singh *et al.* Fl. Maharashtra State Dicot. 2: 593. 2001.

Shrubs, armed, scarcely branched. Leaves oblong elliptic, pinnatifid or toothed, 8–12 × 4–6 cm, acute with sharp spinous teeth on the margin. Flowers sessile in opposite pairs in terminal crowded or interrupted spikes. Calyx glabrous; segments 4, in 2 opposite pairs, outer segments elliptic oblong, 15 mm long, obtuse, mucronulate; inner segments oblong, 12 mm long, rounded, ciliate at apex. Corolla 2.5–4 cm long, blue; 2 lipped; lower lip obovate, nearly 2.5 cm broad, hairy on upper surface, very shortly 3-lobed, the lobes rounded, middle one is much narrower than the lateral. Stamens 4, didynamous; filaments thick, glabrous, slightly striate; anthers densely beard. Style glabrous. Capsules oblong, 2.5 cm long, apiculate, brown, smooth, shining. Seeds 6 mm across, broadly ovoid or suborbicular, much compressed.

Fls. & Frts.: December–June. *Exsiccata*: SUK-1704.

Common, along the coast and tidal swamps.

Localities: Ratnagiri, Sakharppa.

Local uses: Decoction of plant with sugar candy and cumin is given in dyspepsia with acid eructation. Leaves are used to soften an irritated internal part in rheumatism.

Andrographis paniculata (Burm. f.) Wall. ex Nees in Wall. Pl. Asiat. Rar. 3: 116. 1832; C. B. Cl. in Hook. f. Fl. Brit. India 4: 501. 1884; Cooke, Fl. Pres. Bombay 2: 451. 1958 (Repr. ed.); Moorthy in Singh *et al.* Fl. Maharashtra State Dicot. 2: 595. 2001. *Justicia paniculata* Burm. f. Fl. Ind. 9. 1768. KALAMETI, KALPA, KREAT.

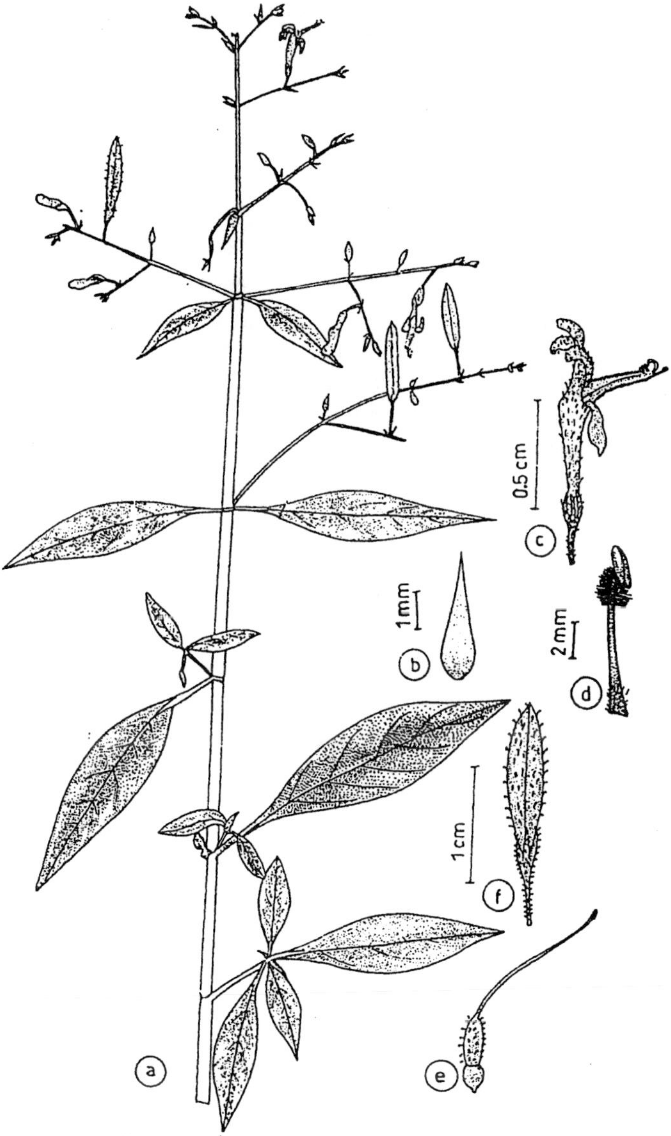

Text fig. 33: Andrographis paniculata a-twig; b-bract; c-flower; d-stamen; e-carpel. f-fruit.

Annual, erect, branched herbs, up to 9 m tall; branches sharply quadrangular, often narrowly winged in upper portions. Leaves lanceolate, 5–7 × 1.2–2.5 cm, acute, glabrous, slightly undulate, pale beneath, tapering. Flowers small, solitary, distant, in lax spreading axillary and terminal racemes or panicles, the whole forming a large pyramidal paniculate inflorescence. Calyx 5 partite, 3 mm long; sepals equal; linear-lanceolate, glandular-pubescent. Corolla rose-colored, tubular, 2-lipped, pubescent 1 cm long; tube 5 mm long, upper lip deeply 3-lobed. Stamens 2; filaments flattened, hairy in upper part; anthers bearded at base. Ovary glabrous, 2-celled. Capsules ca 2 × 0.3 cm, linear oblong, acute at both ends. Seeds numerous, subquadrate, osseous, rugosely pitted, glabrous, yellowish-brown.

Fls. & Frts.: December–April. *Exsiccata*: SUK-1740.
Frequent, as forest undergrowth and in open rocky situations.
Localities: Radhanagari, Shiroli.

Local uses: Decoction of plant is given in low fever to the children. Plant is very useful in general debility, dysentery and certain forms of dyspepsia. The expressed juice of the leaves together with certain spices, such as cardamoms, cloves, cinnamon is dried in sun, and made into small globules, which are prescribed for infants to relieve griping, irregular stools and loss of appetite.

Barleria cristata var. **cristata** L. Sp. Pl. 636. 1753; C. B. Cl. in Hook. f. Fl. Brit. India 4: 488. 1884; Cooke, Fl. Pres. Bombay 2: 460. 1958 (Repr. ed.); Moorthy in Singh *et al.* Fl. Maharashtra State Dicot. 2: 599. 2001. NILI-KORANTI.

Erect, perennial undershrubs, 0.3–1.5 cm tall; stems appressed hairy. Leaves elliptic oblong, 4–15 × 2.5–6 cm, narrowed at base, entire, acute or acuminate, hairy on both surfaces. Flowers in axillary and terminal, short, dense spikes; calyx hairy outside; sepals 4, distinct, the outer membranous, 20–25 × 6–8 mm, long, acuminate mucronate and with spinous-bristle tipped margins, the inner sepals similar but narrower and shorter. Corolla blue, wide; limb 2-lipped, lobes obovate-oblong, 1.5–2 cm long, rounded. Stamens 2 fertile, with 2 staminodes, included. Capsules ellipsoid, 1–1.5 cm long, acute at both ends. Seeds 4, orbicular, 3–4 mm across, hairy.

Fls. & Frts.: November–January. *Exsiccata*: SUK-1710.
Common, in hedges and in gardens.
Localities: Amba, Kolhapur.

Local uses: Dried bark is given in whooping cough. Leaf juice is used in fever and much viscid phlegm. Leaf paste is applied on feet in the rainy season to prevent cracking.

Barleria cristata var. **dichotoma** (Roxb.) Prain, Beng. Pl. 812. 1903. [2:606. 1963 (Repr. ed.); Cooke, Fl. Pres. Bombay 2: 460. 1958 (Repr. ed.); Moorthy in Singh *et al.* Fl. Maharashtra State Dicot. 2: 599. 2001. *B. dichotoma* Roxb. Fl. Ind. 3: 39. 1832. PANDHRI KORANTI.

Very similar to var. *cristata* but with white corolla.

Fls. & Frts.: September–January. *Exsiccata*: SUK-1719.
Frequent, cultivated.
Localities: Hatkanagale, Jaisingpur, Miraj, Shirol,.

Local uses: Dried bark is given in whooping cough. Leaf juice is used in fever and much viscid phlegm. Leaf paste is applied on feet in the rainy season to prevent cracking. Same as *Barleria cristata* var. *cristata*.

Barleria prionitis ssp. **prionitis** L. Sp. Pl. 635. 1753; C. B. Cl. in Hook. f. Fl. Brit. India 4: 482; 1884; Cooke, Fl. Pres. Bombay 2: 457. 1958 (Repr. ed.); Moorthy in Singh *et al.* Fl. Maharashtra State Dicot. 2: 602. 2001. KATE KORANTI, KORANTA.

Much branched shrubs or under shrubs, 0.5–1.5 m tall, armed with white, sharp thorns often forming clusters; stem glabrous. Leaves elliptic-oblong, 9–15 × 4–6 cm, tapering at base, entire, acuminate and often bristle-tipped, glabrous above, pubescent beneath especially when young. Petioles with 2–4 axillary, sharp thorns. Flowers solitary, axillary or in terminal short spikes. Outer sepals unequal, oblong-lanceolate, 8–12 mm long, mucronate, the inner ones narrower and shorter than the outer sepals, mucronate. Corolla orange-yellow; tubc broad, 2–2.5 cm long; limb 2-lipped; lobes oblong-obovate rounded at apex. Stamens 2, perfect with 2 staminodes, included or slightly exserted. Capsules ovoid, 2–2.5 cm long, beaked. Seeds 4, ovoid-orbicular, 6–8 mm across, appressed hairy.

Fls. & Frts.: November–January. *Exsiccata*: SUK-164.
Common, in hedges around fields, also planted in gardens.
Localities: In all districts.

Local uses: Dried bark is given in whooping cough. Leaf juice is used in fever and much viscid phlegm. Leaf paste is applied on feet in the rainy season to prevent cracking.

Carvia callosa (Wall.) Bremek. in Verh. Nederl. Acad. Wet 2, 40: 187. 1944; Moorthy in Singh *et al.* Fl. Maharashtra State Dicot. 2: 609. 2001. *Ruellia callosa* Wall. Cat. 2359. 1830. *Strobilanthes callosus* Nees in Wall. Pl. Asiat. Rar. 3: 85. 1832; C. B. Cl. in Hook. f. Fl. Brit. India 4: 451. 1884; Cooke, Fl. Pres. Bombay 2: 444. 1958 (Repr. ed.). KARVI.

Shrubs, erect, 2–4 m tall. Leaves elliptic-lanceolate, 10–25 × 3.5–7.5 cm, one of each pair often smaller than opposite one, strongly lineolate above, more or less hairy on nerves beneath, apex acute, base tapering into wing like petiole, margins crenate and ciliate. Flowers axillary, in simple or branched ovoid, pedunculate, subtetragonal spikes with rounded edges, 2.5–7 cm long. Bracts 1.2–1.9 cm long, as broad as long, broadly elliptic or obovate, rounded at apex, concave, glabrous green with a pink tinge. Disk bright orange. Calyx 1.2 cm long, much enlarged in fruit, segments leathery, elliptic-lanceolate, 4–5 mm broad, obtuse. Corolla 2–3 cm long, with white tube and purple limb, glabrous outside with yellow hairs in mouth inside. Stamens 2; filaments of the longer stamens bearded. Capsules obovoid, 0.6–2 cm long, acute, apiculate, narrow at base, 2-seeds. Seeds ovate, ca 1 cm in diam., broadly acute, densely clothed with white appressed hairs.

Fls. & Frts.: September–June. Monocarpic. *Exsiccata*: SUK-1720.
Common, in the higher slopes of the Ghats.
Localities: Amba, Amboli, Kas, Mahabaleshwar, Radhanagari.

Local uses: Bark is externally applied to wash ulcers or wounds. Root paste is given in dysuria. Decoction of leaves are rubbed on the body during fever, paste is applied in hemicrania.

Ecbolium ligustrinum (Vahl) Vallesen in Kew Bull. 44: 651. 1989; Moorthy in Singh *et al.* Fl. Maharashtra State Dicot. 2: 620. 2001. *Justicia ligustrina* Vahl, Enum. 1: 118. 1804. *Ecbolium viride* (Forssk.) Alst. var. *laetevirens* (Vahl.) Raiz. in Ind. For. 84: 482. 1950. *E. linneanum* Kurz. var. *laetevirens* (Vahl.) C. B. Cl. in Hook. f. Fl. Brit. India 4: 545. 1885; Cooke, Fl. Pres. Bombay 2: 484. 1958 (Repr. ed.). DHAKTANDULSA; ODU-JATI, RAN-ABOLI.

Undershrubs; stems and branches terete. Leaves lanceolate or elliptic lanceolate, 10–20 × 3.5–7.5 cm, apex acuminate, base acute. Flowers in terminal dense or lax nearly sessile spikes, 6–20 cm long. Bracts foliaceous, ca 22 × 9 mm, elliptic-lanceolate, acute, aristate, entire, calyx ca 6 mm long, hairy, divided to the base; segments linear-lanceolate, very acute. Corolla bluish-green, hairy outside, 2–3 cm long; upper lip narrowly linear from a triangular base, notched at the apex; lower lip ca 1.2 cm long, divided almost to base into 3 lobes; lateral lobes oblong, subacute, the middle lobe broader, oblong obtuse. Stamens 2; filaments of the stamens hairy at base; anthers not spurred not appendaged at base. Ovary pubescent, 2-celled; stigma of 2 short divaricate lobes. Capsules 2.5 cm long, acutely pointed, pubescent all over. Seeds broadly ovoid, 8 mm across, much compressed.

Fls. & Frts.: October–May. *Exsiccata*: SUK-1727.
Frequent, in semi evergreen forests.
Localities: Sawantwadi, Tillari.

Local uses: Leaf decoction is given twice a day with 1/4 teaspoon powder of *Acorus calamus* in cough, bronchitis and asthma. Leaf juice is dropped in nostrils in sinusitis.

Gendarussa vulgaris Nees in Wall. Pl. As. Rar. 3: 104. 1832; Moorthy in Singh *et al.* Fl. Maharashtra State Dicot. 2: 680. 2001. *Justicia gendarussa* Burm. f. Fl. Ind. 10. 1768; C. B. Cl. in Hook. f. Fl. Brit. India 4: 532. 1885; Cooke, Fl. Pres. Bombay 2: 492. 1958 (Repr. ed.). BAAD, JAGAT MADAN, KALA ADULSA TEV.

Erect, much-branched shrubs, 30–120 cm tall; stem violet-brown, glabrous or finely pubescent. Leaves lanceolate, 5–20 × 1–3.5 cm, cuneate at base, subsinuate, acute or acuminate, glabrous, dark green with dark violet nerves. Flowers in axillary spikes or terminal panicles, 3–12 cm long; bracts narrow, 1–3 flowered, violet. Calyx 5-partite to the base, violet green, 3.5–5 mm long; teeth narrow, acute. Corolla white with purple blotch and veins, 1.5–2 cm long; tube pale purple; limb 2-lipped; lobes short, rounded. Stamens 2; filaments glabrous.

Fls. & Frts.: November–April. *Exsiccata*: SUK-1730.
Common, in semi evergreen forests.
Localities: Amba, Amboli, Borbet, Gaganbawada, Radhanagari, Tillari.

Local uses: Leaf paste is applied externally in eczema. Juice is dropped into the nostril in hemicrania.

Haplanthodes tentaculata (L.) R. B. Majumdar in Bull. Bot. Soc. Bengal 25: 76. 1971; Moorthy in Singh *et al.* Fl. Maharashtra State Dicot. 2: 628. 2001. *Ruellia tentaculata* L. in Torner, Cent. Pl. 22. 1756. *Haplanthus tentaculatus* (L.) Nees in DC. Prodr. 11: 513. 1847 p.p.; C. B. Cl. in Hook. f. Fl. Brit. India 4: 507. 1884; Cooke, Fl. Pres. Bombay 2: 453. 1958 (Repr. ed.). JHANKARA.

Herbs, 50–75 cm tall, erect; stems more or less quadrangular. Leaves ovate, 5–10 × 3–5 cm, acuminate, more or less pubescent, decurrent in petiole. Cladodes slender, subquadrangular, 0.6–2 cm long, densely and softly villous. Flowers sessile in the midst of the cladodes; calyx 4–5 mm long, divided almost to the base; lobes 5, linear lanceolate, densely hairy. Corolla 1.2 cm long, blue, lilac, 2-lipped, ca ½ way down. Stamens 2, usually beard on the back. Ovary 2-celled. Capsules 8 mm long, pointed, pubescent.

Fls. & Frts.: December–May. *Exsiccata*: SUK-1741.
Frequent, in moist and semi evergreen forests.
Localities: Ajara, Amboli, Chandgad, Tillari.

Local uses: Decoction of plant is given in fever.

Haplanthodes verticillata (Roxb.) R. B. Majumdar in Bull. Bot. Soc. Bengal 25: 1971; Moorthy in Singh *et al.* Fl. Maharashtra State Dicot. 2: 628. 2001. *Justicia verticillatus* Roxb. Fl. Ind. 1: 135. 1832. *Haplanthus verticillata* (Roxb.) Nees in DC. Prodr. 11: 513. 1847; C. B. Cl. in Hook. f. Fl. Brit. India 4: 506, 1884; Cooke, Fl. Pres. Bombay 2: 452. 1958 (Repr. ed.). JAKARA, KATERI.

Herbs, 50–75 cm tall, erect; stems, glandular. Leaves ovate-lanceolate, apex acute, base cuneate or tapering into a long petiole. Cladodes 2.5–3 cm long, quadrangular, usually with 2 sharp naked spinous teeth at apex, clothed with short glandular and long spreading hairs. Flowers in the midst of the spines, sessile; bracteoles longer than the calyx. Calyx 4 mm long, divided almost to base, lobes 5; glandular pubescent, lanceolate with a subulate point. Corolla 15 mm long, tubular, 2-lipped, ca ¼ way down; limb dark lilac, marked with darker lines; greenish-white inside the throat. Stamens 2. Capsules narrowly oblong, 1 cm long, acute, grooved, brownish-yellow. Seeds 0.25–0.3 × 0.12 cm long, compressed, yellowish-brown, hygroscopically densely hairy.

Fls. & Frts.: December–June. *Exsiccata*: SUK-1744.
Common, in moist and semi evergreen forests.
Localities: Panhala, Radhanagari.

Local uses: Decoction of plant is given in fever.

Hygrophila schulli (Buch. -Ham.) M. R. & S. M. Almeida in J. Bombay Nat. Hist. Soc. (suppl.) 221. 1986; Moorthy in Singh *et al.* Fl. Maharashtra State Dicot. 2: 636. 2001. *Behel schulli* Buch.-Ham. in Trans. Linn. Soc. Lond. 14: 289. 1825. *Barleria longifolia* L. in Torner. Cent. II, Pl. 22 1756. non Nees 1847. *Astercantha longifolia* (L.) Nees in Wall. Pl. As. Rar. 3: 90. 1832; Cooke, Fl. Pres. Bombay 2: 428. 1958 (Repr. ed.). *Hygrophilla spinosa* T. Anders. in Thw. Enum. Pl. Zeyl. 225. 1860; C. B. Cl. in Hook. f. Fl. Brit. India 4: 408. 1884. KOLSHINDA, TALIMKHANA.

Stout herbs with numerous fasciculate, usually unbranched, subquadrangular stems; 60–120 cm tall; nodes tumid, hispid with long hairs. Leaves sessile, oblong-lanceolate, 10–18 × 2–3 cm, in whorls of 6 at each node, the two outer ones much larger than the 4 inner ones. Thorns from the axils of leaves sharp, 2–4 cm long, yellowish-brown. Flowers in axillary clusters of eight at each node; bracts lanceolate, hairy and ciliate, like the leaves. Calyx 4-paritie; sepals unequal, one much longer than the other three, all linear-lanceolate, 1.2–2 cm long, hairy and with hyaline ciliate margins. Corolla purple blue, 2–3 cm long; tube 1.2–1.5 cm long; limb 2-lipped; lobes oblong, truncate or rounded. Stamens 4; filaments glabrous. Capsule linear-oblong, 7–8 mm long, pointed. Seeds 4–8, ovate quadrate, black, compressed, hygroscopically hairy.

Fls. & Frts.: June–February. *Exsiccata*: SUK-271.
Common, in marshes along streams, rivers, ponds, ditches etc.
Localities: Ajara, Kagal, Kolhapur, Narsinghwadi.

Local uses: Plant ash is given as diuretic, used in urinary calculi. Root decoction is given in gonorrhea and in rheumatism. Seed powder is given

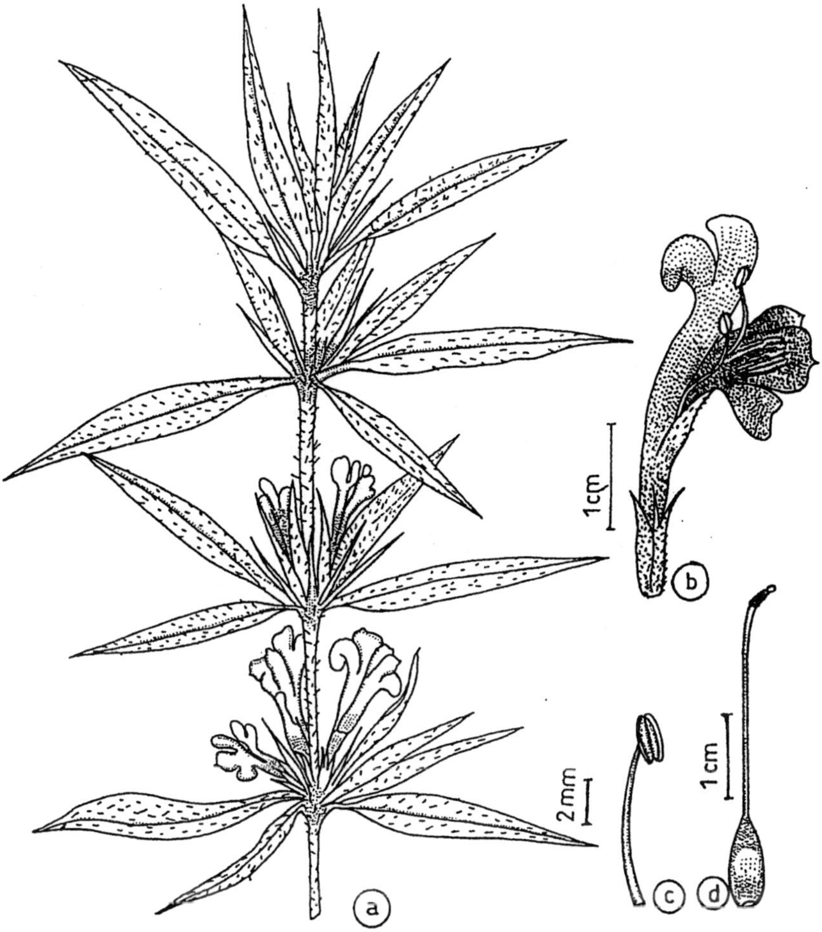

Text fig. 34: Hygrophila schulli a-twig; b-flower; c-stamens; d-carpel.

with cow milk in weakness caused by diarrhoea. Leaves are used as a tonic, hypnotic; useful in diarrhoea and dysentery, thirst, urinary calculi, urinary discharges, inflammations, abdominal troubles, anemia, constipation and sexual debility.

Justicia adhatoda L. Sp. Pl. 15. 1753; Stearn in J. Arnold Arb. 52: 640. 1971; Moorthy in Singh *et al.* Fl. Maharashtra State Dicot. 2: 639. 2001. *Adhatoda zeylanica* Medic. in Hist. & Commentat. Acad. Elect. Sci. Theod. Palast. 6: 393. 1790. *A. vasica* Nees in Wall. Pl. As. Rar. 3: 103. 1832 & in DC. Prodr. 11: 387. 1847; C. B. Cl. in Hook. f. Fl. Brit. India 4: 540. 1885; Cooke, Fl. Pres. Bombay 2: 493. 1958 (Repr. ed.). ADULSA.

Large shrubs, 1–2 m tall; stems terete glabrous. Leaves opposite, elliptic lanceolate, 10–20 × 3–8 cm, tapering at base, entire, acuminate, glabrous or puberulous. Flowers in short, dense, axillary, pedunculate spikes, 2–8 cm long. Calyx 8–10 mm long, divided near to the base; sepals oblong-lanceolate, acute. Corolla white with pinkish tinge in the throat, 2.5–3 cm long, pubescent outside; tube 8–12 mm long, narrowed at base; limb 2-lipped; lobes ovate-oblong, obtuse, 1–2 cm long. Filaments hairy at base; anthers acute but not tailed at base. Capsules clavate, 1.5–2 × 0.6–0.8 cm, shortly and bluntly pointed; solid, stalk flattened. Seeds orbicular-oblong, 5–6 mm long, glabrous.

Fls. & Frts.: August–December. *Exsiccata*: SUK-183.
Common, along villages, mostly grown as hedge plant.
Localities: Hatkangale, Pawas, Ratnagiri, Sangli.

Local uses: The leaves are smoked as a cigarette in asthma, bronchitis and cough. Leaves and root with ginger powder administered in malarial fevers. Leaves are used in jaundice.

Rostellularia procumbens (L.) Nees in Wall. Pl. As. Rar. 3: 101. 1832; Moorthy in Singh *et al.* Fl. Maharashtra State Dicot. 2: 667. 2001. *Justicia procumbens* L. Sp. Pl. 15. 1753; C. B. Cl. in Hook. f. Fl. Brit. India 4: 539. 1885; Cooke, Fl. Pres. Bombay 2: 491. 1958 (Repr. ed.). KALMASHI, KAMRABVL.

Procumbent, annual herbs, 15–25 cm tall; stem much-branched, rooting at lower nodes, pubescent. Leaves elliptic to lanceolate, 2–4.5 × 0.7–1.8 cm, often obtuse at both ends, pubescent. Flowers in dense, thick, axillary and terminal spikes, which are usually 2–3 together, 1–3 cm long. Bracts linear-lanceolate, 4.5–5.5 mm long, with scarious ciliate margins, acute, hairy. Calyx 3.5–4 mm long, 4-partite to the base; sepals linear-subulate, with scarious ciliate margins. Corolla pink, 5–6 mm long; tube equaling the limb; lobes short rounded. Filaments glabrous. Capsules ellipsoid-ovate or oblong, 3–4 mm long, shortly pointed and pubescent at the tip, with 1 mm long solid stalk. Seeds 4, orbicular, 1 mm in diam., finely tuberculate.

Fls. & Frts.: July–November. *Exsiccata*: SUK-1746.
Common, in grasslands and hill slopes.
Localities: Ajara, Chandgad, Mahabaleshwar, Radhanagari, Sawantwadi.

Local uses: Fresh decoction of plant is warmed with tincture of salt and given in fevers and considered as vermifuge.

Rungia repens (L.) Nees in Wall. Pl. As. Rar. 3: 110. 1832; C. B. Cl. in Hook. f. Fl. Brit. India 4: 549. 1885; Cooke, Fl. Pres. Bombay 2: 480. 1958 (Repr. ed.); Moorthy in Singh *et al.* Fl. Maharashtra State Dicot. 2: 672. 2001. *Justicia repens* L. Sp. Pl. 15. 1753. GHATIPITPAPDA.

Decumbent herbs, stem rooting at lower nodes, 15–60 cm long, glabrous or appressed pubescent. Leaves oblong-lanceolate, 2–12 × 1–3 cm, tapering at base, entire, acute, thinly pubescent. Flowers in sessile, terminal spikes, 2–5 cm long; bracts broadly elliptic, 8–10 × 5–6 mm, hairy with broad scarious margins, rounded and cuspidate at apex. Calyx 5 partite nearly to the base, 4–5 mm long; sepals linear-subulate. Corolla blue, 8–10 mm long, pubescent outside, 2-lipped; lobes obtuse. Stamens 2; anthers 2-celled, cells superposed, lower cell often with a white basal appendage. Capsules obovoid, 4–5 mm long, obtuse or truncate, mucronate, pubescent. Seeds 4, suborbicular, 1–1.5 mm in diam., brown glabrous.

Fls. & Frts.: September–January. *Exsiccata*: SUK-1752.
Common, under shade of bushes on gravelly slopes of hills.
Localities: Kolhapur, Satara, Sangli.

Local uses: Leaf decoction is used as tonic and also given as diuretic. Whole plant is dried and pulverized and given in fevers and cough and considered as vermifuge. Fresh decoction of plant is warmed with tincture of salt and given in fevers.

VERBENACEAE J. St. Hil.

Callicarpa tomentosa (L.) Murr. Syst. Veg. ed. 13, 130. 1774; Prasanna in Singh *et al.* Fl. Maharashtra State Dicot. 2: 689. 2001. *Tomex tomentosa* L. Sp. Pl. 118. 1753. *Callicarpa lanata* L. Mant. Alt. 331. 1771; C. B. Cl. in Hook. f. Fl. Brit. India 4: 567. 1885; Cooke, Fl. Pres. Bombay 2: 502. 1958 (Repr. ed.). AISAR.

Small trees, 5–10 m tall. Leaves ovate or elliptic lanceolate, 10–20 × 5–12 cm, acuminate, base rounded, upper surface glabrous, lower surface densely tomentose. Flowers sessile in divaricately tomentose axillary cymes. Calyx 2.5 mm long, campanulate, very faintly 4-lobed, densely stellate tomentose. Corolla 4 mm long, reddish-purple, glabrous, tube 2.5 mm long; lobes subequal, subquadrate, rounded. Stamens much exserted; filaments slender, glabrous. Ovary 2-celled, stigma obscurely 2-fid, glabrous, drupes globose, 2–4 mm across, black, smooth shining.

Fls. & Frts.: December–June. *Exsiccata*: SUK-1735.
Frequent, in semi evergreen forests.
Localities: Amboli, Mahabaleshwar, Pawas, Sawantwadi, Tillari, Vaibhavwadi.

Local uses: The leaves and root decoction is used to less febrile heat and removes hepatic obstruction and hepatic eruptions. Leaves are bruised and applied on wounds and ulcers.

Clerodendrum inerme (L.) Gaertn. Fruct. Sem. 1: 271, t. 75. 1788; C. B. Cl. in Hook. f. Fl. Brit. India 4: 589. 1885; Cooke, Fl. Pres. Bombay 2: 511. 1958 (Repr. ed.); Prasanna is Singh *et al.* Fl. Maharashtra State Dicot. 2: 690. 2001. *Volkameria inermis* L. Sp. Pl. 637. 1753. VANJAI.

Shrubs, 2.5 m tall, much branched, straggling. Leaves opposite, elliptic or obovate, 2–5 × 1–2.5 cm, obtuse, base attenuate. Flowers in axillary pedunculate cymes, 3–9-flowered. Calyx 5 mm long, enlarged, leathery; teeth triangular, acute. Corolla white; tube 2.5 cm or more in length, glabrous, very hairy inside; lobes subequal, 0.6–1 cm long, oblong, hairy at base, much exserted. Stamens 4, didymous, much exserted. Ovary glabrous; style much exserted, stigma bilobed. Drupe pyriform, 1.2–1.5 cm long, seated in the veined persistent enlarged calyx, separating into 4 woody pyrenes.

Fls. & Frts.: May–December. *Exsiccata*: SUK-1734.
Common, in deciduous forests and village outskirts.
Localities: Hatkanagale, Ichalkaranji, Shirol, Shiroli.

Local uses: Juice of leaves and root is used in fever. Leaf is used externally in rheumatism. Leaf paste is warmed and applied externally on fresh wounds.

Clerodendrum multiflorum (Burm. f.) O. Ktze. Rev. Gen. Pl. 3: 526. 1891; Prasanna is Singh *et al.* Fl. Maharashtra State Dicot. 2: 690. 2001. *Volkameria multiflorum* Burm. f. Fl. Ind. 137, t. 45, f. 1. 1786. *Clerodendrum phlomidis* L. f. Suppl. 292. 1781; C. B. Cl. in Hook. f. Fl. Brit. India 4: 590. 1885; Cooke, Fl. Pres. Bombay 2: 511. 1958 (Repr. ed.). IRAN.

Trees or scandent shrubs, ca 9 m tall; branches pubescent. Leaves ovate or subrhomboid, 5–9 × 0.5–4 cm, obtuse or acute, margins crenate-dentate. Flowers fragrant, in small dichotomous axillary cymes arranged so as to form a rounded terminal panicle. Calyx 1 cm long divided about ½ way down, glabrous, not enlarged in fruit; segments ovate, acutely acuminate. Corolla white or pinkish; tube 2–2.5 cm long, pubescent outside, glabrous inside. Stamens 4; filaments slightly pubescent below. Drupe obovoid, 7 mm long, calyx lobes persistent.

Fls. & Frts.: November–March. *Exsiccata*: SUK-252.
Common, in deciduous forests.
Localities: Hatkanagale, Sangli, Shiroli.

Local uses: Leaf paste is applied externally and given orally in rheumatism. Leaves are warmed over flame and made into paste and applied on swellings. Leaf juice is externally applied on piles and smoke of castor oil is given.

Clerodendrum serratum (L.) Moon. Cat. 46. 381. 1824; C. B. Cl. in Hook. f. Fl. Brit. India 4: 592. 1885; Cooke, Fl. Pres. Bombay 2: 512. 1958 (Repr. ed.); Prasanna is Singh *et al.* Fl. Maharashtra State Dicot. 2: 691. 2001. *Volkameria serrata* L. Mant. 90. 1767. BHARANG.

Shrubs, 3 m tall. Leaves elliptic-ovate or oblong, 8–30 × 3–14 cm, acute, coarsely and sharply serrate, glabrous, base acute. Flowers numerous, showy in lax pubescent dichotomous cymes, with a pair of acute bracts at each branching and a flower in the fork. Calyx 2.5 mm long, puberulous, cup-shaped, truncate, not enlarged in fruits, lobes very small, triangular, acute, ciliolate. Corolla glabrous outside, pale blue, the large lower lobe dark bluish-purple; tube 1 cm long, cylindric, hairy, oblique at the mouth; 2 upper and 2 lateral lobes elliptic, obtuse, flat, spreading; 1 cm long. Lower lobe lip like, more than 1.2 cm long, concave, deflexed. Filaments much curved, densely hairy at the base. Ovary and style glabrous. Drupes 6 mm long, succulent, broadly obovoid, normally 4-lobed with 1 pyrene in each lobe.

Fls. & Frts.: July–December. *Exsiccata*: SUK-136.
Common, in deciduous forests.
Localities: In all districts.

Local uses: Decoction of root is given in fever and cough. Leaves are applied externally in tumors and certain skin diseases. Plant is used as a tonic, useful in bronchitis, asthma, ozoena, fevers, tumors inflammations, burning sensation, hiccough, epilepsy and wounds.

Clerodendrum viscosum Vent. Malm. 1: t. 25. 1803; Prasanna is Singh *et al.* Fl. Maharashtra State Dicot. 2: 691. 2001. *C. infortunatum* auct. non L. 1753; C. B. Cl. in Hook. f. Fl. Brit. India 4: 594. 1885; Cooke, Fl. Pres. Bombay 2: 513. 1958 (Repr. ed.). ADULSA, BHANDIRA, KODLI.

Shrubs, 3 cm tall, woody, villous. Leaves broadly ovate, 5–20 × 3.5–15 cm, villous cordate or rounded. Flowers on rather long pubescent pedicels in stalked cymes forming large pubescent panicles. Calyx 1–2 cm long in flower, enlarged in fruit; silky pubescent; segments broadly lanceolate, very acute. Corolla densely pubescent outside, white tinged with pink; tube 2 cm long, slender, lobes 1 cm long, oblong, obtuse. Filaments glabrous. Ovary and style glabrous. Drupes 8 mm across, black.

Fls. & Frts.: June–November. *Exsiccata*: SUK-1711.
Common, in moist deciduous to semi evergreen forests.
Localities: Amba, Amboli, Barki, Dajipur.

Local uses: Fresh juice of leaves is employed as a vermifuge and febrifuge in malarial fevers mostly in children. Plant is used as a tonic, aphrodisiac, antipyretic, anthelmintic, useful in biliousness, thirst, burning sensations, foul odor and diseases of the blood.

Gmelina arborea Roxb. Pl. Cor. 3: 42, t. 246. 1815; C. B. Cl. in Hook. f. Fl. Brit. India 4: 581. 1885; Cooke, Fl. Pres. Bombay 2: 504. 1958 (Repr. ed.). SHIVAN.

Trees, 15 m tall. Leaves broadly ovate, 4–18 × 3–13 cm, acuminate, cordate. Flowers appearing with or before the young leaves, usually in small cymes of about 3 flowers arranged along the branches of densely fulvous-hairy panicle reaching 30 cm long. Calyx 5 mm long, broadly campanulate; densely fulvous hairy; teeth 5, small; triangular acute; corolla brownish-yellow, densely hairy outside, reaching 2–3 cm long, 5-lobed, 2-lipped, upper lip more than 1 cm long, deeply divided into 2 oblong, obtuse lobes; lower lip nearly 2.5 cm long, 3-lobed, middle lobe projecting forward, ovate, subobtuse, with irregularly crenulate margin, much longer and broader than the obovate rounded lateral lobes. Drupes 1–2.5 cm long, ovoid or pyriform, smooth, orange-yellow when ripe.

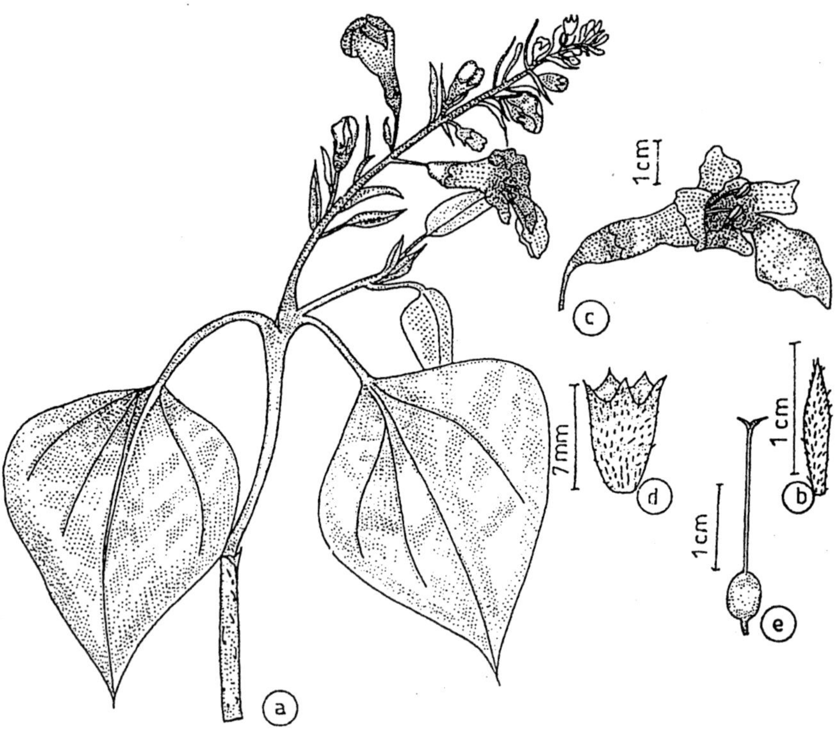

Text fig. 35: Gmelina arborea a-twig; b-bract; c-flower; d-calyx; e-carpel.

Fls. & Frts.: February–July. *Exsiccata*: SUK-333.
Common, in deciduous forests.
Localities: Borbet, Kolhapur, Tillari.

Local uses: Root forms an ingredient of Dashmula. Leaf juice is given in urinary complaints. It is used externally in herpes zoster and given orally also. Bark is given to pregnant women to prevent abortion. Root is useful in hallucinations, thirst, piles, abdominal pains, burning sensations, fever and urinary discharges.

Phyla nodiflora (L.) Greene in Pittonia 4: 46. 1899; Prasanna in Singh *et al.* Fl. Maharashtra State Dicot. 2: 693. 2001. *Verbena nodiflora* L. Sp. Pl. 20. 1753. *Lippia nodiflora* (L.) A. Rich. in Michux, Fl. Bor. Amer. 2: 15. 1803; C. B. Cl. in Hook. f. Fl. Brit. India 4: 563. 1885; Cooke, Fl. Pres. Bombay 2: 499. 1958 (Repr. ed.). GOURMUNDI, RATOLIVA.

Prostrate, perennial herbs; stem much branched, creeping, pubescent, rooting at nodes. Leaves opposite, sessile, obovate or spathulate, 1.5–3 × 1–2.5 cm, cuneate at base, deeply cordate above, base obtuse or rounded at apex, appressed pubescent on both the surfaces. Flowers minute, sessile, in many flowered, densely packed subglobose to oblong, 1.5–2 cm long heads. Calyx spathaceous, 1.5–2 mm long, 2-lobed, pubescent, embracing the fruit. Corolla white to pink, 2–2.5 mm long; the middle lobe the longest; tube as long as the lobes. Fruits oblong, 1–1.5 mm long, flat, splitting into 2, one seeded pyrenes.

Fls. & Frts.: Throughout year. *Exsiccata*: SUK-181.
Common, in wet situations, in marshes and along water courses.
Localities: Kalamba, Kolhapur, Rankala.

Local uses: Poultice of leaf is applied on boils. Leaf paste is applied externally on inflamed swellings. Leaf infusion is given to children suffering from indigestion. Plant is useful in diseases of heart, blood; cure ulcers, wounds, burning sensation, asthma, bronchitis and loss of consciousness.

Premna obtusifolia R. Br. Phytologia 36: 438. 1977. *P. integrifolia* L. Mant. 2: 252. 1771 nom. superfl.; C. B. Cl. in Hook. f. Fl. Brit. India 4: 574. 1885; Cooke, Fl. Pres. Bombay 2: 506. 1958 (Repr. ed.). KHARA-NARVEL.

Shrubs, ca 8 m tall. Leaves ovate, 4–9 × 3–6 cm, acute-apiculate, base rounded, margins serrate. Flowers small, greenish-yellow with disagreeable odor, in terminal pubescent paniculate corymbose cymes. Calyx 2.5 mm long, thick, glabrous, 2-lipped, one lip 2-toothed; the other subentire. Corolla glabrous outside; tube 3 mm long, cylindric, hairy in the throat inside; lobes 4; oblong, rounded. Stamens slightly exserted filaments hairy at base. Ovary and style glabrous; stigma of 2 equal divaricate lobes. Fruits 4 mm long, pear-shaped; endocarp ridged, bony, 4-celled, 4-seeded.

Fls. & Frts.: April–August. *Exsiccata*: SUK-124.
Infrequent, in semi evergreen or moist deciduous forests.
Localities: Pawas, Radhanagari, Ratnagiri.

Local uses: Root forms an ingredient of Dashmula. Decoction of root is given in colds and fevers. Root is useful in anemia, diabetes, chyluria, inflammations, swellings, bronchitis, dyspepsia, piles, constipation and fever. Leaves are good as an external application to piles and tumors.

Tectona grandis L. f. Suppl. 151. 1781; C. B. Cl. in Hook. f. Fl. Brit. India 4: 570. 1885; Cooke, Fl. Pres. Bombay 2: 503. 1958 (Repr. ed.); Prasanna in Singh *et al.* Fl. Maharashtra State Dicot. 2: 697. 2001. SAG, SAGWAN.

Large, deciduous trees, 10–20 m tall; branchlets 4-angled, densely clothed with yellowish gray tomentum. Leaves opposite, subsessile, elliptic, or obovate, 30–50 × 15–20 cm, cuneate at base, entire or crenulate, acute or acuminate, rough and glabrous above, stellate, gray to tawny tomentose beneath. Flowers in large, erect, terminal, branched tomentose panicles, 30–60 cm long. Calyx linear-lanceolate, 2–2.5 mm long; teeth 5–6, subequal, 1 mm long; in fruit enlarged to 2–2.5 cm or more long, bladdery, enclosing the fruit. Corolla white, 5–6 lobed; tube 2 mm long; limb 5–6 mm across; lobes subequal. Stamens 5–6, exserted. Ovary 4-locular. Fruits subglobose, 10–12 mm in diam., shallowly 4-lobed, stellate hairy. Seeds 2–4, oblong, brown, enclosed in body endocarp.

Fls. & Frts.: June–December. *Exsiccata*: SUK-1761.
Common, in hilly tracts and also planted.

Local uses: Oil extracted from seeds is applied externally on skin diseases. Bark is used for the dispersion of inflammatory swellings. Flower paste is given as a diuretic. Roots are given in anria and retention of urine. Flowers are useful in bronchitis and urinary discharges. Oil from the flowers and seeds promotes the growth of hair and useful in scabies.

Vitex altissima L. f. Suppl. Pl. 294. 1781; C. B. Cl. in Hook. f. Fl. Brit. India 4: 584. 1885; Cooke, Fl. Pres. Bombay 2: 510. 1958 (Repr. ed.); Prasanna in Singh *et al.* Fl. Maharashtra State Dicot. 2: 692. 2001. BALAGE, BAVALGE.

Trees, 20 m tall. Leaves 3-foliolate; leaflets 5–14 × 2–5 cm, lanceolate, chartaceous, acuminate, entire, glabrous. Flowers numerous in shortly pedunculate or nearly sessile cymes, clustered along the branches of large lax fulvous pubescent terminal panicles. Calyx 3 mm long, fulvous-villous, teeth small, 1 mm long, oblong, obtuse. Corolla white, tinged with blue, 2-lipped; upper lip 2.5 mm long with 2 triangular subacute lobes; lower lip 4 mm long, 3-lobed, the middle lobe more than twice as large as the lateral ones. Drupes irregularly globose, less than 1 cm in diam., calyx persistent, purple, smooth.

Fls. & Frts.: March–August. *Exsiccata*: SUK-1760.
Common, in semi evergreen forests.
Localities: Ajara, Patgaon, Tillari.

Local uses: Leaf decoction is applied externally in skin diseases and in rheumatism.

Vitex leucoxylon L. f. Suppl. 293. 1781; C. B. Cl. in Hook. f. Fl. Brit. India 4: 587. 1885; Cooke, Fl. Pres. Bombay 2: 510. 1958. (Repr. ed.); Prasanna in Singh *et al.* Fl. Maharashtra State Dicot. 2: 698. 2001. *Wallrothia leucoxylon* (L. f.) Roth, Nov. Pl. Sp. 319. 1821. SHERAS, SHERNI.

Trees, 20 m tall. Leaves 3–5 foliolate; leaflets 2–10 × 0.8–4 cm, petiolate, glabrous above, sparsely pubescent beneath, apex obtuse, acute, base acute. Flowers fragrant in axillary pedunculate lax corymbose cymes. Calyx pubescent, 3 mm long; teeth 5, small, triangular. Corolla 1.2 cm long pubescent, white with purplish hairs; upper divided to the middle into 2 rounded lobes; lower lip 3-lobed, 1 cm long, middle lobe hairy in the center, with purplish hairs and with crisped margins. Filaments hairy at the base. Ovary densely hairy at the top with erect white hairs. Drupes 2 cm long, succulent, obovoid, blunt, dark purple when ripe, calyx persistent; stone large, 4-celled and 4-seeded.

Fls. & Frts.: April–June. *Exsiccata*: SUK-286.
Frequent, on banks of streams and in forests.
Localities: Ajara, Amboli, Phonda, Sawantwadi, Tillari, Vaibhavwadi.

Local uses: Leaves are spread over blanket and patients were allowed to sleep on blanket whole night in rheumatism and muscular pain. Similarly pillow stuffed with the leaves of Nirgudi is placed under the head for relief of headache. Leaf juice is used to remove fetid discharges and worms from ulcers. Similar as *Vitex negundo*.

Vitex negundo L. Sp. Pl. 638. 1753; C. B. Cl. in Hook. f. Fl. Brit. India 4: 583. 1885; Cooke, Fl. Pres. Bombay 2: 508. 1958 (Repr. ed.); Prasanna in Singh *et al.* Fl. Maharashtra State Dicot. 2: 699. 2001. NIRGUDI.

Erect shrubs or small trees with grayish tomentose branches. Leaves 3–5 foliolate; leaflets elliptic-lanceolate, 3–15 × 1–4 cm, very unequal, lateral ones with very short. Petiolules subsessile; the median one the longest with petiolules, 8–12 mm long, margins crenate-serrate, acute, acuminate, densely short hairy beneath. Flowers in grayish-tomentose, pedunculate, terminal or axillary panicles of distantly and laterally disposed cymes. Calyx cup-shaped, 3 mm long, tomentose; teeth triangular, acute. Corolla bluish, 1 cm long, tomentose outside and within at the insertion of the stamens. Stamens 4 included. Fruits subglobose, drupaceous, 2–3 mm across. Seeds 2–4, in bony endocarp.

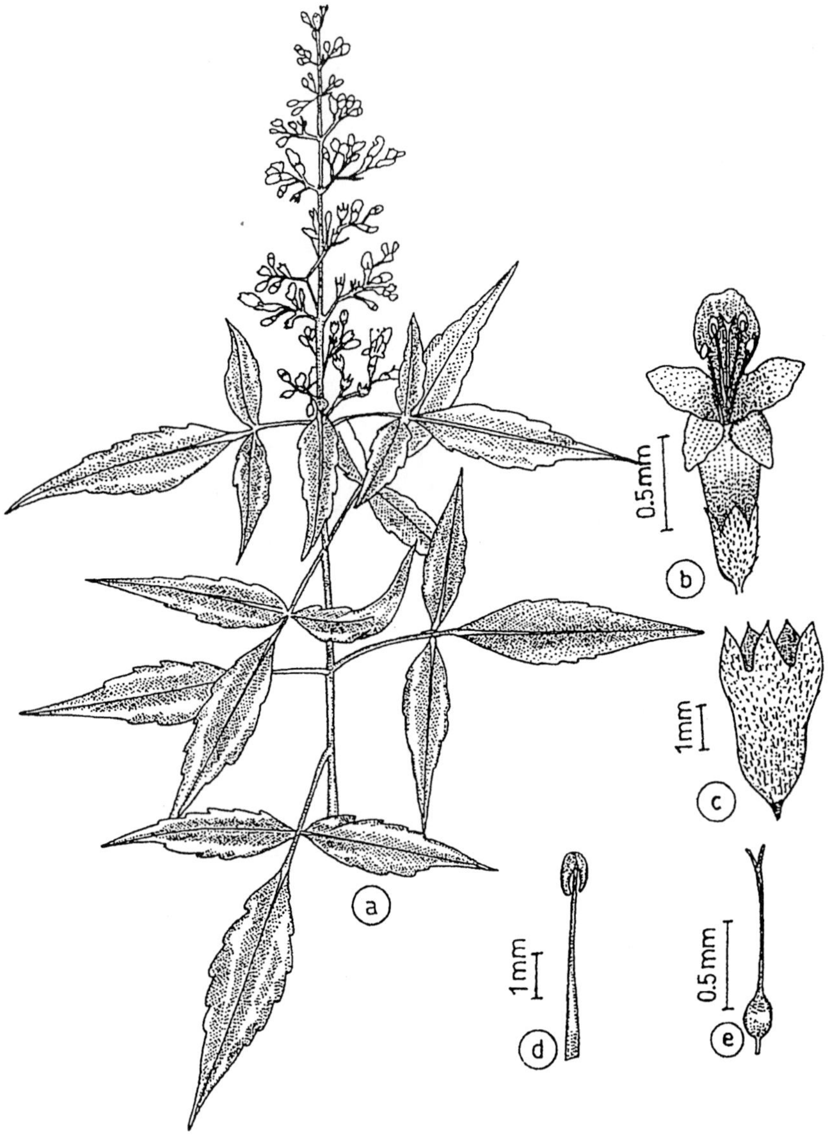

Text fig. 36: Vitex negundo a-twig; b-flower; c-calyx; d-stamen; e-carpel.

Fls. & Frts.: Throughout year. *Exsiccata*: SUK-287.
Common, along riverbanks and along canals.
Localities: Amba, Barki, Gaganbawada, Radhanagari.

Local uses: Leaves are spread over blanket and patients are allowed to sleep on blanket whole night in rheumatism and muscular pain. Similarly pillow stuffed with the leaves are placed under the head for relief of headache. Leaf juice is used to remove fetid discharges and worms from ulcers.

LAMIACEAE Lindl. (*nom. alt.*)
[**LABIATAE** (*nom. cons.*)]

Anisochilus carnosus (L.) Wall. Pl. As. Rar. 2:18. 1830; Hook. f. Fl. Brit. India 4: 627. 1885; Cooke, Fl. Pres. Bombay 2: 532. 1958 (Repr. ed.); Kulkarni & Das Das in Singh *et al.* Fl. Maharashtra State Dicot. 2: 713. 2001. *Lavandula carnosa* L. Amoen Acad. 10: 56, t. 3. 1790. KAPURLI.

Erect, subfleshy herbs, 30–90 cm tall; branches often purplish. Leaves ovate, 4–6 × 2–4.5 cm, truncate or subcordate at base, crenate-serrate, downy on both the surfaces, obtuse or subacute. Flowers in 5–7 cm long, simple to paniculate, cylindrical spikes. Calyx 2-lipped, shortly pubescent and ciliate, upper lip ovate-lanceolate, acute, ciliolate, 3-nerved, often deflexed and in fruits concealing the mouth of calyx tube; lower lip truncate, ciliate. Corolla tubular, 2-lipped, 2–3 mm long, purplish, pubescent outside more or less inflated above; upper lip short, erect shallow; lower lip concave. Stamens 4 didynamous. Nutlets suborbicular, 1–1.5 mm broad, subcompressed, shining, smooth as long as broad.

Fls. & Frts.: September–December. *Exsiccata*: SUK-200.
Occasional, in crevices of rocks, on old walls and along banks of river.
Localities: Kolhapur, Panhala.

Local uses: Plant is externally applied in insect bite. Leaves are externally applied on wounds.

Anisomeles indica (L.) O. Ktze. Rev. Gen. Pl. 512. 1891; Kulkarni and Das Das in Singh *et al.* Fl. Maharashtra State Dicot. 2: 715. 2001. *Nepeta indica* L. Sp. Pl. 571. 1753. *Anisomeles ovata* R. Br. in Ait. Hort. Kew. (ed. 2), 2: 364. 1811; Hook. f. Fl. Brit. India 4: 672; 1885; Cooke, Fl. Pres. Bombay 2: 543. 1958 (Repr. ed.). GOPALI.

Erect, woody herbs, 100–150 cm tall; stems acutely quadrangular, soft pubescent; branches gray tomentose. Leaves broadly ovate, 5–12 × 4.8 cm, subcordate, truncate or rounded at base, crenate-serrate, pubescent on both surfaces, acute. Flowers sessile or nearly so, in densely flowered whorls; whorls distantly arranged in lower part of the axis and closely disposed in upper part to form dense spikes. Calyx tubular, 5–7 mm long, equally 5-toothed, more or less constricted in the middle, hirsute with long stiff hairs; teeth lanceolate, acute as long as the tube, ciliate. Corolla 1–1.5 cm long, pink-purple with deep blue lower lip; upper lip oblong-obtuse; lower lip much longer, 3-lobed; the middle one deeply 2-fid. Stamens 4, filaments beard. Nutlets oblong obovoid, 1.5–2 mm long, smooth, dark-brown.

Fls. & Frts.: September–November. *Exsiccata*: SUK-1716.
Common, along banks of streams, rivers and along gullies in forests.
Localities: Kolhapur, Kagal, Wai.

Local uses: Seeds are soaked in water overnight and given early in the morning in urinary problems. Oil from plant is used in uterine infections.

Lavandula bipinnata O. Ktze. Rev. Gen. Pl. 251. 1891; Kulkarni and Das Das in Singh *et al.* Fl. Maharashtra State Dicot. 2: 721. 2001. *Biseropogon bipinnata* Roth, Nov. Pl. Sp. 255. 1821. *Lavandula burmanni* Benth. Lab. Gen. and Sp. 151. 1883; Hook. f. Fl. Brit. India 4: 631. 1885; Cooke, Fl. Pres. Bombay 2: 534. 1958 (Repr. ed.). GHODEGHUI.

Erect herbs, 20–50 cm high, stem simple or branched, leafy, quadrangular, finely pubescent. Leaves sessile, 2.5–10 cm long and as broad as long, pinnatipartite or very deeply pinnatisect; lobes linear, entire, cut or toothed, obtuse or subacute, glabrous or pubescent above, pale and pubescent beneath. Flowers in spikes which are simple or with few or many branches, sometimes umbellately arranged. Calyx 5 mm long, tubular, slightly curved, clothed with fine hoary pubescent; teeth equal, lanceolate, acute, with penicillate tips. Corolla blue or white, pubescent outside, exceeding 1.2 cm long, lower half very slender, the upper half slightly dilated, upper lip oblong, with 2 short rounded lobes; lower lip 5-partite, the middle lobe oblong; rounded, more than twice as long as the ovate obtuse lateral lobes. Nutlets oblong-ellipsoid, 2 × 2.5 mm, smooth, very mucilaginous when wetted, black; when ripe, with large white oblong areole on the dorsal side.

Fls. & Frts.: October–November. *Exsiccata*: SUK-190.
Common, throughout region.
Localities: Jat, Kagal, Ramling, Wai.

Local uses: Paste of leaves applied externally on wounds and ulcers. Decoction of root and flowers is used repeatedly to wash ulcers.

Leonotis neptiifolia (L.) R. Br. Prodr. 504. 1810. (*neptaefolia*); Hook. f. Fl. Brit. India 4: 691. 1885; Cooke, Fl. Pres. Bombay 2: 555. 1958 (Repr. ed.); Kulkarni and Das Das in Singh *et al.* Fl. Maharashtra State Dicot. 2: 723. 2001. *Phlomis neptiifolia* L. Sp. Pl. 586. 1753. (*nepetaefolia*) DIPMAL.

Erect, woody herbs, 0.5–2 m tall; stems erect, obtusely 4-angled, gray pubescent. Leaves broadly ovate to oblong, 5–15 × 3–10 cm, cuneate at base, membranous, crenate-serrate, gray pubescent on both the surfaces, acute to acuminate; petioles winged in upper part. Flowers in dense globose, many-flowered, head like whorls, 3–6 cm in diam; bractiform leaves lanceolate, deflexed; floral bracts linear, strong spinecent, often deflexed, pubescent. Calyx tubular, 1.1–5 cm long, ribbed, reticulate within, in curved; tube densely pubescent or villous, bristly in upper part; teeth 8–9, unequal, spinous pointed; the upper one the largest and ovate-acute, not decurrent on the tube. Corolla tubular, 2–2.5 cm long, orange-red; tube glabrous below, densely covered with orange-red hairs, annulate within with 3 transverse rings of white hairs; upper lip woolly, with orange-red hairs; lower lip deeply 3-lobed, the middle lobe the longest. Stamens scarcely exserted. Nutlets oblong, obovoid, 3–5.4 mm long, depressed at apex.

Fls. & Frts.: September–January. *Exsiccata*: SUK-269.
Common, on wastelands.
Localities: Jotiba, Kagal, Kolhapur, Panhala, Satara, Wai.

Local uses: Leaf juice in combination with 1 tablespoon of limejuice is mixed with spirit and given in fever. Leaf decoction is used with decoction of *Phyllanthus amarus* in urinary complaints due to fever. Flower ash is mixed with curd and applied externally on elephantiasis and eczema.

Leucas cephalotes (Roth) Spr. Syst. Veg. 2: 743. 1825; Hook. f. Fl. Brit. India 4: 689. 1885; Cooke, Fl. Pres. Bombay 2: 549. 1958 (Repr. ed.); Kulkarni & Das Das in Singh *et al.* Fl. Maharashtra State Dicot. 2: 727. 2001. TUMBA.

Erect, pubescent herbs, 40–70 cm tall. Leaves broadly elliptic or ovate, 3–8 × 1.5–3.5 cm, narrowed at base, crenate-serrate, rather membranous, more or less pubescent, subacute. Flowers sessile in large, terminal globose, dense whorls, 3–5 cm in diam., with often bractiform leaves at the base or crowned at the top of the whorls; floral bract elliptic or ovate lanceolate, 10–15 × 3–4 mm, strongly reticulate, densely ciliate, acuminate, awned, as long as or slightly longer than the calyx. Calyx tubular, ca 1.5 mm long, slightly curved; the lower half quite glabrous, smooth and membranous, the upper half pubescent outside and inside; mouth oblique, densely villous; the villi not exceeding the teeth; teeth small, subulate. Corolla white, ca 2 cm long; upper lip woolly, the lower lip twice as long; tube annulate within near the middle. Nutlets obovoid, 3 mm long, cuneate at base, smooth, brownish.

Fls. & Frts.: August–November. *Exsiccata*: SUK-1728.
Common, along gullies, in hill forests, wastelands of agricultural field and old walls.
Localities: Katyayani, Nesari.

Local uses: Plant is applied externally in inflammatioh. Leaf juice given in urinary complaints. Fresh juice of plant is used externally in scabies. Flowers are given orally in cough. Leaf juice dropped in nostrils in cold. Plant is laxative, anthelmintic, stimulant, diaphoretic; useful in bronchitis, jaundice, inflammations, asthma, paralysis and leucoderma. Leaves are useful in fever and urinary discharges.

Leucas indica (L.) R. Br. ex Vatke in Oestr. Bot. Zeitschr. 25: 95. 1885; Kulkarni & Das Das in Singh *et al.* Fl. Maharashtra State Dicot. 2: 729. 2001. *Leonurus* *indicus* L. Syst. Nat. ed. (ed.) 10. 1101. 1760. *Leucas linifolia* (Roth) Spr. Syst. Veg. 2: 743. 1825; Hook. f. Fl. Brit. India 4: 690. 1885; Cooke, Fl. Pres. Bombay 2: 548. 1958 (Repr. ed.). SHANKROBA.

Erect, pubescent herbs, 15–30 cm tall. Leaves linear-lanceolate, 1.5–5.5 × 0.3–0.5 cm, tapering at both the ends, entire or remotely and shallowly crenate serrate, shortly pubescent, on both the surfaces, acute. Flowers in axillary and terminal whorls, towards the ends of branches; floral bracts linear, setaceous, bristle tipped, shortly pubescent. Calyx 5–7 mm long, minutely pubescent outside, appressed pubescent inside in the upper part, glabrous in the lower part, with a pale tube and a green limb; mouth very oblique, contracted, glabrous; teeth very unequal, the upper tooth large and much produced forward; the others minute. Corolla white, ca 1.5 cm long, annulate inside about the middle; the upper lip woolly, slightly shorter than the lower lip. Nutlets oblong, ca 2 mm long, rounded at apex, brownish.

Fls. & Frts.: September–November. *Exsiccata*: SUK-1729.
Common, in plains and Ghats.
Localities: Appachiwadi, Kagal, Ramling.

Local uses: Plant is applied externally in inflammation. Leaf juice given in urinary complaints. Fresh juice of plant is used externally in scabies. Flowers are given orally in cough. Leaf juice dropped in nostrils in cold.

Ocimum americanum L. Cent. Pl. 1: 15. 1755; Kulkarni & Das Das in Singh *et al.* Fl. Maharashtra State Dicot. 2: 741. 2001. *O. canum* Sims in Bot. Mag. 52: t. 2452. 1823; Hook. f. Fl. Brit. India 4: 607. 1885; Cooke, Fl. Pres. Bombay 2: 521. 1958 (Repr. ed.). RAN-TULSI.

Erect, much branched, hoary pubescent woody herbs, 30–50 cm tall. Leaves ovate or elliptic-lanceolate, 2.5–4 × 1.5–3 cm, narrowed at base; entire or nearly so. Flowers in 5–6 flowered whorls arranged in 7–20 cm long thyrsus. Floral bracts elliptic lanceolate, stalked, ciliate with white hairs. Calyx tubular, 2-lipped, 4–6 mm long; upper lip sub-orbicular, broader than long, entire, reticulate, mucronate; lower lip unequally 4-toothed, the two central teeth longer than the upper. Stamens 4, much exserted, the two upper with a tooth at base. Nutlets ellipsoid, 1–1.5 mm long; black, mucilaginous when wetted.

Fls. & Frts.: September–January. *Exsiccata*: SUK-1731.
Common, on waste lands near fields and around village.
Localities: Kagal, Karad, Wai, Radhanagari.

Local uses: Leaf juice is dropped into the nostrils in headache and cold. Leaf infusion is used to reduce fever and also used in catarrh and bronchitis. Leaf paste is applied externally in various skin diseases. Leaf juice is used with honey in hiccough. Leaf juice mixed with hot water and tincture of salt added and given in sore throat. Seeds are given in urinary complaints.

Ocimum tenuiflorum. L. Sp. Pl. 2: 597. 1753; Kulkarni & Das Das in Singh *et al.* Fl. Maharashtra State Dicot. 2: 743. 2001. *O. sanctum* L. Syst. Nat. ed. 12, 2: 402. 1767; Hook. f. Fl. Brit. India 4: 609. 1885; Cooke, Fl. Pres. Bombay 2: 521. 1958 (Repr. ed.). KALI-TULAS, RAMTULASI, TULSHI.

Erect herbs or undershrubs, 30–60 cm tall; stem and branches pubescent and purplish. Leaves elliptic oblong, 2–8 × 1.5–5 cm, acute or subacute at base, entire or crenate-serrate, obtuse or subacute, minutely gland dotted, pubescent on both surfaces. Flowers in 15–20 cm long, thyrsiform-racemes of 6-flowered whorls closely disposed on pubescent axis. Calyx tubular, 4–6 mm long, slightly accrescent in fruits, pubescent; upper lip broadly obovate, or suborbicular, much reflexed, minutely apiculate; lower lip longer than the upper lip; unequally 4-toothed, central teeth exceeding the upper lip, corolla tubular, 5–7 mm long, purplish, upper lip pubescent on the back. Stamens 4, exserted the upper pair with a small beard appendage at the base. Nutlets broadly ellipsoid, 1–1.3 mm long, yellow with black markings, smooth.

Fls. & Frts.: Almost throughout year. *Exsiccata*: SUK-294.
Cultivated.
Localities: In all districts.

Local uses: Leaf juice is dropped into the nostrils in headache and cold. Leaf infusion is used to reduce fever and also used in catarrh and bronchitis. Leaf paste is applied externally in various skin diseases. Leaf juice is used with honey in hiccough. Leaf juice is used in sore throat. Seeds are given in urinary complaints.

Plectranthus barbatus Andr. Bot. Rep. 9: t. 594. 1809; Kulkarni & Das Das in Singh *et al.* Fl. Maharashtra State Dicot. 2: 747. 2001. *Coleus barbatus* (Andr.) Benth. in Wall. Pl. As. Rar. 2: 15. 1830; Hook. f. Fl. Brit. India 4: 625. 1885; Cooke, Fl. Pres. Bombay 2: 530. 1958 (Repr. ed.). *Coleus forskohlii* (Willd.) Brig. in Engl. & Prantl. Pflanzenf. 4: 39. 359. 1879. GORMAR, MAINMULA.

Perennial herbs, 30–60 cm tall, with thick, subfleshy rootstock. Leaves elliptic-oblong or lanceolate, 2.5–8 × 1.3 cm, slightly narrowed at base, crenate-serrate, pubescent on both the surfaces, obtuse. Flowers in 6–10 flowered whorls, distantly arranged on spike like racemes; rachis glandular hairy; floral bracts broadly ovate, membranous, reticulate, long mucronate, pubescent and ciliate, often 5 toothed, often deflexed in fruits, with white hairs outside and a ring of white hairs in the throat; upper lip ovate entire, reticulate, acuminate, slightly longer than the lower one; lower lip subequally

4-toothed; teeth triangular. Corolla tubular, 1.5–2 cm long, pale purplish or blue, tube often curved; upper lip short erect, 4-lobed; lower lip boat-shaped, and curved upwards, conspicuously produced horizontally beyond the tube from its narrow base. Stamens slightly exserted beyond the corolla tube. Nutlets subglobose, ca 1–5 mm broad, smooth, blackish.

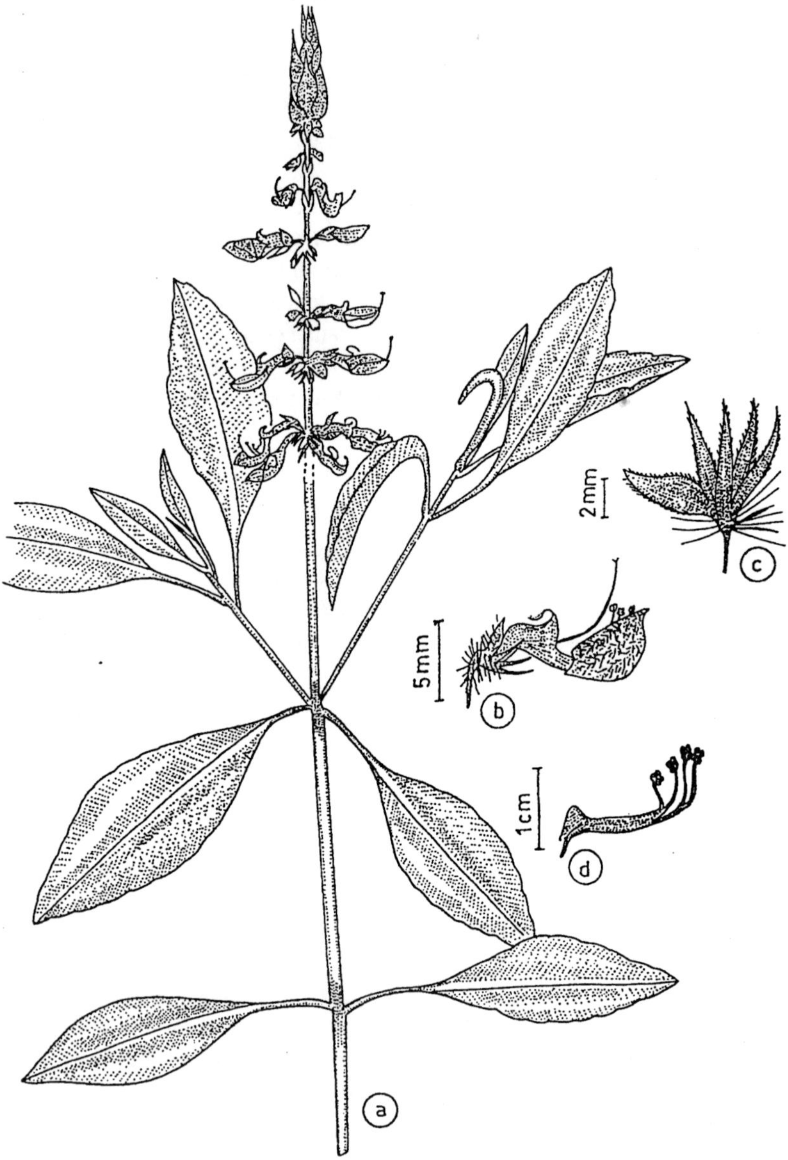

Text fig. 37: Plectranthus barbatus a-twig; b-flower; c-calyx; d-stamen.

Fls. & Frts.: August–October. *Exsiccata*: SUK-290.
Frequent, in crevices of rocks.
Localities: Nesari, Wai Ghat.

Local uses: Leaves are said to have a specific action on the bladder and useful in urinary diseases and vaginal discharges. Roots are used in hypertension.

Pogostemon purpurascens Dalz. in Hook. J. Bot. Misc. 2: 337. 1850; Hook. f. Fl. Brit. India 4: 632. 1885; Cooke, Fl. Pres. Bombay 2: 537. 1958 (Repr. ed.); Kulkarni & Das Das in Singh *et al.* Fl. Maharashtra State Dicot. 2: 757. 2001. PHUDSA.

Herbs, 20–60 cm high; stem and branches quadrangular, furrowed, softly hairy with spreading hairs. Leaves ovate, 5–10 × 2.5 cm, membranous, acute, serrate, hairy, base cuneate. Flowers in long-pedunculate villous spikes, 7–17 cm long; whorls many flowered, globose, the upper whorls approximate, the lower usually separate. Calyx 5 mm long, pubescent; tube pentagonal; teeth nearly equaling the tube, lanceolate, acute, 3-nerved; ciliate and with penicillate tips. Corolla 6 mm long, white, upper lip purple with white margin, 3-lobed; lower lip white, deltoid-ovate, acute. Stamens beard. Nutlets 1 mm long, ovoid, compressed, the inner face angular, the dorsal face rounded, smooth, polished, black when ripe.

Fls. & Frts.: October–January. *Exsiccata*: SUK-1733.
Frequent, as under growth in deciduous forest from ghats to plains.
Localities: Amba, Amboli, Borbet, Gaganbawada, Radhanagari, Tillari, Vaibhavwadi.

Local uses: The roots are used as an antidote for the poison of *Echis carinata* and common snake.

PLANTAGINACEAE Juss.

Plantago erosa Wall. in Roxb. Fl. Ind. 1: 423. 1820; Moorthy in Singh *et al.* Fl. Maharashtra State Dicot. 2: 769. 2001. *P. major* sensu Hook. f. Fl. Brit. India 4: 705. 1885; non. L. 1753. *P. asiatica* auct non L. 1753.

Scapigerous, glabrous, perennial herbs with short, stout rhizome. Leaves in basal rosette, ovate-elliptic to oblong, 5–20 × 3–15 cm, attenuate and decurrent at base, entire or remotely dentate, rounded, obtuse or acute, glabrous; petioles narrowly winged, 4–15 cm long. Flowers in terminal spikes, peduncle 5–20 cm long, terete or ribbed; spike cylindric, 1–30 cm long; bracts ovate, 1–1.5 mm long, scarious margined, acuminate, with a tuft of hairs at the base-within. Calyx 1–1.5 mm long, 4-cleft; sepals ovate-oblong, obtuse. Corolla white or dull-yellow, 3–3.5 mm long, 4-lobed; lobes obovate, acute. Stamens 4; filaments filiform; anthers versatile. Capsules ovoid, 3–4 mm long, transversely dehiscent, with a conical lid. Seeds 4–8, globose angular, minute, rugulose.

Fls. & Frts.: September–January. *Exsiccata*: SUK-1751.

Occasional, along riverbeds and around fields.

Localities: Kolhapur.

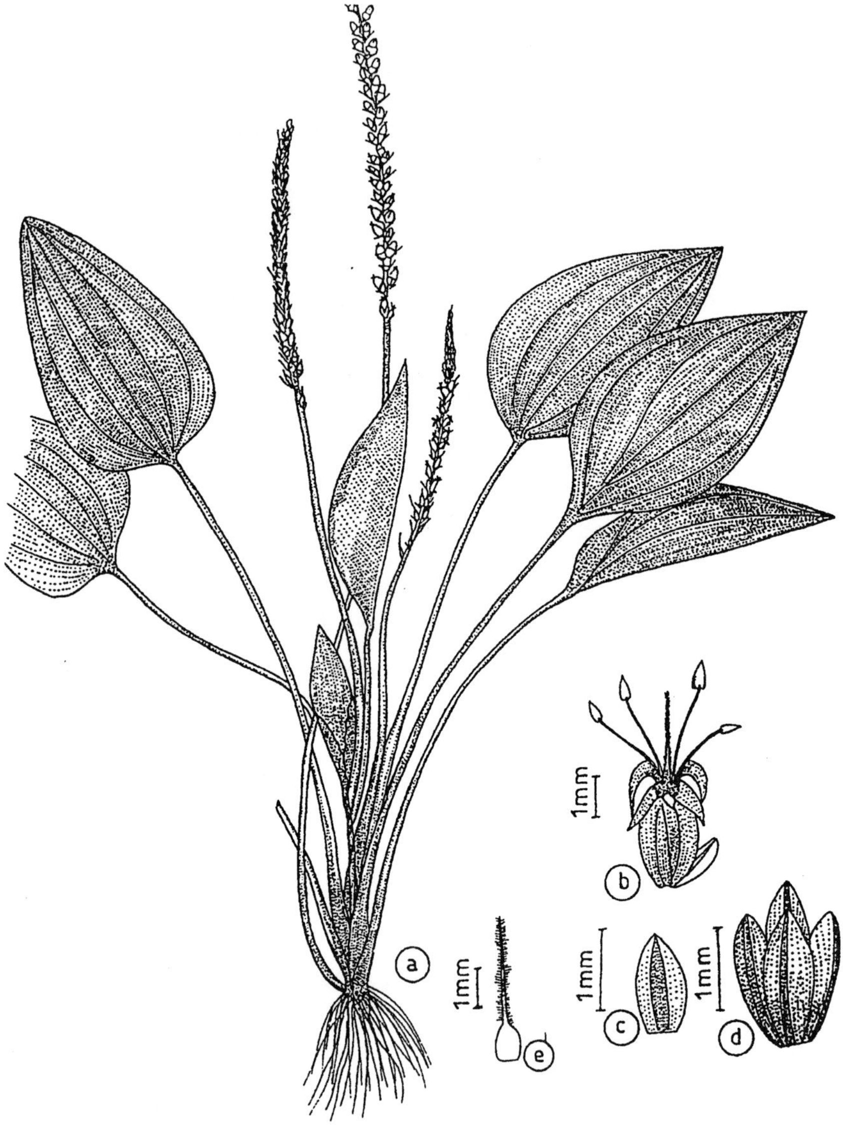

Text fig. 38: Palntago erosa a-twig; b-flower; c-bract; d-calyx; e-carpel.

Local uses: Leaf paste is mixed with sesamum oil and applied to rheumatic swellings. The seeds are used in chronic dysentery. Seeds are astringent to the bowels, tonic, useful in dysentery and leprosy; cause flatulence. Seeds have demulcent and diuretic properties and are generally used in inflammatory conditions of the mucous membrane of gastro-intestinal and genito-urinary tracts.

NYCTAGINACEAE Juss.

Boerhavia erecta L. Sp. Pl. 3. 1753; Moorthy in Singh *et al.* Fl. Maharashtra State Dicot. 2: 772. 2001. *B. punarnava* Saha & Murthy in J. Sci. Ind. Res. 21c: 249. 1962. PUNARNAWA.

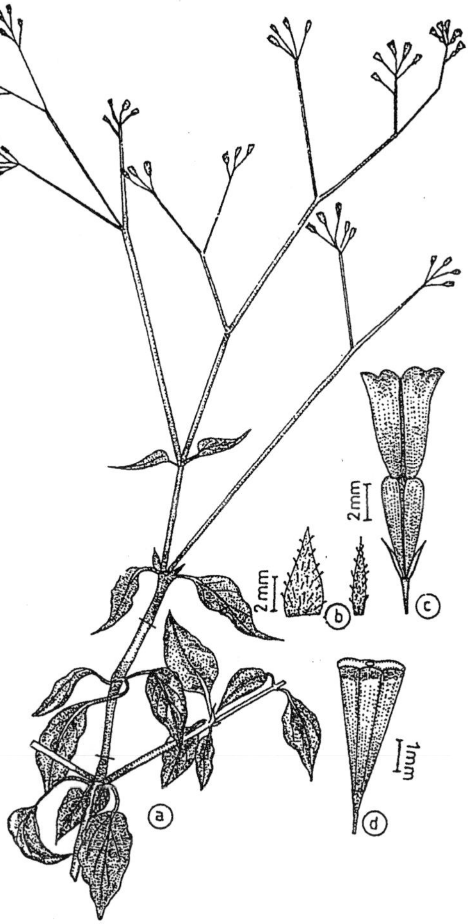

Text fig. 39: Boerhavia erecta a-twig; b-bract; c-flower; d-carpel.

Erect or suberect, pubescent herbs, 20–60 cm tall; stems more or less glandular hairy. Leaves ovate-lanceolate, in unequal pairs, 1.5–5 × 0.8–2.5 cm, acute at base, entire or undulate, glabrous or thinly pubescent. Flowers 2–3 together in lax panicles, 4–5 mm across; bracts minute, ovate, acute. Perianth white or pale pink, 2.5–3 mm long, campanulate, in upper part; lobes deeply notched. Stamens 2–3 slightly exserted. Fruits obconical, 4–5 mm long, 5-angled, truncate at apex, glabrous.

Fls. & Frts.: September–December. *Exsiccata*: SUK-1736.
Native of tropical America, naturalized in some parts.
Localities: Kolhapur, Sangli, Satara, Wai.

Local uses: Whole plant is used with *Acorus calamus* in asthma and in sore throat. It is given orally in jaundice to remove acidity. Root decoction is used to wash eyes in eye complaints. Decoction of whole plant is given as diuretic.

Boerhavia repens L. Sp. Pl. 3. 1753; Hook. f. Fl. Brit. India 4: 709. 1885; Moorthy in Singh *et al.* Fl. Maharashtra State Dicot. 2: 773. 2001. PUNARNAVA.

Prostrate, perennial herbs. Leaves ovate-lanceolate, 0.8–2 × 0.5–1 cm, oblique at base crisped along margins, acute green and glabrous above, paler and thinly pubescent beneath. Flowers very small, shortly stalked or sessile, 4–10 together, in small umbels arranged in slender-stalked corymbose, axillary and terminal panicles. Perianth 3 mm long; contracted above the ovary, glandular, viscid; limb funnel shaped, dark pink with 5 narrow vertical bands outside. Stamens 2–3, slightly exserted. Fruit 3 mm long, clavate, rounded, broadly and bluntly 5-ribbed, very glandular.

Fls. & Frts.: July–March. *Exsiccata*: SUK-1722.
Common, on dry gravelly soil.
Localities: Hatkanagale, Kolhapur.

Local uses: Same as *B. erecta L.*

AMARANTHACEAE Juss.

Achyranthes aspera L. Sp. Pl. 204. 1753. Hook. f. Fl. Brit. India 4: 730. 1885; Cooke, Fl. Pres. Bombay 2: 580. 1958 (Repr. ed.). Lakshminarasimhan & Godbole in Singh *et al.* Fl. Maharashtra State Dicot. 2: 778. 2001.

Erect, herbs, 30–90 cm tall; branches obtusely 4-angled, striate, pubescent. Leaves ovate or broadly elliptic, 2.5–6 × 2–3.5 cm, acute, rounded or subcordate at base, entire or undulate, rounded, subacute or acuminate at apex, finely appressed pubescent on both sides. Flowers numerous on long pubescent rachis of elongating terminal spikes, at first erect, becoming strongly reflexed with age; bracts broadly ovate, 2.5–3 mm long, acuminate aristate; persistent. Perianth 4–6 mm long, glabrous and shining; tepals subequal, ovate-oblong, with narrow scarious margins, finely pointed. Stamens 5; filaments united at base, alternating with truncate, fimbriate, pseudo-staminodes. Fruits oblong-cylindric, 3–4 mm long, truncate at apex, thinly membranous, enclosed in hardened perianth. Seeds subcylindric, truncate at apex, brown.

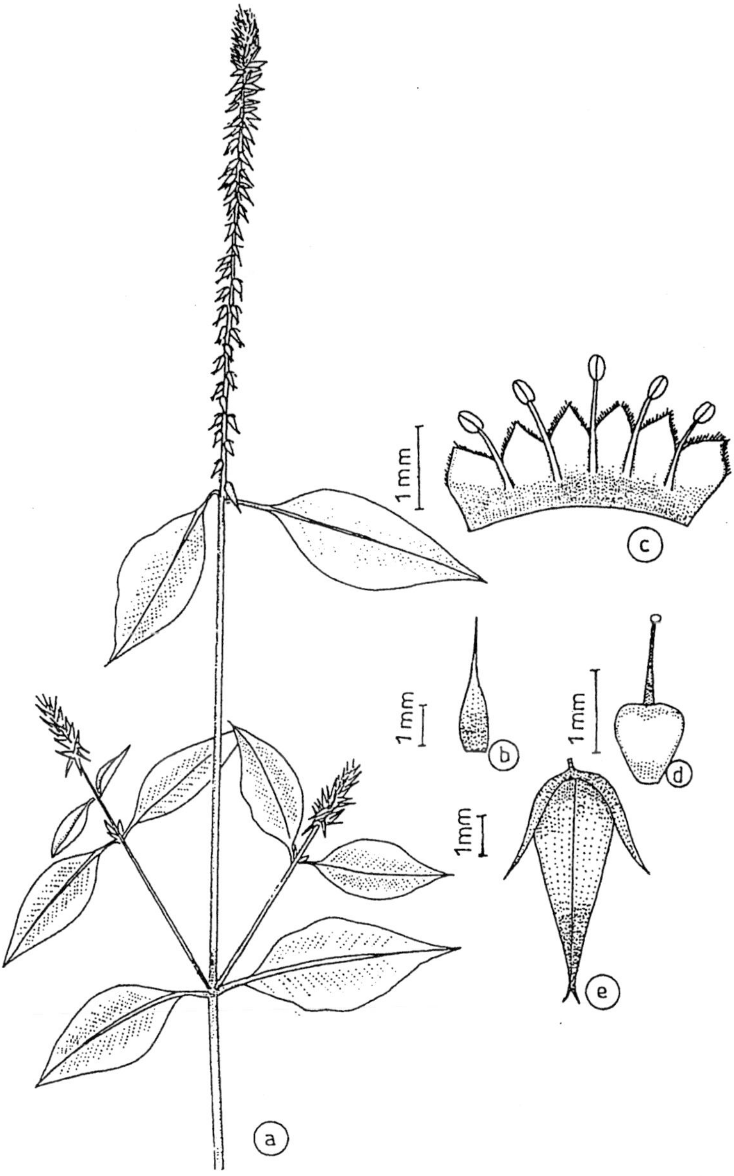

Text fig. 40: Achyranthes aspera a-twig: b-bract: c-dissected flower: d-carpel: e-fruit.

Fls. & Frts.: August–January. *Exsiccata*: SUK-11.
Common, weed of wastelands and around fields.
Localities: Amba, Kolhapur, Karad, Wai.

Local uses: Plant ash or decoction of leaf is given as appetizers and stomachic. Root decoction is used to cure urinary inflammation. Decoction of root is given in backache caused due to excess bleeding in menstrual cycle. Ash of plant is used with sesamum oil and poured in earache. Plant ash yields a large quantity of potash; it is used in cases of asthma and cough. Leaves are useful in dropsy, piles, boils and eruptions of the skin.

Aerva lanata (L.) Juss. ex Schult. Syst. Veg. ed. 15. 5: 564. 1819; Hook. f. Fl. Brit. India 4: 728. 1885; Cooke, Fl. Pres. Bombay 2: 578. 1958 (Repr. ed.); Lakshminarasimhan and Godbole in Singh *et al.* Fl. Maharashtra State Dicot. 2: 781. 2001. *Achyranthes lanata* L. Sp. Pl. 204. 1753. KAPURI-MADHURI.

Erect or climbing herbs, 30–120 cm tall; stem branching from the base; branches densely white woolly. Leaves mostly alternate, broadly elliptic-obovate, 0.5–5 × 0.3–3 cm, cuneate at base, acute and shortly mucronate at apex, densely white woolly on both surfaces. Flowers minute in dense, subsessile, axillary spikes. Tepals oblong, the 3 outer acute, two inner obtuse, all densely white woolly outside. Stamens 5, half as long as tepals; filaments connate at base, alternating with short, subulate, pseudo-staminodes. Fruits membranous, ca 1 mm in diam. Seeds discoid, less than 1 mm in diam., smooth, polished, black.

Fls. & Frts.: September–February. *Exsiccata*: SUK-235.
Common, in hedges, on wall and under the shade of trees.
Localities: Amba, Amboli, Borbet, Dajipur Gaganbawada, Tillari, Vaibhavwadi.

Local uses: Leaf juice is externally applied on headache. Root decoction is given as diuretic.

Alternanthera sessiles (L.) R. Br. ex DC. Cat. Hort. Monsp. 77. 1813; Hook. f. Fl. Brit. India 4: 731. 1885; Lakshminarasimhan & Godbole in Singh *et al.* Fl. Maharashtra State Dicot. 2: 783. 2001. *Gomphrena sessilis* L. Sp. Pl. 225. 1753. *Alternanthera triandra* Lam. Encycl. 1: 95. 1785; Cooke, Fl. Pres. Bombay 2: 584. 1958 (Repr. ed.).

Much branched, erect or ascending herbs. Leaves linear-lanceolate to oblong-obovate, 1–6 × 0.3–1.5 cm, acute at base, entire, obtuse, acute or denticulate at apex, glabrous or thinly pilose. Flowers sessile, in dense, 1–4 axillary spikes, 0.5–1.5 cm long; rachis densely white hairy. Bracts and bracteoles ovate-lanceolate, 0.5–1.5 mm long, acuminate, glabrous. Tepals ovate-lanceolate, 2–3 mm long, shortly acuminate, white or pinkish, 1-nerved or obscurely 3-nerved at base, glabrous. Stamens 3; staminodes 2;

pseudo-staminodes minute or obscure. Fruits obreniform, 1.5–2 × 2.5–3 mm, densely emarginate, glabrous, brown with thickened margins.

Fls. & Frts.: July–March. *Exsiccata*: SUK-1737.

Common, weed in marshes, around ponds, along river banks and in wet fields.

Localities: Karad road, Kolhapur, Shiroli.

Local uses: Plant decoction is given as a diuretic, an inflammation of urinary passage during fever.

Celosia argentea L. Sp. Pl. 205. 1753; Hook. f. Fl. Brit. India 4: 714. 1885; Cooke, Fl. Pres. Bombay 2: 570. 1958 (Repr. ed.); Lakshminarasimhan & Godbole in Singh *et al.* Fl. Maharashtra State Dicot. 2: 789. 2001.

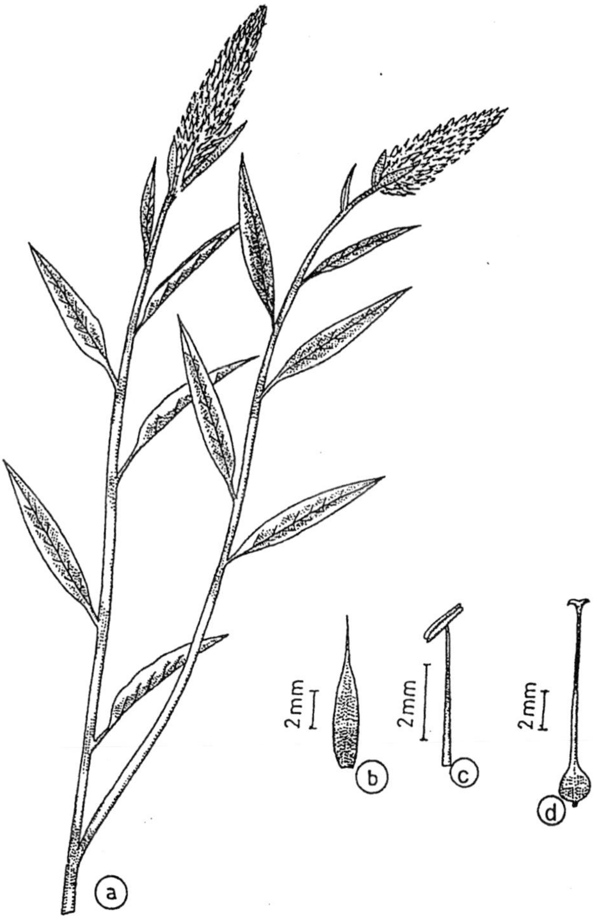

Text fig. 41: Celosia argentea a-twig; b-bract; c-stamen; d-carpel.

Erect, annual herbs, 30–150 cm tall; stem strongly ribbed, green or red. Leaves oblong-lanceolate to linear-lanceolate, 4–15 × 0.7–6 cm, acute or acuminate, green or tinged with red, petioles up to 2 cm long. Flowers in solitary paired, terminal cylindric spikes, 2–20 cm long, with conical apex. Bracts and bracteoles ovate-oblong, 3–7 mm long, pellucid, 1-nerved, persistent. Perianth of 5 distinct tepals, 6–10 mm long, at first pink, withering white. Stamens 5; filaments united at base, pseudo-staminodes minute. Style purple, 3–5 mm long. Fruits obovoid to globose, 3–3.5 mm in diam., rounded at apex. Seeds 2–9, 1–1.5 mm diam., biconvex, shining black, minutely tubercled.

Fls. & Frts.: August–March. *Exsiccata*: SUK-26.

Common, weed of wet and harvested fields.

Localities: Kolhapur, Sangli, Panhala.

Local uses: Seed powder is mixed with sugar and milk and given in aphrodisiac. Roots are given in urinary calculi. Leaves are given in natural sleep.

BASELLACEAE Moq.-Tand.

Basella alba L. Sp. Pl. 272. 1753; Moorthy in Singh *et al.* Fl. Maharashtra State Dicot. 2: 799. 2001. *B. rubra* L. Sp. Pl. 272. 1753; Hook. f. Fl. Brit. India 5: 21. 1886; Cooke, Fl. Pres. Bombay 2: 594. 1958. (Repr. ed.). MAYALU.

Perennial herbs; stem long, slender, twining, succulent, glabrous. Leaves alternate, broadly ovate, 5–15 × 3–12 cm, cordate at base and narrowed into a petiole, entire, acute or acuminate, 2–15 cm long spikes. Perianth white or pink at the tip; divided half way down, 2.5–3 mm long, fleshy; segments elliptic, obtuse. Stamens 5, included. Fruits globose, 6–8 mm in diam., enclosed in fleshy, enlarged, persistent perianth. Seed subglobose, black.

Fls. & Frts.: October–March. *Exsiccata*: SUK-1738.

Cultivated.

Local uses: Herb is used to improve appetite, improves voice and applied to burns.

POLYGONACEAE Juss.

Persicaria glabra (Willd.) Gomer in Ann. Inst. Segunda Ensef. Habana 2: 278. 1896; Jadhav in Singh *et al.* Fl. Maharashtra State Dicot. 2: 805. 2001. *Polygonum glabrùm* Willd Sp. Pl. 2: 447. 1799; Hook. f. Fl. Brit. India 5: 34. 1886; Cooke, Fl. Pres. Bombay 3: 5. 1958 (Repr. ed.). SHERAL.

Procumbent, perennial herbs; stems stout, 40–100 cm tall, striate, green with reddish-brown tinge at base. Leaves lanceolate, 7–25 × 1–3 cm, tapering at base, entire, finely acuminate, glabrous, gland-dotted. Flowers in paniculate slender racemes. Bracts ovate, 3–4 mm long, obtuse with membranous ciliate margins. Perianth pink, 3–5 mm long; segments oblong, obtuse. Stamens 6–8, included. Styles 2, connate at base. Nutlets broadly ovoid or suborbicular, compressed, biconvex, black, shining.

Fls. & Frts.: September–April. *Exsiccata*: SUK-15.
Common, in marshes along stream and riverbanks.
Localities: Bhudergad, Chandgad, Kolhapur, Nesari, Ratnagiri.

Local uses: An infusion of leaves is used to relieve pain in colic. Whole plant paste is used to remove endo and exo-cortosis or tumours.

ARISTOLOCHIACEAE Juss.

Aristolochia bracteolata Lam. Encycl. 1: 258. 1783; Lakshminarasimhan & Godbole in Singh *et al.* Fl. Maharashtra State Dicot. 2: 814. 2001. *A. bracteata* Retz. Obs. Bot. 5: 29. 1789; Hook. f. Fl. Brit. India 5: 75. 1886; Cooke, Fl. Pres. Bombay 3: 16. 1958.(Repr. ed.). GANDHARI, KIDMAR.

Slender, perennial herbs; stem prostrate or trailing, 20–40 cm long, striate, glaucous. Leaves reniform or broadly ovate, 3–7 × 2.5–5 cm, cordate at base with broad sinus, crenulate, obtuse, glabrous above, glaucous beneath. Flowers axillary, solitary; pedicels with a large, sessile, orbicular bract at base. Perianth 2–4 cm long with a sub-globose base and cylindric ash-gray colored tube; mouth trumpet shaped; lip linear, dark purple, with revolute margins, as long as the tube. Stamens 5–6, adnate to the top of the ovary. Ovary inferior, 6-locului. Capsules globose to ovoid, 1.5–2 cm long, 6-grooved and ribbed. Seeds obcordate, 4–5 mm long, black.

Fls. & Frts.: July–November. *Exsiccata*: SUK-1739.
Common, weed of black cotton soil.
Localities: Atpadi, Tasgaon.

Local uses: It is given in the bowel complaints of children and in cholera. It is regarded as a stimulant tonic and is applied externally to the abdomen. Leaves are used with castor oil in elephantiasis. Leaf decoction is applied on foul and neglected ulcers for destroying the larvae of insects. Decoction of plant made in spirit is used with black pepper in Typhoid and also externally applied on body. Plant is useful in joint pains and bowel troubles of children.

Aristolochia indica L. Sp. Pl. 960. 1753; Hook. f. Fl. Brit. India 5: 75. 1886; Cooke, Fl. Pres. Bombay 3: 16. 1958 (Repr. ed.); Lakshminarasimhan and Godbole in Singh *et al.* Fl. Maharashtra State Dicot. 2: 815. 2001. SAPSAN, SAPSUND.

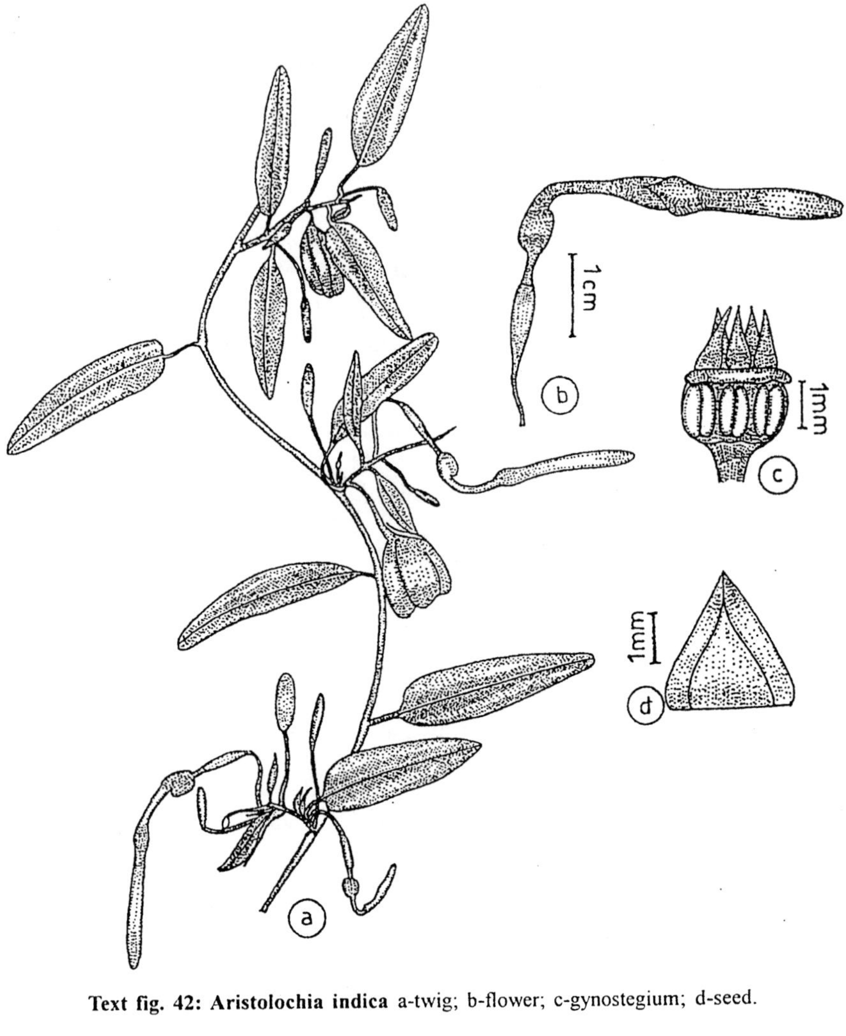

Text fig. 42: Aristolochia indica a-twig; b-flower; c-gynostegium; d-seed.

Woody, twining, glabrous, perennial herbs; stem long, slender, grooved. Leaves variable, linear-oblong to obovate-oblong, 3–15 × 1.5–8 cm, cuneate, subtruncate, rounded or subcordate at base. Flowers in few flowered, axillary racemes. Perianth greenish-white, 3–4 cm long with globose base, bent cylindric tube with oblique mouth and ending in a long, narrow linear-oblong, obtuse, brown lip. Anthers 6, style 6-lobed. Capsule globose, oblong, 3–4 × 2.5–3 cm, 6-valved, grooved and ribbed. Seeds deltoid, ovate, 2–3 mm long, compressed winged.

Fls. & Frts.: September–January. *Exsiccata*: SUK-154.

Infrequent, in deciduous forests.

Localities: Babu Jamal, Hatkanagale, Ichalkaranji, Islampur, Kagal.

Local uses: Decoction of plant is given in common cold, cough and in fever. Root with other combination is given during labor to increase uterine contractions. It is not given to pregnant women in large quantity, it may cause abortion. It is given orally in colic and abdominal complaints. Plant is useful in pains in joints and bowel troubles of children.

PIPERACEAE J. Agardh

Piper betle L. Sp. Pl. 29. 1753; Hook. f. Fl. Brit. India 5: 83. 1886; Cooke, Fl. Pres. Bombay 3: 20. 1998 (Repr. ed.); Godbole & Prasanna in Singh *et al.* Fl. Maharashtra State Dicot. 2: 819. 2001. NAGWEL, PAAN.

Perennial, dioecious climbers; stem producing clinging roots at aerial nodes; glabrous. Leaves alternate, ovate-oblong to broadly ovate, 10–18 × 5–10 cm, cuneate, truncate or sub-cordate at base, entire, acute or broadly acuminate, glabrous. Male flowers in narrow, cylindric, 4–12 cm long spikes, minute, bracts peltate; adnate to the rachis. Perianth absent. Stamens 1–4; anthers 2-celled. Fruiting not seen.

Fls. & Frts.: March–April. *Exsiccata*: SUK-1747.
Cultivated.
Localities: Atigre, Chokak, Mangaon.

Local uses: Leaves are soaked in castor oil and warmed on flame and applied as bandage in cold. Leaf juice is given in sore throat and releases cough. Leaves are chewed orally in anemia. Leaves are fried in ghee and given as a stomachic or as an appetizer. Leaf is used as stomachic, tonic, useful in foul smell in the mouth, ozoena, bronchitis, elephantiasis of the leg and improves appetite.

Piper longum L. Sp. Pl. 29. 1753; Hook. f. Fl. Brit. India 5: 83. 1886; Cooke, Fl. Pres. Fl. Bombay 3: 20. 1958 (Repr. ed.); Godbole and Prasanna in Singh *et al.* Fl. Maharashtra State Dicot. 2: 819. 2001. PIPLI.

Robust scandent shrubs. Leaves 2–7 × 0.5–3 cm, cordate, lower leaves petiolate, upper leaves sessile. Flowers in 3–6 cm long spikes. Bracts peltate. Flowers dioecious, in whitish catkins. Stamens 2. Berries globose, blackish-green when ripe, ca 2.5 mm in diam.

Fls. & Frts.: October–March. *Exsiccata*: SUK-1763.
Cultivated.
Localities: Kagal, Kolhapur.

Local uses: Root is used in bronchitis, abdominal pains, diseases of the spleen and tumors. Ripe fruit is cooling useful in biliousness. Ripe fruits externally used in acute and chronic bronchitis attended with cough. Roasted aments are beaten up with honey and given in rheumatism; powdered with black pepper and rock salt (3:1) used in colic.

Piper nigrum L. Sp. Pl. 28. 1753; Hook. f. Fl. Brit. India 5: 90 1886; Cooke, Fl. Pres. Bombay 3: 19. 1958 (Repr. ed.); Godbole & Prasanna in Singh *et al.* Fl. Maharashtra State Dicot. 2: 818. 2001. KALI-MIRI.

Stout climbers; sparingly rooting, much thickened at the nodes. Leaves broadly ovate, glabrous 5–9 nerved, rounded or oblique at base. Flowers in slightly interrupted glabrous spikes of variable length, dioecious or polygamous. Bracts adnate to rachis, forming a short hemispheric cup beneath the ovary. Stamens 2, stigmas 2–4. Fruit globose, 6 mm in diam., first yellow, afterwards becoming red when fully ripe.

Fls. & Frts.: November–January. *Exsiccata*: SUK-361.
Cultivated.
Localities: Ratnagiri.

Local uses: Fruits are given orally in piles and mixed with ghee and applied externally. Fruits are given as a diuretic. It is employed as an aromatic stimulant in cholera, weakness following fevers. It is used with various other combinations in fever and abdominal complaints. Fruit is used in asthma, diseases of the throat, piles, urinary discharges, toothache, diseases of spleen, chronic fever, paralysis and facilitates menstruation.

MYRISTICACEAE R. Br.

Knema attenuata (Wall. ex Hook. f. & Thoms.) Warb. Monogr. Myrist. 590. 1897; Londhe in Singh *et al.* Fl. Maharashtra State Dicot. 2: 821. 2001. *Myristica attenuata* Wall. ex Hook. f. & Thoms. Fl. Ind. 157. 1855 & in Hook. f. Fl. Brit. India 5: 110. 1886; Cooke, Fl. Pres. Bombay 3: 24. 1958 (Repr. ed.). RANJAYPHAL, RUKT-MARA.

Trees, 8–15 m tall. Leaves elliptic oblong to oblong-lanceolate, 12–22 × 4.5–7.2 cm, thinly coriaceous, dark green above, pale and glaucous beneath, base rounded or shortly cuneate; main nerves 12–20 pairs. Male flowers obovate-globose, 8 mm in diam., in 3–6 flowered, shortly pedunculate axillary umbels. Perianth deeply divided; segments 3, fleshy, concave, densely rufous woolly outside, glabrous inside. Staminal disk stalked, subconcave, glabrous; anthers ca 13, ovate, dehiscing downwards; filaments short. Female flowers in umbels, like males but few-flowered, subsessile or sessile. Perianth cylindric or narrowly ovoid, slightly glabrous at the base, 3-toothed, ovary sessile, ovoid, densely tomentose, style short, fleshy, glabrous, expanding above into 2-lobed fleshy, stigma. Fruit 2.5–3.5 cm long, ellipsoid or subovoid, abruptly and shortly pointed, densely clothed with rusty furaceous tomentose, but ultimately glabrous. Seeds narrowly ovoid; aril thin, uniform, completely covering the seed.

Fls. & Frts.: October–June. *Exsiccata*: SUK-77.

Occasional in semi evergreen and moist deciduous forests.

Localities: Amba, Amboli, Barki, Tillari, Vasota.

Local uses: Bark decoction given orally in diarrhoea. Bark is also given in jaundice to improve appetite. Decoction of bark is used in chronic fever.

Myristica fragrans Houtt. Nat. Hist. 2, 3: 333. 1774; Londhe in Singh *et al.* Fl. Maharashtra State Dicot. 2: 822. 2001. JAYPHAL.

Evergreen trees, 9–12 m tall; main axis of inflorescence is slender and branches dichotomously as top into two equal branches unbranched axis bears only one flower.

Cultivated.

Localities: Dapoli.

Local uses: Fruit paste is given to small children to induce sleep. It is one of ingredient in Balgutti. Other uses are similar as *M. malabarica.*

Myristica malabarica Lam. in Hist. Acad. Sci. Par. 162. 1791; Hook. f. Fl. Brit. India 5: 103. 1886; Cooke, Fl. Pres. Bombay 3:22. 1958 (Repr. ed.); Londhe in Singh *et al.* Fl. Maharashtra State Dicot. 2: 822. 2001. KAYAPHAL, RAN-JAYPHAL.

Trees, 10–15 m tall. Leaves thinly coriaceous, oblong or elliptic lanceolate, 8–12 × 3–5 cm, subacute, shining above, dull beneath, base acute, main nerves 8–14 pairs. Male flowers in slender, open, 2–3-chotomous axillary or lateral subumbellate cymes, 2–5 cm long. Perianth minutely scurfy-puberulous outside, glabrous inside; teeth short, triangular. Staminal column stalk apiculate; anthers 10–14, linear. Female flowers globose, larger than the males, few flowered, axillary umbels scarcely longer than the petioles. Perianth teeth ovate. Ovary ovoid, globose, pointed, densely rufous-tomentose, stigma large, sessile, 2-lobed. Fruit elongate, oblong, obtuse, slightly flattened on one side, 5–5.5 × 2.5–3 cm; aril yellow; irregularly lobed and laciniate extending to the apex of the seed.

Fls. & Frts.: November–March. *Exsiccata*: SUK-257.

Occasional, in semi evergreen forests, along Ghats.

Localities: Amba, Barki, Tillari,Vasota.

Local uses: Seeds are applied to indolent and ill-conditioned ulcers; it allays pain, cleans the surface and establish healthy action. Seeds are given in diabetes. Bark decoction is given in AIDS in primary stages. Seeds are given as sex stimulants.

LAURACEAE Juss.

Actinodaphne angustifolia Nees in Wall. Pl. Asiat. Rar. 3: 31. 1832; Londhe in Singh *et al.* Fl. Maharashtra State Dicot. 2: 823. 2001. *A. hookeri* Meiss. in DC. Prodr. 15 (1): 218. 1864; Hook. f. Fl. Brit. India 5: 149. 1886; Cooke, Fl. Pres. Bombay 3: 31 1958 (Repr. ed.). GULCHAHE, PISA.

Trees, 10–15 m tall. Leaves in two whorls of three each, elliptic lanceolate, 10–18 × 4.5–7 cm, coriaceous, apex finely acuminate, base acute. Flowers dioecious, yellowish, arising from naked branches, males in clusters of about 8, the females umbellate or subracemose on very short, stout peduncles. Perianth densely clothed with tawny hairs; segments elliptic or ovate. Stamens 9 in male flowers, reduced in the female to strap shaped antherless staminodes, filaments thick, covered with whitish hairs, the interior row of 3 with 2 somewhat reniform glands at base of each filament, the glands in the male flowers slightly stalked, those of the female sessile. Style rather thick; stigma large, mushroom shaped, oblique. Berry ellipsoid, 0.8–1.2 cm long, red when ripe.

Fls. & Frts.: October–May. *Exsiccata*: SUK-266, 278.
Common, in semi evergreen and moist deciduous forests.
Localities: Amba, Amboli, Dajipur, Sawantwadi, Tillari, Vaibhavwadi.

Local uses: Leaf powder orally given in night emissions with milk. Leaves are applied to sprains. Oil of the seeds is used in similar case. Cold infusion of the leaves is mucilaginous and is used in urinary disorder and in diabetes.

Cassytha filiformis L. Sp. Pl. 35. 1753; Hook. f. Fl. Brit. India 5: 188. 1886; Cooke, Fl. Pres. Bombay 3: 35. 1958 (Repr. ed.). AMARVEL.

Climbers, parasitic, herbaceous; stems wiry, slender, green, branched, attached to host by suckers, glabrous or hairy. Leaves consisting of minute scales or 0. Flowers small, hermaphrodite, sometimes dimorphic, spicate, capitate or racemose; bracteoles 3. Perianth twice as long as bracteoles, small, tube short; segments 6 in two rows, the 3 outer segments short, rounded, ciliate, the inner oblong, acute, concave, valvate. Perfect stamens 9 or 6; filaments of first row eglandular with introse, 2-celled anthers, those of the fourth row replaced by sessile staminodes. Ovary globose, free. Fruit about 6 mm in diam., white globose, smooth enclosed in the fleshy perianth tube and crowded with the erect perianth segments.

Fl. & Frts.: June–September. *Exsiccata*: SUK-385.
Frequent, in deciduous forests.
Localities: Ratnagiri.

Local uses: Bark powder is applied externally as hair tonic. Root used to wash chronic ulcers. Leaf juice is specifically used in inflamed eyes.

Cinnamomum nitidum (Roxb.) Hook. Exot. Fl. 2: t. 176. 1825; Londhe in Singh *et al.* Fl. Maharashtra State Dicot. 2: 826. 2001. *Laurus nitida* Roxb. Fl. Ind. 2: 300. 1832. *Cinnamomum macrocarpum* Hook. f. Fl. Brit. India 5: 133. 1886; Cooke, Fl. Pres. Bombay 3: 28. 1958 (Repr. ed.). MAWARUNG.

Trees, 10–15 m tall. Leaves elliptic-lanceolate or oblong-lanceolate, 12–15 × 4.5–5 cm, thinly coriaceous, 3–5-nerved, the lateral nerves starting 5 mm above the base. Flowers in panicles, shorter than the leaves. Perianth 5–5.6 mm long; tube 2.5 mm long; segments pubescent on both sides, oblong

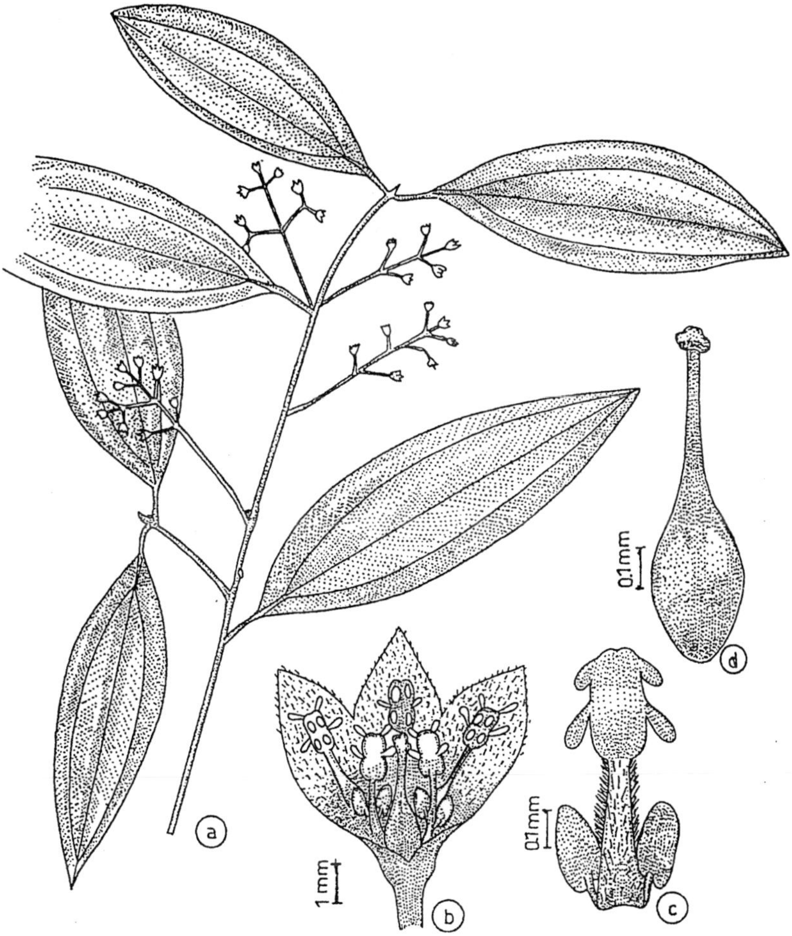

Text fig. 43: Cinnamomum nitidum a-twig; b-flower; c-staminode; d-carpel.

or obovate. Stamens 9 perfect; 2 outer rows with eglandular filaments and introse 4-celled anthers, those of the third row with glandular filaments, fourth row replaced by staminodes. Ovary sessile, free; stigma discoid, or obscurely 3-lobed. Fruit globosely oblong, 2.5 × 0.6 cm.

Fls. & Frts.: March–April. *Exsiccata*: SUK-386.
Rare, in semi evergreen forests.
Localities: Amboli, Barki, Dajipur, Udgiri.

Local uses: Root is applied externally in cold and cough. Roots are prescribed as a brain tonic, used orally to promote hair growth and prevent graying of hairs. Roots are used as vicks vaporub in cold. Decoction is given in fever. Oil from root bark and leaves is used in rheumatic affections as an external medicine.

Cinnamomum verum J. S. Presl in Prir. Rostl. 2: 36, 37–44, t. 7. 1885; Londhe in Singh *et al.* Fl. Maharashtra State Dicot. 2: 826. 2001. *C. zeylanicum* Bl. Bijdr. 568. 1825; Hook. f. Fl. Brit. India 5: 131. 1886; Cooke, Fl. Pres. Bombay 3: 28. 1958 (Repr. ed.). DALCHINI, TAMALPATRA.

Trees, 5–10 m tall. Leaves ovate oblong or elliptic ovate, 3–20 × 3–6 cm, coriaceous, shining above, paler beneath, base acute or rounded; main nerves 3–5 from base or nearly so, strong. Flowers numerous, in silky pubescent panicles, usually longer than the leaves. Perianth 5–6 mm long; tube 2.5 mm long; segments pubescent on both sides, oblong or somewhat obovate, obtuse. Stamens 9, perfect, or few aborted staminodes. 2 outer rows with eglandular filaments, the glands subsessile or stipitate and extrose 4 celled anthers, those of the fourth row replaced by shortly stipitate cordate or sagittate staminodes. Fruits 1.2 cm long, oblong or ovoid oblong, apiculate, dry or slightly fleshy, dark purple, surrounded by the enlarged campanulate perianth.

Fls. & Frts.: February–June. *Exsiccata*: SUK-386.
Common, in moist deciduous and semi evergreen forests.
Localities: Amba, Amboli, Dajipur, Radhanagari, Tillari, Vaibhavwadi, Vasota.

Local uses: Bark is used in weakness of stomach and diarrhoea. Oil is externally used in rheumatic affections. Leaf juice is dropped in nostrils to stop headache. Decoction of bark is given in fever and cough. Bark is useful in parched mouth, bronchitis, itching diseases of heart, rectum and urinary diseases.

Litsea deccanensis Gamble, Fl. Pres. Madras 1235. 1925 [2: 864. 1957 (Repr. ed.)]; Londhe in Singh *et al.* Fl. Maharashtra State Dicot. 2: 829. 2001. *L. tomentosa* Heyne ex Hook. f. Fl. Brit. India 5: 157. 1886; Cooke, Fl. Pres. Bombay 3: 32. 1958 (Repr. ed.). CHIKNA.

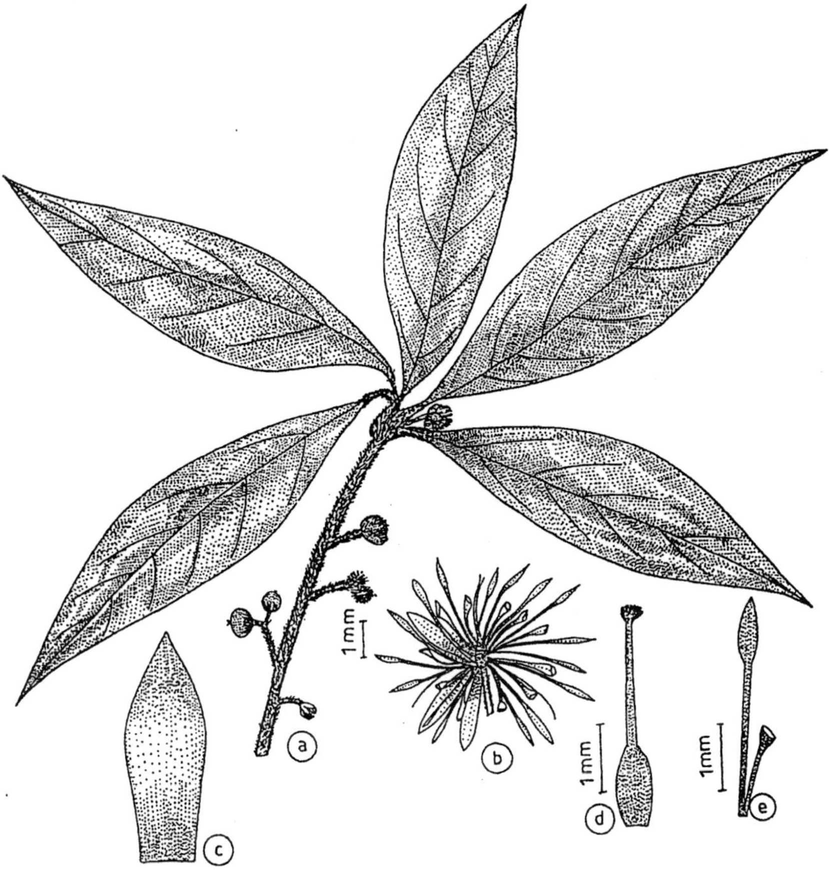

Text fig. 44: Litsea deccanensis a-twig; b-flower; c-tepal; d-carpel; e-staminode.

Shrubs or small trees, 2–10 m high. Leaves crowded towards the ends of the branches, elliptic-lanceolate or obovate, 5–20 × 2.5–8 cm, shortly acuminate, glabrous above, finely tomentose and yellowish-white beneath, base acute. Flowers pale yellow in solitary pedunculate many-flowered umbellate heads, globose before expansion. Perianth tube very short, hairy outside; segments usually 0. Stamens 18–20; filaments long, slender, densely villous, the glands with long stalk. Staminodes of the female long, strap-shaped, numerous. Fruits 1.5 cm, globose, ripens black.

Fls. & Frts.: December–September. *Exsiccata*: SUK-60.

Frequent in moist deciduous forests.

Localities: Amba, Amboli, Barki, Borbet, Dajipur, Lingmala falls, Tosegahar.

Local uses: Stem is used externally to heal wounds. Stem is used as a bandage in bone fracture. Leaves are used in similar cases. Oil from the berries is used in rheumatism. Roots are useful in burning sensations, bronchitis, fever and paralysis.

Neolitsea cassia (L.) Kosterm. in J. Sci. Res. (Jakarta) 1: 85. 1952; Londhe in Singh *et al.* Fl. Maharashtra State Dicot. 2: 832. 2001. *Laurus cassia* L. Sp. Pl. 361. 1753. *Litsea zeylanica* Nees in Amoen. Bot. Bonn. Fasc. 1: 58, t. 5. 1823; Hook. f. Fl. Brit. India 5: 178. 1886 p. p.; Cooke, Fl. Pres. Bombay 3: 34. 1958 (Repr. ed.). CHIRCHIRA.

Trees, 5–8 m tall. Leaves crowded towards the tips of branches, elliptic-lanceolate, 5–15 × 3–6 cm, thinly coriaceous, apex obtuse or acuminate, finely reticulately veined, 3-nerved; flowers glabrous in small, sessile, clustered, 4–5 flowered umbels. Perianth hairy; tube long; segments 4, acute, deciduous. Fruits oblong or globose, 1.2 cm across, dark-purple when ripe, seated on disciform perianth tube.

Fls. & Frts.: September–June. *Exsiccata*: SUK-288.
Occasional, in moist deciduous and semi evergreen forests.
Localities: Anuskara, Barki.

Local uses: Leaf decoction is given in asthma. Dried leaves are smoked as cigarette in cough. Decoction of leaves is given externally in headache and various skin diseases.

Persea macrantha (Nees) Kosterm. in Reniwardtia 6: 193. 1962; Londhe in Singh *et al.* Fl. Maharashtra State Dicot. 2: 833. 2001. *Machilus macrantha* Nees in Wall. Pl. As. Rar. 2: 70. 1831; Hook. f. Fl. Brit. India 5: 140. 1886; Cooke, Fl. Pres. Bombay 3: 29. 1958 (Repr. ed.). GULUM, PALAS.

Trees, 15–20 m tall. Leaves oblong to elliptic lanceolate, 8–18 × 1.5–6.5 cm, coriaceous, glaucous beneath. Flowers yellow, numerous, in panicles near the ends of branches, often several panicles together, 12–20 cm long, stout, puberulous. Perianth 6 mm long, silky pubescent, segments 6, oblong or linear oblong, subacute. Stamens 9 perfect, 2 outer rows with eglandular filaments and introse 4-celled anthers, those of the third row with filaments which have two stipitate glands at the base and extrose 4-celled anthers, those of the forth row replaced by stipitate cordate staminodes. Fruit globosely oblong, 1–2 cm across, smooth, dark green, dotted with white, turning black.

Fls. & Frts.: January–June. *Exsiccata*: SUK-1705.
Infrequent, in moist deciduous and semi evergreen forests.
Localities: Amba.

Local uses: Leaf juice is used to wash wounds and ulcers. Bark is used with bark of *Terminalia elliptica* in 1:1 proportion in rheumatic swellings. Root and fruit decoction is given in asthma and sore throat.

THYMELAEACEAE Juss.

Gnidia glauca (Fresen.) Gilg. in Bot. Jahrb. Syst. 19: 265. 1894; Londhe in Singh *et al.* Fl. Maharashtra State Dicot. 2: 836. 2001. *Lasiosiphon glaucus* Fresen. Fl. 21: 603. 1838. *Lasiosiphon eriocephalus* (Meissn.) Decne. in Jacq. Voy. Ind. Bot. 4(3): 148. 1844; Hook. f. Fl. Brit. India 5: 197. 1886; Cooke, Fl. Pres. Bombay 3: 36. 1958 (Repr. ed.). RAMETHA.

Shrubs, 1.5–4.0 m tall. Leaves linear-oblong to oblong-lanceolate, 4–8 × 1.5–2 cm, subsessile, apex acute, glaucous and glabrous above, glabrous or silky beneath, base acute. Flowers yellow, nearly sessile in erect dense terminal heads, 2.5–3.5 cm in diam., surrounded at the base by involucre of large imbricate elliptic-oblong, acute, silky, villous deciduous bracts. Perianth densely silky-villous, tube 1.8–2 cm long, lobes 2.5–4 mm long, oblong-obtuse, flat with a linear, usually 2-fid scale at each division. Stamens 10, attached to the throat below the scales in 2 rows, almost sessile. Ovary superior, 1-celled, style long, filiform. Fruit ellipsoid, oblong, 0.9–1.2 cm long, pointed, enclosed in the perianth.

Fls. & Frts.: December–May. *Exsiccata*: SUK-1995.
Frequent along the slopes of the Ghats.
Localities: Amba, Amboli, Dajipur, Mahabaleshwar, Panchgani, Radhanagari.

Local uses: One drop of leaf juice is taken on a needle and dropped into ulcerated teeth and teeth are expelled out.

VISCACEAE Miers

Viscum angulatum Heyne ex DC. Prodr. 4: 283. 1830; Hook. f. Fl. Brit. India 5: 225. 1886; Cooke, Fl. Pres. Bombay 3: 47. 1958 (Repr. ed.); Londhe in Singh *et al.* Fl. Maharashtra State Dicot. 2: 847. 2001. BANDA.

Leafless, semi parasitie shrubs on trees. Main stem terete or nearly so; young branches numerous, 4-angled, jointed, the nodes not swollen nor contracted. Flowers unisexual, minute, sessile, solitary or whorled at the nodes. Perianth lobes deciduous. Berries globose, yellowish.

Fls. & Frts.: May–October. *Exsiccata*: SUK-1692.
Frequent, in moist deciduous forests on various trees.
Localities: Bhudergad, Radhanagari.

Local uses: Plant is given on joint pains. Decoction of plant is given orally in bone fracture and applied externally.

Viscum articulatum Burm. f. Fl. Ind. 311. 1768; Hook. f. Fl. Brit. India 5: 226. 1886; Cooke, Fl. Pres. Bombay 3: 47. 1958 (Repr. ed.); Londhe in Singh *et al.* Fl. Maharashtra State Dicot. 2: 847. 2001. JIVANTIKA.

Leafless semi parasitic, pendulous, dichotomously much branched shrubs; branches jointed, the internodes 2.5–5 × 0.25–0.8 mm, linear, flattened, slightly narrowed at each end, radially disarticulating, pale green. Flowers minute, 3–4-merous, very shortly, pedicellate, 1–6 together in fascicles at the nodes, each with a large cup-shaped bract at the base. Berries globose, 4 mm long, smooth, greenish-yellow.

Fls. & Frts.: June–August. *Exsiccata*: SUK-1691.
Localities: Amba, Amboli, Dindewadi, Patgaon.

Local uses: Plant is given in joint pains. Decoction of plant is given orally and applied externally in bone fracture. Used in diseases of blood, ulcers, epilepsy and biliousness.

SANTALACEAE R. Br.

Santalum album L. Sp. Pl. 349. 1753; Hook. f. Fl. Brit. India 5: 231. 1886; Cooke, Fl. Pres. Bombay 3: 49. 1958 (Repr. ed.); Godbole and Lakshminarasimhan in Singh *et al.* Fl. Maharashtra State Dicot. 2: 850. 2001. CHANDAN.

Evergreen trees with slender drooping branches. Leaves elliptic-lanceolate, 1.5–6 × 0.5–0.3 cm, subacute, entire, thin, base acute. Flowers brownish-purple, inodorous in terminal axillary paniculate cymes, shorter than the leaves. Perianth campanulate; limb of 4 valvate triangular segments. Stamens 4, exserted, alternating with the 4 rounded obtuse scales. Ovary free, ½ inferior, ovules 2–3; stigma 2–3 lobed. Drupes globose, 1.2 cm in diam., purple black.

Fls. & Frts.: March–August. *Exsiccata*: SUK-201.
Infrequent, in dry and moist deciduous forests.
Localities: Tillari, Chandgad.

Local uses: Powder of wood is externally applied on various skin diseases and on boils. Paste of wood is commonly applied to local inflammations and in fever. Wood powder is taken orally to reduce heat from the body and given as diuretic. Wood is useful in heart diseases, burning sensation, cold in the head, strangury, bronchitis, vaginal discharges and small pox etc.

EUPHORBIACEAE Juss.

Acalypha indica L. Sp. Pl. 1003. 1753; Hook. f. Fl. Brit. India 5: 416. 1887; Cooke, Fl. Pres. Bombay 3: 108 (Repr. ed.); Londhe in Singh *et al.* Fl. Maharashtra State Dicot. 2: 856. 2001. KOKHALI.

Erect, 60–120 cm tall, sparsely pubescent herbs; stems angular. Leaves broadly ovate-elliptic or rhomboidal, 2–7 × 1.8–5.5 cm, obtuse at base, serrate, acute, pubescent on veins beneath. Flowers in androgynous axillary, erect, 3–9 cm long spikes; the males minute, few towards ends of spikes; the females several at base, bracts of female flowers large 1 or two flowered, broadly ovate, 4–7 × 7–12 mm, dentate at margin, ovary globose; styles long, branched at end. Fruits greenish, 1.5 mm across, hispid. Seeds brown, pear shaped, ca 1.5 mm across, slightly tubercled.

Fls. & Frts.: August–December. *Exsiccata*: SUK-1689.

Common, along banks of watercourses, on waste places, bounds of farms and old walls of houses.

Localities: Kolhapur, Sangli, Shirol.

Local uses: Leaf juice mixed with *Allium cepa* is used as anthelmintic. Dried leaves mixed with common salt are given in constipation. Fresh leaves mixed with lime juice is applied externally on itch, eczema and elephantiasis. A piece of cotton is saturated with expressed juice and inserted into each nostril, relieving head symptoms by causing hemorrhage.

Aporosa lindleyana (Wight) Baill. Etud. Gen. Euph. 645. 1858; Hook. f. Fl. Brit. India 3: 349. 1881; Cooke, Fl. Pres. Bombay 3: 91. 1958 (Repr. ed.); Londhe in Singh *et al.* Fl. Maharashtra State Dicot. 2: 860. 2001. *Scepa lindleyana* Wight, Ic. t. 361. 1840. SALT, BUGDI.

Trees, 8 m tall. Leaves elliptic-oblong, 6–13 × 3.5 cm, coriaceous, acuminate, base acute. Male flowers sessile; spikes 2.5–3.5 cm long, dense. Sepals 4; ovate-oblong, obtuse, 2.5 mm long, ciliate. Stamens 2–3; filaments free, 1.3 mm long, anthers subglobose. Female flowers in very short, hairy, often clustered racemes. Sepals ovate, subacute, 1.3 mm long. Ovary flask-shaped, shortly pedicellate, hairy; style short, with 3 large laciniate curved stigmas. Fruit globose, 1 cm or slightly more in diam., pointed with the style. Seeds planoconvex, broadly ovate, subacute, 6 mm long, as broad as long.

Fls. & Frts.: November–June. *Exsiccata*: SUK-50.

Scarce, in evergreen forests.

Localities: Amba, Amboli, Dajipur, Gaganbawada, Lingmala falls, Radhanagari, Vasota.

Local uses: A decoction of root is given to improve appetite in cases of jaundice. Young fruit juice is dropped into nostril in headache.

Baliospermum montanum (Willd.) Muell.-Arg. in DC. Prodr. 15(2): 1125. 1866; Londhe in Singh *et al.* Fl. Maharashtra State Dicot. 2: 862. 2001. *Jatropha montana* Willd. Sp. Pl. 4: 563. 1805. *Baliospermum axillare* Bl. Bijdr. 604. 1826; Hook. f. Fl. Brit. India 5: 461. 1887; Cooke, Fl. Pres. Bombay 3: 106. 1958 (Repr. ed.). DANTI, JAMALGOTA, SAPIDI.

Stout, monoecious, undershrubs, 1.1–5 m tall; branches herbaceous. Leaves, the lower broadly ovate in outline, 10–20 cm long and as much broad, 3–5 lobed, rounded or cuneate and 2-glandular at base, sinuate-toothed;

lobes acute; the upper ones small, lanceolate; all glabrous above pubescent beneath. Flowers in axillary androgynous racemes, with numerous males and a few females near the base. Male calyx globose, 2–2.5 mm long, 4–5 partite; segments suborbicular, concave, imbricate, membranous. Disk glandular. Stamens 15–20. Female sepals ovate-lanceolate. Disk entire. Ovary globose, hairy; styles 1–1.5 mm long, 2-partite, dull red. Fruits obovoid, 8–12 mm long, pale yellow. Seeds ellipsoid, 7–8 × 5 mm, smooth, mottled.

Fls. & Frts.: September–February. *Exsiccata*: SUK-122.
Common, grows as under growth in moist deciduous forests in Konkan
Localities: Ratnagiri, Sawantwadi.

Local uses: Seeds are used as a drastic purgative. Seeds are used externally as a stimulant and rubefacient. Oil is a powerful hydrogogue cathartic and useful in external application in rheumatism. Root is cathartic and used in dropsy, anasarca and jaundice.

Breynia retusa (Dennst.) Alston in Ann. Roy. Bot. Gard. (Peradeniya) 11: 204. 1929; Londhe in Singh *et al.* Fl. Maharahstra State Dicot. 2: 863. 2001. *Phyllanthus retusus* Dennst. Schluss. Hort. Malab. 31. 1818. *Breynia patens* (Roxb.) Rolfe in J. Bot. 11: 359. 1882; Hook. f. Fl. Brit. India 5: 329. 1887; Cooke, Fl. Pres. Bombay 3: 79. 1958 (Repr. ed.). DOLFODI, KANGLI, RETVND.

Glabrescent, erect shrubs, 2–3 m tall; branches zig-zag, cylindric. Leaves alternate, broadly elliptic-orbicular, 1–2.5 × 1–2 cm, rounded at base, entire, rounded at apex. Flowers: Males few in axillary, 4–5 mm long racemes; females solitary, axillary or in pairs. Male flowers greenish, turning yellow at maturity; bracts compactly set, persistant, greenish-turning pinkish, ovate very minute. Calyx tube obconical, 1.5–2 mm long; lobes six, ovate, minute, curved, rounded, thick at apex. Stamens 3, synandrous. Female flowers green, turning yellow, ca 1 mm long. Calyx tube rather fleshy, obconical, 1–1.5 mm long; lobes 5–6, ovate, minute; style 1 with single stigma. Fruits purple green, globose, ca 5 × 5 mm, glabrous.

Fls. & Frts.: September–March. *Exsiccata*: SUK-1688.
Common, in open forests.
Localities: Amba, Amboli, Radhanagari, Tillari, Vaibhavwadi.

Local uses: Plant is used as astringent to the bowels, useful in inflammations and diseases of blood. Juice of the stem is used in conjunctivitis.

Bridelia retusa (L.) Spreng. Syst. Veg. 3: 48. 1826; Hook. f. Fl. Brit. India 5: 268. 1887; Cooke, Fl. Pres. Bombay 3: 68. 1958 (Repr. ed.); Londhe in Singh *et al.* Fl. Maharashtra State Dicot. 2: 865. 2001. *Cluytia retusa* L. Sp. Pl. 1042. 1753. ASAN.

Shrubs or small trees, 6–10 m tall. Leaves elliptic-oblong, 7–20 × 3.5–10.5 cm, coriaceous, obtuse, subacute or rounded, with rounded or slightly crenulate margins, bright green and glabrous above, glaucous and finely tomentose beneath, base rounded. Flowers dioecious, greenish-yellow, sessile. Crowded in dense axillary clusters or on long axillary or terminal, sometimes paniculate spikes, often exceeding the leaves. Calyx greenish, tinged with red, 4 mm in diam., glabrous; segments fleshy, triangular-ovate, acute, spreading. Petals greenish-white, those of the male flowers obovate, pectinate; female flowers subspathulate. Disk of male flowers thick and pulpy, of the female flowers truncate, enclosing the ovary. Style short. Drupe fleshy, seated on the persistent slightly enlarged calyx, 8 mm in diam., purple black.

Fls. & Frts.: May–October. *Exsiccata*: SUK-90.
Common, in deciduous forests.
Localities: Amba, Borbet, Gaganbawada.

Local uses: Bark is given as a cardiac tonic with water. Bark is given for the removal of urinary concentrations. Bark is used as valuable astringent.

Bridelia scandens (Roxb.) Willd. Sp. Pl. 4: 979. 1805; Londhe in Singh *et al.* Fl. Maharashtra State Dicot. 2: 866. 2001. *Cluytia scandens* Roxb. Pl. Cor. t. 173. 1802. *Bridelia stipularis* sensu Hook. f. Fl. Brit. India 5: 270. 1887 non Bl. 1825 p. p.; Cooke, Fl. Pres. Bombay 3: 68. 1958 (Repr. ed.). CHIKNI, PATHARPHODA.

Shrubs with spines when young. Leaves elliptic-oblong or obovate, 2.5–10 × 1.2–5 cm, subcoriaceous, obtuse, or subacute, glabrous above, pale pubescent beneath. Flowers monoecious, greenish-yellow, in small, dense axillary clusters or long spikes, often subtended by stipular hairy obliquely lanceolate acute bracts. Calyx 5 mm long, hairy; segments triangular-ovate, acute, connate at base. Petals 1.5 mm long, broadly obovate or suborbicular, shortly cuneate at base, crenulate at apex. Staminal column 3 mm long top of anthers. Styles with long slender arms. Drupes 9 mm long or more, ellipsoid-oblong, seated on the slightly enlarged calyx, black and juicy when ripe.

Fls. & Frts.: October–December. *Exsiccata*: SUK-1647.
Common, in dry and moist deciduous forests.
Localities: Katyani, Gajapur.

Local uses: Bark with water is given as a cardiac tonic. Bark is given for the removal of urinary concentrations. Bark is used as valuable astringent.

Chrozophora prostrata Dalz. & Gibs. Bombay Fl. 233. 1861; Hook. f. Fl. Brit. India 5: 410. 1887; Cooke, Fl. Pres. Bombay 3: 105. 1958 (Repr. ed.); Londhe in Singh *et al.* Fl. Maharashtra State Dicot. 2: 866. 2001.

Prostrate or decumbent, monoecious herbs, stellate hairy all over stem, branches red. Leaves broadly ovate, orbicular, 3–6 × 2.5–4 cm, red brown, truncate or rounded at base, undulate margined, rounded or obtuse at apex, plicate, and thinly hairy above, hoary and stellate tomentose beneath. Flowers in few-flowered short axillary racemes. Male calyx 2.5–3 mm long, purple, stellate hairy outside, green and glabrous within. Petals elliptic-lanceolate, 1.5–2.5 mm long, yellow, obtuse. Stamens 15 in 2 whorls. Female sepals linear, 2–2.5 mm long, acute. Petals smaller than the sepals, linear. Styles yellow, 2–2.5 mm long, spreading, bifid up to the middle. Fruits globose, 5–6 mm in diam., stellate hairy, without silvery scales.

Fls. & Frts.: November–May. *Exsiccata*: SUK-1651.
Common, weed in drying ditches, riverbanks, cultivated fields etc.
Localities: Kagal, Kolhapur, Ichalkaranji, Tasgaon.

Local uses: Plant is emetic, drastic and has corrosive properties.

Chrozophora rottleri (Gies.) Juss. Euphorb. Gen. Tent. 28. 1824; Londhe in Singh *et al.* Fl. Maharashtra State Dicot. 2: 868. 2001. *Croton rottleri* Gies. Croton Monogr. 54. 1807. *Chrozophora plicata* Voigt Hort. Sub. Calc. 156. 1846 non Juss. 1824; Hook. f. Fl. Brit. India 5: 408. 1887; Cooke, Fl. Pres. Bombay 3: 104. 1958 (Repr. ed.). SURYAVARTA.

Erect, hoary, annual herbs, 30–60 cm tall. Leaves thick, ovate-orbicular, shallowly 3 lobed, 3–10 × 1.5–9 cm, truncate, cordate or cuneate at base, wrinkled, rugose with deep seated nerves, stellate tomentose on both surfaces. Flowers in terminal, densely stellate hairy racemes. Male calyx 3 mm long; petals greenish-white, ovate lanceolate, shorter than the calyx, thin. Stamens 15 in 2 whorls. Female sepals triangular, acute, 2 mm long; petals shorter and narrower than the sepals. Ovary globose; styles 2–3 mm long, 2-fid up to the middle, red or pink. Fruits globose, 6–8 mm in diam., densely stellate hairy, without silvery scales.

Fls. & Frts.: December–June. *Exsiccata*: SUK-33.
Common, weed on black cotton soil.
Localities: Hatkanangale, Ichalkaranji, Jaisingpur, Kolhapur.

Local uses: Root powder is given to the children in cough and asthma. Seeds are used as purgative.

Croton zeylanicus Muell.-Arg. in Linnaea 34: 107. 1865 and in DC. Prodr. 15(2): 581. 1886; Londhe in Singh *et al.* Fl. Maharashtra State Dicot. 2: 871. 2001. *C. reticulatus* Heyne in Wall. ex Muell-Arg. in DC. op. cit.; Hook. f. Fl. Brit. India 5: 386. 1887; Cooke, Fl. Pres. Bombay 3: 96. 1958 (Repr. ed.). PANDHRI.

Shrubs, 3–5 m tall. Leaves elliptic-lanceolate, 5 × 2.5 cm, apex-acuminate, base acute or rounded, with pair of stipulate glands at base, clothed beneath with silvery scales. Flowers pedicellate, in terminal racemes shorter than the leaves; the females few at the base of the racemes. Male flowers: Sepals ovate-oblong, 2.5 mm long, obtuse, ciliolate. Petals slightly shorter than the sepals, obovate, obtuse. Stamens 14–20, inserted on a hairy receptacle. Female flowers: Calyx 4–6 mm long; sepals oblong-lanceolate, much enlarged in fruit. Ovary with reddish-brown scales and stellate hairs. Style about 6.3 mm long, 3-branched shortly above the base, each branch again 2-branched and again divided into 2 long linear lobes at the apex. Capsules ovoid-oblong, under 1.2 cm long, slightly 3-lobed, apiculate, with reddish-brown, stellate hairs mixed with scales. Seeds purplish-brown mottled with white.

Fls. & Frts.: October–January. *Exsiccata*: SUK-1706.
Infrequent, in semi evergreen forests.
Localities: Amboli, Borbet, Dajipur, Gajapur, Patgaon, Vasota.

Local uses: Seeds are used as a purgative. Root is given in jaundice every morning for 8 days. It is administered in very small quantity.

Drypetes roxburghii (Wall.) Hurusawa in J. Fac. Sci. Univ. Tokyo, Sect. III Bot. 6: 337. 1954; Londhe in Singh *et al.* Fl. Maharashtra State Dicot. 2: 872. 2001. *Putranjiva roxburghii* Wall. Tent. Fl. Nep. 61. 1828; Hook. f. Fl. Brit. India 5: 336. 1887; Cooke, Fl. Pres. Bombay 3: 86. 1958 (Repr. ed.). PUTRANJIV.

Evergreen trees, 8–10 m tall; branches pendent, glabrous or nearly so. Leaves coriaceous, penninerved, leaflets elliptic-oblong, 4–10 × 2–4 cm, acute or rounded at base, entire or distantly serrulate, obtuse or shortly acuminate, shining dark green above, paler beneath, often gland dotted. Flowers axillary, on short pedicles. Male flowers clustered; sepals oblong, 2–2.5 mm long, obtuse, ciliolate. Filaments 1.5–2 mm long, connate at base; anthers 2–4, globose. Females 1–3 in axil; sepals elliptic, 2–2.5 mm long, obtuse ciliolate. Ovary ovoid, pubescent; style recurved; stigmas crescent-shaped. Drupes ellipsoid, 12–18 mm long, white, tomentose. Stone, rugose, hard.

Fls. & Frts.: March–June. *Exsiccata*: SUK-58.
Native of tropical India and Ceylon, cultivated.
Localities: Ichalkaranji, Kolhapur, Tillari.

Local uses: Leaf juice is externally applied and orally given in burning sensation. Leaf juice is dropped in eye complaints. It is given to the women suffering from some infertility problems but it is useless in its action.

Emblica officinalis Gaertn. Fruct. 2: 122, t. 108. 1790; Londhe in Singh *et al.* Fl. Maharashtra State Dicot. 2: 873. 2001. *Phyllanthus emblica* L. Sp. Pl. 982. 1753; Hook. f. Fl. Brit. India 5: 289. 1887; Cooke, Fl. Pres. Bombay 3: 81. 1958 (Repr. ed.). AMALA, AVALA, AWALKANTE.

Middle sized, deciduous trees, 5–8 m tall; branches spreading; branchlets 10–20 cm long, glabrous or finely pubescent, deciduous. Leaves distichous, subsessile, linear-oblong, 10–12 × 2.5–3 mm, entire, obtuse, glabrous. Flowers in axillary fascicles on leaf bearing branchlets. Male flowers numerous on short, slender pedicles. Perianth segments 6, distinct, oblong, 1–1.5 mm long, obtuse. Disk absent. Anthers 3 on short central column. Female flowers few, subsessile; perianth segments as in the males. Disk cupular, lacerate. Ovary 3-locular; styles, 1.2–3.5 cm in diam., with 6, obscure, vertical furrows, pale yellow, with 2-seeded 3 crustaceous cocci. Seeds 6, trigonous.

Fls. & Frts.: October–March. *Exsiccata*: SUK-27.

Common, in hill forests, also cultivated.

Localities: Amba, Amboli, Dajipur, Kolhapur, Mahabaleshwar, Sawantwadi, Vasota.

Local uses: It is used as one of the ingredient in Triphala churna. Juice of fresh bark with honey and turmeric is given in gonorrhea. Unripe fruit is cooling, diuretic and laxative. The exudation from the incision on the fruit is used as an external application in inflammation of the eye. Fruit juice is given to remove general debility. Fruit juice with juice of *Momordica charantia* or turmeric increases insulin and decreases the sugar from the blood. Fruit juice is used in eye complaints. Fruit juice is applied externally for hair growth.

Euphorbia antiquorum L. Sp. Pl. 450. 1753; Hook. f. Fl. Brit. India 5: 255. 1887; Cooke, Fl. Pres. Bombay 3: 57. 1958 (Repr. ed.); Londhe in Singh *et al.* Fl. Maharashtra State Dicot. 2: 876. 2001. NARSEJ.

Much branched, small trees with jointed, terete branches; branchlets with 3–5, thick, sinuate wings; stipular spines sharp in sinuses. Leave small, fleshy, obovate-oblong or spathulate 1–1.5 cm long, rounded at tips, early deciduous, cyathia 3-nate, forming short peduncled cymes in the sinuses. Styles distinct, 2-fid. Fruits trigonous with compressed cocci, glabrous, 12–15 mm wide, yellowish. Seeds globose, greenish.

Fls. & Frts.: February–May. *Exsiccata*: SUK-1686.

Frequent, in hedges around villages.

Localities: Amba, Dajipur, Ramling.

Local uses: Fresh juice is applied to warts, and used as rubefacient embrocation in rheumatism. Juice is given as a purgative. It is useful on cough in children.

Euphorbia hirta L. Sp. Pl. 454. 1754; Londhe in Singh *et al.* Fl. Maharashtra State Dicot. 2: 881. 2001. *E. pilulifera* auct. Pl. non L. 1753; Hook. f. Fl. Brit. India 5: 250. 1887; Cooke, Fl. Pres. Bombay 3: 64. 1958 (Repr. ed.). DUDHANI, DVDHANALI.

Annual, herbs; stems procumbent or ascending, 15–40 cm long, hispid with long, yellowish crisped hairs. Leaves opposite, ovate or ovate-lanceolate, 1–4 × 0.5–1.5 cm, unequal sided and cordate at base, serrulate-dentate, acute or subacute, dark green or reddish-brown on the upper surface, pale beneath. Cyathia many, 1 mm long, crowded in small axillary, subsessile cymes; glands minute, without petaloid appendages. Fruits globose, 1 mm in diam., hispid. Seeds ovoid, trigonous, less than 1 mm long, transversely rugose, reddish-brown.

Fls. & Frts.: Throughout the year. *Exsiccata*: SUK-6.
Very common, in variety of habitats.
Localities: Kolhapur, Satara, Sangli, Wai.

Local uses: Herb is used as an astringent in chronic diarrhoeas and dysentery. Latex is applied externally to ulcers, oedemas and phlegmons. Leaf poultice is applied to abscesses and inflamed glands. Drug has fairly marked action on the respiratory system depressing the respiration and producing well-marked dilations of the bronchioles. Drug produces nausea and vomiting in large doses.

Euphorbia ligularia Roxb. Fl. Ind. 3: 465. 1832; Cooke, Fl. Pres. Bombay 3: 58. 1958 (Repr. ed.); Londhe in Singh *et al.* Fl. Maharashtra State Dicot. 2: 882. 2001. *E. nerifolia* Sec. Boiss in DC. Prodr. 79. 1862; Hook. f. Fl. Brit. India 5: 255. 1887; Cooke, Fl. Pres. Bombay 3: 59. 1958 (Repr. ed.). SABAR, THOR.

Erect, fleshy, glabrous shrubs or small trees, 5–6 m tall; branches ascending, 5-angled, appearing as if twisted; stipular spines arising from thick tubercles, sharp dark, brown; tubercles in conspicuous spirals. Leaves alternate, near the ends of the branches, oblanceolate-oblong or lanceolate, 15–25 × 5–8 cm, narrowed at base, entire or slightly concave, acute or bluntly acuminate, glabrous. Cyathia globose, often reddish, 3-nate, forming small pedunculate cymes, from above the leaf-scars; the central cyathium male, sessile, the lateral ones hermaphrodite, pedicellate; involucres hemispheric, yellowish-red with fimbriate, rounded lobes; glands transversely oblong; bracts many, fimbriate. Stigmas capitate. Fruits globose, 3-lobed, 10–12 mm across; cocci compressed, glabrous.

Fls. & Frts.: February–June. *Exsiccata*: SUK-1685.
Common, on hill slopes and planted in hedges.
Localities: Alas, Alate, Babu Jamal.

Local uses: Fresh decoction is used to remove pimple marks and abscesses from the face. Milky juice is used as a purgative and rubefacient (Reddening the skin, an agent that reddens the skin by producing hyperemia.). Plant is used as laxative, carminative, alexipharmic, improves the appetite; useful in abdominal troubles, bronchitis, tumors, loss of consciousness, delirium, leucoderma, piles, inflammations, enlargement of the spleen, anemia, ulcers and fevers.

Euphorbia nivulia Buch.-Ham. in Trans. Linn. Soc. London 14: 286. 1825; Hook. f. Fl. Brit. India 5: 255. 1887; Londhe in Singh *et al.* Fl. Maharashtra State Dicot. 2: 882. 2001.

Large glabrous trees, 6–10 m tall; branches often whorled, terete; stipular spines from tubercled bases, straight, sharp, black, shining. Leaves obovate-oblong or oblanceolate, 10–25 × 5–12 cm, narrowed at base, entire, obtuse or rounded at apex, very thick, glabrous. Cyathia 3-nate as in *E. nerifolia* but larger and yellowish-green. Fruits globose, trigonous, 10–12 mm in diam., styles simple, connate below. Seeds ovoid-ellipsoid, smooth pale brown or ash brown.

Fls. & Frts.: February–April. *Exsiccata*: SUK-1684.
Occasional, along moist gullies in forests, also planted in hedges.
Localities: Sangavade.

Local uses: Fresh milky juice is applied to warts and used as a rubefacient embrocation in rheumatism. The juice is mixed with jaggery and given as a purgative. The juice is useful in asthma, enlargement of the spleen, dyspepsia, colic and stone in the bladder.

Euphorbia thymifolia L. Sp. Pl. 454. 1753; Hook. F. Fl. Brit. India 5: 252. 1887; Cooke, Fl. Pres. Bombay 3: 64. 1958 (Repr. ed.); Londhe in Singh *et al.* Fl. Maharashtra State Dicot. 2: 885. 2001. DHAKTI-DUDHI, LAHNA-NAYATI.

Prostrate, annual herbs; stems many, spreading, 10–20 cm long, pubescent, often tinged with red. Leaves opposite, obliquely oblong, 3–6 × 2–4 mm, rounded and very unequal sided at base, entire or crenulate, rounded at apex, glabrous and green or reddish above, glaucous and pubescent beneath; stipules fimbriate. Cyathia axillary, 1–3 in short cymes; involucres campanulate, 1 mm long; gland minute or absent, without petaloid appendages. Fruits ovoid-globose, 1–1.5 mm across, 3-lobed; lobes obtusely angled, densely hairy. Seeds obtusely quadrangular, obtuse at both ends, 1 mm long, pale brown transversely rugose.

Fls. & Frts.: Throughout year. *Exsiccata*: SUK-1662.
Common, on gravelly soil and in crevices of rocks, along roadsides.
Localities: Bahubali, Jaisingpur, Sangli.

Local uses: Herb is used as an astringent in chronic diarrhoeas and dysentery. Latex is applied externally to ulcers, oedemas, and phlegmons. Leaf poultice is applied to abscesses and inflamed glands. Drug has fairly marked action on the respiratory system depressing the respiration and producing well marked dilations of the bronchioles. Drug produces nausea and vomiting in large doses.

Euphorbia tirucalli L. Sp. Pl. 452. 1753; Hook. f. Fl. Brit. India 5: 254. 1887; Cooke, Fl. Pres. Bombay 3: 86. 1958 (Repr. ed.); Londhe in Singh *et al.* Fl. Maharashtra State Dicot. 2: 885. 2001. SHEND, SHER.

Much branched shrubs or small trees, 3–6 m tall; branches terete, green, succulent, dichotomously branched. Leaves alternate, linear-lanceolate, 0.7–2.5 cm, obtuse, caducous. Cyathia subsessile, 1–7 in peduncled cymes which are either terminal or in forks of branches, yellowish-green; involucres 2–2.5 mm long, hairy; bracts triangular-semiorbicular, 1 mm long; glands oval, 1.5–2 mm broad, without petaloid appendages. Fruits not developed.

Fls.: March–April. *Exsiccata*: SUK-619.
Native of tropical Africa, planted.
Localities: Jaisingpur, Jat, Sangli.

Local uses: Fresh milky juice is applied to warts and used as a rubefacient embrocation in rheumatism. The juice is mixed with jaggery and given as a purgative. The juice is useful in asthma, enlargement of the spleen, dyspepsia, colic and stone in the bladder.

Homonoia riparia Lour. Fl. Cochinch. 637. 1790; Hook. f. Fl. Brit. India 5: 455. 1887; Cooke, Fl. Pres. Bombay 3: 118. 1958 (Repr. ed.); Londhe in Singh *et al.* Fl. Maharashtra State Dicot. 2: 890. 2001. PASHANBHEDA, SERNI.

Evergreen shrubs, 1–1.5 m tall. Leaves numerous, erect or closely set, linear-oblong or linear-lanceolate, 2–5 × 1–2 cm, acute, apiculate, entire or toothed, glabrous above, clothed beneath with scattered and numerous minute orbicular scales, base acute or rounded. Flowers dioecious, sessile, in axillary bracteate spikes, 5–12 cm long, rachis pubescent. Male flowers: Sepals 3, broadly elliptic, 4 mm long, acute, concave, veined, glabrous. Stamens very numerous, in globose heads; anthers red. Female flowers: Sepals usually 5, narrowly ovate, acuminate, 2.5 mm long. Ovary pubescent, styles 3, papillose. Capsules globose, 3 mm in diam., tomentose. Seeds 1 mm long, rounded on the back, slightly angular on the inner face, smooth, yellowish-brown.

Fls. & Frts.: Throughout year. *Exsiccata*: SUK-312.
Common, mostly in beds of streams in the ravines.
Localities: Amba, Amboli, Radhanagari, Sawantwadi, Tillari, Vaibhavwadi.

Local uses: Root is given in urinary calculi. Flowers are crushed and given in constipation, paste is applied externally on fresh wound to stop bleeding and heal.

Jatropha curcas L. Sp. Pl. 1006. 1753; Hook. f. Fl. Brit. India 5: 383. 1887; Cooke, Fl. Pres. Bombay 3: 95. 1958 (Repr. ed.); Londhe in Singh *et al.* Fl. Maharashtra State Dicot. 2: 891. 2001. MOGALI-YERAND.

Large shrubs, 2–4 m tall; branches terete, glabrous with watery juice. Leaves alternate, broadly ovate to reniform, 8–15 × 8–16 cm, entire or more often 3–5 lobed, cordate at base; lobes acute or acuminate, glabrous. Flowers in terminal, dichotomous cymes with solitary terminal female and rest male flowers. Calyx green, 3–4 mm long; outer segments slightly shorter than the inner ones, ovate, acute. Petals erect or with recurved, apex, oblong, 7–8 mm long, with long white hairs inside near the center. Stamens 8–10; filaments 2–3, inner ones connate. Stigmas sagittate, 2-lobed. Fruits ovoid-globose, 2.5–3 cm long, light yellow turning black, indehiscent. Seeds ellipsoid, carunculate at base.

Fls. & Frts.: October–June. *Exsiccata*: SUK-380.

Native of tropical America, planted in hedges.

Localities: Kagal, Hatkangale, Ramling.

Local uses: Latex is applied externally on wounds and ulcers. One drop of latex is dropped into the ulcerated gum or teeth to stop pain and cure ulcer. Latex is used externally on elephantiasis, eczema and itch. Root is used with black pepper, asafoetida and buttermilk on diarrhoea, dysentery, stomachache, colic and abdominal pain. Oil from the seeds is applied externally on herpes zoster. Latex is slightly warmed and applied externally in fistula.

Jatropha gossipifolia L. Sp. Pl. 1006. 1753; Hook. f. Fl. Brit. India 5: 383. 1887; Cooke, Fl. Pres. Bombay 3: 95. 1958 (Repr. ed.); Londhe in Singh *et al.* Fl. Maharashtra State Dicot. 2: 892. 2001.

Much branched shrubs, 1–1.5 m tall; stem terete, striate, glabrous, with yellow juice. Leaves alternate, orbicular to broadly obovate in outline, 10–15 × 12–16 cm, cordate at base, 3–5-fid, beyond the middle, green or dark purplish-red; lobes obovate, acute or acuminate, ciliate with stalked glands. Flowers in terminal, corymbose panicles. Calyx green with purple tinge, 2–3 mm long; teeth ovate, acute, gland ciliate. Petals obovate, 4 to 6 mm long, dark red with paler base, glabrous. Fruits oblong, 10–12 mm long, trigonous, glabrous, greenish-yellow. Seeds oblong, curunculate.

Fls. & Frts.: September–January. *Exsiccata*: SUK-210.

Native of tropical America, grown in gardens and also grows on wasteland, as an escape.

Localities: Kagal, Karad, Miraj, Sangli, Wai.

Local uses: Decoction of the bark is used as an emmenagogue. Leaves are applied to boils, carbuncles, eczema and itches.

Macaranga peltata (Roxb.) Muell.-Arg. in DC. Prodr. 15 (2): 1010. 1886; Londhe in Singh *et al.* Fl. Maharashtra State Dicot. 2: 892. 2001. *Oxyris peltata* Roxb. Fl. Ind. ed. 2, 3: 755. 1832. *Macaranga roxburghii* Wight, Ic. 5: 23. 1852; & 6: t. 1949, f. 4. 1853; Hook. f. Fl. Brit. India 5: 448. 1887. *M. tomentosa* Wight, Ic. 5: 23. 1852 & 6: t. 1949, f. 1. 1853; Cooke, Fl. Pres. Bombay 3: 117. 1958 (Repr. ed.). CHANDIVA.

Trees, 5 m tall. Leaves ovate or deltoid or orbicular, 6.5–24 × 6–18 cm, cuspidate, broadly peltate, entire or minutely toothed, glabrous above, pubescent and gland dotted beneath, base rounded. Male flowers minute, numerous, in interrupted dense sessile heads, enveloped in bracts and bracteoles, arranged in numerous, much-branched axillary rusty-tomentose panicles shorter or longer than the leaves; sepals 3, minute, obovate, cuneate. Stamens 2–3. Female flowers: Panicles simple than the males. Calyx–limb obsolete. Ovary densely glandular, 1-celled; stigma sessile, often embracing one side of the ovary, thickly papillose. Capsules 6–8 mm in diam., globose, hairy, glandular. Seeds globose, testa brown, crustaceous, rough.

Fls. & Frts.: January–March. *Exsiccata*: SUK-87.

Common, in moist deciduous and evergreen forests.

Localities: Amba, Amboli, Barki, Dajipur, Gaganbawada, Sawantwadi, Vaibhavwadi.

Local uses: The gum powdered and made into a paste is used for external application for venereal sores. Latex is applied externally on wounds.

Mallotus philippensis (Lam.) Muell.-Arg. in Linnaea 34: 196. 1865 (as *philippinensis*); Hook. f. Fl. Brit. India 5: 442. 1887; Cooke, Fl. Pres. Bombay 3: 113. 1958 (Repr. ed.); Londhe in Singh *et al.* Fl. Maharashtra State Dicot. 2: 895. 2001. *Croton philippense* Lam. Encycl. 2: 206. 1876. KAMALA, SHENDRI.

Small, much branched trees, 5–6 m tall. Leaves alternate, ovate-lanceolate, 8–22 × 3–8 cm, rounded or acute and 3-nerved at base, entire or toothed, acuminate, glabrous above, pubescent and with numerous, orbicular red glands beneath. Flowers small; males clustered in erect terminal spikes forming elongated paniculate racemes; the females solitary in short spikes. Male sepals 4–5, lanceolate, 2.5–3 mm long, acute. Stamens numerous. Female calyx 3–4 partite to the base; teeth ovate-lanceolate. Ovary with red glands. Fruits globose, 3-lobed, 8–10 mm in diam., covered with bright red powder consisting of minute stellate hairs. Seeds subglobose, 3–4 mm in diam., black.

Fls. & Frts.: November–February. *Exsiccata*: SUK-1661.
Common, in moist deciduous forests.
Localities: Kagal, Ramling, Mahabaleshwar.

Local uses: Powder prepared from the fruit is used as an anthelmintic, vermifuse. Glandular hairs from the fruits are mixed with oil and used on skin diseases. The glandular brick red powder, consisting of small glands covering the ripe fruit is used with castor oil and given to expel roundworms and thread-worms. It is also given to heal ulcers and wounds, tumors, stone in the bladder, useful in bronchitis, diseases of the abdomen and enlargement of spleen.

Phyllanthus acidus (L.) K. Skeels. in U.S. Dept. Agric. Bur. Pl. Industry Bull. 17. 1909; Londhe in Singh *et al.* Fl. Maharashtra State Dicot. 2: 909. 2001. *Averrhoa acida* L. Sp. Pl. 228. 1753. *Phyllanthus distichous* (L.) Muell.–Arg. in DC. Prodr. 15(2): 413. 1866; Hook. f. Fl. Brit. India 5: 304. 1886; Cooke, Fl. Pres. Bombay 3: 85. 1958 (Repr. ed.). RAI AWLA.

Glabrous, deciduous trees, 3–10 m tall. Leaves distichous, obliquely ovate-oblong, 3–9 × 1.5–4 cm, obtuse or rounded at base, entire, acute, light green above, glaucous beneath. Flowers densely clustered in elongate, simple or branched racemes from old, defoliate branches; males numerous; females 1–2, perianth lobes ovate-orbicular, 1–1.5 mm long, light green or reddish. Disk glands of the males free. Stamens 4, much longer than the perianth. Disk annular in females, crenate. Ovary 3–4 locular. Fruits fleshy, depressed globose, 12–10 mm in diam., 6–8 lobed, pale yellow, acidic.

Fls. & Frts.: October–March. *Exsiccata*: SUK-1660.
Native of Malay Islands and Madagascar, planted in kitchen gardens.

Local uses: It is used as blood purifier. Fruit improves appetite; useful in bronchitis, biliousness, urinary concretions and piles. Fruit and seed powder is given in diarrhoea.

Phyllanthus amarus Schum & Thonn. Kongl. Danske Vidensk. Seslk. Skr. 4: 195–196. 1826; Londhe in Singh *et al.* Fl. Maharashtra State Dicot. 2: 897. 2001. *P. niruri* auct non L. 1753 p. p.; Hook. f. Fl. Brit. India 5: 298. 1887; Cooke, Fl. Pres. Bombay 3: 84. 1958 (Repr. ed.). BHUIAVALI.

Erect, annual herbs, 10–40 cm tall. Leaves distichous, on leafy 4–8 cm long branches, elliptic-oblong, rarely obovate, 5–15 × 2–5.5 mm, rounded at both ends, entire, apiculate, green above, glaucous beneath. Flowers small in leaf axils, 1–2 together. Perianth segments 5 or 6, less than 1 mm long, obovate, oblong, acute, green with broad scarious margins, enlarging in fruit. Disk 5 lobed; lobes orbicular. Stamens 3; filaments entirely connate. Style bifid at apex. Fruits globose-trigonous, depressed at apex, 1–2 mm in diam., smooth. Seeds 3-gonous, 1 mm long, pale brown, with 5–7 subparallel longitudinal ribs.

Fls. & Frts.: July–December. *Exsiccata*: SUK-55.
Common, weed of cultivated fields and waste places.
Localities: Jat, Kolhapur, Miraj, Sangli.

Local uses: Plant decoction is used in chronic dysentery. Seed paste is externally applied on wounds. Leaf paste mixed with lime juice is externally applied on scabies and elephantiasis and itch. Fresh root is chewed with betel leaf in jaundice. Plant decoction is used as diuretic and given in gonorrhea. Leaves are boiled and the liquor drunk to stop acute pains in the stomach. Plant is useful in thirst, bronchitis, leprosy, anemia, urinary discharges, asthma and hiccough.

Phyllanthus debilis Klein ex Willd. Sp. Pl. 4: 582. 1805; Hook. f. Fl. Brit. India 5: 299. 1887; Cooke, Fl. Pres. Bombay 3: 84. 1958 (Repr. ed.); Londhe in Singh *et al.* Fl Maharashtra State Dicot. 2: 898. 2001.

Glabrous, erect, annual herbs, 10–40 cm tall; stem terete, branched above. Leaves elliptic to obovate, 6–15 × 4–10 mm, acute at base, entire, obtuse or acute, membranous, pale green above, glaucous beneath; stipules lanceolate, subulate, scarious. Flowers axillary on short pedicels, males in fascicles, females solitary. Perianth segments obovate-oblong, 1–2 mm long, those of females much longer than the males, with broad, thin scarious margins, often slightly enlarged in fruits. Disk glands of males stellately lobed. Styles short, free, 2-fid. Fruit depressed globose, 2–2.5 mm in diam. Seeds trigonous, 1–1.5 mm long, longitudinally ribbed on the back.

Fls. & Frts.: August–November. *Exsiccata*: SUK-1654.
Occasional, in ditches around fields and in waste places.
Localities: Barki, Kolhapur, Radhanagari, Sangli.

Local uses: Leaf paste mixed with lime juice is externally applied on scabies and elephantiasis and itch. Fresh root is chewed with betel leaf in jaundice. Plant decoction is used as diuretic and given in gonorrhea.

Phyllanthus maderaspatensis L. Sp. Pl. 982. 1753; Hook. f. Fl Brit. India 5: 292. 1887; Cooke, Fl. Pres. Bombay 3: 82. 1958 (Repr. ed.); Londhe in Singh *et al.* Fl. Maharashtra State Dicot. 2: 900. 2001.

Erect, much-branched, glabrous, annual herbs, 15–60 cm tall; branches angular, ascending. Leaves scattered, oblanceolate, 1–4 × 0.5–1.5 cm, sessile, cuneate at base, entire, obtuse, rounded or emarginate at apex, mucronate, glaucous. Stipules lanceolate, acute. Flowers axillary, the males minute in small clusters, subsessile, females large, solitary, on short pedicels. Perianth segments 6, obovate-oblong, 2–2.5 mm long, obtuse, green with white margins. Stamens 3; filaments connate. Styles distinct, bifid at apex. Fruits depressed globose, 2.5–3 mm in diam. 3-lobed, glabrous. Seeds-3-gonous, 1–1.5 mm long, muricate with fine lines on the rounded back.

Fls. & Frts.: June–November. *Exsiccata*: SUK-1653.
Common, weed of agricultural fields and wastelands.
Localities: Hatkanagale, Jat, Ramling, Sangli.

Local uses: Plant decoction is used in chronic dysentery. Seed paste is externally applied on wounds. Leaf paste mixed with limejuice is externally applied on scabies, elephantiasis and itch. Plant is useful in thirst, bronchitis, leprosy, anemia, urinary discharges, asthma and hiccough.

Phyllanthus reticulatus Poir. in Lam. Encycl. 5: 298. 1804; Hook. f. Fl. Brit. India 5: 288. 1887; Cooke Fl. Pres. Bombay 3: 81. 1958 (Repr. ed.); Londhe in Singh *et al.* Fl. Maharashtra State Dicot. 2: 900. 2001. KANGUNI, DATWAN.

Large, glabrous or pubescent, straggling shrubs. Leaves oblong-elliptic or almost rotundate, 10–28 × 8–18 mm, rounded, acute or subcordate at base, entire, obtuse or acute, thin, glabrous, dark green above, paler beneath; stipules ovate, acute, often bristle pointed. Flowers axillary, the males in fascicles of 2–6, the females solitary perianth segments 5–6, oblong, 2–2.5 mm long, obtuse, alternating with disk glands. Stamens 5, the inner 3 with their filament connate into a column, the outer two free, shorter. Ovary 5–10 locular; styles 3, minute. Fruits fleshy, depressed globose, 3.5–4 mm in diam., dark purple or black, shining; seeds 8–10, 3-gonous, 1–1.5 mm long, granulate.

Fls. & Frts.: Almost throughout the year. *Exsiccata*: SUK-1652.
Frequent, in hedges around fields.
Localities: Kolhapur, Sangli, Wai.

Local uses: Juice of the leaves is used with camphor and cubebs made in pill and allowed to dissolve in the mouth as a remedy for bleeding from the gums. Root paste is given in jaundice with milk in morning and night. Juice is used in inflammation of urinary passage and also given in gonorrhea.

Phyllanthus scabrifolius Hook. f. Fl. Brit. India 5: 299. 1887; Cooke, Fl. pres. Bombay 3: 84. 1958 (Repr. ed.); Londhe in Singh *et al.* Fl. Maharashtra State Dicot. 2: 901. 2001.

Erect, branched herbs, 15–60 cm tall; stems distantly winged at least in upper part; wings scabrid. Leaves subsessile, elliptic-obovate, 10–15 × 5–10 mm, narrowed at base, entire, rounded or obtuse at apex, pale green above, glaucous and scaberulous on the nerves beneath; stipules lanceolate-subulate, membranous, ciliate, auricled at base. Flowers shortly pedicellate, axillary, solitary or clustered. Perianth segments oblong or obovate, 2–2.5 mm long with white scarious margins, obtuse. Stamens 3; filaments united into a column. Styles distinct, 2-fid, recurved. Fruits depressed globose, 2.5–3 mm in diam., smooth or faintly granulate. Seeds 3-gonous, 1–1.5 mm long, with longitudinal ribs on the back and concentric ribs on the face.

Fls. & Frts.: September–December. *Exsiccata*: SUK-1701.
Common, on gravelly soil of hill slopes, grasslands.
Localities: Alate, Atigre, Babu Jamal, Chokak, Hatkanagale, Kagal, Ramling, Wai.

Local uses: Plant decoction is used in chronic dysentery. Seed paste is externally applied on wounds. Leaf paste mixed with limejuice is externally applied on scabies, elephantiasis and itch. Fresh roots are chewed with betel leaf in jaundice. Plant decoction is used as diuretic and given in gonorrhea. Leaves are boiled and the liquor drunk to stop acute pains in the stomach.

Phyllanthus urinaria L. Sp. Pl. 982. 1753; Hook. f. Fl. Brit. India 5: 293. 1887; Cooke, Fl. Pres. Bombay 3: 83. 1958 (Repr. ed.); Londhe in Singh *et al.* Fl. Maharashtra State Dicot. 2: 901. 2001.

Erect, annual, glabrous herbs, 15–60 cm tall. Leaves distichous, closely placed, linear oblong, 0.5–2 × 0.1–0.8 cm, broadly rounded and suboblique at base, entire, obtuse or rounded and apiculate at apex, green above, pale beneath; stipules lanceolate, acute. Flowers minute, axillary solitary, subsessile. Perianth segments 0.7–1 mm long, oblong, obtuse, green with white margins. Stamens 3; filaments connate into a column. Fruits depressed globose, 2.5–3 mm in diam., echinulate, scarcely lobed. Seeds 3-gonous, 1–1.5 mm long, rounded on the back, transversely rugose.

Fls. & Frts.: September–January. *Exsiccata*: SUK-1723.
Common, on wet land, in grasslands and along roadsides.
Localities: Amba, Barki, Satara, Kolhapur, Panchgani, Shivaji University campus, Wai.

Local uses: Decoction of root and leaves is a good tonic and given as a diuretic. Decoction of fresh root is used in jaundice. Leaf paste applied externally on scabies. Leaf juice is given in gonorrhea. Fruit and plant is useful in bronchitis, leprosy, anemia, urinary discharges, anuria, biliousness, asthma and hiccough.

Phyllanthus virgatus Forst. f. Inst. Austr. Prodr. 65. 1786; Londhe in Singh *et al.* Fl. Maharashtra State Dicot. 2:902. 2001. *P. simplex* Retz. Obs. Bot. 5: 29. 1788; Hook. f. Fl. Brit. India 5: 95. 1887; Cooke, Fl. Pres. Bombay 3: 83. 1958 (Repr. ed.).

Erect, diffuse, perennial herbs or undershrubs. Leaves distichous, subsessile, narrowly elliptic or linear oblong, 1–2.5 × 0.3–0.6 cm, acute at base, entire, rounded or emarginate at apex, green above, paler beneath; stipules sagittate, brown scarious. Flowers usually solitary on slender axillary pedicels. Perianth segments oblong, ca 1 mm long, obtuse, slightly enlarged in fruit. Stamens 3, distinct, anthers didymous. Style short, 2-fid. Fruits globose, 2.5–3 mm in diam., 3-lobed, slightly rough. Seeds trigonous, 1–1.5 mm long, rounded on the back, covered with minute tubercles in irregular lines, dark brown.

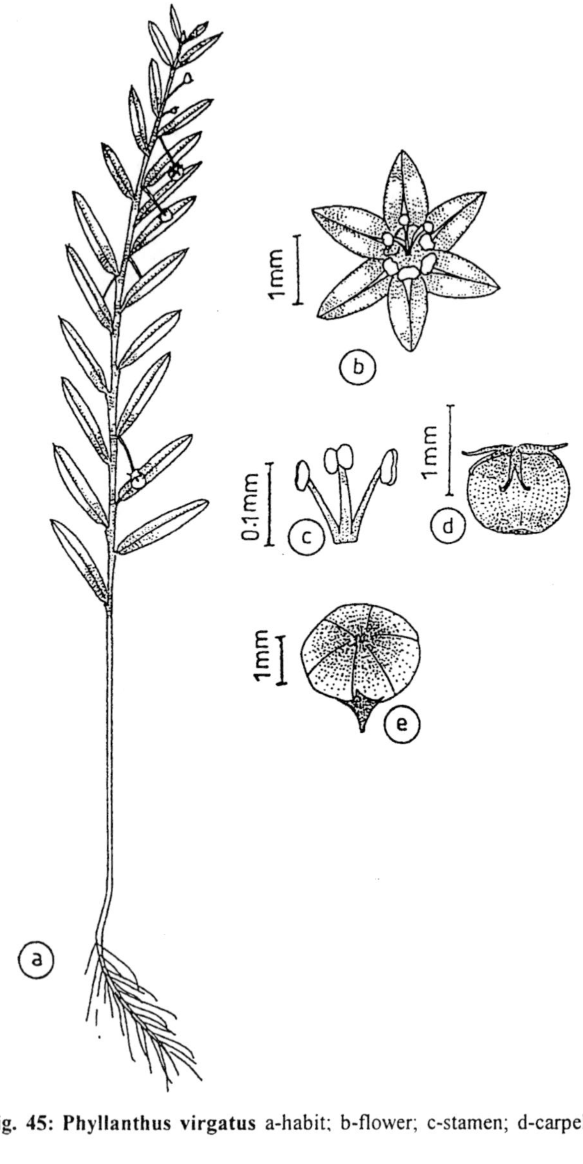

Text fig. 45: Phyllanthus virgatus a-habit; b-flower; c-stamen; d-carpel; e-fruit.

Fls. & Frts.: September–February. *Exsiccata*: SUK-1724.
Common, in crevices of rocks on hill slopes.
Localities: Kagal, Kolhapur, Wai.

Local uses: Decoction of root and leaves is a good tonic and given as a diuretic. Decoction of fresh root is used in jaundice. Leaf paste applied externally on scabies. Leaf juice is given in gonorrhea.

Ricinus communis L. Sp. Pl. 1007. 1753; Hook. f. Fl. Brit. India 5: 457. 1887; Cooke, Fl. Pres. Bombay 3: 125. 1958 (Repr. ed.); Londhe in Singh *et al.* Fl. Maharashtra State Dicot. 2: 1001. 2001. ERAND.

Erect, glaucous shrubs or small trees, 3–5 m tall. Leaves spirally arranged, suborbicular, in outline, 20–40 cm across, peltate, plamately 5–11 lobed, glaucous-green or red-brown; lobes ovate-oblong, crenate-serrate, acuminate. Flowers in narrow, terminal panicles consisting of subsessile cymes of lower male and upper female, up to 30 cm long. Male cymes 3–20 flowered: Perianth segments ovate, acuminate, 6–8 mm long. Stamens many in branched fascicles. Female cymes 1–7 flowered. Perianth as in males, caducous. Stigmas red. Fruits covered with soft prickles, 1.5–2.5 cm long, obtusely trigonous, schizocarpic. Seeds 6, ellipsoid, 8–15 mm long, carunculate at base, mottled with gray-brown, smooth, shining.

Fls. & Frts.: February–June. *Exsiccata*: SUK-1702.
Native of tropical Africa. Cultivated.
Localities: In all districts.

Local uses: 3 to 4 leaves combined and warmed on flame and applied externally to the head to relieve headache, and used as a poultice for boils. Oil is given as a purgative; used with various combination on skin diseases. Seeds paste is applied on rheumatic swellings. Root is useful in inflammation, pains, ascites, fever, glands, asthma, and cructations bronchitis. Leaves are useful in intestinal worms, night blindness, earache and increase biliousness. Flowers are useful in glandular tumors, anal troubles and vaginal pains.

Sapium insigne (Wight) Hook. f. Fl. Brit. India 5: 472. 1888; Londhe in Singh *et al.* Fl. Maharashtra State Dicot. 2: 902. 2001. *Falconeria malabarica* Wight, Ic. t. 1866. 1852. *Sapium insigne* sensu Cooke, Fl. Pres. Bombay 2: 622. 1906. [3: 121. 1958 (Repr. ed.)] non Trim. 1885. HURA, KIR-KIND, SIRAD.

Deciduous trees, 10–15 m tall, with a thick milky poisonous juice. Leaves crowded towards the ends of the branches, elliptic-lanceolate or oblong-lanceolate, 15–22 × 5–8 cm, acuminate, crenate, serrate, often with few small glands on the margins, glabrous, base acute, sometimes unequal sided. Flowers appearing when the tree is bare of leaves, monoecious, in unisexual robust spikes, 15–20 cm long; the males sessile; females sessile or shortly pedicellate. Male flowers in round clusters, 6–8 mm in diam. Stamens 2.

Female flowers: Sepals ovate, acuminate, 1 mm long. Ovary ovoid, glabrous; style 3, short, spreading shortly connate at the base. Capsules 8 mm in diam., 2–3 celled, globosely ovoid, fleshy at first dry and irregularly dehiscent when ripe, packed closely together along the rachis. Seeds 6 mm broad, ovoid, compressed, smooth, pale.

Fls. & Frts.: December–February. *Exsiccata*: SUK-1726.

Few, in semi evergreen and moist deciduous forests.

Localities: Ajara, Amba, Amboli, Barki, Dajipur, Gaganbawada, Phonda, Radhanagari, Tillarinagar.

Local uses: Milky juice is poisonous. Small drop of juice is dropped in chronic ulcerous wounds.

Securinega leucopyrus (Willd.) Muell.-Arg. in DC. Prodr. 15(2): 451. 1866; Londhe in Singh *et al.* Fl. Maharashtra State Dicot. 2: 903. 2001. *Fluggea leucopyrus* Willd. Sp. Pl. 4: 757. 1805; Hook. f. Fl. Brit. India 5: 328. 1887; Cooke, Fl. Pres. Bombay 3: 77. 1958 (Repr. ed.). PANDHARPHALLI.

Rigid shrubs with somewhat straggling branches; branchlets angular, slender, leafy, ending in sharp thorns. Leaves ovate-orbicular, 1–2 × 0.6–1.5 cm, cuneate at base, entire, obcordate or rotundate, green above, glaucous and reticulately veined beneath. Flowers axillary, males crowded in clusters, females 1–3. Stamens 3–5; filaments free, with alternate glands at the base. Ovary ovoid; styles 1–1.5 mm long; deeply bifid. Fruits globose, 3–4 mm in diam., fleshy in the beginning with white epicarp, becoming dry and shining with age. Seeds ovoid, punctate, rounded on the back with acute faces.

Fls. & Frts.: June–December. *Exsiccata*: SUK-128.

Common, on gravelly slopes of hills along gullies.

Localities: Kagal, Kolhapur.

Local uses: Leaf and fruit decoction is given to cure constipation. Leaf juice and fruit juice externally applied on itching and eczema. Plant is used in strangury, biliousness and blood diseases.

Securinega virosa (Roxb. ex Willd.) Baill. in Adansonia 6: 334. 1866; Londhe in Singh *et al.* Fl. Maharashtra State Dicot. 2: 903. 2001. *Phyllanthus virosus* Roxb. ex Willd. Sp. Pl. 4: 578. 1805. *Fluggea microcarpa* Bl. Bijdr. 580. 1825; Hook. f. Fl. Brit. India 5: 328. 1887; Cooke, Fl. Pres. Bombay 3: 77. 1958 (Repr. ed.). PACHURNA.

Large, unarmed, glabrous shrubs, 1.5–2 m tall. Leaves broadly elliptic-obovate, 2.5–7.5 × 1.5–4 cm, acute at base, entire or crenulate, rounded, obtuse or acute at apex, glabrous. Flowers axillary, the males crowded in clusters, females 1–3; bracts minute. Perianth segments 5, obovate-oblong,

Clematis gouriana Roxb. ex DC.

Michelia champaca L.

Diploclisia glaucescens (Blume) Diel.

Tinospora cordifolia (Willd.) Miers.

Nelumbo nucifera Gaertn.

Crateva adansonii (Buch. Ham.) Jacob.

Plate I

Garcinia indica (Thou.) Choisy

Bombax ceiba L.

Adansonia digitata L.

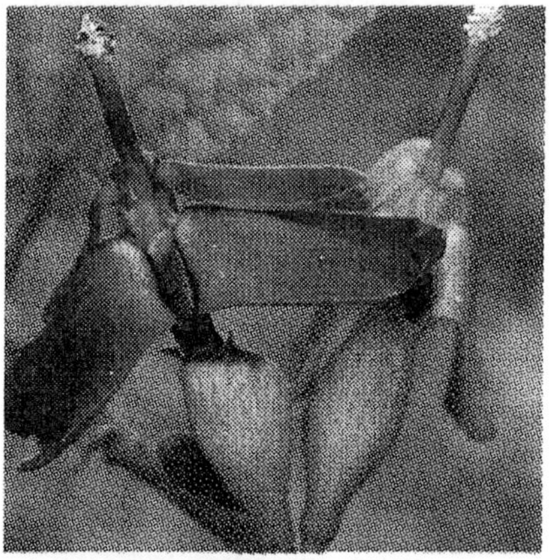

Helicteres isora L.

Fagonia schweinfurthii (Hadidi) Hadidi ex Ghafoor

Ruta chalepensis L.

Plate II

Toddalia asiatica (L.) Lamk.

Naregamia alata Wight. & Arn.

Celastrus paniculatus Willd.

Cissus quadrangularis L.

Cardiospermum helicacabum L.

Azadirachta indica Juss.

Plate III

Abrus precatorius L.

Butea monosperma Lam.

Clitoria ternatea L.

Erythrina variegata L.

Mucuna pruriens (L.) DC.

Pongamia pinnata (L.) Pierre

Plate IV

Cassia alata L.

Moullava spicata (Dalz.) Nicols.

Saraca asoca (Roxb.) de Wilde.

Acacia concinna (Willd) DC.

Mimosa pudica L.

Kalanchoe pinnata (Lam.) Pres.

Plate V

Centella asiatica (L.) Urb.

Catunergam spinosa (Thunb.) Triveng.

Vernonia cinerea (L.) Less

Embelia ribes Burm. f.

Eclipta prostrata (L.) L.

Spianthus calva DC.

Plate VI

Holarrhena pubescens (Buch.-Ham.) Wall. ex. G. Don.

Rauvolfia serpentina (L.) Bth. ex. Kurz.

Calotropis procera (Ait) R. Br.

Pergularia daemia (Forssk.) Choiv.

Tylophora indica (Burm. f.) Merr.

Hemidesmus indicus (L.) Schult

Plate VII

Evolvulus alsinoides (L.)

Withania somnifera (L.) Dunal.

Bacopa monnieri (L.) Penn.

Dolichandrone falcata (Wall. ex. DC.) Seem.

Pedalium murex L.

Clerodendrum serratum

Plate VIII

Celosia argentea L.

Boerhavia repens L.

Santalum album L.

Acalypha indica L.

Drypetes roxburghii (Wall.) Hurusawa

Ficus benjhalensis L.

Plate IX

Eulophia nuda Lindl.

Curculiago orchioides Gaertn.

Smilax zeylanica L.

Gloriosa superba L.

Asparagus racemosus Willd.

Vetiveria zizanioides (L.) Nash.

Plate X

Aegle marmelos (L.) Corr.

Ficus hispida L.

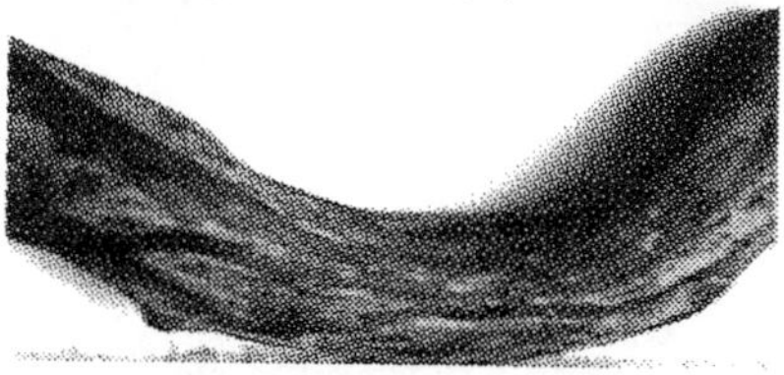

Woodfordia fruticosa (L.) Kurz.

Catunaregam spinosa (Thunb.) Tirveng.

Spermadictyon suaveolens Roxb.

Syzygium cumini (L.) Skeels.

Celastrus paniculatus Willd.

Atalantia racemosa Wight.

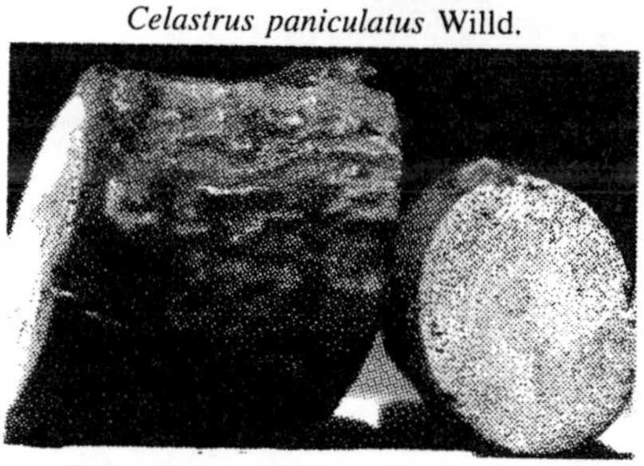

Cinnamomum nitidum (Roxb.) Hook.

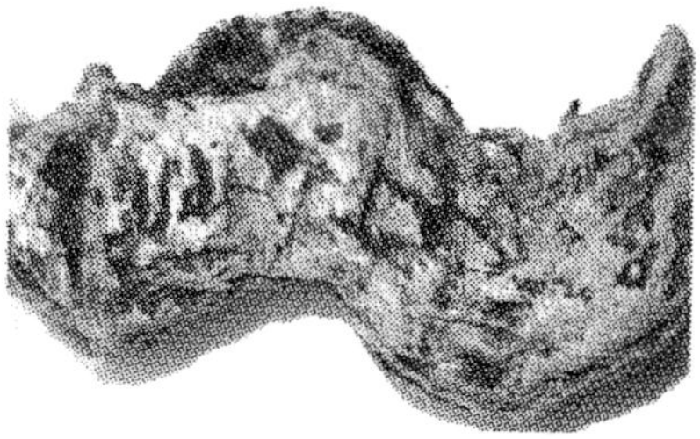

Jasminum malabaricum Wight

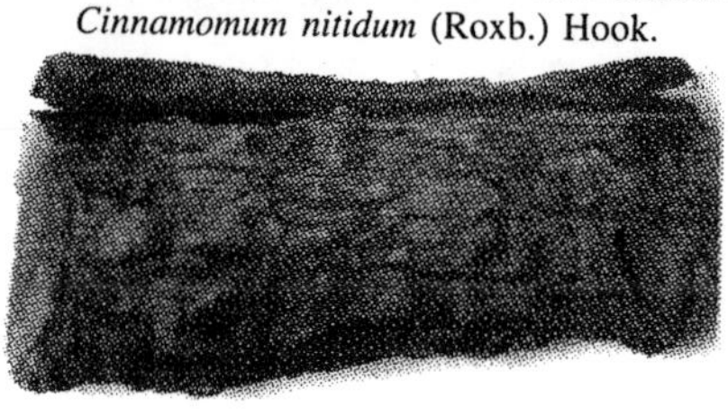

Stereospermum chelenoides (L. f.) DC.

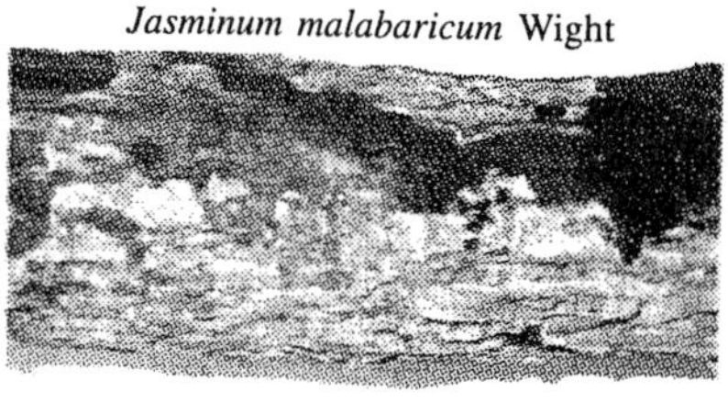

Butea monosperma (Lam.) Taub.

Plate XI

Artocarpus heterophyllus Lam.

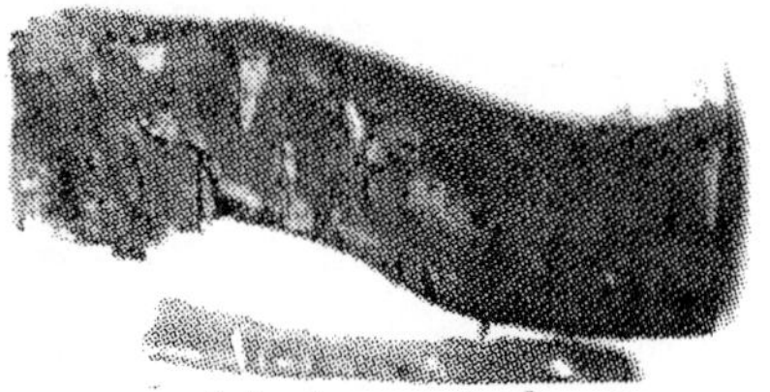

Salvadora persica L.

Cocculus hirsutus (L.) Theob.

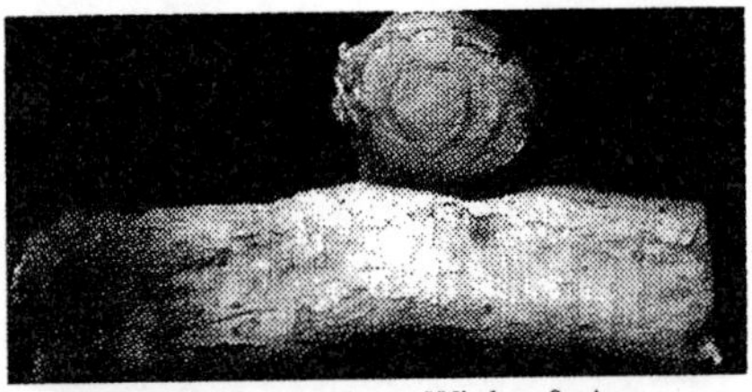

Salacia brunoniana Wight. & Arn.

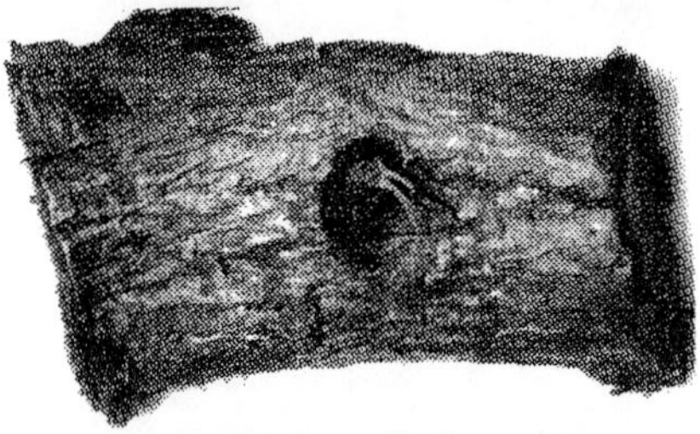

Bombax ceiba L.

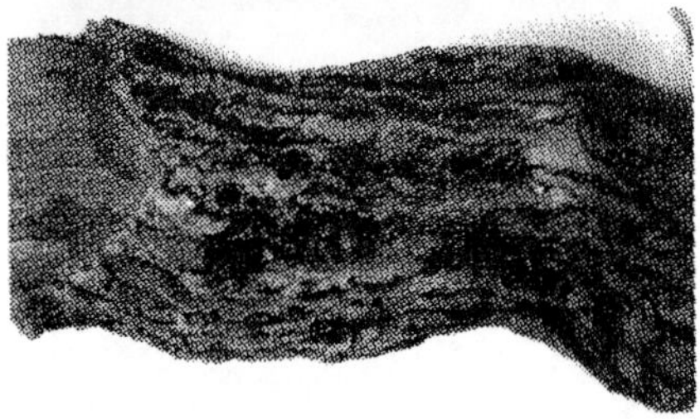

Gmelina arborea Roxb.

Oroxylum indicum (L.) Vent.

Embelia basaal (R & S.) A. DC.

Hemidesmus indicus R. Br.

Plate XII

Aegle marmelos (L.) Corr.

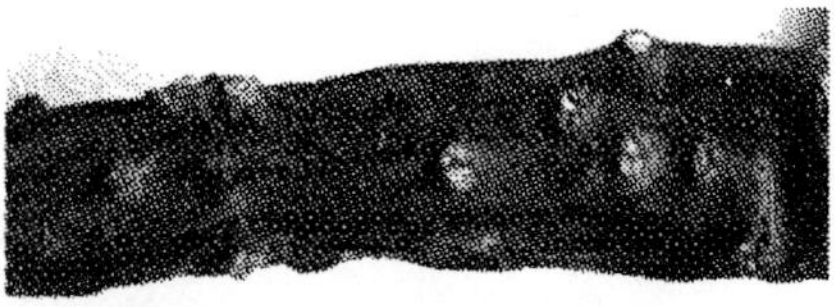

Tinospora cordifolia (Willd.) Mers.

Litsea deccanensis Gamble

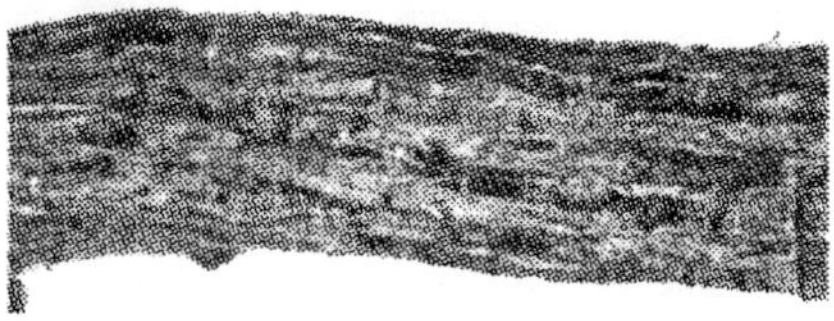

Argyreia boseana Sant. & Patel

Cryptolepis buchanani Roem. & Schult.

Celastrus paniculatus Willd.

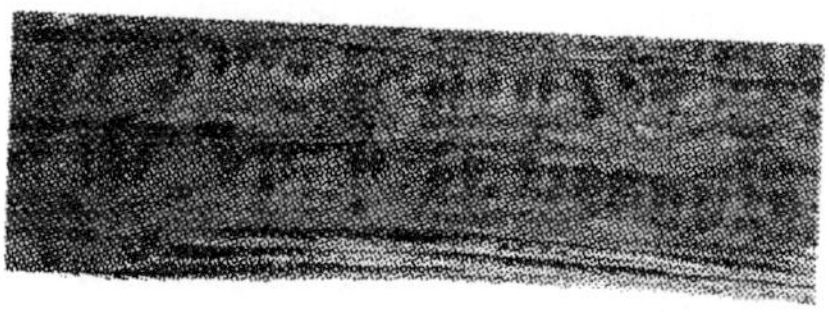

Trichosanthes tricuspidata Lour.

Cissus elongata Roxb.

Calotropis gigantea (L.) Br.

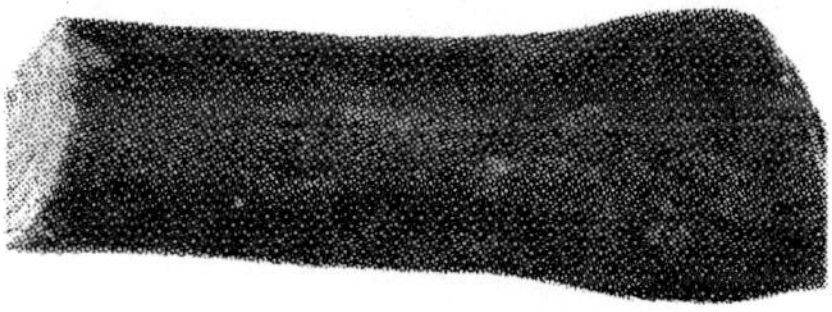

Diploclisia glaucescens (Bl.) Diels.

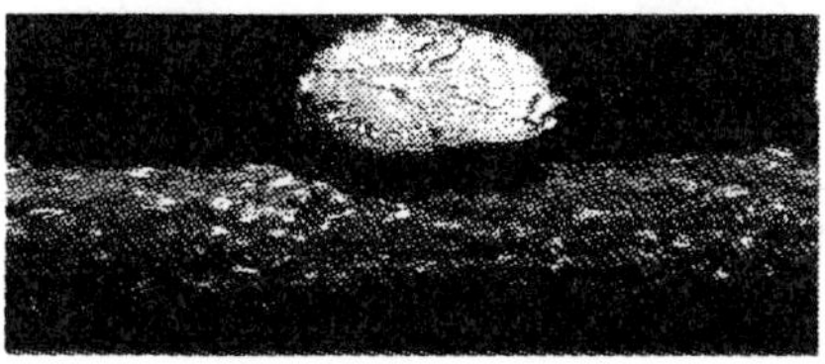

Gnetum scandens Roxb.

Moullava spicata (Dalz.) Nicols.

Plate XIII

Terminalia cuneata Roth

Terminalia elliptica Willd.

Bridelia retusa (L.) Spreng.

Pterocarpus marsupium Roxb.

Pongamia pinnata (L.) Pierre

Ficus exasperata Vahl

Holarrhena pubescens (Buch.-Ham.) Wall ex G. Don.

Wrightia tinctoria ssp. *tinctoria* R.Br.

Cassine paniculata (W. & A.) Lobereau-Collen.

Alstonia scholaris (L.) R. Br.

Crateva adansonii ssp. odora (Buch.-Ham.) Jacob.

Pittosporum floribundum W. & A.

Plate XIV

Semecarpus anacardium L. f.

Catunaregam spinosa (Thunb.) Tirveng.

Terminalia chebula Retz.

Caryota urens L.

Caesalpinia bonduc (L.) Roxb.

Entada rheedei Spreng.

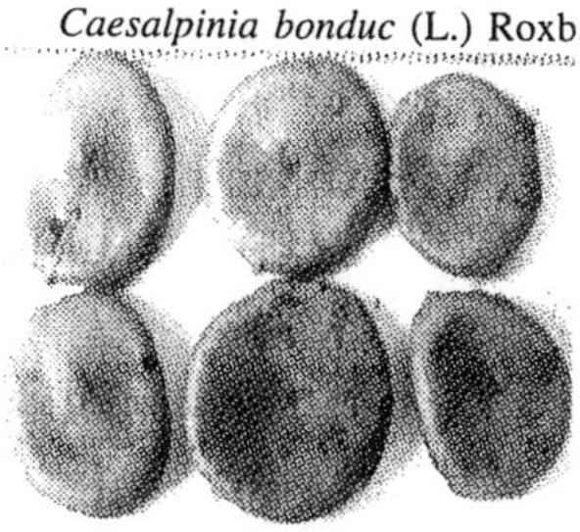

Strychons nux-vomica L.

Butea monosperma (Lam) Taub.

Dioscorea bulbifera L.

Zanthoxylum rhetsa (Roxb.) DC.

Plate XV

1–1.5 mm long, obtuse, concave. Stamens 3–5; filaments free, with alternate glands at base. Ovary ovoid; styles 1–1.5 mm long, deeply bifid. Fruits globose, 3–4 mm in diam., fleshy in the beginning with white epicarp, becoming dry and shining with age. Seeds ovoid, punctate, rounded on the back with acute faces.

Fls. & Frts.: June–December. *Exsiccata*: SUK-1725.
Occasional, on gravelly slopes of hills along gullies.
Localities: Girgaon, Katyayani.

Local uses: Leaf and fruit decoction is given to cure constipation. Leaf juice and fruit juice externally applied on itching and eczema. Plant is used in strangury, biliousness and blood diseases.

Tragia involucrata L. Sp. Pl. 980. 1753; Hook. f. Fl. Brit. India 5: 465. 1888; Cooke, Fl. Pres. Bombay 3: 119. 1958 (Repr. ed.); Londhe in Singh *et al.* Fl. Maharashtra State Dicot. 2: 904. 2001. KAHJKOLTI.

Twining herbs, with scattered stinging hairs. Leaves oblong-lanceolate to broadly ovate, 2–10 × 1.9–5 cm, acuminate, serrate, hairy, base rounded or cordate. Flowers shortly pedicellate in terminal axillary and leaf opposed usually hairy racemes 3–5 cm long; the males in upper parts, yellowish; the females few in the lower part of the racemes. Male flowers: Sepals 3, broadly elliptic to orbicular, concave, 1 mm long, glabrous; stamens 3. Female flowers: Sepals 6, ovate, pinnatifid, very hispid, 3 mm long, much elongating and becoming rigid in fruit; ovary 3-lobed, hispid; styles 3, circinately revolute, united below in stout cylindric column often as long as the branches. Capsules 8 mm in diam., 3-lobed, white, more or less hispid. Seeds globose, smooth, 4 mm in diameter.

Fls. & Frts.: May–June. *Exsiccata*: SUK-235.
Common, in semi evergreen forests.
Localities: Amboli, Barki, Dajipur, Tillari, Vaibhavwadi.

Local uses: Paste of the root is used in stomach complaints. Root with juice of *Ocimum sanctum* is employed as a cure for itchy skin eruptions. Root decoction is given in fever to cause perspiration. A decoction of root is useful in relieving bronchitis and the attendant fever.

CANNABINACEAE Endl.

Cannabis sativa L. Sp. Pl. 1027. 1753; Hook. f. Fl. Brit. India 5: 487. 1888; Cooke, Fl. Pres. Bombay 3: 160. 1958 (Repr. ed.); Jadhav in Singh *et al.* Fl. Maharashtra State Dicot. 2: 928. 2001. APHU, CHARAS, GANJA.

Erect, pubescent, annual, dioecious herbs, 30–90 cm tall. Leaves alternate or opposite, digitate; leaflets in the upper leaves 1–3, in the lower leaves

5–11, linear-lanceolate, 5–15 × 1–2 cm, serrate, acuminate, glabrous above, pubescent beneath. Male flowers in short cymes combined into terminal panicles. Perianth green or white; segments 5, oblong 3–3.5 mm long, obtuse, ciliate, with white margin, appressed pubescent on the back. Stamens 5; female flowers solitary, axillary, forming leafy spikes; bracts glandular-pubescent, accrescent in fruits. Achenes grayish-brown, compressed, 4–5 mm long. Seed flattened.

Fls. & Frts.: September–November. *Exsiccata*: SUK-1709.
Rare.
Localities: Dindewadi, Udgiri.

Local uses: Leaf decoction is given orally and smoke is applied externally on piles. It is given on debility in women. It is given in cold, cough, bronchitis and asthma. It is externally applied in hemicrania. Leaf juice is given orally in urinary discharges. Leaves are used as a tonic, stomachic, useful in cough; causes biliousness, intoxication and hallucinations. Female flowers are abortifacient, useful in convulsions, causes thirst, restlessness, check vomiting and lessen inflammation.

MORACEAE Link

Antiaris toxicaria (Pers.) Lesch. in Ann. Mus. Hist. Nat. Paris 16: 478, t. 22. 1810; Hook. f. Fl. Brit. India 5: 537. 1883; Cooke, Fl. Pres. Bombay 3: 156. 1958 (Repr. ed.); Lakshminarasimhan & Venkanna in Singh *et al.* Fl. Maharashtra State Dicot. 2: 928. 2001. *Ipo toxicaria* Pers. Syn. 2: 566. 1807. JASUND, KARVAT, CHAND KUDA.

Trees, very tall. Leaves often oblong, 5–20 × 2.5–10 cm, acuminate, often mucronate, glabrous, entire or serrulate, polished and shining above, base rounded or subcordate. Flowers monoecious; the male receptacles 3–4 together, orbicular, tomentose, 1.2 cm in diam., peduncles velvety; females solitary, enclosed in pear-shaped axillary subsessile velvety involucre of numerous confluent bracts. Male flowers: Sepals 3–4, spathulate, imbricate; stamens 3–8; filaments erect. Female flowers: Perianth 0; ovary adnate to the involucre. Fruit red, velvety, 2 × 1.5 cm pyriform or subglobose, 1-seeded, on a stalk.

Fls. & Frts.: August–September. *Exsiccata*: SUK
Rare, in evergreen forests.
Localities: Marleshwar.

Local uses: Seeds are poisonous. Seeds are used in diabetes. 'L' shape made on seed with the needles and powder obtained during making of L shape is given to diabetic patients. Seeds are used as a febrifuge and in dysentery. Sap is used as arrow and dart poison.

Artocarpus heterophyllus Lam. Encyl. 3: 210. 1789 (*heterophylla*); Lakshminarasimhan & Venkanna in Singh *et al.* Fl. Maharashtra State Dicot. 2: 930. 2001. *A. integrifolia* sensu Gamble, Fl. Madras Pres. 1369. 1928. [3: 957. 1967 (Repr. ed.)] non L. f. 1781; Hook. f. Fl. Brit. India 5: 541. 1888; Cooke, Fl. Pres. Bombay 3: 158. 1958 (Repr. ed.). PHANAS.

Evergreen trees, cauliflorous with massive syncarps. Leaves elliptic or obovate, acuminate, entire, 5–25 × 3.5–12 cm, dark green and shining above, rough beneath, base narrowed. Stipules large, glabrous soon caducous. Inflorescence solitary-axillary. Male heads 27–70 × 8–28 mm, narrowly clavate. Syncarp 30–100 × 25–50 cm, green, round with conical acute processes; fruiting perianth yellow to light orange. Seeds 25–30 × 15–20 mm.

Fls. & Frts.: March–April. *Exsiccata*: SUK-1703.
Cultivated.
Localities: Kolhapur.

Local uses: Roots applied externally on mouth ulcers. It is also used as wound healing. Juice of plant is applied externally to glandular swelling and abscesses to promote suppuration. Fruit is useful in ulcers. Seeds are diuretic and constipating.

Artocarpus lakoocha Roxb. Fl. Ind. 3: 524. 1832; Hook. f. Fl. Brit. India 5: 543. 1888; Cooke, Fl. Pres. Bombay 3: 158. 1958 (Repr. ed.); Lakshminarasimhan & Venkanna in Singh *et al.* Fl. Maharashtra State Dicot. 2: 931. 2001. UOMT.

Large deciduous trees, 15–20 m high. Leaves oblong-elliptic or subovate, 10–30 cm long, thinly coriaceous, glabrous or softly tomentose above, pubescent beneath, base rounded, apex cuspidate. Flowers in axillary globose, shortly pedunculate heads. Male flowers: Sepals 2 3 together, triangular, truncate, puberulous; stamens 1, filament broad below, tapering upwards; anthers exserted, short, broad, 2-celled. Female flowers: anthocarps completely united. Fruit 5–15 cm, in diam., lobulate, smooth, velvety, yellow, edible. Seeds oblong, few, broad, about 1.2 cm across.

Fls. & Frts.: March–April. *Exsiccata*: SUK-1718.
Occasional in semi evergreen forest.
Localities: Sawantwadi, Tillari.

Local uses: Roots applied externally on mouth ulcers. Juice of plant is applied externally to glandular swelling and abscesses to promote suppuration. Fruit is useful in ulcers. Seeds are diuretic and constipating.

Ficus amplissima J. E. Sm. in Rees, Cyclop. 14. n. 68. 1810; Lakshminarasimhan and Venkanna in Singh *et al.* Fl. Maharashtra State Dicot. 2: 933: 2001. *F. tsiela* Roxb. ex Buch.-Ham. in Trans. Linn. Soc. 15: 149. 1826; King in Hook. f. Fl.

Brit. India 5: 515. 1888; Cooke, Fl. Pres. Bombay 3: 150. 1958 (Repr. ed.). PAYER, PIMPRI.

Trees, 15 m high. Leaves ovate or ovate lanceolate, 5–12 × 2–6 cm, bright green, shining above, base rounded, apex acute or cuspidate, entire along margins, 3-nerved, membranous. Receptacles in clusters of 2–6, on very short, crowded tubercles, in the axils of the leaves or more frequently in the axils or the scars of fallen leaves, sessile, depressed globose, 5 mm diam., whitish-yellow, dotted when ripe. Male flowers; few, near the mouth of receptacle, sessile; sepals 3, ovate, shorter than the single stamen; filaments as long as or longer than the ovate anthers. Female flowers: Sepals 3, ovate; achenes long; stigma very long, cylindric.

Fls. & Frts.: April–September. *Exsiccata*: SUK-1717.
Common, in forests areas.
Localities: Dajipur, Panhala, Radhanagari.

Local uses: Fruits are chewed in mouth ulcers. Leaf juice applied externally in case of old chronic wounds. Bark is boiled with salt and boiled rice water in equal proportion and lukewarm solution is given in colic.

Ficus arnottiana (Miq.) Miq. in Ann. Mus. Bot. Lugd. Bat. 3: 287. 1867; King in Ann. Roy. Bot. Gard. Calcutta 1: 56, t. 68 A and 84 V. 1887 and Hook. f. Fl. Brit. India 5: 513. 1888; Cooke, Fl. Pres. Bombay 3: 149. 1958 (Repr. ed.); Lakshminarasimhan and Venkanna in Singh *et al.* Fl. Maharashtra State Dicot. 2: 933. 2001. *Urostigma arnottiana* Miq. in London J. Bot. 6: 564. 1847. KADAK PAYER, PAIR.

Trees, 10 m high. Leaves spirally arranged; veins and petioles bright pink, ovate-cordate or elliptical, 6–14 × 5–9 cm, base deeply cordate, apex caudate-acuminate with ca 20 mm long, acuminate, entire along margins, coriaceous. Figs in axillary pairs and raised below leaves, ripening yellowish-brown red or purple with green spots; peduncles 2–5 mm long, tepals red, fleshy. Male flowers sessile; stamens 1, female flowers sessile or shortly stalked, ovary sessile.

Fls. & Frts.: February–June. *Exsiccata*: SUK-1708.
Common, in open forests in dry zones or on hillsides and rocky places.
Localities: Amboli, Dajipur, Radhanagari, Vaibhavwadi, Vasota.

Local uses: A paste of bark is used as an absorbent in inflammatory swellings. In cracked feet the juice is employed. Bark decoction is used in gonorrhea. Fruits are used orally for days with empty stomach commencing on last day of menstruation helps in conception.

Ficus benghalensis L. Sp. Pl. 1059. 1753; King in Ann. Roy. Bot. Gard. Calcutta 1:18, t. 13. 1887 & in Hook. f. Fl. Brit. India 5: 499. 1888; Cooke, Fl. Pres.

Bombay 3: 145. 1958 (Repr. ed.); Lakshminarasimhan & Venkanna in Singh *et al.* Fl. Maharashtra State Dicot. 2: 934. 2001. WAD.

Trees with numerous aerial roots from spreading branches; twigs, stipules, petioles and leaves puberulous at early stage, late glabrescent. Leaves ovate, coriaceous, 10–30 × 7–20 cm, base cordate, or rarely cuneate; stipules stout. Figs sessile, axillary, paired, ripening orange to red, depressed-globose; basal bracts obtuse; internal bristles none. Tepals free, 2–3 in male, 3–4 in galls and female flowers; anthers shortly mucronate.

Receptacles: April–June. *Exsiccata*: SUK-178.
Planted along roadsides and near villages.
Localities: In all districts.

Local uses: It is applied to the soles of the feet cracked or inflamed and is also applied to the teeth and gums as a remedy for toothache. An infusion of bark is used as tonic. Fruits are used in treatment of diabetes. Tender leaves and bark is used to stop bleeding. Young tender roots are dried in shadow and given every night after menstrual cycle to overcome infertility. Bark decoction of *F. benghalensis* and *F. religiosa* is used in leucorrhoea. It is said that tender ends of hanging roots given with *Withania somnifera* invoke the chances of male baby would be become much greater. Decoction of mature leaves are given in inflammation of urinary bladder and also given in prostatitis.

Ficus exasperata Vahl, Enum. Pl. 2: 197, 402. 1806; Lakshminarasimhan & Venkanna in Singh *et al.* Fl. Maharashtra State Dicot. 2: 936. 2001. *F. asperrima* Roxb. Fl. Ind. 3: 554. 1832; King in Ann. Roy. Bot. Gard. Culcutta 2: 80, t. 100. 1888 and in Hook. f. Fl. Brit. India 5: 522. 1888; Cooke, Fl. Pres. Bombay 3: 158. 1958 (Repr. ed.). BUIVMBAR, KARYAT, KHARVATI.

Shrubs to small trees, ca 8 m high. Leaves variable in size and shape, ovate-elliptic or more or less 3-lobed, scabrid, 6–19 × 3–9 cm, base cuneate, more or less symmetric, apex acute to acuminate with a tip, 15 mm long, denticulate or sinuate-crenate to serrate along margins. Receptacles solitary, axillary ca 1.5 cm across, greenish to yellow and reddish at maturity. Tepals white, free, rarely 1–2-jonited, puberulous or hispidulous upwards. Internal bristles copious, white. Male flowers ostiolar, sessile; tepals 3–4 or 6; stamens 1, female and gall flowers sessile to pedicellate; tepals 4–7, linear-spathulate. Seeds 1.2–1.5 × 0.8–1.0 mm, shortly oblong, slightly keeled.

Receptacles: April–June. January–June. *Exsiccata*: SUK-1707.
Common, along streams.
Localities: Radhanagari, Panhala, Sawantwadi, Tillari.

Local uses: Bark is applied externally immediately after scorpion sting. Bark is externally applied on wounds and ulcers. Juice of bark is used for glandular enlargement of the abdomen, such as liver and spleen.

Ficus heterophylla L. f. Suppl. 432. 1781; King in Ann. Roy. Bot. Gard. Calcutta 2: 75, t. 94. 1888 and in Hook. f. Fl. Brit. India 5: 518. 1888; Cooke, Fl. Pres. Bombay 3: 152. 1958 (Repr. ed.); Lakshminarasimhan & Venkanna in Singh *et al.* Fl. Maharashtra State Dicot. 2: 937. 2001. DATIR.

Creeping or often prostrate shrubs, ca 3 m long; twigs reddish-brown, thinly pubescent with short, white, spreading hairs, subscabrid. Leaves distichous, oblong or narrowly elliptic to subovate, often lanceolate, 5–15 × 2–5 cm, base widely cuneate or rounded, apex acute to acuminate, dentate to subentire along margins, membranous to subcoriaceous; lateral veins 4–7 pairs. Figs solitary, axillary, rarely paired, ripening orange to dark red; basal bracts 3, ovate-sub-acute, small. Tepals white, free, entire; internal bristles none. Male flowers in 2–3 rings, tepals 3–4, linear-oblong. Females and gall flowers pedicellate, tepals 4–5, lanceolate. Seeds slightly keeled, smooth.

Receptacles: May–June. *Exsiccata*: SUK-267.
Occasional, in dry zones.
Localities: Ichalkaranji, Kolhapur.

Local uses: Juice of root is internally administered in colic pains. Juice of the leaves mixed with milk is given in dysentery. Juice is externally applied on scorpion sting.

Ficus hispida L. f. Suppl. 442. 1781; King in Ann. Ray. Bot. Gard. Calcutta 2: 116, t. 154. 1888; & Hook. f. Fl. Brit. India 5: 522. 1888; Cooke, Fl. Pres. Bombay 3: 154. 1958 (Repr. ed.); Lakshminarasimhan in Singh *et al.* Fl. Maharashtra State Dicot. 2: 937. 2001. BOKEDA, KALA-UMBAR.

Shrubs to small trees, 5 m high. Leaves opposite, ovate, oblong or subovate, base cordate to cuneate, apex acuminate, crenate-serrate along margins. Figs obovate, axillary to cauliflorous on short, tubercled branches, ripening pale yellow; basal bracts 3, subtriangular, internal bristles 0. Male flowers in 1–2 rows; stamen 1. Galls and female flowers sessile or pedicellate; style hispid. Seed slightly keeled, tuberculate.

Receptacles: January–July. *Exsiccata*: SUK-227.
Common, in forest and along streams or in moist localities.
Localities: Amba, Amboli, Barki, Borbet, Dajipur, Gaganbawada.

Local uses: Fruits heated with water to form a poultice is applied to buboes. Stem is externally applied on wounds. Bark is used as emetic. All parts are useful in dysentery; psoriasis, anemia, piles, jaundice, haemorrhage of the nose and mouth and disease of the blood. Fruits are used as emetic and constipation. Latex from the fruit the roots is used in tuberculosis.

Ficus racemosa L. Sp. Pl. 1060. 1753; Lakshminarasimhan & Venkanna in Singh *et al.* Fl. Maharashtra State Dicot. 2: 939. 2001. *F. glomerata* Roxb. Pl.

Corom. 2: 13, t. 1. 23. 1798; King in Ann. Roy. Bot. Gard. Calcutta 2: 173, tt. 218, 219. 1888 & in Hook. f. Fl. Brit. India 5: 535. 1888; Cooke, Fl. Pres. Bombay 3: 154. 1958 (Repr. ed.). UMBAR.

Trees, 4–10 m high, much branched, often with well developed canopy. Leaves ovate-oblong, elliptic-lanceolate, 3–6 × 6.2 × 2–4 cm, base cuneate or rarely subcordate, apex subacute or subacuminate, entire, subcoriaceous. Receptacles in large clusters on the main branches and trunks, obovoid, pyriform, pale green to red to purplish, on ripening. Perianth 3–4, dentate-lacerate, lobes joined below, red. Male flowers in 2–3 rings, sessile, much compressed; stamens 1–2. Galls long stalked. Females sessile or shortly stalked, ovary sessile or substipitate. Seed 1 mm long, lenticular, smooth.

Receptacles: February–June. *Exsiccata*: SUK-179.
Common, along streams and rocky hill slopes.
Localities: Kolhapur, Sangli, Satara.

Local uses: Ripe fruits are used in abortion, given orally in diarrhoea. Fruit are given orally during menstrual cycle in excess flow. Decoction of fruit mixed with decoction of *Cuscuta* sp is given to weak children. In dry bronchitis, latex is given with other species of *Ficus* sp. Decoction of bark is given in inflammation of feet. Latex is useful in all kinds of urinary complaints.

Ficus religiosa L. Sp. Pl. 1059. 1753; King in Ann. Roy. Bot. Gard. Calcutta 1: 55, t. 67. 1887 & in Hook. f. Fl. Brit. India 5: 513. 1883; Cooke, Fl. Pres. Bombay 3: 149. 1958 (Repr. ed.). Lakshminarasimhan & Venkanna in Singh *et al.* Fl. Maharashtra State Dicot. 2: 940. 2001. PIMPAL.

Trees, 10 m high, occasionally epiphytic. Leaves coriaceous, ovate, rotund, base truncate; apex caudate-acuminate, 1–5 cm long; margins often sinuous long petiolate. Receptacles axillary, paired, sessile, pink, purple or black when ripens; basal bracts 3, silky puberulous, internal bristles none. Male flowers sessile in 1 ring; tepals 2, red, free, ovate-lanceolate. Galls and females sessile; tepals 3–4, free.

Receptacles: March–August. *Exsiccata*: SUK-1721.
Planted near temples.
Localities: Sangli, Satara.

Local uses: A paste of bark is used as an absorbent in inflammatory swellings. In cracked feet the juice is employed. Bark decoction is used in gonorrhea. Fruits taken orally for days with empty stomach commencing on last day of menstruation helps in conception. Juice of leaves is good in urinary tract infections. Bark is one of the constituents of panch-walklya. Bark is useful in diseases of the blood, vagina, uterus and given in leucorrhoea, burning sensations and urinary discharges.

SALICACEAE Mirbel

Salix tetrasperma Roxb. Pl. Cor. t. 97. 1798; Hook. f. Fl. Brit. India 5: 626. 1888; Cooke, Fl. Pres. Bombay 3: 162. 1958 (Repr. ed.); Venkanna & Kothari in Singh *et al.* Fl. Maharashtra State Dicot. 2: 950. 2001. WALUNJ.

Small, much-branched trees, 5–8 m tall. Leaves alternate, lanceolate, 5–15 × 1.5–3 cm, drooping, acute or rounded at base, entire or serrulate, acute or acuminate, green and glabrous above, white and more or less silky beneath when young. Flowers unisexual, sessile, in axillary 5–10 cm long catkins, bracts ovate, concave, silky villous on the back. Disk of 2 yellow glands. Stamens 5–10. Female flowers pedicellate, in the axils of oblong or obovate, hairy bracts. Ovary glabrous; stigma 2-branched. Capsules ovoid, 3–4 mm long, 2-valved. Seeds 4–6, oblong, with a pencil of long, silky deciduous hairs.

Fls. & Frts.: October–December. *Exsiccata*: SUK-627.
Common, along river banks.
Localities: Barki, Kolhapur, Radhanagari, Vaibhavwadi.

Local uses: Bark decoction is given in Typhoid fever, fever due to acidity. Flower juice is given to improve appetite.

CERATOPHYLLACEAE S. F. Gray

Ceratophyllum demersum L. Sp. Pl. 992. 1750; Hook. f. Fl. Brit. India 5: 639. 1888; Cooke, Fl. Pres. Bombay 3:165. 1958 (Repr. ed.); Venkanna & Kothari in Singh *et al.* Fl. Maharashtra State Dicot. 2: 951. 2001.

Submerged water plants; stems 20–90 cm long, forming a much-branched tangle of slender branches and leaves. Leaves whorled, dichotomously divided into filiform, 2–2.5 cm long, toothed segments. Flowers minute, axillary solitary. Male flowers with 6–12 narrow, subclavate, 2-fid tepals. Stamens 20–30; filaments very short, anthers with terminally produced, 2-fid connectives. Female perianth as in males, 1–1.5 mm long. Ovary sessile; style subulate. Fruits ellipsoid, 3.5 × 4 × 2 mm excluding spines, slightly compressed, biconvex with narrowly winged margins; spines 3, unequal, 3–6 mm long, lateral ones usually half as long as the terminal one.

Fls. & Frts.: October–May. *Exsiccata*: SUK-1696.
Frequent, in ponds and stagnant waters.
Localities: Kolhapur, Shiroli lake, Rankala lake.

Local uses: Plant is used as cooling, oleaginous, laxative, antipyretic, useful in diseases of the blood and acidity.

ORCHIDACEAE Juss.

Acampe praemorsa (Roxb.) Blatt. & McCann. in J. Bombay. nat. Hist. Soc. 35: 495. 1932. Lakshminarasimhan in Sharma *et al.* Fl. Maharashtra State Monocot. 11. 1996. *Epidendrum praemorsum* Roxb. Pl. Cor. 1: 34, t. 43. 1795. *Acampe wightinan* Lindl. Fol. Orchid. 2. 1853. Cooke, Fl. Pres. Bombay 3: 209. 1958 (Repr. ed.); *Saccolabium praemorsum* (Roxb.) Hook. f. Fl. Brit. India 6: 762. 1890.

Stem thick, sheathed. Leaves thick coriaceous, chanelled. Flowers yellow in corymbose racemes. Capsule 6–7 × 0.6–0.8 cm in size, subcylindric, longitudinally ribbed.

Fls. & Frts.: February–August. *Exsiccata*: SUK-1792.
Localities: Amba, Amboli, Radhanagari, Sawantwadi.

Local uses: Leaf juice is used for earache.

Bulbophyllum neilgherrense Wight, Ic. t. 1650. 1851; Hook. f. Fl. Brit. India 6: 761. 1890; Cooke, Fl. Pres. Bombay 3: 189. 1958 (Repr. ed.); Lakshminarasimhan in Sharma *et al.* Fl. Maharashtra State Monocot. 14. 1996.

Small herbs, with creeping slender stem; pseudobulbs ovoid, 1.5–2.5 × 1.0–3 cm. Leaf one from the top of the pseudobulb, oblong-elliptic, 4–10 × 1.5–3 cm, obtuse-emarginate. Flowers dense, dull yellow in racemes, 5–8 cm long, drooping, arising from the base of the pseudobulb, scapes covered by sheaths at base. Sepals dull brownish-yellow, suffused with red at their bases. Petals smaller than the dorsal sepals, ovate, acuminate, pale yellow. Column with lanceolate spurs.

Fls.: July–December. *Exsiccata*: SUK-1792.
Localities: Amboli, Koyananagar, Mahabaleshwar, Patgaon.

Local uses: Stem is given orally in acidity. It also eaten as a tonic in general debility.

Cymbidium aloifolium (L.) Swartz in Nov. Act. Reg. Soc. Sci. Upsal. 2, 6: 73. 1779; Hook. f. Fl. Brit. India 6: 10. 1890; Cooke, Fl. Pres. Bombay 3: 199. 1958 (Repr. ed.); *Epidendrum aloifolium* L. Sp. Pl. 953. 1753. Lakshminarasimhan in Sharma *et al.* Fl. Maharashtra State Monocot. 17. 1996.

Epiphytes, pseudo-stem short. Leaves linear-oblong, curved, 20–55 × 1–3 cm, coriaceous, obtuse, fleshy, slightly and obliquely notched at the apex, somewhat sheathing and slightly expanded at the base. Flower yellowish-red in many-flowered drooping racemes, 22–40 cm long. Bracts ovate, acute. Sepals subequal, oblanceolate-oblong, obtuse; lateral pair somewhat falcate. Petals as long as the sepals, oblanceolate-ovate, obtuse.

Lip purplish as along as the sepals, oblong, 3-lobed, its upper surface with 2 lamellae broken and disconnected in the middle. Capsules 5–6 cm long, elliptic, ribbed.

Fls.: April–May. *Exsiccata*: SUK-1791.
Occasional, in semi evergreen forests.
Localities: Amboli, Patgaon, Sawantwadi.

Local uses: Plant juice is given in acidity.

Dendrobium barbatulum Lindl. Gen. Sp. Orchid. 84. 1830; Hook. f. Fl. Brit. India 5: 179. 1890; Cooke, Fl. Pres. Bombay 3: 184. 1958 (Repr. ed.); Lakshminarasimhan in Sharma *et al.* Fl. Maharashtra State Monocot. 19. 1996. NALKAND.

Herbs, caespitose, stem 9–16 cm long, slender, plurinodal, leafless at the time of flowering. Leaves lanceolate, 4.5–13 × 1.5–3.5 cm, acute, deciduous. Flowers white, often suffused with pink, glistening when fresh, in lateral and terminal racemes, 7–10 cm long. Ovary with pedicel, 1.1–1.3 cm long.

Fls & Frts.: January–July. *Exsiccata*: SUK-1790.
Common, epiphyte on trees in open moist deciduous forests.
Localities: Amba, Amboli, Borbet, Radhanagari, Sawantwadi, Tillari, Vaibhavwadi.

Local uses: The entire plant is given in problems of urine passage. The entire plant is given internally for all kinds of stomachache and in gas troubles, excites bile and acts as a laxative to the intestines.

Eulophia nuda Lindl. Gen. Sp. Orchid. 180. 1833; Hook. f. Fl. Brit. India 6: 5. 1890; Cooke, Fl. pres. Bombay 3: 197. 1958 (Repr. ed.); Lakshminarasimhan in Sharma *et al.* Fl. Maharashtra State Monocot. 29. 1996. AMARKAND.

Terrestrial herbs with creeping, under ground tubers. Leaves from the sides of the tuber, 20–60 × 5–6 cm, oblong-lanceolate, acute, entire, plicate. Flowers 9–20, in lax racemes from the base of the pseudo-stem; scape ca 50 cm tall. Flowers greenish-purple, ca 4.5 cm across. Petals white oblong, obtuse, many nerved white or yellow flushed with pink or purple. Capsules 2.5 cm long, fusiform, conspicuously ribbed.

Fls.: June. *Exsiccata*: SUK-335, 336.
Common, in moist deciduous forests in shady places.
Localities: Amboli, Dajipur, Gaganbawada, Kolhapur, Panchgani, Radhanagari.

Local uses: The tuber is given in gas troubles and in acidity. It is also given internally in backaches and also used as a food. The tuber is used as appetizer, useful to tuberculous glands in neck, tumors and in bronchitis.

Eulophia ochreata Lindl. in J. Linn. Soc. 3: 24. 1858; Hook. f. Fl. Brit. India 6: 2. 1890; Cooke, Fl. Pres. Bombay 3: 196. 1958 (Repr. ed.); Lakshminarasimhan in Sharma *et al.* Fl. Maharashtra State Monocot. 29. 1996. AMARKAND.

Terrestrial, tuberous herbs. Leaves 2–5, arising from the base of the pseudobulb, forming a pseudostem, 8–23 cm long. Leaves oblong-lanceolate to ovate or ovate-elliptic, 13–28 × 4–10 cm, sheathing at base, acute, entire, glabrous. Scape appearing with the leaves from the pseudostem, 25–45 cm long, with 2–3 sheaths; the latter are about 2–4 cm long, purplish-brown, acute, somewhat inflated. Flowers in dense racemes, clustered at the top of the scape, about 1–1.5 cm across. Bracts 4–13 × 1–2 mm, narrowly linear, acute, entire, 3-nerved, pale green; pedicles with ovary 10–13 mm long, slightly curved, twisted, faintly yellowish-green. Sepals 10–14 × 2–3 mm, yellowish-green with purple or brown spots on the inner surface near the base, linear-lanceolate, subacuminate or apiculate, entire, 1-nerved, the dorsal slightly broader. Petals broadly ovate-oblong. Capsules 28–12 mm long; broadly ovoid, deflexed, strongly ridged, green.

Fls.: June–July; *Frts*: August–October. *Exsiccata*: SUK-1789.
Occasional on hill slopes.
Localities: Kartikswami Ghat, Panhala.

Local uses: The tuber is given in gas troubles and in acidity. It is also given internally in backaches and also used as a food. The tuber is used as appetizer, useful to tuberculous glands in neck, tumors and in bronchitis.

Flickingeria macraei (Lindl.) Seid. in Dansk. Bot. Ark. 34: 39. 1980; Lakshminarasimhan in Sharma *et al.* Fl. Maharashtra State Monocot. 31. 1996. *Denbrobium macraei* Lindl. Gen. Sp. Orchid. 75. 1830; Hook. f. Fl. Brit. India 5: 714. 1890 p.p.; Cooke, Fl. Pres. Bombay 3: 182. 1958 (Repr. ed.). JIVANTI.

Stems long, creeping, annulate, profusely branching; branches polynodal, more or less trigonous or terete, smooth, shining, yellowish-brown, ending in pseudobulb. Leaf, solitary sessile, ovate-lanceolate or oblong-elliptic, 4–10 × 1–2.5 cm, acute, obtuse or rarely subretuse. Flowers one or two, ovary with short pedicel 5–6 mm long, subclavate, pale yellow, ridged, separated from the stalk by the bract. Sepals white, very slightly suffused with pink and yellow, ovate-oblong, acute. Capsules 3–5 cm long, oblong, yellow, ridged.

Fls.: July–August. *Exsiccata*: SUK-1788.
Occasional, in moist deciduous forests
Localities: Amboli, Patgoan, Ramtirtha.

Local uses: The plant is given as tonic and as an aphrodisiac and expectorant. It is useful in asthma and bronchitis. The plant is prescribed as a blood purifier. It is given internally in low fevers. Fruit is used as an aphrodisiac.

ZINGIBERACEAE Lindl.

Amomum pterocarpum Thw. Enum. Pl. Zeyl. 317. 1861; Hook. f. Fl. Brit. India 6: 241. 1892; Lakshminarasimhan in Sharma *et al.* Fl. Maharashtra State Monocot 68. 1996. *A. microstephanum* Baker in Hook. f. op. cit. 239; Cooke, Fl. Pres. Bombay 3: 239. 1958 (Repr. ed.).

Perennial herbs, 60–100 cm in height. Leaves linear-lanceolate; 50–60 × 5–10 cm, acuminate, glabrous above, silky pubescent beneath, base tapering. Flowers white, in globose shortly pedunculate spikes, 5 cm in diam. Calyx tubular, 1.9–2.2 cm long, membranous, 3 toothed, split down one side. Corolla tube shorter than the calyx; lobes linear-oblong, half as long as the tube. Lip 2.5 cm long, obovate. Anthers glabrous, 1.5 cm long, crest small, orbicular. Capsules 9-ribbed, echinate.

Fls. & Frts.: May–June. *Exsiccata*: SUK-1787.
Rare, in moist situations along streams in semi evergreen and moist deciduous forests.
Localities: Amboli, Sawantwadi.

Local uses: The seeds act as a stomachic. Decoction of the seeds is used as a gargle in affections of the teeth and gums. It is used as a diuretic in cases of gravel of the kidneys.

Costus speciosus (Koen.) J. E. Sm in Trans. Linn. Soc. 1: 249. 1800; Baker in Hook. f. Fl. Brit. India 6: 249. 1892; Cooke, Fl. Pres. Bombay 3: 243. 1958 (Repr. ed.); Lakshminarasimhan in Sharma *et al.* Fl. Maharashtra State 70. 1996. *Banksia speciosa* Koen. *apud.* Retz. Obs. Bot. 3: 75. 1783. KOSHT, PIWAY, RATANJYOT.

Perennial herbs with long leafy stems; 2–3 m tall; rootstock tuberous. Leaves oblong or oblanceolate, 15–30 × 3–7 cm, subsessile, acute or acuminate, often cuspidate, glabrous above, silky pubescent beneath, base rounded. Flowers white, numerous, in very dense spikes, 5–12.5 × 3–7.5 cm; bracts ovate, acuminate, bright-red. Calyx lobes 1.9 cm long, deltoid-ovate, cuspidate. Corolla tube as long as the calyx; lobes ovate-oblong, apiculate. Lip suborbicular, white with yellow center, concave, plicate, crisped. Stamens with a tuft of hairs at the base of the filament; connective petaloid, ca 1.2 cm broad, pubescent, produced into a glabrous appendage as long as the linear anther cells. Capsules globosely 3-gonous ca 2 cm in diam., red. Seeds black with a white aril.

Fls. & Frts.: August–February. *Exsiccata*: SUK-104.
Common, in moist deciduous and semi evergreen forests.

Localities: Amboli, Dajipur, Gaganbawada, Radhanagari, Tillari, Vaibhavwadi.

Local uses: Rhizome is given as a tonic. Decoction of rhizome is given in inflammations. It is also used as an anthelmintic. The rhizome paste is applied externally on skin to remove excess pigmentation on skin after pregnancy. The rhizome is used as adulterant of *Gloriosa superba*.

Curcuma longa L. Sp. Pl. 2. 1753, pro max parte; Baker in Hook. f. Fl. Brit. India 6: 214. 1890; Cooke, Fl. Pres. Bombay 3: 238. 1958 (Repr. ed.); Lakshminarasimhan in Singh *et al.* Fl. Maharashtra State 77. 1996. HALAD, HALDI.

Herbs tall; rootstock large, ovoid, with sessile cylindric tubers orange inside. Leaves oblong-lanceolate, apex caudate-acuminate, tapering at base. Flowers in terminal spikes, 10–15 × 5 cm; peduncles ca 15 cm long; flowering bracts pale green, bracts of the coma tinged with pink. Corolla white.

Fls. & Frts.: June–November. *Exsiccata*: SUK-1783.

Cultivated.

Local uses: The rhizome is used in various skin diseases and used as blood purifier. It is applied to promote the eruption of exanthematous fevers. It is also used externally in scabies and the eruption of small pox. The rhizome in combination with the *Ocimum santum* leaf is applied on forehead to stop the headache. It is also applied as a face pack and used to remove boils and to soften the skin. Rhizome powder is applied externally in muscle pain or in contraction of muscles.

Hitchenia caulina (Grah.) Baker in Hook. f. Fl. Brit. India 6: 224. 1890; Cooke, Fl. Pres. Bombay 3: 233. 1958 (Repr. ed.); Lakshminarasimhan in Sharma *et al.* Fl. Maharashtra State Monocot. 79. 1996. *Curcuma caulina* Grah. Cat. Pl. Bombay 210. 1839. CHAVA, CHOWAR.

Annual herbs; rhizomes tuberiferous fibers; tubers white inside; stem leafy, 1–1½ m high. Leaves oblong-lanceolate, 30–50 × 7.5–15 cm, acuminate, narrowed at base, waved, strongly nerved with sheathing petioles. Flowers yellow or white in terminal spikes 12–22 cm long, and ca 6.5 cm in diam., bracts greenish and greenish-white, sometimes tinged with red, obovate-cuneate. Calyx thinly membranous, tubular, 3-fid. Corolla tube 3.5 cm long, slender; lobes thinly membranous, one larger than the other two. Lateral staminodes petioled, 1.2 cm long. Lip 2-fid with a yellow band at the division. Anther cells oblong, spurred at the base. Ovary 3-celled; stigma 2-lipped, funnel shaped, ciliate at the top, projecting beyond the anther. Seeds ovoid, with lacerate aril.

Text fig. 46: Hitchenia caulina a-twig; b-flower; c-front view of flower.

Fls. & Frts.: July–October. *Exsiccata*: SUK-1784.
Common or abundant, in crevices of rocks on plateaus of higher altitudes.
Localities: Kas, Mahabaleshwar.

Local uses: Tubers are given in general debility with milk.

Zingiber neesanum (Grah.) Ramam. in Sald. & Nicols. Fl. Hassan Dt. 769. 1976; Lakshminarasimhan in Sharma *et al.* Fl. Maharashtra State Monocot 85. 1996. *Alpinia neesana* Grah. Cat. Pl. Bombay 207. 1839. *Zingiber macrostachyum* Dalz. in Hook. Kew J. Bot. 4: 342. 1852; Baker or Hook. f. Fl. Brit. India 6: 247. 1892; Cooke, Fl. Pres. Bombay 3: 241. 1958 (Repr. ed.). NISAM.

Herbs; stem lefty reaching 50 cm high, red, pubescent. Leaves lanceolate, 12–30 × 2.5–5 cm, acuminate, dark green and glabrous above, paler and pubescent beneath. Flowers in long cylindric spikes, 15–30 cm long; bracts obovate, acute, reddish, 3.5 cm long. Corolla white or greenish-white; tube as long as the bracts; lobes lanceolate, 2.5 cm long. Lip obovate, 3-lobed; midlobe rounded, emarginate, pale yellow marked with diverging purple lines. Capsules obovoid, pubescent red. Seeds dark purple with large white aril.

Fls. & Frts.: July–September. *Exsiccata*: SUK-398.

Common, in moist deciduous or semi evergreen forests.

Localities: Amboli, Gaganbawada, Mahabaleshwar, Panchgani, Radhanagari, Tillari.

Local uses: It is given in cough and asthma. The rhizome is externally applied in itching. It is applied externally after leech bite. It is also used as diuretic and in piles.

Zingiber officinale. Rosc. in Trans. Linn. Soc. 8: 348. 1807; Hook. f. Fl. Brit. India 6: 246. 1892; Cooke, Fl. Pres. Bombay 3: 242. 1958 (Repr. ed.); Lakshminarasimhan in Sharma *et al.* Fl. Maharashtra State Monocot. 86. 1996. ADRARK, ALE, SUNT.

Herbs with slender lefty stems 1–1.2 cm high. Rhizome horizontal, branched creeping with brown scale leaves. Leaves sessile, lanceolate-linear, 10–25 × 1.5–3 cm, narrowed to the base, acute or acuminate; sheaths 10–15 cm long, rather loose, glabrous. Flowers not seen.

Fls. & Frts.: Not seen. *Exsiccata*: SUK-1785.

Native to tropical Asia, cultivated on a small scale.

Local uses: Bruised rhizome is mixed with jaggery and *Allium sativum* and given to the patients suffering from low blood pressure. Rhizome made into paste with honey or with sugar and small pills made from the mixture is given in cough. The fresh juice is given as a diuretic. Juice is made little hot and poured into ear to stop the pain.

Zingiber purpureum Rosc. in Trans. Linn. Soc. 8: 348. 1807; Lakshminarasimhan in Sharma *et al.* Fl. Maharashtra State Monocot. 85. 1996. *Z. cassumunar* Roxb. in Asiat. Res. 11: 347, t. 5. 1810; Baker in Hook. f. Fl.

Brit. India 6: 248. 1890; Cooke, Fl. Pres. Bombay 3: 242. 1958 (Repr. ed.). MALBARI HALAD.

Herbs; root stock perennial, yellow inside; stem leafy, 1.2–1.5 m tall. Leaves lanceolate, 22–35 × 2–4 cm, acute, base lightly rounded, glabrous above, pubescent beneath. Flowers in dense fusiform or oblong-ellipsoid spikes, 10–16 × 3–3.5 cm. Bracts 2.5–3.5 cm long and as broad as long; broadly ovate, subacute, bright red or greenish-red, or green pubescent and with narrow membranous margins. Calyx 1.9 cm long, membranous, truncate, glabrous, split halfway down. Corolla tube 2.5 cm long, slender glabrous; lobes lanceolate, yellow, with crisped margins. Stamens yellowish-white, shorter than the lip, the appendix of the connective long, flexuous. Style glabrous; stigma obconic, ciliate. Capsules subglobose, 1.5 cm in diam., membranous. Seeds many, minute, purple.

Fls. & Frts.: July–August. *Exsiccata*: SUK-1781.
Rare, in semi evergreen forests.
Localities: Sawantwadi.

Local uses: It has similar reputation to the official ginger. It is given internally in case of diarrhoea.

Zingiber zerumbet (L.) Rosc. ex J. E. Sm. Exot. Bot. 2: 103, t. 112. 1804; Baker in Hook. f. Fl. Brit. India 6: 247. 1892; Cooke, Fl. Pres. Bombay 3: 241. 1958 (Repr. ed.); Lakshminarasimhan in Sharma *et al.* Fl. Maharashtra State Monocot. 86. 1996. *Amomum zerumbet* L. Sp. Pl. 1. 1753.

Perennial herbs, root stock large, yellow inside, with strong aromatic ginger like taste; stem leafy, 1–1.5 m tall. Leaves oblong-lanceolate or oblanceolate, acuminate, glabrous, base narrowed. Flowering stems 30–45 cm long, flexuous, clothed with long appressed obtuse sheaths. Flowers pale sulphur-yellow, in conical-oblong or ovoid-obtuse spikes, 7.5–10 × 5 cm. Bracts closely imbricate, ovate-oblong or obovate, with rounded, red in fruit. Calyx tube 2.5 cm long, appressed to the corolla tube, 3-toothed, glabrous. Corolla tube 3 cm long; lobes ovate-lanceolate, acuminate, Lip shorter than the corolla-lobes and of a dark yellow, 3-fid; lobes obtuse. Anther glabrous, style glabrous; stigma minute, funnel shaped with ciliate mouth. Capsules ellipsoid, 2.5 cm long. Seed 4 mm long, oblong, black.

Fls. & Frts.: July–October. *Exsiccata*: SUK-359.
Rare, along edges of forests and also cultivated.
Localities: Amboli, Gaganbawada, Sawantwadi, Tillari, Vaibhavwadi.

Local uses: The rhizome is used like the official ginger. It is given in cough, asthma. The rhizome is externally applied in itching. It is applied externally after leech bite. It is also used as diuretic and in piles.

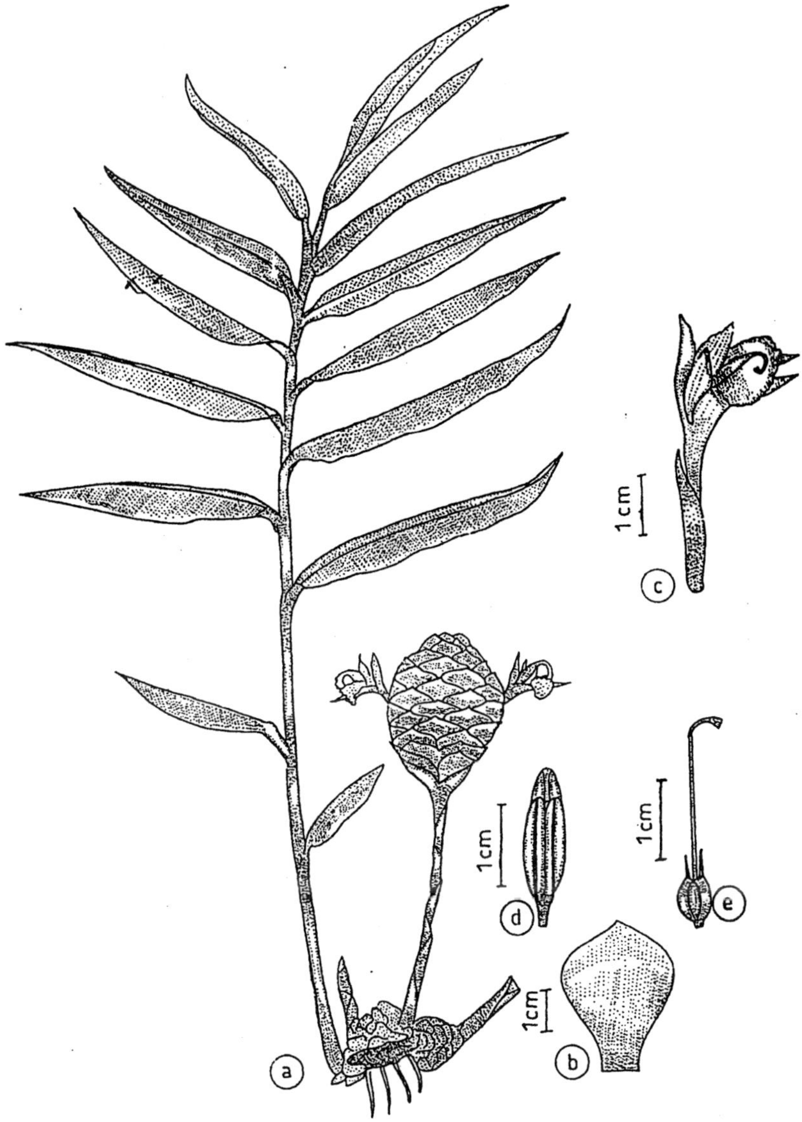

Text fig. 47: Zingiber zerumbet a-habit; b-bract; c-flower; d-stamen; e-carpel.

MUSACEAE Juss.

Ensete superbum (Roxb.) Cheesm. in Kew Bull. 1947; 100. 1958; Lakshminarasimhan in Sharma *et al.* Fl. Maharashtra State Monocot. 89. 1996.

Musa superba Roxb. Pl. Cor. t. 223. 1811 & Pl. Ind. 1: 667. 1832; Baker in Hook. f. Fl. Brit. India 6: 261. 1892; Cooke, Fl. Pres. Bombay 3: 247. 1958 (Repr. ed.). RANKEL.

Stout, perennial herbs, 1–3 m tall; trunk not stoloniferous. Leaves oblong, 0.5–2 × 0.2–0.3 m, narrowed at base, suddenly acuminate; petioles 20–40 cm long on long sheaths forming pseudostem. Flowers in 2 dense rows of 10–15 each, arranged in spikes which are at first globose, 30 cm in diam., drooping and 1/3 the length of the trunk. Bracts orbicular, reaching 30 cm long and broad, dull claret brown. Calyx lobes 3, linear, loosely cohering. Petals 1.2 cm long, pellucid, apex 3-lobed. Fruit oblong, subcoriaceous, 7.6 × 3.5 cm. Seeds numerous, subglobose, angled by pressure, 0.8–1.2 cm in diam., smooth brown or black.

Fls. & Frts.: July–January. *Exsiccata*: SUK-700.

Frequent, on steep rocky slopes in semi evergreen forests.

Localities: Amba, Amboli, Gaganbawada, Radhanagari, Sawantwadi, Vaibhavwadi.

Local uses: Ash of stem is mixed with 4–5 drops of castor oil and given to expel worms from the stomach.

HYPOXIDACEAE R. Br.

Curculigo orchioides Gaertn. Fruct. 1: 63; t. 13. 1788; Hook. f. Fl. Brit. India 6: 279. 1882; Cooke, Fl. Pres. Bombay 3: 255. 1958 (Repr. ed.); Lakshminarasimhan in Sharma *et al.* 102. 1996. KALIMUSALI.

Perennial herbs with elongate, cylindric rootstock up to 30 cm long, root fleshy fibers. Leaves basal, sessile or narrowed into short or long petioles, linear-lanceolate, 15–100 × 1.5–4.5 cm, plicate, acuminate; sheaths 5–10 cm long, membranous. Flowers distichous on a short, clavate, flattened scape; the lowest flowers in the racemes bisexual, upper ones male; bracts lanceolate, 1.5–4 cm long, acuminate. Perianth yellow; segments elliptic oblong, united into a short tube, free parts 10–20 × 3–6 mm, ciliate at apex, otherwise glabrous. Stamens 6 shorter than the perianth-segments. Ovary densely hairy; stigmas 3. Capsules oblong, 10–12 mm long, hairy. Seeds black, ovoid with an appendage above the funicle vertically striated.

Fls. & Frts.: June–November. *Exsiccata*: SUK-214.

Common, in hill forests, on rocky soils.

Localities: Gaganbawada, Kas, Mahabaleshwar.

Local uses: Decoction of roots with milk is given in gonorrhea and dysuria. It is also used as an aphrodisiac, useful in piles and pains in joints. The roots are given in biliousness, fatigue, diseases of the blood.

AMARYLLIDACEAE J. St. Hill

Crinum asiaticum L. Sp. Pl. 292. 1753; Hook. f. Fl. Brit. India 6:280. 1893; Cooke, Fl. Pres. Bombay 3: 256. 1958 (Repr. ed.); Lakshminarasimhan in Sharma *et al.* Fl. Maharashtra State Monocot. 94. 1996. NAGDAVANA.

Stout perennial herbs with under ground bulbs, 5–15 cm in diam., narrowed into 15–30 cm long neck and clothed with old leaf sheaths. Leaves radical, many, linear-lanceolate, 6–120 × 12–20 cm, narrowed at sheathing base, shortly acuminate, bright green; sheaths membranous, 15–20 cm long. Flowers in simple, 15–20 flowered umbels; scapes stout, 45–60 cm long; bracts membranous, ovate-lanceolate, 6–10 cm long; perianth tube greenish-white, cylindric, slender; lobes white, nearly as long as the tube, deflexed. Stamens 6, shorter than the perianth anthers linear, 1.2–2 cm long, red brown, fruits subglobose, 2.4 cm in diam., beaked. Seeds solitary or 2, rounded, black.

Fls. & Frts.: August–December. *Exsiccata*: SUK-699.
Cultivated in gardens.
Localities: Kolhapur.

Local uses: The bulb is used in night blindness and defective visions. It is also used to wash the eyes. The seeds are used as purgative, diuretic, emenagogue. The leaves are applied in local inflammations. Bulb antipyretic, anthelmintic, useful in biliousness, strangury, vomiting, urinary discharges, tumors, diseases of the vagina, abdomen and blood complaints.

Crinum latifolium L. Sp. Pl. 291. 1753. Hook. f. Fl. Brit. India 6: 283. 1892; Cooke, Fl. Pres. Bombay 3: 258 (Repr. ed.); Lakshminarasimhan in Sharma *et al.* Fl. Maharashtra State Monocot. 95. 1996.

Stout perennial herbs with underground bulbs 7–12 cm in diam., neck short, stout. Leaves numerous, radical, lorate, 60–90 × 7–12 cm, acuminate, bright green, margins scabrous. Scape inserted on the neck of the bulb as long as the leaves, stout, tinged with purple down the middle, in 10–20 flowered umbels; bracts oblong lanceolate, 7.5–10 cm long. Perianth tube 7.6–15 cm long, curved, cylindric; lobes oblong-lanceolate, 7–10 × 2.5 cm, acute, much longer than the stamens. Stamens declinate, much shorter than the perianth lobes; filaments 5–7.5 cm long; anthers 1.2–2 cm long, gray. Style longer than the stamens. Ovary with 5–6 superposed ovules in each cell. Fruit subglobose, 3.5–2.5 cm in diam.

Fls. & Frts.: April–June. *Exsiccata*: SUK-698.
Rare, in forests and also cultivated.
Localities: Ratnagiri, Ramghat.

Local uses: The bulbs are used in bronchitis and inflammation. It is used externally in rheumatism. The juice of the leaf is used in earache. The crushed and roasted bulb is applied to piles and abscesses to cause suppuration.

TACCACEAE R. Br.

Tacca leontopetaloides (L.) O. Ktze. Gen. Pl. 2: 704. 1891; Lakshminarasimhan in Sharma *et al.* Fl. Maharashtra State Monocot. 108. 1996. *Leontice leontopetaloides* L. Sp. Pl. 313. 1753. *Tacca pinnatifida* J. R. & J. G. A. Forst. Char. Gen. 35 t. 35. 1776; Hook. f. Fl. Brit. India 6: 287. 1892; Cooke, Fl. Pres. Bombay 3: 262. 1958 (Repr. ed.). DEVA-KANDA.

Perennial herbs with subterranean, depressed-globose, white or brownish corm, 5–10 cm in diam. Leaf solitary, 30–90 cm long. Petiole, blade circular in outline, 3 partite, 30–90 cm across; segments variously pinnately divided into linear-lanceolate lobes. Flowers long pedicelled, drooping, 12–15 mm across, at the end of 60–100 cm long, terete, green or brown-mottled scape. Involucral bracts 6–12, oblong-lanceolate, acuminate, white with purple stripes; bracteoles filiform, much longer than the bracts. Perianth subglobose, 1–1.5 cm long, greenish-yellow; lobes connivent, with purple margin. Stamens 6, included. Ovary 3-gonous, 1-locular; fruit, globose, 3 × 2 cm, yellow, 6-ribbed. Seeds many, angular.

Fls. & Frts.: July–September. *Exsiccata*: SUK-166.

Common in open and moist shady places.

Localities: Ajara, Amboli, Dajipur, Sawantwadi, Shiroda, Tillari, Vaibhavwadi.

Local uses: Fruit paste is applied externally as a poultice in ulcerated wounds. Leaf paste made with little salt is warmed over flame and applied externally in plies and smoke of *Embelia basaal* seeds given to affected area.

DIOSCOREACEAE R. Br.

Dioscorea alata L. Sp. Pl. 1033. 1753; Hook. f. Fl. Brit. India 6: 296. 1892; Lakshminarasimhan in Sharma *et al.* Fl. Maharashtra State Monocot. 109. 1996. *D. alata* var. *globosa* (Roxb.) Prain, Bengal Pl. 2: 1067. 1903 [803. 1963 (Repr. ed.)]; Cooke, Fl. Pres. Bombay 3: 267. 1958 (Repr. ed.). GORADV.

Twining, perennial herbs with large subterranean tubers. Stem unarmed; with 4, longitudinal, 0.5–2 mm wide, crispy, membranous wings, glabrous, bearing axially, globose bulbils. Lower leaves alternate, upper often subopposite or opposite, ovate, 6–25 × 3–15 cm, cordate at base, herbaceous, long acuminate, glabrous; male flowers in 1–2-nate, 1–3 cm long, axillary spikes. Perianth creamy white, 1.5–2 mm long; stamens 6, shorter than the perianth. Female flowers in axially, solitary, 12–30 m long, pendulous spikes. Perianth 2–2.5 mm long; staminodes absent; ovary oblong, glabrous. Capsules quadrate-oblong, 2–2.5 cm long, 3-winged.

Fls. & Frts.: September–December. *Exsiccata*: SUK-697.
Commonly cultivated.
Localities: Doddamarg, Sawantwadi.

Local uses: The tuber is used to expel worms from the stomach; it is externally applied in leprosy. It is given orally in piles.

Dioscorea bulbifera L. Sp. Pl. 1033. 1753; Hook f. Fl. Brit. India 6: 296. 1892; Cooke, Fl. Pres. Bombay 3: 265. 1958 (Repr. ed.); Lakshminarasimhan in Sharma *et al.* Fl. Maharashtra State Monocot. 111. 1996. KADVKARANDA, GORKAN.

Twining, perennial herbs with deeply penetrating, hairy root tubers. Stems terete or obscurely winged, with axillary bulbs. Leaves alternate, or the lower opposite, ovate, 5–25 × 3–20 cm, cordate at base, membranous and thin margined, acuminate, cuspidate or caudate. Male flowers in short, slender, axillary, solitary or 2–4-nate, 5–10 cm long spikes; bracts ovate, acuminate, minute. Perianth white or purplish; segments lanceolate, 2–2.5 mm long. Stamens 6; filaments short, pistillode 3-lobed, minute. Female flowers in axially solitary or fascicled, 10–20 cm long spikes. Perianth white; segments 2–2.5 mm long, acute; staminodes 6. Style short; stigmas 3. Capsules quadrately oblong, 2–2.5 cm long, reflexed, 3-winged. Seeds winged at one end only.

Fls. & Frts.: September–January. *Exsiccata*: SUK-59.
Common, in bushes in hill forests.
Localities: Amba, Amboli, Dajipur, Radhanagari, Ratnagiri, Mahabaleshwar.

Local uses: Tuber powder mixed with seeds of *Cuminum cyminum* and sugar or with honey and given in indigestion in stomachache. Bulbils are given orally in acidity. Tuber is used as aphrodisiac, stomachic, anthelmintic, improves appetite and complexion, useful in dyspepsia, urinary discharges, leucoderma, bronchitis, biliousness, piles, tumors, strangury.

Dioscorea hispida Dennst. Schluess. Hort. Malab. 15. 1818; Lakshminarasimhan in Sharma *et al.* Fl. Maharashtra State Monocot. 112. 1996. *D. daemona* non Roxb. 1892; Hook. f. Fl. Brit. India 6: 289. 1892; Cooke, Fl. Pres. Bombay 3: 264. 1958 (Repr. ed.). VASKAND.

Dioecious, twining, perennial herbs with lobed root tubers. Stems prickly leaves 3-foliolate; leaflets obliquely obovate, 7.5–27 × 4–18 cm, cuneate at base, cuspidately caudate-acuminate, villous when young, glabrous with age. Male flowers in dense, cylindric, pedunculate, 5–10 mm long spikes, arranged in simple or branched, 15–20 cm long racemes; bracts ovate-deltoid. Perianth pale green, 2–3 mm long; tube short or none; lobes broadly oblong or orbicular, the 3 outer shorter than the 3 inner. Stamens 6, shorter than the perianth. Female flowers in solitary, distant spikes. Perianth as in male. Staminodes 6. Ovary obovoid. Capsules oblong, 3.5–4.2 × 2.2 × 2.7 cm, 3 winged, subcordate at base, rounded at apex. Seeds winged at base, ca 1.5–1.8 cm long including wing; nucleus oblique.

Fls. & Frts.: July–October. *Exsiccata*: SUK-1793.
Frequent, in hill forests.
Localities: Gaganbawada, Nesari.

Local uses: Tubers are given orally with tincture of sugar in piles. Tubers are applied externally in glandular swellings.

Dioscorea oppositifolia L. Sp. Pl. 1033. 1753; Hook. f. Fl. Brit. India 6: 292. 1892; Cooke, Fl. Pres. Bombay 3: 266. 1958 (Repr. ed.); Lakshminarasimhan in Sharma *et al.* Fl. Maharashtra State Monocot. 112. 1996. PASPOLI.

Dioecious, twining herbs with cylindric root tubers. Stem slender, smooth, glabrous. Leaves coriaceous, opposite, subopposite or alternate, ovate-lanceolate or broadly elliptic to orbicular, 6–15 × 3.5–9 cm, rounded or sub cordate at base, with well defined cartilaginous margins, acute or caudate-acuminate, glabrous. Male flowers in short, dense, 1–3 cm long axillary spikes which are fascicled or scattered along slender 10–20 cm long axillary rachis. Perianth segments broadly ovate or orbicular, white spotted with brown, the inner slightly smaller than the outer. Female flowers distant in solitary or fascicled, 15–20 cm long, axillary spikes. Capsules sub orbicular, 2.5–3.5 cm in diam., 3 winged, retuse at apex with rounded lobes of wings, cuneate at base, glabrous. Seeds orbicular, 2–2.5 cm in diam., broadly winged with a central nucleus.

Fls. & Frts.: August–October. *Exsiccata*: SUK-146.
Common in hill forests.
Localities: Ajara, Amboli, Borbet, Gaganbawada, Kas, Sawantwadi, Vaibhavwadi.

Local uses: The roots are made into paste, heated and applied to reduce swelling. It is also externally applied in itching.

Dioscorea pentaphylla L. Sp. Pl. 1032. 1753; Hook. f. Fl. Brit. India 6: 289. 1892; Cooke, Fl. Pres. Bombay 3: 264. 1958 (Repr. ed.); Lakshminarasimhan in Sharma *et al.* Fl. Maharashtra State Monocot. 113. 1996. SHENDON-VEL.

Twining, perennial herbs with oblong root tubers. Stems slender, prickly, glabrous. Bulbils often present in the leaf axils. Leaves alternate, 3–5 foliolate; leaflets obovate-elliptic at base, 5.5–17.5 × 2.5–7.5 cm, oblique and acute at base, acuminate or cuspidate, glabrous; petiolules very short. Male flowers in slender, solitary, axillary 2–4 cm long racemes or fascicled along branches 15–30 cm long panicles. Bracts broadly ovate or orbicular, 2–2.5 mm across, membranous, with brown spots. Perianth greenish-white, 1.5–2 mm long; segments ovate; subacute, subequal. Stamens 3 perfect; staminodes 3, minute. Female flowers in axillary, 5–15 cm long, pendulous spikes. Bracts and perianth as in males. Capsules quadrately oblong, 2–2.5 cm long, retuse at both ends, 3 winged. Seeds 10–20 cm long including the basal wings.

Fls. & Frts.: September–December. *Exsiccata*: SUK-147.
Common, in hill forests.
Localities: Gaganbawada, Ramling.

Local uses: The tubers are given orally in debility with milk. Tubers are externally applied in swelling, itch and in other skin diseases.

LILIACEAE Juss. *sensu lato*

Aloe vera (L.) Burm. f. Fl. Ind. 83. 1768; Cooke, Fl. Pres. Bombay 3: 283. 1958 (Repr. ed.); Lakshminarasimhan in Sharma *et al.* Fl. Maharashtra State Monocot. 115. 1996. *A. perfoliata* var. *vera* L. Sp. Pl. 320. 1753. KORPHAD. KUMARI.

Scapigerous herbs, root fibers slender. Leaves fleshy, ensiform, dense aggregate, 20–60 × 5–10 cm, broad and sheathing at base with distant horny prickles along margins, glaucous green, plane on the upper surface, convex beneath. Scapes 60–120 cm tall; simple or branched; branches ascending; racemes 15–40 cm long, dense, bracts ovate, 8–15 mm long, finally reflexed, perianth orange red, 2.5–3.5 cm long, divided half way or little more from the tip, erect, nodding afterwards. Filaments shortly exserted. Capsules ovoid-ellipsoid, 1.5–2 cm long, brown, obtusely trigonous. Seeds many, trigonous black, winged.

Fls. & Frts.: September–January. *Exsiccata*: SUK-314.
Native of N. America, cultivated. Common in dry region of shady area.
Localities: Hatkanagale, Ramling.

Local uses: Leaf with salt is given for 10 days in Hepatitis (Jaundice). Leaves are given in cough. Leaves are also given orally in menstrual problems for 15 days. Dried leaves are given in piles, constipation and in cough. The plant is used as tonic, aphrodisiac, anthelmintic, alexiteric, useful in eye disease, tumors, enlargement of spleen, liver complaints, vomiting, fever due to bronchitis, erysipelas, skin diseases, biliousness, asthma, leprosy, jaundice, strangury and ulcers.

Asparagus racemosus Willd. Sp. Pl. 2: 152. 1799; Hook. f. Fl. Brit. India 6: 317. 1892; Cooke, Fl. Pres. Bombay 3: 270. 1958. (Repr. ed.); Lakshminarasimhan in Sharma *et al.* Fl. Maharashtra State Monocot. 119. 1996. SHATAVARI, ASHWAL.

Extensive, scandent, undershrubs; roots tuberous; branches angular. Leaves reduced to spines, linear-subulate, 4–6 mm long, conical at base, straight or curved. Cladodes slender, spinous pointed. Flowers fragrant, in simple, 5–15 cm long, axillary racemes, rachis triquetrous, crowned with a tuft of cladodes. Pedicels filiform, 3–5 mm long, jointed above the middle, solitary or in pairs along the rachis; bracts ovate, 1–1.5 mm long acute. Perianth white; segments distinct, linear-oblong, 2–3 mm long, obtuse. Anther subglobose. Style short; stigmas spreading. Fruits fleshy, globose, 3–4 mm in diam., red. Seed usually solitary.

Fls. & Frts.: June–November. *Exsiccata*: SUK-86.

Common, on barren, rocky soil under the shade of trees.

Localities: Kolhapur, Jaisingpur, Jat, Panhala.

Local uses: The roots are given to the women's suffering from leucorrhoea and general debility. The tuberous roots are applied externally in pains. The decoction of roots is given as tonic in any kind of disease. Seeds are given to 1–1.1/2 year's children possessing stunted or slow growth. Seeds are also given in foul smell of mouth. The roots are given as stomachic, tonic, aphrodisiac, glactagogue, astringent to bowels; useful in dysentery, tumors, inflammations, biliousness, diseases of the blood and the eye, throat complaints, tuberculosis, leprosy, epilepsy, night blindness.

Chlorophytum bharuchae Ansari, Raghvan & Hemadri in Indian Forester 96: 304; f. 1–10. 1970; Lakshminarasimhan in Sharma *et al.* Fl. Maharashtra State Monocot. 125: 1996. SAFED MUSALI.

Erect, perennial herbs with fleshy, cylindric root, root fibers without tubers. Leaves radical, sessile, linear-lanceolate, 15–50 × 1.5–3 cm, slightly narrowed at base, wavy along margins, acute-acuminate, glaucous green; sheaths membranous, 1–2.5 cm long, scapes usually solitary, 40–90 cm long, naked below, branched; branches ascending. Racemes rather lax, 10–30 cm long; bracts ovate-lanceolate, 5–30 × 2–7 mm, acuminate, with hyaline margins; pedicels solitary or 4–8 together, 5–10 mm long, jointed near the top, elongating in fruit. Perianth white; lobes distinct, oblong-lanceolate, 8–12 × 2–3 mm, with brown midrib prominent on the back. Stamens longer than the perianth; filaments glabrous. Style exserted; stigma papillose. Capsules obcordate, 8–10 mm long, acutely 3-lobed, transversely veined, brown. Seeds discoid, 2–4 mm across, black.

Fls. & Frts.: June–September. *Exsiccata*: SUK-202.

Rare, on rocky hill slopes.
Localities: Appachiwadi, Kagal.

Local uses: The roots are given as stomachic, tonic, aphrodisiac, glactagogue, astringent to bowels; useful in dysentery, tumors, inflammations, biliousness, diseases of the blood and eye, throat complaints, tuberculosis, leprosy, epilepsy and night blindness.

Chlorophytum borivilianum Sant. & Fernand. in J. Bombay nat. Hist. Soc. 52: 898. 1955; Lakshminarasimhan in Sharma *et al.* Fl. Maharashtra State Monocot. 125. 1996. SAFED MUSALI.

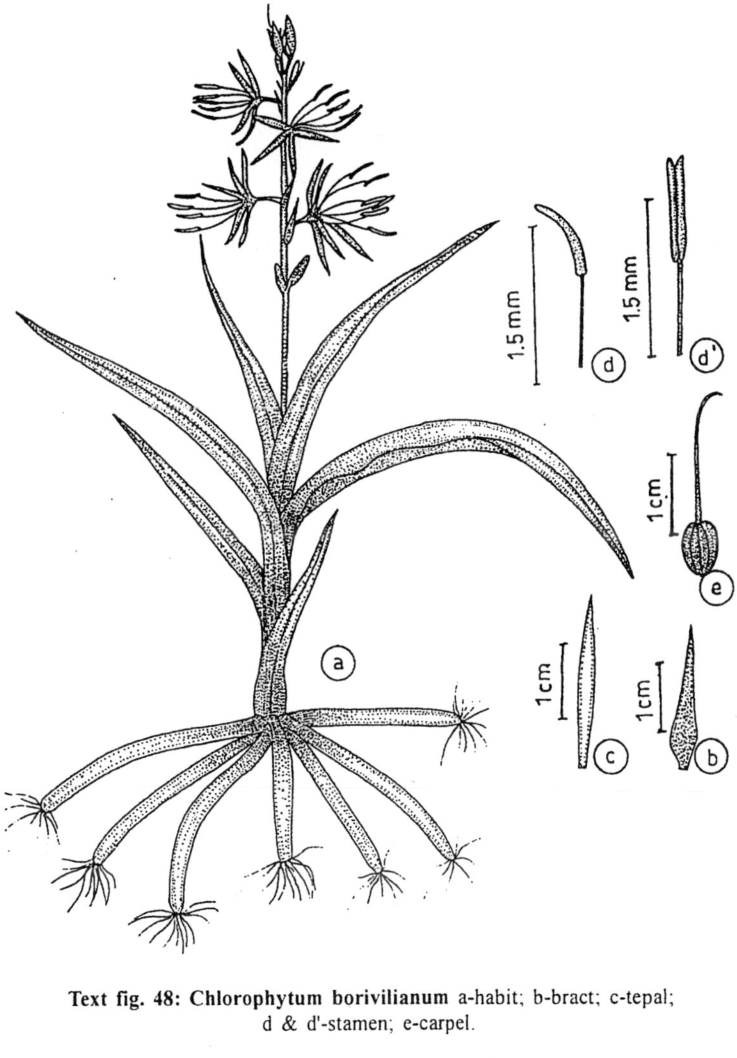

Text fig. 48: Chlorophytum borivilianum a-habit; b-bract; c-tepal; d & d'-stamen; e-carpel.

Herbs, rhizomatous; sessile, root tubers 1–9, up to 5 × 1 cm. Leaves radical, coriaceous, spirally imbricate at base, sessile, 12–18 × 1–5 cm, linear, lorate or ensiform, flat, apex acute. Scape 16–30 cm long. Flowers white, 2.5 cm across, in racemes, bracteate, pedicellate, in cluster of 3; bractes linear, papery, purplish. Capsules loculicidal, greenish-yellow, trigonous, obcordate. Seeds discoid, orbicular, black.

Fls. & Frts.: July–August. *Exsiccata*: SUK-339.
Cultivated, in Kolhapur.

Local uses: The roots are given as stomachic, tonic, aphrodisiac, glactagogue, astringent to bowels; useful in dysentery, tumors, inflammations, biliousness, diseases of the blood and the eye, throat complaints, tuberculosis, leprosy, epilepsy and night blindness.

Drimia indica (Roxb.) Gessop in J. S. Afr. Bot. 43: 272. 1977; Lakshminarasimhan in Sharma *et al.* Fl. Maharashtra State Monocot. 134. 1996. *Scilla indica* Roxb. Fl. Ind. 2: 147. 1824. *Urginea indica* (Roxb.) Kunth, Enum. Pl. 4: 333. 1843; Hook. f. Fl. Brit. India 6: 347. 1892; Cooke, Fl. Pres. Bombay 3: 277. 1958. (Repr. ed.). JANGALI KAND, JANGALI PIAZ.

Herbs; bulbs 6–8 × 4–6 cm, ovoid. Leaves linear lanceolate or lorate, ensiform, hysteranthus, apex acute. Scapes 17–100 cm long, purplish brown. Racemes 4–30 flowered. Flowers reddish-green or variable. Capsule ellipsoid. Seeds 4–10 per locule, black, wing scarious.

Fls. & Frts.: February–May. *Exsiccata*: SUK-1665.
Common, in open grasslands and sandy soils.
Localities: Phonda, Ratnagiri.

Local uses: The bulbs are used in cough and asthma. It is given in very small quantity. In large quantity they may cause vomiting. It is given as cardial exhilerant. The bulb is used as stomachic, diuretic, emmenagogue, anthelmintic, bronchitis, asthma, dropsy, rheumatism, renal calculi, leprosy, skin diseases, headache, diseases of the nose and internal pains and scabies.

Drimia polyantha (Blatt. & McC.) Ansari & Raghavan in J. Bombay nat. Hist. Soc. 77: 174. 1980; Lakshminarasimhan in Sharma *et al.* Fl. Maharashtra State Monocot. 135. 1996. *Urginea polyantha* Blatt. & McC. Ibid. 32: 735. 1928. PACHANKANDA.

Perennial herbs with underground tunicated bulbs and ephemeral aerial shoots. Bulbs ovoid, up to 5 × 4.5 cm, with long neck. Leaves linear, 3–30 × 0.6–2 cm, glaucous green, often curled. Scapes 15–60 cm long, slender; racemes dense, many flowered, 10–25 cm long; bracts deltoid, 1 mm long, persistent; pedicles filiform, spreading, perianth brownish-purple, 5–6 mm long; segments oblong, obtuse, 1-nerved. Stamens about half as long as the perianth. Ovary oblong, style short; stigma 3-lobed. Capsules ellipsoid,

8–10 × 5–6 mm, trigonous. Seeds often only 5–6, ellipsoid, 6–8 mm long, winged shining black.

Fls. & Frts.: March–April. *Exsiccata*: SUK-1665.
Rare, on gravelly slopes of hills.
Localities: Kas, Panchgani, Thoseghar.

Local uses: The bulbs are used in cough and asthma. It is given in very small quantity. In large quantity they may cause vomiting. It is given as cardial exhilerant. The bulb is used as stomachic, diuretic, emmenagogue, anthelmintic, bronchitis, asthma, dropsy, rheumatism, renal calculi, leprosy, skin diseases, headache, diseases of the nose, internal pains and scabies.

Gloriosa superba L. Sp. Pl. 305. 1753; Hook. f. Fl. Brit. India 6: 358. 1892; Cooke, Fl. Pres. Bombay 3: 274. 1958 (Repr. ed.). Lakshminarasimhan in Sharma *et al.* Fl. Maharashtra State Monocot. 137. 1996. BACHNAG, KALLAVI.

Perennial, glabrous, climbers; root stock fleshy, white, arched, often v-shaped; roots fibrous; aerial stems annual, 2–5 m long, terete. Leaves alternate, opposite or in whorls of 3, ovate-lanceolate, 10–15 × 2–4 cm, rounded or cordate at base, entire, tip ending in a tendril like spiral. Flowers large, axillary, solitary or subcorymbose towards the ends of branches; pedicels 5–15 cm long, tips deflexed. Perianth segments distinct, linear-lanceolate, 5–6 × 0.8–1.2 cm, greenish-yellow turning red, with crispy waved margins, strongly reflexed. Filaments 3–4 cm long, spreading; anthers linear 1–1.2 cm long. Style long, trifid. Capsules fusifrom or oblong, 4–5 × 1.5–2 cm. Seeds many, subglobose, 1–1.5 mm in diam., gray.

Fls. & Frts.: September–November. *Exsiccata*: SUK-25.
Common, in hill forests admist bushes.
Localities: Amba, Amboli, Borbet, Dajipur Gaganbawada, Sawantwadi.

Local uses: The tuberous root are reduced to a paste and applied to the naval, suprapubic region and vagina with the object of promoting labour. In retained placenta, a paste of the root is applied to the palms and soles. While entire tubers are tied around the back during labour. The tuber is used as anthelmintic, laxative, alexiteric, abortifacient; used in chronic ulcers, leprosy, inflammation, piles, abdominal pain, itching, thirst; used to remove placenta from the uterus.

Iphigenia indica (L.) A. Gray *apud* Kunth, Enum. PL. 4: 213. 1843; Hook. f. Fl. India 6: 357. 1892: Cooke, Fl. Pres. Bombay 3: 275. 1958 (Repr. ed.); Lakshminarasimhan in Sharma *et al.* Fl. Maharashtra State Monocot. 139. 1996. *Melanthium indicum* L. Mant. 226. 1771. MARKALI

Erect herbs, 7–20 cm tall; under ground corm subglobose, 1–2 cm in diam. Leaves few, sessile, linear, grass like, 8–15 cm long, 2–3 mm wide, sheathing at base, acute or acuminate. Flowers few in terminal erect racemes;

bracts linear, the lower leafy; pedicels 1–2.5 cm long. Perianth segments distinct, purple brown narrow, acute, spreading. Stamens 3–4 mm long; filaments flattened, ovary hairy oblong; styles 3, recurved. Capsules oblong-ellipsoid, 10–15 × 6–7 mm, 3-grooved, obtuse. Seeds many, ellipsoid or subglobose, 2 mm in diam., brown.

Fls. & Frts.: June–September. *Exsiccata*: SUK-355.
Occasional, on hill slopes amidst grasses.
Localities: Gaganbawada, Panhala, Radhanagari.

Local uses: Tubers are externally applied in skin diseases such as itch, elephantiasis. Diluted juice with limewater is given in colic or abdominal pain.

Iphigenia pallida Baker in J. Linn. Soc. Bot. 17: 451. 1879; Hook. f. Fl. Brit. India 6: 357. 1892; Cooke, Fl. Pres. Bombay 3: 275. 1958 (Repr. ed.); Lakshminarasimhan in Sharma *et al.* Fl. Maharashtra State Monocot. 139. 1996.

Erect herbs, 7–12 cm tall; corm globose, 1.2 cm in diam. Leave usually 4, linear, grass like, 7–10 × 0.3–0.6 cm, upper smaller. Flowers 1–4 white or tinged with purple, corymbose. Perianth–segments 4–6 × 2.5 mm, oblanceolate, acute, many-nerved. Stamens 2.5–3 mm long; filaments flattened; anthers 2.5 mm long, verstaile. Ovary obovoid; styles 3, recurved, 1 mm long. Capsules obovoid, 8 mm long. Seeds many, ellipsoid.

Fls. & Frts.: June–September. *Exsiccata*: SUK-345, 356.
Common, in grasslands with rocky substratum.
Localities: Kagal, Kas, Panhala, Satara, Shivaji University campus.

Local uses: Tubers are crushed and mixed with one tablespoon of lime-juice and boiled in half cup of water for 1/2 hours and then applied externally in leprosy.

Iphigenia stellata Blatt. in J. Bombay nat. Hist. Soc. 32: 734. 1928 *emend.* Ansari & Rolla Rao in Bull. Bot. Surv. India 15: 120, f. 1–7b. 1972; Lakshminarasimhan in Sharma *et al.* Fl. Maharashtra State Monocot. 140. 1996.

Herbs with tunicated bulb, 7–10 cm tall. Bulb globose. Leaves linear, 7–10 × 0.2–0.3 cm, apex acuminate, glabrous. Flowers white or tinged with purple in corymbose inflorescence; bracts linear, foliaceous, pedicels ascending, 0.6–3.5 cm long. Perianth lobes 6, oblanceolate, acute. Stamens 6; filaments flattened; anthers oblong, versatile. Ovary obovoid; style 3, recurved. Capsule obovoid with persistent style. Seed subglobose brown.

Fls. & Frts.: July–August *Exsiccata*: SUK-357.
Common, on plateaus of higher altitude.
Localities: Kas, Kolhapur, Panhala, Thoseghar.

Local uses: Tubers are crushed and mixed with one tablespoon of lime juice and boiled in half cup of water for 1 and ½ hours and then applied externally in leprosy.

Scilla hyacinthina (Roth) McBride in Contrib. Gray herb. 2, 56: 14. 1918; Lakshminarasimhan in Sharma *et al.* Fl. Maharashtra State Monocot. 141. 1996. *Ledbouria hyacinthina* Roth, Nov. Pl. Sp. 195. 1821. *Scilla indica* (Wight) Baker in Saund. Refug. Bot. 3: App. 12. 1870 and J. Linn. Soc. Bot. 13: 250. 1872 *non* Roxb. 1832 (Which is *Urginea indica* Roxb.); Hook. f. Fl. Brit. India 6: 348. 1892; Cooke, Fl. Pres. Bombay 3: 276. 1958 (Repr. ed.).

Perennial herbs, 10–20 cm tall; bulbs ovoid, 1.5–5 × 3 cm, bitter to taste. Leaves few very variable, linear to lanceolate, ovate obovate or oblanceolate, narrowed at base, entire or undulate, obtuse, green or with brown spots on the upper surface. Scapes 6–20 cm tall, branched; bracts obovate, 1 mm long, serrate. Perianth persistent, greenish-pink, 4–5 mm long; segments oblong, obtuse. Stamens shorter than the perianth; filaments deep purple; anthers oblong. Ovary globose on short stipe; style short; stigma curved. Capsules ovoid-globose, 5 mm in diam., unequally 3-lobed. Seeds 3, obovoid, 3 mm long.

Fls. & Frts.: May–July. *Exsiccata*: SUK-82, 384.
Common, in rocky soil of hill slopes and grasslands of plains.
Localities: Kas, Kartikswami Ghat, Nesari, Pachgani, Toseghar.

Local uses: Decoction of bulb is given as cardiac tonic. Paste of bulb is given in urinary complaints.

SMILACACEAE Vent.

Smilax zeylanica L. Sp. Pl. 1429. 1753; Hook. f. Fl. Brit. India 6: 309. 1892; Lakshminarasimhan in Sharma *et al.* Fl. Maharashtra State Monocot. 144. 1996. GHOTVEL.

Deciduous, scandent, prickly climber. Leaves alternate oblong, broadly elliptic or ovate-lanceolate, 5–10 × 2–5 cm, coriaceous, 3–5 nerved, glabrous, base obtuse rotund or subcordate, apex obtuse subacute; petiole 1.5 cm long, often twisted; leaf sheath narrow; stipular tendril long and slender. Flowers dioecious, pedicellate, white in umbels; bracts ovate, acute. Male flowers: Perianth lobes 6, free, oblong, greenish-white, obtuse. Stamens 6; filaments ca 5 mm long, flat, anthers oblong. Female flowers: Perianth lobes 6, oblong, shorter than the males, reflexed. Ovary 3-celled; style 3-fid, recurved. Staminodes 3–6, linear. Berries globose, turning red after ripening.

Fls. & Frts.: November–April. *Exsiccata*: SUK-93.
Common, in hilly forests.

Localities: Ajara, Amboli, Dajipur, Dapoli, Gaganbawada, Sawantwadi, Tillari, Vasota.

Local uses: Roots are used in psoriasis.

ARECACEAE Sch.-Sch. (*nom. alt.*)
[**PALMAE** Juss. (*nom. cons.*)]

Areca catechu L. Sp. Pl. 1189. 1753. Becc. & Hook. f. in Hook. f. Fl. Brit. India 6: 405. 1892; Cooke, Fl. Pres. Bombay 3: 319. 1958 (Repr. ed.); Lakshminarasimhan in Sharma *et al.* Fl. Maharashtra State Monocot. 198. 1996. SUPARI.

Trunk solitary, straight 12–30 m high, uniformly thick. Leaves 1.2–1.8 m long; leaflets numerous, 30–60 cm long, upper confluent. Spathes double, compressed. Spadix much branched bearing male and female flowers. Male flowers sessile, numerous. Female flower solitary or 2–3, sessile. Fruits smooth, orange and scarlet, supported by persistent perianth. Seeds in upper part.

Fls. & Frts.: Throughout year.
Cultivated.
Localities: Kolhapur, Malwan, Ratnagiri, Sawantwadi.

Local uses: Nuts are used as aphrodisiac, useful in urinary disorders and given as nervine tonic.

Caryota urens L. Sp. Pl. 1189. 1753; Becc. & Hook. f. in Hook. f. Fl. Brit. India 6: 422. 1892; Cooke, Fl. Pres. Bombay 3: 315. 1958 (Repr. ed.); Lakshminarasimhan in Sharma *et al.* Fl. Maharashtra State Monocot. 189. 1996. BHERLI-MAD.

Trees, 10–12 m tall; trunk stout, annulate, up to 30 cm in diam. Leaves pinnatisect, 5–6 × 2–3 m; leaflets 1.5–2 m long, drooping; ultimate segments broadly cuneate, 10–20 cm long, very obliquely truncate, acutely serrate, toothed, the outer margin caudate; petiole very stout. Flowers greenish in 2–3 m long, slender drooping spadices forming a bunch subtended by 3–5, tubular, 40–50 cm long spathe. Male flowers 1–1.5 cm long; perianth segments orbicular to linear-oblong, valvate. Stamens many; filaments short. Female flowers in between the two lateral males, globose with orbicular perianth segments. Ovary globose, 1.5–2 cm in diam., reddish.

Fls. & Frts.: March–June. *Exsiccata*: SUK-1799.
Common in semi evergreen forests and moist deciduous forests.
Localities: Borbet, Panhala, Thoseghar.

Local uses: The nut is used as an application to the head in cases of hemicrania. Nut is used to allay thirst and fatigue, causes biliousness, flatulence. A glass of the freshly drawn 'toddy' taken early in the morning acts as a laxative.

Phoenix sylvestris (L.) Roxb. Fl. Ind. 3: 787. 1832; Becc. & Hook. f. in Hook. f. Fl. Brit. India 6: 425. 1892; Cooke, Fl. Pres. Bombay 3: 311. 1958 (Repr. ed.); Lakshminarasimhan in Sharma *et al.* Fl. Maharashtra State Monocot. 195. 1996. *Elate sylvestris* L. Sp. Pl. 1189. 1753. SHINDI.

Dioecious trees, 2–10 m tall; trunk clothed with persistent bases of petioles. Leaves crowned at the top of the trunks, pinnatisect, 1–3 m long, grayish-green, glabrous; segments fasciculate, 2–4-farious, rigid, linear, 15–40 × 2–2.5 cm, pungent at apex; petioles 15–30 cm long, compressed towards apex, spinous margined towards base; spines triangular, up to 10 cm long. Spadix 60–90 cm long, splitting into 2, boat-shaped. Flowers 6–8 mm long, numerous, angular, oblique. Outer perianth segments united into a cup with 3 rounded teeth; inner ones 3–4 times longer than the outer, concave. Filaments short, female flowers rather distinct, roundish, in nodding, 60–90 cm long spadices. Fruits oblong-ellipsoid, 2.5–3 cm long, orange-yellow. Seeds oblong, 1–1.5 cm long, pale brown.

Fls. & Frts.: January–May. *Exsiccata*: SUK-1800.
Frequent, in low lying areas along stream banks.
Localities: Jat, Kartikswami, Kolhapur.

Local uses: The fruit is used as cardiotonic, fattening, constipating, aphrodisiac; good in heart complaints, abdominal complaints, fevers, vomiting, wandering of the mind, loss of consciousness. The root is used in toothache.

PANDANACEAE R. Br.

Pandanus furcatus Roxb. Fl. Ind. 3: 744. 1832; Hook. f. Fl. Brit. India 6: 484. 1893; Cooke, Fl. Pres. Bombay 3: 324. 1958 (Repr. ed.); Lakshminarasimhan in Sharma *et al.* Fl. Maharashtra State Monocot. 206. 1996. KEVADA.

Trees; trunk 3–9 m high, branched, with aerial roots from the lower part of the trunk. Leaves 200–450 × 7.5–15.0 cm, linear-lanceolate, apex acute, base caudate-acuminate, margins and midrib armed with short stout spines. Male flowers in compound cylindric spikes, 10–15 × 2–3 cm, densely floriferous; female flowers in solitary spadix; syncarpium ripens orange-red, 15–22 cm long. Drupes obconic with convex crown.

Fls. & Frts.: October–November. *Exsiccata*: SUK-1185.
Common, along streams.
Localities: Sawantwadi, Sindhudurg, Tillari.

Local uses: Oil from the bracts extracted is applied externally in headache. Bracts are inhaled during hemicrania. Paste is applied externally in case of leucoderma.

Pandanus odoratissimus L. f. Suppl. 64. 1781; Lakshminarasimhan in Sharma *et al.* Fl. Maharashtra State Monocot. 206. 1996. *P. tectorius* auct. Plur. non Soland. ex Parkinson, J. Voy. H. M. S. End. 46. 1773; Cooke, Fl. Pres. Bombay 3: 324. 1958 (Repr. ed.). *P. fascicularis* Lam. Encycl. 1: 372, t. 1. 1785; Hook. f. Fl. Brit. India 6: 484. 1893; KEVADA.

Shrubby, up to 6 m high; Stem supported by aerial roots. Leaves glaucous green, 90–150 cm long, ensiform, caudate acuminate, coriacious, marginal spines pointing forward, those on midrib pointing forward or backward. Male flowers: spadix with numerous, subsessile, cylindric spike, 5–10 × 2.5–3.7 cm enclosed in long white fragrant, caudate acuminate spathes. Female flowers: spadix solitary, 5 cm in diam. Syncarpium oblong or globose, yellow or red. Drupes numerous each consisting 5–12 carpels.

Fls. & Frts.: March–December. *Exsiccata*: SUK-909.
Common, along streams.
Localities: Kolhapur, Nerurpar, Ratnagiri.

Local uses: Oil from the bracts extracted is applied externally in headache. Bracts are inhaled during hemicrania. Paste is applied externally in case of leucoderma.

TYPHACEAE Juss.

Typha angustifolia L. Sp. Pl. 971. 1753; Lakshminarasimhan in Sharma *et al.* Fl. Maharashtra State Monocot. 207. 1996. *T. angustata* Bory & Chaub. Exped. Sci. Moree Bot. 2, 1: 338. 1832; Hook. f. Fl. Brit. India 6: 489. 1893; Cooke, Fl. Pres. Bombay 3: 326. 1958 (Repr. ed.). PANKANIS.

Robust, erect, perennial herbs, 1–3 m tall. Leaves linear, 1–3 m long, 1–2 cm wide, planoconvex, entire, acute or obtuse, glaucous; sheaths compressed, 15–20 cm long. Flowers minute, densely crowded in cylindric spikes, male and females. Spikes separated by 2–4 cm long naked stem, males in the upper, yellow spikes, females in the lower, brown spike, subtended by leafy spathes. Stamens 2–3 on a short common stock, inter mixed with linear, spathulate hairs; anthers linear; connective shortly produced. Female flowers intermixed with filiform bracteoles, with ovary raised on a long, slender gynophore, style long; stigma flattened. Fruits minute, obovate, obtuse.

Fls. & Frts.: October–May. *Exsiccata*: SUK-270.
Common, forming extensive beds along river banks and streams.

Localities: Kolhapur, Rankala lake, Kalamba lake.

Local uses: Dried inflorescence of the plant is stuffed in wounds to stop contamination of water and dust and microorganisms and to heal wound. Rootstock is astringent and diuretic.

ARACEAE Juss.

Acorus calamus L. Sp. Pl. 324. 1753; Hook. f. Fl. Brit. India 6: 555. 1893; Cooke, Fl. Pres. Bombay 3: 341. 1958 (Repr. ed.); Lakshminarasimhan in Sharma *et al.* Fl. Maharashtra State Monocot. 227. 1996. VEKHAND.

Herbs, perennial, up to 80 cm tall; rootstock stout, 1.0–1.5 cm broad, creeping with fibrous roots from the lower surface. Leaves 55–100 × 0.8–1.5 cm, spathe leaf like, up 46 cm long. Spadix 5–6.5 cm long, cylindrical, 1.0–1.4 cm broad. Flowers densely arranged. Berries ellipsoid. Seeds few, obconical, 0.2 cm long.

Fls. & Frts.: February–June. *Exsiccata*: SUK-1795.
Cultivated.
Localities: Amboli, Choukul, Sawantwadi.

Local uses: A small piece of the dried rootstock kept in the mouth acts better than many cough lozenges. In small doses is given in asthma. Rhizome in small doses acts as stomachic and in large doses it is emetic, improves appetite, voice throat; good for mouth disease, useful in abdominal pains, inflammations, fevers, epilepsy bronchitis, hysteria, dysentery and tumors.

·**Amorphophallus paeonifolius** (Dennst.) Nicols. var. **campanulatus** (Dence.) Sivadasan in Taxon 32: 130. 1983; Lakshminarasimhan in Sharma *et al.* Fl. Maharashtra State Monocot. 210. 1996. *A. campanulatus* Decene. in Nouv. Ann. Mus. Hist. Paris 3: 366. 1834; Hook. f. Fl. Brit. India 6: 513. 1893; Cooke, Fl. Pres. Bombay 3: 336. 1958 (Repr. ed.). JANGLI SURAN.

Tubers depressed-globose, 20–25 cm in segments spreading, simple or forked, petioles 60–90 cm long, stout; leaflets 5–12 cm long, obovate or oblong, acute, peduncles short stout, elongating in fruit. Spathes campanulate, 15–25 cm broad, greenish-pink externally with pale ocellated blotches, base purple within. Spadix as long as the spathes; appendages varying in size, up to 20 × 12 cm, globose, conoid or amorphous, sinuately lobulate, dark red-purple, spongy within. Male inflorescence subturbinate, ca 7.6 long, 2.5–5 cm in diam. Anthers densely crowded, pale yellow, female inflorescence densely crowded, sessile, depressed globose; style 1.2 cm long, stout, ascending, purple; stigma large 2–3 lobed. Berries red, 2–3 seeded, obovoid.

Fls. & Frts.: June--September. *Exsiccata*: SUK-655.
Common, in shady places or open places.

Localities: Amboli, Doddamarg, Sawantwadi, Tillari, Vaibhavwadi.

Local uses: The corm and the seeds are used as irritants and relieve the pain or rheumatic swellings when applied externally. It is also applied externally in piles. Plant is used as a stomachic, constipating, useful in piles, enlargement of the spleen, tumors, asthma, bronchitis, vomiting, abdominal pain, blood disease, elephantiasis; causes itching sensation; harmful in leprosy and leucoderma.

Arisaema tortuosum (Wall.) Schott & Endl. Meletem. Bot. 1: 17. 1832; Hook. f. Fl. Brit. India 6: 502. 1893; Cooke, Fl. Pres. Bombay 3: 331. 1958 (Repr. ed.); Lakshminarasimhan in Sharma *et al.* Fl. Maharashtra State Monocot. 214. 1996. *Arum tortuosum* Wall. Pl. As. Rar. 2: 10, t. 114. 1830. SARDACHA-JAD.

Herbs, reaching 90 cm high; tubers up to 10 cm in diam., spheroid. Leaves 2–3; petioles 30–90 cm long; leaflets sessile or petiolate, 10–20 cm long, ovate-lanceolate or linear-lanceolate, subcaudately acuminate, distant or crowded or almost radiately arranged. Spathes 10–15 cm long, green outside; tube as long as the limb, subcylindric, gaping, gradually dilating into the limb, pale purplish inside; limb ovate-oblong, acuminate, broadly cymbiform, curved. Spadix 1 or 2 sexual; male flowers stalked; appendage very long, much exserted, tapering like a rat's tail, smooth, erect at 1st, then porrect and again erect. Ovaries ovoid, attenuated into a short style. Berry 4–5 seeded

Fls. & Frts.: May–September. *Exsiccata*: SUK-657.
Frequent, in hilly forests along roadsides.
Localities: Amboli, Dajipur, Kas, Radhanagari, Sawantwadi.

Local uses: It is stated to have poisonous properties. The tubers are used to kill the worms which infect cattle in the rains. Tuber is also applied externally in wounds in cattle.

Cryptocoryne spiralis (Retz.) Fischer ex Wydler in Linnaea 5: 428. 1830; Hook. f. Fl. Brit. India 6: 494. 1893; Cooke, Fl. Pres. Bombay 3: 329. 1958 (Repr. ed.); Lakshminarasimhan in Sharma *et al.* Fl. Maharashtra State Monocot. 217. 1996. *Arum spirale* Retz. Obs. Bot. 1: 30. 1779.

Herbs, rootstock tuberous, sotoloniferous; roots vermiform. Leaves linear-lanceolate, 7.5–20 × 0.15–1.75 cm, acute or acuminate, narrowed from the middle to both ends; spathes subsessile, 7.5–12.5 cm long, linear-lanceolate, greenish outside, dark purple and transversely lamellate within. Fruits subglobose. Seeds 0.5 cm long, spindle shaped, irregularly verrucose rugose.

Fls. & Frts.: August–December. *Exsiccata*: SUK-1770.

Frequent, partially submerged in pools or in margins of ponds and rivers.
Localities: Gaganbawada, Malkapur, Thoseghar.

Local uses: It is given as a remedy for infantile vomiting and cough and in case of adults for abdominal complaints and fever.

Lagenandra ovata (L.) Thw. Enum. Pl. Zeyl. 334. 1864; Lakshminarasimhan in Sharma *et al.* Fl. Maharashtra State Monocot. 217. 1996. *Arum ovatum* L. Sp. Pl. 967. 1753. *Lagenandra toxicaria* Dalz. in Hook. J. Bot. Kew Gard. Misc. India 6: 495. 1893; Cooke, Fl. Pres. Bombay 3: 330. 1958 (Repr. ed.). VATSANAB.

Root stocks reaching 5 cm thick, creeping, simple; root fibers vermiform. Leaves elliptic-oblong, 15–37 × 5–12.5 cm, obtuse or acute, margins undulate, entire, base acute or rounded. Spathes 7.5–22.5 cm long, tubular below; the limb ovate-lanceolate, 2–3.5 cm broad, caudate acuminate, slightly twisted; tube much shorter than the long-tailed limb. Male inflorescence cylindric; anthers crowded, yellow, the cells with tubular tips. Female inflorescence of many ovaries in many cycles, crowded in globose head; stigma sessile, pulvinate, 5 angled; ovule 4–8 on a basal placenta. Syncarpium on short decurved peduncle, globose, 2.5–5 cm in diam., carpels 8 mm long, partially dehiscent. Seeds 9 mm long, narrowly oblong, furrowed.

Fls. & Frts.: February–May. *Exsiccata*: SUK-1773.
Rare, in marshy places along streams.
Localities: Ajara, Borbet, Dhamapur, Patgaon.

Local uses: The rootstock is used in Rheumatism. The plant is said to have insecticidal properties.

Remusatia vivipara (Roxb.) Schott & Endl. Meletem. Bot. 18. 1832; Hook. f. Fl. Brit. India 6: 521. 1893; Cooke, Fl. Pres. Bombay 3: 339. 1958 (Repr. ed.); Lakshminarasimhan in Sharma *et al.* Fl. Maharashtra State Monocot. 220. 1996. *Arum viviparum* Roxb. Fl. Ind. 3: 496. 1832. RUKHAL KAND.

Tubers 2.5–3.7 cm in diam., clustered, depressed, rooting from the crown; bulbiferous shoots 15–30 cm long, simple or shortly branched, bearing at the nodes clusters of oblong, squarrosely scaly bulbils. Spathe 10–12 cm long, coriaceous; tube 2.5–5 cm long, oblong or ovoid, green; limb 5–7.6 cm long, broadly orbicular ovate or ovate-cordate, 5–7.5 cm broad, golden yellow. Spadix 2.5–3.5 cm long; male inflorescence 6.3 mm long, 2–3 stamens; female flowers of densely packed ovaries; neutral flowers mixed with male. Berries obovoid, minute.

Fls. & Frts.: April–September. *Exsiccata*: SUK-388.
Common, in the clefts of trees in Ghats and in jungles.

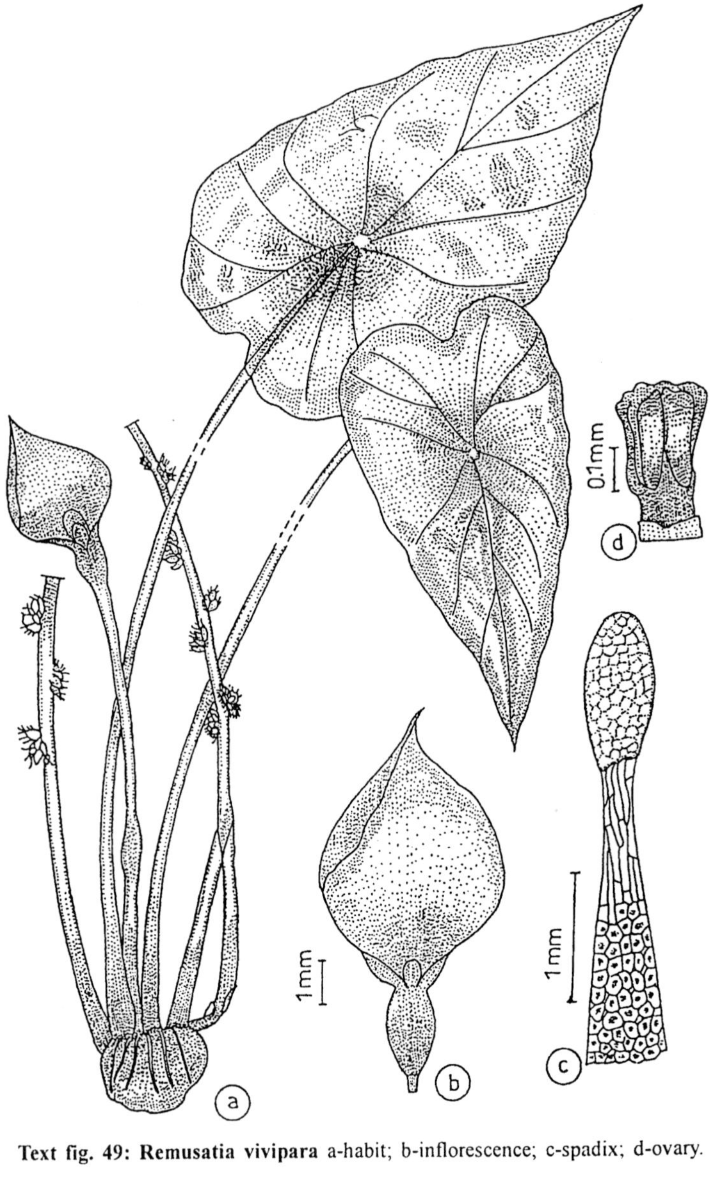

Text fig. 49: Remusatia vivipara a-habit; b-inflorescence; c-spadix; d-ovary.

Localities: Amboli, Barki, Bhudergad, Borbet, Dajipur, Phonda, Tillari, Vaibhavwadi.

Local uses: Tuber is given for external application in piles with the mixture of *Dioscorea bulbifera* bulbils and bark of *Pittosporum dacycaulon*. The root is made into ointment with turmeric and used as a remedy for itch and the juice with cow's urine is considered to be alexipharmic.

Sauromatum venosum (Ait.) Schott, Prodr. Aroid. 71. 1860; Lakshminarasimhan in Sharma *et al.* Fl. Maharashtra State Monocot. 220. 1996. *Arum venosum* Ait. in Hort. Kew ed. 10. 3: 315. 1789. *Sauromatum guttatum* (Wall.) Schott in Schott & Endl. Meletem. Bot. 1: 17. 1832; Hook. f. Fl. Brit. India 6: 508. 1893; Cooke, Fl. Pres. Bombay 3: 335. 1958 (Repr. ed.). LOTH, NURKI, AGYASVRAN.

Tubers large. Leaf pedatisect, solitary; 15–30 cm broad; segments 7–15 variable, 5.0–37. 5 × 2.5–7.5 cm, elliptic, acuminate; petiole stout, 30–45 cm long. Spathe 30–70 cm long, tube ovoid or subglobose; limb linear-lanceolate, thick, green or yellowish, usually with dark, spots or blotches margins generally purple and wavy. Spadix shorter than the spathe; appendage 5–20 cm long, cylindric and obtuse or narrowed from the base to a fine point, green or dark purple. Male inflorescence 1.2 cm long, about 7.5 cm above the female; anthers densely packed. Female inflorescence 1.2 cm long, neuters 1.2 cm long, spreading, close above the female. Berries ob-pyramidal, 1-seeded; seeds 0.18 cm broad ovoid.

Fls. & Frts.: March–May. *Exsiccata*: SUK-1774.
Very common, in shady places along roadsides.
Localities: Ichalkaranji, Kagal, Kartikswami Ghat, Kolhapur, Ramling, Wai.

Local uses: Tubers are externally applied in piles with the castor oil. The tubers are used as a stimulating poultice.

CYPERACEAE Juss.

Cyperus esculentus L. Sp. Pl. 76. 1753; C. B. Cl. in Hook. f. Fl. Brit. India 6: 616. 1893; Cooke, Fl. Pres. Bombay 3: 390. 1958 (Repr. ed.); Lakshminarasimhan in Sharma *et al.* Fl. Maharashtra State Monocot. 285. 1996. NAGARMOTHA.

Perennials, 10–30 cm tall; stolons slender, wiry, often ending in horizontally zoned, ovoid, blackish tubers, 0.8–1.5 cm in diam.; stems triquetrous, 1–1.5 wide, slender. Leaves sheath glabrous, closely distichous; blades linear, 1.5–2 mm wide, coriaceous, glaucous beneath, shorter than or as long as the stems, scabrid, long acuminate. Umbels simple, or sub compound. Involucral bracts 4, 2 often foliaceous; rays 3–6 × 4–6 cm long. Spikes ovoid with 3–8 spikelets; rachis 3–5 mm long, smooth, flexous. Spikelets linear, 1.5–2.5 by 1.5–2 mm, subcompressed, erect, more or less clustered on short rachis or alternately disposed, reddish-brown, bright, obtuse or subacute; rachilla broadly winged. Nuts obovoid, grayish-black, 0.15 × 0.1 cm.

Fls. & Frts.: September–October. *Exsiccata*: SUK-1731.
Occasional, in rice-fields, wet-fields, etc.

Localities: Kolhapur.

Local uses: The bulbous roots powdered taken early in the morning in case of dysuria. Tubers powder is also applied externally to stop hairloss and in graying of hairs and applied externally to the breast as a galactogogue. In pregnancy it is given orally as milk purifier.

Cyperus rotundus ssp. **rotundus** L. Sp. Pl. 45. 1758 C. B. Cl. in Hook. f. Fl. Brit. India 6: 614. 1893; Cooke, Fl. Pres. Bombay 3: 385. 1958 (Repr. ed.); Lakshminarasimhan in Sharma *et al.* Fl. Maharashtra State Monocot. 293. 1996. NAGARMOTHA.

Perennial, 30–50 cm tall; stem slender, trigonous, 1.5–2 mm wide, arising from tuberous base. Leaves sheaths glabrous, stramineous or pale yellowish-brown, loosely distichous; umbels compound or subcompound; involucral bracts 4–5; primary rays 7–10 spikelets. Spikelets linear, 1–2 cm by 2 mm, compressed, erect, pale, to dark reddish-brown, acute; rachilla broadly winged. Glumes elliptic-oblong, 3.5–4 × 2 mm, 5–7 nerved, closely imbricated; keel 3-nerved. Nuts acutely trigonous, ellipsoid, ca 1.5 × 0.4 mm, smooth or faintly reticulate, grayish, apiculate, sessile.

Fls. & Frts.: September–December. *Exsiccata*: SUK-1797.
Common, weed of wet agricultural fields.
Localities: Kagal, Satara, Sindhudurg.

Local uses: The bulbous roots powdered and taken early in the morning in case of dysuria. Tubers powder is also applied externally to stop hairloss and in graying of hairs and externally to the breast as a galactogogue. In pregnancy it is given orally as milk purifier. Tubers are used as appetizer, stomachic, anthelmintic, useful in leprosy, thirst, fevers, blood diseases, biliousness, dysentery, pruritus, pain, vomiting and epilepsy.

POACEAE Barnhart (*nom. alt.*)
[**GRAMINEAE** Juss. (*nom. cons.*)]

Coix gigantea Koen. ex Roxb. Fl. Ind. 3: 570. 1832; Bor, Grass. Ind. 264. 1960; Lakshminarasimhan in Sharma *et al.* Fl. Maharashtra State Monocot. 437. 1996. KASAI.

Annual herbs. Culms terete, 60–150 cm, erect or ascending; nodes glabrous. Leaves sheaths compressed, 4–7 cm long, glabrous; ligule membranous, 1 mm long; blades flat, linear-lanceolate, 8–45 × 0.8–2.5 cm, cordate at base. Racemes spiciform, consisting of one basal female spikelet enclosed in a globose basal bracts; male racemes above the female spikelets or female spikelet absent. Male spikelets or 2–3-nate. Ovary ellipsoid. Grains flat, broadly ovoid.

Fls. & Frts.: October–February. *Exsiccata*: SUK-186.
Common, along stream banks and margins of ponds.
Localities: Kagal, Kolhapur, Rankala lake, Shivaji University campus.

Local uses: Root decoction is given as diuretic. Fruit are boiled in water and given to reduce excess fats from the body.

Cymbopogon citratus (DC.) Stapf in Kew Bull. 1906. 357. 1906; Blatt & McC. Bombay Grass, 103. 1935; Bor, Grass. Ind. 126. 1960; Lakshminarasimhan in Sharma *et al.* Fl. Maharashtra State Monocot. 441. 1996. *Andropogon citratus* DC. Cat. Hort. Monse. 78. 1813. GAVATICHAHA.

Perennial herbs. Culms 50–200 cm high, erect; nodes glabrous. Leaves linear-lanceolate, 15–60 × 0.8–1.5 cm, narrowed at base, glaucous. Panicles lax, 10–50 cm long, Racemes 1–1.5 cm long; joints slender, villous. Sessile spikelets linear-lanceolate, 5–6 mm long. Pedicelled spikelets linear-lanceolate or lanceolate, 4–6 mm long. Occasional cultivated in the region.

Fl. & Frts.: September–December. *Exsiccata*: SUK-276.
Cultivated.
Localities: Amba, Shirol, Kandala.

Local uses: The leaf with that of citrus lemon and tea powder are boiled and taken in dysentery and also in fever. Oil extracted from the leaves is applied on body during fever and common cold.

Cynodon dactylon (L.) Pers. Syn. Pl. 1:85. 1805; Hook. f. Fl. Brit. India 7: 288. 1896; Cooke, Fl. Pres. Bombay 3: 554. 1958 (Repr. ed.); Blatt & McC. Bombay Grass. 249. 1935; Bor. Grass. Ind. 469 f. 52. 1960; Lakshminarasimhan in Sharma *et al.* Fl. Maharashtra State Monocot. 442. 1996. *Panicum dactylon* L. Sp. Pl. 58. 1753. DURVA, HARALI.

Perennial herbs, culms terete, 5–40 cm tall from widely creeping stolons, rooting at the nodes; nodes glabrous. Leaves: sheath compressed; sometime terete, up to 6 cm long, glabrous; ligule a ciliate rim; blades flat, linear-lanceolate, 2–15 × 0.1–0.3 cm, glabrous, glaucous, usually distichous, acute or pungent at apex. Spikes 2–8, whorled, sometimes in two whorls one above the other, 2–10 cm long, erect or spreading, curved; rachis slender. Spikelets sessile, oblong, 1.5–2.5 mm long, laterally compressed, 1-flowered, green or purple. Lower glume thinly membranous, narrowly oblong, 1–1.5 mm long, keeled, acute. Upper glume similar, 1.5–2 mm long. Lemma membranous, boat-shaped, 2.3–2.5 mm long, 3-nerved; keels and margins hispid with white hairs; apex acute. Lodicules 2, small. Stamens 3, small. Grain up to 1.5 mm long.

Fls. & Frts.: Throughout year. *Exsiccata*: SUK-312.
Very common, in irrigated fields on wet lands.

Localities: Kolhapur, Sangli, Satara.

Local uses: Decoction of plant is taken early in the morning in acidity it also acts as a diuretic leaves are brushed and applied externally in itch and wounds. It is used in burning sensation, bad taste in the mouth, hallucinations, epileptic fits, leprosy, scabies skin diseases, dysentery and fever.

Paspalum scrobiculatum L. Mant. Pl. 1: 29. 1767; Blatt & McC. Bombay Grass. 136. 1935; Bor, Grass. Ind. 340. 1960; Lakshminarasimhan in Sharma *et al.* Fl. Maharashtra State Monocot. 561. 1996.

Annual or perennial herbs. Culms tufted, terete, 10–60 cm tall, erect or decumbent or creeping, rooting at lower nodes, simple or branched, glabrous; nodes glabrous. Leaves sheaths compressed, 4–10 cm long, glabrous; ligules membranous, short, blades flat, linear-lanceolate, 4–16 × 0.2–0.7 cm, glabrous; apex acuminate. Racemes 2 or 3 erect or shortly spreading, 2–6 cm long; rachis flattened, keeled on the abaxial side. Grain elliptic.

Fls. & Frts.: August–December. *Exsiccata*: SUK-188.
Frequent, in harvested rice fields and wet lands.
Localities: Amba, Gaganbawada, Vaibhavwadi.

Local uses: Plant is used as tonic, alexiteric, useful for ulcers, causes constipation, flatulence, hallucinations and dysuria. Plant is useful in inflammation, disease of the liver, causes constipation, and heats the body.

Vetiveria zizaniodes (L.) Nash in Small, Fl. S.E. U. S. 67. 1903; Blatt & McC. Bombay Grass. 65. 1935; Bor, Grass. Ind. 258. 1960; Lakshminarasimhan in Sharma *et al.* Fl. Maharashtra State Monocot. 632.1996. *Andropogon squarrosus* L. f. Suppl. Pl. 433. 1781. (nomen confusum). VALA.

Perennial herbs, culms stout, up to 2 m tall; rhizome aromatic. Leaves: sheaths compressed, sharply keeled; blades flat, linear-lanceolate, 15–60 × 0.5–0.7 cm glabrous, acuminate. Panicles oblong, 10–30 cm long, usually contracted; rachis stout, smooth; racemes simple, whorled, 5–10 cm long; joints and pedicels slender, glabrous. Sessile spikelets linear-lanceolate, 4–4.5 mm long, yellowish, violet brown or purplish-black. Glumes coriaceous, acute. Lower glume muriculate all over the back, upper glume muriculate on the keel. Grain oblong.

Fls. & Frts.: September–December. *Exsiccata*: SUK-258.
Rare, in moist, low-lying areas in open grasslands, often cultivated.
Localities: Kolhapur, Amboli.

Local uses: Roots are placed in water during hot season. Root placed in one full glass of water over night and given early in the morning in gonorrhoea. It is used as laxative, appetizer, alexipharmic, an aphrodisiac, anthelmintic, useful in bronchitis, leprosy, epileptic fits, it causes burning sensation.

PHARMACOGNOSTIC STUDIES

Roots used in Folk Medicine

BELMUL [Photoplate-XI; Text fig.: 50a-i]
Botanical name: **Aegle marmelos** (L.) Corr.

Family: Rutaceae

Vernacular name: Bel, Bilwa.
Parts used: Root.
Organoleptic characters: Taste slightly bitter, odorless.
Macroscopic characters: The drug occurs in the form of cylindrical, pale yellow colored pieces of roots of ca 6–8 × 1–1.5 cm in size [photoplate-XI]. Bark of root is 1–3 mm thick and is rough. Outer bark is pale yellow and contains thick layer of rhytidome. Inner bark is yellowish brown in color. Root is longitudinally wrinkled; wrinkles are very shallow. Wood is pale yellow, hard. Fractures are fibrous. A transversely cut surface of root shows central wood with distinct numerous vessels and radiating medullary rays.

Microscopic characters: Transverse section of root is circular in outline. It shows outer 2–3 layers of rhytidome. It consists of alternate layers of thin walled rectangular parenchymatous cells and thick walled cells. Phelloderm is compactly arranged and consists of islands of stone cells. Secondary cortex is made up of loosely arranged parenchymatous cells containing starch grains. Small, 2–3 celled bunches of sclerenchyma cells are irregularly scattered in secondary cortex. Phloem is wide and consists of 10–16 celled thick rectangular patches of sclerenchyma. Rays are uniseriate. Wood consists of vessels, xylem parenchyma and fibers [Text fig. 50a-c]. Vessels are ca 259–370 × 55–92 μ in size; perforation plates simple and horizontal; intervascular pitting is alternate [Text fig. 50a]. Pits to xylem parenchyma are similar as vessels [Text fig. 50c]. Fibers are of two types. Thin walled fibers are ca 722–805 μ in length, slightly pitted and unseptate [Text fig. 50b]. Thick walled fibers are 1000–1330 μ in length, slightly pitted and unseptate [Text fig. 50b'].

Powder characters: Powder is fresh yellow in color containing pieces of vessels, fibers, stone cells, crystals of calcium oxalate and starch grains [Text fig. 50e-i].

Stone cells are 46–77 × 22–37 μ, rectangular to polygonal in shape with blunt ends narrow lumen [Text fig. 50g]. Crystals of oxalate solitary, 18–12 μ in size [Text fig. 50h]. Starch grains are abundant in powder, simple, ca 6–12 μ in diameter [Text fig. 50i].

Local uses: The root is given as one of the *Dashmula* and *Mahapanchmula*. Decoction of root is given thrice a day in fever. Root decoction is given in epilepsy. Root is dipped in neem oil and root is burned over flame and that oil is dropped in ear to cure all kinds of ear problems. Root is boiled with rice and water obtained from rice is given to the children suffering from diarrhoea.

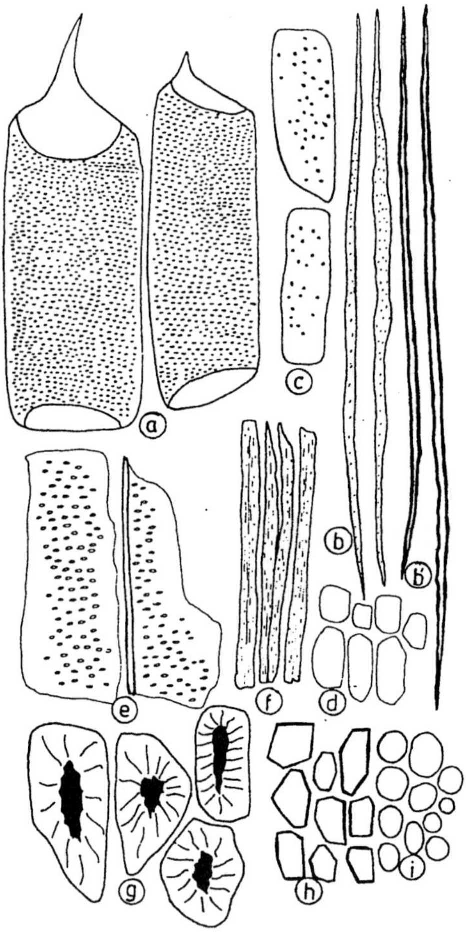

Text fig. 50: Aegle marmelos a-Vessels (x 225), b-Thin walled fibers (x 150), b'-Thick Walled fibers (x 100), c-Xylem parenchyma (x 450), d-Parenchyma cells (x 150), e-Pieces of vessel (x 450), f-Pieces of fibers (x 100), g-Stone cells (x 450), h-Prism type of Crystals of Calcium oxalate (x 450), i-Starch grains (x 450).

BOKIDA [Photoplate-XI; Text figs.: 51a–f; 52g–l]

Botanical name: **Ficus hispida** L.

Family: Moraceae

Vernacular names: Bhuiumbar, Bokeda, Gandy–umbar, Kalaumber.
Part used: Root.
Organoleptic characters: Tasteless, odorless

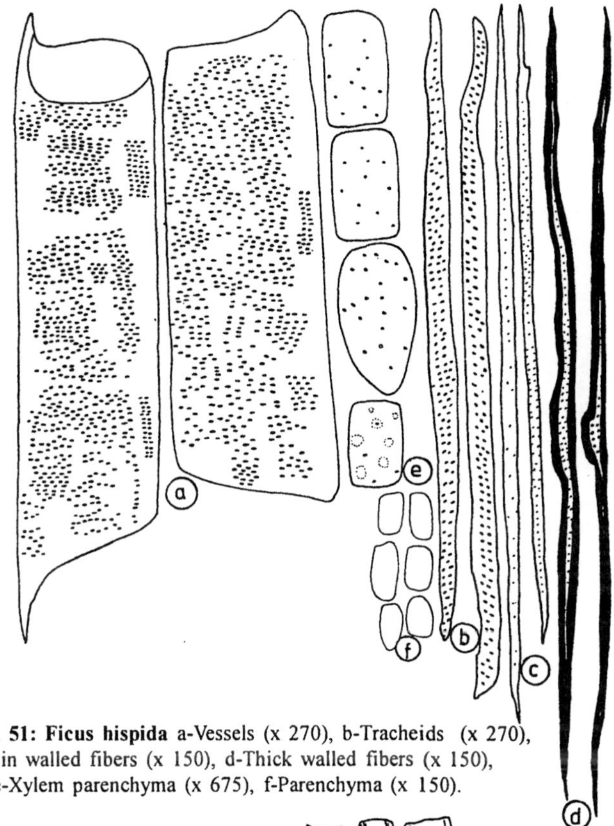

Text fig. 51: Ficus hispida a-Vessels (x 270), b-Tracheids (x 270), c-Thin walled fibers (x 150), d-Thick walled fibers (x 150), e-Xylem parenchyma (x 675), f-Parenchyma (x 150).

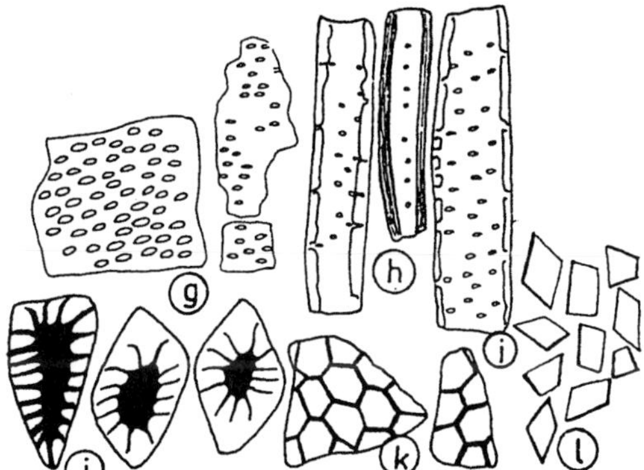

Text fig. 52: Ficus hispida g-Pieces of vessel (x 450), h-Pieces of fibers (x 450), i-Pieces of tracheids (x 450), j-Stone cells (x 450), k-cork cells (x 450), l-Crystals of Calcium oxalate (x 450).

Macroscopic characters: The drug occurs in the form of cylindrical dark brown or black or reddish-brown colored pieces of roots of ca 6–7 × 1.5–2 cm in size [Photoplate-XI]. Bark of root is ca 1–2 cm thick, rough. Outer cork is thin papery, flanking off in the form of sheets. Inner bark is longitudinally striated. Lenticels are large, rounded and fractures are short. Wood is pale yellow or pale brown in color. A transversely cut portions of the root shows thick bark and central wood having wavy outline.

Microscopic characters: Transverse section of root is circular in outline. Cork consists of outer thick walled, 3–5 layered phellem and inner 5–8 layers of phelloderm made up of compactly arranged cells. Phelloderm consists of numerous solitary crystals of calcium oxalate. Secondary cortex is wide. Sclerenchymatous cells in the form of small or large islands are scattered in secondary cortex. The secondary phloem is crushed and consists of uniseriate or biseriate rays . Wood consists of vessels, xylem parenchyma, tracheids and fibers [Text fig. 51 a-d]. Vessels are ca 250–468 × 93–125 μ; perforation plates are simple and horizontal; intervascular pitting is alternate; small and bordered [Text fig. 51 a]. Pits to xylem parenchyma are large and rounded [Text fig. 51 e]. Tracheids are ca 405–500 μ in length, pitted and tapering towards ends [Text fig. 51 d]. Thin walled fibers are ca 833–916 μ in length, pitted and unseptate [Text fig. 51 c]. Thick walled fibers are ca 890–1100 μ in length, thick walled with or without lumen, pitted and unseptate [Text fig. 51 d].

Powder characters: Powder is pale yellow containing small black or reddish colored thin pieces of cork, pieces of vessels, tracheids, fibers, stone cells, crystals of calcium oxalate and starch grains [Text fig. 52 g-i].

Stone cells are ca 37–55 × 22–37 μ in size, square to rectangular in shape with blunt ends [Text fig. 52 j]. Crystals of calcium oxalate are abundant in powder, solitary and 18–22 μ in size [Text fig. 52 i].

Local uses: Paste of root is applied externally on wounds. Milky latex is filled in ulcers to promote suppuration. Root paste is externally applied on boils and itch.

DHAYATIMUL [Photoplate-XI; Text fig.: 53 a-k]
Botanical name: **Woodfordia fruticosa** (L.) Kurz.

Synonyms: *W. floribunda* Salisb.

Family: Lythraceae

Vernacular names: Dhayati, Dhauri etc.

Part used: Root.

Organoleptic characters: Taste astringent, odorless.

Macroscopic characters: The drug occurs in the form of cylindrical pale pink or reddish-colored pieces of roots of ca 5–6 × 1–2.5 cm. Bark of root is ca 1–2 mm thick, hard. Outer bark is pale pink and inner bark is reddish brown and smooth. Root is without any fissures. Wood is pale pink or brown in color, hard, fractures short. A transversely cut surface of stem shows prominent annular growth rings and very small or minute vessels.

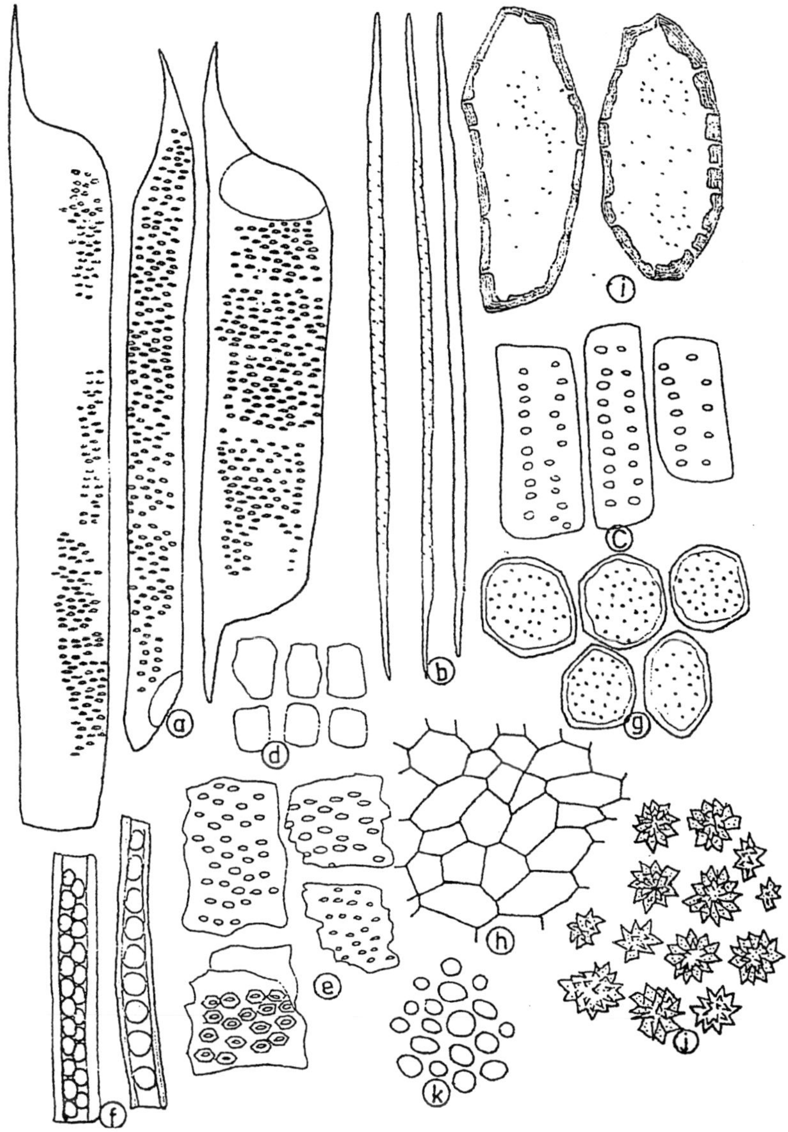

Text fig. 53: Woodfordia fruticosa a-Vessels (220x), b-Fibers (x 150), c-Xylem parenchyma (x 675), d-Parenchyma (x 150), e-Pieces of vessel (x 675), f-Pieces of fibers (x 450), g-Stone cells (x 675), h-Cork cells (450), j-Crystals of Calcium oxalate (x 450), k-Starch grains (x 450).

Microscopic characters: Transverse section of root is circular in outline. Outer cork is 3–4 layered made up of thick walled cells while inner cork consists of wide zone of phelloderm. Cells of secondary cortex are filled with tannin and starch grains. Crystals of calcium oxalate are observed in secondary cortex and in phloem. Secondary phloem is wide and consists of uniserate rays. Wood consists of vessels, xylem parenchyma and fibers [Text fig. 53 a-c]. Vessels are ca 518–670 × 55–81 μ; perforation plates are simple, oblique to horizontal; intervascular pitting is alternate with small to moderate sized bordered pits [Text fig. 53 a]. Pits on xylem parenchyma are large and rounded [Text fig. 53 c]. Fibers are 611–972 μ in length, thin walled, pitted and unseptate [Text fig. 53 b].

Powder characters: Powder is reddish-brown in color, containing pieces of vessels and fibers, stone cells, thick walled pitted stone cells, crystals of calcium oxalate and starch grains [Text fig. 53 e–k].

Stone cells are 83–129 × 37–55 μ, rectangular to polygonal in shape, thin walled with broad and pitted lumen [Text fig. 53 g]. Pieces of thick walled stone cells are also present [Text fig. 53 h]. Crystals of calcium oxalate are clustered, 12–25 μ in diameter [Text fig. 53 j]. Starch grains are abundant in powder, simple, rounded to oblong in shape, 5–11 μ in diameter [Text fig. 53 k].

Chemical taste:

Tests	Reagents	Results
Starch	Iodine	+ ve bark
Tannins	1% $FeCl_3$	+ ve Phelloderm, secondary cortex, secondary phloem. Powder extract shows pale green color.

Local uses: The root powder is given twice a day with water in diabetes. One teaspoon of powder is given every night to the women suffering from excess bleeding and in leucorrhoea.

DHORGUNJ/MAHAGUNJ [Text fig.: 54 a-i]

Botanical name: **Withania somnifera** (L.) Dunal.

Family: Solanaceae

Vernacular names: Ashwagandha, Askand.

Part used: Root.

Organoleptic characters: Taste initially sweetish, later on bitter and acrid, odor sweetish, nauseous.

Macroscopic characters: The drug occurs in the form of cylindrical or slightly torturous gray or fawn colored pieces of roots of ca 3–5 × 1–2 cm. Bark is ca 1–2 mm thick, rough, possessing transversely elongated lenticels. Root is longitudinally wrinkled; wrinkles deep or shallow. Wood is dirty white, starchy and soft. Fractures are short. A transversely cut portion of root shows a creamy starchy surface with a closely adhering bark.

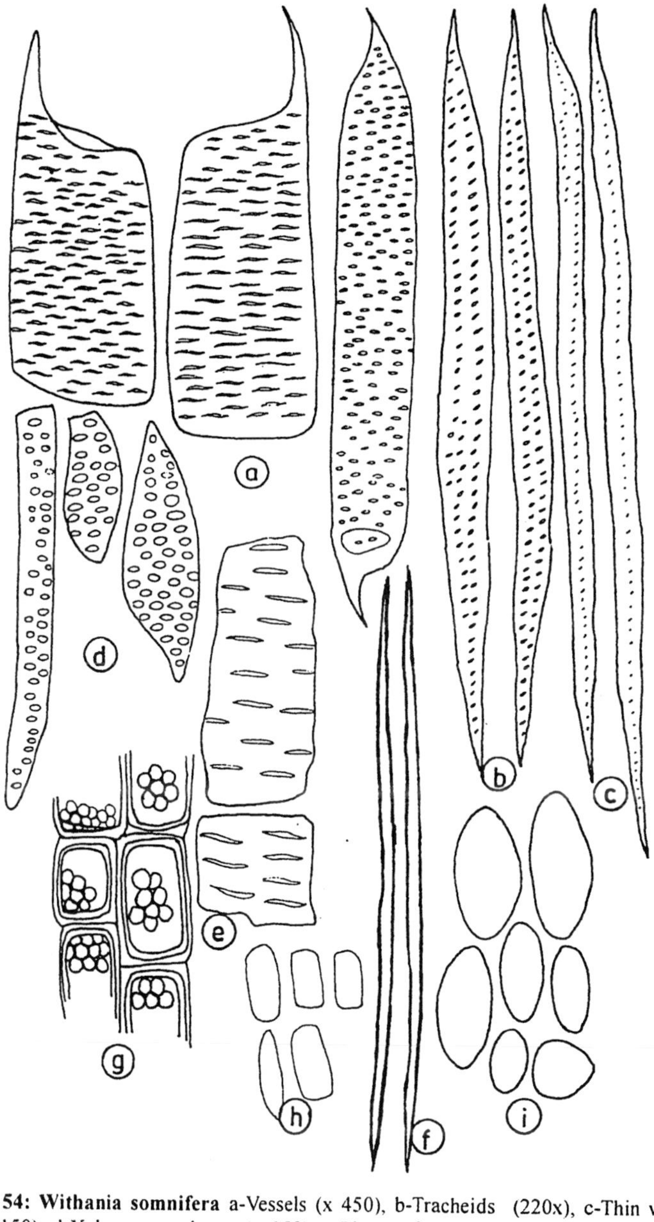

Text fig. 54: Withania somnifera a-Vessels (x 450), b-Tracheids (220x), c-Thin walled fibers (x 150), d-Xylem parenchyma (x 150), e-Pieces of vessel (x 675), f-Thick walled fibers (220x), g-Thick walled cells containing Starch grains (x 675), h-Parenchyma (x 150), i-Starch grains (x 675).

Microscopic characters: Transverse section of root is wavy in outline. It shows outer 2–3 layered thick walled cork cells. Secondary cortex is made up of thin walled loosely arranged parenchymatous cells. The cortical cells are filled with starch grains. Ground tissue is parenchymatous and loosely arranged. Vascular bundles are scattered irregularly more towards the outer region of ground tissue. Wood consists of vessels, xylem parenchyma, tracheids and fibers [Text fig. 54]. Vessels are ca 148–100 × 37–50 μ, with simple horizontal perforation plate. Intervascular pitting is alternate to scalariform [Text fig. 54 a-d]. Pits to xylem parenchyma are large and rounded [Text fig. 54 d]. Tracheids are ca 250–296 μ in length with alternate pitting [Text fig. 54 b]. Tracheid fibers are ca 200–235μ in length [Text fig. 54 c]. Fibers are ca 138–220 μ in length, thick walled and unseptate [Text fig. 54 f].

Powder characters: Powder is dirty white or gray in color containing pieces of vessels, parenchymatous cells filled with starch grains and starch grains [Text fig. 54 e-i]. Starch grains are abundant in powder. They are simple, elliptic oblong, ca 12–36 μ in diameter [Text fig. 54 i].

Chemical tests:

Tests	Reagents	Results
Starch	Iodine	+ve. Powder extract gives deep blue color with iodine showing presence of abundant starch grains.

Local uses: Half teaspoon of powder given every night with the milk in menorrhoea, in night emissions and in case of general debility. Tincture of powder is given as a stimulant.

GELAMUL [Photoplate–XI; Text fig.: 55 a-g]

Botanical name: **Catunaregam spinosa** (Thunb.) Tirveng.

Synonyms: *Randia dumatorum* (Retz.) Poir.

Family: Rubiaceae.

Vernacular names: Gehela, Gelphal, Madanphal.

Part used: Root.

Organoleptic characters: Taste none, odorless.

Macroscopic characters: The drug occurs in the form of cylindrical gray or pale pink colored pieces of roots of 5–8 × 1–2 cm [Photoplate-11]. Bark of root is ca 2 mm thick, hard, light pink colored without any fissures. Fractures are short. Wood is pale brown and hard. A transversely cut surface of root shows thick bark, central pale brown wood with very small vessels and rays.

Microscopic characters: Transverse section of root is circular in outline. It shows 2–4 layered thin walled cells of cork. Phelloderm is completely

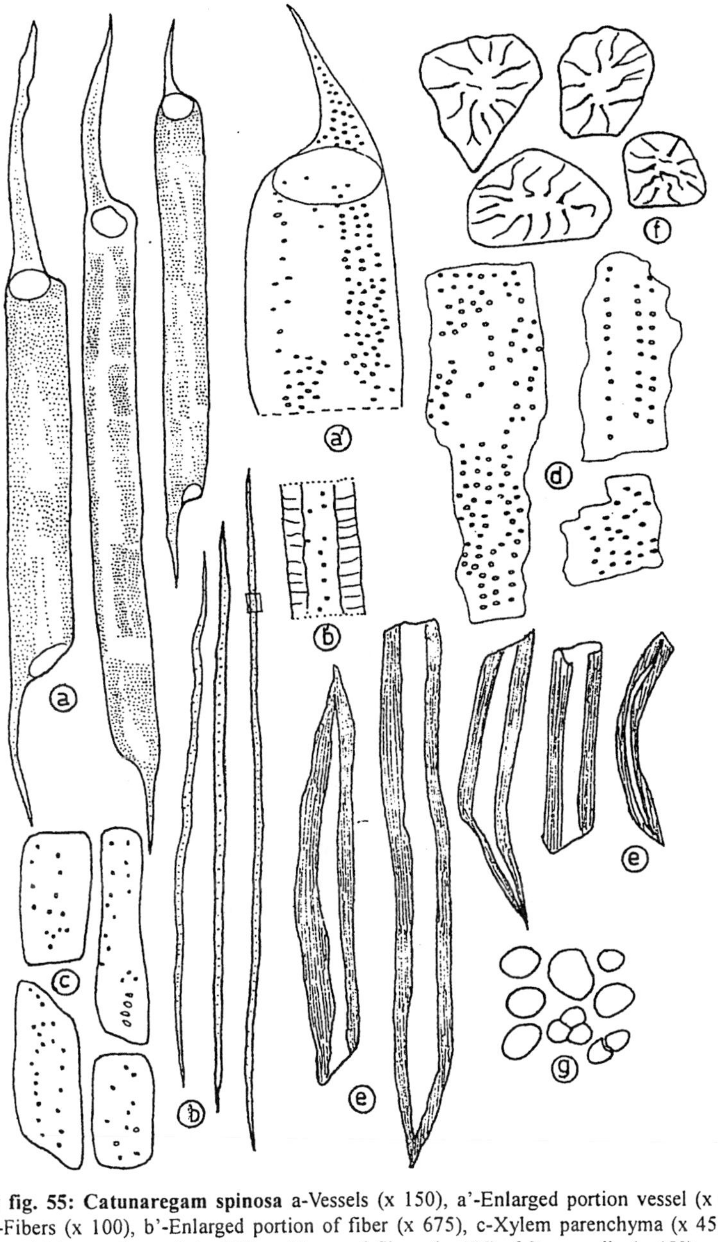

Text fig. 55: Catunaregam spinosa a-Vessels (x 150), a'-Enlarged portion vessel (x 450), b-Fibers (x 100), b'-Enlarged portion of fiber (x 675), c-Xylem parenchyma (x 450), d-Pieces of vessel (x 675), e-Pieces of fibers (x 675), f-Stone cells (x 450), g-Starch grains (x 450).

sclerosed and consists of broad bands of stone cells. Secondary cortex consists of thin walled cells possessing abundant starch grains and band of stone cells. Secondary phloem is crushed and consists of uni or biseriate rays and rounded islands of stone cells. Wood consists of vessels, xylem parenchyma and fibers [Text fig. 55 a–c]. Vessels are 694–944 × 38–55 μ in size; perforation plates are simple, horizontal or slightly oblique. Intervascular pitting is alternate, very small to minute. Vessels are tapering towards both ends [Text fig. 55 a]. Pits to xylem parenchyma are larger than pits on the vessels [Text fig. 55 c]. Fibers are 916–1208 μ in length, thick walled, pitted and unseptate [Text fig. 55 b].

Powder characters: Powder is pale gray in color containing pieces of vessels and fibers, stone cells and starch grains [Text fig. 55 d–g]. Stone cells are abundant in powder, 37–55 × 20–59 μ in size, triangular to polygonal in shape, thick walled, with or without lumens and with blunt end walls [Text fig. 55 f]. Starch grains are abundant in powder. They are simple or compound, 6–24 μ in diam. and rounded in shape [Text fig. 55 g].

Local uses: Root paste is used to remove post pregnancy marks from the face due to the excess secretion of melanin. Root decoction is used to wash the teeth and to stop bad smell from the mouth. Root decoction is used externally to wash the wounds and ulcers.

GITSAYA [Photoplate–XI; Text fig.: 56 a-i]
Botanical name: **Spermadictyon suaveolens** Roxb.

Synonyms: *Hamiltonia suaveolens* Roxb.

Family: Rubiaceae

Vernacular names: Gidsawwa, Gides, Gidas.

Part used: Root.

Organoleptic characters: Tasteless or slightly bitter, odor pungent.

Macroscopic characters: The drug occurs in the form of cylindrical silvery gray colored pieces of roots, 5–6 × 1.7–2 cm in size. Bark is thin and rough. Outer surface of bark flanks off in thin silvery gray colored sheets. Root is longitudinally fissured. Fractures are short. Wood is gray and soft with prominent radiating rays. Vessels are few and sparsely distributed.

Microscopic characters: Transverse section of root is circular in outline. It shows outer lamillated cork containing thick walled compactly arranged layers of cork cells. Secondary cortex is made up of thin walled parenchymatous cells. Ring of raphid sacs is present in secondary phloem encircling the secondary xylem. Rays are 6–9 cell wide. Wood consists of vessels, xylem parenchyma and fibers [Text fig. 56 a-i]. Vessels are ca 296–425 × 40–50 μ in size; perforation plates are simple, slightly oblique; intervascular pitting is alternate or scalariform; pits are bordered [Text fig. 56 a]. Pits to xylem parenchyma are present in straight lines on lateral walls or in the center and smaller than vessel pits [Text fig. 56 c].

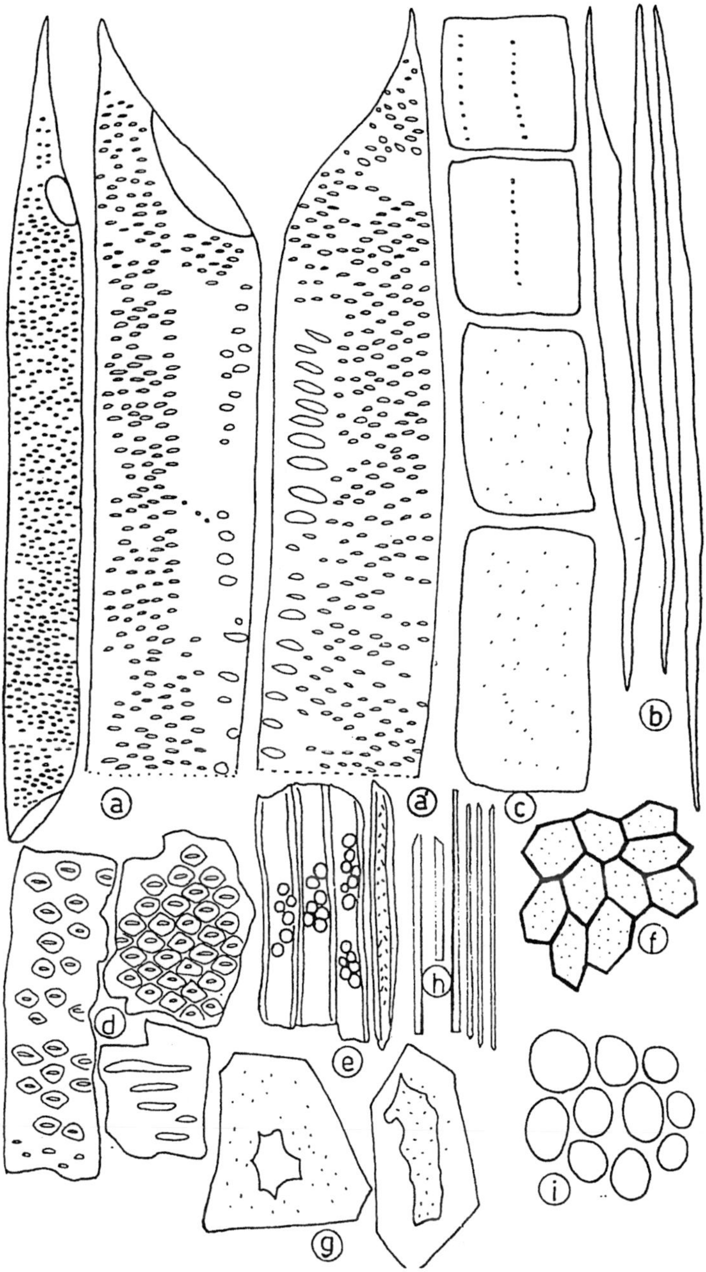

Text fig. 56: Spermadictyon suaveolens a-Vessels (450, × 225), b-Fibers (x 150), c-Xylem parenchyma (x 450), d-Pieces of vessel (x 675), e-Pieces of fibers (x 150), f-Cork cells (x 150), g-Stone cells (x 450), h-Raphides (x 675), i-Starch grains (x 675).

Fibers are 440–861 μ in length, thin walled and tapering at both the ends [Text fig. 56 b].

Powder characters: Powder is silvery gray in color containing pieces of vessels, fibers, cortical cells, raphides, starch grains and unidentified thick walled pitted stone cell like structures [Text fig. 56 c-i].

Cortical cells are thick walled, pitted [Text fig. 56 f]. Raphides are abundant in powder, 30–90 μ in length, acicular and pointed at both ends [Text fig. 56 h]. Starch grains are abundant in powder. They are simple, 12–30 μ in diameter, rounded to oblong in shape [Text fig. 56 i].

Local uses: Root powder is externally applied on Herpes zoster virus. Paste of root is prepared in boiled rice water with dried ginger and given in peritonitis. Root is mixed with the honey and given orally twice a day in stomach complaints of children.

JAMBHOOL MUL [Photoplate-XI; Text fig.: 57a-h]
Botanical name: **Syzygium cumini** (L.) Skeels

Synonyms: *Eugenia jambolana* Lam.

Family: Myrtaceae

Part used: Root

Vernacular names: Jambhal, Jambhul, Jambul.

Organoleptic characters: Taste astringent slightly sweet, odorless.

Macroscopic characters: The drug occurs in the form of cylindrical pale pink or reddish colored pieces of roots, 4–6 × 1.5–2 cm. Bark of root is ca 1–3 mm thick, smooth, pale pink outside and reddish inside. The root is without any wrinkles or fissures [Photoplate-XI]. Fractures are short. Wood is hard and pale pink in color. A transversely cut surface of root shows thick reddish colored bark and central wood. Wood consists of annual large brown colored alternating rings.

Microscopic characters: Transverse section of root is circular in outline. It shows outer 3–4 layered thick walled cork cells, inner cork consists of 1–3 layers of phellogen and 5–6 layers of phelloderm. Phelloderm consists of small islands of stone cells. Secondary cortex is wide and consists of abundant starch grains, tannin sacs and sclerenchyma. Secondary phloem is wide and consists of 2–3 celled small islands of sclerenchyma cells and uniseriate rays. Rays uniseriate. Wood consists of vessels, xylem parenchyma and fibers [Text fig. 57a–c]. Vessels are ca 185–259 × 37–53 μ in size; perforation plates simple and oblique; intervascular pitting is alternate and pits bordered [Text fig. 57 a–a"]. Pits to xylem parenchyma are similar to that of vessels [Text fig. 57 c]. Fibers are of two types, thin walled and thick walled. Thin walled fibers are ca 833–1166 μ in length and unseptate [Text fig. 57 b]. Thick walled fibers are ca 500–830 μ in length with or without lumen [Text fig. 57 b'].

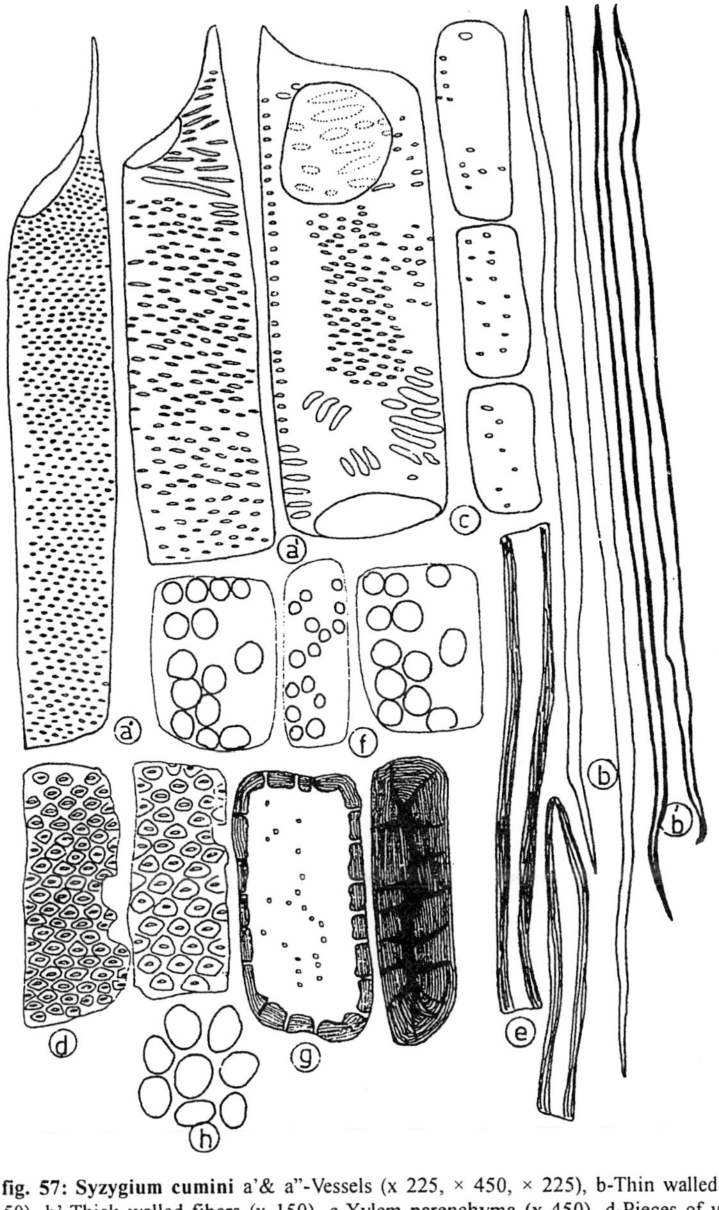

Text fig. 57: Syzygium cumini a'& a"-Vessels (x 225, × 450, × 225), b-Thin walled fibers (x 150), b'-Thick walled fibers (x 150), c-Xylem parenchyma (x 450), d-Pieces of vessel (x 675), e-Pieces of fibers (2x 50), f-Thin walled parenchymatous cells containing Starch grains (x 450), g-Stone cells (x 675), h-Starch grains (x 675).

Powder characters: Powder is reddish-pink in color containing pieces of vessels, fibers, stone cells, parenchymatous cells containing starch grains and starch grains [Text fig. 57 d-h]. Stone cells ca 72–55 × 14–30 μ in size, mostly rectangular in shape, thick walled, striated, pitted and with or without lumen [Text fig. 57 g]. Thin walled parenchymatous cells filled with starch grains are very abundant in powder [Text fig. 57 f]. Starch grains are simple, elliptic oblong, 8–12 μ in size [Text fig. 57 h].

Chemical tests:

Tests	Reagents	Results
Starch	Iodine	+ ve
Tannins	1% $FeCl_3$	+ ve Phelloderm, secondary cortex, secondary phloem. Powder extract is shows pale green color.

Local uses: Half teaspoon of powder and one teaspoon of *Gymnema sylvestre* leaf powder is given daily to the diabetic patients after meal. Two teaspoon of root powder is given early in the morning with water in urinary calculi. Root decoction is used in mouth ulcers. Root paste is applied externally on pimples.

KANGONIMUL [Photoplate–XI; Text fig.: 58 a-h]

Botanical name: **Celastrus paniculatus** Willd.

Family: Celastraceae.

Vernacular names: Dhirmarbel, Jyotishmati, Kangli, Kanguni, Karadkanguni.

Part used: Root.

Organoleptic characters: Taste astringent, odorless.

Macroscopic characters: The drug occurs in the form of cylindrical yellow colored pieces of roots, 5–6 × 1.5–2.5 cm [Photoplate-XI]. Bark of root is ca 3–5 mm thick, yellow, rough and longitudinally fissured. Inner bark is reddish brown. Lenticels are transversely elongated. Wood is hard and pale pink in color. Fractures are short

Microscopic characters: Transverse section of root is circular in outline. It shows multilayered thick walled cork cells. Phellogen is 5–6 layered made up of compactly arranged parenchymatous cells. Secondary cortex is wide in which solitary or groups of large stone cells are present. Solitary crystals of calcium oxalate are irregularly scattered in secondary cortex. Tannin is filled in cortical cells. Secondary phloem is interrupted by wide medullary rays. Wood consists of vessels, xylem parenchyma and fibers [Text fig. 58 a-d]. Vessels are ca 370–556 × 74–259 μ in size. Perforation plates are simple and horizontal; intervascular pitting is alternate and pits are bordered [Text fig. 58 a]. Pits to xylem parenchyma are similar to that of vessels [Text fig. 58 d]. Fibers are 278–370 μ in length, septate and pitted [Text fig. 58 b].

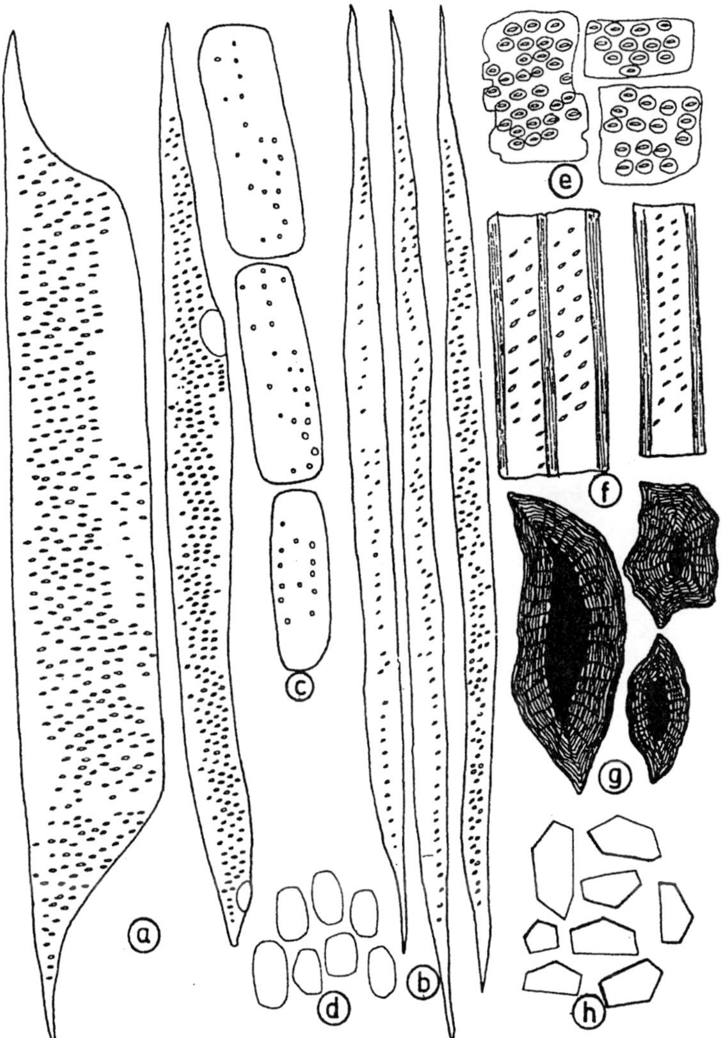

Text fig. 58: Celastrus paniculatus a-Vessels (x 450), b-Fibers (x 450), c-Xylem parenchyma (x 450), d-Parenchyma (150'x), e-Pieces of vessel (x 450), f-Pieces of fibers (x 450), g-Stone cells (220x), h-Crystals of Calcium oxalate (x 450).

Powder characters: Powder is reddish pink in color containing pieces of vessels and fibers, stone cells and crystals of calcium oxalate [Text fig. 58 a–h]. Stone cells are variously shaped, irregular, ca 222–74 × 55–81 μ, thick walled, striated and have blunt ends and pitted lumen [Text fig. 58 g]. Crystals of calcium oxalate are solitary, ca 24–14 cm in size [Text fig. 58 h].

Chemical tests:

Tests	Reagents	Results
Starch	Iodine	+ ve
Tannins	1% $FeCl_3$	+ ve Phelloderm, secondary cortex, secondary phloem. Powder extract shows pale green color.

Local uses: Root paste is externally applied in rheumatism and is given orally. Root is also given as brain tonic. Root is crushed with the leaves of *Litsea chinensis* and applied externally on chronic ulcers, wounds and in bone fractures.

MAKADLIMBU/GHANSARMUL [Photoplate-XI; Text fig.: 59 a–h]
Botanical name: **Atalantia racemosa** Wight.

Family: Rutaceae

Vernacular names: Makadlimbu, Ranlimbu.

Part used: Root.

Organoleptic characters: Taste bitter and biting, odor fresh, aromatic like orange which remains even in dry samples.

Macroscopic characters: The drug occurs in the form of cylindrical gray or greenish-yellow colored pieces of root of ca 4–6 × 2–4 cm [Photoplate-XI]. Bark of root is ca 1–2 mm thick, rough, gray or greenish-yellow or sometimes brown colored. Root is longitudinally wrinkled with small protuberances. Longitudinal cracks are present on the roots. Wood is lemon yellow and hard. Fractures are short. A transversely cut surface of stem shows central wood possessing annular growth rings with alternate dark brown bands.

Microscopic characters: A transverse section of root is circular in outline. It shows outer thin walled, multilayered cork. Phelloderm is wide in which islands of stone cells are rarely present. Secondary cortex consists of discontinuous bands of sclerenchyma cells. The secondary phloem is a narrow zone which consists of thin walled parenchymatous cells and uniseriate or biseriate rays. Wood consists of vessel elements, xylem parenchyma and fibers [Text fig. 59 a–c]. Vessels elements are ca 268–283 × 58–83 μ in size, tapering towards both the ends with blunt ends; perforation plates are oblique and intervascular pitting is alternate [Text fig. 59 a]. Pits to xylem parenchyma are similar to that of vessel elements [Text fig. 59 c]. Fibers are ca 583–638 μ in length, thin walled and unseptate [Text fig. 59 b].

Powder characters: Powder is lemon yellow in color containing pieces of vessels and fibers, crystals of calcium oxalate and starch grains. Stone cells are mostly absent or very rare in powder [Text fig. 59 e–h].

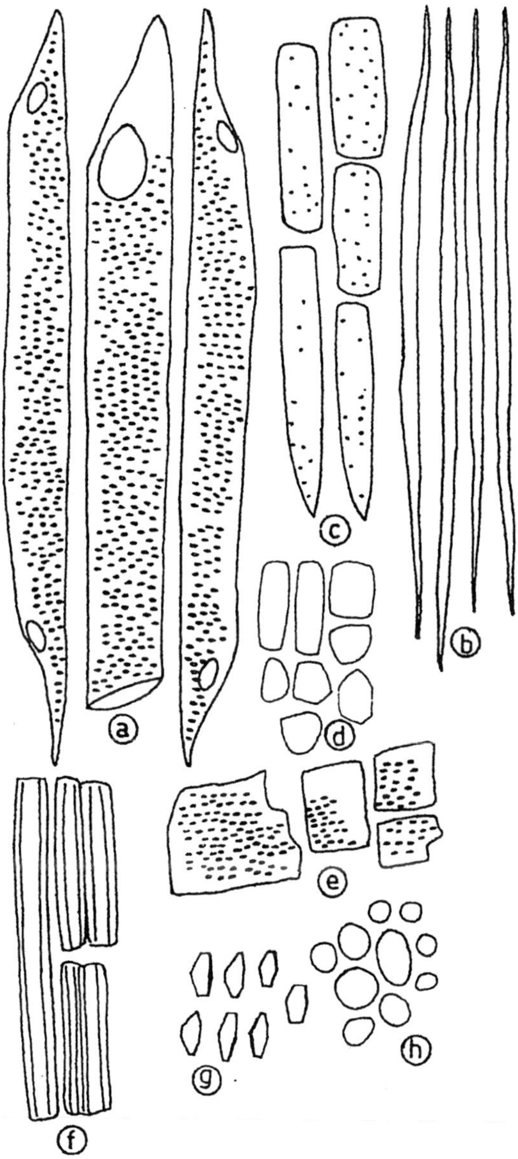

Text fig. 59: Atalantia racemosa a-Vessels (220x), b-Fibers (x 150), c-Xylem parenchyma (x 450), d-Parenchyma (x 150), e-Pieces of vessel (x 450), f-Pieces of fibers (x 150), g-Crystals of Calcium oxalate (x 450), h-Starch grains (x 675).

Local uses: One teaspoon of root powder is given in cough and sore throat after every 6 hours. Similarly, a teaspoon of root powder is given every night as blood purifier.

MAWARUNG MUL [Photoplate-XI; Text fig.: 60 a–h]
Botanical name: **Cinnamomum nitidum** (Roxb.) Hook.

Synonyms: *C. macrocarpum* Hook. f.

Family: Lauraceae

Vernacular names: Mawarungmul.
Part used: Root.

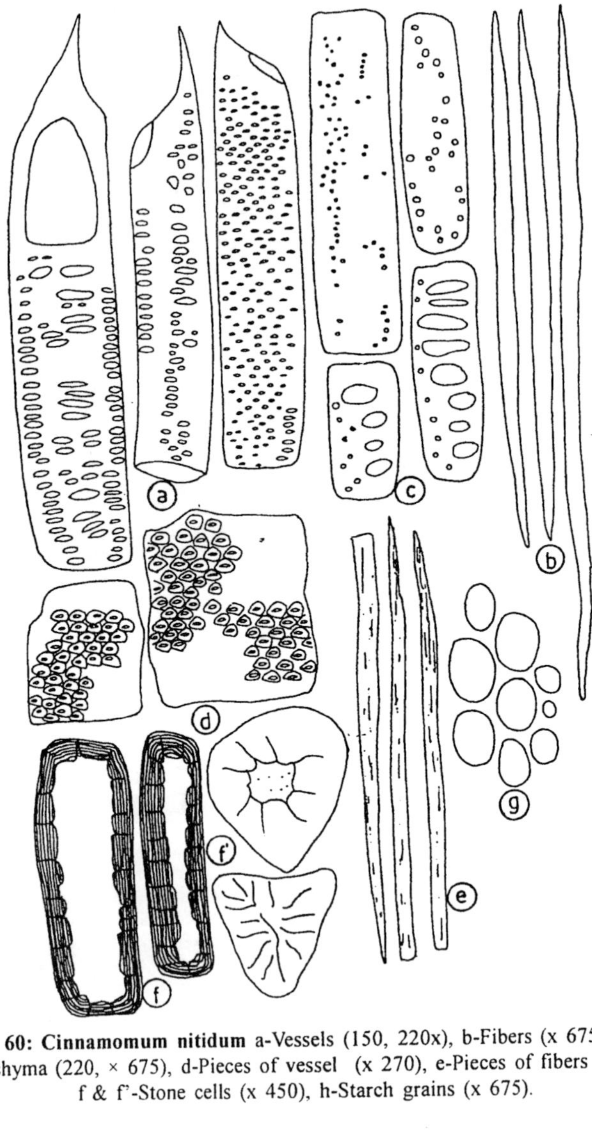

Text fig. 60: Cinnamomum nitidum a-Vessels (150, 220x), b-Fibers (x 675), c-Xylem parenchyma (220, × 675), d-Pieces of vessel (x 270), e-Pieces of fibers (x 150), f & f'-Stone cells (x 450), h-Starch grains (x 675).

Organoleptic characters: Taste mucilaginous, odor aromatic like Vicks-vaporub which remain even in dried samples.

Macroscopic characters: The drug occurs in the form of cylindrical reddish-brown colored pieces of root of ca 5–3 × 1.5–3 cm. Bark of root is ca 2–5 mm thick which is reddish-brown in color. Surface of bark is smooth without wrinkles or fissures. Wood is soft and reddish-brown or pale pink in color. Fractures are short. Vessels are large and prominent on wood surface.

Microscopic characters: Transverse section of root is circular in outline. It shows 3–6 layered cork made up of thick walled cells. Phelloderm is sclerosed. Islands of stone cells and oil containing cells are found in secondary cortex. Secondary phloem consists of uniseriate or biseriate rays. Wood consists of vessels, xylem parenchyma, and fibers [Text fig. 60 a-c]. Vessels are ca 370–462 × 62–118 μ in size; perforation plates are simple and oblique; intervascular pitting is alternate to scalariform and pits are bordered [Text fig. 60 a]. Pits to xylem parenchyma are large, circular to scalariform [Text fig. 60 c]. Fibers are ca 556–889 μ in length, pitted, unseptate and thin walled [Text fig. 60 b].

Powder characters: Powder is reddish-brown in color containing pieces of vessels and fibers, stone cells and starch grains [Text fig. 60 d-h].

Stone cells are of two types; thin walled stone cells are 69–120 × 42–55 μ in size rectangular to square in shape [Text fig. 60 f']. Thick walled stone cells are ca 55–83 × 31–46 μ in size, rectangular to triangular in shape with blunt ends and without lumen [Text fig. 60 f']. Starch grains are simple, rounded, ca 7–18 μ in diameter [Text fig. 60 h].

Chemical tests:

Tests	Reagents	Results
Starch	Iodine	+ ve. Starch grains are very few in powder. Powder extract shows pale blue color with iodine.
Oil	Sudan III	+ ve in bark region

Note: Powder swells in water and becomes mucilaginous.

Local uses: The paste of root is rubbed on head to stop headache. Tincture of powder is taken orally in cough and asthma. Root decoction is given as brain tonic and as an appetizer. Root powder is smoked as a *Bidi* in seasonal asthma.

PADMUL [Photoplate-XI; Text fig.: 61a–i]

Botanical name: **Jasminum malabaricum** Wight

Family: Apocynaceae

Vernacular name: Kasur.

Part used: Root.

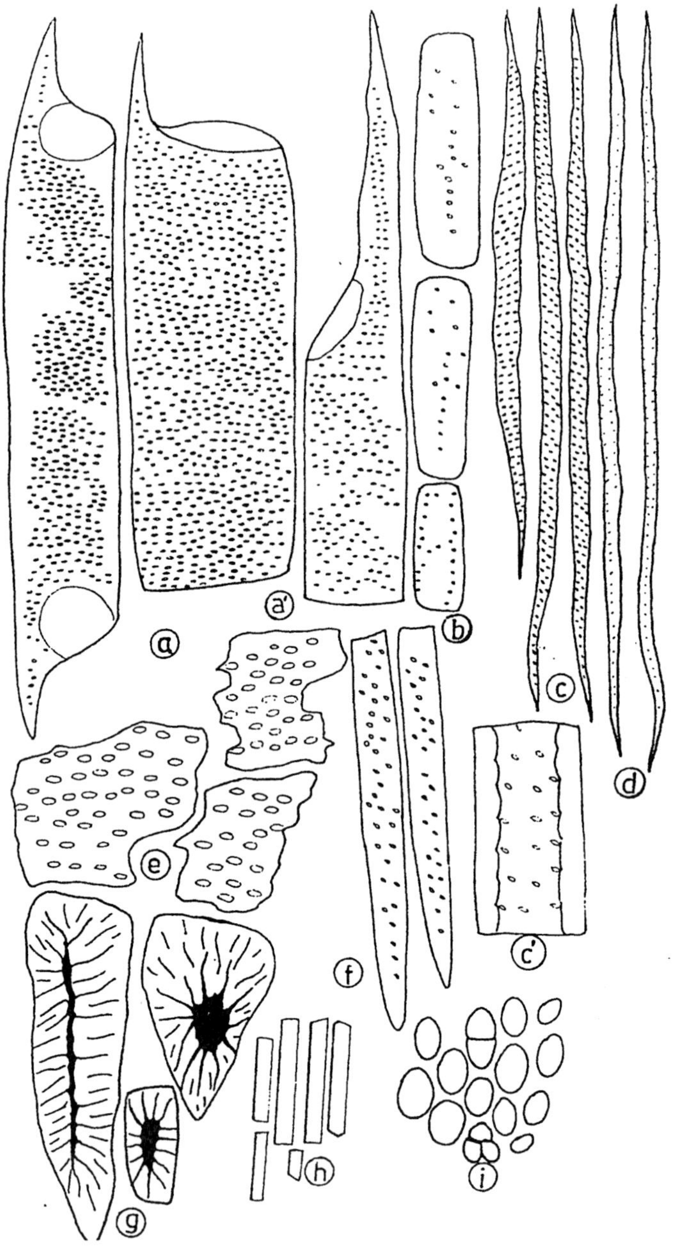

Text fig. 61: Jasminum malabaricum a-Vessels (x 270); a'-Enlarged portion of vessel (x 450), b-Xylem parenchyma (x 270), c-Tracheids (x 150), c'-Enlarged portion of tracheid (x 450), d-Thick walled fibers (x 150), e-Pieces of vessel (x 675), f-Pieces of fibers (x 450), g-Stone cells (x 450), h-Crystals of Calcium oxalate (x 450), i-Starch grains (x 675).

Vernacular name: Kasur.

Part used: Root.

Organoleptic characters: Tasteless, odor disagreeable like bug.

Macroscopic characters: The drug occurs in the form of cylindrical pale brown colored pieces of roots of ca 6–9 × 1–1.3 cm [Photoplate-XI]. Bark of root is ca 1–2 mm thick, rough and longitudinally wrinkled, wrinkles are shallow. Fractures are short. Wood is hard and pale yellow in color. A transversely cut surface shows rounded islands of stone cells in bark region.

Microscopic characters: Transverse section of root is circular in out line. It shows outer thin walled multilayered cork. Cork consists of inner 6–8 layers of phelloderm in which large rounded islands of stone cells are found. They are also present in secondary cortex. Rays in secondary phloem are uniseriate. Wood consists of vessels, xylem parenchyma, tracheid and fibers [Text fig. 61 a-d]. Vessels are ca 328–406 × 46–125 μ in size; perforation plates are simple, horizontal or oblique; intervascular pitting is alternate [Text fig. 61 a]. Tracheids are ca 556–830 μ in length, thick walled and pitted [Text fig. 61 c]. Fibers are ca 722–940 μ in length, thick walled, pitted and unseptate [Text fig. 61 d].

Powder characters: Powder is pale brown in color containing pieces of vessels, tracheids fibers, stone cells and crystals of calcium oxalate [Text fig. 61 e-i].

Stone cells are ca 55–129 × 18–37 μ in size, elongated oblong to triangular to polygonal in shape, thick walled and without lumen [Text fig. 61 g]. Crystals of calcium oxalate are solitary, ca 12–46 μ in length, rectangular or rod shaped [Text fig. 61 h]. Starch grains are simple, ca 7–12 μ in diameter and oblong to round in shape [Text fig. 61 h].

Chemical tests:

Tests	Reagents	Results
Starch	Iodine	+ ve, bark, phelloderm, phloem, medullary rays. Powder extract shows pale blue color with iodine.

Local uses: It is given as a tonic. One teaspoon of powder is given with milk in general debility and also in menorrhoea.

PADALMUL [Photoplate–XI; Text fig.: 62 a-f]

Botanical name: **Stereospermum chelenoides** (L. f.) DC.

Synonyms: *Bigonia chelenoides* L. f., *S. suaveolens* (Robx.) DC.

Family: Bignoniaceae

Vernacular names: Padal, Paral, Taamrapushpi.

Organoleptic characters: Taste none at first later on bitter, odorless.

Macroscopic characters: The drug occurs in the form of cylindrical, earthy brown or pale brown colored pieces of roots of ca 5–7 × 1–2 cm.

Bark of root is ca 5 mm thick, rough, and lamillated. Rhytidome is soft and longitudinally shriveled, breaking in the form of powder. Lenticels are rounded and raised. Fractures are tough and fibrous. Wood is hard and brown or pale gray in color. A transversely cut portion of root shows gray surface with distinct layers of bark and cork tissue. Wood shows light brown colored portion having dark brown concentric rings and radiating medullary rays.

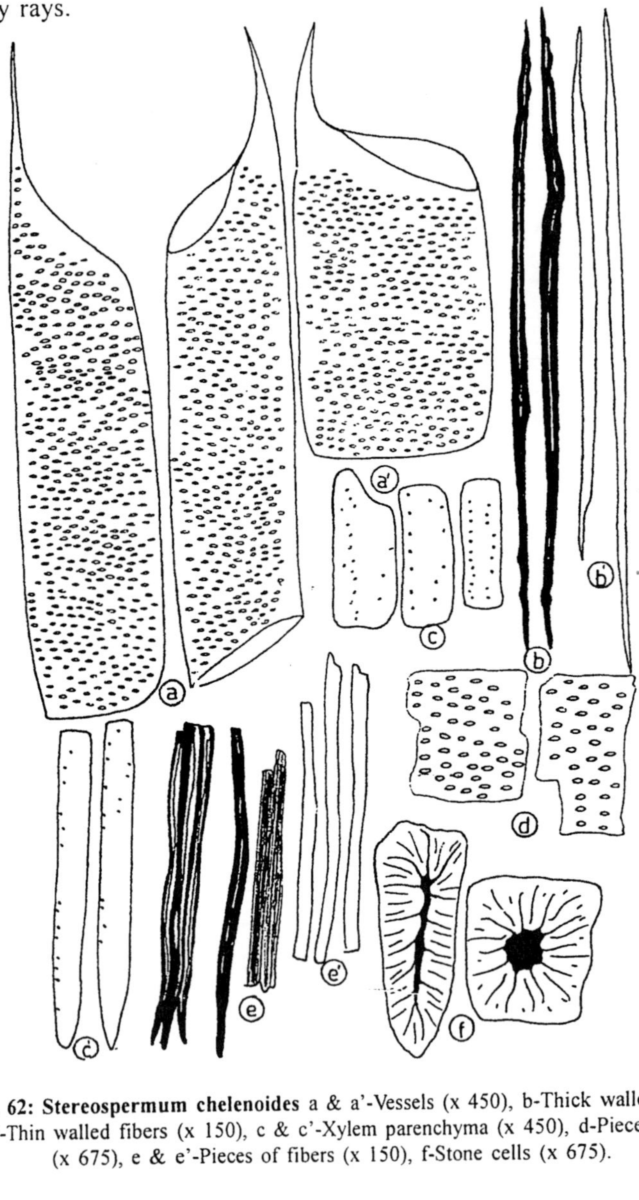

Text fig. 62: Stereospermum chelenoides a & a'-Vessels (x 450), b-Thick walled Fibers (x 150), b'-Thin walled fibers (x 150), c & c'-Xylem parenchyma (x 450), d-Pieces of vessel (x 675), e & e'-Pieces of fibers (x 150), f-Stone cells (x 675).

Microscopic characters: Transverse section of root is circular in outline. It shows 7–8 layers of thin walled parenchymatous tissue of rhytidome. Cork consists of 5–6 layered compactly arranged thin walled parenchymatous cells. Phelloderm sometimes possesses patches of stone cells. Secondary cortex is wide and made up of alternate bands of thick walled (1–2 layers) and thin walled (4–6 layers) parenchymatous cells. Secondary cortex consists of patches of sclerenchymatous cells forming a mesh like structure. Medullary rays are 1 to 2 cells wide. Parenchyma is paratracheal, vasicentric in narrow terminal bands. Wood consists of vessels, xylem parenchyma and fibers [Text fig. 62 a–c']. Vessels are ca 185–296 × 44–50 μ in size; perforation plates are simple and oblique; intervascular pitting is alternate [Text fig. 62 a & a']. Pits to xylem parenchyma similar as that of vessels [Text fig. 62 c & c']. Fibers are of two types; thin walled and thick walled. Thin walled fibers are ca 500–890 μ in length and unseptate [Text fig. 62 b'] while thick walled fibers are ca 500–1056 μ in length, without lumen and are unseptate [Text fig. 62 b].

Powder characters: Powder is brown in color containing pieces of fibers as a major component. In addition to this, pieces of vessels and stone cell are found in the powder. [Text fig. 62 d–f].

Stone cells are scarce in the powder. They are ca 30–48 × 24–42 μ, rectangular, square or triangular in shape, unpitted and have blunt ends and walls thick with narrow lumen [Text fig. 62 f].

Local uses: It is used as one of *Dashmula* drug in Ayurveda and also in folk medicines. A teaspoon of root decoction is given in cough, bronchitis and in fever. Root powder is given thrice in a day in general debility.

PALASMUL [Photoplate-XI; Text fig.: 63 a-k]
Botanical name: **Butea monosperma** (Lam.) Taub.

Synonyms: *Butea frondosa* Roxb. ex Willd.

Family: Fabaceae

Vernacular names: Palas.

Part used: Root.

Organoleptic characters: Taste slightly bitter, odorless.

Macroscopic characters: The drug occurs in the form of cylindrical pieces of roots of ca 6–7 × 1.5–3 cm. Bark of root is ca 1 mm thick, rough. Outer surface of bark consists of thin, silvery brown colored sheets of cork which are easily detachable. Root is longitudinally deeply fissured. Fractures are fibrous. Wood is pale yellow in color, soft and fibrous. A transversely cut surface shows annual growth rings and dark brown colored alternate rings.

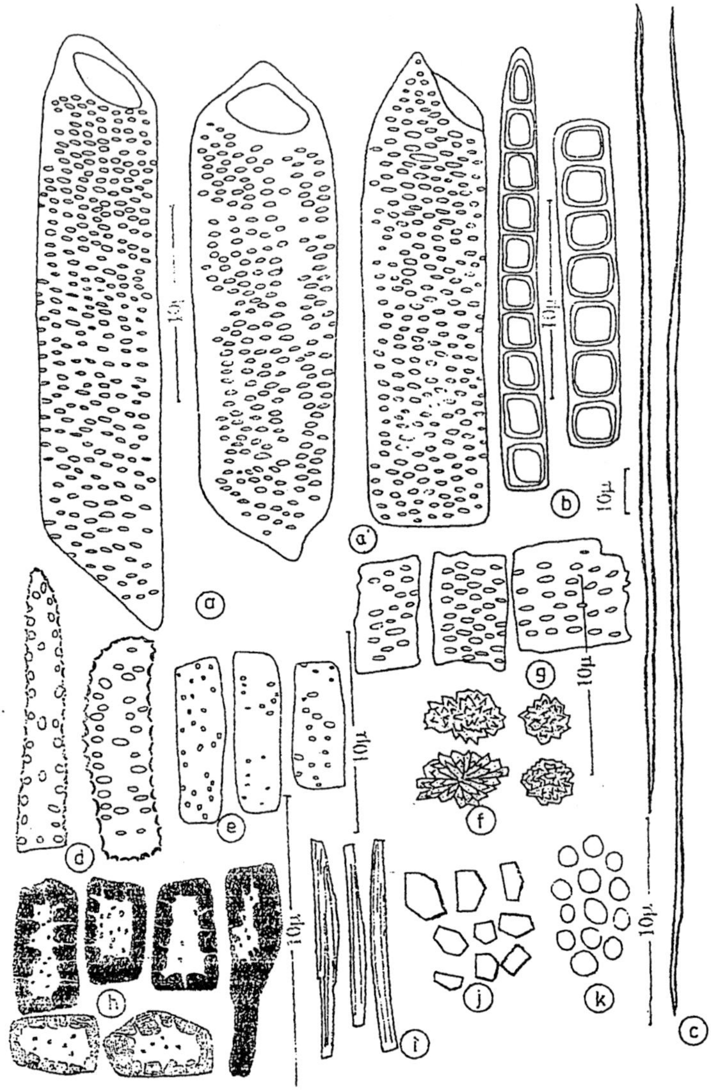

Text fig. 63: Butea monosperma a–a'-Vessels (x 450), b-Phloem Fibers (x 450), c-Fibers (x 100), d & e-Xylem parenchyma (x 450), f-Clustered Crystals of Calcium oxalate (x 450), g-pieces of vessel (x 450), h-Stone cells (x 675), i-Pieces of fibers (x 450), j-Prism type of Crystals of Calcium oxalate (x 450), k-Starch grains (x 450).

Microscopic characters: Transverse section of root is circular in outline. It shows outer thin walled multilayered cork. Phellogen is 2–3 layered. Phelloderm is 2–5 layered. Secondary cortex is wide and made up of thin walled parenchymatous cells in which islands of sclerenchyma cells are scattered. Secondary phloem is small and crushed. It consists of islands of phloem fibers. Parenchyma is very abundant in broad moderately regular bands of 4–6 cells wide. Rays 3–8 cell in width. The phloem fibers form small islands on the cross section. Wood consists of vessels, xylem parenchyma, fibers and phloem fibers [Text fig. 63 a-e]. Vessels are ca 203–240 × 37–55 μ in size; perforation plates are simple; intervascular pitting is alternate and pits bordered [Text fig. 63 a-a']. Pits to xylem parenchyma are large and rounded [Text fig. 63 d & e]. Fibers are ca 1166–1850 μ in length, thick walled and unseptate [Text fig. 63 c].

Powder characters: Powder is dark brown in color containing pieces of vessels, fibers, stone cells, crystals of calcium oxalate and starch grains [Text fig. 63 f-k]. Stone cells are ca 46–61 × 27–35 μ in size, rectangular in shape, walls striated or plain, lumen broad and pitted [Text fig. 63 h]. Crystals of calcium oxalate are clustered or solitary, clustered crystals are 12–22 μ in size, [Text fig. 63 f]. Solitary crystals are prism type [Text fig. 63 j]. Starch grains are simple, rounded and ca 7–12 μ in diameter [Text fig. 63 k].

Local uses: The paste of root is applied externally on wounds as well as on skin diseases such as itch.

PANDHARACHITRAK [Text fig.: 64 a-g]
Botanical name: **Plumbago zeylanica** L.

Family: Plumbaginaceae

Vernacular names: Chitrak.

Part used: Root.

Organoleptic characters: Root has acrid and biting taste, odourless.

Macroscopic characters: The drug occurs in the form of cylindrical dark brown or black colored pieces of roots of ca 5–6 × 1–1.5 cm. Bark of root is ca 1–3 mm thick. Bark is longitudinally fissured, fissures shallow. Wood is pale brown to dark brown in color, hard. Fractures are short. A transversely cut surface shows thick dark brown or black colored bark and central wood with prominent growth rings.

Microscopic characters: Transverse section of root is circular in outline. It shows outer 4–5 layers of thick walled cells. Phelloderm is wide ca 5–10 cell wide, cells of phelloderm contain yellow amorphous contents. Secondary cortex is wide and consists of strands of sclerenchymatous cells. Phloem consists of rays of 2–3 cell wide. The wood consists of vessels, xylem parenchyma and fibers [Text fig. 64]. Vessels are ca 185–203 × 24–37 μ in size; perforation plates simple and horizontal; intervascular pitting is alternate, pits small and bordered [Text fig. 64 a]. Pits to xylem parenchyma similar as vessels [Text fig. 64 c]. Fibers are of two types. Thin walled and thick walled fibers [Text fig. 64 b & b']. Thin walled fibers are ca 281–375 μ in length, pitted and unseptate [Text fig. 64 b]. Thick walled fibers are pitted, ca 440–610 μ in length and with or without lumen [Text fig. 64 b'].

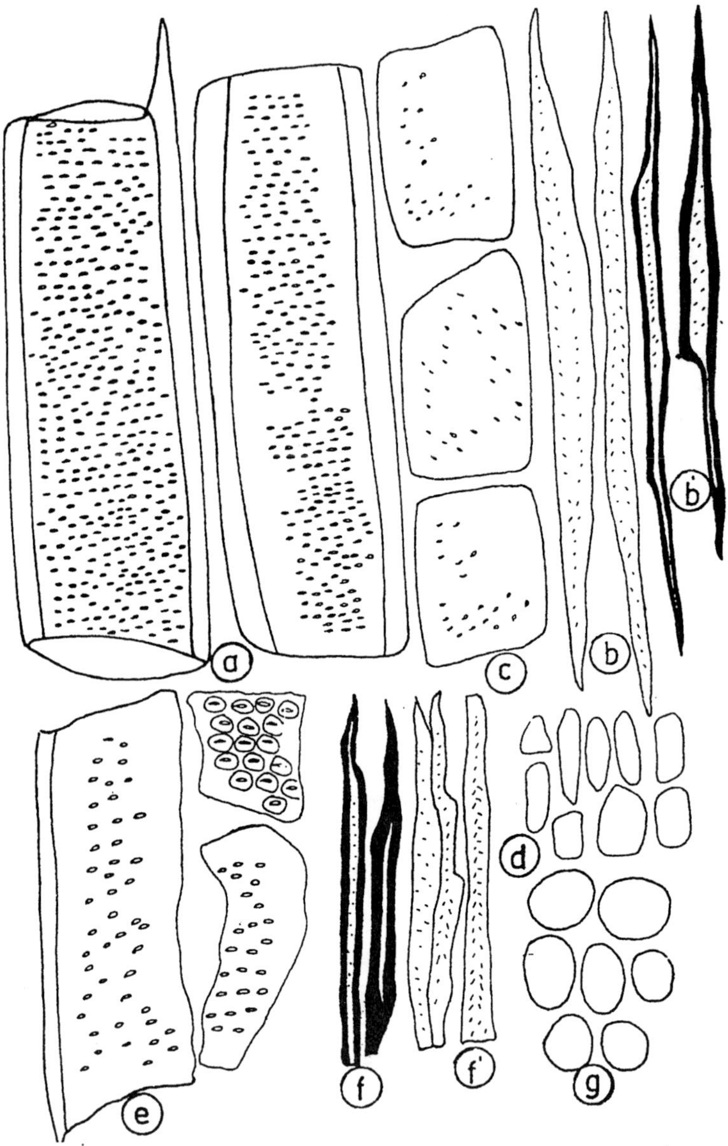

Text fig. 64: Plumbago zeylanica a-Vessels (x 450), b-Thin walled fibers (x 270), b'-Thick Walled fibers (x 150), c-Xylem parenchyma (x 675), d-Parenchyma cells (x 150), e-Pieces of vessel (x 450), f & f'-Pieces of fibers (f- 1500, f'- x 270), g-Starch grains (x 450).

Powder characters: Powder is pale brown in color containing pieces of vessels, fibers and starch grains [Text fig. 64 c-g]. Starch grains are abundant in powder. They are simple, ca 8–12 μ in diameter and oblong to rounded inshape [Text fig. 64 g].

Local uses: A tincture of root powder is given as antiperiodic. The root powder is given as enema to cure piles. It is not given to women's during pregnancy because it may cause abortion.

PHANASMUL [Photoplate-XII; Text fig.: 65 a–i]
Botanical name: **Artocarpus heterophyllus** Lam.

Synonyms: *A. integrifolia* Gamble

Family: Moraceae

Vernacular name: Phanas.

Part used: Root.

Organoleptic characters: Taste sweetish, odorless.

Macroscopic characters: The drug occurs in the form of cylindrical, reddish-brown or orange colored pieces of roots of ca 4–6 × 1–1.5 cm. Outer bark is reddish-brown or orange colored, rough and irregularly fissured. Lenticels are large, prominent, fresh orange colored and rounded or elongated. Outer cork is lamilatted breaking off into thin sheets. Inner bark is pale brown, ca 1–2 mm thick and fractures are fibrous and hard. A transversely cut surface of root shows orange cork, ring of latex, vessels and pale brown wood possessing dark brown concentric rings and radiating medullary rays.

Microscopic characters: Transverse section of root is circular in outline. It shows outer multilayered cork. Phellogen is 3–4 layered. Phelloderm contains latex tubes and solitary crystals of calcium oxalate. Cortex is wide made up of parenchymatous cells with numerous sclerenchyma cells. Secondary phloem is crushed and consists of sclerenchyma cells. Secondary xylem contains paratracheal confluent parenchyma. Rays are 2–3 cell wide. Wood consists of vessels, xylem parenchyma and fibers [Text fig. 65 a-c]. Vessels are ca 175–246 × 55–62 μ in size; perforation plates are simple and horizontal; intervascular pitting is alternate and pits are bordered [Text fig. 65 a]. Pits to xylem parenchyma are usually larger than the pits of vessels [Text fig. 65 c]. Fibers are ca 166–750 μ in length, thick walled and unseptate [Text fig. 65 b].

Powder characters: Powder is pale yellow in color containing pieces of vessels and fibers, stone cells, crystals of calcium oxalate and starch grains [Text fig. 65 d–i].

Stone cells are very common in powdered material. Stone cells are ca 37–74 × 35–42 μ in size, triangular to oblong or square in shape, thick walled and with broad lumen [Text fig. 65 f]. Crystals of calcium oxalate are of two types, clustered and solitary. Clustered crystals are ca 14–18 μ in diameter where as solitary crystals are 12–24 μ in size [Text fig. 65 g & h]. Starch grains are found to be abundant in powder. Starch grains are simple, rounded and ca 6–18 μ in diameter [Text fig. 65 i].

Local uses: Root paste applied twice a day on mouth ulcers. Similarly root latex is externally filled in ulcerated wounds.

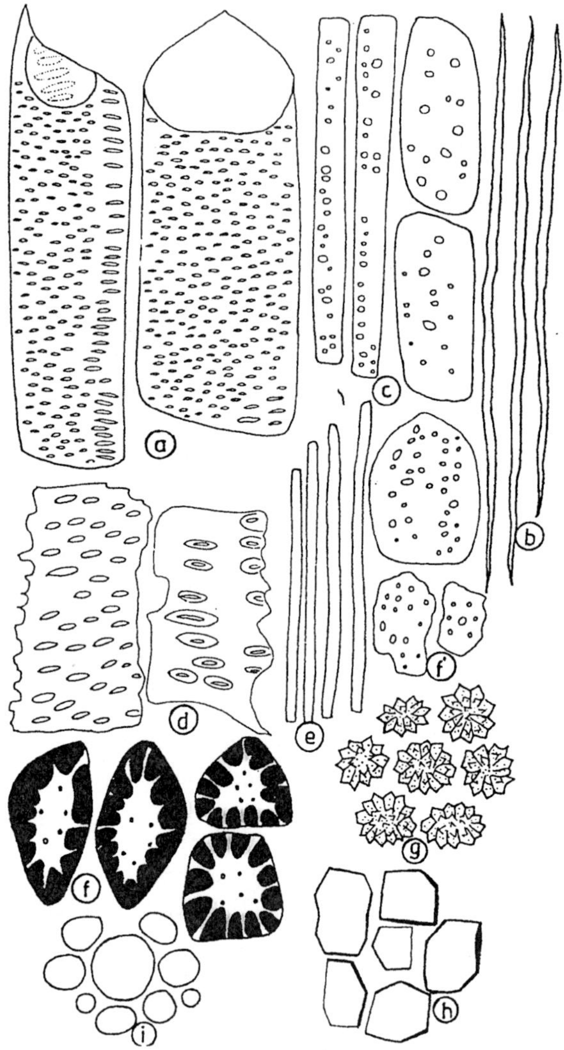

Text fig. 65: Artocarpus heterophyllus a-Vessels (x 450), b-Fibers (x 150), c-Xylem parenchyma (x 675), d-Pieces of vessel (x 675), e-Pieces of fibers (x 150), f-Stone cells (x 675), f'-Thick walled pitted pieces of Stone cells (x 675), g-Clustered Crystals of Calcium oxalate (x 675), h-Prism type of Crystals of Calcium oxalate (x 675), i-Starch grains (x 675).

PILUMUL [Photoplate-XII; Text fig.: 66 a–f]
Botanical name: **Salvadora persica** L.

Family: Salvadoraceae.

Vernacular name: Pilu.

Part used: Root.

Organoleptic characters: Taste sweetish and slightly biting, odor slightly pungent when fresh.

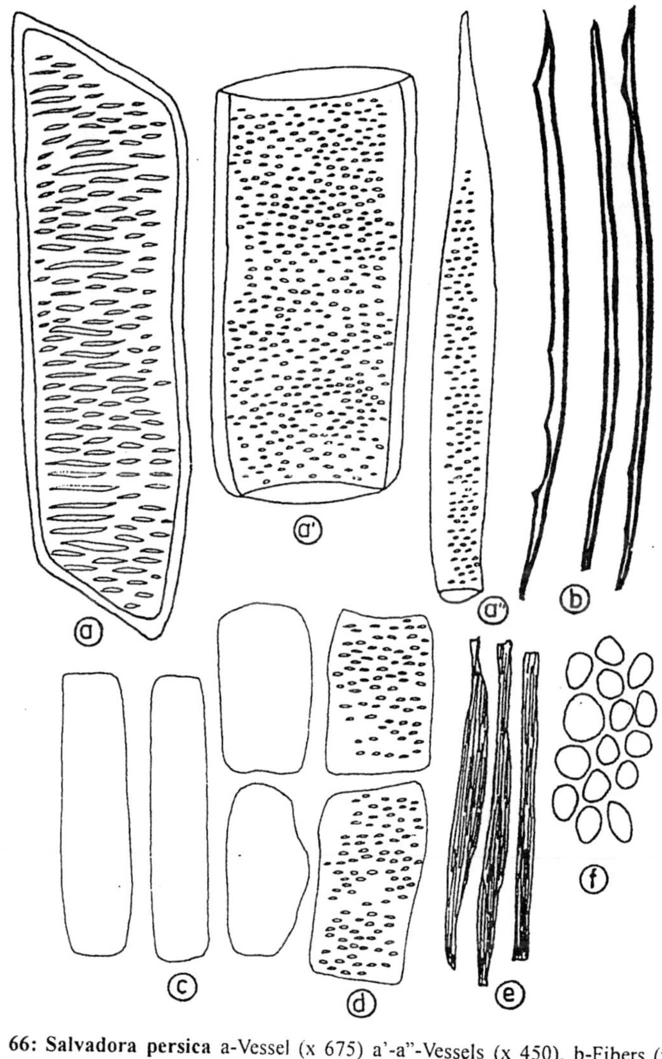

Text fig. 66: Salvadora persica a-Vessel (x 675) a'-a"-Vessels (x 450), b-Fibers (x 150), c-Xylem parenchyma (x 450), d-Pieces of vessel (x 450), e-Pieces of fibers (x 150), f-Starch grains (x 675).

Macroscopic characters: The drug occurs in the form of cylindrical, fawn or pale brown colored pieces of root of ca 9–15 × 1–8 cm. Bark of root is pale brown in color, ca 1–2 mm thick. It is smooth and easily separate from the wood. Root is without wrinkles or fissures. Wood is pale white and soft. Fractures are short. A transversely cut surface shows dark brown concentric rings.

Microscopic characters: Transverse section of root is circular in outline. It shows outer thin walled, 6–8 layered cork cells. Phellogen is 1–2 layered. Phelloderm is 5–6 layered made up of rectangular parenchymatous cells. Starch grains are abundant in phelloderm. Secondary cortex is made up of thin walled parenchymatous cells containing abundant starch grains. Transverse section of root shows secondary anomalous structure. Phloem is included, interxylary and of foraminate type with round to oval strands of included phloem surrounded and commonly linked together by parenchyma around vessels and the phloem strands and often lining the latter in broad bands. Rays are up to 1–2 cell wide. Wood consists of vessels, xylem parenchyma and fibers [Text fig. 66 a-c]. Vessels are ca 100–132 × 11–80 μ in size. Perforation plates are simple and horizontal. Intervascular pitting is alternate to scalariform [Text fig. 66 a–a"]. Pits to xylem parenchyma are similar as vessels [Text fig. 66 c]. Fibers are ca 440–670 μ in length and thick walled [Text fig. 66 b].

Powder characters: Powder is pale yellow in color, mainly containing pieces of vessels, fibers and starch grains [Text fig. 66 d–f]. Starch grains are found to be abundant in powder. Starch grains are simple, ca 8–12 μ in diameter and elliptic oblong to rounded in shape [Text fig. 66 f].

Local uses: Root is used as a toothbrush. It strengthens the gums, keeps them from becoming spongy and improves digestion. Root decoction is given 3 times in a day in mild fever.

SAPSHINDI/SPASAN [Photoplate-XII; Text fig.: 67 a–h]

Botanical name: **Cocculus hirsutus** (L.) Theob.

Synonyms: *Menispermum hirsutum* L., *Cocculus villosus* (Lam.) DC.

Family: Menispermaceae.

Vernacular name: Jaljamni, Karvom, Paravel, Vasan vel.

Part used: Root.

Organoleptic characters: Taste very bitter, odorless.

Macroscopic characters: The drug occurs in the form of cylindrical, dark brown color pieces of root of ca 4–7 × 1–1.5 cm in size. Bark of root is ca 1–2 mm thick, dark brown and irregularly fissured; fissures are deep. Transverse cracks are present on root. Wood is soft and yellowish-gray in color. Fractures are short. Transversely cut surface of root shows anomalous secondary growth. The xylem is wedge shaped and stellate.

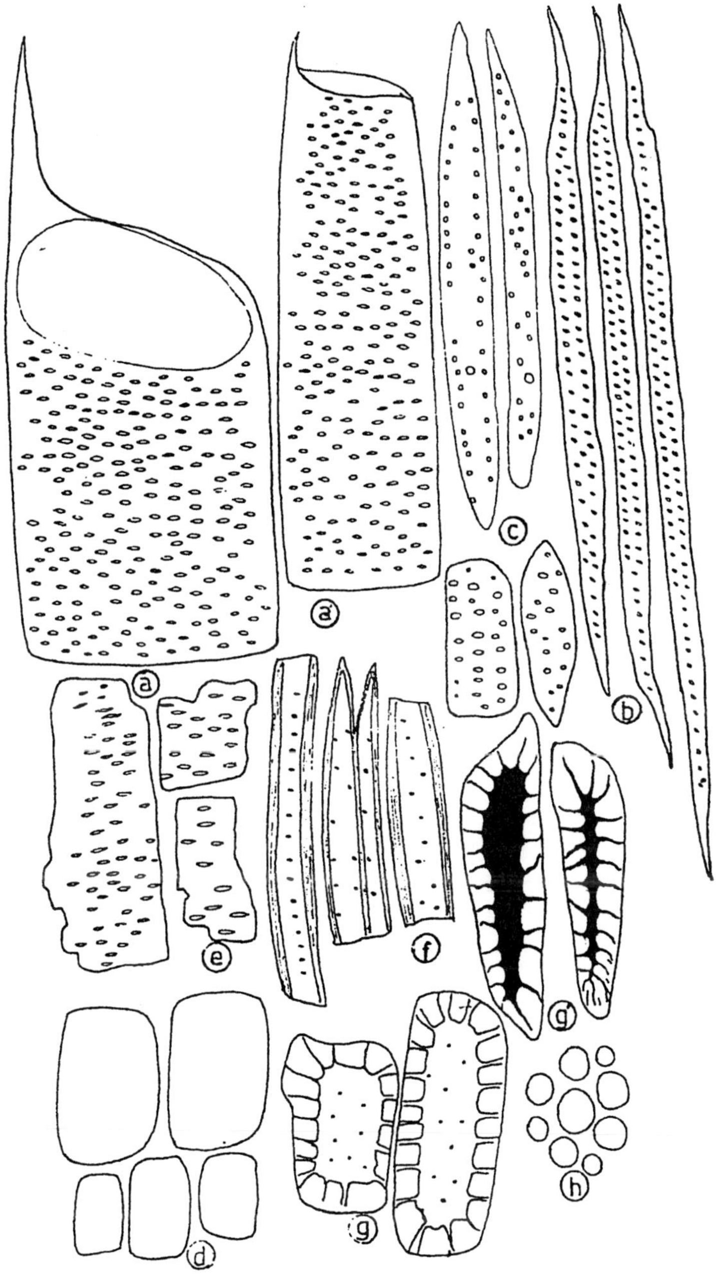

Text fig. 67: Cocculus hirsutus a & a'-Vessels (x 450), b-Fibers (x 450), c-Xylem parenchyma (x 450), d-Parenchyma (x 450), e-Pieces of vessel (x 450), f-Pieces of fibers (x 450), g & g'-Stone cells (x 450), h-Starch grains (x 450).

Microscopic characters: Transverse section of root is circular in outline. It shows outer 6–8 layered cork. The cork consists of 1–2 layered phellogen and 4–5 layers of phelloderm in which bands of stone cells are present. Secondary cortex is wide in which stone cells are scattered. Transverse section of root shows anomalous secondary growth. Xylem is star shaped situated at the center by partially or completely obliterating pith. Phloem is situated at the ends of radii. Rays are ca 10–23 cell wide and high. Wood consists of vessels, fibers and xylem parenchyma [Text fig. 67 a–c]. Vessels are ca 125–333 × 58–100 μ in size; perforation plates are simple and slightly oblique; intervascular pitting is alternate [Text fig. 67 a–a']. Pits on xylem parenchyma are rounded with slit like apertures [Text fig. 67 c]. Fibers are fibrotracheids ca 240–296 μ in length, moderately thick walled with numerous distinct bordered pits on both radial and tangential walls [Text fig. 67 b].

Powder characters: Powder is chocolate brown in color containing pieces of vessels and fibers, stone cells and starch grains [Text fig. 67 c–h]. Stone cells are ca 55–124 × 22–55 μ in size, triangular to rectangular in shape. They have moderately thick walls, and broad lumen [Text fig. 67 g-g']. Starch grains are found to be abundant. They are simple and ca 9–16 μ in diam. [Text fig. 67 h].

Local uses: Teaspoon of root powder is given with buttermilk to cure gas trouble and stomachache.

SAPTRANGI [Photoplate-XII; Text fig.: 68 a-i]
Botanical name: **Salacia brunoniana** Wight & Arn.

Family: Celastraceae.

Vernacular name: Saptrangi, Nisulbondi.

Part used: Root.

Organoleptic characters: Taste astringent, odorless.

Macroscopic characters: The drug occurs in the form of cylindrical yellow colored pieces of root of ca 3–6 × 1.5–2.5 cm in size. Bark is ca 2–4 mm thick. Outer part of bark consists of thin yellow colored 4–5 sheets of cork, which are crumpled. Inner part of bark is longitudinally fissured or is without fissures. Transverse cracks are present on root. Wood is pale pink or reddish-brown and hard. Fractures are short. Transversely cut surface of root shows secondary anomalous growth containing alternating rings of secondary xylem and phloem.

Microscopic characters: Transverse section of root is circular in outline. It shows multilayered cork cells. Cork consists of inner 1–2 layers of phellogen and 5–6 layers of phelloderm. Secondary cortex is wide consisting of irregularly scattered stone cells, crystals of calcium oxalate and tannin sacs. Secondary growth is abnormal consists of successive rings of interxylary

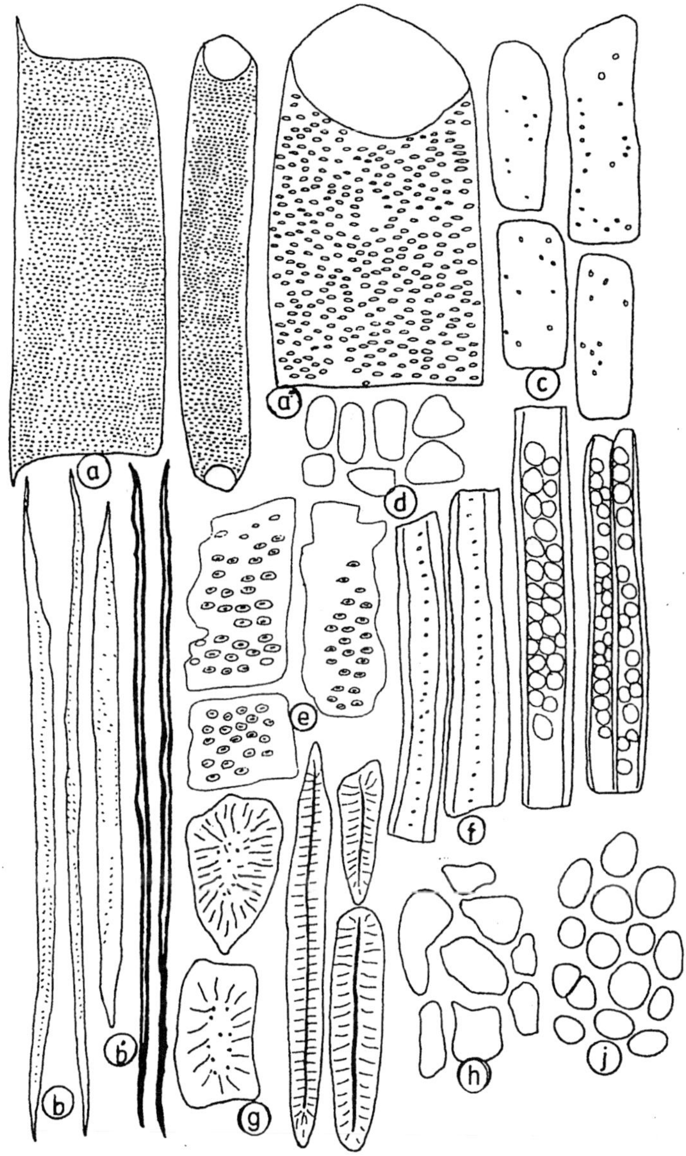

Text fig. 68: Salacia brunoniana a-Vessel (x 150) & a'-Vessel (x 450), b-Thin walled fibers (x 150), b'-Thick walled fibers (x 150), c-Xylem parenchyma (x 450), d-Parenchyma cells (x 150), e-Pieces of vessel (x 450), f-Pieces of fibers (x 450), g-Stone cells (x 450), h-Tannin (x 675), i-Starch grains (x 675).

phloem. Rays are exclusively uniseriate. Wood consist of vessels, xylem parenchyma, tracheids, tracheid fibers and fibers [Text fig. 68 a–c]. Vessels are ca 444–527 × 55–150 μ in size; perforation plates are simple and horizontal; intervascular pitting is alternate. Pits are very small and bordered [Text fig. 68 a–a'] Pits to xylem parenchyma are small [Text fig. 68 c]. Tracheids are ca 722–780 μ in length, tapering at both ends [Text fig 68 b]. Tracheid fibers are ca 611–1055 μ in length, slightly pitted [Text fig. 68 b']. Fibers are ca 527–861 μ in length, thick walled with or without lumen [Text fig. 68 b'].

Powder characters: Powder is reddish brown or pink in color with small yellow colored pieces of outer bark. Powder consists of pieces of vessels, tracheids, fibers, stone cells, crystals of calcium oxalate, thin walled cells containing starch grains and starch grains [Text fig. 68 e–i].

Stone cells are pitted or unpitted, ca 51–166 × 18–46 μ in size, rectangular to oblong or linear lanceolate to triangular in shape,. They have thick walls and narrow or broad lumen [Text fig. 68 g]. Crystals of calcium oxalate are solitary, ca 8–10 μ in size [Text fig. 68]. Starch grains abundant in powder. They are simple or compound, ca 7–12 μ in diameter [Text fig. 68 i]. Yellow colored tannin contents are very common and irregular in shape [Text fig. 68 h].

Chemical tests:

Tests	Reagents	Results
Starch	Iodine	+ ve bark
Tannins	1% $FeCl_3$	+ ve Phelloderm, secondary cortex, secondary phloem. Rings of secondary phloem. Powder extract shows pale green color.

Local uses: Tincture of root powder is given thrice a day as a blood purifier. The root powder is given early in the morning with water in urinary calculi. The powder is used as antidiabetic. The decoction of root is used to clean the eyes in eye complaints.

SAVARMUL [Photoplate-XII; Text fig.: 69 a-j]

Botanical name: **Bombax ceiba** L.

Family: Bombacaceae.

Vernacular names: Kantesavar, Samar, Savári, Shevari.

Part used: Root.

Organoleptic characters: Taste sweetish and mucilaginous, odorless.

Macroscopic characters: The drug occurs in the form of cylindrical, silvery pale brown colored pieces of roots of ca 6–7 × 3.5–2 cm. Bark of root is ca 3–4 mm thick and rough. Outer bark is thin, papery, shining pale brown colored and flanking off in the form of thin pieces [Photoplate-XII] Inner bark is dark reddish brown or black colored, wrinkled and mucilaginous. Root is longitudinally wrinkled. Wood is pale brown and soft. Fractures are fibrous. Vessels are sparsely distributed.

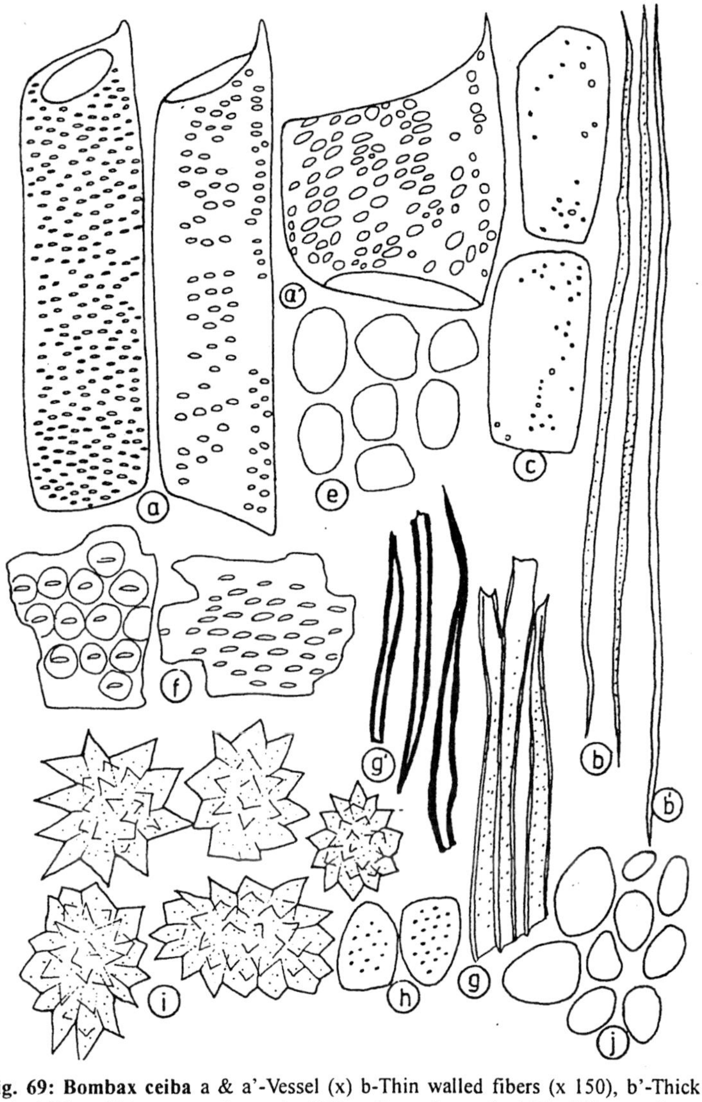

Text fig. 69: Bombax ceiba a & a'-Vessel (x) b-Thin walled fibers (x 150), b'-Thick walled fibers (x 150), c-Xylem parenchyma (x 450), e-Parenchyma (x 450), f-Pieces of vessel (x 675), g & g'-Pieces of fibers (x 450), h-Pitted thick walled cells (x 675), i-Clustered crystals of Calcium oxalate (x 675), j-Starch grains (x 675).

Microscopic characters: Transverse section of root is circular in outline and shows outer multilayered cork containing thin walled and thick walled cells. Inner phelloderm is 3–5 layered. Secondary cortex is wide and consists of crystals of calcium oxalate and sclerenchyma patches. Secondary phloem consists of rectangular, incomplete or discontinuous rings of sclerenchyma. Parenchyma is abundant and diffuse. Rays are 2–7 cell wide. Wood consists of vessels, xylem parenchyma and fibers [Text fig. 69 a–c]. Vessels are ca 120–200 × 74–37 μ in size; perforation plates are horizontal or oblique; intervascular pitting is alternate and large [Text fig. 69 a & a']. Pits to xylem parenchyma are smaller than pit on vessels [Text fig. 69 c]. Fibers are ca 833–890 μ in length and are thin walled, pitted and unseptate [Text fig. 69 b]. Thick walled fibers are ca 890–1278 μ in length, with or without lumen and unseptate [Text fig. 69 b'].

Powder characters: Powder is dark brown in color containing pieces of vessels and fibers, crystals of calcium oxalate and starch grains [Text fig. 69 f–j]. Fibers are found to be major component of powder. Powder becomes mucilaginous in water. Crystals of calcium oxalate are clustered, ca 30–48 μ in diameter, abundant in powder [Text fig. 69 i]. Starch grains are abundant in powder. They are simple and ca 4–14 μ in diameter [Text fig. 69 j].

Local uses: One teaspoon of powder is given twice a day with milk to women suffering from leucorrhoea. Roots are crushed with water and filtered and filtrate is used for ulcerated wounds. It is also given as a tonic.

SHIVANMUL [Photoplate-XII; Text fig.: 70 a–h]
Botanical name: **Gmelina arborea** Roxb.

Family: Verbenaceae

Vernacular names: Gambari, Kasmari, Khamar, Shivan.

Part used: Root.

Organoleptic characters: Taste bitter and astringent, odorless.

Macroscopic characters: The drug occurs in the form of cylindrical pale brown pieces of roots of ca 5–6 × 1–1.5 cm [Photoplate-XII]. Bark of root is 1–2 mm thick, pale brown and rough. Root is longitudinally striated and striations are deep. Wood is pale yellow in color and hard. Fractures are short. A transversely cut surface of root shows central wood containing annual growth rings and prominent vessels.

Microscopic characters: Transverse section of root is circular in outline. It shows outer multilayered cork consists of outer 4–5 layers of thick walled cells, inner 2–3 layers of phellogen and 5–6 layers of phelloderm. Secondary cortex is wide in which small islands of stone cells are scattered. Secondary phloem is crushed and possesses bands of yellow colored crystalline substances. Wood consists of vessels, xylem parenchyma and fibers [Text fig. 70 a-c]. Vessels are ca 250–312 × 71–140 μ in size; perforation plates are simple and horizontal; intervascular pitting is alternate and pits are bordered [Text fig. 70 a]. Pits to xylem parenchyma are similar as in vessels [Text fig. 70 c]. Fibers are ca 583–1050 μ in length, thin walled and septate [Text fig. 70 b].

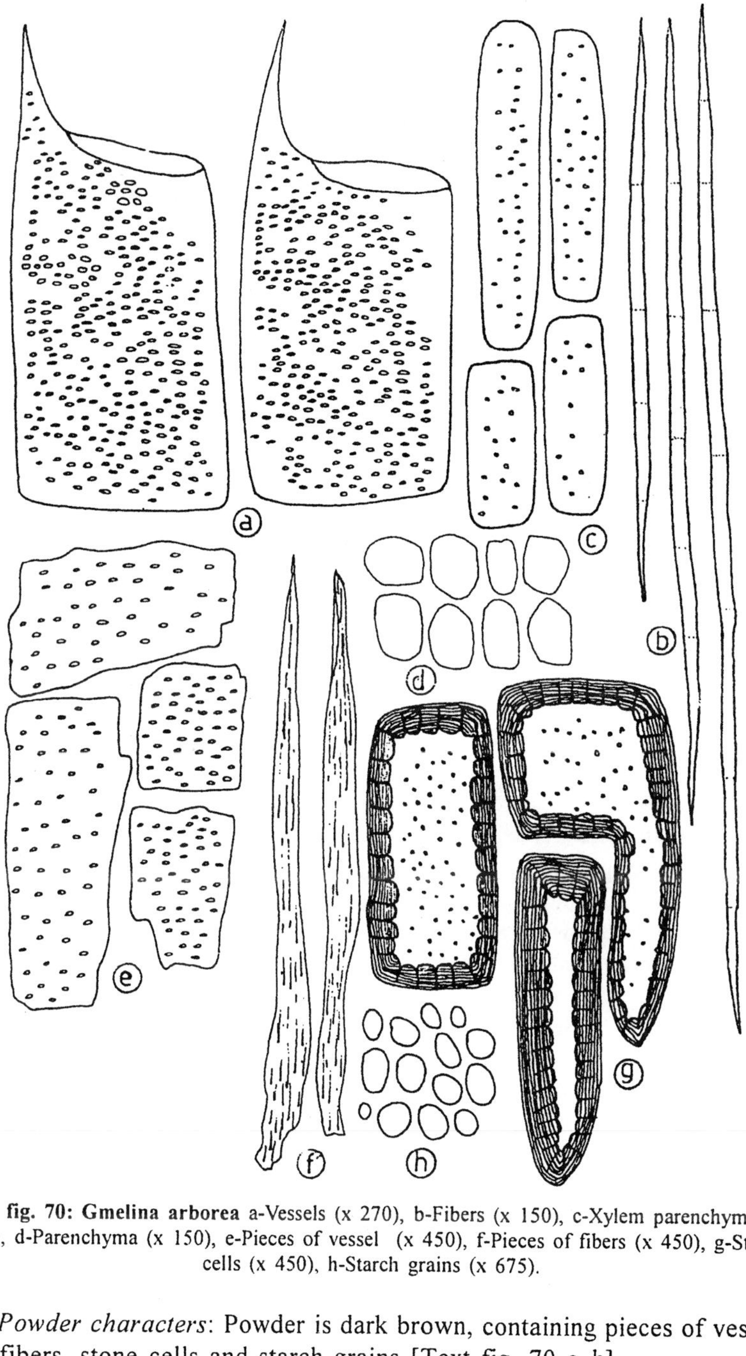

Text fig. 70: Gmelina arborea a-Vessels (x 270), b-Fibers (x 150), c-Xylem parenchyma (x 450), d-Parenchyma (x 150), e-Pieces of vessel (x 450), f-Pieces of fibers (x 450), g-Stone cells (x 450), h-Starch grains (x 675).

Powder characters: Powder is dark brown, containing pieces of vessels and fibers, stone cells and starch grains [Text fig. 70 e–h].

Stone cells are ca 74–111 × 27–59 μ in size, rectangular in shape, thick walled, striated, pitted and have lumen [Text fig. 70 g]. Starch grains are abundant in powder. They are simple, ca 6–12 μ in diameter and rounded in shape [Text fig. 70 h].

Local uses: The root is used as one of the *Dashmula* in Ayurveda. The decoction of root is given in abdominal pains. Decoction of root with leaf decoction of *Andrographis paniculata* is given in low fever.

TETUMUL [Photoplate–XII; Text fig.: 71 a–h]
Botanical name: **Oroxylum indicum** (L.) Vent.

Synonyms: *Bignonia indica* L.

Family: Bignoniaceae.

Vernacular names: Shivan, Tetu.

Parts used: Root.

Organoleptic characters: Taste bitter, odorless.

Macroscopic characters: The drug occurs in the form of pale brown colored cylindrical pieces of root of ca 5–7 × 2–4 cm [Photoplate-XII]. Bark of root is ca 5–8 mm thick and rough. Outer surface is pale brown corky and longitudinally fissured. Fissures are deep due to presence of rhytidom. Lenticels are transversely elongated. Wood is pale yellow and hard.

Microscopic characters: Transverse section of root is circular in outline. It shows outer multilayered thin walled parenchymatous cells of rhytidome. Inner bark consists of compactly arranged phelloderm and stone cells. Secondary cortex is wide consisting of patches of sclerenchyma cells. Yellow colored crystalline content is present in cortical cells. Parenchyma is paratracheal and confluent. Rays are 2–3 cell wide. Wood consists of vessels, xylem parenchyma and fibers [Text fig. 71 a–c]. Vessels are ca 229–351 × 111–129 μ in size; perforation plates are simple and oblique; intervascular pitting is alternate and pits are border [Text fig. 71 a]. Pits on xylem parenchyma are similar as that of vessels [Text fig. 71 c]. Fibers are of two types, thin walled fibers are ca 791–1083 μ in length, unseptate and without pits [Text fig. 71 b]. Thick walled fibers are ca 916–1125 μ in length and are without lumen [Text fig. 71 b'].

Powder characters: Powder is brown, containing pieces of vessels and fibers, stone cells and starch grains [Text fig. 71 e-h].

Stone cells are scarcely found in the powder. They are ca 74–83 × 35–55 μ, rectangular to hexagonal or polygonal in shape and have blunt ends, thin walls, broad striated lumen and pits [Text fig. 71 g]. Starch grains are simple, ca 12–18 μ in diameter. They are scarcely found in the powder [Text fig. 71 h].

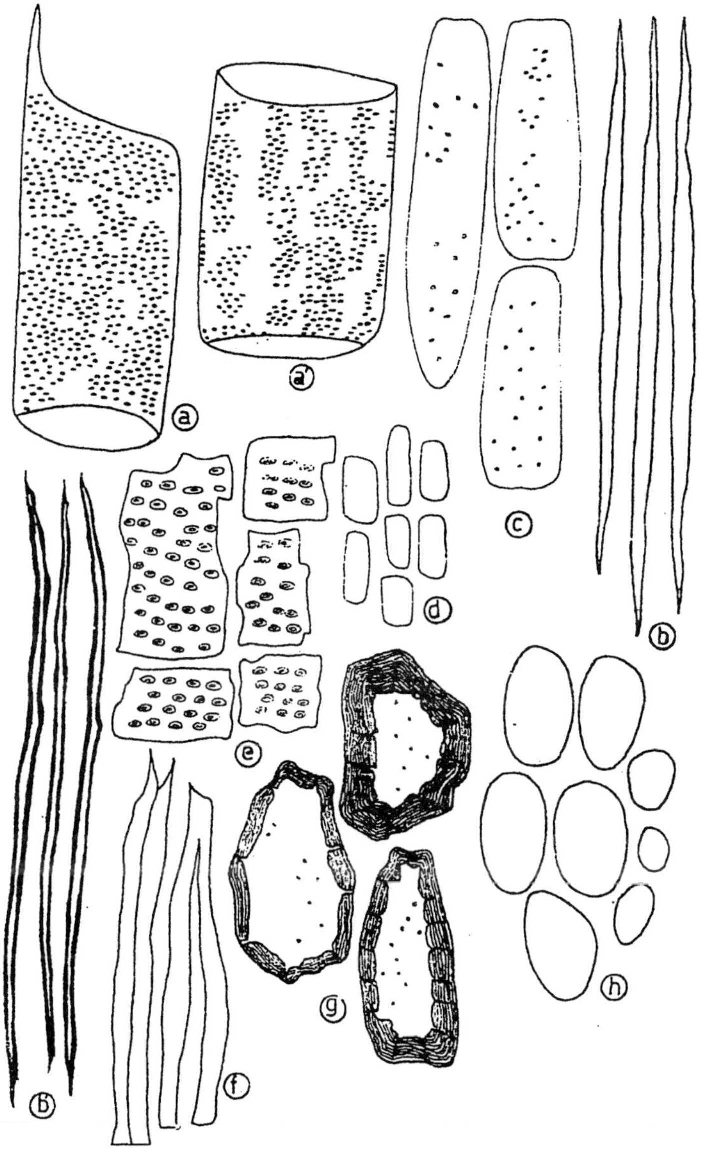

Text fig. 71: Oroxylum indicum a-Vessels (220x), b-Thin walled fibers (x 100), b'-Thick walled fibers (x 100), c-Xylem parenchyma (x 450), d-Parenchyma cells (x 150), e-Pieces of vessel (x 675), f-Pieces of fibers (x 150), g-Stone cells (x 450), h-Starch grains (x 675).

Local uses: It is one of the *Dashmula* drug of Ayurveda. Root decoction is given in cough and fever.

VAVDINGMUL [Photoplate-XII; Text fig.: 72 a–i]
Botanical name: **Embelia basaal** (R. & S.) A. DC.

Synonyms: *E. tsjerium-cottam* Sant., *E. robusta* C. B. Cl.

Family: Myrsinaceae

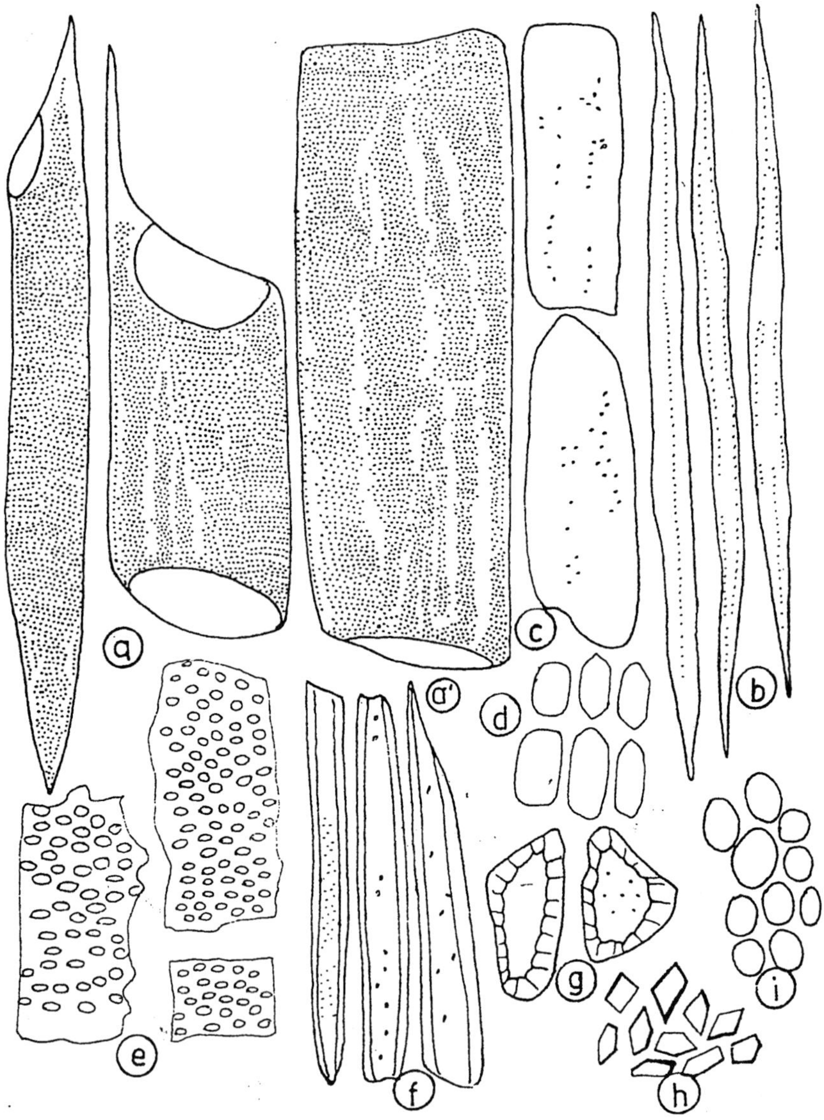

Text fig. 72: Embelia basaal a & a'-Vessels (x 270), b-Fibers (x 270), c-Xylem parenchyma (x 270), d-Parenchyma (x 150), e-Pieces of vessel (x 450), f-Pieces of fibers (x 450), g-Stone cells (220x), h-Crystals of Calcium oxalate (x 450), i-Starch grains (x 675).

Vernacular name: Ambaldange, Ambati, Elang, Vowding.

Part used: Root.

Organoleptic characters: Taste bitter, odorless.

Macroscopic characters: The drug occurs in the form of pale pink, cylindrical pieces of roots of ca 5–6 × 1.9–1 cm [Photoplate-XII]. Bark of root is 2–5 mm thick, pale pink, smooth. Root slightly fissured. Wood is pale pink and hard, fractures are hard and fibrous. A transversely cut surface of stem shows broad white colored radiating rays in central wood and thick bark.

Microscopic characters: Transverse section of root is circular in outline. Transverse section of root shows cork made up of 2–5 layered thick walled cells. Secondary cortex is wide and consists of tannin cells and sclerenchyma patches. The secondary phloem is interrupted by broad wedge shaped bands of stone cells directly encircling in ray cells. Rays are 6–13 cell wide. Wood consists of vessels, xylem parenchyma and fibers [Text fig. 72 a-c]. Vessels are ca 234–312 × 93–125 μ in size; perforation plates are simple and horizontal and intervascular pitting is alternate and small [Text fig. 72 a-a']. Pits to xylem parenchyma are similar as to that of vessels [Text fig. 72 c]. Fibers are ca 218–406 μ in length. They are pitted and either septate or unseptate [Text fig. 72 b].

Powder characters: Powder is pale pink in color which contains pieces of vessels and fibers, stone cells, crystals of calcium oxalate and starch grains [Text fig. 72 e-i].

Stone cells are scarcely found in powder. Stone cells are pitted, variously shaped, ca 62–93 × 46–31 μ in size, and with broad lumen [Text fig. 72 g]. Crystals of calcium oxalate are abundant in powder, solitary and ca 20–31 μ in size, prism shaped [Text fig. 72 h]. Starch grains are abundant in powder. They are simple, rounded to oblong and ca 6–12 μ in diameter [Text fig.72 i].

Chemical tests:

Tests	Reagents	Results
starch	Iodine	+ ve
Tannin	1% $FeCl_3$	+ ve Phelloderm, secondary cortex, secondary phloem. Medullary rays. Powder extract shows pale green color.

Local uses: The root powder is given every morning with water to reduce heat from body. It is also applied externally in earache. Paste of root is applied externally on chronic ulcers and on wounds.

UAPLSARI [Photoplate-XII; Text fig.: 73 a–i]

Botanical name: **Hemidesmus indicus** R. Br.

Family: Periplocaceae.

Vernacular name: Ananta, Anantmul, Sariva, Upalsari.

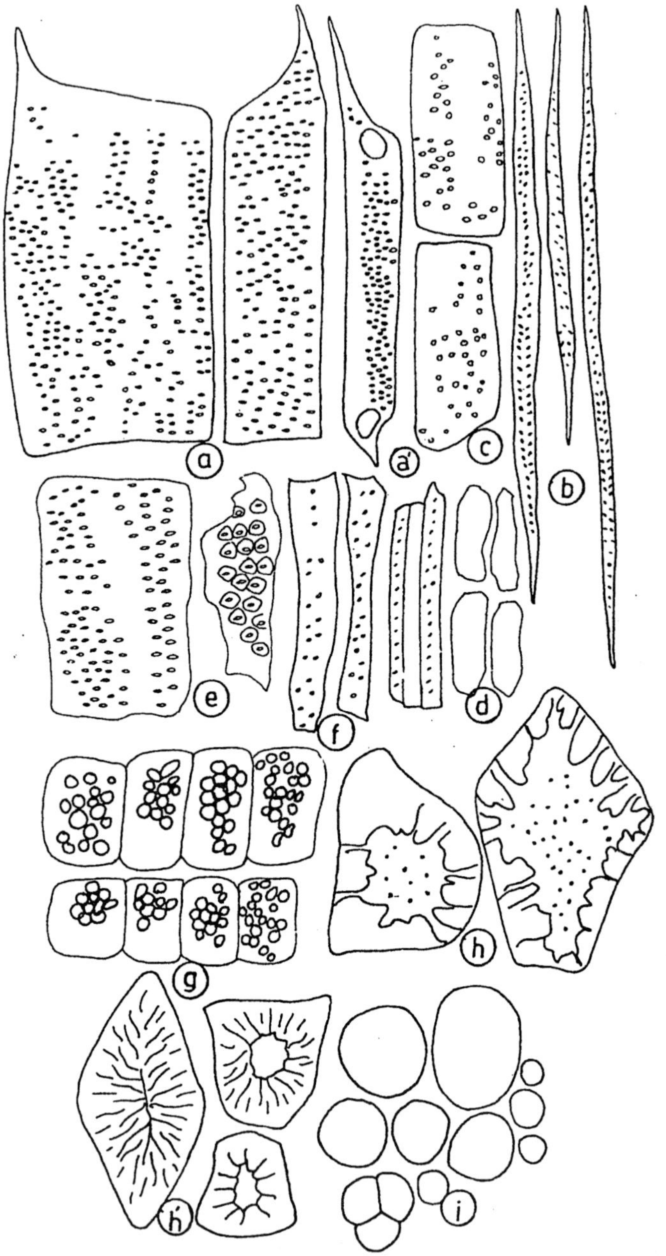

Text fig. 73: Hemidesmus indicus a & a'-Vessels (x 450, × 270), b-Fibers (x 150), c-Xylem parenchyma (x 675), d-Parenchyma (x 450), e-Pieces of vessel (x 450), f-Pieces of fibers (x 150), g-Thin walled cells containing starch grains (x 450), h & h'-Stone cells (x 450), i-Starch grains (x 450).

Part used: Root.

Organoleptic characters: Taste bitter, odor agreeable.

Macroscopic characters: The drug occurs in the form of cylindrical, tortuous, grayish purple or reddish-brown colored pieces of roots of ca 6–10 × 3–7 cm [Photoplate-XII]. Bark is thin, rough. Root is longitudinally fissured and transversely cracked. Wood is soft and yellow colored. Fractures are short.

Microscopic characters: Transverse section of root is circular in outline. It shows outer 6–10 layers of thick walled crushed cells of cork. Phellogen is 1–2 layered. Phelloderm is 6–8 layered. Secondary phloem consists of uniseriate rays. Laticiferous canals are present in phelloderm and in phloem. The wood consists of vessels, xylem parenchyma and fibers [Text fig. 73 a–c]. Vessels are ca 92–157 μ × 27–55 μ in size; perforation plates are simple and horizontal; intervascular pitting is alternate. Pits are small and bordered [Text fig. 73 a & a']. Pits to xylem parenchyma are similar as that of vessels [Text fig. 73 c]. Fibers are ca 400–670 μ in length, thin walled and pitted [Text fig. 73 b].

Powder characters: Powder is buff or yellow colored containing pieces of vessels and fibers, thin walled parenchymatous cells filled with starch grains, stone cells and starch grains [Text fig. 73 e–i].

Stone cells are scarcely found in the powder. They are ca 24–60 × 18–42 μ in size rectangular to polygonal in shape, thick or thin walled and have pitted lumen or lumen is absent [Text fig. 73 h & h']. Starch grains are abundant. They are simple or compound, ca 14–20 μ in diameter, oval to spherical in shape and striated [Text fig. 73 i].

Local uses: The root is given as a blood purifier. Root powder is used with root powder of *Rubia cordifolia* to remove black marks created due to pimples. Root powder is given twice a day to cure inflammation of urinary passage. Powder is used on various skin diseases.

TUBERS, CORMS AND RHIZOMES USED IN FOLK MEDICINES:

AMARKAND

Botanical name: **Eulophia nuda** Lindl.

Family: Orchidaceae

Vernacular names: Amarkand.

Part used: Rhizome.

Organoleptic characters: Taste slightly bitter, odorless.

Macroscopic characters: The drug occurs in the form of rhizome. Rhizomes are ca 5–10 × 5–3 cm in size which are unbranched, rounded to oblong or spherical in shape and taper towards both ends. Rhizome color varies from pale yellow to green. Rhizome surface is smooth and possess transverse ridges and thick roots. The other rhizomes of other species i.e. *E. ochreata* are also sold under name of *Amarkand*. The rhizome of *E. ochreata* can be distinguished from rhizome of *E. nuda* by its conical shape and longitudinal wrinkles.

Powder characters: Powder is gray or black in color which contains only parenchyma cells and starch grains.

Local uses: Rhizomes are used as anthelmintic, vermifuge, to remove scrofulous glands in the neck, in bronchitis and disease of the blood. It is used as nutrient source. The rhizome is appetizer, stomachache, tonic and in aphrodisiac. Powdered rhizomes are given with the milk in acidity. Rhizomes are given twice a day orally to the women in backache. Rhizomes are given orally in piles.

BHUI KOHALA [Text fig.: 74 a–g]
Botanical name: **Pueraria tuberosa** (Roxb. ex Willd.) DC.

Family: Fabaceae

Vernacular names: Bhumikusmand, Vidarika, Vidarikand.

Parts used: Tubers.

Organoleptic characters: Taste mucilaginous, slightly bitter, odorless.

Macroscopic characters: The drug occurs in the form of globular or pear shaped, brown colored tubers or sliced tubers of ca 20–40 × 15–25 cm size. Bark of tuber is thin and light brown. The surface of tuber is smooth and glabrous in fresh tubers while wrinkled in dried samples with beady spots. Fractures are fibrous. A transversely cut surface shows a white spongy, highly fibrous inner tissue. Sliced pieces are rough, creamy in color and are composed of fibrous pulp which exfoliates in papery flakes.

Powder characters: Powder is dirty white containing vessels, phloem fibers, thin walled parenchymatous cells, crystals of calcium oxalate, stone cells and starch grains [Text fig. 74 a–g]. Vessel pieces are ca 10–60 × 5–20 μ; intervascular pitting is alternate and pits are bordered [Text fig. 74 a]. Fragment of phloem fibers are ca 330–470 μ in length which are encircled by a parenchymatous sheath. Fibers possesses rows of crystals of calcium oxalate encircled by sheath. [Text fig. 74 b–c]. Stone cells are ca 3–9 × 1–6 μ in size, thick walled and are without lumen [Text fig. 74 e]. Thin walled parenchymatous transparent cells are present [Text fig. 74 d]. Crystals are ca 10–30 μ in length [Text fig. 74 f]. Starch grains are ca 3–7 μ in diameter, and are rounded in shape [Text fig. 74 g].

Local uses: Tuber is used as a tonic. Pieces of tubers are given in menorrhoea and leucorrhoea. Tubers are given orally in every night and applied externally in case of rheumatism. Tubers are given orally with water early in the morning in case of epilepsy. It is given in debility caused by fever.

CHAMBHAR GONDA [Text fig.: 74 h-k]
Botanical name: **Bryonia dioica** Jacq.

Family: Cucurbitaceae.

Vernacular names: Chambargonda

Part used: Tubers.

Organoleptic characters: Taste bitter, (very) mucilaginous, odor disagreeable fetid.

Macroscopic characters: The drug occurs in the form of tubers. Tubers are ca 3–7 to 1.5–3 cm in size, unbranched, and rounded to oblong in shape. Apical portion of tuber or towards shoot of tuber is hard stem like. Bark is thin. Lenticels are small, circular or elongated. Fractures are short. Transversely cut surface of tuber is mealy white. Tuber consists of yellowish watery juice which comes out when surface is ruptured.

Powder characters: Powder is dirty white in color containing pieces of vessels, stone cells, parenchyma cells and starch grains [Text fig. 74 h–k]. Pieces of vessels are ca 46–157 × 37–55 μ in size. Intervascular pits are alternate and pits are bordered [Text fig. 74 h]. Parenchyma cells are rounded to oblong in shape containing starch grains [Text fig. 74 j]. Stone cells are ca 46–59 × 27–46 μ in size, thick walled, square to oblong, narrow at one end with narrow lumen which is sometimes pitted [Text fig. 74 i]. Starch grains are abundant, simple, ca 7–18 μ in diameter, elliptic oblong in shape [Text fig. 74 k].

Local uses: A small tincture of powder of tuber is given to stop the stomachache. The tubers are not given to pregnant women. It may cause abortion even to the 7 month fetus. It is given as abortifacient.

KOMBADNKHI [Text fig.: 74 l–p]
Botanical name: **Tectaria cicutaria** L.

Family: Aspidiaceae

Vernacular names: Waghchawadi.

Part used: Rhizome.

Organoleptic characters: Initially taste sweetish and astringent, later on bitter, odor disagreeable in powdered material.

Macroscopic characters: Drug occurs in the form of irregularly shaped, dark brown or black colored rhizome of ca 3–9 × 2–3 cm. Surface of rhizome is rough with many hard pointed leaf bases, bearing many hard, dark brown membranous ramenta. Rhizome is hard without any wrinkles or fissures. Fractures are short and hard. A transversely cut surface of stem shows irregular outline with inner yellowish-green colored surface. The surface remains as it is in color for long time even after drying.

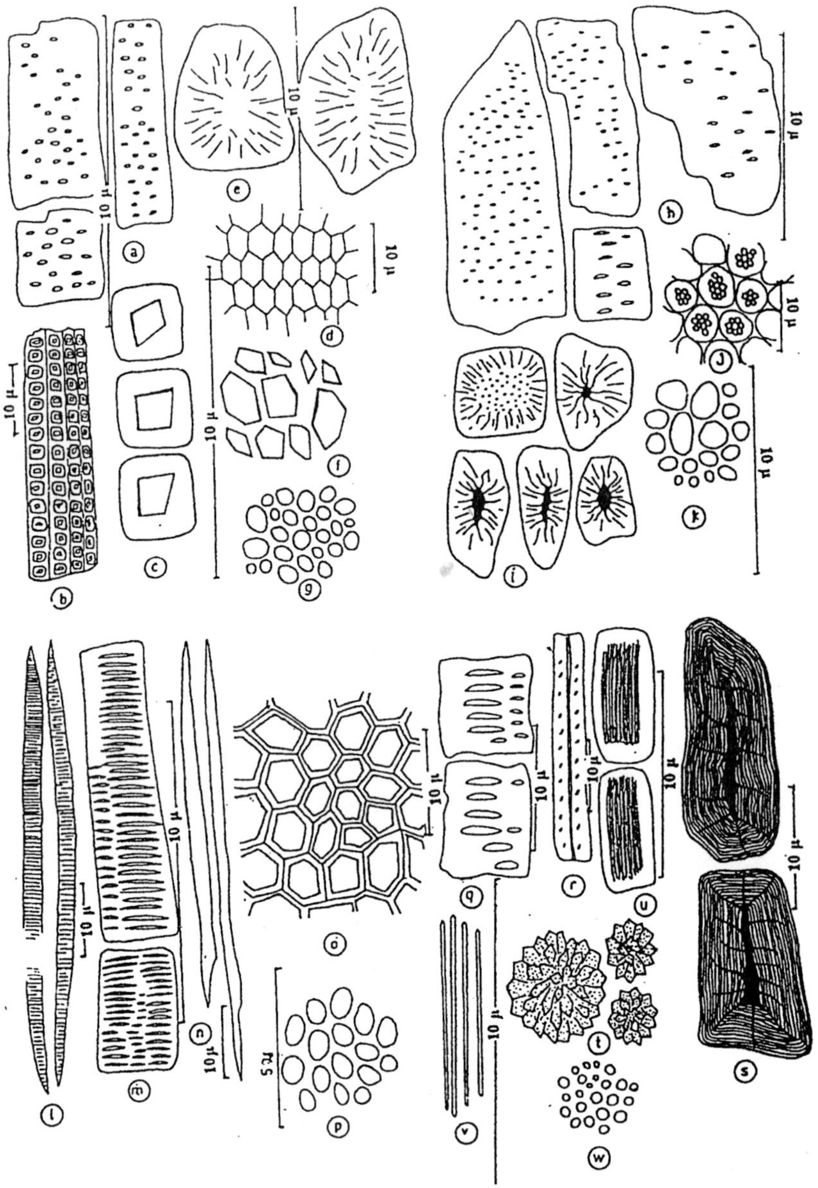

Text fig. 74: Pueraria tuberosa a-Pieces of vessels, b–c-Pieces of phloem fibers, d-Parenchymatous thin walled cells, e-Stone cells, f-Crystals of Calcium oxalate, g-Starch grains; **Bryonia dioica** h-Pieces of vessels, i-Stone cells, j-Parenchymatous cells filled with starch grains, k-Starch grains; **Tectaria cicutaria** l–m-Pieces of trachieds, n-Fibers, o-Epidermal cells, p-Starch grains; **Leea setuligera** q-pieces of vessels, r-fibers, s-Stone cells, t-Crystals of Calcium oxalate, u-Raphid sac, v-Raphides, w-Starch grains.

Powder characters: Powder is dark brown in color containing pieces of vessels, fibers, epidermal cells and starch grain [Text fig. 74 l–p]. Vessels are ca 638–833 × 24–30 μ in size; intervascular thickening is spiral [Text fig. 74 l–m]. Fibers are ca 500–638 μ in length, thin walled tapering towards both the ends [Text fig. 74 n]. Pieces of thick walled, brown colored epidermal cells are present in powder [Text fig. 74 o]. Starch grains are abundant in powder, simple, ca 3–8 μ in diameter and oblong to rounded in shape [Text fig. 74 p].

Local uses: Rhizome powder is filled in ear to stop earache and pus oozing out from the ear. It is also used to stop the pain in teeth.

LAHAN DINDA/TAMBADA DINDA [Text fig.: 74 q–w]
Botanical name: **Leea setuligera** C. B. Cl.

Family: Leeaceae

Vernacular names: Lahandinda or Tambada dinda.

Part used: Tuber.

Organoleptic characters: Taste slightly bitter, mucilaginous. Tubers if chewed causes irritation in throat. Odourless.

Macroscopic characters: Drug occurs in the form of cylindrical, or dumbbell shaped, reddish-brown colored tubers of ca 6–12 × 1–4 cm. Bark of tuber is thin, reddish-brown and smooth. Lenticels are small and transversely elongated. Tubers are longitudinally wrinkled in dried samples; wrinkles are shallow; fractures are short. A transversely cut surface of tuber shows central mealy white inner tissue.

Powder characters: Powder is pale pink in color. Powder shows pieces of vessels and fibers, parenchymatous cells, stone cells, raphid sacs, raphides, crystals of calcium oxalate and starch grains [Text fig. 74 q–w]. Vessel pieces are ca 10–40 × 20–40 μ; intervascular pitting is alternate and pits are bordered [Text fig. 74 q]. Fibers are ca 250–410 μ in length, thin walled and pitted [Text fig. 74 r]. Stone cells are ca 150–200 × 70–90 μ in size, oblong to rectangular in shape and have striated walls, blunt ends and narrow lumen [Text fig. 74 s]. Raphid sacs are ca 40–50 μ in length containing bunches of raphides [Text fig. 74 u]. Raphides are ca 60–100 μ in length with needle like ends [Text fig. 74 v]. Crystals of calcium oxalate are clustered, ca 20–30 μ in diameter [Text fig. 74 t]. Starch grains are abundant in powder. They are simple, rounded and ca 1–5 μ in diameter [Text fig. 74 w].

Local uses: Tubers are used to cure skin diseases. Paste of tuber is applied externally in elephantiasis and itch. Tubers are also applied to heal wounds. Tubers are also used on affected areas of skin due to biting of leech to stop bleeding.

RANKARLE [Text fig.: 75 a–c]
Botanical name: **Momordica dioica** Roxb.

Family: Cucurbitaceae

Vernacular names: Kartola, Karantoli.
Part used: Tuber.
Organoleptic characters: Taste slightly bitter, odorless.
Macroscopic characters: Drug occurs in the form of entire tuber. Tubers are cylindrical, elliptic-oblong or fusiform, ca 8–30 × 2–10 cm in size. Apical region of tuber or portion towards stem becomes hard and woody. Bark is ca 1–2 mm thick, thin in young tubers. It is yellowish-brown in color with rough, lenticellate surface. Bark has deep longitudinal wrinkles. Lenticels are numerous, prominent, circular and dirty white in color. Fractures are short and granular.
Powder characters: Powder is pale yellow or yellow in color, contains pieces of vessels, stone cells and starch grains [Text fig. 75 a-c]. Pieces of vessels are ca 30–70 × 25–30 μ; intervascular pitting is alternate, pits are small and rounded [Text fig. 75 a]. Stone cells are ca 60–80 × 30–50 μ in size, square to rectangular in shape. They are thick walled, pitted and without lumen [Text fig. 75 b]. Starch grains found in bunches in powdered materials. Starch grains are simple, ca 12–6 μ in diameter, rounded in shape [Text fig. 75 c].

Local uses: Root powder is recommended in diabetes. One teaspoon of powder given thrice a day to reduce sugar percentage in diabetic patient suffering from urinary calculi. A teaspoon of powder is given early in the morning empty stomach.

RATANJYOAT/RATANPURUSH [Text fig.: 75 d–j]
Botanical name: **Costus speciosus** (J. Koenigh) Smith

Family: Costaceae

Vernacular names: Kemuka, Kobee, Kushta, Penva, Pinnga etc.
Parts used: Rhizome.
Organoleptic characters: Initially taste sweetish, later on bitter aftertaste, odorless.
Macroscopic characters: The drug occurs in the form of sub-cylindrical, unbranched, pale brown colored rhizome of ca 5–12 × 1–1.2 cm. Bark is thin and hairy. A transversely cut portion of rhizome shows creamy white surface. Rhizome is longitudinally wrinkled. Wrinkles are deep. At intervals on the upper and lower surface are found small scars of ca 4–6 mm in diameter.
Powder characters: Rhizome powder is dirty white in color containing pieces of vessels and fibers, epidermal cork cells, hairs, crystals of calcium

oxalate and starch grains [Text fig. 75 d–j]. Vessels members with spiral thickening are ca 130–70 μ in length [Text fig. 75 d]. Vessels members with scalariform or opposite pitting are ca 110–200 μ in length [Text fig. 75 e]. Fibers are ca 410–580 μ in length, thin walled, pitted [Text fig. 75 g]. Hairs are ca 4100–5800 μ in length brown, colored, 2–3 celled, uniseriate and pointed towards tip [Text fig. 75 f]. Epidermal cork cells are thick walled found in bunches [Text fig. 75 h]. Crystals of calcium oxalate are clustered, ca 10–20 μ in diameter and are found in parenchymatous cells [Text fig. 75 i]. Starch grains are abundant in powder. They are simple, ca 10–20 μ in diameter and elliptic oblong in shape [Text fig. 75 j].

Local uses: The paste of rhizome is applied on skin to remove marks remained due to the pimple and acne. The rhizome paste is applied on skin to remove mark formed due to excess secretion of melanin after pregnancy. It is also used to remove black circle below the eyes. Rhizome is given orally with buttermilk in gas trouble and used as a tonic.

RUKHALKAND [Text fig.: 75 k–o]
Botanical name: **Remusatia vivipara** Schott.

Family: Araceae

Vernacular names: Rukhalu.

Parts used: Corms.

Organoleptic characters: Taste mucilaginous, cause irritation to throat, odorless.

Macroscopic characters: Drug occurs in the form of corms. Corms are rounded planoconvex, ca 2–5 cm in diameter. Red when fresh. Dried corms are gray to brown in color with ridges of fallen leaf scars. Bark is smooth in fresh material, white dirty in dried materials. Transverse cut of sample shows mucilaginous white surface with red colored ring of bark.

Microscopic characters: Powder is pale pink in color containing pieces of vessels, raphides, crystals of calcium oxalate and starch grains [Text fig. 75 k–o]. Pieces of vessels are ca 300–690 μ in length; intervascular thickening is spiral [Text fig. 75 k–i]. Raphides are ca 50–60 μ in length, needle like pointed at both ends [Text fig. 75 n]. Crystals of calcium oxalate are ca 10–20 μ in length [Text fig. 75 m]. Starch grains are abundant in powder. They are simple, ca 6–10 μ in diameter, oblong elliptic in shape [Text fig. 75 o].

Local uses: Tubers are used in pile. Paste of tubers with bark paste of *Pittosporum floribundum* and root paste of *Salacia brunoniana* is applied externally in piles. Paste of tubers is also applied in itch.

SHATAVARIMUL [Text fig.: 75 t–y]
Botanical name: **Asparagus racemosus** Baker

Family: Liliaceae

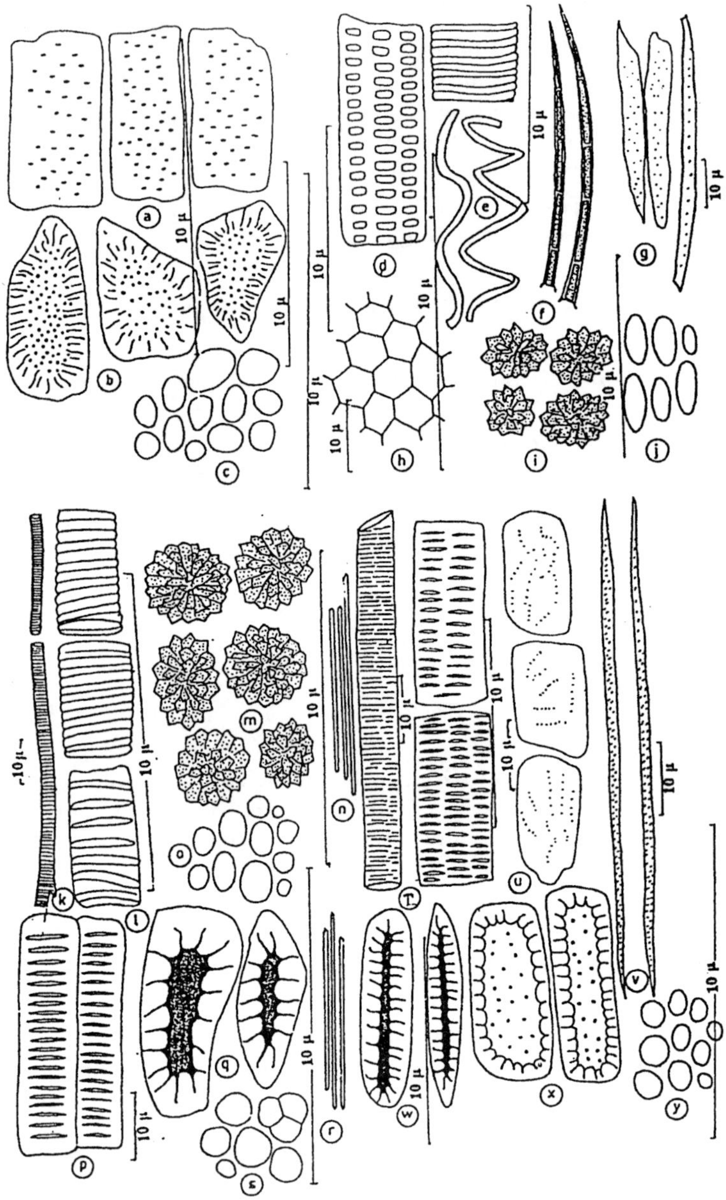

Text fig. 75: Momordica dioica a-Pieces of vessels, b-Stone cells, c-Starch grains; **Costus speciosus** d–e-Pieces of vessels, f-Hairs, g-Pieces of fibers, h-Epidermal cells, i-Crystals of Calcium oxalate, j-starch grains; **Remusatia vivipara** k–l-Pieces of vessels, m-Crystals of Calcium oxalate, n-Raphides, o-Starch grains; **Sauromatum venosum** p-Pieces of vessels, q-Stone cells, r-raphides, s-Starch grains. **Asparagus racemosus** t-Pieces of vessels, u-Parenchymatous cells, v-Fibers, w-Stone cells, x-Stone cells, y-Starch grains.

Vernacular names: Asvel, Indivari, Narayani, Satavari, Shatmul.

Part used: Tuber

Organoleptic characters: Taste mucilaginous sweet later on slightly bitter.

Macroscopic characters: The drug occurs in the form of cylindrical, fusiform, fawn or flesh colored entire roots. Roots are tuberous, ca 15–40 × 1–1.5 cm in size, tapering towards both the ends. Bark of root is thin leathery, waxy, smooth with shallow transverse wrinkles when fresh while deep longitudinal wrinkled in dried samples. Sometimes longitudinal wrinkles are interrupted by transverse cut. Fractures are short. A transversely cut surface of root shows peripheral mealy white region with central yellowish strand.

Microscopic characters: Root powder is dirty white in color contains pieces of vessels and fibers, pitted parenchyma cells, stone cells and starch grains [Text fig. 75 t–y]. Pieces of vessels are ca 110–150 × 20–30 μ in size and with scalariform intervascular pitting [Text fig. 2 t]. Pieces of fibers are ca 250–300 μ in length, pitted and unseptate [Text fig. 75 v]. Stone cells are found in two forms; stone cells with thick wall are ca 70–90 × 20–30 μ in size and oblong lanceolate with tapering ends and without pits [Text fig. 75 w], thin walled stone cells are ca 90–100 × 10–30 μ in size, rectangular in shape, central lumen is broad and pitted [Text fig. 75 x]. Starch grains are abundant in powder. They are simple, ca 10–20 μ in diameter and oblong to circular in shape [Text fig. 75 y].

Chemical tests:

Tests	Reagents	Results
Starch	Iodine	+ ve
Saponin	Shake vigorously with distilled water.	+ ve Honeycomb froth more than 3 cm above the surface, persists for 30 min.

Local uses: One tablespoon of root powder given with milk in every night to women in leucorrhoea, it is also given in excess bleeding. The root powder is orally given to women as a galactogogue and also as a tonic. The roots are given to the cows to increase milk yield. The oil extracted from roots is used to stop hair loss.

SURAN/RANSURAN/AGYASURAN [Text fig.: 75 p–s]

Botanical name: **Sauromatum venosum** Schott.

Synonyms: *S. guttatum* (Wall.) Schott.; *S. pedatum* Schott.

Family: Araceae

Vernacular names: Nurki.

Parts used: Corms.

Organoleptic characters: Taste is not detected, odorless.

Macroscopic characters: Drug occurs in the form of corms. Corms are planoconvex, large, ca 7–12 cm in diameter and brown in color. Dried corms are gray in color, rough containing small protuberances with leafy scars at apex. A transversely cut surface of tuber shows mucilaginous white surface.

Powder characters: Powder of tuber is gray or pale brown or dirty white in color containing pieces of vessels, stone cells, raphides and starch grains [Text fig. 75 p–s]. Pieces of vessels are ca 222–360 × 55–83 μ. Intervascular thickening is spiral [Text fig. 75 p]. Stone cells are rarely present in powder, ca 66–83 × 30–37 μ, irregular in shape, walls thick [Text fig. 75 q]. Raphides are very abundant in powder. They are needle shaped, pointed and ca 12–36 μ in length [Text fig. 75 r]. Starch grains are also abundant in powder. They are simple or compound, ca 12–15 μ in diameter and rounded or oblong in shape.

Local uses: Tuber paste is applied externally on piles with castor oil and smoke of castor seed is given for 15 minutes to affected area.

Stems used in Folk Medicines

BELMUL [Photoplate–XIII; Text fig.: 76 a-j]
Botanical name: **Aegle marmelos** (L.) Corr.

Family: Rutaceae

Vernacular names: Bel, Bilwa.

Organoleptic characters: Taste is bitter, odorless.

Macroscopic characters: Drug occurs in the form of cylindrical pale yellow colored pieces of stem of ca 5–7 × 3–2.5 cm in size. Bark of stem is ca 1–2 mm thick, rough due to the presence of rhytidom. Stem is without any fissures or wrinkles, central wood is pale yellow in color and hard. Fractures are short. A transversely cut surface of stem shows normal secondary growth.

Microscopic characters: Transverse section of stem is circular in outline. It shows outer thin walled parenchymatous cells of rhytidom. Inner bark consists of 3–6 layers of phelloderm which contain large yellow colored islands of stone cells. Secondary cortex contains isolated strands of sclerenchyma cells. Secondary phloem is crushed and contain uniseriate rays. Parenchyma is paratracheal and scanty. Rays are 1–2 cell wide. Wood consists of vessels, xylem parenchyma and fibers [Text fig. 76 a–c]. Vessels are ca 259–388 × 22–77 μ in size, perforation plates are simple with horizontal or oblique end walls; intervascular pitting is alternate and pits are very small [Text fig. 76 a–a']. Pits to xylem parenchyma are similar to vessels [Text fig.76 c]. Fibers are either thick walled or thin walled, ca 916–1625 μ in length, pitted and unseptate [Text fig. 76 b].

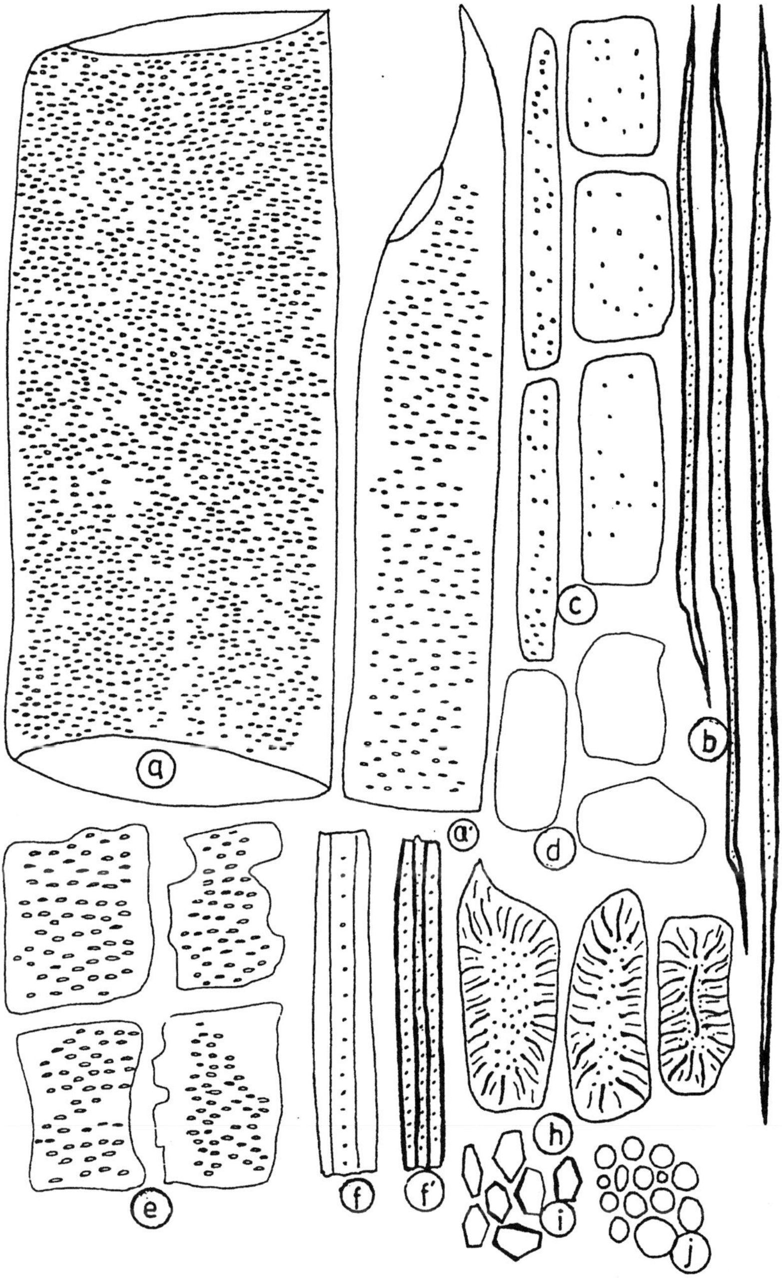

Text fig. 76: Aegle marmelos a & a'-Vessels (x 450), b-Fibers (x 100), c-Xylem parenchyma (450, × 675), d-Parenchyma cells (x 450), e-Pieces of vessel (x 450), f & f'-Pieces of fibers (x 100), h-Stone cells (x 450), i-Crystals of Calcium oxalate (x 450), j-Starch grains (x 450).

Powder characters: Powder is pale yellow colored containing pieces of vessels and fibers, stone cells, crystals of calcium oxalate and starch grains [Text fig. 76 c-j].

Stone cells are ca 50–129 × 21–22 μ in size, oblong to rectangular in shape, thick walled and with pitted lumen [Text fig. 76 h]. Crystals of calcium oxalate are solitary and ca 6–12 μ in size [Text fig. 76 i]. Starch grains are simple, ca 4–8 μ in diameter and circular in shape [Text fig. 76 j].

Local uses: It is used as one of the 'Dashmula' drug of Ayurveda. One teaspoon of root decoction is given in fever.

GULVEL [Photoplate-XIII; Text fig.: 77 a–m]
Botanical name: **Tinospora cordifolia** (Willd.) Mers.

Family: Menispermaceae

Vernacular names: Ambarvel, Gaudichi, Jivantika, Giroli.

Organoleptic characters: Taste is slightly bitter and mucilaginous, odorless.

Macroscopic characters: Drug occurs in the form of cylindrical, gray colored pieces of stems of ca 4–6 × 0.5–1.5 cm. Bark of stem is gray colored, rough, ca 3–5 mm thick. Lenticels are circular and prominent. Outer bark is thin and papery. Stem is warty longitudinally wrinkled, wrinkles are shallow. Wood is gray and soft. Fractures are fibrous. A transversely cut surface of stem shows secondary anomalous growth. Xylem is stellate in structure.

Microscopic characters: Transverse section of stem is circular in outline. It shows 2–3 layers of cork. Phelloderm is 4–5 layered. Secondary cortex is wide zone contain large columnar type of cells filled with mucilage. Transverse section of stem shows secondary anomalous growth. Xylem is united at the center, thereby completely obliterating the pith and giving xylem a stellate appearance with the phloem at the ends of radii. Rays are large, ca 6–15 cell wide. Xylem consists of vessels, xylem parenchyma, tracheids, tracheid fibers, fibers and phloem fibers [Text fig. 77 a–f]. Vessels are ca 150–250 × 20–160 μ, perforation plates simple horizontal or oblique; intervascular pitting is alternate, pits are bordered [Text. fig.77 a]. Sometimes vessel elements are with spiral thickening, ca 120–220 × 27–10 μ in size [Text fig. 77 b]. Tracheids are ca 610–880 μ in length with small pits [Text fig. 77 c]. Tracheid fibers are ca 410–1030 μ in length, tapering at both the ends [Text fig. 77 d–d']. Fibers are ca 520–1000 μ in length, thin walled, slightly pitted and unseptate [Text fig. 77 d-d']. Phloem fibers are ca 110–660 μ in length. Phloem fibers are encircled by a thick parenchymatous sheath of cell possessing rows of solitary crystals of calcium oxalate [Text fig. 77 e].

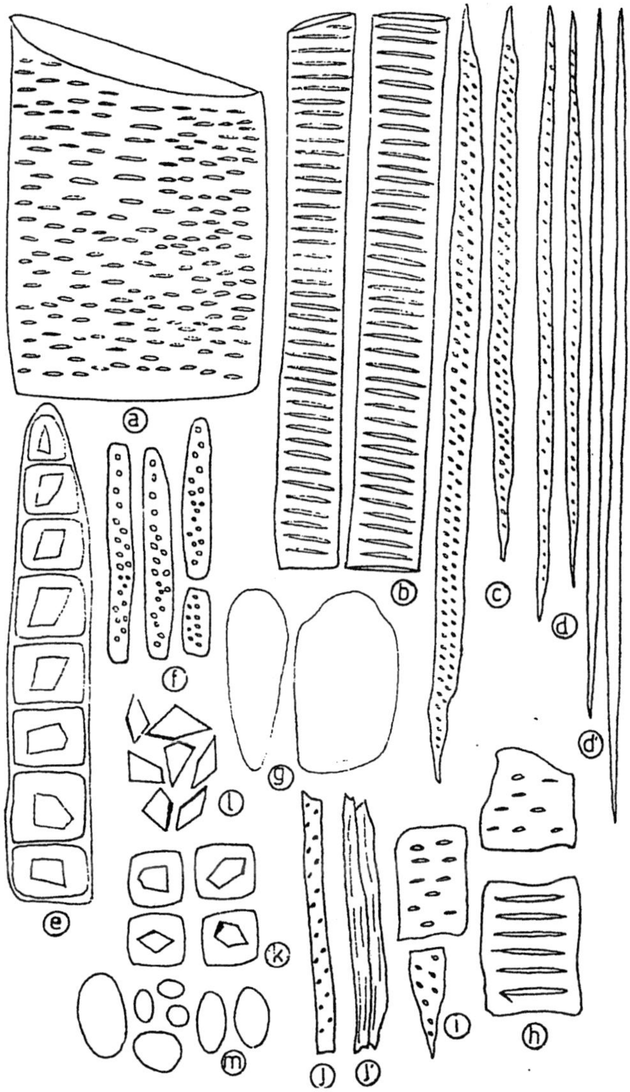

Text fig. 77: Tinospora cordifolia a-Vessels (x 450), b-Vessels (x 450), c-Tracheids (x 225), d & d'-Fibers (x 150), e-Phloem fibers (x 675), f-Xylem parenchyma (x 150), g-Parenchyma (x 450), h-Pieces of vessel (x), i-Pieces of tracheids (x 225), j & j'-Pieces of fibers (x 150), k-Pieces of phloem fibers (x 675), l-Crystals of calcium oxalate (x 450), m-Starch grains (x 675).

Powder characters: Powder is gray in color containing pieces of vessels, fibers and phloem fibers, crystals of calcium oxalate and starch grains [Text fig. 77 h–m].

Crystals of calcium oxalate are ca 8–10 μ, solitary and prism type [Text fig. 77 l]. Starch grains are simple or compound, ca 10–25 μ in diameter, ovoid elliptic in shape and with faintly marked striations [Text fig. 77 m].

Local uses: The stem is used in jaundice. Small pieces of stems are tied around the neck of women after delivery. It is used to reduce heat from the body. Decoction of stem is given thrice in a day in fever.

HADSANDHI [Photoplate-XIII; Text fig.: 78 a-i]
Botanical name: **Litsea deccanensis** Gamble

Synonyms: *L. tomentosa* Heyne ex Hook

Family: Lauraceae

Vernacular names: Chikana.

Organoleptic characters: Taste is slightly sweet and later on bitter, mucilaginous, odor is aromatic.

Macroscopic characters: Drug occurs in the form of reddish-brown or black colored, cylindrical pieces of stem of ca 5–6 × 1.3–2 cm in size. Bark of stem is thin, reddish or black in color and rough with few small oblong to rounded lenticels. Stem is longitudinally striated. Wood is hard and gray or pale brown colored. Fractures are short. Transversely cut portion shows normal secondary growth. Annual growth rings are observed in wood. Vessels are very small and dot like.

Microscopic characters: Transverse section is circular in outline. Cork is 4–5 layered, made up of thick walled cells. Cork consists of 1–2 layers of phellogen and 3–5 layers of phelloderm. Secondary cortex is interrupted by continuous bands of stone cells. Secondary phloem is crushed containing uniseriate ray cells. Secondary growth is normal. Rays are usually uniseriate or biseriate and prominent. Wood is diffuse porous. Wood consists of vessels, xylem parenchyma and fibers [Text fig. 78 a–e]. Vessels are ca 492–678 × 44–55 μ; perforation plate are simple and oblique; intervascular pitting is alternate. Scalariform pitting is also observed [Text fig. 78 a–a']. Pits to xylem parenchyma are large, elongated and mostly scalariform [Text fig.78 c]. Fibers are ca 722–955 μ in length, thick walled, unseptate without pits and taper towards both the ends [Text fig. 78 b].

Powder characters: Powder is dirty white. Powder consists of pieces vessel and fibers, stone cells and starch grains [Text fig. 78 f–i].

Stone cells are of two types viz. thin walled pitted and thick walled unpitted. Pitted thin walled stone cells are ca 85–92 × 55–62 μ in size, square to rectangular or hexagonal to pentagonal in shape and with striated walls [Text fig. 78 f]. Thick walled stone cells are ca 85–92 × 122–129 μ in size, mostly triangular to oblong in shape and with striated walls [Text fig.78 f']. Starch grains are simple, ca 8–6 μ in diameter and rounded in shape [Text fig. 78 i].

Local uses: Paste of stem is applied on cut wounds. Stem is crushed and it is tied around the fractured bone.

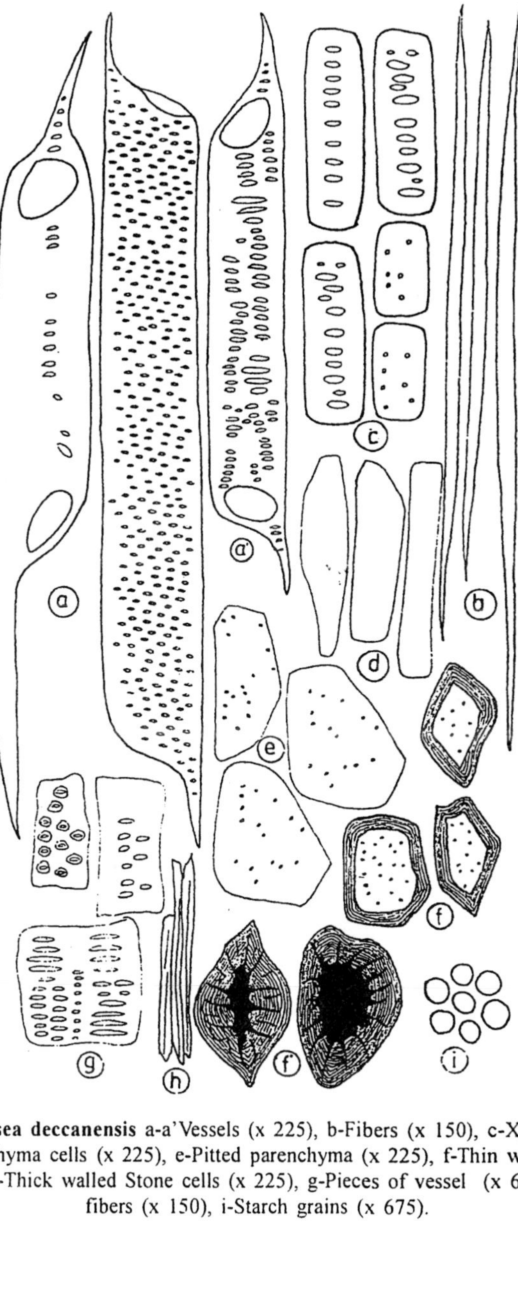

Text fig. 78: Litsea deccanensis a-a'Vessels (x 225), b-Fibers (x 150), c-Xylem parenchyma (x 450), d-Parenchyma cells (x 225), e-Pitted parenchyma (x 225), f-Thin walled pitted Stone cells (x 225), f'-Thick walled Stone cells (x 225), g-Pieces of vessel (x 675), h-Pieces of fibers (x 150), i-Starch grains (x 675).

MARALMATHANGI [Photoplate–XIII; Text fig.: 79 a–k]
Botanical name: **Argyreia boseana** Sant. & Patel

Family: Convolvulaceae

Vernacular names: Maralmathangi.
Organoleptic characters: Taste is bitter, odorless.

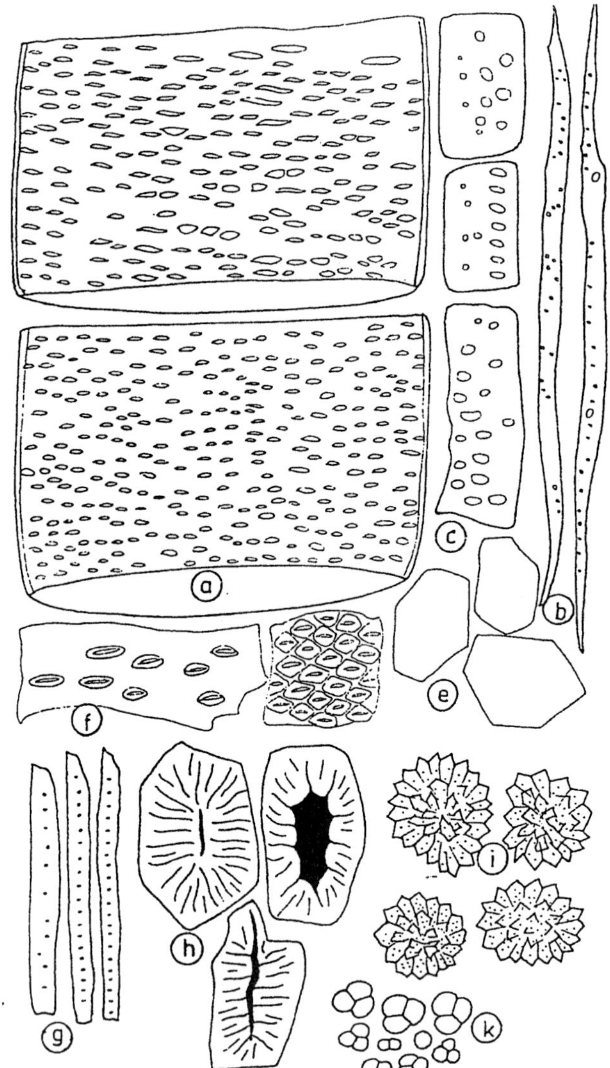

Text fig. 79: Argyreia boseana a-Vessels (x 225), b-Fibers (x 150), c-Xylem parenchyma (x 450), e-Parenchyma cells (x 450), f-Pieces of vessel (x 450), g-Pieces of fibers (x 150), h-Stone cells (x 675), i-Crystals of Calcium oxalate (x 675), k-Starch grains (x 675).

Macroscopic characters: The drug occurs in the form of pale gray or greenish-gray colored cylindrical pieces of stem of ca 5–6 × 2–3 cm in size. Bark of stem is thin and rough. Lenticels are small, white dot like. Stem is longitudinally wrinkled. Wrinkles are shallow (Deep in mature stem). Inner wood is soft and gray in color. Fractures are fibrous. A transversely cut surface of stem shows secondary anomalous growth.

Microscopic characters: Transverse section is circular in outline. It shows 2–3 layered thick walled cork cells. Secondary cortex is occupied by rounded islands of stone cells. Secondary anomalous growth shows intraxylary, concentric phloem. Xylem groups occasionally developed in association with the intraxylary phloem. Clustered crystals of calcium oxalate are present in phloem. Secondary phloem is interrupted by large rounded islands of stone cells. Wood consists of vessels, xylem parenchyma and fibers [Text fig.79 a-c]. Vessels are drum shaped, ca 92–120 × 140–185 μ in size; perforation plates simple and horizontal; intervascular pitting alternate, pits are bordered and large [Text fig.79 a]. Pits to xylem parenchyma are large and rounded [Text fig. 79 c]. Fibers are ca 660–1027 μ in length, pitted and unseptate [Text fig.79 b].

Powder characters: Powder is yellowish-brown in color containing pieces of vessels, fibers, stone cells, crystals of calcium oxalate and starch grains [Text fig.79 f-k].

Stone cells are abundant in powder ca 45–49 × 24–26 μ in size, rectangular-polygonal in shape, thick walled with narrow lumen [Text fig.79 h]. Crystals of calcium oxalate are abundant in powder, clustered, ca 26–30 μ in diameter [Text fig.79 i]. Starch grains are abundant in powder, compound or simple, ca 6–12 μ in diameter [Text fig. 79 k].

Local uses: Stem is used as a wound healing. Paste of stem is applied on a wound thrice a day after clean washing of wounds till recovery.

KAJARICHAVEL [Photoplate-XV; Text fig.: 80 a–i]
Botanical name: **Entada rheedei** Spreng.

Synonyms: *Entada pusaetha* DC., *E. scandens* Benth.

Family: Mimosaceae

Vernacular names: Garambi, Garbi, Gardul.

Organoleptic characters: Taste is slightly bitter, odorless.

Macroscopic characters: Drug occurs in the form of cylindrical gray colored pieces of stems of ca 6–8 × 2–4 cm. Bark of stem is ca 2–3 mm thick and rough. Lenticels are large and rounded. Stem is longitudinally and transversely fissured. Longitudinal fissures are deep. Large transverse cracks present on stem. Wood is brown colored, soft and fibrous. Fractures are fibrous. A transversely cut surface of stem shows large prominent vessels and secondary anomalous growth.

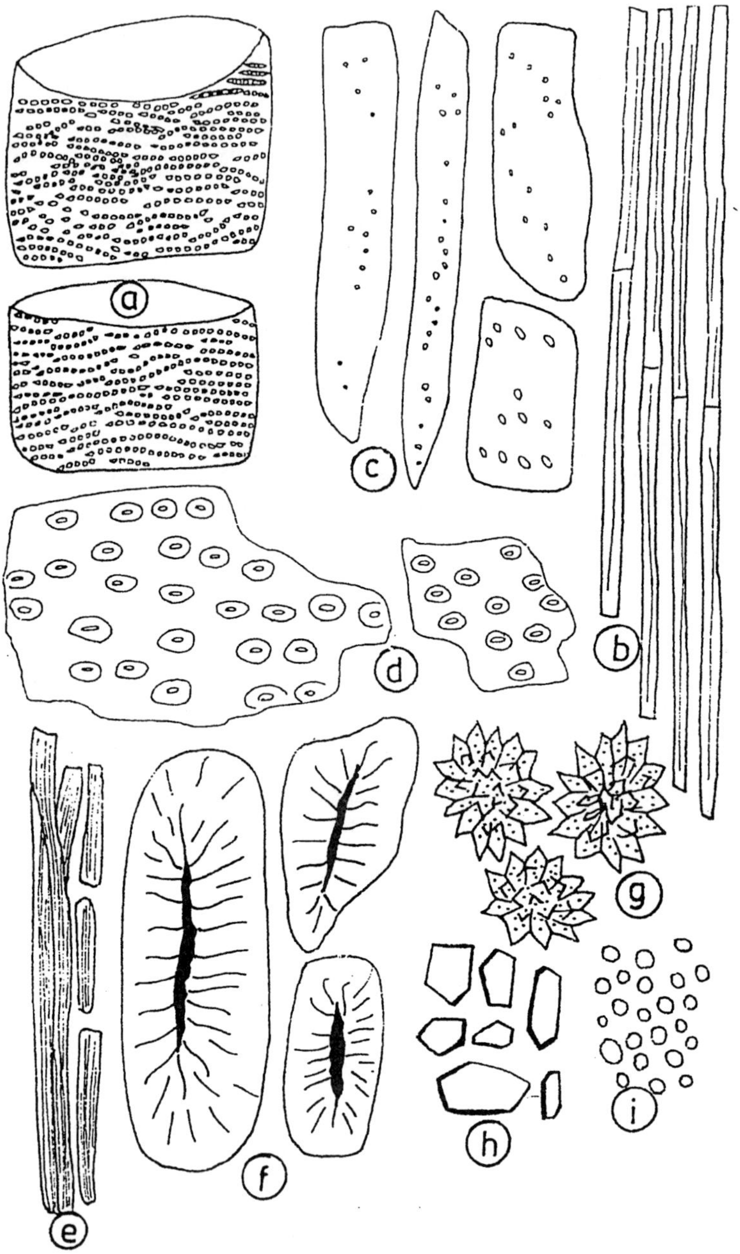

Text fig. 80: Entada rheedei a-Vessels (x 50), b-Fibers (x 150), c-Xylem parenchyma (x 450), d-Pieces of vessel (x 675), e-Pieces of fibers (x 150), f-Stone cells (x 675), g-Clustered Crystals of Calcium oxalate (x 675), h-Prism type of Crystals of Calcium oxalate (x 675), i-Starch grains (x 675).

Microscopic characters: Transverse section of stem is oblong to circular in outline. It shows multilayered thick walled cork cells alternating with thin walled cells. Secondary cortex is wide containing bands of stone cells and crystals of calcium oxalate. Anomalous secondary growth shows secondary phloem embedded in parenchymatous ground tissue of the xylem. Parenchyma is paratracheal. Wood consists of vessels, xylem parenchyma, phloem fibers and fibers [Text fig. 80 a–c]. Vessels are ca 111–129 × 129–722 μ; perforation plate is simple and horizontal; intervascular pitting is alternate, larger in size and pits are bordered [Text fig. 80 a]. Pits to xylem parenchyma are similar as vessels [Text fig. 80 c]. Fibers are ca 550–780 μ in length, thick walled, without lumen and septate [Text fig. 80 b]. Fragments of phloem fibers are ca 380–440 μ in length encircled by parenchymatous sheath, possessing rows of crystals of calcium oxalate [Text fig. 80 c].

Powder characters: Powder is reddish–brown in color containing pieces of vessels, fibers, phloem fibers, stone cells and crystals of calcium oxalate [Text fig.80 d-i].

Stone cells are ca 36–78 × 18–27 μ in size, oblong to rectangular in shape, thick walled and without lumen [Text fig. 80 f]. Crystals of calcium oxalate are clustered, ca 18–24 μ in diameter and [Text fig. 80 g]. Solitary crystals of calcium oxalate are ca 7–18 μ in size [Text fig. 80 h]. Starch grains are simple, ca 2–3 μ in diameter, small, rounded in shape [Text fig. 80 i].

Local uses: Paste of stem is applied on wound.

KALITAKALI [KAWALI] [Photoplate-XIII; Text fig.: 81a–i]
Botanical name: **Cryptolepis buchanani** Roem. & Schult.

Family: Periplocaceae

Vernacular names: Karanta, Setakavali.

Organoleptic characters: Taste is bitter, odorless.

Macroscopic characters: The drug occurs in the form of brown or reddish-black colored cylindrical pieces of stem of ca 5–6 × 1–2 cm in size. Bark of stem is ca 2–4 mm thick, rough. Outer bark is thin, papery, reddish-brown in color, flanking off in the form of sheets. Stem is longitudinally fissured or wrinkled; wrinkles are shallow, fractures are short. Wood is yellow colored and hard. A transversely cut surface shows normal secondary growth.

Microscopic characters: Transverse section of stem is circular in outline. Transverse section of stem shows outer cork containing thick walled, 3–4 layers of cork, 1–2 layers of phellogen and 7–8 layers of phelloderm. Phelloderm consists of solitary crystals of calcium oxalate. Secondary cortex contains thick walled sclerenchyma cells scattered irregularly. Latex canals are present in secondary cortex. Secondary phloem is crushed. Rays are

2–3 cell wide. Wood consist of vessels, tracheids, tracheid fibers and fibers [Text fig. 81 a-d]. Vessels are cylindrical, ca 277–370 × 55–105 μ in size; perforation plates are simple, slightly oblique; intervascular pitting is alternate [Text fig. 81 a-a']. Pits to xylem parenchyma similar as vessels [Text fig. 81 d]. Tracheids are ca 500–592 μ in length, with alternate and minute pits [Text fig. 81 b]. Tracheid fibers are ca 833–958 μ in length, slightly pitted and unseptate [Text fig.81 c]. Fibers are thick walled, ca 625–1250 μ in length, tapering at both the ends [Text fig. 81 c'].

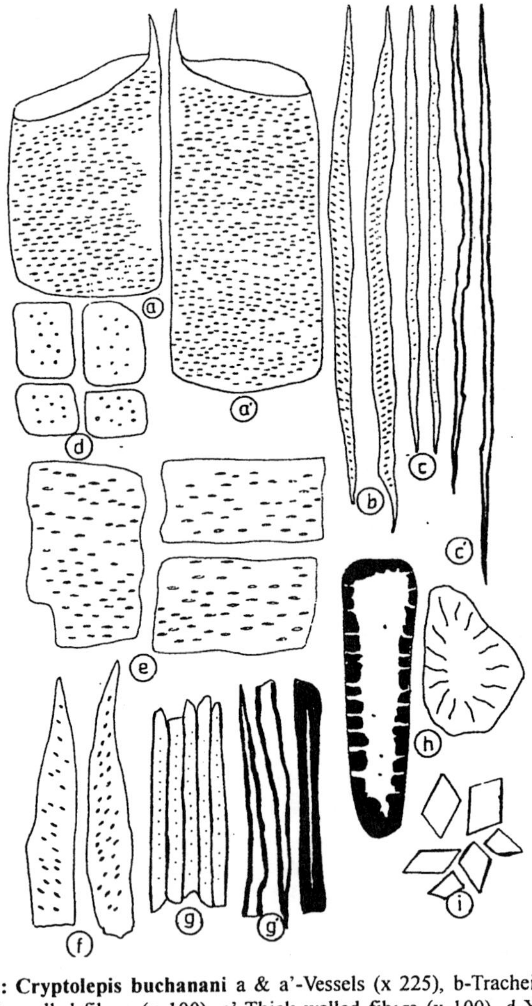

Text fig. 81: Cryptolepis buchanani a & a'-Vessels (x 225), b-Tracheids (x 225), c-Thin walled fibers (x 100), c'-Thick walled fibers (x 100), d-Xylem parenchyma (x 225), e-Pieces of vessel (x 675), f-Pieces of tracheids (x 450), g & g'-Pieces of fibers (x 150), h-Stone cells (x 450), i-Crystals of Calcium oxalate (x 675).

Powder characters: Powder is lemon yellow in color containing brown or reddish-brown colored pieces of outer bark, pieces of vessels, tracheids, fibers, stone cells and crystals of calcium oxalate [Text fig. 81 e–i].

Stone cells are not abundant in powder, ca 64–84 μ in size, thick walled slightly pitted [Text fig. 81 h]. Solitary crystals of calcium oxalate are ca 18–20 μ in size [Text fig. 81 i]. Starch grains absent.

Local uses: Stem paste is externally applied to cure the wounds. Fresh stems are dipped in hot water and bath is given to children suffering from stomach complaints.

KANGONIMUL [Photoplate-XIII; Text fig.: 82 a-l]

Botanical name: **Celastrus paniculatus** Willd.

Family: Celastraceae

Vernacular names: Dhimar-bel, Jyotishmati, Kanglikangun, Karkanganvel, Kazad-kangoni, Kiri, Malkangoni, Piwalwel.

Organoleptic characters: Taste is astringent, odorless.

Macroscopic characters: Drug occurs in the form of whitish-gray colored pieces of stem of ca 5–8 × 1.5–2.5 cm in size. Bark is ca 1–2 mm thick. Stem is without any wrinkles or fissures. The wood is pale brown in color, wavy in outline and hard. Fractures are short. Transversely cut surface of stem shows normal secondary growth.

Microscopic characters: Transverse section of stem is circular in outline. It shows thick walled 2–5 layered cork. Secondary cortex consists of patches of stone cells. Rays are usually 2–4 cell wide. Wood consists of vessels, xylem parenchyma, tracheids, tracheid fibers and fibers [Text fig. 82 a-d]. Vessels are cylindrical, ca 437–459 × 55–148 μ; perforation plate are simple, horizontal or slightly oblique; intervascular pitting is alternate [Text fig. 82 a]. Tracheids are ca 500–592 μ in length, tapering at both the ends [Text fig. 82 b]. Tracheid fibers are ca 518–600 μ in length, pitted and septate [Text fig. 82 c]. Fibers are ca 1000–1250 μ in length, thick walled and without lumen [Text fig. 82 c'].

Powder characters: Powder is reddish-pink in color containing pieces of vessels, tracheids fibers, stone cells, crystals of calcium oxalate and starch grains [Text fig. 82 f–l].

Stone cells are ca 100–157 × 27–64 μ in sizes, irregular in shape, thick walled and striated [Text fig. 82 j]. Crystals of calcium oxalate are abundant in powder. They are solitary, ca 14–27 μ in size [Text fig. 82 k]. Starch grains are very few, ca 3–7 μ in diameter and rounded [Text fig. 82 l].

Local uses: Stem paste is applied externally in rheumatism. It is also given orally twice a day in rheumatism.

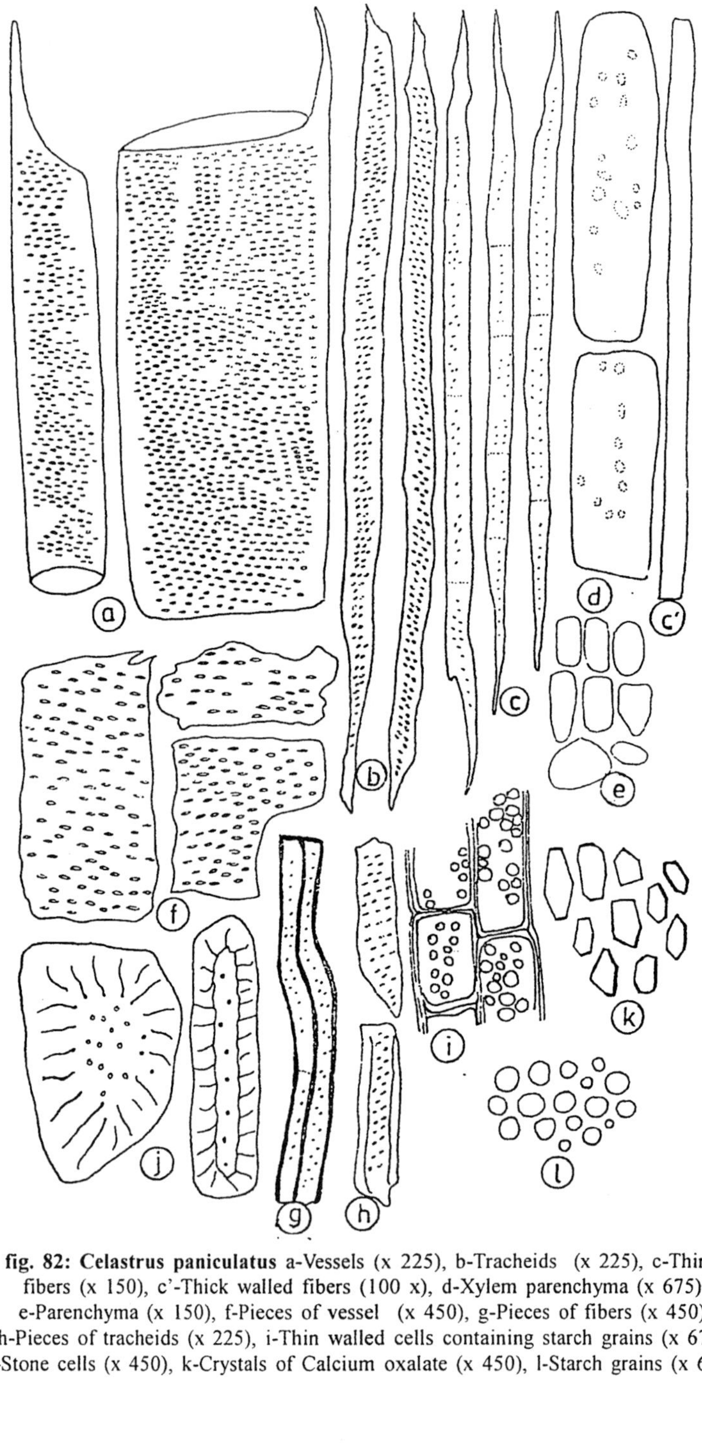

Text fig. 82: Celastrus paniculatus a-Vessels (x 225), b-Tracheids (x 225), c-Thin walled fibers (x 150), c'-Thick walled fibers (100 x), d-Xylem parenchyma (x 675), e-Parenchyma (x 150), f-Pieces of vessel (x 450), g-Pieces of fibers (x 450), h-Pieces of tracheids (x 225), i-Thin walled cells containing starch grains (x 675), j-Stone cells (x 450), k-Crystals of Calcium oxalate (x 450), l-Starch grains (x 675).

KAUNDALKHOD [Photoplate–XIII; Text fig.: 83 a–h]
Botanical name: **Trichosanthus tricuspidata** Lour.

Synonyms: *T. palmata* Roxb., *T. palmata* var. *tomentosa* Heyne ex C.B.Cl. *T. bracteata* (Lam.) Voigt, *T. bracteata* var. *tomentosa* (Heyne ex C.B.Cl) Chakr.

Vernacular names: Kaundal, Kavandala, Kodmdol, Indrayeum, Mukul.

Organoleptic characters: Taste is slightly bitter, odorless.

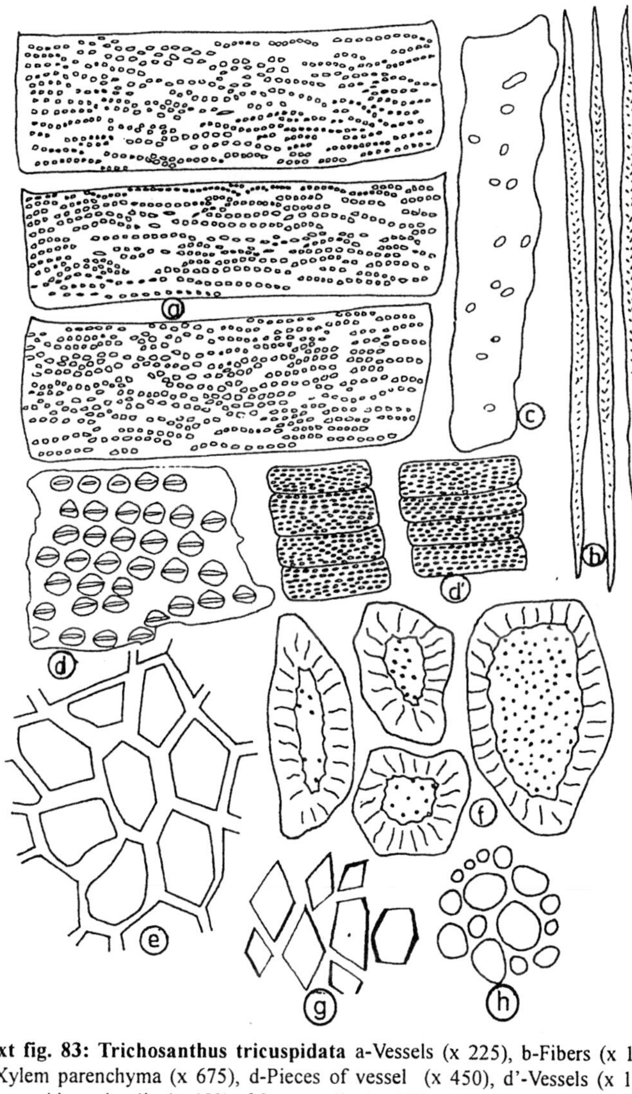

Text fig. 83: Trichosanthus tricuspidata a-Vessels (x 225), b-Fibers (x 100), c-Xylem parenchyma (x 675), d-Pieces of vessel (x 450), d'-Vessels (x 100), e-epidermal cells (x 450), f-Stone cells (x 450), g-Crystals of Calcium oxalate (x 450), h-Starch grains (x 450).

Macroscopic Characters: The drug occurs in the form of cylindrical, angular, furrowed, gray colored pieces of stems of ca 5–7 × 2–2.5 cm in size. Bark is thin and rough. Lenticels are small, rounded and arranged on ridges. Stem is deeply furrowed. Wood is soft and fibrous. Fractures are fibrous. A transversely cut surface of stem shows secondary anomalous growth and stellately arranged vascular strands.

Microscopic characters: Transverse section of stem is wavy in outline. It shows outer 2–3 layered thick walled cork. Secondary cortex containing large oblong islands of stone cells or bands of stone cells. Transverse section of stem shows secondary anomalous growth. Xylem is formed in between the portions into which the normal xylem becomes separated. Rays are broad, ca 5–20 cell wide. Xylem groups of the bundles united at the center, thereby practically or completely obliterating the pith and giving xylem stellate appearance with phloem at the ends of radii. Wood consists of vessels, xylem parenchyma and fibers [Text fig. 83]. Vessels are ca 92–111 × 333–307 μ in size; perforation plates are simple, horizontal; intervascular pitting is alternate, pits are bordered [Text fig. 83 a]. Pits to xylem parenchyma are similar as of vessels [Text fig. 83 d–d']. Fibers are thin walled ca 700–1083 μ in length, pitted and unseptate [Text fig. 83 b].

Powder characters: Powder is lemon yellow in color containing pieces of vessels, fibers, outer cork cells, stone cells, crystals of calcium oxalate and starch grains [Text fig. 83 d–h].

Stone cells are ca 46–92 μ × 40–46 μ in size, variously shaped with broad, pitted lumen [Text fig. 83 f]. Crystals of calcium oxalate are abundant in powder, ca 14–22 μ in size, solitary [Text fig. 83 g]. Starch grains are abundant, simple, rounded in shape, ca 9–18 μ in diameter [Text fig. 83 h].

Local uses: Stem paste is applied on back pain.

PANDHRI GIRNOLI [Photoplate–XIII; Text fig.: 84 a–k]
Botanical name: **Cissus elongata** Roxb.

Synonyms: *Vitis elogata* (Roxb.) Wall., *Cayratia elongata* (Roxb.) Suesseng.

Family: Vitaceae

Vernacular names: Pandhari girnoli.

Organoleptic characters: Taste is astringent, later on slightly bitter, odorless.

Macroscopic characters: Drug occurs in the form of pale yellow colored cylindrical pieces of stem of ca 5–6 × 1.5–2 cm in size. Bark of stem is ca 1–2 mm thick, pale yellow and rough due to presence of rhytidome. Stem is longitudinally fissured, fissures deep. Wood is pale pink in color and soft. Fractures are fibrous. Transversely cut surface of stem shows secondary anomalous growth.

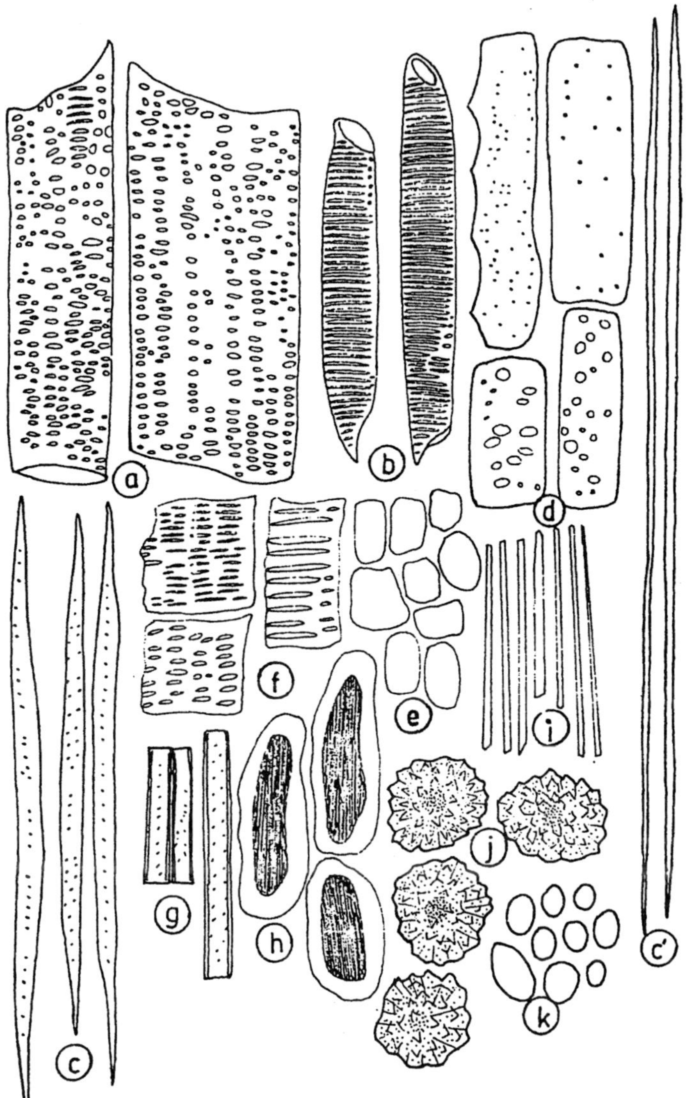

Text fig. 84: Cissus elongata a-Vessels (x 150), b-Vessels (x 150), c & c'-Fibers (x 150), d, d'-Xylem parenchyma (d-150, d'-x 225), e-Parenchyma, f-Pieces of vessel (x 225), g-Pieces of fibers (x 150), h-Raphid sacs (x 675), i-Raphides (x 675), j-Crystals of Calcium oxalate (x 675), k-Starch grains (x 675).

Microscopic characters: Transverse section of stem is circular in outline. It shows outer thin walled 5–8 layered parenchymatous cells of rhytidome. Inner bark consists of thick walled multilayered cork. Cork consists of

1–2 layers of phellogen followed by 4–5 layers of phelloderm. Secondary cortex consists of discontinuous ring of sclerenchyma cells. Transverse section of stem shows secondary anomalous growth. It is *Aristolochia* type. Secondary xylem is formed between the portions into which the normal xylem becomes separated. Rays are usually broad, very high and 5–10 cell wide. Central wood consists of vessels, xylem parenchyma, phloem fibers and fibers [Text fig. 84 a-d]. Vessels are of two types viz. vessels with alternate or scalariform pits and vessels with spiral thickening. Vessels with scalariform pitting are ca 333–472 × 83–222 μ in size; perforation plate are simple, horizontal [Text fig. 84 a]. Vessels with spiral thickening are ca 416–472 × 27–55 μ in size, perforation plate oblique [Text fig. 84 b]. Pits to xylem parenchyma are rounded [Text fig. 84 d-d']. Fibers are of two types. Thin walled fibers are ca 527–611 μ in length pointed at both ends, pitted and unseptate [Text fig. 84 c]. Thick walled fibers are ca 777–955 μ in length, with or without lumen [Text fig. 84 c'].

Powder characters: Powder is reddish-black in color and consists of vessels, fibers pieces, raphide sacs, raphides, crystals of calcium oxalate and starch grains [Text fig. 84 f–k].

Raphide sacs are ca 26–48 × 16–18 μ in size, containing group of raphides [Text fig. 84 h]. Raphides are ca 45–54 μ in length, pointed at one end [Text fig. 84 i]. Crystals of calcium oxalate are abundant in powder, clustered, ca 20–24 μ in diameter [Text fig. 84 j]. Starch grains are abundant in powder, simple, ca 4–8 μ in diameter and elliptic oblong in shape [Text fig. 84 k].

Local uses: Stem paste is applied externally on wounds. Stem is used on various skin diseases such as elephantiasis, scabies, itch, swelling etc.

RUIKHOD [Photoplate-XIII; Text fig.: 85 a–h]
Botanical name: **Calotropis gigantea** (L.) Br.

Family: Asclepiadaceae

Vernacular names: Arka, Mandar, Kalimandur Rui.

Organoleptic characters: Taste is bitter and acrid, odorless.

Macroscopic characters: The drug occurs in the form of gray or dirty white or brown colored pieces of stem of ca 7–8 × 1–2.5 cm. Bark of stem is thin, rough due to presence of rhytidome, gray in color. Stem is longitudinally wrinkled, wrinkles are shallow. Transverse cracks are present on bark. Wood is lemon yellow or buff colored and soft. Fractures are short. The outer rhytidom breaks in the form of powder. Secondary growth is normal.

Microscopic characters: Transverse section of stem is circular in outline. It shows outer thin walled, multilayered parenchymatous cells of rhytidome and soft phellem. Inner bark consists of 4–5 layered phelloderm. Secondary cortex consists of sclerenchyma. Rays are uniseriate and parenchyma is scanty or absent. Wood consists of vessels, xylem parenchyma and fibers [Text fig. 85 a-c]. Vessels are ca 289–407 × 29–44 μ in size, perforation plates simple and oblique; intervascular pitting alternate. Pits are bordered [Text fig. 85 a-a']. Vessels with scalariform pits were also observed. Pits to xylem parenchyma are similar to as vessels [Text fig. 85 c]. Fibers are ca 550–665 μ in length, slightly pitted and unseptate [Text fig. 85 b].

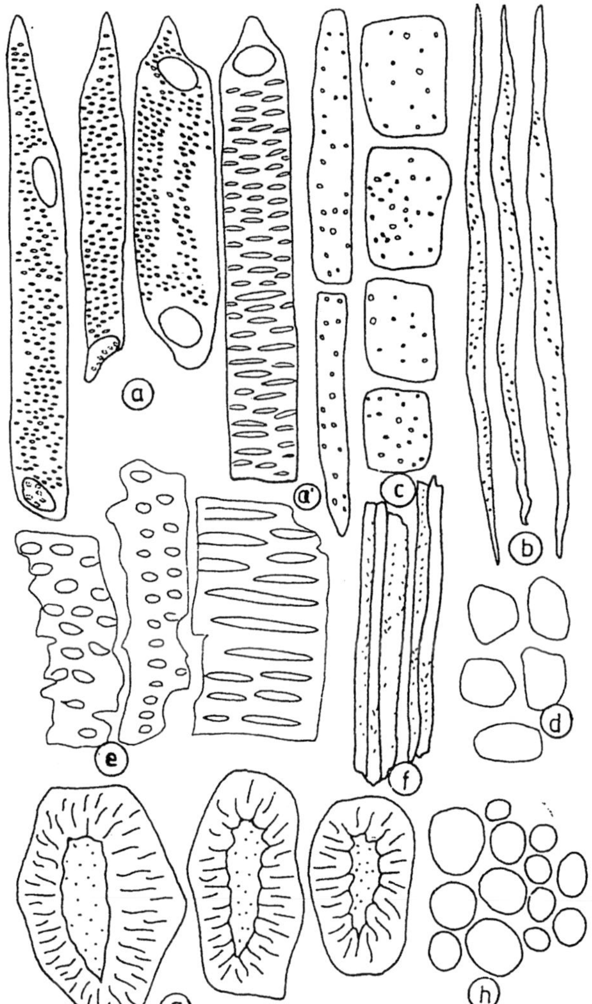

Text fig. 85: Calotropis gigantea a-a'-Vessels (x 225), b-Fibers (x 150), c-Xylem parenchyma (x 450), d-parenchyma (x 450), e-Pieces of vessel (x 675), f-Pieces of fibers (x 150), g-Stone cells (x 450), h-Starch grains (x 675).

Powder characters: Powder is pale yellow or buff colored containing pieces of vessels fibers, stone cells and starch grains [Text fig. 85 c–h].

Stone cells are ca 6–10 × 6–8 μ in size, oval in shape, thick walled and with pitted lumen [Text fig. 85 g]. Starch grains are rare in powder material, ca 7–14 μ in diameter, rounded in shape [Text fig. 85 h].

Local uses: Powder of stem is used on various skin diseases. The fresh stem is warmed on flame and tied around the wound made due to the thorn. The thorn is expelled out from the wounds.

SAPTAKAPPI/CHAKRUMBI/WARDHARA [Photoplate–XIII; Text fig.: 86 a–k]

Botanical name: **Diploclisia glaucescens** (Bl.) Diels.

Synonyms: *Cocculus glaucescens* Bl., *C. macrocarpus* Wight & Arn.

Family: Menispermaceae

Vernacular names: Ramrakh, Vardhara.

Organoleptic characters: Initially taste none, later on bitter, odorless.

Macroscopic characters: The drug occurs in the form of gray colored cylindrical pieces of stem of ca 5–6 × 2–2.5 in size. Bark of stem is thin, smooth or slightly rough with small, elongated to circular lenticels. Stem is without wrinkles or fissures. Small transverse cracks are present on the bark. Inner wood is hard and gray in color. Fractures are short. Cut surface of stem shows concentric rings of secondary anomalous growth.

Microscopic characters: Transverse section of stem is circular in outline. Cork is 2–5 layered made up of thick walled rectangular cells. Secondary cortex consists of 4–5 layers of parenchymatous cells. Transverse section of stem shows anomalous secondary growth. It consists of successive bundles of xylem and phloem repeating the structure of the young stem. Phloem is interxylary of the concentric type interrupted by broad bands of conjunctive parenchyma containing irregular groups of stone cells. The bundles of xylem and phloem are separated by tangential bands of parenchyma and large interfasicular rays. Wood consists of vessels, xylem parenchyma, tracheids, tracheid fibers and fibers [Text fig. 86]. Vessels are cylindrical, ca 583 μ -500μ × 94–122 μ; perforation plates are simple, slightly oblique; intervascular pitting is alternate [Text fig. 86 a]. Pits to xylem parenchyma usually rounded [Text fig. 86 d-d']. Tracheids are ca 770–1027 μ in length, pits are small and tapers at both ends [Text. fig.86 b]. Tracheid fibers are ca 638–916 μ in length, tapering at both the ends with small pits [Text fig. 86 c]. Thick walled fibers are ca 650–1000 μ in length [Text fig. 86 c']. Parenchymatous cells are with or without pits [Text fig. 86 e].

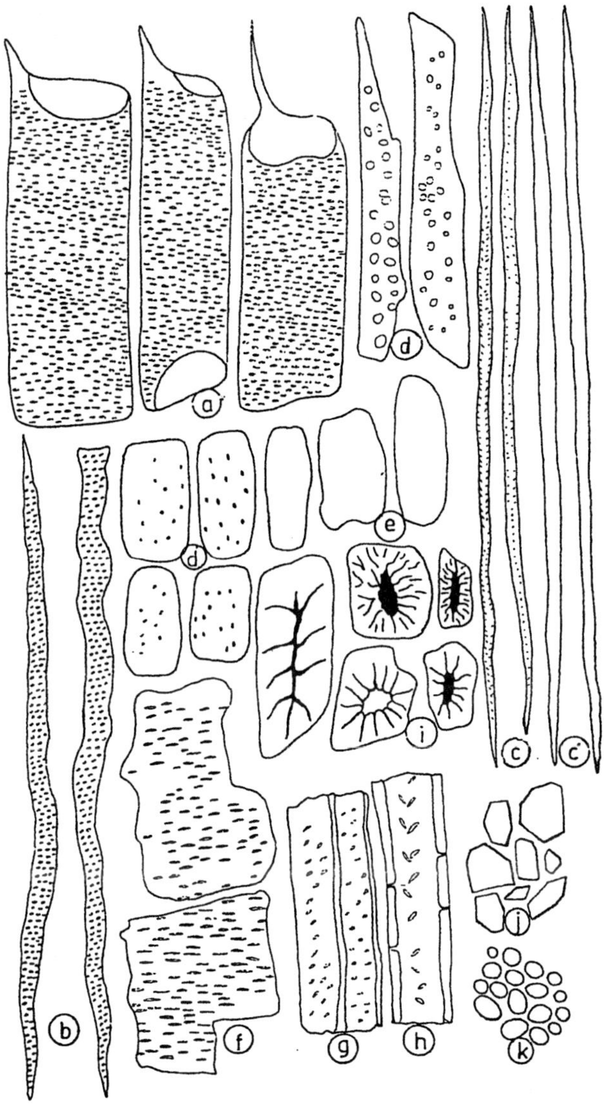

Text fig. 86: Diploclisia glaucescens a-Vessels (x 150), b-Tracheids (x 450), c-Thin walled fibers (x 150), c'-Thick walled fibers (x 150), d & d'-Xylem parenchyma (x 450), e-Parenchyma (x 450), f-Pieces of vessel (x 450), g-Pieces of tracheids (x 450), h-Pieces of fibers (x 450), i-Stone cells (x 675), j-Crystals of Calcium oxalate (x 450), k- Starch grains (x 450).

Powder characters: Powder is pale black or gray in color containing pieces of vessels, tracheids, fibers, stone cells, crystals of calcium oxalate and starch grains [Text fig. 86 f–k].

Stone cells ca 54–37 × 20–12 μ, mostly rectangular to square in shape, thick walled with narrow lumen [Text fig. 86 i]. Crystals of calcium oxalate are solitary, abundant in powder, ca 3–18 μ in size [Text fig. 86 j]. Starch grains are abundant in powder, simple, ca 8–12 μ in diameter, elliptic-oblong in shape [Text. fig. 86 k].

Local uses: Stem powder is given thrice a day orally to diabetic patient. Stem powder in combination with *Gnetum scandens* (Umbalvel) is given orally every night with water in pile troubles. It is also given in kidney stone. Stem in combination with *Cinnamomum nitidum* (Mawarung) powder and root of *Salacia brunoniana* (Saptrangi) is given in asthma.

UMBALVEL [Photoplate–XIII; Text fig.: 87 a–j]
Botanical name: **Gnetum scandens** Roxb.

Family: Gnetaceae

Vernacular names: Umbal.

Organoleptic characters: Taste is slightly bitter, odorless.

Macroscopic characters: The drug occurs in the form of dark gray or black colored cylindrical pieces of stems of ca 3–6 × 1–25 cm. Bark of stem is ca 1–2 mm thick, rough with small circular dirty white colored and rounded lenticels. Stem is longitudinally fissured. Fissures are deep. Transverse cracks are present on stem. Fractures are fibrous. Wood is gray and pale brown in color. A transversely cut surface of stem shows secondary anomalous growth.

Microscopic characters: Transverse section of stem is circular in outline. Cork is multilayered made of thick walled cells. Cork consists of 2–3 layers of phellogen and 5–6 layers of phelloderm. Secondary cortex consists of bands of stone cells. Transverse section of stem shows secondary anomalous growth. Phloem is intraxylary. Secondary anomalous growth is eccentric. Interfasicular rays are broad 3–6 cell wide. Wood consists of vessels, tracheids and fibers [Text fig. 87 a-c]. Vessels are ca 472–583 × 375–166 μ; perforation plates are simple, oblique; intervascular pitting is alternate and pits are bordered [Text fig. 87 a]. Pits to xylem parenchyma similar as vessels [Text fig. 87 c]. Tracheids are ca 555–1277 μ in length, thick walled with circular and bordered pits [Text fig. 87 b]. Fibers are thick walled, ca 667–808 μ in length, without lumen [Text fig. 87 c].

Powder characters: Powder is blackish-gray in color containing pieces of vessels, tracheids, fibers, stone cells, crystals of calcium oxalate and starch grains [Text fig. 87 d–j].

Stone cells are ca 55–83 × 12–40 μ in size, rectangular to square in shape, thick walled and with broad lumen [Text fig. 87 d–j]. Crystals of calcium oxalate are ca 4–10 μ in size and solitary [Text fig. 87 i]. Starch grains are abundant in powder. They are simple or compound, ca 4–10 μ in diameter and rounded in shape [Text fig. 87 j].

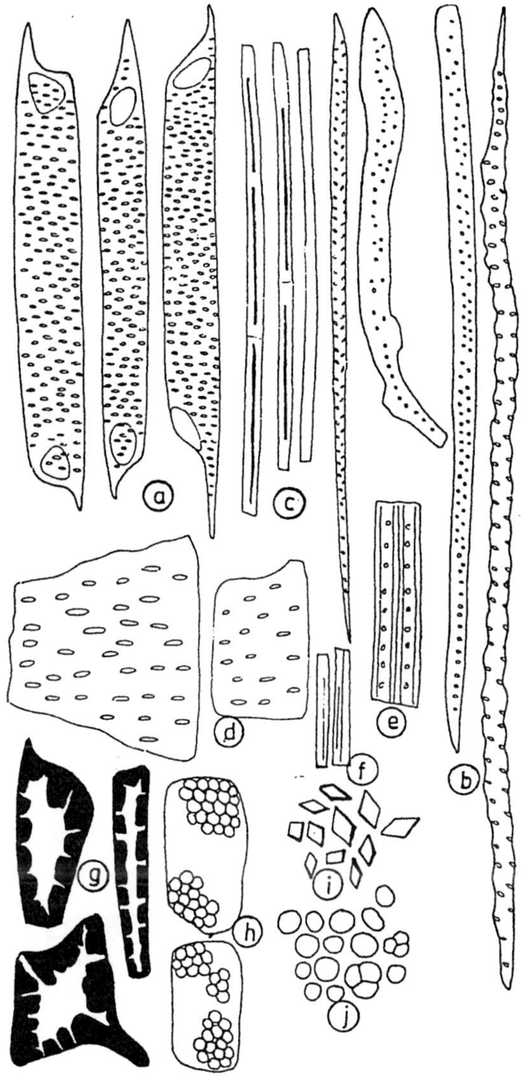

Text fig. 87: Gnetum scandens a-Vessels (x 150), b-Tracheids (x 150), c-Fibers (x 100), d-Pieces of vessel (x 450), e-Pieces of tracheids (x 450), f-Pieces of fibers (x 100), g-Stone cells (x 450), h-Parenchyma cells filled with starch grains (x 450), i-Crystals of Calcium oxalate (x 675), j-Starch grains (x 675).

Local uses: Paste of stem in combination with bark of *Pittosporum floribundum* ('Vikari') and root of *Salacia brunoniana* ('Saptrangi') is given for external application on piles. The stem powder is orally given in fever. Paste of stem is applied on forehead in hemicrania.

VAKERICHE BAL BHATE [Photoplate-XIII; Text fig.: 88 a–j]
Botanical name: **Moullava spicata** (Dalz.) Nicols.

Synonyms: *Caesalpinia spicata* Dalz., *Wagatea spicata* (Dalz.) Wight

Family: Caesalpiniaceae

Vernacular names: Vakeri, Waghati.
Organoleptic characters: Taste is slightly bitter, odorless.

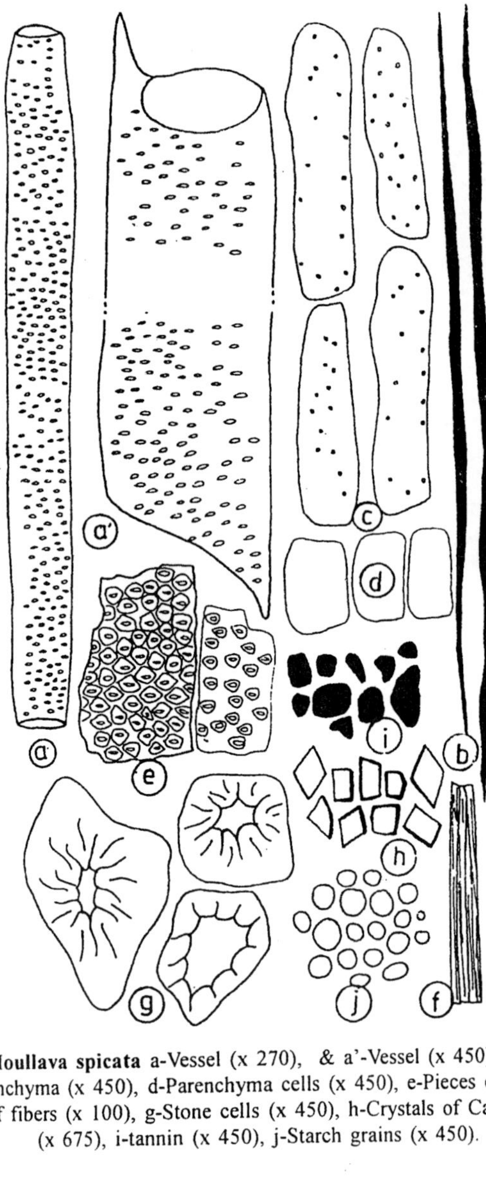

Text fig. 88: Moullava spicata a-Vessel (x 270), & a'-Vessel (x 450), b-Fibers (x 100), c-Xylem parenchyma (x 450), d-Parenchyma cells (x 450), e-Pieces of vessel (x 450), f-Pieces of fibers (x 100), g-Stone cells (x 450), h-Crystals of Calcium oxalate (x 675), i-tannin (x 450), j-Starch grains (x 450).

Macroscopic characters: The drug occurs in the form of pale pink or brown colored cylindrical pieces of stem of about 6–8 × 2–3 cm. Bark of stem is ca 2–4 mm thick, rough with small elongated or circular lenticels. Stem is transversely fissured or sometimes without fissures. Wood is pale pink or dark brown in color and hard. Fractures are short.

Microscopic characters: Transverse section of stem is circular in outline. It shows multilayered cork made up of alternate layers of thick walled and thin walled cells. Phelloderm region of cork contains band of stone cells and crystals of calcium oxalate. Wide zone of secondary phloem contains biseriate medullary rays in bark. Parenchyma is usually abundant, paratracheal, confluent. Rays are 1–2 cell wide. Wood consists of vessels, xylem parenchyma and fibers [Text fig. 88 a–c]. Vessels are ca 296–500 × 37–292 μ; perforation plates simple, horizontal; intervascular pitting is alternate, pits are bordered [Text fig. 88 a]. Pits to xylem parenchyma are similar as of vessels [Text fig. 88 c]. Fibers are thick walled without lumen, ca 116–141 μ in length, tapering at both ends and unseptate [Text fig. 88 b].

Powder characters: Powder is reddish-brown in color containing pieces of vessels and fibers, stone cells, crystals of calcium oxalate, starch grains and brown colored content [Text fig. 88 e–j].

Stone cells are ca 26–54 × 24–83 μ in size, mostly rectangular to square in shape, thick walled, with narrow or broad lumen [Text fig. 88 g]. Crystals of calcium oxalate are abundant, 6–13 μ in size, prism type [Text fig. 88 h]. Starch grains are simple, ca 3–6 μ in diameter and rounded in shape [Text fig. 88 j].

Local uses: The stem powder is given every morning with milk in night emissions. Tincture of powder is also given with water early morning to reduce heat from the body. The stem paste is applied on any type of endo or exocortosis. The stem powder is also given in diabetes.

ARJUNASAL [*Arjunsada*] [Photoplate-XIV; Text fig. 89 a–c]
Botanical name: **Terminalia cuneata** Roth

Synonyms: *T. arjuna* (Roxb. ex DC.) Wight & Arn.

Family: Combretaceae

Vernacular names: Arjuna, Arjunasadada, Dhavala, Sanmadat.

Organoleptic characters: Taste gritty or astringent, odorless.

Macroscopic characters: Drug occurs in the form of flat or slightly curved pieces of bark, ca 3–5 × 0.5–1 cm size. Outer surface of bark is pink or flesh colored with mealy coating. It is smooth and glabrous. Bark is without wrinkles or fissures. Inner surface of bark is reddish-brown, finely striated and peels out in thin flakes. Fractures are short.

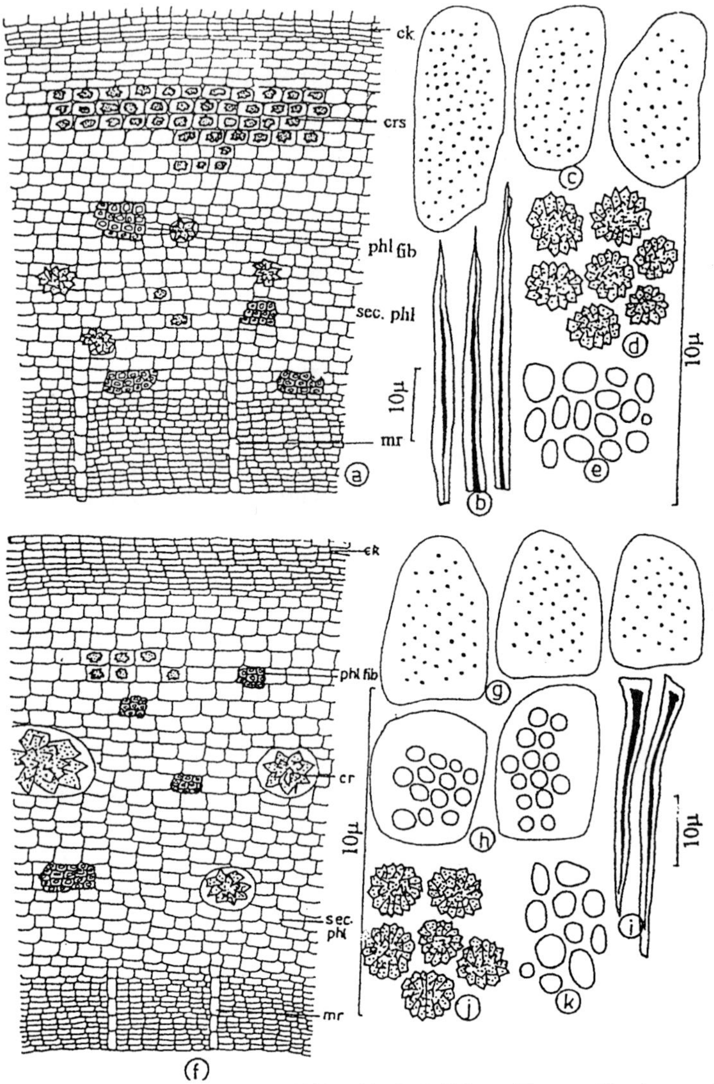

Text fig. 89: Terminalia cuneata a-T. S. of bark, b-Phloem fibers, c-Pitted parenchyma cells, d-Crystals of Calcium oxalate, e-Starch grains; **Terminalia elliptica** f-T. S. of Bark, g-Pitted parenchyma cells, h-Parenchyma cells containing starch grains, i-phloem fibers, j-Crystals of Calcium oxalate, k-Starch grains.

Microscopic characters: Transverse section of bark shows cork and wide zone of secondary phloem [Text fig. 89 a]. Cork cells are 3–5 layered, thin walled, radialy arranged and filled with tannin. Secondary phloem consists of phloem fibers, crystals of calcium oxalate. The secondary phloem is deep and traversed by uniseriate ray cells [Text fig. 89 a].

Powder characters: Powder of bark is pink in color containing pitted parenchyma, pieces of phloem fibers, crystals of calcium oxalate and starch grains [Text fig. 89 b–e].

Pieces of phloem fibers are ca 330–440 μ in length and thick walled [Text fig. 89 b]. Parenchyma cells are pitted [Text fig. 89 c]. Crystals of calcium oxalate are ca 12–43 μ in diameter and clustered [Text fig. 89 d]. Starch grains are abundant, simple, ca 6–10 μ in diameter and rounded to oblong in shape [Text fig. 89 e].

Local uses: The paste of bark is applied externally to cure wounds. Decoction of bark is given orally early in the morning in bone fracture for one month. It is also given as a cardiac tonic. Though it is given as cardiac tonic, according to Mukerji (1943) it is found to be negative in its effect and cause fall of blood pressure in intact animals even in very small doses. Large dose causes death of animal from stoppage of heart. On isolated heart, beats are recorded weaker and eventually heart stops contrary to general belief.

In market the bark of both *T. cuneata* [Photoplate-XIV] and *T. elliptica* [Photoplate-XIV] is sold under the same name of "*Arjuna*". Bark of *T. elliptica* is used as substitutes to bark of *T. cuneata*. The following macroscopic and microscopic characters are useful in distinguishing both the barks.

Table I: Showing comparison of Barks of *T. cuneata* Roth and *T. elliptica* Willd.

Terminalia cuneata Roth (*T. arjuna* (Roxb. ex DC.) W. & A.) [Photoplate-XIV;Text fig. 89 a–e]	**Terminalia elliptica** Willd. (*T. tomentosa* (Roxb. ex DC.) W. & A.) [Photoplate-XIV; Text fig. 89 f–k]
1. Bark is flat, ca 0.5–1 cm thick, and flesh in colour with mealy coating. Surface is smooth without fissures. T. S. of bark shows 1–2 layered, thin walled cork. Secondary phloem is deep and traversed by uniseriate ray of cells [Text fig. 89 a].	1. Bark is curved, ca 1–1.5 cm thick due to rhytidome. Pale brown to dark brown in color with may cracks. Surface is rough & lamellated. T. S. of bark shows outer 2–3 layers of rhytidome composed of alternate layers of periderm and crushed secondary phloem. Cork is multilayered and suberised [Text fig. 89 f].
2. Powder is pale pink in color containing abundant crystals of calcium oxalate.	2. Powder is reddish-brown in colour containing few crystals of calcium oxalate as compared to *T. cuneata*.

ASANSAL [Photoplate–XIV; Text fig. 90 a–e]
Botanical name: **Bridelia retusa** (L) Spreng.

Family: Euphorbiaceae.

Vernacular names: Asan, Ashind, Kutgi, Pashanpheda.
Organoleptic characters: Taste astringent, odorless.

Macroscopic characters: Drug occurs in the form of slightly curved pieces of bark of ca 5–6 × 0.3–0.5cm size. Outer surface of bark is reddish-brown to pale brown in color. Inner surface of bark is reddish in color and finely striated, pilling out in thin flakes, fractures are fibrous. Cut pieces of bark show red colored tannins exuding on their edges which is easily soluble in water giving red color to the water

Microscopic characters: Transverse section of bark shows cork layers, secondary phloem and phloem fibers [Text fig. 90 a]. Bark consists of 1–2 layers of phellogen and 4–6 layers of phelloderm. Secondary phloem consists of phloem fibers. Phloem fibers are in rectangular patches. T. S. of bark looks like a mesh possessing phloem fibers and secondary phloem. Solitary crystals of calcium oxalate are found in cells of bark. Medullary rays are 2–3 seriate [Text fig. 90 a].

Powder characters: Powder is reddish-pink in color containing pieces of phloem fibers, crystals of calcium oxalate and starch grains [Text fig. 90 d-e].

Pieces of phloem fibers are ca 38–66 μ in length, which contains crystals of calcium oxalate encircled by parenchymatous sheath [Text fig. 90 b-c]. Crystals of calcium oxalate are solitary. They are abundant in powder and ca 9–10 μ in size [Text fig. 90 d]. Starch grains are absent or very few, simple, ca 5–10 μ in diameter and elliptic to rounded in shape [Text fig. 90 e].

Local uses: Bark is used as a blood purifier. One teaspoon of powder of bark is mixed with water and taken early in the morning with empty stomach. Same way, it is given in urinary calculi. It is also given as cardiac tonic.

BIWALASAL [Photoplate–XIV; Text fig. 90 f–j]
Botanical name: **Pterocarpus marsupium** Roxb.

Family: Fabaceae

Vernacular names: Bilwa, Biwala

Organoleptic characters: Taste astringent, odorless.

Macroscopic characters: Drug occurs in the form of curved pieces of bark of ca 5–6 × 0.3–1cm size [Photoplate-20]. Outer surface of bark is reddish brown to pale brown in color, rough due to presence of rhytidome. Inner surface of bark is pale brown in color and finely striated, fractures are short.

Microscopic characters: Transverse section of bark shows cork layers and secondary phloem [Text fig. 90 f]. Bark consists of multilayered cork cells. Secondary phloem consists of phloem fibers. Phloem fibers are in small islands. Stone cells are also present in the form of islands [Text fig. 90 f].

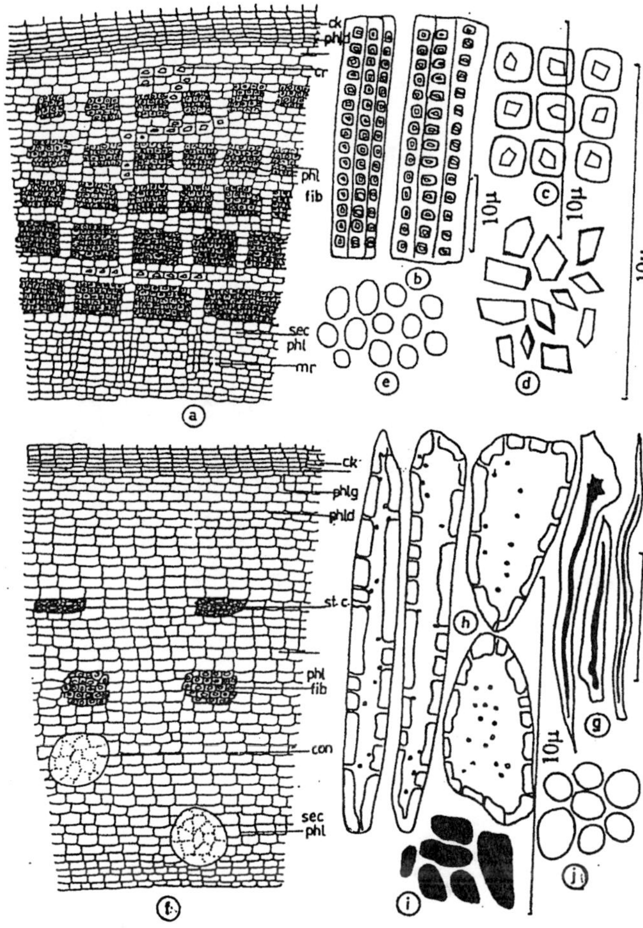

Text fig. 90: Bridelia retusa a-T. S. of bark, b-Pieces of phloem fibers, c-fragments of phloem fibers containing crystals of Calcium oxalate, d-Crystals of Calcium oxalate, e-Starch grains; **Pterocarpus marsupium** f-T. S. of bark, g-Phloem fibers, h-Stone cells, i-Orange colored tannin content, j-Starch grains.

Powder characters: Powder is reddish-brown in color containing pieces of phloem fibers, stone cells, orange colored contents and starch grains [Text fig. 90 g–j].

Pieces of phloem fiber are ca 300–400 μ in length and thick walled [Text fig. 90 g]. Stone cells are 80–100 × 18–50 μ in size, linear lanceolate to triangular in shape. They are pitted, thin walled and with broad lumen [Text fig. 90 h]. Orange colored content is abundant in powder [Text fig. 90 I]. Starch grains simple, ca 7–10 μ in diameter, rounded to oblong in shape [Text fig. 90 j].

Local uses: The paste of bark is applied over gum to stop toothache. Bark is soaked whole night in water and one full glass of water is given in diabetes early in the morning.

KARANJSAL [Photoplate–XIV; Text fig. 91 a–d]
Botanical name: **Pongamia pinnata** (L.) Pierre.

Synonyms: *Pongamia glabra* Vent.

Family: Fabaceae

Vernacular names: Karanj, Kirmal, Naktamal.

Organoleptic characters: Taste very bitter, pungent odor noxious, and remain permanent in dry samples.

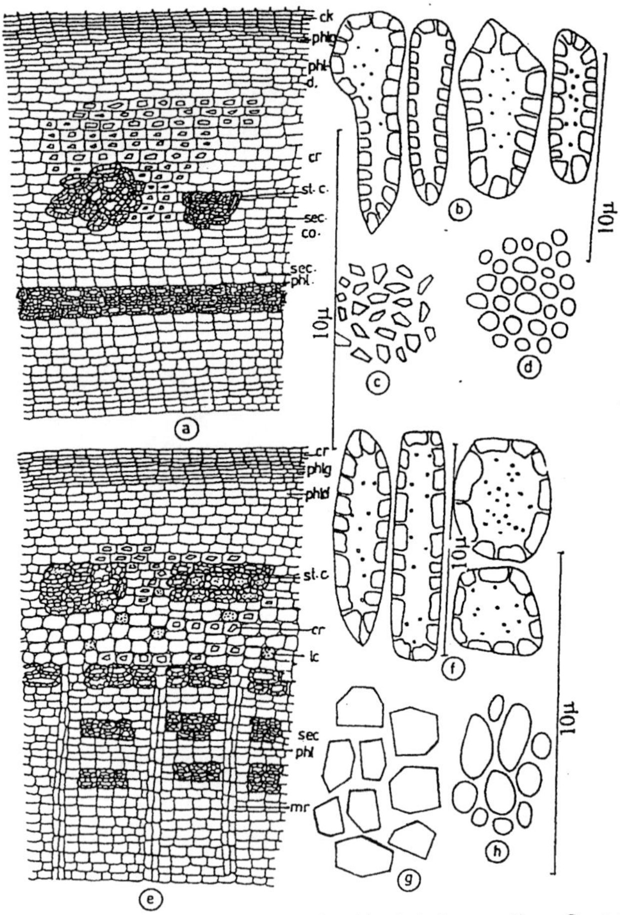

Text fig. 91: Pongamia pinnata a-T. S. of bark, b-Stone cells, c-Crystals of Calcium oxalate, d-Starch grains; **Ficus exasperata** e-T. S. of bark, f-Stone cells, g-Crystals of Calcium oxalate, h-Starch grains.

Macroscopic characters: Drug occurs in the form of incurved or channeled pieces of bark of ca 3–6 × 0.3–0.7 cm size. Bark is gray or brown in color and rough. Lenticels are numerous and transversely elongated or rounded in shape. Inner surface of bark is pale brown or off white in color, smooth and longitudinally striated; fractures are short.

Microscopic character: Transverse section of bark shows cork and secondary phloem [Text fig. 91 a]. Cork is thick walled, 5–6 layered. Cork consists of thin walled, 1–2 layered phellogen and 4–6 layers of phelloderm. Secondary phloem is deep and contains islands of stone cells or rows of stone cells [Text fig. 91 a].

Powder characters: Powder is pale yellow in color and consists of stone cells, crystals of calcium oxalate and starch grains [Text fig. 91 b–d].

Stone cells are ca 64–96 × 18–37 μ in sizes and elliptic oblong in shape, with thick walled, narrow lumen and are pitted [Text fig. 91 b]. Crystals of calcium oxalate solitary ca 4–7 μ in size [Text fig. 91 c]. Starch grains are abundant in powder. They are simple, ca 4–7 μ in diameter and rounded in shape [Text fig. 91 d].

Local uses: The paste of bark is applied externally on itch. The bark is chewed to clean the teeth. Bark is used in wounds and external injuries.

KHARVATSAL [Photoplate–XIV; Text fig. 91 e–h]

Botanical name: **Ficus exasperata** Vahl

Synonyms: *Ficus asperrima* Roxb.

Family: Moraceae

Vernacular name: Bhuimbar, Karyat, Kharmate, Kharvati.

Organoleptic characters: Tasteless, odorless.

Macroscopic characters: Drug occurs in the form of curved pieces of bark of ca 6–8 × 0.3–0.5 cm. The outer surface of bark is yellow or greenish-brown in color with numerous brown lenticels. Lenticels are rounded and scattered irregularly. Inner surface of the bark is brown or pale brown in color with fine striations, fractures short.

Microscopic character: Transverse section of bark shows cork, secondary cortex and secondary phloem [Text fig. 91 e]. Cork is 3–4 layered, made up of thick walled cells. It consists of 1–2 layered phellogen and 4–6 layered phelloderm. Phelloderm consists of islands of stone cells and secretary canals. Secondary cortex consists of rectangular parenchymatous cells interrupted by stone cells. The crystals of calcium oxalate are present in parenchymatous cells of cortex. Phloem is interrupted by bands of stone cells [Text fig. 3 e].

Powder characters: Powder is yellowish-brown in color containing stone cells, crystals of calcium oxalate and starch grains [Text fig. 91 f–h].

Stone cells are rectangular to elliptic-oblong in shape, ca 37–111 × 20–46 μ in size. They are thick walled, Pitted and possesses broad [Text fig. 91 f].

Crystals of calcium oxalate are abundant in powder, solitary, ca 8–18 μ in length [Text fig. 91 g]. Starch grains abundant in powder. They are simple, rounded to elliptic oblong in shape, ca 8–18 μ in diameter [Text fig. 91 h].

Local uses: Paste of bark is immediately applied externally after scorpion sting. Paste of bark is also applied externally to heal wounds.

KUDASAL [Photoplate-XIV; Text fig. 92 a–d]

Botanical name: **Holarrhena pubescens** (Buch.-Ham.) Wall. ex G. Don

Synonyms: *H. antidysentrica* (Roth) Wall. ex A. DC.

Family: Apocynaceae

Vernacular names: Kuda, Indrajaw, Pandhrakuda.

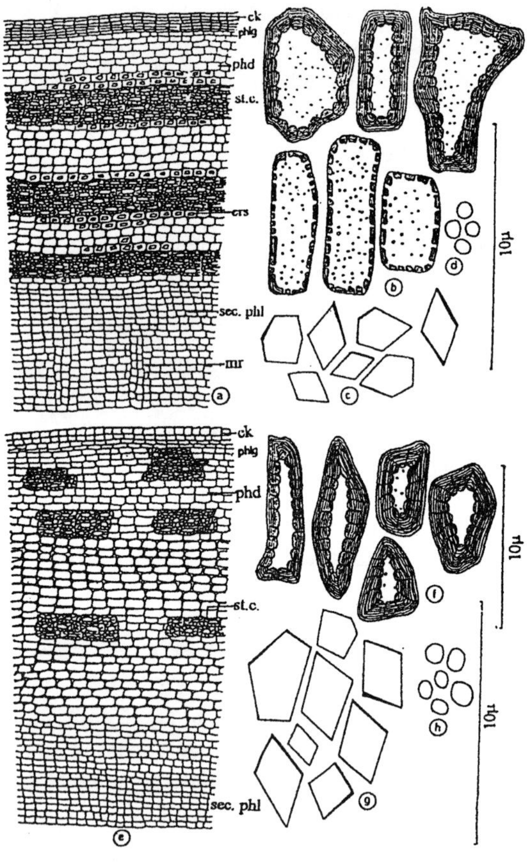

Text fig. 92: Holarrhena pubescens a-T. S. of bark, b-Stone cells, c-Crystals of Calcium oxalate, d-Starch grains; **Wrightia tinctoria** ssp. **tinctoria** e-T. S. of bark, f-Stone cells, g-Crystals of Calcium oxalate, h-Starch grains.

Organoleptic characters: Taste very bitter, odor very faint, agreeable.

Macroscopic characters: Drug occurs in the form of recurved pieces of bark of ca 3–4 × 0.1–0.4 cm in size [Photoplate-XIV]. Outer surface of bark is gray, rough and warty, longitudinally wrinkled with occasional transverse and longitudinal cracks. Lenticels are small rounded to transversely elongated, irregularly scattered. Inner surface of bark is dark, buff colored, stratified, rough and scaly. Fractures are brittle.

Microscopic characters: Transverse section of bark shows cork, secondary cortex and secondary phloem [Text fig. 92 a]. Cork cells are 5–8 layered, thick walled and rectangular. Cork consists of 1–2 layered phellogen and 6–10 layered phelloderm. The cells are thin walled containing rhomboidal crystals of calcium oxalate. Cortex is wide consisting of groups or bands of pitted, lignified stone cells. Cortical parenchyma surrounding the stone cells contains rhomboidal crystals of calcium oxalate. It is a peculiar character of bark of *H. pubescens*. Secondary phloem consists of phloem parenchyma and medullary rays. Medullary rays are 2–3 seriate, wider towards outside [Text fig. 92 a].

Powder characters: Powder is gray to buff colored containing abundant stone cells, crystals of calcium oxalate and very few starch grains [Text fig. 92 b-d].

Stone cells are in groups or solitary in powder. Stone cells are ca 59–101 × 27–55 μ in size and rectangular to elongated in shape with thin or thick walls, striated, lumen broad and pitted [Text fig. 92 b]. Crystals of calcium oxalate are abundant. They are ca 18–32 μ in size, solitary and prism shaped [Text fig. 92 c]. Starch grains are very few or absent. They are simple and ca 6–10 μ in diameter [Text fig. 92 d].

Local uses: The bark is given in dysentery. Tincture of bark powder is given thrice in a day during dysentery and also given to stop stomach pain.

KALAKUDA [Photoplate–XIV; Text fig. 92 e–h]

Botanical name: **Wrightia tinctoria** ssp. **tinctoria** R.Br.

Synonyms: *Wrightia tinctoria* R. Br.

Family: Apocynaceae

Vernacular names: Kalakuda, khirni, kuda. Godaindrajaw.

Organoleptic characters: Taste slightly bitter, odorless.

Macroscopic characters: Drug occurs in the form of recurved pieces of bark of ca 3–5 × 0.3–0.5 cm in size. The outer surface of bark is gray or greenish-black, rough and possesses lenticels. Lenticels are pale brown or gray, small, rounded and present in groups. The outer surface of bark is without fissures. Inner surface of bark is pale brown or greenish-yellow, smooth and longitudinally striated. Fractures are short.

Microscopic characters: Transverse section of bark shows cork, secondary cortex and secondary phloem [Text fig. 92 e]. Outer layer of cork is 2–3 layered made up of thick walled square shaped cells. Cork consists of 2–3 layers of phellogen and 4–5 layers of phelloderm. Stone cells are distributed in phelloderm. Secondary cortex consists of crystals of calcium oxalate and latex cavities. Secondary cortex is followed by secondary phloem [Text fig. 92 e].

Powder characters: Powder is gray or greenish-black in color containing abundant stone cells, crystals of calcium oxalate and few starch grains [Text fig. 92 f–h].

Stone cells are ca 55–111 × 40–45 μ in size, rectangular to oblong in shape. They have thick walls with fine striations [Text fig. 92 f]. Crystals of calcium oxalate are ca 14–22 μ in size. They are solitary, prism shaped, very few in number as compared to *Holarrhena pubescens* ("Kudasal") [Text fig. 92 g]. Starch grains are found to be very less in number or are sometimes absent. They are ca 6–18 μ in diameter in size, simple and elliptic oblong in shape [Text fig. 92 h].

Local uses: Bark is applied externally in piles and to cure wounds. It is also given to the animals suffering from wounds.

RAGATWAD SAL [Photoplate–XIV; Text fig. 93 a–i]

Botanical name: **Cassine paniculata** (Wight & Arn.) Lobereau-Collen.

Synonyms: *Elaeodendron paniculatum* Wight & Arn. *E. glaucum* Law.

Family: Celastraceae

Vernacular names: Ragatwad.

Organoleptic characters: Initially taste astringent and later on slightly bitter, odorless.

Macroscopic characters: Drug occurs in the form of curved pieces of bark of ca 4–6 × 0.3–1 cm in size [Photoplate-XIV]. Outer surface of bark is green, rough or smooth without any wrinkles or fissures. Lenticels are numerous and scarlet or orange in color. Inner surface of bark is creamish-brown in color, smooth, without any striations. Fractures are short.

Microscopic characters: Transverse section of bark shows cork, secondary cortex and secondary phloem [Text fig. 93 a]. Cork is made up of 3–6 layered thick-walled cells containing 1–2 layers of phellogen and 3–5 layers of phelloderm. Secondary cortex is wide, parenchymatous. It is interrupted by large, oblong, patches of stone cells. Secondary phloem is deep and rays are biseriate [Text fig. 93 a].

Powder characters: Powder of bark is reddish-brown in color containing crystals of calcium oxalate, abundant stone cell and very few starch grains [Text fig. 93 b–e].

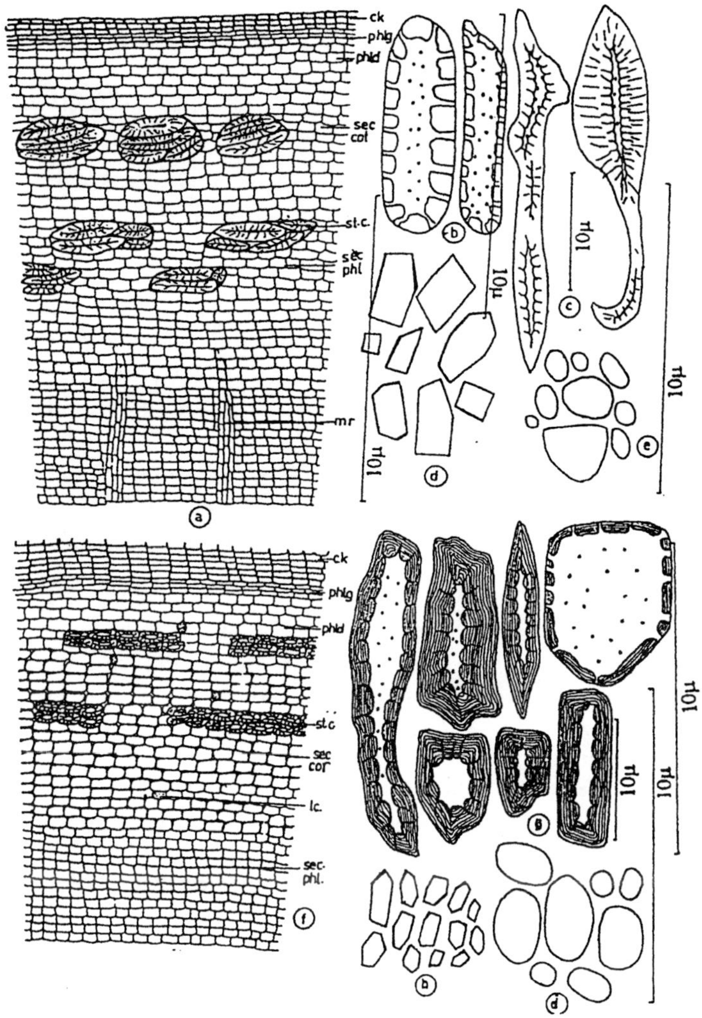

Text fig. 93: Cassine paniculata a-T. S. of bark, b-Stone cells with broad lumen, c-Stone cells with narrow lumen, d-Crystals of Calcium oxalate, e-Starch grains; **Alstonia scholaris** f-T. S. of bark, g-Stone cells, h-Crystals of Calcium oxalate, i-Starch grains.

Stone cells are of two types. First type of stone cells are ca 50–60 × 10–20 μ in size, rectangular to oblong in shape. They are moderately thick walled with broad pitted lumen [Text fig. 93 b]. However, second type of stone cells are 28–31 × 3–6 μ in size, thick walled with narrow lumen and are pitted [Text fig. 93 c]. Crystals of calcium oxalate are solitary, ca 10–20 μ in size [Text fig. 93 d]. Starch grains are absent or very few. They are simple, ca 6–20 μ in diameter and rounded in shape [Text fig. 93 e].

Local uses: Paste of bark is given to women in anemia every night with milk. Tincture of bark powder is given as a blood purifier.

SATVIN SAL [Photoplate–XIV; Text fig. 93 f–i]
Botanical name: **Alstonia scholaris** (L.) R. Br.

Family: Apocynaceae

Vernacular names: Saptaparni, Satvin, Shalmali, Purna.

Organoleptic charters: Taste persistently bitter, odor mild and pleasant.

Macroscopic characters: Drug occurs in the form of channeled or quilled pieces of bark of ca 5–7 × 3–4 cm in size [Photoplate–XIV]. The bark is dark gray to greenish in color, rough, uneven fissured. Lenticels are prominent, transversely elongated to rounded in shape, buff colored. Inner surface of bark is brownish buff to grayish brown in color. Fractures are short.

Microscopic characters: Transverse section of bark shows multilayered cork, broad zone of secondary cortex and secondary phloem [Text fig. 93 f]. Cork is multilayered, 6–12 cell in thickness, brick shaped. The secondary cortex is made up of thin walled parenchymatous cells. The region consists of many rounded latex cavities. Solitary crystals of calcium oxalate are also present in secondary cortex. Stone cells are in the form of non continuous layers of 4–8 cell in width. The cells of secondary phloem are smaller in dimension than the cortical cells [Text fig. 93 f].

Powder characters: Powder is gray in color containing stone cells, crystals of calcium oxalate and starch grains [Text fig. 93 g–i].

Stone cells are ca 26–108 × 12–42 μ in size, variously shaped, rounded, linear fiber like, rectangular. Stone cells are blunt, thick or thin walled and possess striated lumen broad or narrow and are pitted or unpitted [Text fig. 93 g]. Crystals of calcium oxalate are solitary, ca 6–18 μ in size [Text fig. 93 h]. Starch grains are found to be abundant in powder. They are simple, ca 7–15 μ in diameter and rounded or irregular in shape [Text fig. 93 i].

Local uses: The paste of bark is applied on wounds. Tincture of bark powder is given twice a day with buttermilk in dysentery.

VAIVARNA SAL [Photoplate–XIV; Text fig. 94 f–j]
Botanical name: **Crateva adansonii** spp. **odora** (Buch.-Ham.) Jacob.

Synonyms: *C. odora* Buch.-Ham. *C. religiosa* var. *roxburghii* (R.Br.) Hook. f. & Thoms.

Family: Capparaceae

Vernacular names: Vaivarna, Varuna.

Organoleptic characters: Taste slightly bitter, odor disagreeable when fresh whereas dried samples are odorless.

Macroscopic characters: The drug occurs in the form of curved pieces of bark of ca 3–5 × 0.4–0.8 cm in size [Photoplate–XIV]. Outer surface of bark is greenish-black in color, rough lenticellate and minutely fissured. Fissures are longitudinal, sometimes interrupted by longitudinal cuts. Lenticels are small, numerous, white and rounded to elongate in shape. Inner surface of bark is lemon yellow or buff in color, rough with small granular tubercles. It is thinly striated. Fractures are short.

Microscopic characters: Transverse section of bark shows cork, secondary cortex and secondary phloem [Text fig. 94 f]. Cork is brown in color and made up of 4–5 layered phelloderm. Cortex is wide and consists of groups or bands of stone cells [Text fig. 94 f]. Phloem is thin and, medullary rays are biseriate [Text fig. 94 f].

Powder characters: Powder is lemon yellow to buff in color containing stone cells, starch grains and cut part of stone cells [Text fig. 94 g–j].

Stone cells are of two types. Pitted moderately thick walled stone cells are ca 40–60 × 20–50 μ in size [Text fig. 94 g] and thick walled stone cells with narrow, unpitted lumen are ca 5–7 × 2–4 μ in size [Text fig. 94 h]. Shape of stone cells varies from rounded to rectangular. Pieces of pitted stone cells are also present in powder [Text fig. 94 i]. Starch grains are abundant in powder. They are simple, ca 3–7 μ in diameter and rounded to oblong in shape [Text fig. 94 j].

Local uses: Powder of bark is taken early in the morning with empty stomach to remove excess fats (Vata) remained after pregnancy. During course of treatment, heavy and spicy food is not allowed. Paste of bark is also applied in rheumatism.

VIKARISAL [Photoplate-XIV; Text fig. 91 a–e]
Botanical name: **Pittosporum floribundum** Wight & Arn.

Family: Pittosporaceae

Vernacular names: Gapsendi, Ikali, Vikari.

Organoleptic characters: Taste bitter, odor aromatic, disagreeable, remains even in dry simples.

Macroscopic characters: Drug occurs in the form of channeled or quilled pieces of bark of ca 5–15 × 0.1–0.3 cm size. Outer surface of the bark is gray or greenish-black in color, smooth, and lenticellate; lenticels are rounded to elongated and pale brown in color. Outer surface of bark is without fissures. Inner surface of bark is pale brown or yellowish in color, smooth and is without striations. Fractures are short [Photoplate-XIV].

Microscopic characters: Transverse section of bark shows cork, secondary cortex and secondary phloem [Text fig. 94 f]. Cork is made up of thin walled, rectangular, 2–3 layered cells. Cork consists of 1–2 layered phellogen and 6–10 layered phelloderm. Secretary canals are present in

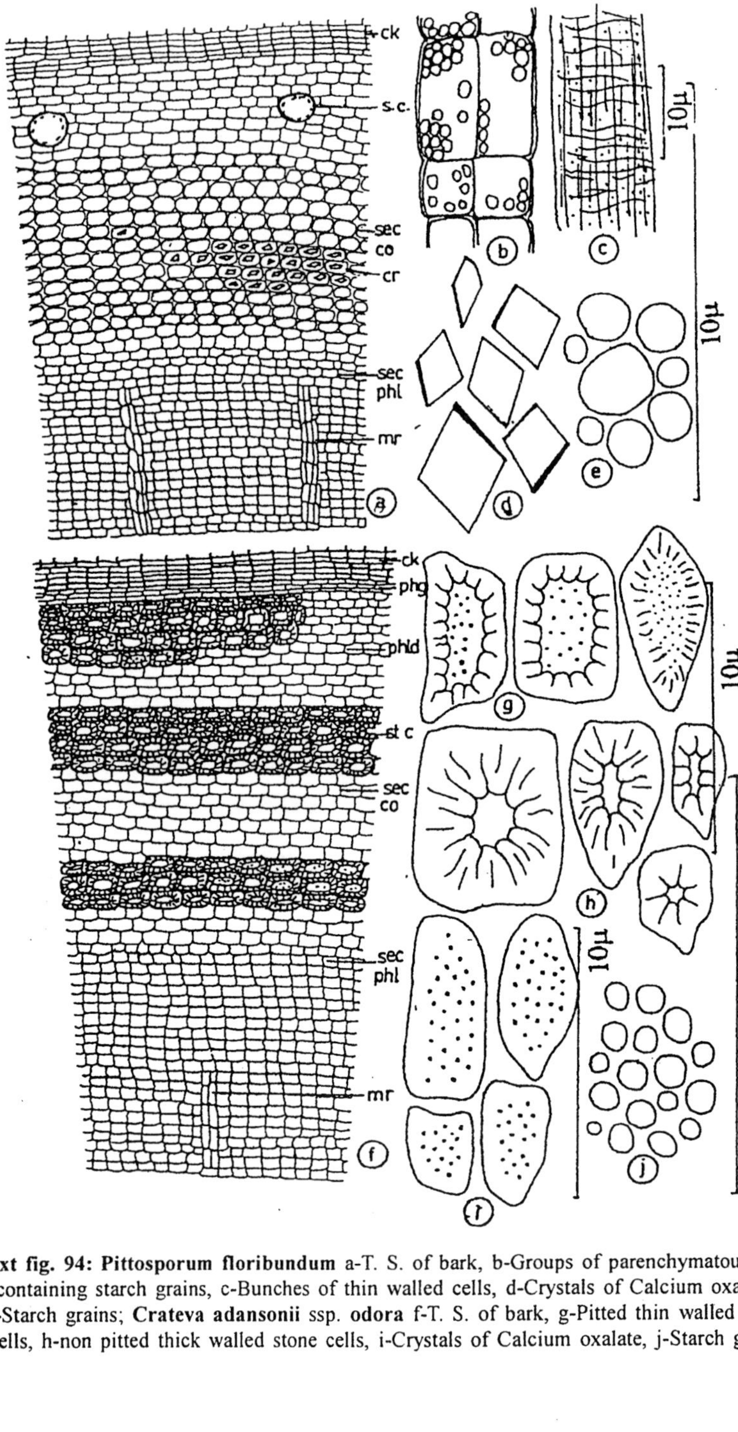

Text fig. 94: Pittosporum floribundum a-T. S. of bark, b-Groups of parenchymatous cells containing starch grains, c-Bunches of thin walled cells, d-Crystals of Calcium oxalate, e-Starch grains; **Crateva adansonii** ssp. **odora** f-T. S. of bark, g-Pitted thin walled stone cells, h-non pitted thick walled stone cells, i-Crystals of Calcium oxalate, j-Starch grains phelloderm region. Secondary cortex consists of thin walled parenchymatous cells. Parenchymatous cells contain crystals of calcium oxalate. Secondary phloem consists of phloem parenchyma and medullary rays. Medullary rays are uni-biseriate, wider towards outside [Text fig. 94 f].

Powder characters: Powder is gray in color containing abundant crystals of calcium oxalate and starch grains. Thin walled parenchymatous cells contains starch grains [Text fig. 94 b–e].

Crystals of calcium oxalate are solitary, ca 18–33 μ in size [Text fig. 94 d]. Starch grains are simple, ca 6–18 μ in diameter and rounded to oblong in shape [Text fig. 94 e].

Local uses: It is used with *Gnetum scandens* stem (Umbal vel), root of *Salacia brunoniana* (Saptrangi) and fruit of *Terminalia chebula* (Hirda) for external application in piles. Bark is externally applied in itch and in many skin diseases.

Leaves used in Folk Medicines

ADULSA [Text fig.: 95 a–f]
Botanical name: **Justicia adhatoda** L.

Synonyms: *Adhatoda zeylanica* Medic., *A. vasica* Nees

Family: Acanthaceae

Vernacular names: Adulsa, Malabar nut, Vasaka.

Organoleptic characters: Taste bitter, odor characteristic, disagreeable.

Macroscopic characters: The drug occurs in market in the form of leaves. The leaves are ca 7–20 × 3–5 cm, simple, petiolate, dark green above and paler beneath; lamina emarginate, elliptic to lanceolate in shape, apex acuminate, base tapering, margin entire or crenate, minutely puberulous when young, venation reticulate.

Microscopic characters: Leaf is dorsiventral and amphistomatic. Transverse section of leaf shows single layered epidermis made up of rectangular parenchymatous cells covered with cuticle [Text fig. 95 f]. Cells of upper and lower epidermis are more or less similar. Stomata's are diacytic [Text fig. 95 b]. Glandular and non-glandular trichomes are present on both surfaces of the leaf [Text fig. 95 d & c]. Non-glandular trichomes are uniseriate 2 to 4 celled, thick walled and pointed [Text fig. 95 c]. Mesophyll tissue is differentiated into two layered vertically elongated compact palisade cells and 3 to 5 layered spongy parenchyma cells with intercellular spaces. Midrib region shows continuous layer of epidermis, however, more trichomes appear on the epidermal layer of the midrib. 2–4 layers of collenchyma cells are present below the upper and lower epidermis in midrib region. The rest of the midrib is occupied by the parenchyma cells. Vascular bundle is arc

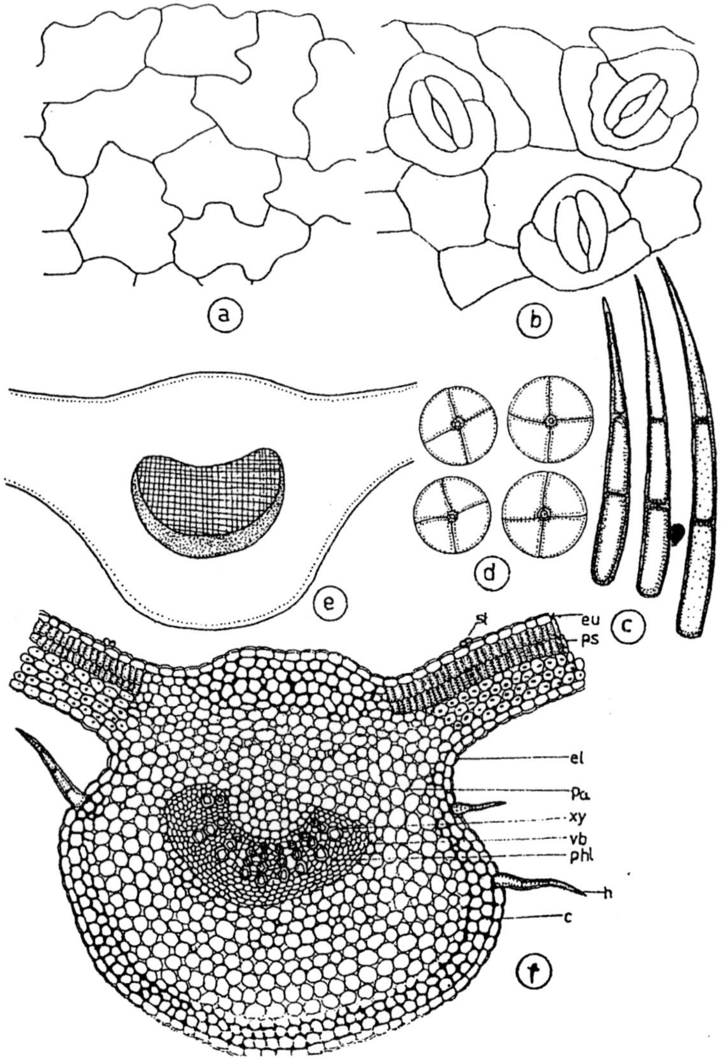

Text fig. 95: Justicia adhatoda a-Upper epidermis (x 675), b-Lower epidermis with stomata (x 675), c-Non glandular trichomes d-Glandular trichomes (x 675), e-Outline of transverse section of leaf (x 50), f-Transverse section of leaf passing through the mid vein (x 150).

shaped and collateral. Xylem is situated towards lower epidermis [Text fig. 95 f].

Powder characters: Leaf powder is dark green in color containing 2–4 celled uniseriate thick walled epidermal hairs and cuticle with diacytic stomata [Text fig. 95 b–d].

Local uses: One tea spoon of leaf powder mixed in hot water is given to patient suffering form cough or bronchitis after every six hours. Decoction of leaves is given in fever every 6 hrs. Entire leaves are smoked as bidi in asthma.

BEDKI CHAPALA [Text fig.: 96 a–g]
Botanical name: **Gymnema sylvestre** (Retz) R. Br. ex Schult.

Family: Asclepiadaceae

Vernacular names: Medhasingi, Bedukuli, Parapatraha.

Organoleptic characters: Taste slightly bitter when fresh, odor none. Leaves when chewed paralyze the taste buds and tongue which cannot determine sweet taste for a period.

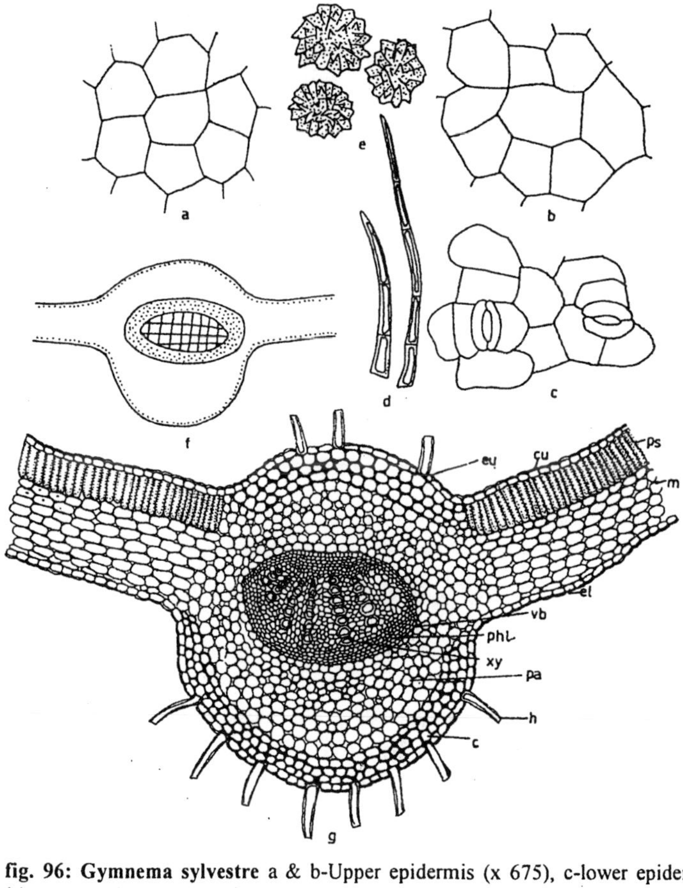

Text fig. 96: Gymnema sylvestre a & b-Upper epidermis (x 675), c-lower epidermis with stomata (x 675), d-Hairs (x 150), e-Crystals of Calcium oxalate (x 675), f-Outline of transverse section of leaf (x 150), g-Transverse section of leaf passing through the mid vein (x 675).

Macroscopic characters: The drug occurs in market in the form of leaves or leaf powder. Leaves simple and petiolate; lamina simple, dull green, ca 3–6 × 1.5–2.7 cm, ovate, elliptic or ovate lanceolate, acute or shortly acuminate, base rounded or cordate, sometimes narrow, lamina pubescent on both surfaces, densely beneath especially on nerves, pennineved, midrib prominent on lower surface, lateral nerves obscure, 3–5 nerved, reticulately veined.

Microscopic characters: Leaf is dorsiventral and hypostomatic. Transverse section made of leaf shows single layered epidermis made up of rectangular cells with thick cuticle [Text fig. 96 g]. Epidermal cells are straight walled, upper and lower epidermal cells more or less same in size and shape [Text fig. 96 a & b]. Stomata are paracytic [Text fig. 96 c]. Trichomes present on both epidermis, trichomes ca 22–460 μ, uniseriate and non glandular [Text fig. d]. Mesophyll tissue is differentiated into two layered compactly arranged vertically elongated palisade cells and 3 to 5 layered, loosely arranged spongy parenchyma cells. Crystals of calcium oxalate are present in parenchyma cells [Text fig. 96 g]. The midrib region shows 2–3 layers of collenchyma below upper and above lower epidermis, remaining part is occupied by cortical parenchyma. Vascular bundle consists of xylem and phloem, xylem encircled by phloem [Text fig. 96 b].

Powder characters: Leaf powder contains uniseriate trichomes, ca 220–466 μ in length, [Text fig. 96 d] and crystals of calcium oxalate of ca 12–20 μ in diameter [Text fig. 96 e]. Cuticle with straight wall and paracytic stomata [Text fig. 96 a–c].

Local uses: One teaspoon of leaf powder is given 2 times in a day after meal with water to diabetic patient.

CHIKANA PAN [Text fig.: 97 a–g]
Botanical name: **Sida acuta** Burm. f.

Family: Malvaceae

Vernacular names: Bala, Chikana, Janglimethi, Pata, Rajbala.

Organoleptic characters: Taste slight bitter, mucilaginous, odorless.

Macroscopic characters: The drug occurs in market in the form of leaves. Leaves are simple, petiolate; pale green, ca 2–7 × 1–2.3 cm, upper and lower surface hairy, glandular, lamina lanceolate, acute towards apex and rounded towards base, margin sharply serrate, penninerved, recticulately veined, midrib prominent, lateral nerves 5–11 pairs, obscure.

Microscopic characters: Leaf is dorsiventral and amphistomatic. Transverse section of leaf shows single layered epidermis made up of rectangular parenchymatous cells, covered with cuticle [Text fig. 97 g]. Cells of upper epidermis are larger than cells of lower epidermis [Text fig. 97 a & b]. Stomata are ranunculaceous type [Text fig. 97 b]. Epidermal

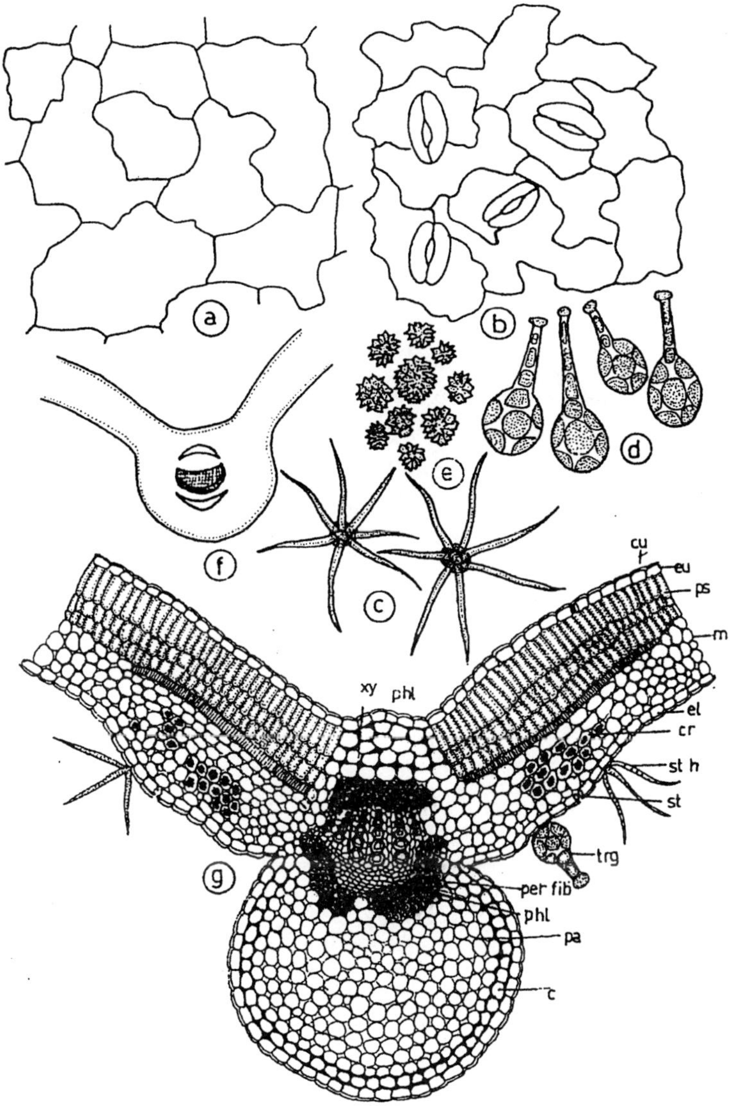

Text fig. 97: Sida acuta a-Upper epidermis, b-Lower epidermis with stomata (x 675), c-Stellate non glandular trichomes (x 150), d-Bulbous based glandular trichomes (x 225), e-Crystals of Calcium oxalate (x 675), f-Outline of transverse section of leaf (x 50), g-Transverse section of leaf passing through the mid vein (x 150).

cells are wavy in outline. Glandular and non-glandular trichomes are present on lower surface of the lamina. Non-glandular trichomes are stellate, thick walled and pointed [Text fig. 97 c]. Glandular trichomes are sessile, bulbous based with capitate head [Text fig. 97 d]. Mesophyll is differentiated into 2–3 layered vertically elongated compact palisade cells towards upper epidermis and 2–3 layered loosely arranged spongy parenchyma towards lower epidermis. Parenchyma cells contains crystals of calcium oxalate [Text fig. 97 g]. Midrib region shows 1–2 layers of collenchyma cells. The rest of the midrib region occupied by the parenchyma cells. Vascular bundle is arc shaped, collateral with xylem towards upper epidermis and phloem towards lower epidermis. Pericyclic fibrous sheath encircles the vascular bundle [Text fig. 97 g].

Powder characters: Leaf powder is pale green in color. It contains stellate hair of ca 0.18–0.3 μ in breadth [Text fig. 97 c]. Glandular hairs of bulbous base and capitate head [Text fig. 97 d], pieces of epidermis with wavy outline, ranuncuelaceous stomata [Text fig. 97 b] and crystals of calcium oxalate of ca 0.60–1 μ in diameter [Text fig. 97 e].

Local uses: Leaves are used in healing of wounds and in bone fracture. Poultice of leaves applied or tied around the injured part. The crushed leaves are tied with the help of clean cloth around the fractured part of body and after 2 days it is removed and fresh leaves are again used and the process is repeated till recovery.

MAKAD-LIMBU [Text fig.: 98 a–d]
Botanical name: **Atalantia racemosa** Wight & Arn.

Family: Rutaceae

Vernacular names: Makad-limbu, Ran-limbu.

Organoleptic characters: Taste biting, astringent, odor aromatic, when fresh as that of orange fruit.

Macroscopic characters: The drug occurs in market in the form of leaves or leaf. Leaves simple, petiolate, ca 6–10 × 2–3 cm; lamina glabrous, gland dotted, dark green above and paler below, elliptic, rounded at apex, emarginate acute or tapering towards base, margin entire or crenulate, penninerved, reticulately veined, midrib prominent, lateral nerves obscure.

Microscopic characters: Leaf is dorsiventral and hypostomatic. Transverse section of leaf shows single layered epidermis made up of rectangular parenchymatous cells covered with thick cuticle [Text fig. 98 d]. Cells of upper epidermis are slightly larger than the lower epidermis [Text fig. 98 a & b]. Stomata are paracytic and rubiaceous [Text fig. 98 b]. Mesophyll tissue is differentiated into single layered compactly arranged vertically elongated palisade cells towards lower epidermis and 2–5 layered loosely arranged spongy parenchyma towards upper epidermis. Palisade layer is interrupted by circular secretory canal [Text fig. 98 d]. Midrib region is occupied by cortical parenchyma cells distributed towards upper and lower

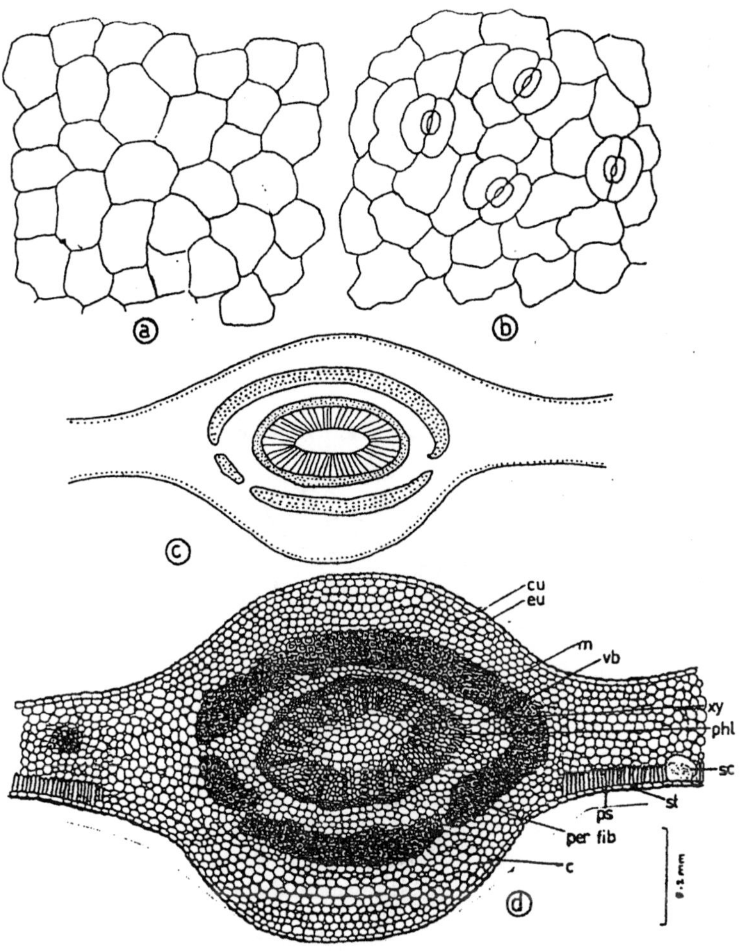

Text fig. 98: Atlantia racemosa a-Upper epidermis (x 675), b-Lower epidermis with stomata (x 675), c-Outline of transverse section of leaf (x 50), d-Transverse section of leaf passing through the mid vein (x 150).

epidermis. It is followed by bundle sheath. Vascular bundle is bicollateral [Text fig. 98 d].

Powder characters: Leaf powder is dark green in color. It contains epidermal cells with wavy outline and paracytic stomata [Text fig. 98 a & b]. Crystals are absent.

Local uses: Decoction of leaf is given in bronchitis, asthma and in cough. The leaves are chewed in bronchitis. Leaf powder is used as blood purifier. Decoction of leaf is given every night as a blood purifier.

MALAWA [Text fig.: 99 a–f]

Botanical name: **Actinodaphne angustifolia** Nees.

Synonyms: *Actinodaphne hookeri* Meissin.

Family: Lauraceae

Vernacular names: Pisa.

Organoleptic characters: Taste none, odor aromatic

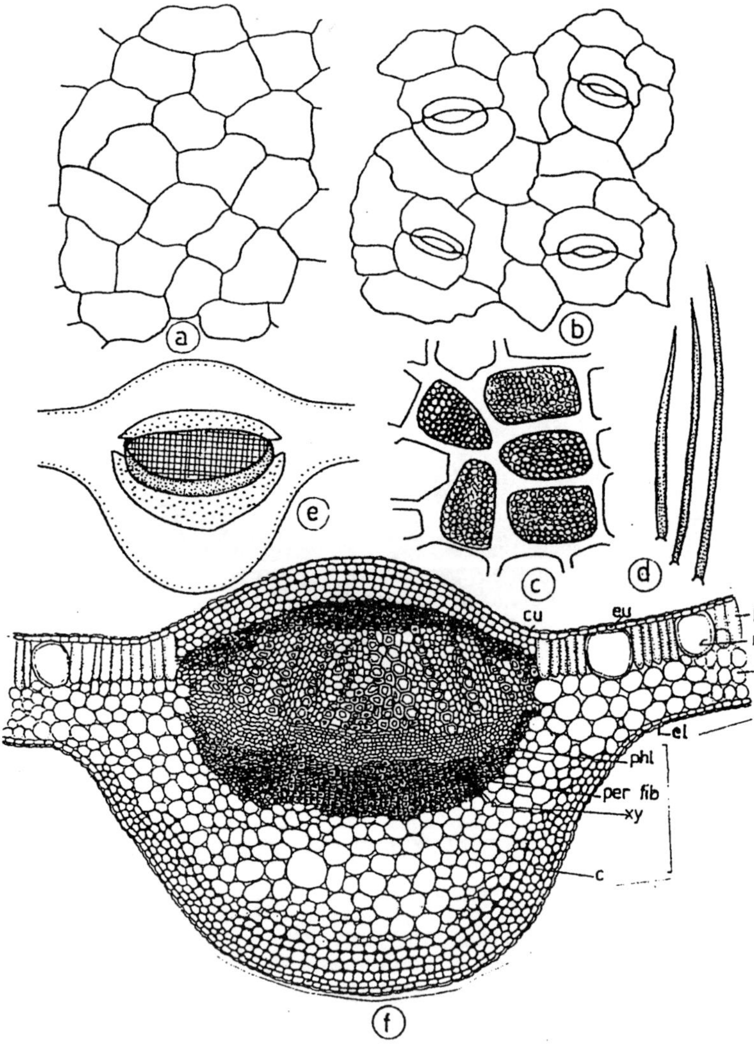

Text fig. 99: Actinodaphne angustifolia a-Upper epidermis (x 675), b-Lower epidermis with stomata (x 675), c-Surface view of leaf, d-hairs (x 675), e-Outline of transverse section of leaf (x 50), f-Transverse section of leaf passing through the mid vein (x 150).

Macroscopic characters: Leaves simple and petiolate; lamina pale green above and thinly white tomentose below, coriaceous, elliptic to lanceolate in shape, young leaves densely silky, white, mature leaves glabrous, penninerved and main nerves 6–10 pairs.

Microscopic characters: Leaf is dorsiventral and hypostomatic. Transverse section of leaf shows single layered epidermis covered with cuticle. Cells of upper and lower epidermis are more or less similar in structure. Epidermal hairs are unicellular, ca 1–2 mm long, thick and brown in color [Text fig. 99 d]. Stomata are paracytic [Text fig. 99 b]. Mesophyll tissue is differentiated into upper palisade cells and lower spongy parenchyma cells. Palisade cells are vertically elongated, compactly arranged in single layer, interrupted by spherical suberised mucilage cavities. Spongy parenchyma is 3 to 4 layered, loosely arranged with intercellular space. Midrib shows continuous layer of epidermis. 3–5 layered collenchyma cells present below upper and lower epidermis. Vascular bundle with xylem present towards upper epidermis and phloem towards lower epidermis, surrounded by incomplete sheath of pericyclic fibres. Cortex consists of 2–5 layers of parenchyma cells towards lower epidermis containing mucilage cavities [Text fig. 99 f].

Powder characters: Powder is pale green in color. It contains many unicellular, thick walled epidermal hairs [Text fig. 99 d], pieces of midrib, cuticle with paracytic stomata [Text fig. 99 b] and epidermal pieces with transparent rounded dots [Text fig. c].

Local uses: The leaf powder is given to men suffering from menorrhoea or in premature ejaculation. Powder is given in every night with milk. Leaf powder with water is given early morning in urinary disorder.

Flowers used in Folk Medicines

DHAYATI

Botanical name: **Woodfordia fruticosa** (L.) Kurz.

Synonyms: *Woodfordia floribunda* Salisb.

Family: Lythraceae

Vernacular names: Dhayati.

Parts used: Flower powder.

Organoleptic characters: Taste astringent, odorless.

Macroscopic characters: Flower powder is red in color. Observations of powder through hand lens shows bulbous based glandular hairs, unicellular simple hairs and small, many cuneate, obovoid, brown, smooth seeds.

Local uses: Flower powder is given during the excess menstrual discharges. Decoction of flower is given for one week to overcome menstrual complaints.

Fruits used in Folk Medicines

BAHAWA SHENG
Botanical name: **Cassia fistula** L.

Family: Caesalpiniaceae

Vernacular name: Amaltaash, Bahawa.
Parts used: Fruit.
Organoleptic characters. Taste sweet. Odor disagreeable.
Macroscopic characters: The drug occurs in the form of cylindrical, chocolate brown or black colored pieces of pod of ca 4–6 × 1–1.5 cm in size. Pod is divided into number of cellular compartments by thin woody transverse dissepiments. These compartments contains flat, oval, reddish-brown colored seeds imbedded in soft, black gray or brown colored greasy pulp.

Local uses: One teaspoon of pulp of pod is given twice a day in piles.
[Note: Entire pod is ca 40–70 cm in length and 2–2.5 cm in diameter, straight, cylindrical with short woody stalk].

BEHDA
Botanical name: **Terminalia bellirica** (Gaertn.) Roxb.

Family: Combretaceae

Vernacular names: Behda, Bellirina, Hela.
Parts used: Fruit.
Organoleptic characters: Taste astringent. Odor sweet but disagreeable.
Macroscopic characters: The drug occurs in the form of ovoid or obovoid more often pentagonal, light amber colored fruits of ca 2–5 × 2 cm size. Fruit is velvety covered with minute pale brown tomentum. Drupe is suddenly narrowed into a short stalk or a scar of pedicel. Dry fruits are obscurely angled. Fruit is one seeded; seed is stony light yellow with fine longitudinal ridges on surface.

Local uses: Paste of fruit is applied on painful swellings or in inflammation. The fruit decoction is used to wash the eyes. The decoction of fruit is given to the patient suffering sore throat.

BIBBA [Photoplate-XV]

Botanical name: **Semecarpus anacardium** L. f.

Family: Anacardiaceae

Vernacular names: Bibba, Bhilaava, Marking nut.

Part used: Fruits.

Organoleptic characters: Taste acrid. Odorless.

Macroscopic characters: The drug occurs in the form of heart shaped obliquely ovoid, drupaceous nut with a fleshy pear shaped receptacle. Fruit is brownish-black glossy ca 2.5 × 2 cm in size. Internally the nut contains a viscous oily liquid of black color.

Local uses: Fruit is ruptured by needle and one drop of black colored liquid is taken on finger tip and mixed with other contents and applied externally on larynx early in the morning in bronchitis.

GELPHAL [Photoplate-XV]

Botanical name: **Catunaregam spinosa** (Thunb.) Tirveng.

Randia dumatorum (Retz.) Poir., *Xermophis spinosa* Thunb.

Family: Rubiaceae

Vernacular name: Gehela, Gela, Madanphal.

Part used: Fruits.

Organoleptic characters: Taste nauseous. Odour mild and reminiscent.

Macroscopic characters: The drug occurs in the form of globular, pear shaped, reddish-brown or black colored fruit. Fruits are ca 4–5 × 2–4 cm in size. Fruit is compound with rim of persistent calyx at top of fruit which appears like a disk, surface smooth, glabrous minutely or obscurely longitudinally ribbed. The fruit is two celled and consists of numerous seeds in gray gelatinous pulp.

Local uses: Paste of fruit is externally applied on face to remove post pregnancy marks formed due to excess secretion of melanin.

HIRDA [Photoplate-XV]

Botanical name: **Terminalia chebula** Retz.

Family: Combretaceae

Vernacular name: Amrita, Haritaki, Hirda, Rohini.

Part used: Fruits.

Organoleptic characters: Taste astringent or bitter. Odorless.

Macroscopic characters: The drug occurs in the form of ellipsoid, obovoid, club shaped, greenish-yellow or greenish colored fruits. Fruit is 3–4 × 2–3 cm, drupe, glabrous and more or less 5 ribbed, ribs running from apex to base. The fruit consists of single stony seed; seed is light yellow, oblong, very thick and obscurely angled.

Local uses: The decoction of fruit is applied in eye infection. Also decoction of fruit is given to wash the mouth. Fruit wall is chewed as appetizer.

MURUDSHENG

Botanical name: **Helicters isora** L.

Family: Sterculiaceae

Vernacular name: Kewan, Murud sheng.

Part used: Fruits.

Organoleptic characters: Taste mucilaginous. Odorless.

Macroscopic characters: The drug occurs in the form of spirally twisted, gray or green colored fruit of ca 6–7 × 0.5–0.6 in size. Fruits are follicle type, 5 beaked, linear, twisted together in the form of a screw, stellate tomentose. Seeds numerous, black colored, angular and tubercled, testa loose wrinkled.

Local uses: Paste of fruit is given to children suffering from stomach complaints.

VAVDING

Botanical name: **Embelia basaal** (R. & S.) A. DC.

Synonyms: *E. robusta* Clarke, *E. tsjerium-cottam* Sant.

Family: Myrsinaceae

Vernacular names: Ambaldange, Ambati, Elang, Vavding.

Organoleptic characters: Taste astringent. Odorless.

Macroscopic characters: The drug occurs in the form of fruits. Fruits are 0.4–0.5 cm in diameter, red to reddish-black or brown in color, surface longitudinally striated. Calyx is persistent; sepals 4–5, shortly connate at the base, ovate, subacute, gland dotted and with minutely glandular margins. Pedicels are 1–2 mm in length. Pericarp is followed by membranous sheath covering a single globular to sub globular seed. Outer surface of seed is brownish-black, speckle with yellowish-brown or white spot.

Local uses: Fruit are placed on burning coal and smoke is given to women after pregnancy as a laxative, antiseptic and fruits are given twice a day for a child to expel worms.

VAVDING

Botanical name: **Embelia drupacea** (Dennest.) M.R. & S.M. Almeida.

Family: Myrsinaceae.

Synonyms: *E. viridiflora* Scheft.

Vernacular names: Ambat.

Organoleptic characters: Taste astringent. Odorless.

Macroscopic characters: The drug occurs in the form of fruits. Fruits are 0.6–0.9 cm in diameter, globose or slightly oblong, succulent red to reddish-black when ripe, surface warty in dried samples, style persistent. Pericarp is longitudinally striated brittle followed by a membranous sheath covering a single globular to sub globular seed. Outer surface of seed is brownish-black which is speckle with yellowish-brown or white spot.

Local uses: Fruit are placed on burning coal and smoke is given to women after pregnancy as a laxative, antiseptic and fruits are given twice a day for a child to expel worms.

VAVDING

Botanical name: **Embelia ribes** Burm. f.

Family: Myrsinaceae

Vernacular names: Vavding.

Organoleptic characters: Taste astringent. Odorless.

Macroscopic characters: The drug occurs in the form of fruits. Fruits are ca 0.5 cm in diameter, globose, grayish-black when ripe, surface warty in dried samples, style persistent. Pericarp is warty brittle followed by a membranous sheath covering a single globular to sub globular seed. Outer surface of seed is brownish-black, speckle with yellowish brown or white spot.

Local uses: Fruits are placed on burning coal and smoke is given to women after pregnancy as a laxative, antiseptic and fruits are given twice a day for a child to expel worms.

Table I: Distinguishing characters of 'Vavding' samples

Embelia basaal (R. & S.) A. DC. (*E. robusta* Clarke, *E. tsjerium-cottam* Sant.)	***Embelia drupacea*** (Dennest.) M.R. & S.M. Almeida. (*E. viridiflora* Scheft.).	***Embelia ribes*** Burm. f.
Fruits are 0.4–0.5 cm in diameter, black or brown colored, with longitudinally striated surface.	Fruits are 0.6–0.9 cm in diameter, black colored, with warty surface.	Fruits are 0.2–0.4 cm in diameter, grayish-black colored, with warty surface.
Pericarp longitudinally striated, brittle.	Pericarp longitudinally striated, brittle.	Pericarp warty and brittle.

Seeds used in Folk Medicines

ARDH–SUPARI [Photoplate-XV]
Botanical name: **Caryota urens** L.

Family: Arecaceae

Vernacular name: Ardhi supari.
Part used: Seeds.
Organoleptic characters: Taste astringent. Odorless.
Macroscopic characters: The drug occurs in the form of plano-convex, black colored seeds. The seeds are ca 1–1.5 cm × 1–1.5 cm in size, shining black, possessing small reticulately veined outer thin papery covering. Seed is longitudinally ribbed. Seeds consist of ruminate endosperm and small rudimentary embryo.

Local uses: The seeds are rubbed on copper coin and paste collected is applied in hemicrania.

GAJAGA [Photoplate-XV]
Botanical name: **Caesalpinia bonduc** (L.) Roxb.

Synonyms: *C. bonducella* (L.) Felming

Family: Caesalpiniaceae

Vernacular name: Gaja, Gajaga, Sagargota.
Parts used: Seeds.
Organoleptic characters: Taste is very bitter. Odorless.
Macroscopic characters: The drug occurs in the form of rounded slightly oblong, marble like seeds. Seeds are ca 1.2–1.8 cm in diameter. Surface of testa is very hard, glossy and bluish-gray or lead colored. It shows very fine horizontal cracks. The kernel consists of yellowish-white cotyledons.

Local uses: Fruit is soaked in water for 12 hours and paste prepared from it is given for 3–4 days once in day to the ladies suffering from menstrual problems. Similarly tincture of powder is given to diabetic patients once in a day.

GARAMBI [Photoplate-XV]
Botanical name: **Entada rheedei** Spreng.

Synonyms: *E. pusaetha* DC., *E. scandens* Benth.

Family: Mimosaceae

Vernacular name: Garambi, Garbi, Garbul.
Parts used: Seeds.
Organoleptic characters: Taste bitter. Odorless.

Macroscopic characters: The drug occurs in the form of compressed, ovoid-orbicular, stony seeds. Seeds are ca 3–5 in diameter, 1–1.5 cm in thickness, dark brown to reddish-black. Testa is shiny and pericarp is hard. Cotyledons are starchy.

Local uses: Paste of seeds is immediately applied after scorpion sting. The seeds paste in small quantity is given in stomachache. Seed paste is applied on rheumatic pains and also on headache.

JAMBHULBEE
Botanical name: **Syzygium cumini** (L.) Skeels.

Synonyms: *Eugenia jambolana* Lam

Family: Myrtaceae

Vernacular name: Jambhul, Jamunguthi.
Parts used: Seeds.
Organoleptic characters: Taste is slightly bitter astringent. Odorless.
Macroscopic characters: The drug occurs in the form of ovoid brown colored seeds. The seeds are ca 2.5 × 1 cm in size. The seeds are covered with brownish papery shell enclosing a dark brown kernel.

Local uses: Teaspoon of powder is mixed with buttermilk and given in diarrhoea of children, dysentery and loss of appetite. Seed powder is given to women suffering from leucorrhoea. Similarly seed powder is also used in diabetes.

KAJRA [Photoplate-XV]
Botanical name: **Strychons nux-vomica** L.

Family: Loganiaceae

Vernacular name: Kajra, Kasarkamora.
Parts used: Seeds.
Organoleptic characters: Taste extremely bitter. Odorless.
Macroscopic characters: The drug occurs in the form of flat, irregularly bent, disk shaped, compressed, ash-gray colored seeds. Seeds are ca 1–3 cm in diameter and ca 6 mm thick. Surface of seeds is covered by numerous closely appressed silky hairs radiating from the center. There is distinct hilum in the center of one of the flattened side and on the circumference. Besides this, there is small projection, which makes the position of the micropyle. Cotyledons are heart shaped embedded in gray horny endosperm.

Local uses: The very small piece of seed is placed in mouth in diabetic patient. Seeds are reported to be poisonous.

PALASBI [Photoplate-XV]
Botanical name: **Butea monosperma** (Lam) Taub.

Synonyms: *Butea frondosa* Roxb. ex Willd.

Family: Fabaceae

Vernacular names: Palas, Palash.
Parts used: Seeds.
Organoleptic characters: Taste acrid, oily. Odorless.
Macroscopic characters: The drug occurs in the form of flat, kidney shaped seeds of ca 3–3.5 × 2–2.5 cm. The seeds are ca 2 mm thick. Outer surface of seed is leathery, glossy veined and wrinkled, deep reddish-brown in color. The seed has conspicuous hilum which is situated near the middle of its concave edge.

Local uses: The seeds are soaked in cow urine for 24 hrs, then dried in shade and thoroughly washed with water. Process is repeated for 2 to 3 times and given to the patient suffering from menorrhoea.

Prickle used in Folk Medicines

CHIRPHALICHAKATA [Photoplate-XV]
Botanical name: **Zanthoxylum rhetsa** (Roxb.) DC.

Family: Rutaceae

Vernacular names: Chirphal.
Parts used: Prickles.
Organoleptic characters: Tasteless. Odorless.
Macroscopic characters: Prickle is sharp, pointed 2–5 cm in length, yellow colored and have conical solid base. Mature prickles are without sharp points.

Local uses: Prickle is used to stop bleeding from the fresh cut wounds.

Bulbil used in Folk Medicines

DUKKAR-KAND [Photoplate-XV]
Botanical name: **Dioscorea bulbifera** L.

Family: Dioscoreaceae.

Vernacular names: Dukkar-kand.
Parts used: Bulbils.
Organoleptic characters: Taste mucilaginous slightly bitter. Odorless.
Macroscopic characters: Bulbils are globular to ovoid in shape, ca 4 × 3 cm in size, brown in color with many lenticels all over the surface. It is white inside and is mucilaginous inside.

Local uses: Paste of bulbils mixed with powder of *Salacia brunoniana* and *Gnetum scandens* is applied externally in piles and also given orally in similar case.

Result and Discussions

Rural communities in most of the world continue to rely on locally produced, plant-based remedies (Balee and Gely, 1989; Leaman, 1996; Leaman, 2001; Shankar and Majumdar, 1997; Caniago and Siebert, 1998) and will continue to rely on their indigenous medicines. According to WHO, the herbal medicine serves the health need of about 80% of the worlds population. It is reported that around 20,000 plant species are used medicinally in third world countries (Sharma, 1999).

Indian system of medicines are predominantly dependent on plant drugs. Presently about 1000 single drugs and 8,000 compound formulations of recognized merit are in vogue in Ayurvedic systems of medicines. Similarly, Siddha, Unani and Amachi systems of medicines utilize as many as 600, 700, 600 medicinal plants respectively (Sushil Kumar *et al.,* 1999).

The tribal and ethno or folk medicines are also quite prevalent in the remote areas. It is reported that these tribal people in India utilize over 7,500 species of plants for medicinal purpose (Pushpangadhan undated report). Similarly, it is estimated that there could be 8–9 million vendors across the country selling herbs (Natesh, 2000). Once again, medicinal plants are gaining importance because there have been number of spectacular successes with herbal medicines.

Loss of biodiversity is worldwide problem. According to IUCN by the end of this century about 20,000 plant species used in medicines and about 60,000 higher plant species may extinct. Presently, it is estimated that roughly 10,000 species of medicinal plants are threatened with extinction worldwide (Leaman, 1998). The international trade in medicinal plants is major force in the global economy and demand is increasing in both developing and industrial nations. Destruction of forests, unrestricted over exploitation, fragmentation and modification of habitats, pollution, climatic changes etc. are responsible for worldwide loss of biodiversity. In this scenario, medicinal plants are proven to be extinct due to over exploitation from wild for increased demands all over the world.

Presently, the need to study medicinal plants in detail, from various points of view, is generally well recognized. Traditional knowledge provide hints for development of new plant based drugs. India is one of the twelve megacentres of biodiversity (Nayar, 1987, 1988, 1989). It is considered as a veritable emporium of medicinal plants. The need for detailed and systematic study of medicinal plants has been discussed by many workers (Farnsworth and Bingel, 1977, van Reis Altschul, 1973; Balandran *et al*

1985; Naranjio, 1995; Kameswara Rao, 2000; Leaman 2001; Susan *et al.*, 2001). Traditional remedies in India provide an affordable health care system and also hold immense potential for the international market in this age of the patents and intellectual property rights.

Taking into account an urgency and need of studies on Indian medicinal plants, an attempt has been made to survey medicinal plant wealth of south western Maharashtra. Efforts, have been made to collect data on their field status, local-vernacular names and local uses. Similarly special efforts have been made to determine correct botanical identity of folk medicines sold in local bazaars by comparing external and internal morphology of plant part sold with that of authentic samples. The work is discussed here under two parts viz. A] Taxonomic account and B] Correct botanical identity of folk medicines.

A] Taxonomic Account of Medicinal Plants of South Western Maharashtra:

Maharashtra is one of the important states of our country which possesses rich biodiversity. Several floristic work are available on the state and fairly good amount of work has been done on flowering plants of the region (Graham, 1837; Dalzell and Gibson, 1861; Nairne, 1894; Woodrow, 1897–1901; Birdwood, 1896 and 1897; Cooke, 1901–1908; Talbot, 1894; Haines, 1912–1914, 1916; Gammie, 1905–1912; Blatter, 1926–1935; Santapau, 1952, 1953, 1958; Santapau and Irani, 1962; Santapau and Henry, 1973; Vartak, 1966; Patel, 1968; Almeida, 1990; Kambale and Pradhan, 1988; Kulkarni, 1988; Lakshminarasimhan and Sharma, 1991; Kothari and Moorthy, 1993; Deshpande *et al.*, 1993–1995; Sharma *et al.*, 1996; Singh and Kartikeyan, 2000; Singh *et al.*, 2001; Naik, 1998; Yadav and Sardesai, 2002). Recent publication on flora of Maharashtra State reports 3869 species of flowering plants for the region. General references have been made to medicinal plants of the region in different floristic works but detailed account on medicinal plants is not available for the state.

South western Maharashtra, a study area of present investigation includes five districts. Of the five districts, 3 are floristically well explored viz. Sindhudurg (Kulkarni, 1988; Almeida 1990), Satara (Deshpande *et al.*, 1993–95) and Kolhapur (Yadav and Sardesai, 2002). Floristic account of these districts clearly indicates the richness of the region in flowering plant diversity. South western Maharashtra also possesses high number of medicinal plants. About 565 plant species of some therapeutic value have been collected and documented in present survey from south western Maharashtra. A comparative account of medicinal plants recorded by different authors for India, and Maharashtra is presented in following table.

Table I: Comparative account of medicinal plants reported for India and Maharashtra

Sl. No.	*Name of the author*	*Year of publication*	*Name of the region*	*Number of species*
1.	Kirtikar & Basu	1918	India	22305
2.	Chopra *et al.*	1956	India	3500
3.	Dymock	1891	India	1225
4.	Nadkarni	1954	India	2671
5.	Agharkar	1953	Maharashtra	350
6.	Sharma & Singh	1997	Maharashtra	443
7.	Kumbhojkar & Upadhya	1997	Maharashtra	424
8.	Naik	1998	Marathwada	350
9.	Present study	—	South Western Maharashtra	565

Comprehensive account on medicinal plants, their local names, distribution and uses was first published by Agharkar (1915) reporting 350 medicinal plant species for the state of Maharashtra. Kumbhojkar and Upadhya (1997) recorded 424 medicinal plants for Maharashtra while Sharma and Singh (1997) enumerated 443 species of medicinal value for the region. Naik (1998) described 350 plant species of medicinal value from Marathwada region.

Present investigation on medicinal plants of south western Maharashtra (5 districts) resulted into collection and documentation of 565 species. Out of 565 species of medicinal plants, 508 are dicot species and 57 are monocots. Out of total 565 species of medicinal plants, 474 are wild and 91 are garden species. About 47 species of medicinal value and endemic to peninsular India are recorded from south western Maharashtra.

Similarly, about 20 species recorded to be rare (CAMP, 2001) are also found to occur in the region viz., *Capparis moonii*, *Cassine paniculata*, *Cadaba fruticosa*, *Chlorphytum arundinaceum*, *Chlorophytum borivilianum*, *Embelia ribes*, *Eulophia nuda*, *Eulophia ochreata*, *Holostemma annulare*, *Iphigenia stellata*, *Lamprachaenium microcephalum*, *Nothapodytes nimmoniana*, *Operculina turphetum*, *Oroxylum indicum*, *Pueraria tuberosa*, *Rauvolfia serpentina*, *Terminalia cuneata*, *Salacia brunoniana*, *Santalum album*, *Saraca asoca*.

During present investigation some medicinal plant species viz. *Ancistrocladus heyneanus*, *Argyreia boseana*, *Cassine paniculata*, *Diploclisia glaucescens*, *Eulophia ochreata*, *Jasminum malabaricum*, *Leea setuligera* and *Parthenium hysterophorus* which are not reported in earlier literature on medicinal plants of Maharashtra are reported from the region.

Field studies revealed that some of the medicinal plant species viz. *Holostemma annulare*, *Iphigenia stellata*, *Nothapodytes nimmoniana* and

Rauvolfia serpentina are exploited on large scale by local people, trader, pharmaceutical industries, local practitioners and are under the threat of extinction. They are reported to be critically endangered (CAMP, 2001). *Nothapodytes nimmoniana* was reported to be very common and abundant plant in the region, (Cooke, 1958), however, due to its over exploitation during last decade for the foreign market, it has become a rare plant in the region.

Some of the plants viz. *Bryonia dioica*, *Chlorphytum arundinaceum*, *Chlorophytum borivilianum*, *Eulophia nuda*, *Eulophia ochreata*, *Iphigenia stellata*, *Oroxylum indicum*, *Pueraria tuberosa* and *Salacia brunoniana* are used as root drugs. These root drugs are largely exploited from their wild sources. Many of the root drug yielding plant species viz. *Asparagus racemosus*, *Baliospermum montanum*, *Chlorophytum borivilianum*, *Eulophia nuda*, *E. ochreata* and *Iphigenia stellata* are of rare occurrence in the region and some of them may become locally extinct. Some of species of *Chlorophytum*, *Eulophia*, *Iphigenia* and *Salacia*, are already listed in Red Data book (Nayar and Sastry, 1987, 1988, 1990).

Considerable amount of information has been gathered on local uses and methods of administration on medicinal plants of the region. It has been revealed that number of local medicinal uses on these medicinal plants have not been recorded in literature. For example, *Toddalia asiatica* is used to treat any kind of urinary trouble, *Ancistrocladus heyneanus*, *Castanospermum australe* and *Saraca asoca* are used to treat HIV patients, roots of *Smilax zeylanica* are used to cure psoriasis. Roots of *Spermadictyon suaveolens* are used to treat Herpes zoster infection. Root or stem of *Moullava spicata* is used to cure any kind of exo or endocortosis. Seeds of *Antiaris toxicaria* are used in diabetes. Leaf of *Abutilon indicum* is used to relive mental stress.

Some of the species viz. *Catharanthus roseus*, *Hardwiekia binata*, *Nothapodytes nimmoniana*, and *Vernonia cinerea* are used in curing major diseases like cancer. *Catharanthus roseus*, *Gymnema sylvestre*, *Helicters isora*, *Salacia* sp., *Syzygium cumini*, *Woodfordia fruticosa* are commonly used to treat diabetic patient. *Phyllanthus amarus*, *P. urinaria* and *Tinospora cordifolia* are regularly used in treatment of Hepatitis and found to be very effective. *Tylophora indica* is effectively used to reduce asthma.

Fruits of *Embelia ribes* locally known as 'Vavding' are usually used in medicine, however, it is very rare plant in the region. Market survey showed that the fruits of *Embelia basaal* and *E. drupacea* are sold in market under the name of 'Vavding'. The fruits of these two species are used as substitute and both the species are fairly common in the area.

South western Maharashtra has rich biodiversity as well as good wealth of medicinal plants. There is further need for studies on various aspects of these medicinal plants for sustainable harvest and utilization. From present study, it is revealed that there is still need to collect information on local medicinal uses of various plant species which will orm basis for further scientific studies on drug formulations and discovery of new drugs.

B] Correct Botanical identity of folk medicines of South Western Maharashtra:

Folk medicines consist of independent practices of primitive nomadic tribes almost in every country of the developing world. During present investigation over 100 samples of folk medicines have been collected from south western Maharashtra. Out of 100 samples of folk medicines, botanical identity of 81 samples has been confirmed. An account of folk medicines collected and studied is presented in following table.

Table II: Showing number of plant sample collected & identified from different groups of plants

Total samples collected	100
Total samples identified	81
Samples belonging to Dicots	71
Samples belonging to Monocots	8
Samples belonging to Gymnosperm	1
Samples belonging to Pteridophyte	1

Angiosperms constitute a major part of folk medicines in south western Maharashtra. Other than Angiosperms, *Gnetum scandens* Roxb. (Gymnosperms) and *Tectaria cicutaria* L. (Pteridophytes) are used in folk medicines. Plant part wise number of folk medicines collected, studied and identified during present investigation are presented in table below.

Table III: Showing plant part wise number of folk medicines collected & identified during present studies

Sr. No.	*Folk form collected*	*No. of samples collected and identified*
1	Roots	22
2	Corm	2
3	Tubers	5
4	Rhizomes	4
5	Stems	13
6	Barks	12
7	Leaves	5
8	Flower	1
9	Fruits	8
10	Seeds	7
11	Bulbil	1
12	Prickle	1
	Total folk medicines identified	81

Roots

Comparative account of plant parts collected during present investigation shows that underground plant parts such as root, corm, rhizome and tuberous roots constitute major share in folk medicines. Kirtikar and Basu (1918) reported 300 root drugs while Agarwal and Ghosh (1985) reported about 650 root drugs used in Indian medicines. Work on pharmacognosy of indigenous drugs have been complied and published into 3 volumes (Raghunathan and Mitra, 1999). Sarin (1996) described 66 root drugs used in Ayurveda. During present study, total 33 underground plant parts have been collected and identified botanically by comparative study on external and internal morphology of folk and authentic samples. Underground plant samples of folk medicines consists of 22 roots, 2 corms, 5 tubers or tuberous roots and 4 rhizomes. Their major uses and diagnostic external and internal features are described in following paragraphs.

'Jaribooty' is used as trade name for root drugs. Dymock (1889–93) and Nadkarni (1954) reported the 10 root drugs viz. *Aegle marmelos*, *Desmodium gangeticum*, *Ficus religiosa*, *Gmelina arborea*, *Oroxylum indicum*, *Premna obtusifolia*, *Solanum anguivi*, *Solanum virginianum*, *Stereospermum sauveolens* and *Uraria picta* used as Dashmula in Indian system of medicines. Out of these 10 root drugs four (viz. *Aegle marmelos*, *Stereospermum sauveolens*, *Gmelina arborea* and *Oroxylum indicum*) are found to be used in folk medicines in the region.

Roots of plant species such as *Atalantia racemaosa*, *Catunaregam spinosa*, *Celastrus paniculatus*, *Cinnamomum nitidum*, *Embelia basaal* and *Jasminum malabaricum* are used in folk medicines of the region, however, this is not reported in earlier literature by Kirtikar and Basu (1918) Agarwal and Ghosh (1985) and Jain (1991).

Root of *Aegle marmelos* is given under the name of 'Belmul' which is one of the constituents of Dashmula. In folk medicines, it is given in fever same use is reported by Kirtikar and Basu (1918) and Nadkarni, (1954). The root pieces can be easily diagnosed externally by pale yellow or lemon colored root pieces with thin rhytidome and by internal feature such as presence of lemon yellow colored islands of stone cells in cortex, cylindrical vessels with alternate small pits, thin walled wood fibers, thick walled phloem fibers and presence of abundant starch grains.

Root of *Ficus hispida* is given under the name of 'Bokida' in folk medicines used to treat ulcers, boils and itch. The roots can be identified externally by their dark brown or reddish-brown or black color, thin papery bark flanking off in the form of thin papery sheets and internally by longitudinally striated inner bark, soft pale yellow wood, phelloderm with abundant crystals of calcium oxalate, phloem fibers and latex canals;

cylindrical vessels with alternate small bordered pits and root powder with solitary crystals of calcium oxalate.

Root of *Woodfordia fruticosa* is given under the name 'Dhayatimul' in folk medicine to treat excess bleeding in leucorrhoea and diabetes. Flowers of species are reported to have same uses Nadkarni (1954). The root can be identified by their astringent taste, pale or reddish colored hard bark with tannin and abundant starch grains, cylindrical vessels with alternate bordered pits and reddish-brown colored root powder containing abundant clustered crystals of calcium oxalate and starch grains.

Root of *Withania somnifera* is given under the name of 'Dhorgunj' or 'Mahagunj' which is called as 'Ashwagandha' (Kirtikar & Basu, 1918 and Cooke, 1958). It is given in menorrhoea, in night emissions and in case of general debility. These uses are well known for the species (Nadkarni, 1954). Roots are tuberous. The roots can be externally identified by their cylindrical, slightly tortuous gray or fawn colored tuberous, longitudinally wrinkled nature and internally by loosely arranged ground tissue with abundant starch grains, drum shaped vessels with alternate or scalariform pitting. Mukerji (1953) also reported same character for the root of the species.

Root of *Catunaregam spinosa* is called as 'Gelmul' in folk medicines. In folk medicines it is given on skin complaints, foul breath, on wounds and ulcers. Usually fruits are used in medicine but even roots are also used in medicine in South Western Maharashtra. Roots are very distinct in external characters by presence of pale pink or gray colored pieces of roots. Bark is thick and hard. Transverse section of root shows broad bands of stone cells. Starch is abundant in secondary cortex. Vessels are small, cylindrical, longer in length, tapering towards both ends. Ends of vessels are pointed, pits to vessels are very small and minute. Powder is dominated by abundant triangular or polygonal thick walled stone cells and starch grains.

Root of *Spermadictyon sauveolens* is given under the name of 'Gitsaya' in stomach complaints and also used in Herpes zoster virus. Root is characterized by its silvery gray color, thin rough bark, outer bark flanking off in thin sheets, longitudinal fissures and internally by ring of raphid sac in secondary phloem, cylindrical vessels with alternate or scalariform bordered pits and root powder with abundant raphides and starch grains.

Root of *Syzygium cumini* is given under the name 'Jambhoolmul' in diabetes, urinary problem and ulcers. Same uses are also given by Kirtikar and Basu (1918) and Nadkarni, (1954). Root can be identified by its astringent taste, pale pink or reddish color, thick, soft, smooth bark without wrinkles and internally by presence of abundant starch grains and tannin in secondary cortex, vessels with alternate and bordered pits and pale pink colored root powder with simple, elliptic, oblong starch grains.

Root of *Celastrus paniculatus* is given under the name of 'Kangonimul' in rheumatism and as a brain tonic, in chronic ulcers and in bone fracture.

Seeds and seed oil is used in rheumatism and also as brain tonic (Kirtikar and Basu, 1918 and Nadkarni, 1954). Roots are characterized by golden yellow color, reddish-brown or pale pink colored wood bark with abundant tannin and islands of stone cells, cylindrical vessels with alternate bordered pits, septate fibers and root powder with abundant crystals of calcium oxalate, striated stone cells and absence of starch grains.

Root of *Atalantia racemosa* is given under the name of 'Ghansarmul' or 'Makad limbu'. In literature, the word 'Ghansar' is used for *Croton oblongifolius* and *Croton aromaticus* root (Kirtikar and Basu, 1918; Nadkarni, 1954 and Cooke, 1958). The root can be identified by its bitter and biting taste, aromatic odor, gray color, thick and longitudinally wrinkled bark, lemon yellow hardwood and internally by secondary cortex which is interrupted by discontinuous bands of phloem fibers, vessels with tapering blunt ends and alternate pits and root powder dominated by crystals of calcium oxalate and starch grains. Root of *A. racemosa* is reported to be used in cough and in sore throat and as blood purifier but in literature no reference to these uses is found. Nadkarni (1954) reported roots of *A. monophylla* to be used as antispasmodic and stimulant and given in snake bite.

Root of *Cinnamomum nitidum* called as 'Mawarung mul' in folk medicine is used as brain tonic, hair tonic and to cure asthma, headache and cough. These uses of the root of the species are not recorded in any of the previous literature. The root is very distinctive in its Vicks vaporub like odor and mucilaginous taste. It is characterized by dark reddish or reddish-brown pieces, thick non wrinkled bark, pale pink or brown colored soft wood, sclerosed phelloderm with oil canals and cylindrical vessels with alternate to scalariform bordered pits.

Root of *Stereospermum suaveolens* is given under the name of 'Padalmul' used as one of 'Dashmula' in Ayurveda. Root is available in the form of brown colored cylindrical pieces. Root is characterized by thick and lamilatted rough bark, soft and longitudinally shriveled rhytidome of alternate layers of thick walled loosely arranged and thick walled compactly arranged layers of cells, cortex with sclerenchyma forming mesh, paratracheal vasicentric wood parenchyma in terminal narrow bands (Metcalf and Chalk 1950), cylindrical vessels with alternate pits and root powder with abundant pieces of fibers.

Root of *Plumbago zeylanica* is given under the name of 'Padhra chitrak' in folk medicines is given in fever, as an enema. Same uses are reported by Kirtikar and Basu (1918) and Nadkarni (1954). Root is characterized by dark brown or black color, longitudinally wrinkled pieces, hard and pale brown wood with yellow colored amorphous contents, vessels simple, cylindrical with alternate small and bordered pits, pitted phloem fibers and root powder with abundant starch grains.

Root of *Jasminum malabaricum* is given under the name of 'Padmul' as tonic in general debility and in menorrhoea. In literature it is called as 'Kachor' (Cooke, 1958; Kirtikar and Basu, 1918). Root is characterized by insect (Bug) like odor, thick, pale brown color, longitudinally wrinkled bark, large rounded islands of stone cells, vessels with alternate intervascular pitting and root powder is dominated by thick walled stone cells, solitary rectangular rod shaped crystals of calcium oxalate and abundant starch grains.

Root of *Butea monosperma* is given under the name 'Palasmul' in skin diseases. Kirtikar and Basu (1918) reported that the roots are used in night blindness and in elephantiasis. Root is characterized by reddish-brown color, thin silvery brown colored ,outer surface of bark, longitudinally fissured, soft, fibrous wood, presence of phloem fibers, parenchyma is abundant in broad moderately regular bands of 4–6 cells, cylindrical vessels with alternate bordered pits and root powder is dominated by phloem fibers and crystals of calcium oxalate.

Root of *Artocarpus heterophyllus* is given under the name of 'Phanasmul' in mouth ulcers and on wounds. Nadkarni (1954) reported that roots are given in diarrhoea and useful in skin diseases. Root is very distinct by its scarlet or orange color, large prominent fresh orange colored lenticels, lamelatted bark breaking off into thin sheets; pale brown wood, presence of latex canals in secondary cortex and in phloem region, paratracheal confluent parenchyma, cylindrical vessels with alternate bordered pits and root powder dominated by stone cells and crystals of calcium oxalate.

Root of *Salvadora persica* is given under the name of 'Pilu' (Kirtikar and Basu, 1918; Nadkarni, 1954 and Cooke, 1958) and used as toothbrush, in teeth complaints and in mild fever. Same is reported in various literature. Root is characterized by sweetish and biting taste, pungent odor, pale brown or fawn color thick bark with abundant starch grains, easily separates from the wood, soft wood with anomalous structure, included interxylary phloem of foraminate type with round to oval strands (Metcalf and Chalk, 1950), vessels with alternate to scalariform pitting and thick walled fibers and root powder with abundant starch grains.

Root of *Cocculus hirsutus* is given under the name of 'Sapasan' or 'Sapshindi' in gas trouble and stomachache. Cooke (1958) reported same local name i.e. 'Sapasan' for the roots of *Aristolochia indica*. Roots of both the plant shows similar anatomical characters but roots of *Aristolochia indica* differ in its bitter taste and slight camphor like odor. Root of *Cocculus hirsutus* is bitter and odorless. Nadkarni (1954) reported that the roots of *Aristolochia indica* are used in snake bite, scorpion sting, in leprosy (white), diarrhoea and bowel complaints. Pharmacognostic studies on *Cocculus hirsutus* is carried out by Singh (1956). Root of *Cocculus hirsutus* is characterized by cylindrical, dark brown, deeply fissured root pieces with yellowish-gray softwood, star shaped xylem situated at the center by partially or completely obliterating pith, rays are 10–23 cell wide, drum shaped broad vessels and root powder dominated by abundant starch grains. Same characters have been reported for *Cissampelos pareira* by Mukerji (1953) which is used under the name of "Cissampelos" or "Venivel".

Root of *Salacia brunoniana* is given under the name of 'Saptrangi' as blood purifier, in urinary calculi and as antidiabetic. Agarwal and Ghosh (1985) also cited same local name for the species but the same local name is referred also for *Caesaria esculenta* (Desai, 1927 and Naik, 1998). During present studies, it is confirmed that the roots of *Salacia* species show rings of 7 colors. Same has also been confirmed by Handa *et al.* (1968). Root of *Salacia brunoniana* is very distinct in its golden yellow color, thick bark exfoliating in the form of thin sheets, pale pink hard wood, secondary anomalous growth of successive rings of interxylary phloem and abundant tannin and crystals of calcium oxalate in bark and rings of phloem (Metcalf and Chalk, 1950).

Root of *Bombax ceiba* is given under the name of 'Savarmul' in leucorrhoea and on wounds. Same uses are reported by Kirtikar and Basu (1918) and Nadkarni (1954). Mehra and Karnik (1968) studied pharmcognosy of *Bombax ceiba* while Rai (1971) studied stem bark of the species. Root is characterized by its mucilaginous taste; silvery pale brown colored outer bark, reddish-brown or black colored wrinkled inner bark, softwood, fibrous fractures, presence of abundant crystals of calcium oxalate and sclerenchymatous patches in secondary cortex, abundant diffuse parenchyma, 2–7 cell wide rays (Metcalf and Chalk, 1950), vessels with large, bordered pits and root powder dominated by abundant clustered crystals of calcium oxalate and starch grains.

Root of *Gmelina arborea* is given under the name of 'Shivanmul' as one of 'Dashmula'(Kirtikar and Basu, 1918 and Nadkarni, 1954). Root is characterized by bitter and astringent taste, pale brown and longitudinally striated hard wood, secondary cortex with islands of stone cells, crushed phloem with bands of yellow colored crystalline substances, cylindrical vessels with alternate and bordered pits and root powder dominated by thick walled striated, rectangular pitted stone cells and simple starch grains.

Root of *Oroxylum indicum* is given under the name 'Tetumul' as one of 'Dashmula' (Nadkarni, 1954). Root is characterized by pale brown color, longitudinal fissures, pale yellow hardwood, secondary cortex with patches of sclerenchyma cells containing yellow colored crystalline content, paratracheal confluent parenchyma (Metcalf and Chalk, 1950) and root powder is dominated by fibers, vessel pieces with alternate bordered pits, stone cells and absence of starch grains.

Root of *Embelia basaal* is given under the name of 'Vavdingmul' in gonorrhoea, in earache, chronic ulcers and on wounds. Nadkarni (1954) reported that the root bark of *Embelia ribes* is used in toothache and in pneumonia. Root is characterized by bitter taste pale pink color, thick bark, hardwood, secondary phloem with wedge shaped patches of stone cells, 6–13 cell wide rays, cylindrical vessels with alternate pits and root powder dominated by crystals of calcium oxalate and starch grains.

Root of *Hemidesmus indicus* is given under the name of 'Upalsari' as blood purifier, on skin diseases and urinary complaints. Same is reported by Kirtikar and Basu (1918) and Nadkarni (1954). Root is characterized by bitter taste, cylindrical and tortuous purple or reddish-brown root pieces, thin, rough bark, soft and yellow wood, latex canals in phelloderm and in phloem, cylindrical vessels with alternate, small, bordered pits and root powder dominated by abundant simple or compound striated starch grains. Pharmacognosy of the species worked by Datta and Mukerji (1950); Mukerji (1953); Nayar *et al.* (1978) and Raghunathan and Roma Mitra, (1999).

During present investigations 10 tuberous and rhizomatous species have been collected and studied.

Rhizome of *Eulophia nuda* is given under the name 'Amarkand' in folk medicines of South Western Maharashtra. The same local name is used by Kirtikar and Basu (1918) and Chopra (1956). In folk medicines it is given orally to stop backache and in piles. In south western Maharashtra rhizome of *E. ochreata* is also given under the name 'Amarkand' however, it is distinct by its conical shape from that of elliptic oblong rhizomes of *E. nuda* (Santapau and Kapadia, 1966).

In south western Maharashtra, tubers of *Peuraria tuberosa* are given under the name 'Bhuikohla' used on various diseases such as menorrhoea, leucorrhoea, rheumatism and epilepsy. Similar uses are reported in literature by Kirtikar and Basu (1918), Nadkarni (1954), Chopra, (1956) and Naik, (1998). Powder of *Peuraria tuberosa* is characterized by presence of phloem fibers.

Tuber of *Bryonia dioica* is given under the name 'Chambar gonda' in stomach complaints. This is also reported by Wallis (1967) and Chakravarty (1982), however, it is also given as abortive in folk medicines. In field, it is very easily identified by its fetid odor, bitter taste and presence of yellowish watery juice (Wallis, 1968). Powder of tubers of *Bryonia dioica* is characterized by abundant starch grains, thick walled, square to oblong stone cells tapering at one end and narrow pitted lumen.

Rhizome of *Tectaria cicutaria* is given under the name 'Kombadnakhi' or 'Waghchawadi' in toothache and in gum complaints. Rhizome can be easily identified by its dark brown or black color, distinctive hard, thick leaf bases which look like paws of tiger, bitter taste, nauseous disagreeable odor, yellowish-green internal tissue, tracheids with spiral thickenings and rhizome powder with abundant, oblong, rounded starch grains. Similar characters are also described for *Dryopteris filis-mas* is used as anthelmintic (Iyengar, 1985).

Tuber of *Leea setuligera* is given as 'Tambada dinda' or 'Lahan dinda' to cure skin diseases, in elephantiasis, itch and to heal wounds. These uses are not reported in previous literature. Similar uses are reported for *Leea macrophylla* by Kirtikar and Basu (1918) and Chopra (1956) which is called as a 'Dinda' or 'Pandhra dinda' in literatures. Tuber is very distinct in its cylindrical, oblong or dumbbell shape, reddish-brown color and tuber powder with striated stone cells, raphid sacs, raphides, crystals of calcium oxalate and very small simple starch grains.

Tuber of *Momordica dioica* is sold under the name 'Rankarle'. It is used in urinary complaints in folk medicines. Similar uses are reported by Kirtikar and Basu (1918); Nadkarni (1954) and Agarwal and Ghosh (1985). The tuber is very distinct in its external yellowish-brown color, presence of large prominent lenticels, slightly bitter taste and fetid odor and tuber powder containing vessels with alternate, bordered pits, abundant starch grains and square to rectangular, pitted stone cells.

Rhizome of *Costus speciosus* is called as 'Ratanjyoti' or 'Ratanpurush' in folk medicines. In literature, it is called as a 'Penva', 'Kamuka', 'Kobee' and it is reported to be used as purgative, depurative, anthelmintic, stimulant and tonic (Nadkarni, 1954 and Agarwal and Ghosh, 1982). In folk medicines, it is used on skin to remove marks remained due to the pimple and acne, mark formed due to excess secretion of melanin, and black circle below the eyes, which is not reported in earlier literatures. Rhizomes are characterized by presence of simple, 2–3 celled hairs, subcylindric shape, fleshy nature, pale brown color and elliptic oblong starch grains. Phrmaocgnosy of rhizome is worked in detail by Datta and Mukerji (1950) and Raghunathan and Roma Mitra (1999).

Corms of *Remusatia vivipara* are called as a 'Rukhalkand' used on piles as an external application with other combinations. Tubers are very distinct in external characters. Tubers are rounded, planoconvex, smaller in size, red or pink in color with ridges of fallen leaf scars. Powder is characterized by spirally thickened vessel elements, abundant raphides, starch grains and crystals of calcium oxalate.

Corms of *Sauromatum venosum* are called as 'Suran' or 'Janglisuran' or 'Agyasuran' mostly used on piles as an external application with other combinations. Tubers are very distinct in shape and size. Tubers of *Sauromatum venosum* are planoconvex, large, 7–12 cm in diam., brown in color. Powder is characterized by presence of abundant crystals of calcium oxalate and starch grains.

Tubers of *Asparagus racemosus* are given under the name of 'Shatavarimul'. It is easily identified in field by its cylindrical, fusiform, fawn or flesh colored entire roots tapering towards both the ends (Iyengar, 1985 and Sarin, 1996). Powder is also very distinct by presence of stone cells and abundant starch grains. It is widely used in debility, in case of leucorrhoea, as glalactogoguge. These uses are also reported in earlier literature by Kirtikar and Basu (1918), Chopra (1956) and Agarwal and Ghosh (1982).

During present investigation 13 stem drugs have been collected and identified on the basis of external and internal characters. Out of 13 stem drugs 5 stem drugs are given as substitute to root viz. *Aegle marmelos*, *Cryptolepis buchananii*, *Celastrus paniculatus*, *Calotropis gigantea*, *Moullava spicata*. Stem of *Trichosanthus tricuspidata*, *Cissus elongata*, *Diploclisia glaucescens*, *Argyreia boseana* and *Entada rheedei* are reported newly which are used as medicine in south western Maharashtra.

Stems

During present investigation it was noted that the stem of *Aegle marmelos* is given under the name 'Belmul' for same medicinal purpose as given by Kirtikar and Basu (1918), Nadkarni (1954) and Chopra (1956). It is one of the 'Dashmula' used in Ayurveda (Kirtikar and Basu, 1918 and Nadkarni, 1954). Stem can be easily identify by its thick, pale brown or cream colored rhytidome, few fibers, normal secondary growth, lemon yellow colored islands of stone cells, secondary cortex with phloem fibers, cylindrical vessels with alternate, very small pits and stem powder with thick walled pitted lumen and solitary crystals of calcium oxalate.

Stem of *Tinospora cordifolia* is given under the name 'Gulvel' jaundice and in fever. Same uses are also reported in literature by Kirtikar and Basu (1918), Cooke (1958), Chopra, (1956) and Nadkarni (1954). Stem is very distinct in its gray colored pieces, thin outer bark, flanking off from the stem in dried samples, warty and longitudinally wrinkled stem (Mukerjee, 1953; Deshmuk *et al.*, 1957 and Aiyer and Kolammal, 1963). Anatomically stem is characterized by presence of secondary anomalous growth, xylem completely obliterating the pith and giving xylem a stellate appearance (Aiyer and Kolammal, 1963 and Deshmukh *et al.*, 1957), drum shaped vessels with alternate bordered pits or with spiral thickening, phloem fibers encircled by a parenchymatous sheath of cells possessing rows of prisms of calcium oxalate (Mukerjee, 1953; Deshmukh *et al.*, 1957 and Aiyer and Kolammal, 1963) and stem powder is gray in color with crystals of calcium oxalate and starch grains. Pharmacognosy of the species is described in detail by Raghunathan and Roma Mitra (1999).

Stem of *Litsea deccanensis* is called as 'Hadsandhi' in folk medicines of south western Maharashtra. In literature it is called as 'Chikna' (Cooke, 1953). It is given to cure wounds and bone fractures. In literature only 3 species viz. *L. chinensis*, *L. polyantha* and *L. stocksii* are reported as medicinal (Kirtikar and Basu, 1918). Dymock (1889–93) reported two species of *Litsea* viz. *L. sebifera* and *L. stocksii* as medicinal but their use in bone fracture is not recorded. Agarwal and Ghosh (1985) reported that root of *L. monopetala* is used in bone fractures in animals and same is reported for the stem of *Litsea deccanensis*. Pharmacognostic studies on stem of *L. deccanensis* have not so far been worked out. Stem is characterized by its cylindrical, longitudinally striated, reddish-brown stem pieces, hard, pale brown wood, aromatic odor, bitter taste and anatomically by normal secondary growth, secondary cortex with continuous band of stone cells, cylindrical vessels with alternate and bordered or sometimes scalariform pits, thick walled unseptate fibers and stem powder with abundant two types of stones cells, some with thin pitted walls and others with thick non pitted striated walls.

Stem of *Argyreia boseana* is given under the name 'Maral Mathangi' for wound healing in folk medicines. It is not reported in earlier literature on medicinal plants. Another species which is used as medicines is *A. speciosa* (Dymock, 1890; Kirtikar and Basu, 1918; Nadkarni, 1954 and Chopra, 1956). Root of *A. speciosa* is used in rheumatism, diseases of nervous systems and as tonic and alterative (Kirtikar and Basu, 1918 and Agarwal and Ghosh, 1985). Stem is very distinct in its external morphology. It is characterized by gray colored cylindrical pieces with longitudinally wrinkled rough surface and anatomically by presence of anomalous growth, intraxylary phloem (Metcalf and Chalk, 1950), drum shaped vessels with alternate, bordered and large pits and yellowish-brown stem powder containing abundant rectangular or polygonal thick walled stone cells, clustered crystals of calcium oxalate and simple or compound starch grains.

Stem of *Celastrus paniculatus* is given as substitute to the roots of *Celastrus paniculatus* in folk medicines of south western Maharashtra. It is given in rheumatism and also as a brain tonic. Same uses are reported for seeds of *Celastrus paniculatus* by Kirtikar and Basu (1918), Nadkarni (1954) and Chopra (1956). Use of stem in medicine is not recorded in literature. Stem of *Celastrus paniculatus* is very distinct in external as well as anatomical characters. Stem is available in gray or pale pink colored, cylindrical pieces. Stem is characterized by thick bark containing tannin in inner layers of bark (Metcalf and chalk, 1950), cylindrical vessels with alternate small pits, septate thick walled fibers and stem powder with abundant crystals of calcium oxalate, few starch grains and striated stone cells.

Stem of *Trichosanthus tricuspidata* is given under the name 'Kaundalkhod' in folk medicines of south western Maharashtra. Same local name is reported by Kirtikar and Basu (1918), Nadkarni, (1954) and Cooke, (1958). In literature, root and fruits are reported to be mainly used on asthma and carbuncles (Kirtikar and Basu, 1918; Nadkarni, 1954 and Chopra 1956). During present investigation use of stem in backache, strains and bruises is reported for the first time. Stem is characterized by presence of brown lenticels, furrowed brown color pieces with secondary anomalous growth, xylem stellate and phloem at the ends of radii, vessels broad, drum shaped with alternate and bordered pits and stem powder by stone cells and abundant starch grains.

Stem of *Entada rheedei* is given under the name of 'Kajarichavel' in folk medicines. It is also called as 'Garabi' or 'Garambi' (Cooke, 1953 and Kirtikar and Basu, 1918). Seeds are used as anthelmintic, tonic, used in leprosy, dysentery, bronchitis, asthma and piles. During present study, stem is used in wound healing. Kirtikar and Basu (1918) reported that bark is used for scorpion sting, while Dymcok (1889–93) reported that spongy fibers of trunk is used with various advantage for various skin infection in the Philippine. Use of stem in wound healing is recorded for the first time. Stem is available in the form of cylindrical, gray colored pieces with thick rough bark covered with lenticels, fibrous fractures, anomalous secondary growth phloem embedded in parenchymatous ground tissue, short barrel shaped vessels with bordered pits, septate fibers (Metcalf and Chalk, 1950) and powder with solitary and clustered crystals of calcium oxalate.

Stem of *Cryptolepis buchananii* is given under the name 'Kalitakali' or 'Kawali'. Same local name is reported for the species by Kirtikar and Basu (1918) and Cooke (1958). Singh (1955) reported that root of *Cryptolepis buchananii* is used as a substitute for the roots of *Hemidesmus indicus*. In folk medicines of south western Maharashtra, stem is available in reddish-brown or black colored cylindrical pieces characterized by longitudinally wrinkled thick bark, outer bark flanking of in thin papery sheets, buff or lemon-yellow wood, cylindrical vessels with alternate pitting, thick walled fibers, presence of few stone cells and absence of starch grains. Pharmacognosy of *Cryptolepis buchananii* has been studied by Raghunathan and Roma mitra (1999).

Stem of *Cissus elongata* is given under the name of 'Pandhrigirnoli' in skin diseases (Kirtikar and Basu, 1918). Stem is characterized by cylindrical, pale yellow colored pieces, soft rhyitidome, deeply wrinkled or fissured bark, abnormal secondary growth (reported to be 'aristolochia' type by Metcalf and Chalk, 1950), cylindrical vessels with alternate or scalariform pits and stem powder with raphid sacs and clustered crystals of calcium oxalate.

Stem of *Calotropis gigantea* is given under the name of 'Ruikhod' in skin diseases and for the early suppuration of ulcers or wounds. Same uses are reported by Kirtikar and Basu (1918), Nadkarni (1954) and Agarwal and Ghosh (1985). Studies on root and leaf drugs were carried out by various workers but studies on stem are not reported in earlier literatures. During present investigations following diagnostic characters of stem of *Calotropis gigantea* were reported. Stem is gray colored, cylindrical with thin rhytidome. Rhytidome shows transverse cracks. Wood is soft, pale yellow or lemon yellow in color. Vessels are cylindrical with alternate or scalariform pitting. Powder is pale yellow or buff in color and dominated by starch grains.

Pharmacognosy of the species is described in detail by Raghunathan and Roma Mitra, (1999).

Stem of *Diploclisia glaucescens* is given under the name of 'Saptkappi', 'Chakrumbi' and 'Vardhara' in folk medicines of south western Maharashtra. 'Ramrakh' or 'Vardhara' are the local names given by Kirtikar and Basu (1918) and Cooke (1958). 'Vardhara' is also reffered for *Raourea santaloides* by the Cooke (1958). Leaves of *D. glaucescens* are given in syphilis, biliousness and in gonorrhoea (Kirtikar and Basu, 1918). During present investigation, stem of *D. glaucescens* is noted to be used on diabetes, kidney stone and in asthma. These uses of stem are reported for first time. During present investigation, external and internal characters were studied. Stem is very distinct in its external morphology by gray color of cylindrical, smooth pieces. Secondary growth is abnormal and shows successive bundles of xylem and phloem repeating the structure of young stem. Phloem is interxylary (Metcalf and Chalk, 1950). Stem powder is black or gray in color which contains abundant starch grains and prism type of crystals of calcium oxalate.

Stem of *Gnetum scandens* is called as 'Umbelvel' in folk medicines of South western Maharashtra. In previous literatures, Kirtikar and Basu (1918) reported that in Andaman the roots and stems are used as an antiperiodic. Stem is given for penetrating wounds of the abdomen caused by horn thrust. During present investigation, it is noted that the stem is used on piles and in hemicrania. These uses are not reported in earlier literature. Stem of *Gnetum* is anatomically studied by various workers. Externally stem is characterized by the cylindrical, dark gray or black colored pieces of stem. Stem is longitudinally fissured, contains pale yellow or brown colored lenticels, longitudinal fissures, long, cylindrical vessels with alternate bordered pits and thick walled fiber tracheids without lumen.

Stem of *Moullava spicata* is given under the name 'Vakerichebalbhate' in folk medicines of south western Maharashtra. Kirtikar and Basu (1918) reported that root is used on pneumonia and in skin diseases. Nadkarni (1954) and Agarwal and Ghosh (1985) reported that root is used in pneumonia and in tuberculosis. But during present investigation, it is noted that the stem is used in night emission, in gonorrhoea and to remove any type of endo or exocortosis. These uses are not reported in earlier literature. Stem is characterized by bitter taste, reddish brown thick bark, paratracheal confluent parenchyma, cylindrical vessels with bordered pits and abundant crystals of calcium oxalate.

Bark

Bark drugs are playing important role in herbal medicines. Sarin (1996) described external morphology of 14 barks used in Ayurveda. During present investigations 12 bark samples have been collected and identified botanically.

Bark of *Terminalia cuneata*, *T. elliptica*, *Bridelia retusa*, *Pterocarpus marsupium*, *Pongamia pinnata*, *Ficus exasperata*, *Holarrhena pubescens*, *Wrightia tinctoria* ssp. *tinctoria*, and *Crateva adansonii*, are commonly used in various Ayurvedic preparations by local peoples and Traditional practitioners, vaidyas, and doctors. Bark of *Pittosporum floribundum* and *Cassine paniculata* is not used very frequently in medicines.

During present investigations bark of *Terminalia cuneata* and *T. elliptica* were collected under the same name i. e 'Arjuna sal'. Cooke (1958), Nadkarni (1958) and Chopra, (1956) reported same local name for only *T. cuneata* and 'Ain' for the *T. elliptica*. Thus bark of *T. elliptica* is used as substitute to the bark of *T. cuneata*. Same is reported by Chopra (1958). Both the barks are very distinct in external and anatomical attributes. Bark of *T. cuneata* is flat, covered with flesh mealy coating, taste is astringent, powder is pale pink in color and shows presence of abundant crystals of calcium oxalate. Same characters are reported for the species by Shaha (1956). Bark of *T. elliptica* characterized by thick, rhytidome composed of alternate layers of periderm and crushed phloem and bark powder with crystals of calcium oxalate. Both the barks are given in bone fracture and as wound healing in folk medicines of South Western Maharashtra. Same uses are reported by Kirtikar and Basu (1918), Nadkarni (1954) and Chopra (1956).

Bark of *Bridelia retusa* is given under the name of 'Asan' in folk medicines. It is used in treatment of urinary calculi. Same use is reported by Kirtikar and Basu (1918) and Nadkarni (1954). Bark is described as one of the most astringent barks in India (Dymock, 1889–93). Bark is externally characterized by reddish or pale brown color finely striated inner surface and fibrous fractures. Anatomically it is characterized by mesh like arrangements of phloem fibers and presence of solitary crystals of calcium oxalate. Fresh bark when wounded exudes a blood-red juice. Tanninic acid is present in bark (Dymock, 1891).

Bark of *Pterocarpus marsupium* is given under the name of 'Biwalasal' in folk medicines. Kirtikar and Basu (1918), Dymock (1890) and Nadkarni (1954) reported that it is used in teeth complaints. In folk medicines of south western Maharashtra, it is used for similar purpose. In addition, it is also given in diabetes. Bark is characterized by sweetish astringent taste, reddish-brown or pale or brown color and orange colored rounded patches in the bark.

Bark of *Pongamia pinnata* is given under the name 'Karanj sal' in folk medicines. Bark of *Pongamia pinnata* is used on various skin diseases, chewed to clean the teeth and is used on wounds (Kirtikar and Basu, 1918 and Nadkarni, 1954). Bark is very distinct in its very bitter taste and nauseous odor. Bark is externally incurved, brown or gray in color, lenticels are numerous and powder is characterized by abundant crystals of calcium oxalate and stone cells.

Bark of *Ficus exasperata* is given under the name 'Kharvat sal' in scorpion sting and to heal wounds. Bark can be identified by yellowish-brown or greenish-brown color, presence of abundant lenticels and latex canals and bark powder with abundant stone cells, solitary crystals of calcium oxalate and starch grains. Dymock (1889–93) and Kirtikar and Basu (1918) reported that it is given in glandular enlargement of liver and spleen.

Bark of *Holarrhena pubescens* is given under the name 'Kuda Sal' in amoeboid dysentery. Bark is characterized by its bitter taste, hard, thick and recurved pieces, anatomically by presence of continuous bands of rectangular, striated and pitted stone cells, crystals of calcium oxalate and absence of starch grains. All these characters are also reported by Iyengar and Nayak (1984).

Bark of *Crateva adansonii* ssp. *odora* is given under the name of 'Vaivarna Sal' in folk medicines. Similar local name is reported by Kirtikar and Basu (1918) and Nadkarni, (1954). Bark is used to reduce excess fats from the body (Kirtikar and Basu, 1918). Stem bark shows diagnostic characters such as bitter taste, disagreeable odor in fresh material, greenish-black with small, white, rounded to elongate lenticels, lemon yellow or buff colored inner bark, abundant, thick walled or thin walled, pitted or unpitted stone cells and abundant starch grains.

Bark of *Wrightia tinctoria* is given under the name of 'Kala kuda' on piles and to cure wounds of animals. In literature, it is reported as stomachic, tonic and febrifuge by Nadkarni (1954). It is used as adulterant to the bark of *Holarrhena pubescens* and given in stomach complaints. But during present investigation, it is reported for different uses and not as adulterant to the bark of *Holarrhena pubescens*. Bark of *Wrightia tinctoria* is characterized by thin, rough gray or greenish-black pieces, with small lenticels, slightly bitter taste and presence of few crystals of calcium oxalate and striated, rectangular stone cells and absence of starch grains.

Bark of *Cassine paniculata* is given under the name 'Ragat wad' as blood purifier. It is also given in anemia. Same use of bark is reported for the bark of *Cassine glaucum* by Nadkarni (1954). Bark of *Cassine paniculata* is very distinct in external morphology. It can be identified by its fresh green color, scarlet or orange colored lenticels (Dymock, 1889–93), astringent taste, abundant stone cells and solitary crystals of calcium oxalate.

Bark of *Alstonia scholaris* is given under the name 'Satvin Sal' as wound healing and in dysentery in folk medicines. Same uses are reported by Kirtikar and Basu (1918); Nadkarni (1954) and Chopra (1956). Bark is very distinct with bitter taste. Bark is gray to greenish in color. Bark powder is characterized by abundant stone cells of various size and shape. Walls of the stone cells are striated, pitted or unpitted. Crystals of calcium oxalate are solitary. Same characters have been reported for the species by lyengar and Nayak (1984). Nadkarni, (1954) reported that it is valuable in debility and after effects of fever, also in chronic diarrhoea, dysentery and in catarrhal fever.

Bark of *Pittosporum floribundum* is given under the name 'Vikari Sal' on skin diseases, on piles and in itch with other combinations. Same uses are reported by the Kirtikar and Basu, (1918) and Nadkarni, (1954). Bark is very distinct in its aromatic odor, bitter taste, gray or greenish-black color, quilled or channeled pieces, presence of oil glands and bark powder with abundant crystals of calcium oxalate and starch grains (Metcalf and Chalk, 1950).

Leaf

During present investigation 5 leaf samples of folk medicines were collected from south western Maharashtra. Studies on Indian leaf drugs are carried out by Datta and Mukerji (1952). *Actinodaphne hookeri* is given in night emission and *Atlantia racemosa* is used in cough and bronchitis which is not reported in previous literature. During present investigation, 3 leaf drugs are reported which are well-known to medicinal science viz. *Justicia adhatoda*, *Gymnema sylvestre* and *Sida acuta*.

In folk medicines, leaf of *Justicia adhatoda* is given under the name 'Adulsa' in cough, bronchitis and in asthma (Kirtikar and Basu, 1918; Chopra, 1956; Nadkarni, 1954 and Naik, 1998). Leaves are very distinct and characterized by dark green upper surface and paler hairy on lower surface. Taste is bitter, odor is disagreeable. Same is reported by Aiyer and Kolammal (1963) and Sarin (1996). Leaf is dorsiventral and amphistomatic. Stomata are diacytic. Epidermal cells are wavy in outline and possesses glandular and non glandular trichomes.

In folk medicines, leaf of *Gymnema sylvestre* is given under the name of 'Bedkichapala' in diabetes which is also reported earlier by Mahaskar and Cauis (1930). Leaves are very easy to identify by its slightly bitter taste. It paralyzes the taste buds and tongue cannot determine the sweet taste for a period (Nadkarni, 1954). Leaves are dorisventral, hypostomatic and characterized by straight walled epidermal cells, paracytic stomata, 1–2 celled uniseriate non glandular trichomes and presence of abundant crystals of calcium oxalate. Same characters reported by Wahi and Chunekar (1964) for the species.

Leaf of *Sida acuta* is called as 'Chikana pan' given as wound healing and also used in bone fractures in folk medicines. It is also called as 'Bala', 'Jangli methi' and 'Lahan-Chikana' (Kirtikar and Basu, 1918 and Cooke, 1958). Dymock (1889–93) reported that leaves of *S. cordifolia* are applied in opthalmia. Jain (1998) reported that leaves of *S. acuta* are used to cure boils and swelling of scrotum. Leaves are simple, hairy on both surface, glandular, bitter, mucilaginous, dorsiventral and amphistomatic. Stomata are of ranunculaceous type. Epidermal cells are wavy in outline. Trichomes are stellate. Glandular trichomes are sessile and bulbous based with capitate head. The tissue shows presence of abundant crystals of calcium oxalate.

Leaf of *Atalantia racemosa* is given under the name of 'Makdicha pala' in bronchitis, asthma, cough and as a blood purifier in folk medicines of south western Maharashtra. Kirtikar and Basu (1918) reported that leaf of *A. racemosa* is used as an ingredient in a compound liniment used in hemiplegia. The leaves of *A. racemosa* are simple, elliptic, dark green, glabrous and gland dotted, crenulate, dorsiventral and hypostomatic. Stomata are paracytic.

In folk medicines leaf of *Actinodaphne hookeri* is given under the name of 'Malawa' in night emissions, urinary disorder and in diabetes. It is mentioned as 'Pisa' by Cooke (1958) and Kirtikar and Basu (1918). Leaf is very distinct in it's external and anatomical characters. Leaf of *A. hookeri* is simple, white tomentose below, coriaceous and acuminate at apex. Young leaves are densely silky. Leaf is dorsiventral and hypostomatic. Stomata are paracytic and hairs are unicellular.

Flower

During present investigation one flower powder is collected in folk medicines of south western Maharashtra. Flower powder of *Woodfordia fruticosa* is given in excess menstrual discharges in folk medicines of south western Maharashtra. Same use is reported by Kirtikar and Basu, 1918. Flower powder is characterized by astringent taste, red color, bulbous based glandular hairs, unicellular simple hairs and small, cuneate, obovoid, brown, smooth seeds. Pharmacognosy of the species worked by Raghunathan and Roma Mitra, (1999).

Fruits

During present investigations 8 fruit samples have been collected and identified on the basis of external characters. All the fruits are very common in south western Maharashtra. All of them are widely used in various systems of medicines. Some of them are commercially exploited in large scale viz. *Cassia fistula*, *Catunaregam spinosa*, *Terminalia bellirica* and *Terminalia chebula*.

Fruit of *Cassia fistula* is given under the name 'Bahava' in folk medicines of south western Maharashtra. It is used on piles in folk medicines. Similar uses are reported by Kirtikar and Basu (1918) and Nadkarni (1954). Pod is divided into cellular compartments by thin woody transverse dissipiments. Fruit is cylindrical and chocolate brown colored with soft black brown colored greasy pulp (Sarin, 1996).

Fruits of *Terminalia bellirica* is given under the name 'Behda' in folk medicines. It is given in inflammation in eye complaints and sore throat. It is used as one of the ingredient in 'Triphala churna' (Kirtikar and Basu, 1918 and Nadkarni, 1954). Fruit is ovoid or obovoid, velvety with minute pale brown tomentum (Sarin, 1996).

Fruit of *Semecarpus anacardium* is called as 'Bibba' in folk. Dymock (1889–93) and Chopra, (1956) reported that it may cause skin allergies, however, in folk medicines, it is given in bronchitis. Fruit is drupaceous nut with pear shaped receptacle, containing viscous oily liquid of black color (Sarin, 1996).

Fruit of *Catunaregam spinosa* is given under the name 'Gelphal.' It is used to remove post pregnancy marks remained on face due to excess secretion of melanin. Dymock (1889–93) and Nadkarni (1954) reported it as powerful emetic. Fruits are pear shaped, black or reddish brown in color.

Fruit of *Terminalia chebula* is called as 'Hirda' in folk medicines. It is one of the ingredient of 'Triphala churna' (Nadkarni, 1954). Fruit is ellipsoid, obovoid, club shaped, glabrous, ribbed single seeded drupe. In folk medicines it is used in eye infection and as an appetizer.

Fruit of *Helicters isora* is given in stomach complaints of children. Fruit is follicle type, 5 follicles twist together in the form of screw. Fruits are stellate tomentose (Cooke 1958).

Fruits of 'Vavding' are being used in Indian system of medicine on various disorders. Due to rarity of *Embelia ribes*, which is an official source of 'Vavding'. It is substituted by other source such as *E. basaal* and *E. drupacea*. The fruits of *E. ribes* are smaller in size as compared to the fruits of *E. basaal* and *E. drupacea*. The fruits of *E. ribes* and *E. drupacea* are warty, while fruits of *E. basaal* are longitudinally striated (Cooke, 1958; Shah and Khanna, 1961 and Sarin, 1996). Pericarp of *E. ribes* is warty while pericarp of *E. basaal* and *E. drupacea* is longitudinally striated. *Embelia ribes* is very rare while other two species are common in Maharashtra. Studies showed that fruits of *E. basaal* and *E. drupacea* are given as substitutes to the fruits of *E. ribes* in Maharashtra. These three species can easily be identified on the basis of external characters of fruits by their size, shape and pericarp.

Seeds

During present investigation about 7 seed have been collected. All of them are commonly used in herbal medicines. During present investigation 6 seeds are reported from dicot and one is from monocot. Seeds of *Caryota urens* are given under the name 'Ardh-supari' in folk medicines. It is given in hemicrania. Same local name and uses are reported by Dymock (1889–93), Kirtikar and Basu (1918) and Nadkarni (1954). Seeds are shining black, plano convex, longitudinally ribbed. Seeds of *Caesalpinia bonduc* are given under the name 'Gajaga' in folk medicines. They are very distinct in its bitter taste, hard and glossy testa of bluish-gray color (Sarin, 1996). Seeds of *Entada rheedei* are given under the name of 'Garambi' in folk medicines. They are given in stomachache and in scorpion sting. The seeds are compressed, ovoid orbicular and stony (Dymock, 1889–93 and Cooke, 1958). Seeds of *Syzygium cumini* is given under the name of 'Jambhulbi', in folk medicines. They are given in diarrhoea, dysentery, leucorrhoea and in diabetes. Seeds are characterized by ovoid shape, covered with brownish-papery shell (Sarin, 1996). Seed of *Strychnos nux-vomica* are given under the name of 'Kajra' in folk medicines. The seeds are given to diabetic patients. The seeds are distinct in its flat, ash-gray colored seeds. Seeds are covered by numerous closely appressed silky hairs radiating from the center. Seeds of *Butea monosperma* are given under the name 'Palasbi' in folk medicines. They are given in menorrhoea while Dymock (1889–93) and Kirtikar and Basu (1918) reported the use of seeds as anthelmintic. The seeds are flat, kidney shaped, reddish-brown colored, leathery, glossy veined and wrinkled (Sarin, 1996).

Prickle

During present investigation prickle of *Zanthoxylum rhetsa* is given under the name of 'Chirphal' used stop bleeding from the fresh cut wounds in folk medicines of south western Maharashtra. Prickle is sharp, pointed, yellow colored, and with conical solid base. It is not reported in previous literature.

Bulbil

Bulbils of *Dioscorea bulbifera* is given under the name of Dukkar-kand in piles in folk medicines of south western Maharashtra. Bulbils are characterized by mucilaginous slightly bitter taste, globular to ovoid in shape, brown in color with many lenticels all over the surface.

These studies have revealed that the plant parts used in folk medicines can be identified on basis of external and internal microscopic characters. It is hoped that this will help in bringing some of the folk medicines in limelight opening new doors for further scientific studies on folk medicine.

Summary and Conclusions

A. Taxonomic Account of Medicinal Plants of South Western Maharashtra

All the medicinal plants that occur in south western Maharashtra are studied critically from taxonomic point of view with special emphasis on their availability in natural habitats, present status, local uses and methods of administration.

Following is a brief account of medicinal plants of the region.

- Extensive fieldwork and careful plant collections resulted into collection of 565 flowering plant species of medicinal potential from the south western Maharashtra. Out of 565 species of medicinal plants, 508 are dicot species and 57 are monocots species.
- Among 565 species of medicinal plants, 474 species are wild and 91 are mostly garden species. About 47 species endemic to Peninsular India which occur in south western Maharashtra have some medicinal uses. Similarly about 22 species recorded to be rare (CAMP, 2001) are also found to occur in the region.
- *Rauvolfia serpentina* (L.) Benth. ex Kurz. and *Holostemma anuulare* (Roxb.) K. Schum. are of very rare occurrence in the region and recorded to be critically endangered (CAMP, 2001). There is need for cultivation of both the species.
- *Nothapodytes nimmoniana* reported to be common in the region has become rare plant due to its over and unrestricted exploitation for the foreign markets.
- Plants yielding root drugs viz. *Bryonia dioica* Jacq., *Chlorophytum arundinaceum* Baker., *Cinnamomum nitidum* (Roxb.) Hook., *Embelia basaal* (R. & S.) DC., *Eulophia nuda* Lindl., *Eulophia ochreata* Lindl, *Hemidesmus indicus* R. Br., *Iphigenia stellata* Blatt., *Kedrostris rostrata* (Rottl.) Cogn., *Operculina turpethum* (L.) Manso, *Oroxylum indicum* (L.) Vent, *Pueraria tuberosa* DC., *Salacia chinensis* L., *Salacia brunoniana* Wt. & Arn. have also become rare plants of the region. Uprooting of entire plant has resulted in decline of populations. Therefore, there is immediate need to conserve them and establish methods for their cultivation.
- Usually fruits of *Embelia ribes* Burm. f. (Vavding) are used in medicine, however, it is very rare plant in the region. Market survey showed that the fruits of *Embelia basaal* (R. & S.) DC. and *E. drupacea* (Dennest.) M.R. & S.M. Almeida are used as substitute to *Embelia ribes*. *Embelia basaal* and *E. drupacea* are fairly common plants in the region and seeds of these species are sold under the name of Vavding.
- During present investigation some medicinal plant species viz. *Ancistrocladus heyneanus*, *Argyreia boseana, Cassine paniculata, Diploclisia glaucascens, Eulophia ochreata, Jasminum malabaricum, Leea setuligera,* and *Parthenium hysterophorus* which are not reported in earlier literature on medicinal plants of Maharashtra are reported from the region.

- During present investigation some medicinal uses of plants species have been recorded which are not reported in earlier literatures. They are as follows:
  - Root and fruit of *Toddalia asiatica* var. *asiatica* (L.) Lamk. is found to be effective in curing urinary complaints.
  - Root of *Smilax zeylanica* L. is used to cure psoriasis.
  - Root and stem of *Moullava spicata* (Dalz.) Nicol. is used to cure any kind of exo or endocortosis and also used to heal diabetic wounds.
  - Root of *Parthenium hysterophorus* L. is used in chronic fever and in sinusitis.
  - Tubers of *Eulophia nuda* Lindl. are used to cure piles.
  - Tuber of *Pueraria tuberosa* DC. is used in epilepsy.
  - Bark of *Saraca asoca* (Roxb.) Willd. is used to treat HIV patients.
  - *Asclepias curassavica* L. is used in asthmatic troubles.
  - Leaf of *Abutilon indicum* (L.) Sweet is used to relive mental stress used to heal any kind of ulcerous wounds in animals.
  - Leaf decoction of *Eclipta prostrata* (L.) L. is given in jaundice.
  - Leaves of *Gmelina arborea* Roxb. are used in Herpes zoster.
  - Leaves of *Lawsonia inermis* L. are used in bronchitis.
  - Leaves of *Pongamia pinnata* (L.) Pierre. are used in fissures.
  - Seed of *Antiaris toxicaria* (Pers.) Lesch. is used in diabetes.
  - Seed of *Thespesia populnea* (L.). Soland. ex Corr. is given in sex debility.

B. Folk Medicines and their Correct Botanical Identity

Over 100 samples including roots, rhizomes, bulbs, tubers, corms, stems, barks, leaves, fruit and seeds were collected from local bazaars, Dhangars and traditional practitioners. Relevant information on their source, local names, specific uses for each of the samples was gathered during collection. By using this information and literature survey, samples were tentatively identified. Authentic plant species were collected and botanical identity of folk medicines was confirmed by comparing morphological and anatomical features of authentic known samples.

- In present investigations, botanical identity of about 81 samples have been confirmed. Following is the list of folk medicines which have been collected and their botanical identity has been confirmed.
- *Actinodaphne angustifolia* Ness., *Adhatoda zeylanica* Medicus., *Aegle marmelos* (L.) Corr., *Alstonia scholaris* R. Br., *Argyreia boseana* Sant. & Patel, *Artocarpus heterophyllus* Lam., *Asparagus racemosus* var. *javanica* Baker, *Atalantia racemosa* Wight & Arn., *Bombax ceiba* L., *Bridelia retusa* Spreng., *Bryonia dioica* Jacq., *Butea monosperma*

(Lam).Taub., *Caesalpinia bonduc* (L.) Roxb., *Calotropis gigantea* (L.) Br., *Caryota urens* L., *Cassia fistula* L., *Cassine paniculata* (W. & A.) loberau-Collen., *Catunaregam spinosa* (Thub.) Tirveng, *Celastrus paniculatus* Willd., *Cinnamomum nitidum* (Roxb.) Hook, *Cissus elongata* Roxb., *Cocculus hirsutus* (L.) Theob., *Costus speciosus* (J. Koenigh) Smith, *Crateva adansonii* DC., *Cryptolepis buchanani* R. & S., *Dioscorea bulbifera* L., *Diploclisia glaucascence* (Bl.) Diels., *Embelia basaal* (R. & S.) DC., *Embelia drupacea* (Dennest.) M. R. & S. M. Almeida., *Embelia ribs* Burm. f., *Entada rheedei* Spreng., *Eulophia nuda* Lindl., *Eulophia ochreata* Lindl., *Ficus exasperata* Vahl., *Ficus hispida* L. f., *Gmelina arborea* Roxb., *Gnetum scandens* Roxb., *Gymnema sylvestre* R. Br., *Helicteres isora* L., *Hemidesmus indicus* R. Br., *Holarrhena antidysenterica* Wall., *Jasminum malabaricum* Wight., *Leea setuligera* (L.) Ribd., *Litsea deccanensis* Gamble, *Momordica dioica* L., *Moullava spicata* (Dalz.) Nicol., *Oroxylum indicum* Vent., *Pittosporum wightii* A. K. Mukherjee, *Plumbago zeylanica* L., *Pongamia glabra* Vent., *Pterocarpus marsupium* var. *marsupium*. Roxb., *Pueraria tuberosa* DC. *Remusatia vivipara* Schott., *Salacia brunoniana* Wt. & Arn., *Salvadora persica* L., *Sauromatum venosum* (Ait.) Schott., *Semecarpus anacardium* var. *anacardium* L. f., *Sida acuta* Burm. f., *Spermadictyon suaveolens* Roxb., *Stereospermum chelenoides* (L. f.) DC., *Strychnos nux-vomica* L., *Syzygium cumini* (L.) Skeels, *Tectaria cicutaria* L., *Terminalia bellerica* Roxb., *Terminalia chebula* Retz., *Terminalia cuneata* Roth., *Terminalia elliptica* Willd., *Tinospora cordifolia* (Willd.) Miers., *Trichosanthus tricuspidata* Lour., *Withania somnifera* (L.) Dunal., *Woodfordia fruticosa* (L.) Kurz., *Wrightia tinctoria* R. Br. and *Zanthoxylum rhetsa* (Roxb.) DC.

- During present investigation 100 samples of folk medicines was collected from south western Maharashtra. Out of 100 folk medicines 81 are correctly identified. Out of 81 samples 71 are dicot samples and 8 are monocots one is Gymnosperm *Gnetum scandens* Roxb. and one is Pteridophytes *Tectaria cicutaria* L.
- During present investigation 22 roots, 2 corms, 5 tubers, 4 rhizomes, 13 stems, 12 barks, 5 leaves, 1 flower, 8 fruits, 7 seeds, one bulbil and one prickle is collected and identified.
- *Aegle marmelos, Desmodium gangeticum, Ficus religiosa, Gmelina arborea, Oroxylum indicum, Premna obtusifolia, Solanum anguivi, Solanum virginianum, Stereospermum suaveolens* and *Uraria picta* used as Dashmula in Indian system of medicines. Out of these 10 root drugs four (viz. *Aegle marmelos, Stereospermum sauveolens, Gmelina arborea and Oroxylum indicum*) are found to be used in folk medicines in the region.

- According to literature, 'Saptarangi' is either *Casearia esculenta* or *Salacia brunoniana*. The root sample available under the name 'Saptrangi' in the region has been confirmed as *Salacia brunoniana* and show colored rings in transverse section, however, *Casearia esculenta* root do not possesses such rings.
- Under the name 'Amarkand' usually rhizomes of *Eulophia nuda* are sold, however, rhizomes of *E. ochreata* are also sold under the same name and purpose.
- The roots sold under the 'Ghansarmul' in the region are roots of *Atalantia racemosa*, however, in literature roots of *Croton oblongifolius* are called as 'Ghansarmul'.
- Usually bark of *Cinnamomum nitidum* is given as brain tonic, however, roots of the species are also administered for the same purpose.
- Roots of *Jasminum malabaricum* is used in aphrodisiac which is not reported in earlier literature.
- The roots of *Catunaregam spinosa* are sold under the name of 'Gelamul' used in skin diseases and foul breath is not reported in previous literature.
- *Aegle marmelos* (Bel), *Calotropis gigantea* (Rui), *Cryptolepis buchanani* (Kawali), *Celastrus paniculatus* (Karad kanguni, Malkanguni), *Moullava spicata* (Vakeriche bhate) are stem drugs given as substitutes to the root drugs of the respective species.
- *Argyreia boseana*, *Jasminum malabaricum*, *Moullava spicata* and *Salacia brunoniana* endemic to peninsular India are used in folk medicine.
- Rare plant species such as *Bryonia dioica*, *Cassine paniculata*, *Eulophia ochreata* are also used as folk medicines in South Western Maharashtra.
- *Aegle marmelos* (Bel), *Gmelina arborea* (Shivan), *Oroxylum indicum* (Tetu), and *Stereospermum suaveolens* (Padal) which are constituents of 'Dashmula' have been collected and identified. They are used in folk medicines.
- *Argyreia boseana* (Maral mathangi), *Cassine paniculata* (Ragatwad), *Diploclisia glaucescens* (Vardhara), *Eulophia ochreata* (Amarkand), *Jasminum malabaricum* (Padmul), *Leea setuligera* (Lahan dinda) are not recorded in literature but these species are used in South Western Maharashtra on various disorders.
- During present investigation two different samples of fruits *Embelia* sp. were collected and identified botanically as *Embelia drupacea* and *E. basaal*. Both are used as substitutes for the fruits of *Embelia ribes* in the region.
- During present investigation some uses of plant species are recorded for the first time which are not reported in earlier literatures. They are as follows:
    - Root of *Artocarpus heterophyllus* (Phanas) is given in mouth ulcers.

- Root of *Catunaregam spinosa* (Gelmul) is given to remove pimples and post pregnancy marks from the face. Same is reported for the *Costus speciosus* (Ratanjyot or Ratanpurush).
- Roots of *Cinnamomum nitidum* (Mawarung) are given as brain tonic.
- Root of *Embelia basaal* (Vavding) is given to cure old and chronic wounds.
- Root of *Jasminum malabaricum* (Pad mul) is given as an aphrodisiac.
- Root of *Spermadictyon suaveolens* (Jitsaya) is used in Herpes zoster.
- Tubers of *Eulophia nuda* and *E. ochreata* are given in acidity and in piles and stomach complaints.
- Tuber of *Leea setuligera* (Lahandinda) is given in fissures and piles.
- Rhizome of *Tectaria cicutaria* (Kombadnakhi) is given in teeth complaints, in earache, and in gum complaints.
- Stem of *Diploclisia glaucescens* (Vardhara) is given in kidney stone and in diabetes.
- Stem of *Moullava spicata* (Wagati) is given to cure any kinds of exo or endocortosis and to treat diabetic wounds.
- Leaves of *Actinodaphne angustifolia* (Malawa) are given in night emission.
- Prickle of *Zanthoxylum rhetsa* (Chirphal) is given to stop bleeding from the cut wounds and in eye complaints.

Acknowledgements

I am thankful to Smt. L.J. Bhosale Professor and Head, Botany Department, Shivaji University, Kolhapur and, (Dr.) P.D. Chavan, (Former Head) for providing laboratory facilities and constant encouragement. It is my pleasant duty to express my thanks to the Department of Science and Technology for the award of Research Fellowship.

BIBLIOGRAPHY

Aalbersberg, B. 1999. Kava boom hits the pacific, Medicinal plant conservation. *Newsletter* of the IUCN species Survival commission-Medicinal plant specialist. Group, **5**:20.

Abdallah, M.S. and Sa'ad, F.M. 1977. Medicinal plants of Egypt: **1.** A preliminary survey. *Notes Agric. Res. Centre Herb.* (Egypt) 4. 18p.

Acosta, C. 1578. Tractado de las drogas Y medicinas de las Indias orientales. Burgos.

Agarwal, V.S. and Ghosh Barin. 1985. Drug Plants of India: Root Drugs. Kalyani Publishers, New Delhi.

Agarwal, V.S. 1986. Economic plants of India. Kailash Prakashan, Calcutta.

Agharkar, S.P. 1953. Medicinal plants of Bombay. *Bombay State Gazeteer, Sec.* A, Pt. **I.**

Ahmedullah, M. and Nayar, M.P. 1987. Endemic Plants of the Indian Region. Fl. India Ser. 4, BSI.

Alami, R. and Macksad, A.S. 1974. Medicinal plants in Kuwait. Kuwait.

Almeida, M.R. 1996. Flora of Maharashtra. St. Xaviers College, Mumbai, Vol. **I** (Ranunculaceae-Sabiaceae).

Almeida, M.R. 1998. Flora of Maharashtra. St. Xaviers College, Mumbai, Vol. **II**. (Fabaceae-Caprifoliaceae).

Almeida, S.M. 1990. Flora of Sawantwadi, Maharashtra, India. *J. Econ. Tax. Bot.* Addl. Ser. **2** vols.

Ambasta, S.P. (ed.) 1986. The useful plants of India. CSIR, New Delhi.

Amico, A. 1980. Plants used in traditional medicines in Lower Zambesia. *J. Ethnopharmac.* **2(1)**:33–34.

Anislie, W. 1813. Materia Medica of Hindustan and Artisan's and Agriculturist's being in the English, Tamool, Kukhanie, Hindotanie, Tetingoo, Arabic, Persian, Sanskrit and Latin Languages, Madras Govt. Press, Madras. **IV, VIII**. 301 p.

Anislie, W. 1826. Materia Medica or some account of those articles which are employed by the Hindus and other eastern nations in their medicine, arts and agriculture. Longmans, Reewss, Orme, Brown and Gerrn, London. **2V**.

Anonymous. 1923. Nos Plantes Medicinales de France. Comite interministeriel del Plantes Medicinales *et al.,* Essences; Paris 8 Pl.

Anonymous. 1997. Indian Medicinal Plants: A sector study occasional Paper No. 54. Export Import Bank of India: Mumbai, India.

Anonymous. 1999. Asian region inventory of medicinal and aromatic plants and polyherbal formulations. CIMAP, Lucknow, India.

Anonymous. 1948–1976. The Wealth of India-A Dictionary of Indian raw Material and Natural Products: **I-II.** CSIR and New Delhi.

Anonymous. 1962. Atlas of Medicinal Plants of the USSR. Government Pharmaceutical editions, Moscow.

Anonymous. 1975. Pharmacognosy and Phytochemistry edited by Wagner, H. and Hoerhammer, L. **VII**.

Anonymous. 1976. Medicinal plants of India: **1**. ICMR, New Delhi.

Anonymous. 1981. West African medicinal plants 1970–1980. Farnham Royal, Commonwealth Agricultural Bureaux.

Anonymous. (1948–76). Wealth of India, Rev. Edn. 1985, 1992: **1–11**. New Delhi: Council of Scientific & Industrial Research, New Delhi.

Awasthi, A.K. and Goel, A.K. 1999. Ethnobotanical studies on the Onge of Little Andaman Island. *J. Econ. Taxo. Bot.* **23**:569–574.

Ayensu, E.S. 1978. Medicinal plants of West Africa. Reference publication, Inc. 218 st clair, River Drive, Algonal Michigam-48001.

Ayensu, E.S. 1981. Medicinal plants of the West Indies. Algonac, Reference Publications, Algonal Michigam-48001.

Ayier, K.N. and Kolammal, M. 1951–53. Pharmacognosy of Ayurvedic Drugs of Travencore-cochin. University of Kerala, Trivendrum.

Aynehchi, Y.; Salehi Sormaghi, M.H.; Amin, G.H.; Khoshkhow, M. and Shabani, A. 1985. Survey of Iranian Plants for Saponins, Alkaloids, Flavonoids and Tannins. III. *Int. J. Crude Drug Res.* **23(1)**:33–41.

Bailey, L.W. 1944. The development of vessels in and its significance in morphological research. *Amer. J. Botany.* **31**:421–428.

Balandran, M.F.; Kloche, J.A.; Wurtele, E.S. and Bollinger, H. 1985. Natural plant chemicals. *Science*, **228**:1154.

Balasubramaniam, P. and Prasad, S.N. 1996. Medicinal plants among the Nilgiri Biosphere Reserve. *J. Econ. Tax. Bot. Addl. ser.* **12**:253–226.

Balasubramaniam, P.; Rajasekaran, A. and Prasad, S.N. 1997. Folk Medicines of Coimbatore forests. *Anci. Sci. Life.* **16**:122–226.

Balee, W. 1994. Footprints of the forest, Ka'apor ethnobotany — the historical ecology of plant utilization by an Amazonian people. New York: Columbia University Press.

Bapalal, G. 1970. Some controversial drugs of Indian medicine-II. *J. Res. Indian Med.* **5(1)**:140–56.

Bapalal, G. 1970. Some controversial drugs of Indian medicine-II. *J. Res. Indian Med.* **6(1)**:95–104.

Bapalal, G. 1972. Some controversial drugs of Indian medicine-III. *J. Res. Indian Med.* **6(2)**:178–96.

Bapalal, G. 1972. Some controversial drugs of Indian medicine-IV. *J. Res. Indian Med.* **7(1)**:76–90.

Bapalal, G. 1972. Some controversial drugs of Indian medicine-VI. *J. Res. Indian Med.*, **7(3)**:64–77.

Bapalal, G. 1974. Some controversial drugs of Indian medicine-VII. *J. Res. Indian Med.* **9(2)**:72–80.

Bapalal, G. 1975. Some controversial drugs of Indian medicine-IX. *J. Res. Indian Med.* **10(4)**:127–40.

Bapalal, G. 1975. Some controversial drugs of Indian medicine-VIII. *J. Res. Indian Med.* **10(3)**:90–103.

Baquar, S.R. 1967. Medicinal Plants of southern West Pakistan PSIR. *Bull. Monogr.* **3**:1–108.

Barnett, R. 2000. Traditional Practitioners in Kenya: putting theory into practice, *TRAFFIC. Bull.* **18**:87–89.

Bentham, G. and Hooker, J.D. 1862–1883. Genera Plantarum I-III. Reeve & Co., Williams and Norgate.

Bharatiya Vidya Bhavan's, Swami Prakashananda, Ayurveda Research, Centre 1992. Selected medicinal plants of India (a monograph of identity, safety, and clinical usage). CHEMEXCIL Bombay.

Bhat, R.B.; Eterjere, E.O. and Oladipo, V.T. 1990. Ethnobotanical studies from Central Nigeria. *Econ. Bot.* **44(3)**:382–390.

Bhattacharjii, S.K. 1998. Hand book of medicinal plants. Pointer Publishers, Jaipur, India.

Bhide, B.V.; Phalinkar, N.L. and Prenjpe, K. 1941. Chemical investigation of *Tinospora cordifolia* Miers. *J. Univ. Bombay*, **89**:10.

Bhutkar, A.V. 2000. Traditional use of medicinal plants of Hingoli District. Proceeding of seminar in plant resource development, held at Marathwada University, Aurnagabad. 146–152.

Billore, K.V. 1889. Some threatened medicinal plants of Rajasthan and their conservation. *Ind. J. For.* **115(8)**:59–599.

Birdwood, H.M. 1886. A catalogue of the flora of Matheran. *J. Bombay Nat. Hist. Soc.* **1**:203–214.

Birdwood, H.M. 1887. A catalogue of the flora of Mahabaleshwar and Matheran. *J. Bombay nat. Hist. Soc.* **2**:107–132.

Birdwood, H.M. 1896. A catalogue of the flora of Mahabaleshwar and Matheran. *J. Bombay nat. Hist. Soc.* **10**:349–439.

Birdwood, H.M. 1897. A catalogue of the flora of Mahabaleshwar and Matheran. (Revised edition by H.M. Birdwood with a note by T. Cooke) *J. Bombay nat. Hist. Soc.* Vol. 10.

Birdwood, H.M. 1899. The hill forests of western India. *J. Bombay nat. Hist. Soc.* **12**:659–674.

Bisset, N.G. 1994. Herbal drugs and pharmaceuticals. CRC Press, Boca Raton.

Blatter, E. 1926–35. Revision of flora of the Bombay Presidency. *J. Bombay nat. Hist. Soc.* **31**:547–557, 897–917, **32**:14–33*,281–298*, 408–435*, 622–649*, **33**:7–25*, 229–243*, 480–496*, 753–775*, **34**:12–268, 291–306*, 623–237, 877–900; **35**:13–318, 254–275*, 484–495*, 722–736*, **36**:13–28*, 307–320*, 524–537*, 781–795*, **37**:15–358, 255–277, 532–548*, 764–779*, **38**:6–18. (* marked jointly with C. McCann.)

Bocco, M.E.; Vischi, N. and Montani, N. 1997. Relevamiento de las plantas medicinales espontaneas del departamento Rio Cuarto (Cordoba, Argentina). (Survey of spontaneous medicinal plants in the department of Rio Cuarto (Cordoba, Argentina)). *Parodiana.* **10(1–2)**:11–18.

Boulos, L. 1983. Medicinal plants of North Africa. Algonac, Reference Publications. Michigan.

Button, P.N. 1933. Notes on herbalism in Southern Spain. *Jl. R. Nav. Med. Serv.* **19(2)**:11–19.

Cambie, R.C. and Ash, J.E. 1994. Fijian medicinal plants. CSIRO, Melbourne.

Cantoria, M. 1976. Aromatic and medicinal herbs of the Phillippines. *Q. Jl. Crude Drug. Res.* **14**:97–128.

Castetter, E.F. 1944. The domain of Ethnobiology in *Am. Nat.* **78**:148.

Chadha, Y.R. and Kashyapa, K. 1980. Information sources and services on medicinal plants. *Indian Sci. Congr. Assoc. Proc.* **67(4)**:53–56.

Chakravarty, H.I. 1982. Cucurbitaceae Fasc. Fl. Ind. **11**:1–136.

Champion, H.G. and Seth, S.K. 1968. A Revised survey of the forest types of India. Manger of Publication, Delhi.

Chatterjee, A. and Pakrashi, S.C. 1991; 1992, 1994; 1995; 1997. The treatise on Indian Medicinal Plants.: **1–5**. Publication and Information Directorate, New Delhi.

Chauhan, N.S. 1999. Medicinal and aromatic plants of Himachal Pradesh. Indus Publishing Company, New Delhi, India.

Chevallier, A. 1996. Encyclopedia of medicinal plants: [a practical reference guide to over 550 key herbs and their medicinal uses]. Dorling Kindersley, London.

Chopra, R.N., Nayar, S.L. and Chopra, I.C. 1956. Glossary of Indian Medicinal Plants. CSIR, New Delhi.

Chopra, R.N.; Chopra, I.C.; Handa, K.L. and Kapoor, L.D. 1958. Indigenous Drugs of India. U.N. Dhur and Sons, Calcutta.

Chopra, R.N.; Chopra, I.C. and Verma, B.S. 1969. Supplement to the Glossary of Indian Medicinal Plants. CSIR, New Delhi.

Chuakul, W.; Saralamp, P.; Paonil, W.; Temsiririrkkul R. and Clayton, T. 1997. Medicinal plants in Thailand: **2**. Mahidol University, Department of Pharmaceutical Botany, Bangkok.

Clute, W.N. 1919. Official Drugs of the United State. *Am. Bot.* **25**:47–50.

Coee, F.G. and Anderson, G.J. 1996. Ethnobotany of the Garifuna of Eastern Nicaragua. *Econ. Bot.* **50(1)**:71–107.

Cooke, T. 1901–1908. The Flora of the Presidency of Bombay. London. (BSI. Reprint). Calcutta, Vols.:**I-III**, 1958.

Correa, Junior C.; Lin Chau Ming and Scheffer, M.C. 1994. Cultivo de plantas medicinais, condimentares e aromaticas. Jabotical: FUNEP.

Cotton, M. 1996. Ethnobotany: principles & applications. John Wiley & Sons, Chichester, England, UK.

Cronquist, A. 1988. The evolution and classification of flowering plants. ed. **2.** Nelson, London, Edinburgh.

Cunningham, A.B. 1993. African medicinal plants: setting priorities at the interface between conservation and primary health care. Peoples and plant working paper **1.** United Nations Educational, Scientific, and cultural organization, Paris, France.

Cunningham, A.B. 2001. Applied Ethnobotany. People, wild Plant use and conservation. Earth Scan Publication, London and Sterling, USA.

Dagar, H.S. 1989. Plants in folk medicines of Nicobarese Bompok Islands. *J. Andaman Sci. Assoc.* **5:**69–71.

Dalzell, N.A. and Gibson, A. 1861. The Bombay flora. Bombay.

Dastur, J.F. 1951. Medicinal plants of India and Pakistan. Taraporveala Sons and Co. Ltd., Bombay.

Datta, S.C. and Mukerji, B. 1950. Pharmacognosy of Indian Rhizomes and Root Drugs. Vol.: I-II Ministry of Health, Government of India, Pharmcognosy Laboratory, Calcutta.

Datta, S.C. and Mukerji B. 1952. Pharmacognosy of Indian Leaf Drugs. Ministry of Health. Government of India, Pharmcognosy Laboratory, Calcutta.

Datta, S.K. and Datta, P.C. 1980. Study of Bark Drugs of Laticiferous Family I (Apocynaceae) *Proc.* 3rd *All Indian Botanical Conference, Sec. IX. Ethnobotany.* 167.

David, M. Bates. 2001. Ethnobotany: Mindsets: Externalities and challenges. *Phytomorphology.* (Golden Jubilee Issue). 31–38.

De Mello, J.F. 1980. Plants in traditional medicine in Brazil. *J. Ethnopharmac.* **2(1):**49–55.

Deshpande, S.; Sharma B.D. and Nayar M.P. 1993–95. Flora of Mahabaleshwar and it's adjoining Maharashtra. **BSI.** Fl. India Ser. 3, vol. pp. 431 pl. 4, f. 38. 1993; vol. 2. pp. 433–776, pl. 4. f. 18. 1995.

Deshpande, S.; Sharma, B.D. and Nayar, M.P. 1993–95. Flora of Mahabaleshwar and its adjoinings, Maharashtra. **BSI.** Fl. India ser. 3, vol. p. 431, pl. 4, f. 38. 1993; vol. 2. pp. 433–776, pl. 4. F. 18. 1995.

Desmukh, V.K.; Godbole, S.R. and Mahabale, T.S. 1957. Studies on pharmacognosy of some Indian medicinal plants. Indian Drug Research Association, Poona. 9–10.

Dev, S. 1999. Ancient-modern concordance in Ayurvedic plants: Some examples. *Environ Health Prospect.* **107:**783–789.

Dey, K.L. 1896. Indigenous Drugs of India. Thacker Spink and Co., Calcutta.

Diwakar, P.G. and Sharma, B.D. 1999. Flora of Buldhana District, Maharashtra Sate. BSI. Calcutta.

Doleres, L. and Latorrre, F.A. 1977. Plants used by Mexican Kickapor Indian. *Econ. Bot.* **31**:340–357.

Duke, J.A. 1985. Medicinal Plant science, **229**:1036.

Dutt, U.C. 1877. Materia Medica of Hindus, with a Glossary of Indian Plants by King and author. Thacker, Spink & Co., Calcutta.

Dutta, U.C. 1877. Materia Medica of Hindus. Calcutta.

Dymock, W. 1883. The Vegetable Materia Medica of Western India. Education Society's Press, Byculla, Bombay.

Dymock, W. and Gadgil, N.K. 1883. A glossary of the vernacular names of the principal plants and drugs (Indigenous and imported) found in Bombay and on the Western Coast of India. Bombay, pp. 104.

Dymock, W.; Warden, C.J.H. and Hooper, D. 1989–93. Pharmacographia Indica. Vols. **3**: London.

Eisenberg, D.M.; Davis, R.B.; Ettner, S.L.; Apple, S.; Wilkey, S.; Van Rompey, M. and Kessler, R.C. 1998. Trends in alternative medicine use in the United States. 1990–1997: Results of follow up national survey. *JAMA*. **280**:1569–1575.

El, Ghazali, G.E.B. 1986. Medicinal plants of the Sudan. Part **1**. Medicinal plants of Erkowit. Khartoum: National Council for Research, Khartoum.

El, Ghazali, G.E.B.; El, Tohami, M.S. and El, Egami, A.A.B. 1994. Medicinal plants of the Sudan. Part **3**. Medicinal plants of the White Nile provinces. Khartoum: National Centre for Research, Khartoum.

Esau, K. 1965. Plant anatomy (2nd ed.). John Wiley & Sons, New York.

Fadzil, M.F.B.A. 2000. Current scenario of the Malaysian herbal/natural products industry. *J. Trop. Med. Plants*. pp. 36–42.

Farnsworth, N.R. and Soejarto, D.D. 1991. Global importance of Medicinal plants In: The conservation of medicinal Plants. Cambridge University Press, Cambridge, UK.

Farnsworth, N.R. and Bingle, A. 1977. New natural products and plant drugs with pharmaceutical, biological or therapeutically activity (eds.) Wagner, H. and Wolff. P. Springer–verlag, New York.

Farnsworth, N.R. and Bunyapraphatsara, N. (eds.) 1992. Thai medicinal plants recommended for primary health care system. Medicinal Plant Information Center, Mahidol University, Bangkok .

Felming, J. 1810. A catalogue of Indian Medicinal plants and Drugs. Printed at the Hindustani Press by A.H. Hubbard, Calcutta.

Feng, P.C.; Haynes, L.J.; Magnus, K.E. and Plimmer, J.R. 1964. Further pharmacological screening of some West Indian medicinal plants. *J. Pharm. Pharmacol.* **16**:115.

Fielder, Mildred. 1975. Plant medicine and folklore. Winchester Press. New York.

Fluckiger, F.A. and Hanbury, D. 1879. Pharmacographia a history of Principle drugs of vegetable origin, met in Great Britain and British India. Macmillan and Co., London, p. XX+803.

Fuller, D. 1991. Medicines from the wild. *TRAFFIC* USA: Washington DC. USA.

Gammie, G.A. The orchids of Bombay Presidency. *J. Bombay nat. Hist. Soc.* **16**:429–433, 562–569; **17**:31–37, 940–942; **18**:88–91, 586–590, 833–834; **19**:139–144, 624–626; **20**:126–129, 597–602; **21**:171–174, 1129–1130. 1905–1912.

Garcia De Orta. 1563. Collquios dos simples e Drogas e Cousas medicinaes da India e Assi de Algumas Fructas Achadas Nella, Goa.

Ghazanfar, S.A. 1994. Handbook of Arabian medicinal plants. CRC Press, Boca Raton, Fla.

Ghosh, B.N. 1951. Need for pharmacognostic study of Indian plants of medicinal value. *Proc. Symposium on Indian Medicinal plants.* Bengal Immunity Research Institute, Calcutta. p. 15–16.

Giberti, G.C. 1983. Herbal folk medicine in Northwestern Argentina: Compositae. *J. Ethnopharmacol.* **7(3)**:321–341.

Gibson, A. 1852. A Report on the teak and other plantation and forests in the Bombay Presidency. Bombay.

Gibson, A. 1863. A handbook to the forests flora of Bombay Presidency. Bombay.

Gilg, E.; Brandt, W. and Schurhoff, P.N. 1927. Text book of pharmacognosy (4th ed.). Julis Springer, pp. XIV+530.

Giron, L.M.; Freire, V.; Alonzo, A. and Caceres, A. 1991. Ethnobotanical survey of the medicinal flora used by the Caribs of Guatemala. *J. Ethnopharmacol.* **34(2/3)**:173–187.

Githens, S.T. 1949. Drug Plants of Africa, African handbook. University of Pennsylvania Press, Philadelphia.

Graham, J. 1839. A Catalogue of the plants growing in Bombay and it's vicinity, spontaneous, cultivated or introduced as far as have been ascertained. Bombay.

Gram, B. 1927. Text book of pharmacognosy. Published by author, Kobenhav. pp. 369, 338 fig.

Gramling, L.G. 1942. Some methods of presentation in a course in pharmacognosy. *Indian J. Pharm.* **4**:116–18.

Gupta, M.P.; Arias, T.D.; Correya, M. and Lamba, S.S. 1979. Ethnopharmacognostic observations on Panamanian medicinal plants. 1. *Q. Jl. Crude, Drug Res.* **17(3/4)**:115–30.

Gupta, M.P.; *et al.*, 1979. Ethnopharmacognostic observations on Panamanian medicinal plants. Part I. *Q. Jl. Crude, Drug Res.* **17(3–4)**:115–130.

Haines, H.H. 1912–1914. List of the trees, shrubs and economic herbs of the Southern forest circle of the Central provinces. *Indian Forester*. **38**:495–509. 1912; **39**:49–69. 1913; **40**:194–229, 264–283, 330–335, 392–403, 429–449, 472–502. 1914.

Haines, H.H. 1916. Descriptive list of trees, shrubs and economic herbs of the southern Circle, Central Provinces. Allahabad.

Hajra, P.K.; Rao, R.R.; Singh, D.K. and Uniyal, B.P. 1995. *Flora of India*. BSI, Calcutta Asteraceae (Anthemideae-Heliantheae): **I-VII**, 1–454.

Hajra, P.K.; Rao, R.R.; Singh, D.K. and Uniyal, B.P. 1995. *Flora of India*. BSI, Calcutta Asteraceae (Inuleae-Vernoieae): **I-VII**, 1–411.

Hajra, P.K.; Sharma, B.D.; Sanjappa, M. and A. Sastry, R.K. 1996. Flora of India Introductory volume Part-1. BSI. **I-XVI**, 1–538.

Hajra, P.K.; V.J. Nair and Daniel, P. 1997. Flora of India. BSI, Calcutta (Malpighiaceae-Dichapetalaceae): **I-XII**, 1–561.

Halberstein, R.A. and Saunders, A.B. 1978. Traditional medical practices and medicinal plant usage on a Bahamian Island. *Cul. Med. Psychiat.* **2**:177–203.

Handa, S.S. and Kapoor, V.K. 1990. Pharmacognosy. Vallabha Prakashana, New Delhi.

Handa, S.S. 1993. Standardization quality of medicinal plant products. *G-15. GEBMAP. Newsletter*. **3 & 4**:13–17.

Handa, S.S.; Surange, S.R. and Kanitkar, U.K. 1968. Saptrangi from Kolhapur area and its identification. *J. Univ. Poona Sci. Technol.* **34(suppl.)**:51–54.

Harnischfeger, G. 2000. Proposed guidelines for commercial collection of medicinal plant material. *J. Herbs, Spices, & Med. plants*. **7**:43–50.

Hart, J.A. 1979. The ethnobotany of the Flathead Indians of Western Montana. *Botanical Museum Leaflet.* Harvard University. **27(10)**.

Hartwell, J.L. 1967–71. Plants used against cancer. A survey. *Lloydia*. pp. 30–34.

Heilbronn, E. 1953. Pharmakognostische monographien turkischer arzneipflanzen. *Pharmazie*. **8(10)**:94–98.

Heizer, Fleming and Elsasser, and Albert B. 1980. *The Natural World of the California Indians*. University of California Press. Berkeley.

Heywood, V. 1991. Botanic Gardens and the conservation of medicinal plants. Conservation of medicinal plants. Cambridge University press, Cambridge, UK. pp. 213–228.

Heywood, V. 1992. Botanic gardens and conservation: New perspectives. *Opera. Bot.* **113**:9–13.

Hocking, G.M. 1955. A Dictionary of terms in pharmacognosy and other divisions of Economic Botany. Charles C. Thomas, XXVI+284.

Hodge, W.H. and Taylor, D. 1956. The Ethnobotany of the Island Caribs of Dominica. *Webbia.* **12**:513–644.

Holdsworth, D. and Lacanienta, E. 1981. Traditional medicinal plants of the central providence of Popua, New Guinea in *Q.J. Crude Drug Res.* **19**:147–167.

Holdsworth, D. 1974. A phytochemical survey of medicinal plants in Papua New Guinea. *Sci. New Guinea.* **2(2)**:142–154.

Holdsworth, D. 1975. Medicinal plants from Milne Bay, eastern Papua. *Sci. New Guinea* **3(2)**:98–102.

Holdsworth, D. 1984. Phytomedicine of the Madang Province, Papua New Guinea Part I. Karkar Island. *Int. J. Crude Drug Res.* **22(3)**:111–119.

Holdsworth, D.K. 1977. Medicinal plants of Papua New Guinea. Techn. Pap. S. Pacific Commis.

Holdsworth, D.K. 1991. Traditional medicinal plants of Rarotonga, Cook Islands. Part **II**. *Int. J. Pharmacog.* **29(1)**:71–79.

Hooker, J.D. *et al.,* 1872–1897. The Flora of British India London. Vols. **7**.

Hooper, David. 1937. Useful plants and drugs of Iran and Iraq. Publication **9(3),** 387, Field Museum of Natural History, Chicago.

Houghton, P.J. 1994. Herbal authentication. Paper presented at the Meeting "Towards the safer use of traditional Remedies". Abstract. p. 18. (National poisons unit: The royal Botanic Gardens, Kew, UK).

Husain, A. 1993. Medicinal plants and their cultivation. CIMAP, Lucknow.

Hutchens, Alma R. 1992. A Handbook of Native American Herbs. Shambhala. Boston & London

Hutchinson, J. 1973. The families of Flowering Plants. (3rd ed.). Clarendon press, Oxford.

Ichimura, T. 1932. Important medicinal plants of Japan. Kanazawa, Japan.

Iglesias, G. 1985. Hierbas medicinales de los Quichaus del Napo. Quito: Ediciones Abya-yala.

Inamdar, A.C. and Mahabale, T.S. 1980. Comparison between Shatavar and *Asparagus* spp. *Biovigynam.* **6(1)**:27–35.

Institute of Chinese Materia Medica, China Academy of Traditional Chinese Medicine 1992. Medicinal plants in China: a selection of 150 commonly used species. Manila: World Health Organization, WHO regional publications, Western Pacific Ser.: **2**.

Iwu, M.M. 1993. Handbook of African medicinal plants. Fla: CRC Press, Boca Raton.

Iyengar, M.A. 1977. Bibliography of investigated medicinal plants (1950–75). College of Pharmacy, Kasturba Medical college, Manipal.

Iyengar, M.A. and Nayak, S.G. 1984. Anatomy of crude drugs. Manipal Power Press, Manipal.

Jackson, B.P. and Snowdon, D.W. 1969. Powdered vegetable drugs: An Atlas of Microscopy for use in identification and authentication of some plant material employed as medicinal agents. American Elsevier Publishing company, Inc: New York, N.Y., VIII+203p.

Jain S.K. 1968. India the land and people: Medicinal plants. National Book Trust, New Delhi. India, XII+176 pp.

Jain, S.K. (ed.) 1981. Glimpses of Indian ethnobotany. Oxford and IBH Publishing Co., New Delhi.

Jain, S.K. 1963. Studies in Indian ethnobotany-less known uses of fifty common plants for the tribal areas of Madhya Pradesh. *Bull. Bot. Surv. India.* **5(324)**:223–26.

Jain, S.K. 1963a. Studies in Indian ethnobotany-plants used in medicine by tribal of Madhya Pradesh in *Bull. Rec. Rec. Lab., Jammu.* **1**:126–128.

Jain, S.K. 1965. Medicinal plant lore of the Tribal of Bastar. *Economic Botany.* **19**:236–250.

Jain, S.K. 1968. *Medicinal plants*. National Book trust, New Delhi.

Jain, S.K. 1991a. Contribution to Indian ethnobotany. Scientific Publisher, Jodhpur.

Jain, S.K. 1991b. Dictionary of Indian folk medicine and ethnobotany. Deep Publications, New Delhi.

Jain, S.K. 1997. Cultural dimensions of biodiversity. *Proc. Indian natn. Sci. Acad. B.* **62**:449–466.

Jain, S.K. 2000. Human aspects of Plant diversity. *Econ. Bot.* **54**:459–470.

Jain, S.K. 2001. Ethnobotany in modern India. *Phytomorphology.* (Golden Jubilee Issue). pp. 39–54.

Jain, S.K. and De Filipps, R.A. 1991. Medicinal plants of India. 2 vols. Reference Publications, Algonac Michigan.

Jain, S.K. and Rao, R. 1960. A Handbook of field and herbarium methods. Today and Tomorrow's publishers, New Delhi.

Jain, S.K. and Saklani, A. 1992. Cross-cultural ethnobotanical studies in Northeast India. *Ethnobotany.* **4**:25–38.

Jain, S.K.; Banerjee, D.K. and Pal, D.C. 1976. Medicinal plants among certain Adibasis in India. *Bull. Bot. Surv. India.* **15(1–2)**:85–91.

Jain, S.K.; Vera Froes Fernandes; Sneh Lata and Ayub, A. 1995. Indo-Amazonian ethnobotanic connections-similar uses of some common plants. *Ethnobotany.* **7**:29–37.

Janardhanan, K.P. (1963) 1964. An Enumeration of medicinal plants of Khed taluka (Maharashtra State). *Bull. Bot. Surv.* India. **5**:363–374, map 1.

Janardhanan, K.P. (1963) 1965. An Enumeration of medicinal plants of Khed taluka (Maharashtra State). *Poona Agri. Coll. Mag.* **55**:38–46, map 1.

Johanson, D.A. 1940. Plant microtechnique. MCGraw Hill, New York.

Jones, V.H. 1941. The Nature and Status of Ethnobotany. *Chron. Bot.* **6**:219.

Kadens, Joseph, M. 1970. Modern Encyclopedia of herbs. Parker Publishing Co. Inc. West.

Kamble, S.Y. and Pradhan, S.G. 1988. Flora of Akola district, Maharashtra. **BSI**. Fl. India ser. **3**, pp. xix+320, pl. 4, fig. 2.

Kameswara Rao C. 2000. Material for the database of medicinal plants. Karnataka State Council for Science and technology.

Kantikar, U.K. 1970. Identification of *Asparagus racemosus* Willd. J. Univ. Poona, Sci. *Technol. Sec.* **38**:5–11.

Kantikar, U.K. and Pendse, G.S. 1969. The source of commercial Shatavar from southern India. *Indian J. Pharm.* **31**:180.

Kantikar, U.K.; Dange, P.S. and Pendse, G.S. 1969. A Pharmacognostic studies on hitherto unrecorded form of *Asparagus racemosus* Willd. (Shatavar*). J. Res. Indian Med.* **3**:123.

Kapoor, L.D. and Chopra, I.C. 1961. The Ancient History and Evolution of the Present Indigenous Drugs in India.

Karnick, C.R. 1966. Some medicinal plants from Satpura Mountains. *Indian Forester* **92**:173–183.

Karnick, C.R. 1968. True identity of soma plant the famous Ayurvedic rejuvenating and longivity promoting drug. *Q. Jl. Crude Drug Res.* **9(4)**:1473–77.

Karstem, G. and Weber, U. 1949. Text book of pharmacognosy. Gustav Fischer, Jena, 414p., 622 illus.

Kaushik, P. and Dhiman, A.K. 2000. Medicinal plants and raw drugs of India. Bishen Singh Mahendra Pal Singh, Dehradun, India.

Kay, L.A. 1939. The microscopical study of crude drugs. Williams and Wilkines Co., Baltimore, pp. Viii+228.

Khosa, R.L. and Prasad, S. 1971. Pharmacognostical Studies on Guduchi (*Tinospora cordifolia* Miers.). *J. Res. Indian Med.* **6**(3): 261–69.

Kim, T.J. 1994. Korean medicinal plants in the wild. GUK-IL Publishing, Seoul.

Kimura, K. and Noro, Y. 1965a. Pharmacognostical studies on Chinese drug 'Gui-Sui-bu. **II.** Anatomical studies on Commercial 'Gui-Sui-bu. *Syoyakugku Zasshi.* **19(2)**:85–98.

Kimura, K. and Noro, Y. 1965b. Pharmacognostical studies on Chinese drug 'Gui-Sui-bu. *Syoyakugku Zasshi.* **19(2)**:99–110.

Kindscher, K. 1992. Medicinal wild plants of the Prairie. University of Kansas Press, Lawrence Kansas. p. 340. [EB 633.88(13.4)]

King and Dutt, U.C. 1877. Materia Medica of the Hindus with a glossary of Indian plants. Thacker, Spink & Co., Calcutta.

Kinghorn, A.D. and Seo, E.K. 1996. Cultivating the pharmacopoeia. *Chemtech.* **26**:46–54.

Kirtikar, K.R. and Basu, B.D. 1935. Indian Medicinal Plants Vol.: **I-IV**. Lalit Mohan Basu, Allahabad.

Krishnamurthy, K.H. 1971. Botanical identification of ayurvedic medicinal plants: A new method of pharmacolinguistics. A preliminary account. *Indian J. Med. Res.* **59(1)**:90–103.

Kulkarni, B.G. 1988. Flora of Sindhudurg. BSI. Fl. India ser. **3,** pp. xx+320, pl. 11, fig. 6.

Kulkarni, D.K. and Kumbhojkar, M.S. 1992. Ethnobotanical studies on Mahadeokoli tribe in Western Maharashtra-Part **I,** Cordage Plants; Part-**II** Fodder Plants; **III,** on Conventional wild edible fruits. *J. Econ. Taxo. Bot.* Addl. Ser. **10:**11–115, 123–128; 151–158.

Kundu, B.C. and Gupta, B. 1967. Importance of taxonomic studies of medicinal plants. *Indian J. Pharm.* **29**:340.

Kuo-Chin, H.; Ling-Wen, C.; Yung-Luan, C.; Jing-Han, L. and Luo-San, H. 1976. The study of microscopical identification of powdered Chinese drugs. IV. *Acta Bot. sin.* **16(4)**:342–53.

Kurup, R.N.V.; Ramdas, V.N.K. and Joshi, P. 1979. Handbook of medicinal plants. New Delhi.

Lambert, J.; Srivastava, J. and Vietmeyer, N. 1997. Medicinal Plants: Rescuing a global heritage. World Bank technical Paper No. 355. The World bank, Washington, D.C.

Lange, D. 1998. Europe's medicinal and aromatic plants: Their use, trade and conservation. *TRAFFIC* International, Cambridge, UK.

Lange, O. 1996. Medicinal plant market study in Germany-state of Project. Medicinal plant conservation *Newsletter* of the IUCN species survival commission-Medicinal Plant specialist group. **2**:9–10.

Langley, L. (ed). 1995. Royal horticultural society Encyclopedia of herbs and their uses. Dorling Kindersley, London.

Lassak, E.V. and McCarthy, T. 1983. Australian medicinal plants. North Ryde Methuen, Australia.

Launert, E. 1981. The Hamlyn guide to edible and medicinal plants of Britain and northern Europe. London.

Leaman, D.J. 2001. Conservation, trade, sustainability and exploitation of medicinal plant species in Development of plant based: Conservation, efficacy and safety (ed.) Saxena, P.K. Kluwer Academic Publ. Boston. pp. 1–15.

Li, M.; Feng, Y.X. and Xiao, P.G. 1991. Comparative anatomical studies on Chinese medicinal plants of the genus *Aconitum*. *Acta Phytotax. Sin.* **29(2)**:97–112.

Lien, W.Y. 1975. A preliminary study of menispermaceous medicinal plants in China. *Acta Phytotax. Sin.* **13(1)**:53–55.

Lin, C.C. and Kan, W.S. 1990. Medicinal plants used for the treatment of hepatitis in Taiwan. *Amer. J. Chinese Med.* **18 (1/2)**:35–43.

Lindley, J. 1838. Flora Medica. A Botanical account of all the more important Plants used in Medicine in different parts of the world, Longma, Orme, Brown, Green and Logman, pp. 655+XIII.

Ling, Y.R. 1995. A new compendium of Materia Medica: pharmaceutical botany and China medicinal plants. Science Press, Beijing.

Linington, S.H. and Pritchard, H.W. 2001. *Gene Banks in encyclopedia of Biodiversity*. Vol **3**. pp. 165–181. Academic Press, San Diego, California, USA.

Lipp, F.J. 1971. Ethnobotany of the Chinantech Indians Oxaca Mexico *Econo. Bot.*

Lishihichen. 1590. The Plants of China and their usefulness to man Smithsonia Institution, Washington.

Lorenzi, H. 1991. Plantas Daninhas do Brasil. Brazil: Nova Odessa.

Madge, A.G.M. 1987. Alternative complementary medicine. Marshell Publications, Devon, UK.

Maheshwari, J.K. (ed). 2000. Ethnobotany and medicinal plants of (the) Indian subcontinent. Scientific Publishers, Jodhpur, India.

Maheshwari, J.K. (ed.) 1996. Ethnobotany in South Asia. Scientific Publishers, Jodhpur, India.

Mahunnah, R.L.A. 1992. Some medicinal plants of the southern highlands of Tanzania. **15(3):**557–563.

Malhotra, S.K. and Moorthy, S. (1973) 1976. Some useful and medicinal plants of Chandrapur district (Maharashtra State) *Bull. Bot. Surv. India* **15:**13–21.

Martin, G.J. 1995. Ethnobotany: A methods manual. Chapman and Hall, London. UK.

Mehra, P.N. and Handa, S.S. 1969. Pharmacognosy of Saptrangi, antibiotic drug of Indian origin. *Res. Bull. Punjab Univ.,* **20(3–4):**487.

Mehra, P.N.; Bhatnagar, J.K. and Handa, S.S. 1969. Researches in Pharmacognosy in India. *Res. Bull. Punjab Univ.* (India) Vol. **20 (Pt. III and IV):**261–337.

Metcalfe, C.R. and Chalk. 1950. Anatomy of the Dicotyledons. Vol.: **I-II**, Oxford University Press, Oxford.

Meyer, Joseph E. 1960. The Herbalist. Publisher not listed. Copyright by Clarence Meyer.

Micallef, H. 1977. Medicinal plants in Malta. In Haslam, S.M., Sell, P.D. and Wolseley, P.A. A flora of the Maltese Islands. Msida, Malta University Press.

Mitra, Roma. 1979. Pharmacognostical Studies on Some indigenous Drugs. Progress in Plant Research Silver Jubilee Publications, Vol.: I NBRI, Lucknow.

Monachino, J. 1956. Chinese herbal medicine-recent studies. *Econ. Bot.* **10(1):**42–48.

Moore, Michael. 1979. Medicinal plants of the Mountain. Museum of New Mexico Press. Santa Fe, New Mexico.

Moore, Michael. 1989. Medicinal plants of the Desert and Canyon West. Museum of New Mexico Press. Santa Fe, New Mexico.

Morton, J. 1975. Current folk remedies of Northern Venezuela. *Q. Jl. Crude Drug Res.* **13(3–4)**:97–121.

Morton, J. 1977. Some folk medicines plants of Central American Markets. *Q. Jl. Crude Drugs Res.* **15(4)**:165–92.

Morton, J.F. 1977. Major medicinal plants: Botany, Culture, and Uses. Charles C. Thomas, Illinois. North America.

Morton, J.F. 1981. Atlas of medicinal plants of Middle America, Bahamas to Yucatan. Charles C. Thomas Publ, Springfield.

Mukerji, B. 1953. Indian Pharmaceutical Codex Vol.: I-II. CSIR, New Delhi.

Mukerji, B. and Datta, S.C. 1951. Common adulterants and substitutes of crude drugs of the Indian market. *Proc. symposium on Indian Medicinal plants*. Bengal Immunity Research institute, Calcutta. pp.22–26.

Nadkarni, A.K. 1954. K.M. Nadkarni's Indian Materia Medica Vol.: I-II Popular Book Depo. Bombay.

Naegele, T.A. 1996. *Edible and medicinal plants of the Great Lakes region.* Davisburg, Mich.: ilderness Adventure Books.

Naik, V.N. 1998. Flora of Marathwada, Vol.: I-II. Amrut Prakashan, Aurangabad.

Naik, V.N. 1998. Taxonomy of Angiosperms. Tata Mc-graw-Hill Publications, New Delhi.

Naik, V.N. 1998. Vanaushadhi (Marathi). Amrut Prakashan, Aurangabad.

Nairne, A.K. 1894. The flowering plants of Western India. Bombay and London.

Namba, T.; Tada, I. and Takahashi, S. 1967. Pharmacognostic studies on water soluble seed gums. *Syoyakugaku Zasshi*, **21(10)**:47–56.

Naranjiro. 1995. The urgent need for the study of medicinal plants. In Ethnobotany: Evolution of discipline (eds.) Schultes, R.E. and Rej. S. Vo, Chapman and Hall, London. pp. 362–368.

Narayana Aiyer, K. 1953. Patala, Pharmacognosy of Ayurvedic drugs of Travancore-Cochin. Central Research Institute, Trivandrum. **1(2)**:49–51.

Narayana Aiyer, K. 1953. Syonakh. Pharmacognosy of Ayurvedic Drugs of Travancore Cochin. Central Research Institute. Trivandrum. **1(2)**:39–49.

Narayana Aiyer, K. and Kolammal, M. 1963. Guduchi: *Tinospora cordifolia* Miers. Pharmacognosy of Ayurvedic *Drugs*, Kerala, Trivandrum. **1(7)**:13–31.

Natesh, S. 1999. Conservation of medicinal and aromatic plants in India — an overview. Conservation of medicinal and aromatic plants: Strategies and technologies. pp. 1–11. Forest research Institute: Malaysia.

Natesh, S. 2000. Biotechnology in the conservation of medicinal and aromatic plants. Biotechnology in Horticulture and plantation crops. pp. 548–561 Malhotra publishing House, New Delhi, India.

Natesh, S. 2001. The changing Scenario of herbal drugs: Role of botanists. *Phytomorphology.* (Golden Jubilee Issue). pp. 75–93.

Natesh, S. and Mohan Ram, H.Y. 1999. An Update of green medicine. *J. Indian Bot. Soc.* **78:**13–33.

Nayar, M.P. and Sastry, A.R.K. 1887. Red Data Book of Indian Plants. Vol. **I-III. BSI** Calcutta.

Nayar, R.C.; Pattan Shetty, J.K.; Mary, Z. and Yoganarasimhan, S.N. 1978. Pharmacognostical studies on the roots of *Decalepis hamiltonii* Wt. & Arn. and comparison with *Hemidesmus indicus* R. Br. *Proc. Indian Acad. Sci.* **87B:**37–48.

Nguyen Van Duong. 1993. Medicinal plants of Vietnam. Cambodia and Laos.

Noire, P.D. and Ma, T.S. 1975. Microchemical Investigation of Medicinal Plants: **XII** the Chinese Herbal drug. *San ch I Mikrochin. Acta.* **2(415):**455–61.

Nwosu, M.O. and Omaliko, C.G. 1993. An anatomic-systematic study of medicinal plants of Nigeria: **1.** *Ocimum gratissimum* Linn. (Lamiaceae). *J. Pl. Anat. Morphol.* **6(2):**107–121.

Oliver, and Bever, B. 1982. Medicinal plants in tropical West Africa: **1.** Plants acting on the cardiovascular system. *J. Ethnopharmacol.* **5(1):**1–72 .

Pal, D.C. and Jain, S.K. 1998. Tribal Medicine. Naya Prakashan, Calcutta.

Patel, R.I. 1968. Forest flora of Melghat. Bishen Singh Mahendra Pal Singh, Dehradun. pp. xlviii+380.

Pereira; Sheteolu, A.O. 1992. Taxonomy of medicinal plants: Foliar epidermal characters in the genus *Monodora* (Annonaceae). *Feddes Repert.* **103(5–6):**375–379.

Perry, L.M. 1980. Medicinal plants of east and Southeast Asia: attributed properties and uses. etc., The Massachusetts Institute of Technology Press, Cambridge.

Pongparayoon, U. 1991. Pharmacognostic studies on Thai Medicinal plants *Ipomoea pes-caprae* (L.) R. Br. (Pak Bung Ta Lae). *Acta Pharm. Nordica.* pp. **3(3):**186.

Powers, Stephen. 1976. Tribes of California. University of California Press, Berkeley.

Prasad, S. 1950. Role of pharmacognosy in the development of drug industry. *Chemical age Series* **I:**135–138.

Prasad, S. and Wahi, S.P. 1965. Pharmacognostic investigations on Indian Sarasaparilla Part-I root and root stock of *Hemidesmus indicus* R. Br. *Indian J. Pharm.* **27(2):**35–39.

Pravin K. Saxena. 2001. Development of Plant based medicines: Conservation, efficacy and safety. Kluwer Academic publishers Dordrecht/Boston/London.

Puri, H.S. 1971. Anatomical Studies on Indian Medicinal plants. Review. *Herba hung.* **10(10)**:47–62.

Puri, H.S. 1972. Quantitative Microscopy of Herbaceous Stems. Q. *Jl. Crude Drugs. Res.* **12(4)**:1633.

Pushpangadan, P. (undated) Ethnobiology in India: A status report. Ministry of Environment and forests. Government of India, New Delhi, India.

Rai, L. and Sharma, E. 1994. Medicinal plants of the Sikkim Himalaya: status, usage and potential. Bishen Singh Mahendra Pal Singh, Dehradun.

Rao, R.R. and Sharma, B.D. 1990. A manual for herbarium collections. BSI. Calcutta.

Rastogi, P.R. and Mehrotra. 1993. Compendium of Indian Medicinal Plants Vol.: I-III CDRI and Publication and Information Directorate, New Delhi.

Rawls, R. 1996. Europe's strong herbal brew. Chem & Eng News pp. 53–60.

Rheede, H.A. van. 1678–1693. Hortus Malabaricus vol. 1–12. Amsterdam.

Riaz, M.; Qamar, S. and Chaudhary, F.M. 1995. Chemistry of the medicinal plants of the genus *Acorus* (family Araceae). *Hamdard Med.* **38(2)**:50–62.

Rizvi, M.A., Ahmed, L. and Sarwar, G.R. 1996. Wild medicinal plants of Madinatal Hikmah and its adjacent areas: part 2. *Hamdard Med.* **39(4)**:8–10.

Roxburgh, W. 1820 and 1824. Flora indica, Vol. 1–2. Serampore.

Santapau, H. 1952. The Acanthaceae of Bombay. *Univ. Bombay Bot. Mem.* **2**:iii+104, fig. 16.

Santapau, H. 1953. The Flora of Khandala on Western Ghats of *India Rec. Bot. Surv. India.* **16(1)**:xxvii+396, map, fig. 3. (Rev. ed. 1967).

Santapau, H. 1955. Botanical collectors manual. Govt. Press, New Delhi.

Santapau, H. 1958d. The flora of Purandhar or an enumeration of all Phanerogamic plants discovered in Purandhar during the years 1944–56. New Delhi. pp. 158.

Santapau, H. 1962. The Asclepiadaceae and Periplocaceae of Bombay. *Univ. Bombay Bot. Mem.* **4**:iv+118, tab.12.

Santapau, H. and Henry. 1973. A dictionary of flowering plants of India. CSIR. New Delhi. pp. viii+198.

Santapau, H. and Kapadia. 1966. The Orchids of Bombay. pl. 54. Manager of Publication, Delhi. pp. vi+239.

Sanyal, M.N. and Namhata, D. 1995. A census of medicinal plants from Bankura district, West Bengal. *J. Econ. Taxon. Bot.* **19(2)**:435–442.

Sanyal, P.K. 1951. Adulteration and spuriousness of crude drugs sold in Indian Market and detection of the same. *Proc. Symposium on Indian Medicinal plants.* Bengal Immunity Research Institute, Calcutta. pp. 30–31.

Sarin, Y.K. 1996. Illustrated Manual of Herbal Drugs Used in Ayurveda. CSIR and ICMR New Delhi.

Satyavati, G.V. and Gupta, A.K. 1976–1987. Medicinal Plants of India. Vol.: **I-II.** ICMR, New Delhi.

Schultes, R.E. 1941. Ethnobotany on alcancey sus objects. *Caldesia* **3**:7–12.

Schultes, R.E. 1956. The strange narcotic snuffs of eastern Colombia: Their sources, preparation and effect on American Botanist. *London News.* **229**:520–521.

Schultes, R.E. 1960. Tapping our heritage of ethnobotanical lore. *Econ. Bot.* **14**:257–262.

Schultes, R.E. 1962. The Role of Ethnobotanist on search for new medicinal plants. *Lloydia.* **25(4)**:257–266.

Schultes, R.E. 1963. The virgin field in the psychoactive plant research. *Ethnobotany.* **5**:5–61.

Schultes, R.E. 1963a. Hallucinogenic plants of the new World. *Harward Rev.* **I**:18–32.

Schultes, R.E. 1963b. Plantae Colombianae-13 plants as oral contraceptives in the West Amazon. *Lloydia.* **26**:67–74.

Schultes, R.E. 1987a. Ethnopharmacological conservation's: A key to progress in medicine *Opera Botanica*, **92**:217–224.

Schultes, R.E. 1987b. Coca and other psychoactive plants. Magico-religeous roles in primitive societies of the New World in clinical and behavioural aspects. I. Fisher, A. Raskin and E.H. Uhlenhunth (Ed.). pp. 212–249.

Schultes, R.E. 1990. De Plantis toxiariis e Mundo Novo Tropicale commentations XL. *Justicia* as a source of a hallucinogenic snuff in *Econ. Bot.* **44**:61–70.

Schultes, R.E. 1992. Unusual and promising aboriginal food Plants of Northeastern Amazon. *Ethnobotany* **4**:7–29.

Schultes, R.E. and Raffauf, R.F. 1990. The healing forest. Oregon: Dioscorides Press, Portland.

Schultes, R.E. and Ven Reis, S. 1995. Preface in Ethnobotany: Evolution of Discipline. Dioscorides Press, Portland, USA.

Seychelles. Ministry of Education and Culture 1994. *Medicinal plants of Seychelles*. National Arts Council, Seychelles.

Sharma, B.D.; Kartikeyan, S. and Singh, N.P. (eds.) 1996. Flora of Maharashtra State: Monocotyledones. **BSI**, Calcutta.

Sharma, P.C.; Yelne, M.B. and Dennis, T.J. 2000–2001. *Database on Medicinal plants*. Central Council of Research in Ayurveda and Siddha, Vol. **1–3**. New Delhi.

Sharma, S. 1979. Realms of Ayurveda. Arnold-Heinemann, New Delhi.

Sharma, U. 1992. Complimentary medicine today. Tavistock/Routledge. London, UK.

Sheriff, M. 1978. A catalogue of Indian synonyms of the medicinal plants, products, inorganic substances. Periodical Experts Book Agency and Dehradun International Book Distributors, Delhi.

Shimomura, T. 1952. Microscopical anatomy of powdered vegetable drugs in the Japanese Pharmacopoeia IV. *J. Jap. Bot.*, **27(12)**:362–67.

Shimomura, T. and Kurokawa, H. 1955. Microscopical anatomy of powdered vegetable drugs in Japan (5). *J. Jap. Bot.* **31(2)**:51–55.

Shipek, Florence Connolly. 1991. Delfina Cuero Her Auto-biography an account of her last years and Her Ethnobotanic Contributions. Balena Press, Menlo Park, CA.

Sing, N.P.; Lakshminarasimhan, Karthikeyan S. and Prasanna P.V. (eds.) 2000. Flora of Maharashtra State: Dicotlydones: **I** (Ranunculaceae-Rhizophoraceae) 1–898. **BSI**, Calcutta.

Sing, N.P.; Lakshminarasimhan, P.; Karthikeyan, S. and Prasanna, P.V. (eds.) 2001. Flora of Maharashtra State: Dicotlydones:**II** (Combretaceae-Ceratophyllaceae) 1–1080. **BSI**, Calcutta.

Singh, B. 1955. Bihar ki Vanaspatiyan, 86 and 88, Sri Baidyanath Ayurveda Bhawan, Calcutta.

Singh, H.B. and Vishwanathan, M.V. 2000. Need for authentication of samples of crude drugs. *JMAPS*. **22/4A** & **23/1A**:612–618.

Singh, N.P. and Karthikeyan, S. (eds.) 2000. *Flora of Maharashtra State*: *Dicotyledones*: **I** (Ranunculaceae-Rhizophoraceae) **BSI**, Calcutta. 1–898.

Singh, P. 1956. Pharmacognosy of Root and Rhizomes of *Cocculus villosus*. *Indian J. Pharm.* **18(7)**: 288–91.

Singh, Y.N. 1986. Traditional medicine in Fiji: Some Herbal Folk Cures used by Fiji Indians. *J. Ethnopharmacol.* **15(1)**:57–88.

Singh,V. and Pandey, R.P. 1983. Economic and medicinal plants of Indian desert. Desert Resources and Technology **1**:307–368.

Sivrajan, V.V. and Balchandran, I. 1994. Ayurvedic Drugs and their plant resources. Oxford and IBH Bombay, Calcutta, New Delhi.

Small, E. and Catling, P.M. 1999. *Canadian Medicinal Crops*. National Research Council Press, Ottawa, Ontario.

Sofowora, A. 1982. Medicinal plants and traditional medicine in Africa. Chichester, J. Wiley.

Subudhi, H.N.; Choudhury, B.P. and Acharya, B.C. 1992. Some potential medicinal plants of Mahanadi delta in the state of Orissa. *J. Econ. Taxon. Bot.* **16(2)**:479–487.

Sukh Dev. 1997. Ethnotherapeutics and modern drug development: the potential of Ayurveda. *Curr. Sci.* **73**:909–928.

Surange, S.R. and Pendse, G.S. 1972. Pharmacognostical studies of *Boerhavia diffusa* Linn. *Q. Jl. Crude Drug Res.* **12(3)**:1937–56.

Surange, S.R.; Patiskar, R.D. and Pendse, G.S. 1973. Comparative pharmacognostic studies on genuine and commercial samples of *Boerhavia* spp. Punarnava species Nov. (Sweta Punarnava). *J. Res. Indian Med.* **3(4)**:42.

Susan, J.; Kristen, L. and Saxena, P.K. 2001. The unique challenge of plant based medicines in Development of plant based: Conservation, efficacy and safety (ed.) Saxena, P.K. Kluwer Academic Publ. Boston. pp. 107–116.

Sushil Kumar; Hasan, S.A. and Misra, N.P. Cultivation of Aromatic Plants in India Lucknow, Central Institute of Medicinal and Aromatic Plants.

Talbot, W.A. 1894 (3rd ed., 1949). The Trees shrubs and woody climbers of the Bombay Presidency, Bombay.

Tanaka, T.; Sakai, Y. and Sakai, E. 1992. Medicinal plants of Mt. Ibuki: **1.** Folk medicine Ibuki-Hyakuso. *Bull. Jap. Ass. Bot. Gard.* **26**:26–33.

Tenorio, M.A.R.; de Oliveria, M.E.; Van den Berg, O.F. de Menezes and Salles, P. 1990. Fitoterapia: Uma strategia terapeutica natural do Amapa. Macapa: Museo de Plantes Medicinais Waldemiro Gomes.

Thakur, R.S.; Puri, H.S. and Husain, A. 1989. *Major medicinal plants of India*. CIMAP. Lucknow.

Tiwari, K.C.; Majumder, R. and Bhattachrjee, S. 1979. Folklore medicines from Assam and Arunachal Pradesh (Distric Tirap). *Q. Jl. Crude Drug Res.,* **17(2)**:61–67.

Tiwari, V.J. and Padhye, M.P. 1993. Ethnobotanical study of Gond tribe of Chandrapur and Gadchiroli districts of Maharashtra State, India. *Fitoterapia,* **64**:58–61.

Toyota, A.; Ninomiya, R.; Kobayashi, H.; Kawanishi, K.; Uhara, Y.; Kato, A. and Hashimoto, Y. 1979. *Studies of Brazilian crude drugs*. **1.** Muira-Puama. *Shoyakugaku Zasshi.* **33**:57–64.

Trease, G.E. 1952. Historica Introduction. *A text Book of Pharmacognosy*, 6th edn., Bailliere, Tindall & Cox, London. pp. 1–15.

Trease, G.E. and W.C. Evans 1972. *Pharmacognosy*. Bailliere Tindall, London.

Trivedi, V.P.; Dixit, R.S. and Lal, V.K. 1980. Orchids in the drug markets of Bareilly, Kanpur and near by districts. *Nagarjun.* **23(8)**:157–63.

Tyler, V.E. 1989. Potential hazards of herbal medication in plants and society, pp. 312–327. MacMillan Publishers: London, UK.

Tyler, V.E. 1996. "Pharmacognosy" What's that ? you spell it how ? *Econ. Bot.* **50**:3–9.

Tyler, V.E. 1997. The herbal remedies market. *Chemtech.* **27**:52–57.

Tyler, V.E. Jr. and Schwarting, A.E. 1958. *Experimental pharmacognosy*, Revised edn. Burgess Publishing Co. Minneapolis 15, Minnesota, p. iii+81.

Tyler, V.E.; Bbrady, L.R. and Robbers, J.E. 1976. *Pharmacognosy*, (7th ed.). Lea and Febiger, Philadelphia, pa, USA, p. 537.

Ugemuge, N.R. and Padhye, M.D. 1979. Some interesting reports of herbal medicines from Nagpur districts (Abstr.). *J. Indian Bot. Soc.* **58:** (Supplement). pp. 115–116.

Upadhye, A.S.; Kumbhojkar, M.S. and Vartak, V.D. 1986. Observations on wild plants used in folk medicine in the rural areas of the Kolhapur district. *Ancient Sci. Life*. **6(2):**119–121.

Upadhye, A.S.; Vartak, V.D. and Kumbhojkar, M.S. 1990. Cultivation of medicinal plants along sea costs in Maharashtra. *Bull. Med. Eth. Bot. Res.* **12:**32–39.

Upadhye, A.S.; Vartak, V.D. and Kumbhojkar, M.S. 1994. Ethno-Medico Botanical Studies in Western Maharashtra. *Ethnobotany.* **6:**25–31.

Van Reiss, Altschuls. 1973. Drugs and foods from little known plants. Harvard Univ. Press. Cambridge, MA.

Van, Wyk, B.E.; Oudtshoorn, B. van and Gericke, N. 1997. Medicinal plants of South Africa. Briza Publications, Pretoria.

Vartak V.D. 1959. Medicinal plants from Ferguson College Campus. *Ferg. Coll. Mag.*, **51(2):**22–25.

Vartak, V.D. 1957. Medicinal plants from Katraj Hills, Poona Districts. *Ayurvidya Mag.* p. 9.

Vartak, V.D. 1959. Medicinal Plants from the hilly region of Poona and Satara district, Bombay State. Part-A. *Ayurvidya Mag.* **22:**11–12

Vartak, V.D. 1959. Medicinal Plants from the hilly region of Poona and Satara district, Bombay State. Part-B. *Ayurvidya Mag.* **23:**129–135.

Vartak, V.D. 1979. Contributions of botany towards progress of ayurvedic medicine of western Maharashtra and Goa during British regime (1812–1947) (Absr.). *J. Indian Bot. Soc.* **58:** (supplement). p. 116.

Vartak, V.D. 1982. Studies on ethnobotany of Maharashtra and Goa. *Trib. Res. Bull.* **4(2):**9–10.

Vartak, V.D. and Kumbhojkar, M.S. 1985. Identification of Ramkand or Chakrakand. *Ind. For.* **111(2):**242–243.

Vartak, V.D. and Mandavgane. 1981. Enumeration of medicinal plants from Karnala tribal area, Kolaba district, (Maharashtra State) *J. Univ. Poona* (Sci. & Tec.) **54:**91–99.

Ved Prakash and Mehrotra, B.N. 1987. Ethnobotanical studies on the Flora of Khandala, Maharashtra State. *J. Econ. Tax. Bot.* **9:**205–208.

Vestal, A and Schultes, R.E. 1939. The Economic Botany of Kiwa Indians as it relates to the history of the Tribes. Harward University, USA.

Wahi, S.P.; Ansari, M.S. and Prasad, S. 1971. Pharmacognostical investigation on Indian Sarasaparilla pt. IV. Root of *Cryptolepis buchananii*. *J. Res. Indian Med.* **6(2):**159–66.

Wallis, T.E. 1920. Analytical microscopy. *Pharm. J.* **105:**159–60, 283–84.

Wallis, T.E. 1920. Analytical microscopy. *Pharm. J.* **105:**247–51, 395–97, 541–42, 578–79.

Wallis, T.E. 1920. Analytical microscopy. *Pharm. J.* **105:**528–31.

Wallis, T.E. 1920. Analytical microscopy. *Pharm. J.* **105:**376–78.

Wallis, T.E. 1921. Analytical microscopy. *Pharm. J.* **105:**48–50.

Wallis, T.E. 1925. Practical pharmacognosy. Churchill, London, pp. 125, 81 illus.

Wallis, T.E. 1967. Text book of pharmacognosy. (5th ed.) J. & A. Churchill Ltd. London.

Wallis, T.E. and Saber, A.H. 1933. Determination of mixed leaf powders e.g. Senna, *Ailanthus. Yb. Pharm*: 655.

Wanner, H. and Hoerhammer, L. 1971. Pharmacognosy and phytochemistry. pp. **VII**+386.

Warrier, P.K. *et al.,* 1996. Indian Medicinal Plants: a compendium of 500 species: **I–V.** Indcom Press, Madras.

Watt, G. 1889–96. The dictionary of the economic products of India. Repr. ed.: **I–VI** Superintendent Govt. Printing, Calcutta, India.

Wen, W. 1978. Herbal medicine in China. *Herbalcure.* **3(4):**4 & 5.

Whistler, W.A. 1985. Traditional and Herbal Medicine in the Cook Islands. *J. Ethnopharmacol.* **13(3):**239–280.

Willard, T. 1992. Edible and medicinal plants of the Rocky Mountains and neighboring territories. Calgary: Wild Rose College of Natural Healing, Calgary.

Williamson, E.M. and Evans, F.J. Potter's Encyclopaedia of botanical drugs and preparations. Revised ed., Saffron Walden, the C.W. Daniel Co. Ltd., Essex UK.

Woodley, E. (ed.) 1991. Medicinal plants of Papua New Guinea: part 1. Morobe Province. Weikersheim: Verlag Josef Margraf.

Yadav, S.R. and Sardesai, M.M. Flora of Kolhapur District. Shivaji University, Kolhapur, India.

Yogaanarasimhan, S.N.; Venkataram, B.S. and Togunashi, V.S. 1974. Standard specimen-Its need and advantage for ayurvedic drugs. *J. Res. Indian Med.* **9(3):**11–14.

Yoganarasimhan, S.N. 1996. Medicinal plants of India: **1.** Interline Publishing, Bangalore, Karnataka.

Youngken, H.W. 1933. A new field of investigation in pharmacognosy: The microscopy of glandular products. *J. Am. Pharm. Ass.* **22(12):**1215–18.

Youngken, H.W. 1947. Pharmacognostical studies of four Americans arnicas. *J. am. Pharm. Ass. Sci.* edn. **36(3):**85–96.

Youngken, H.W. 1950. A text book of pharmacognosy (5th ed.) Philadelphia, USA.

Zafar, R. 1994. Medicinal plants of India. CBS Publishers and Distributors, Delhi

Zagari, A. 1992. Medicinal Plants: **4**. (5th ed.). Tehran University Publications, Dept. Pharmacognosy Coll Pharmacy Tehran Univ. Med. Sci. Tehran. Iran.

Zalzala, A.H. and Al, Shamma, A. 1979. Medicinal plants of Iraq: **1.** Phytochemical investigation of *Haplophyllum tuberculatum*: **2.** Volatile oil components. *Bull. Biol. Res. Centre* (Baghdad). **11**:25–31 (1979).

Zedler, Joy B. 1982. The ecology of southern California coastal salt marshes: A community profile. U.S. Fish and Wildlife Service, Biological Services Program, Washington, D.C.

Zhang, X. 1998. Significance of traditional medicine in human health care. pp. WIPO *Asian Regional seminar on Intellectual Property issues in the field of Traditional Medicines*. New Delhi, India.

Index to Families and Species*

* Synonyms are in Italics